MUNICIPAL WASTEWATER TREATMENT

MUNICIPAL WASTEWATER TREATMENT:

EVALUATING IMPROVEMENTS IN NATIONAL WATER QUALITY

Andrew Stoddard
Andrew Stoddard & Associates, Hamilton, VA

Jon B. Harcum
Tetra Tech, Inc., Fairfax, VA

Jonathan T. Simpson
Tetra Tech, Inc., Fairfax, VA

James R. Pagenkopf
Tetra Tech, Inc., Fairfax, VA

Robert K. Bastian
U.S. Environmental Protection Agency,
Office of Wastewater Management,
Washington, DC

JOHN WILEY & SONS, INC.

This book is printed on acid-free paper. 

Published simultaneously in Canada.

Library of Congress Cataloging-in-Publication Data:
JK
Municipal Wastewater Treatment : Evaluating Improvements in National Water Quality / by Andrew Stoddard . . . [et al.].
p. cm.
Includes bibliographical references and index.
ISBN 0-471-24360-4 (cloth : alk. paper)
1. Water quality—United States. 2. Sewage disposal plants—United States—Design and construction—Finance. 3. Water quality management—United States—Case studies. 4. Federal aid to water quality management—United States. 5. United States. Federal Water Pollution Control Act. I. Stoddard, Andrew.

TD223 .P763 2002
363.739′45′0973—dc21

2001046658

Printed in the United States of America.

10 9 8 7 6 5 4 3 2 1

IN MEMORIAM

Donald J. O'Connor (1922–1997)

Francis J. Stoddard (1920–1999)

Judith K. Patterson (1951–1997)

DEDICATION

This book attempts to document the water quality benefits associated with the federal funding provided through the Construction Grants Program and Clean Water State Revolving Fund (CWSRF) Program to help plan, design, and construct publicly owned treatment works (POTWs). The effort was initiated at the request of Michael J. Quigley while he served as Director of the Office of Municipal Pollution Control. It is dedicated to the many hardworking and conscientious individuals—including the program advocates and critics alike—who help manage, direct (or in some cases redirect), and implement the Construction Grants and CWSRF Programs, which are among the nation's largest public works programs, in a highly professional and effective manner. They include many program managers and staff at the U.S. Environmental Protection Agency (USEPA), and at state levels, as well as local wastewater authority managers and staff, not to mention the many highly qualified consultants and contractors who help the local authorities conduct the necessary studies, develop the required facilities plans and project design documents, and construct and operate the treatment facilities that were established or upgraded with funding from these highly successful public works programs.

This book could not have been written without the extensive water quality monitoring efforts across the country undertaken by a legion of highly qualified field staff and researchers for many local authorities, state and federal agencies, and colleges and universities. Their efforts produced the extensive water quality data available in the USEPA's STORET database system and in local reports, as well as the water quality models and local assessments that served as the basis for the analyses undertaken and reported on in this book.

CONTENTS

Preface xv

Acknowledgments xix

Acronyms xxiii

1. Introduction 1

Background 2

Study Approach 4

The First Leg: An Examination of BOD Loadings Before and After the CWA 5

The Second Leg: An Examination of "Worst-Case" DO in Waterways Below Point Sources Before and After the CWA 6

The Third Leg: Case Study Assessments of Water Quality 12

The Target Audience for this Book 13

References 13

2. An Examination of BOD Loadings Before and After the CWA 17

A. History of Water Supply and Its Effects on Public Health 17

Impacts on Water Supply Users and "The Great Sanitary Awakening" 19

Impacts on Water Resources Users 21

B. Evolution of Wastewater Treatment 23

Primary Treatment 23

Dissolved Oxygen as an Indicator of Water Quality 23

Secondary Treatment 25

Biochemical Oxygen Demand (BOD) as a Measure of Organic Wasteload Strength 26

C. The Federal Role in Implementing Secondary Treatment in the Nation's POTWs 27

The Federal Role in Secondary Treatment Before the Clean Water Act 27

The Federal Role in Secondary Treatment After the Clean Water Act 33

D. Nationwide Trends in BOD Loading from POTWs 42

Types of BOD Reported in this Trends Analysis 43
Trends in POTW Inventory 48
Trends in Population and Influent Wastewater Flow to POTWs 50
Trends in Influent BOD Loading to POTWs 52
Trends in Effluent BOD Loading from POTWs 57
Trends in BOD Removal Efficiency 63
Future Trends in BOD Effluent Loading 64

E. BOD_5 Loadings from Point and Nonpoint Sources 70

Pollutant Loading from Sources Other than POTWs 71
BOD_5 Loads from the National Water Pollution Control Assessment Model (NWPCAM) 76
Comparison of Point and Nonpoint Sources of BOD_5 at the National Level 85

F. Investment Costs for Water Pollution Control Infrastructure 85

The Construction Grants Program 85
Other Investment Costs for Water Pollution Control Infrastructure 90
Future Infrastructure Needs 93

G. Summary, Conclusions, and Future Trends 95

Key Points of the Background Sections 95
Key Points of the BOD Loading Analysis Sections 96
Key Points of the Investment Costs Section 98
Conclusions and Future Trends 98

References 100

3. An Examination of "Worst-Case" DO in Waterways Below Point Sources Before and After the CWA 105

A. Background 106

Sources of DO Data 107
"Worst Case" Conditions as a Screening Tool 107
The Role of Spatial Scale in this Analysis 119

B. Data Mining 119

Step 1—Data Selection Rules 119
Step 2—Data Aggregation Rules from a Temporal Perspective 120
Step 3—Calculation of the Worst-Case DO Summary Statistic Rules 121
Step 4—Spatial Assessment Rules 122
Step 5—Data Aggregation Rules from a Spatial Perspective 125

Step 6—Development of the Paired Data Sets (at Each Spatial Scale) 126

C. Comparison of Worst-Case DO in Waterways Below Point Source Discharges Before and After the CWA at Three Spatial Scales 126

Before and After DO at Reach Scale 127
Before and After DO at Catalog Unit Scale 130
Comparison of the Change in Signal Between the Reach and Catalog Unit Scales Using the Upper White River Basin (Indiana) as an Example 154
Before and After DO at Major River Basin Scale 162

D. Summary and Conclusions 171

Key Points of the Background Section 172
Key Points of the Data Mining Section 173
Key Points of the Comparison Analysis Section 174
Conclusions 176

References 179

4. Case Study Assessments of Water Quality **181**

A. Background 181

B. Selection of Case Study Waterways 183

C. Before and After CWA 184

D. Policy Scenarios for Municipal Effluent Discharges 189

E. Discussion and Conclusions 194

References 195

5. Connecticut River Case Study **199**

Background 199

Physical Setting and Hydrology 200

Population, Water, and Land Use Trends 203

Historical Water Quality Issues 204

Legislative and Regulatory History 205

Impacts of Wastewater Treatment 206
Pollutant Loading and Water Quality Trends 206
Recreational and Living Resources Trends 207

Summary and Conclusions 210

References 211

6. Hudson-Raritan Estuary Case Study **213**

Background 213

Physical Setting and Hydrology 215
Population, Water, and Land Use Trends 215

Historical Water Quality Issues 220

Legislative and Regulatory History 223

Impacts of Wastewater Treatment 225
Pollutant Loading and Water Quality Trends 225
Recreational and Living Resources Trends 242

Summary and Conclusions 248

References 249

7. Delaware Estuary Case Study **255**

Physical Setting and Hydrology 256

Population, Water, and Land Use Trends 259

Historical Water Quality Issues 261

Legislative and Regulatory History 262

Impacts of Wastewater Treatment 263
Pollutant Loading and Water Quality Trends 263
Evaluation of Water Quality Benefits Following Treatment Plant Upgrade 270
Recreational and Living Resources Trends 273

Summary and Conclusions 279

References 281

8. Potomac Estuary Case Study **285**

Physical Setting and Hydrology 286

Population, Water, and Land Use Trends 288

Historical Water Quality Issues 290

Legislative and Regulatory History 290

Impacts of Wastewater Treatment 291
Pollutant Loading and Water Quality Trends 291
Evaluation of Water Quality Benefits Following Treatment Plant Upgrades 294
Recreational and Living Resources Trends 298

Summary and Conclusions 308
References 309

9. James River Estuary Case Study 311
Physical Setting and Hydrology 311
Population Trends 314
Historical Water Quality Issues 315
Legislative and Regulatory History 316
Impacts of Wastewater Treatment 317
Pollutant Loading and Water Quality Trends 317
Evaluation of Water Quality Benefits Following Treatment Plant Upgrades 321
Recreational and Living Resources Trends 324
Summary and Conclusions 325
References 325

10. Upper Chattahoochee River Case Study 327
Physical Setting and Hydrology 328
Population, Water, and Land Use Trends 331
Historical Water Quality Issues 333
Legislative and Regulatory History 335
Impacts of Wastewater Treatment 336
Pollutant Loading and Water Quality Trends 336
Recreational and Living Resources Trends 340
Summary and Conclusions 342
References 342

11. Ohio River Case Study 345
Physical Setting and Hydrology 346
Population, Water, and Land Use Trends 347
Historical Water Quality Issues 350
Legislative and Regulatory History 350
Impacts of Wastewater Treatment 352
Pollutant Loading and Water Quality Trends 352

Recreational and Living Resources Trends 356
Summary and Conclusions 358
References 358

12. Upper Mississippi River Case Study 361
Physical Setting and Hydrology 362
Population, Water, and Land Use Trends 366
Historical Water Quality Issues 367
Legislative and Regulatory History 372
Impacts of Wastewater Treatment 373
Pollutant Loading and Water Quality Trends 373
Evaluation of Water Quality Benefits Following Treatment Plant Upgrades 383
Recreational and Living Resources Trends 386
Summary and Conclusions 390
References 393

13. Willamette River Case Study 399
Physical Setting and Hydrology 400
Population, Water, and Land Use Trends 403
Historical Water Quality Issues 404
Legislative and Regulatory History 405
Impacts of Wastewater Treatment 406
Pollutant Loading and Water Quality Trends 406
Recreational and Living Resources Trends 408
Summary and Conclusions 410
References 412

Appendix A. United States Waterways Identified with Water Pollution Problems Before the 1972 Clean Water Act 415

Appendix B. National Municipal Wastewater Inventory and Infrastructure, 1940–2016 449

Appendix C. National Public and Private Sector Investment in Water Pollution Control 501

Appendix D. **Before and After CWA Changes in Tenth Percentile Dissolved Oxygen and Ninetieth Percentile BOD_5 at the Catalog Unit Level** **507**

Appendix E. **Before and After CWA Changes in Tenth Percentile Dissolved Oxygen at the RF1 Reach Level** **523**

Appendix F. **Hydrologic Conditions of the 48 Contiguous States, Summer (July–September) from 1961 through 1995** **537**

Appendix G. **Municipal and Industrial Wastewater Loads by Major River Basin Before and After the Clean Water Act: 1950, 1973, and *ca.* 1995** **583**

Appendix H. **Municipal and Industrial Water Withdrawals by Major River Basin: 1940–1995** **593**

Glossary **601**

Index **615**

PREFACE

This study was prepared under the sponsorship of several programs in the USEPA Office of Water, primarily to document the water quality benefits associated with the more than 16,000 publicly owned treatment works (POTWs) across the country. This study emphasizes the role of USEPA's Construction Grants Program, which provided $61.1 billion in federal grants to local authorities from 1972 through 1995 to help support the planning, design, and construction of POTWs to meet the minimum treatment technology requirements established by the secondary treatment regulations or water quality standards (where applicable). The program has also provided more than $16 billion under the Clean Water State Revolving Fund (CWSRF) Loan Programs as capitalization grants to the states since 1988 to support a wide range of water quality improvement projects. The study was subjected to a formal peer review process that included detailed reviews and input from NOAA, USGS, AMSA, NRDC, NRC/NAS, NWRI, University of North Carolina, Johns Hopkins University, University of Alabama, states, consultants, local authorities, and others.

The book contains 13 chapters, including a background chapter, and chapters addressing biochemical oxygen demand (BOD) loadings before and after the Clean Water Act (CWA), the "worst case" dissolved oxygen (DO) levels in waterways downstream of point sources before and after the CWA, and nine case study assessments of water quality changes associated with POTW discharges.

The book presents the results of a unique, three-way approach for addressing such frequently asked questions as:

1. Has the CWA regulation of POTW discharges been a success?
2. How does the nation's water quality before the 1972 Federal Water Pollution Control Act (FWPCA) Amendments compare with the water quality conditions after secondary and better treatment was implemented?
3. Has the reduction of BOD loadings to surface waters from POTWs resulted in improved water quality in the nation's waterways? If so, to what extent?

By examining the numbers and characteristics of POTWs, their populations served, and BOD loadings on a nationwide basis before and after the CWA, we were able to document changes in the number of people served by POTWs and the level of treatment provided, the amount of BOD discharged to the nation's waterways, and the aggregate BOD removal efficiencies of the POTWs, while providing insight into the likely impact of future discharges if treatment efficiencies aren't improved to accommodate economic growth and expansions in service population.

The authors examined the "worst-case" historical DO levels in waterways located downstream of point sources before and after the CWA in a systematic manner. By identifying water quality station records that related to the water quality impact of point source discharges from the "noise" of millions of historical records archived in the USEPA's STORET database, and by using DO as the study's indicator of water quality responses to long-term changes in BOD loadings from POTWs, the authors evaluated changes in DO for only those stations on receiving waters affected by point sources over time under comparable worst-case low-flow conditions (during July–September in 1961–1965 for before CWA and 1986–1990 for after CWA) using only surface (within 2 meters of the surface) DO data. The authors documented statistically significant improvements in worst-case summer DO conditions at three different spatial scales, in two-thirds of the reaches, catalog units, and major river basins.

Case study assessments were also completed on nine urban waterways with historically documented water pollution problems. These case study sites included the Connecticut River, Hudson-Raritan estuary, Delaware estuary, Potomac estuary, James estuary, Chattahoochee River, Ohio River, Upper Mississippi River, and Willamette River. Most of the these waterways were sites of interstate enforcement cases from 1957 to 1972, were listed as potential waterways for which state-federal enforcement conferences were convened in 1963, or were the subjects of water quality evaluation reports prepared for the National Commission on Water Quality. Two sites were on a 1970 list of the top 10 most polluted rivers. The case study sites did not include, however, any of the 25 river reaches with the greatest before versus after CWA improvements in DO found in our study. The case studies characterized long-term trends in population, point source loadings, ambient water quality, environmental resources, and recreational uses. Validated water quality models for the Delaware, Potomac, and James estuaries and the Upper Mississippi River were used to quantify the water quality improvements that have been achieved by upgrading POTWs to secondary and higher levels of treatment. The case study assessments document that tremendous progress has been made in improving water quality, restoring valuable fisheries and other biological resources, and creating extensive recreational opportunities (angling, hunting, boating, bird-watching, etc.) in all nine case study sites. At many of the sites, there have been significant increases in species diversity and abundance-returned or enhanced populations of valuable gamefish (e.g., bass, bluegill, catfish, perch, crappies, and sturgeon) and migratory fish populations, waterfowl and fish-eating bird populations, opened shellfish beds, and more. Some of the sites have seen a return of abundant mayflies and other pollution-sensitive species, as well as dramatic increases in recreational boating and fishing. Water quality improvements associated with BOD, suspended solids, coliform bacteria, heavy metals, nutrients, and algal biomass have been linked to reductions in municipal and industrial point source loads for many of the case studies.

The unique, three-way approach undertaken by this study quantitatively supports the hypothesis that the 1972 CWA's regulation of wastewater treatment processes at POTWs and industrial facilities has achieved significant success—success in terms of reduction of effluent BOD from POTWs, worst-case (summertime, low-flow) DO improvement in waterways, and overall water quality improvements in urban case study areas with historically documented water pollution problems. It is important to emphasize that the water quality improvements documented in this book have resulted from the combined efforts of state, local, and federal government public funding, and investments by private industry, to upgrade the nation's infrastructure of municipal and industrial wastewater treatment facilities. However, the study also points out that without continued investments and improvements in our municipal wastewater treatment infrastructure, future population growth will erode away many of the CWA achievements in effluent loading reduction.

ROBERT K. BASTIAN

Senior Environmental Scientist
U.S. Environmental Protection Agency
Office of Wastewater Management (4204)
Washington, DC

ACKNOWLEDGMENTS

The authors gratefully acknowledge the support of Bob Bastian, our USEPA Project Officer and coauthor for this book, and our other USEPA Project Officers for this study. Bob Bastian, Karen Klima, Virginia Kibler, and Dr. Mahesh Podar of USEPA's Office of Water all contributed to this research effort with their vision that, a quarter century after enactment of the 1972 Clean Water Act, an evaluation of the water quality improvements that could be attributed to federal funding by USEPA's Construction Grants Program and the Clean Water State Revolving Fund (CWSRF) was a feasible undertaking. Their guidance, encouragement, and challenging questions helped to shape the study documented in this book. This project was funded by the U.S. Environmental Protection Agency under the following contracts with Tetra Tech, Inc.: EPA-68-C3-0303, EPA-68-C1-0008, and EPA Purchase Order No. 7W-0763-NASA.

The authors gratefully acknowledge the efforts of many of our colleagues who assisted us in identifying, compiling, analyzing, and visualizing an enormous amount of data for this study. We acknowledge Alexander Trounov of Tetra Tech, Inc., for his expert assistance in the extraction and processing of data from USEPA's mainframe databases (STORET, Reach File Version 1, Permit Compliance System, Clean Water Needs Survey) and USGS streamflow databases. Patrick Solomon of Tetra Tech, Inc., is acknowledged for his expert assistance in transforming geographically based data sets into maps that are works of art. Timothy Bondelid of Research Triangle Institute (RTI) is acknowledged for his invaluable contributions of point and nonpoint source pollutant loading data, including the Reach File Version 1 transport routing database, which was developed as part of RTI's National Water Pollution Control Assessment Model (NWPCAM). With a professional career in water pollution investigations that began in Chicago, Illinois, with the U.S. Public Health Service during the early 1960s, Phill Taylor, our colleague at Tetra Tech, Inc., is acknowledged for his invaluable knowledge of the "early years" of water pollution control investigations and the insight stimulated by our frequent questions about historical water quality data archived in STORET. Phill's "corporate memory" and his personal library of reports documenting water pollution conditions during the 1950s and 1960s were instrumental in guiding and completing this research effort. The authors gratefully acknowledge the significant contributions of the late Ralph Sullivan in preparing material documenting the legislative and regulatory history of the Federal Water Pollution Control Act.

The authors wish to acknowledge the efforts of the Peer Review Team, whose insight and often critical observations undoubtedly increased the value and credibility of the study's results. The members of the Peer Review Team included:

- Mr. Leon Billings
- Mr. Tom Brosnan, National Oceanic and Atmospheric Administration
- Mr. Michael Cook, U.S. Environmental Protection Agency
- Mr. John Dunn, U.S. Environmental Protection Agency
- Dr. Mohammad Habibian, Washington Suburban Sanitation Commission
- Dr. Leo Hetling, Public Health and Environmental Engineering, New York State Department of Environmental Conservation (retired)
- Dr. Russell Isaacs, Massachusetts Department of Environmental Protection
- Dr. Norbert Jaworski, U.S. Environmental Protection Agency (retired)
- Dr. William Jobin, Blue Nile Associates
- Mr. Ken Kirk, American Metropolitan Sewerage Association
- Mr. John Kosco, U.S. Environmental Protection Agency
- Mr. Rich Kuhlman, U.S. Environmental Protection Agency
- Mr. Joseph Lagnese
- Ms. Jessica Landman, Natural Resource Defense Council
- Mr. Kris Lindstrom, K. P. Lindstrom, Inc.
- Mr. Ronald Linsky, National Water Research Institute
- Dr. Berry Lyons, University of Alabama
- Dr. Alan Mearns, National Oceanic and Atmospheric Administration
- Dr. Daniel Okun, University of North Carolina
- Mr. Steve Parker, National Research Council
- Mr. Richard Smith, U.S. Geological Survey
- Mr. Phill Taylor, U.S. Environmental Protection Agency and Tetra Tech, Inc. (retired)
- Dr. Red Wolman, Johns Hopkins University

Many state and local agency officials reviewed the case study chapters for accuracy and completeness. The authors want to specifically acknowledge the invaluable contributions and data sets provided by Alan Stubin and Tom Brosnan for the Hudson-Raritan estuary case study; by Cathy Larson for the Upper Mississippi River case study; by Tyler Richards for the Chattahoochee River case study; by Ed Santoro and Richard Albert for the Delaware estuary case study; and by Dr. Virginia Carter and Dr. Nancy Rybicki for the Potomac estuary case study. Jim Fitzpatrick (Hydro Qual, Inc.) and Dr. Wu-Seng Lung (Enviro Tech, Inc.) are acknowledged for providing the results of water quality model simulations for case studies of the Potomac, Delaware, James, and Upper Mississippi Rivers. The late Bob Reimold and his colleagues at Metcalf & Eddy Engineers, Inc., are acknowledged for their contributions to the case study of the Connecticut River.

The members of the Case Study Peer Review Team included:

- Mr. Richard Albert, Delaware River Basin Commission
- Mr. Tom Brosnan, National Oceanic and Atmospheric Administration
- Dr. Virginia Carter, U.S. Geological Survey
- Ms. Linda Henning, St. Paul Metropolitan Council Environmental Services
- Ms. Cathy Larson, St. Paul Metropolitan Council Environmental Services
- Dr. Nancy Rybicki, U.S. Geological Survey
- Mr. Alan Stubin, New York City Department of Environmental Protection
- Mr. Ed Santoro, Delaware River Basin Commission
- Ms. Pat Stevens, Atlanta Regional Commission
- Mr. Peter Tennant, Ohio River Valley Sanitation Commission

Finally, we recognize the cheerful cooperation and expert editing, graphic arts, and document production efforts of Marti Martin, Robert Johnson, Kelly Gathers, Krista Carlson, Emily Faalasli, Elizabeth Kailey, Melissa DeSantis, and Debby Lewis of Tetra Tech, Inc., in Fairfax, Virginia. Thanks for a great job!

ACRONYMS

7Q10 10-year, 7-day minimum flow

AMSA American Metropolitan Sewerage Association

ARC Atlanta Regional Commission (Georgia)

ASIWPCA Association of State and Interstate Water Pollution Control Administration

AWT Advanced wastewater treatment

BOD Biochemical oxygen demand

BOD_5 5-day biochemical oxygen demand

BODu Ultimate biochemical oxygen demand

C:DW Carbon-to-dry weight ratio

CBOD Carbonaceous biochemical oxygen demand

CCMUA Camden County Municipal Utility Authority (New Jersey)

CMSA Combined Metropolitan Statistical Area

CSO Combined sewer overflow

CTDEP Connecticut Department of Environmental Protection

CU Catalog unit

CWA Clean Water Act

CWNS Clean Water Needs Survey

CWSRF Clean Water State Revolving Fund

DDT 2, 2-bis (p-chlorophenyl)-1,1,1-trichlorethane

DECS Delaware Estuary Comprehensive Study

DEL USA Delaware Estuary Use Attainability Study

DEM Dynamic Estuary Model

DNR Department of Natural Resources (Georgia)

DMR Discharge monitoring report

DO Dissolved oxygen

DRBC Delaware River Basin Commission

EPD Environmental Protection Division (Georgia)

FR Federal Register

FWPCA Federal Water Pollution Control Act/Administration

FWQA Federal Water Quality Administration

FY Fiscal year

GAO General Accounting Office

GICS Grants Information and Control System

gpcd Gallons per capita per day

HEP Harbor Estuary Program (NY, NJ)

HUC Hydrologic unit code

IBI Index of Biotic Integrity

ICPRB Interstate Commission on Potomac River Basin

IFD Industrial Facilities Discharge File

INCODEL Interstate Commission on the Delaware River Basin (NJ, DE, PA)

ISC Interstate Sanitation Commission (NJ, NY, CT)

JEM James Estuary Model

JMSRV James River Model

LTI Limno-Tech, Inc. (Ann Arbor, Michigan)

MCES Metropolitan Council Environmental Services (Minneapolis-St. Paul, MN)

mgd Million gallons per day

MPCA Minnesota Pollution Control Agency

MPN Most probable number

MRPA Metropolitan River Protection Act (Georgia)

MSA Metropolitan Statistical Area

MSX Parasitic protozoan *(Haplosporidium nelsoni)* responsible for massive oyster mortalities in Delaware Bay and Chesapeake Bay; MSX is popular name for the disease

mt/day Metric tons per day (1,000 kg per day)

MUA Municipal Utility Authority

MWCC Metropolitan Waste Control Commission (Minneapolis-St. Paul, MN)

MWCOG Metropolitan Washington Council of Governments

N Nitrogen

NAS National Academy of Sciences

NBOD Nitrogenous biochemical oxygen demand

NCWQ National Commission on Water Quality

NH_3—N Ammonia nitrogen

NO_2—N Nitrite nitrogen

NO_3—N Nitrate nitrogen

NOAA National Oceanic and Atmospheric Administration

NPDES National Pollutant Discharge Elimination System

NPS Nonpoint source; also National Park Service

NRC National Research Council

NRDC Natural Resources Defense Council

NURP National Urban Runoff Project
NWPCAM National Water Pollution Control Assessment Model
NWRI National Water Research Institute
NYCDEP New York City Department of Environmental Protection
O Oxygen
O&M Operation and maintenance
ODEQ Oregon Department of Environmental Quality
OMB Office of Management and Budget
ORSANCO Ohio River Valley Sanitation Commission
OTA Office of Technology Assessment
OWM USEPA Office of Wastewater Management
P Phosphorus
PAH Polynuclear aromatic hydrocarbons
PCB Polychlorinated biphenyls
PCS Permit Compliance System
PE Population equivalent
PEM Potomac Eutrophication Model
PL Public Law
PMSA Primary Metropolitan Statistical Area
PO_4—P Phosphate phosphorus
POC Particulate organic carbon
POM Particulate organic matter
POTW Publicly owned treatment works
ppm parts per million (concentration)
ppt parts per thousand (concentration)
QA/QC Quality assurance/quality control
RF1 Reach File Version 1
RF3 Reach File Version 3
SAV Submersed aquatic vegetation
SIC Standard Industrial Classification
SRP Soluble reactive phosphorus
STORET USEPA's STOrage and RETrieval database
SWCB State Water Control Board (Virginia)
SWEM System Wide Eutrophication Model (New York Harbor)
TKN Total Kjeldahl nitrogen
TMDL Total maximum daily load
TN Total nitrogen
TOC Total organic carbon

TP Total phosphorus

TPC Typical pollutant concentration

TSS Total suspended solids

UM Upper Mississippi River milepoint measured from confluence of Ohio River and Upper Mississippi River

USCOE U.S. Army Corps of Engineers

USCB U.S. Census Bureau

USDA U.S. Department of Agriculture

USDOC U.S. Department of Commerce

USDOI U.S. Department of Interior

USEPA U.S. Environmental Protection Agency

USGS U.S. Geological Survey

USPHS U.S. Public Health Service

VIMS Virginia Institute of Marine Science

WEF Water Environment Federation

WIN Water Infrastructure Network

WPCF Water Pollution Control Federation

WQS Water Quality Standard

WRBWQS Willamette River Basin Water Quality Study

WRE Water Resources Engineers (Walnut Creek, California)

CHAPTER 1

Introduction

> I think there is no sense in forming an opinion when there is no evidence to form it on. If you build a person without any bones in him he may look fair enough to the eye, but he will be limber and cannot stand up; and I consider that evidence is the bones of an opinion.
>
> —Attributed to Mark Twain in "Personal Recollections of Joan of Arc."

Today, a student writing a paper on the Federal Water Pollution Control Act Amendments of 1972 (Public Law 92-500, later to be known as the Clean Water Act or CWA) would be hard-pressed to find a public official who would say the legislation was not a success. Vice President Gore's remarks in October 1997, celebrating the twenty-fifth anniversary of the act, are representative of the good feelings people have about the CWA (USEPA, 1997a; WEF, 1997).

In his speech, the Vice President lauded the cooperative efforts of federal, state, tribal, and local governments in implementing the act's pollution control provisions. He reported that the quality of rivers, lakes, and bays has "improved dramatically." He related success stories involving water-based commerce, agriculture, tourism, fisheries, and quality of life for a variety of locations, including Alaska's St. Paul Harbor, the Chesapeake Bay, Cleveland's Cuyahoga River, the Long Island Sound, and the Houston Ship Channel. With cheers like that ringing in people's ears, it's no wonder that the prevailing public opinion is that the act has been a success. But what if the paper-writing student were to inquire skeptically about the "bones" of this opinion? What scientific evidence could she cite to back up this claim? Was the nation indeed able to buy water quality success with the approximately $200.6 billion in capital costs and $210.1 billion in operation and maintenance costs (current year dollars) invested from 1972 to 1994 by public and private authorities in point source water pollution control?

A centerpiece of the CWA was a dramatic increase in federal support for upgrading publicly owned treatment works (POTWs). From 1970 to 1999, $77.2 billion in federal grants and contributions through the U.S. Environmental Protection Agency's (USEPA's) Construction Grants and Clean Water State Revolving Fund (CWSRF) programs was distributed to municipalities and states for this activity. A 1995 editorial in the Water Environment Federation's research journal noted that no comprehensive national study has ever been done to document whether this investment has paid off in terms of improved water quality (Mearns, 1995). Who could blame the student, then, if she applied Mark Twain's logic and concluded that the public's opinion concerning the success of the CWA was "limber" and could not "stand up."

The purpose of this book is to provide that student with the "bones" to form an opinion that will stand up. Specifically, it was designed to examine whether "significant" water quality improvements [in the form of increased dissolved oxygen (DO) levels] have occurred downstream from POTW discharges since the enactment of the CWA.

BACKGROUND

The framers of the CWA, drawing on the experience of the Ohio River Valley Sanitation Commission (ORSANCO), recognized that two basic sets of users depend on the chemical, physical, and biological integrity of the nation's waterways:

1. Water supply users, people who take delivery of and use water drawn from various surface water and groundwater sources. Whether intentionally or not, these users usually contaminate the water they receive with pollutants such as organic matter, sediments, nutrients, pathogens, and heavy metals. Contaminated water (wastewater) is then collected, transported away from the site, treated, and returned back to a natural waterbody, where it can be withdrawn and cycled again by the same or another water supply system. Figure 1-1 illustrates this process, known as the urban water cycle.
2. Water resource users, people such as fishermen, boaters, and swimmers, who use water in its natural settings—lakes, streams, rivers, and estuaries. This category might even be assumed to encompass the fish, waterfowl, and other living things that depend on clean water to live, reproduce, and thrive. These users can be directly affected by the return flow of wastewater from water supply users.

Meeting the needs of water supply and water resource users has been a problem that has vexed public officials for centuries. Only in the latter part of the twentieth century did it become clear that the secret for keeping both sets of users satisfied is to have all components of the cycle in place and functioning properly. This fundamental concept played a pivotal role in the development of the CWA.

By the mid-1900s, it was becoming more and more apparent that the weak link in the urban water cycle was the wastewater treatment component. Many communities were effectively short-circuiting the cycle by allowing raw or nearly raw sewage to flow directly into lakes, streams, rivers, estuaries, and marine waters. The organic matter contained in this effluent triggered increased growths of bacteria and corresponding decreases in DO levels. This situation, in turn, negatively affected the life functions of fish, shellfish, and other aquatic organisms. In addition, pathogens, nutrients, and other pollutants present in wastewater made body contact unsafe, increased the growth of algae and rooted aquatic plants, and reduced the potential for recreation and other uses. In sum, this weak link in the urban water cycle was greatly affecting the lives and livelihoods of water resource users downstream from POTWs.

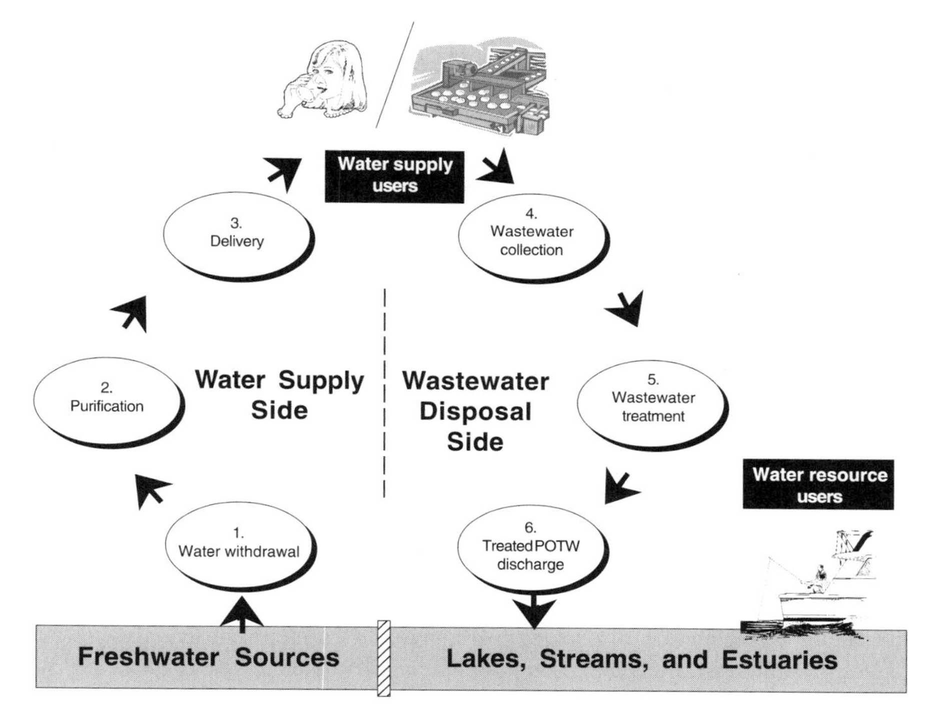

Figure 1-1 Simplified urban water cycle.

Through the 1972 CWA, Congress aimed to remedy this situation by establishing a national policy requiring secondary treatment of municipal wastewater as the minimum acceptable technology, supplemented by more stringent water quality–based effluent controls on a site-specific, as-needed basis. At that time, approximately 4,859 systems in the country serving 56.8 million people were providing only raw discharge or primary treatment of wastewater, a method that uses physical processes of gravitational settling to separate settleable and floatable solids from raw sewage. Secondary treatment, in contrast, yields a much cleaner effluent because it uses biological processes to break down much of the organic matter contained in the wastewater before allowing the wastewater to leave the facility.

Between 1970 and 1995, a total of $61.1 billion (in current year dollars, equivalent to $96.5 billion as constant 1995 dollars) was allocated by Congress through USEPA's Construction Grants Program for the purpose of building new, and upgrading old, POTWs. An additional $16.1 billion in federal contributions was also distributed to states through the CWSRF from 1988 through 1999. In addition to this federal expenditure, state and local governments and private industry made significant investments to comply with regulations of the CWA and other state and local environmental legislation. On a nationwide basis, actual expenditure data compiled by the U.S. Department of Commerce, Bureau of Economic Analysis, in the annual Pollution Abatement Cost Expenditures, documents a cumulative public and private sector capital expenditure of approximately $200.6 billion and an additional $210.1 billion as operating expenditures (current year dollars) for water pollution control activities during the period from 1972 through 1994 (Vogan, 1996). In this context, the Construction Grants Program provided federal grant support to local municipalities that amounted to almost one-half of the public sector costs and about one-third of the total public and private sector capital investment for water pollution control.

STUDY APPROACH

For years, members of Congress, as well as citizens and special interest, environmental, and business groups, have been quizzing the USEPA about the benefits gained from the nation's extraordinary public and private investment in wastewater treatment (GAO, 1986a, 1986b, 1986c; USEPA, 1988). Addressing their questions is a difficult task because environmental systems are very complex—so complex, in fact, that researchers can't even agree what "stick" to use to measure success. Consequently, a number of tools have been applied in an attempt to measure the success of water pollution control efforts. These include:

- Reporting the number of discharge permits issued, enforcement actions taken, and other administrative actions and programmatic evaluations (Adler et al., 1993).
- Reporting on the number of POTWs built or upgraded, population served by various treatment levels, effluent loading rates, and other trends in the construction and use of wastewater infrastructure (USEPA, 1997b).
- Inventorying state and national waterways meeting designated uses (e.g., reports prepared by states to comply with CWA section 305(b), USEPA's 305(b) summary reports to Congress) (ASIWPCA, 1984; USEPA, 1995a, 1995b).
- Investigating changes in specific waterways following wastewater treatment plant upgrades (GAO, 1978, 1986c; Leo et al., 1984; Patrick et al., 1992).
- Investigating the statistical significance of national-scale changes in water quality following the 1972 CWA (GAO, 1981; Knopman and Smith, 1993; Smith et al., 1987a, 1987b).

Although each of the above approaches provides some evidence of the accomplishments of municipal wastewater treatment under the CWA, none could be considered a comprehensive assessment of national progress in meeting the CWA's main goal of maintaining, or restoring, fishable and swimmable waters. Clearly, a fresh measuring stick is needed—one that is simple enough to provide nonscientists with evidence of the overall success or failure of the act, yet rigorous enough to stand up to the scrutiny of people who make their living analyzing water quality data trends.

This book takes a unique, three-pronged approach for answering the *prima facie* question: Has the Clean Water Act's regulation of wastewater treatment processes at POTWs been a success? Or posed more directly: How have the nation's water quality conditions changed since implementation of the 1972 CWA's mandate for secondary treatment as the minimum acceptable technology for POTWs? The three-pronged approach described below was developed so that each study phase could provide cumulative support regarding the success, or failure, of the CWA-mandated POTW upgrades to at least secondary treatment. Using the analogy of a three-legged stool, the study authors believed that each leg must contribute support to the premise of CWA success. If one or more legs fail in this objective, the stool will, in the words of Mark Twain, be "limber" and unable to "stand up."

THE FIRST LEG: AN EXAMINATION OF BOD LOADINGS BEFORE AND AFTER THE CWA (CHAPTER 2)

Biochemical oxygen demand (BOD) is a measurement that allows scientists to compare the relative polluting strength of different organic substances. The widest application of the BOD test, however, is for measuring wasteload concentrations to (influent load) and discharged from (effluent load) POTWs and other facilities and evaluating the BOD-removal efficiency of these treatment systems. From 1970 to 1999, $77.2 billion (as current year dollars) in federal grants and contributions through USEPA's Construction Grants and CWSRF programs was distributed to municipalities and states to upgrade POTWs and, among other objectives, to increase their BOD-removal efficiency. Did this investment pay off in terms of decreasing BOD effluent loadings to the nation's waterways? The purpose of the first leg of this study is to examine nationwide trends in both influent and effluent BOD loadings before and after the CWA.

Chapter 2 begins with some background discussions to help the reader better understand the significance of the wastewater component of the urban water cycle and the pivotal role the 1972 CWA played in establishing the national policy requiring secondary treatment as the minimum acceptable technology for this component. Specifically, Sections A and B trace some historical consequences of not incorporating the wastewater treatment component of the urban water cycle. Beginning with ancient Athenians and moving through time, societies around the world suffered the results of releasing raw or inadequately treated sewage into waterways, including outbreaks of disease and the destruction of fragile aquatic ecosystems. Sparked by the Lawrence (Massachusetts) Experiment Station's discovery of the trickling filter method in 1892 and the development of the BOD test in the 1920s, many states subsequently adopted water quality standards and encouraged the use of secondary treatment for the purpose of protecting their waterways and water supply and water resource users. Unfortunately, rapidly growing urban populations and uneven applications of wastewater treatment funding and technology caused conditions to deteriorate in many highly populated watersheds in the first two-thirds of the twentieth century. Section C of Chapter 2 traces the evolution of the federal government's role in water pollution control during this time period. Key legislation is highlighted to document its movement from passive advisor through to the passage of the 1972 CWA, the decisive legislation that transferred authority for directing and defining water pollution control policy and initiatives from the states to the USEPA. Post-1972 legislation and regulations continue to refine water pollution control goals and objectives and authorize the funding and policies necessary to meet them.

Twenty-five years after the passage of the CWA, the number of people served by POTWs has increased from about 140 million in 1968 to 189.7 million in 1996. In spite of this population increase (and corresponding increases in the amount of BOD flowing into these facilities), has there been a significant decline in BOD loading to the nation's waterways? Section D examines trends in influent and effluent BOD loading from 1940 to 1996, based on population served and BOD removal rates associated with various treatment levels.

Section E helps put POTW effluent BOD loading into national perspective by examining rates and spatial distribution of BOD loadings associated with other point and nonpoint sources of BOD in addition to municipal loadings. Using USEPA's National Water Pollution Control Assessment Model (NWPCAM) (Bondelid et al., 2000), loading estimates were derived for urban and rural runoff, combined sewer overflows, and industrial wastewater discharges, in addition to municipal discharges. Comparison of these sources at a national level provides insight on how total BOD loading is distributed among sources in various regions of the United States. Section F presents a discussion of the investment costs associated with water pollution control infrastructure over the time period 1970 to 1999 and summarizes projections of future wastewater infrastructure needs into the twenty-first century.

THE SECOND LEG: AN EXAMINATION OF "WORST-CASE" DO IN WATERWAYS BELOW POINT SOURCES BEFORE AND AFTER THE CWA (CHAPTER 3)

Professionals in the water resource field use many different parameters to characterize water quality. If one's interest centers on protecting fish and other aquatic organisms, however, DO concentration is a key parameter on which to focus. This interest is articulated in section 101 of Title I of the Clean Water Act in the form of a national goal for fishable waters. Fish kills are the most visible symptom of critically low levels of DO. Some species of fish can handle low levels of oxygen better than others. Cold-water fish (salmon, trout) require higher DO concentrations than warm-water fish (bass, catfish). Early life stages usually require higher DO concentrations than adult stages. Table 1-1 presents USEPA's water quality criteria for DO for cold-water and warm-water biota for four temporal categories. The reader should note that a DO concentration of 5 mg/L has been adopted in this study as a general benchmark threshold for defining desirable versus undesirable levels of DO (i.e., the minimum concentration to be achieved at all times for early life stages of warm-water biota).

The concentration of DO in a stream fluctuates according to many natural factors, including water temperature, respiration by algae and other plants, nitrification by autotrophic nitrifying bacteria, and atmospheric reaeration. By far the biggest factor in determining DO levels in most waterbodies receiving wastewater discharges, however, is the amount of organic matter being decomposed by bacteria and fungi. Twenty-five years after the passage of the CWA, the nation's investment in upgrading POTWs to secondary or greater levels of treatment resulted in significant reductions in BOD loadings. Has the CWA's push to reduce BOD loading resulted in improved water quality in the nation's waterways?

TABLE 1-1 USEPA Water Quality Criteria for Dissolved Oxygen Concentration

	Cold-Water Biota		Warm-Water Biota	
	Early Life Stages[a,b]	Other Life Stages	Early Life Stages[b]	Other Life Stages
30-day mean	NA[c]	6.5	NA	5.5
7-day mean	9.5 (6.5)	NA	6.0	NA
7-day mean minimum	NA	5.0	NA	4.0
1-day minimum[d]	8.0 (5.0)	4.0	5.0	3.0

[a]Recommended water column concentrations to achieve the required intergravel dissolved oxygen concentrations shown in parentheses. The figures in parentheses apply to species that have early life stages exposed directly to the water column.
[b]Includes all embryonic and larval stages and all juvenile forms to 30 days following hatching.
[c]NA = not applicable.
[d]All minima should be considered instantaneous concentrations to be achieved at all times. Further restrictions apply for highly manipulative discharges.

The challenge in evaluating the effectiveness of point source BOD loading reductions is the need to isolate their impacts on downstream DO from impacts caused by urban stormwater runoff and rural nonpoint sources and the natural seasonal influences of streamflow and water temperature. An innovative approach was developed to reduce these confounding factors and screen for water quality station records that inherently contain a "signal" linking point source discharges with downstream DO. It includes the following steps:

- Developing before- and after-CWA data sets of DO summary statistics derived from monitoring stations that were screened for worst-case conditions (i.e., conditions that inherently contain the sharpest signal).
- Assigning the worst-case DO summary statistic to each station for each before- and after-CWA time period and then aggregating station data at sequentially larger spatial scales.
- Conducting a "paired" analysis of spatial units that have both a before- and an after-CWA worst-case DO summary statistic and then documenting the direction (improvement or degradation) and magnitude of the change.
- Assessing how the point source discharge/downstream worst-case DO signal changes over progressively larger spatial scales.

The hierarchy of spatial scale plays an especially important role in this second leg of the three-legged stool approach for examining water quality conditions before and after the CWA. Three spatial scales are addressed in this portion of the study: reach, catalog unit, and major river basin.

Reaches are segments of streams, rivers, lakes, estuaries, and coastlines identified in USEPA's Reach File Version 1 (RF1) and Reach File Version 3 (RF3). In this system, a reach is defined by the confluence of a tributary upstream and a tributary downstream. Reaches in RF1 average about 10 miles in length and have a mean drainage area of 115 square miles. Created in 1982, RF1 contains information for 64,902 reaches in the 48 contiguous states, covering 632,552 miles of streams. Figure 1-2 is a map of the stream reach network in the Chesapeake Bay drainage area.

An individual reach in the RF1 system is identified by an 11-digit number. This number carries much spatial information. It identifies not only the reach itself, but also the hierarchy of watersheds to which the reach belongs. The first eight digits of the identification number are the Hydrologic Unit Code (HUC). Originally developed by the U.S. Geological Survey (USGS), the HUC number identifies four scales of watershed hierarchy. The highest scale, coded in the first two digits of the identification number, is the hydrologic region (commonly referred to as a major river basin). Hydrologic regions represent the largest river basins in the country (e.g., the Missouri River Basin and the Tennessee River Basin). Subregions are identified by the next

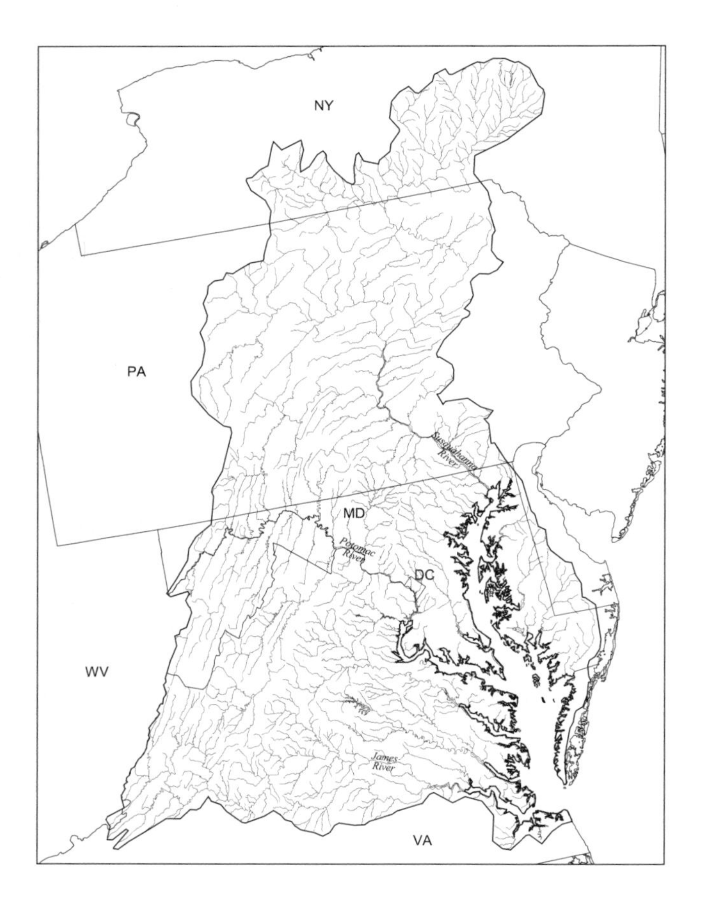

Figure 1-2 Reach File Version 1 stream reach network in the Chesapeake Bay drainage area.

two numbers. These are followed by the accounting unit and the cataloging unit, the smallest scale in the hierarchy. Figures 1-3 and 1-4 display the 18 hydrologic regions and the 2,111 cataloging units in the contiguous 48 states.

Developers of RF1 extended the eight-digit HUC code by three digits for the purpose of identifying the reaches within the cataloging unit. Table 1-2 is an example of the RF1 identification codes for a reach of the Upper Mississippi River near Hastings, Minnesota. This 33.1-mile reach is defined by the confluence of the Minnesota River (upstream) and the St. Croix River (downstream).

Many engineering studies have documented the impact of BOD loading on the DO budget in reaches immediately below municipal outfalls. Consequently, one would expect to find a sharp signal linking point source discharges with worst-case DO in those reaches. The key aspect of this investigation, therefore, was to see how the signal changed (or if it could be detected at all) as one aggregated worst-case DO data at increasingly larger spatial scales and then compared summary statistics associated with time periods before and after the CWA. Detection of a statistically significant signal at the catalog unit and major river basin scales would provide evidence that the CWA mandates to upgrade to secondary treatment and greater levels of wastewater treatment yielded broad, as well as localized, benefits.

Figure 1-5 illustrates signal and noise relationships over the range of spatial scales (reach, catalog unit, and major river basin), using the Upper Mississippi River near Hastings, Minnesota, as an example. The line graphs on the left side of the figure display DO data collected at monitoring stations from 1953 to 1997 aggregated by spatial unit. The bar graphs on the right side of the figure compare worst-case DO (mean tenth percentile) for designated time periods before and after the CWA and are produced as the final step of the comparison analysis process described in Chapter 3. The

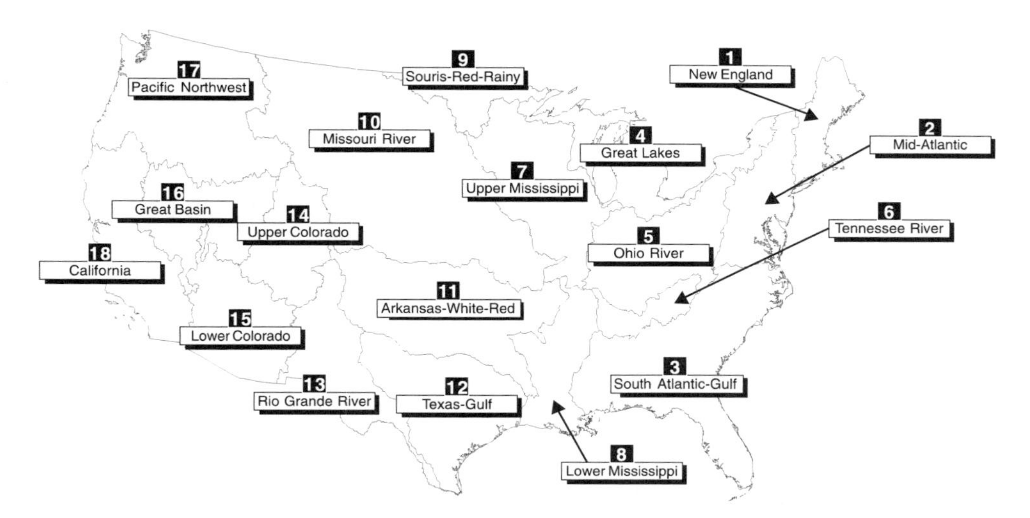

Figure 1-3 The 18 major river basins (hydrologic regions) of the 48 contiguous states.

Figure 1-4 The 2,111 hydrologic catalog units of the 48 contiguous states.

summary statistics they present are derived from station data that have been selected, aggregated, and spatially assessed so that they might have the best chance of inherently containing a "signal" linking point source discharges with downstream DO.

Examining the line graphs in Figure 1-5, one can see that each broader spatial scale aggregation of station data yields a "noisier" data pattern. The bar chart for the reach scale (the finest scale) displays the greatest improvement in worst-case DO, in-

TABLE 1-2 Station and Reach Identification Codes: Reach File Version 1 (RF1)

Agency ID:	21MINN
Station ID:	MSU-815-BB15E58
Station location:	Mississippi River at Lock & Dam No. 2 at Hastings
Major river basin name:	Upper Mississippi River
Major river basin ID:	07
Subbasin ID:	0701
Accounting unit ID:	070102
Catalog unit ID:	07010206
Reach ID:	07010206001
Station milepoint on reach:	UM 815.5
Reach length (miles):	33.1
Upstream milepoint of reach (Minnesota River):	UM 844.7
Downstream milepoint of reach (St. Croix River):	UM 811.6

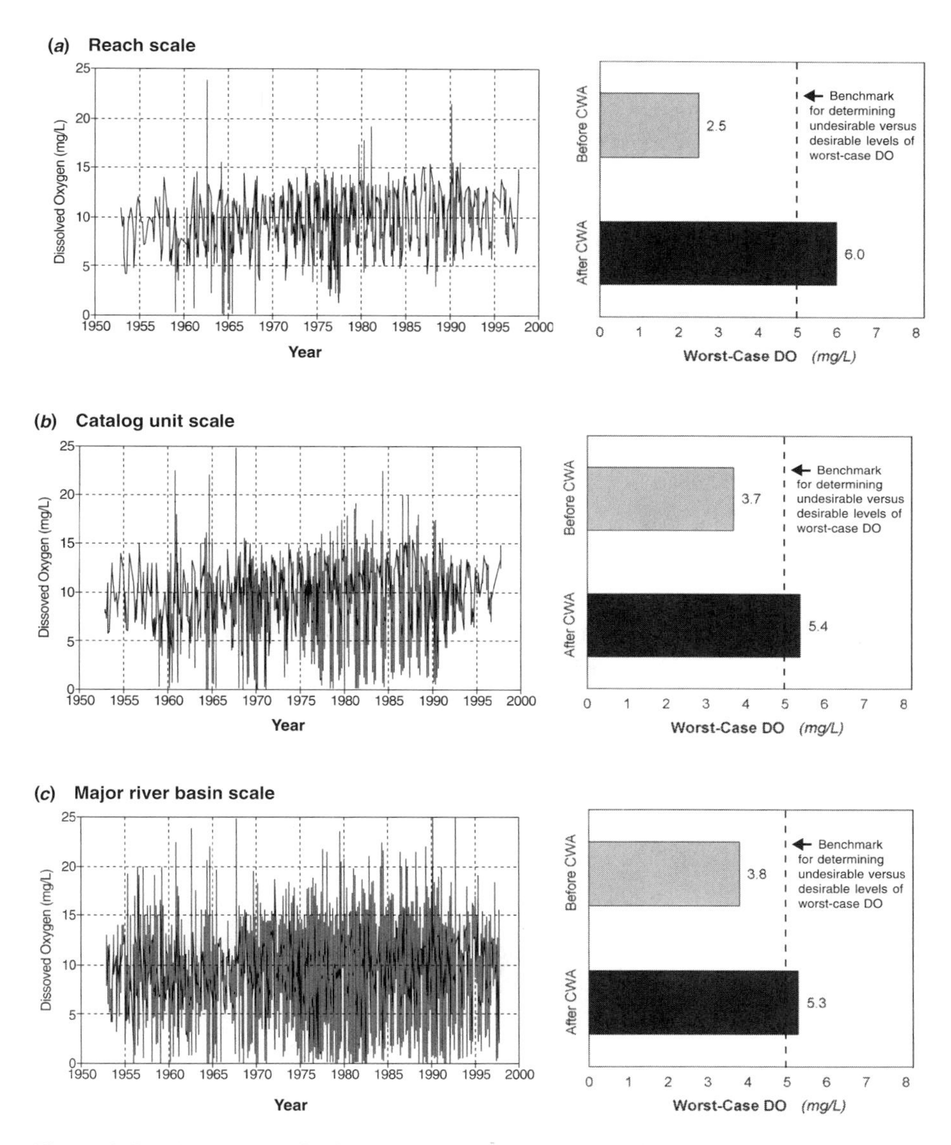

Figure 1-5 Line graphs of DO observations for the Upper Mississippi River from 1953 to 1997 and bar charts of worst-case DO before (1961–1965) and after (1986–1990) the CWA for (*a*) reach scale, (*b*) catalog unit scale, and (*c*) major river basin scale. *Source:* USEPA STORET. (*a*) For Rf1 reach 07010206001 (UM 811.6-844.7). (*b*) For catalog unit 07010206 (UM 811.6-879.8). (*c*) For major river basin (07).

creasing 3.5 mg/L from before to after the CWA. At the broader scales, an improvement is detected, but it is not as large (a before and after difference of 1.7 mg/L at the catalog unit scale and 1.5 mg/L at the major river basin scale). This is because the larger spatial units contain stations both near and far from point source outfalls. In spite of the unavoidable introduction of data noise, however, the signal linking point source discharge to downstream DO is still detectable at the broader scales using the data mining and statistical methodology developed by the study authors. Readers should note that, in this example, the worst-case DO concentration was below the benchmark threshold of 5.0 mg/L at all three scales before the CWA and above the threshold at all three scales after the CWA.

Section A of Chapter 3 provides background on the relationship between BOD loading and stream water quality and discusses the two key physical conditions (high temperature and low flow) that create "worst-case" conditions for DO. Section B describes the development and application of a set of screening rules to select, aggregate, and spatially assess before- and after-CWA worst-case DO data drawn from USEPA's STORET database. Section C presents the results of the comparison analysis of worst-case DO from before and after the CWA for reach, catalog unit, and major river basin scales.

THE THIRD LEG: CASE STUDY ASSESSMENTS OF WATER QUALITY (CHAPTERS 4 THROUGH 13)

The second leg of this study focused on the use of large national databases and statistical methods to examine temporal and spatial trends in DO conditions nationwide. However, the uniqueness of each waterway and the activities surrounding it requires an investigation to go beyond STORET to identify, quantify, and document in detail the specific actions that have resulted in water quality improvements and associated benefits to water resource users.

In the third and final leg of this study, nine urban waterways have been selected to characterize changes in population, point source effluent loading, water quality, and environmental resources before and after the CWA:

- Connecticut River
- Hudson-Raritan estuary
- Delaware estuary
- Potomac estuary
- James estuary
- Chattahoochee River
- Ohio River
- Upper Mississippi River
- Willamette River

These waterways were selected to represent heavily urbanized areas with historically documented water pollution problems. A variety of data sources, including the scientific literature, USEPA's national water quality database (STORET), and federal, state, and local agency reports, were used to characterize long-term trends in population, point source effluent loading rates, ambient water quality, environmental resources, and recreational uses. Additional information was obtained from validated water quality models for the Delaware, Potomac, and James estuaries and Upper Mississippi River case studies to quantify the water quality improvements achieved by upgrading municipal facilities to secondary and better levels of treatment as mandated by the 1972 CWA.

Chapter 4 presents an overview of the case study assessment approach and provides background on previous efforts that have used case studies to examine long-term changes in water quality conditions in the United States. Chapter 4 also summarizes the overall findings for the nine urban waterways; detailed assessments are provided for each case study in Chapters 5 through 13.

THE TARGET AUDIENCE FOR THIS BOOK

This study was designed with two broad groups in mind. The primary audiences are the technical scientists and engineers who try to understand and evaluate cause–effect relationships of pollutants, their sources, and the fate of these pollutants in receiving waters. Understanding these relationships is crucial for developing appropriate (cost-effective and environmentally protective) pollution control measures. This same audience is often tasked with the responsibility of developing and carrying out large-scale monitoring programs whose purpose is to track the performance of various policy decisions related to pollution source control.

The secondary audience is Congress, regulatory/policy professionals, and the informed public, who have often questioned the effectiveness of major pollution control programs directed at the national level. It may benefit future decision-makers to know if major public works programs (i.e., the CWA Construction Grants and CWSRF programs) accomplished what they were designed to do—namely reduce effluent BOD loads from municipal and industrial sources and improve dissolved oxygen in many previously degraded waterways of the nation. These same groups also need to understand that water pollution control efforts are neverending. The 1972 CWA did not "solve" the problem. In fact, waste materials are generated continuously, and effluent removal efficiencies must increase in the future to compensate for economic growth and population growth. Planning for operation and maintenance (O&M) expenditures, as well as for capital expenditures for replacement of obsolete facilities and upgrades to maintain adequate levels/efficiency of wastewater removal, is an ongoing requirement. A projection analysis presented in Chapter 2 demonstrates that many of the gains in national water quality improvements may be lost if future wastewater infrastructure investments and capacity do not keep pace with expected urban population growth.

REFERENCES

Adler, R. W., J. C. Landman, and D. M. Cameron. 1993. The Clean Water Act: 20 years later. Island Press, Washington, DC.

ASIWPCA. 1984. America's clean water: The states' evaluation of progress 1972–1982. Executive Summary and Technical Appendix. Association of State and Interstate Water Pollution Control Administrators, Washington, DC.

Bondelid, T., S. Unger, and A. Stoddard. 2000. National water pollution control assessment model (NWPCAM) Version 1.1. Final report prepared by Research Triangle Institute, Research Triangle Park, NC for U.S. Environmental Protection Agency, Office of Policy, Economics and Innovation, Washington, DC, November, RTI Project Number 92U-7640-031.

GAO. 1978. Secondary treatment of municipal wastewater in the St. Louis Area: Minimal impact expected. GAO/CED-78-76. U.S. General Accounting Office, Program, Evaluation and Methodology Division, Washington, DC.

GAO. 1981. Better monitoring techniques are needed to assess the quality of rivers and streams. Vol. 1. Report to Congress. GAO/CED-81-30. U.S. General Accounting Office, Program, Evaluation and Methodology Division, Washington, DC.

GAO. 1986a. The nation's water: Key unanswered questions about the quality of rivers and streams. Vol. 1. GAO/PMED-86-6. U.S. General Accounting Office, Program, Evaluation and Methodology Division, Washington, DC.

GAO. 1986b. Water quality: An evaluation method for the Construction Grants Program—methodology. Vol. 1. Report to the Administrator. GAO/PMED-87-4A. U.S. General Accounting Office, Program, Evaluation and Methodology Division, Washington, DC.

GAO. 1986c. Water quality: An evaluation method for the construction grants program—case studies. Vol. 1. Report to the Administrator. GAO/PMED-87-4B. U.S. General Accounting Office, Program, Evaluation and Methodology Division, Washington, DC.

Knopman, D. S. and R. A. Smith. 1993. Twenty years of the Clean Water Act: Has U.S. water quality improved? Environment 35(1):17–41.

Leo, W. M., R. V. Thomann, and T. W. Gallagher. 1984. Before and after case studies: Comparisons of water quality following municipal treatment plant improvements. EPA430/9-007. Technical report prepared by HydroQual, Inc., for U.S. Environmental Protection Agency, Office of Water Programs, Washington, DC.

Mearns, A. 1995. Ready . . . shoot . . . aim! The future of water. Editorial. WEF Water Env. Res. 67(7):1019.

Patrick, R., F. Douglass, D. M. Palavage, and P. M. Stewart. 1992. Surface water quality: Have the laws been successful? Princeton University Press, Princeton, NJ.

Smith, R. A., R. B. Alexander, and M. G. Wolman. 1987a. Analysis and interpretation of water quality trends in major U.S. rivers, 1974–81. Water-Supply Paper 2307. U.S. Geological Survey, Reston, VA.

Smith, R. A., R. B. Alexander, and M. G. Wolman. 1987b. Water quality trends in the nation's rivers. Science 235 (27 March): 1607–1615.

USEPA. 1988. POTW's and water quality: In search of the big picture: A status report on EPA's ability to address several questions of ongoing importance to the nation's municipal pollution control program. U.S. Environmental Protection Agency, Office of Water, Office of Municipal Pollution Control, Washington, DC.

USEPA. 1995a. National water quality inventory: 1994 report to Congress. EPA841-R-95-005. U.S. Environmental Protection Agency, Office of Water, Washington, DC.

USEPA. 1995b. National water quality inventory: 1994 report to Congress. Appendices. EPA841-R-95-006. U.S. Environmental Protection Agency, Office of Water, Washington, DC.

USEPA. 1997a. The Clean Water Act: A snapshot of progress in protecting America's waters. Vice President Al Gore's remarks on the 25th Anniversary of the CWA. U.S. Environmental Protection Agency, Washington, DC.

USEPA. 1997b. 1996 Clean Water Needs Survey: Conveyance, treatment, and control of municipal wastewater, combined sewer overflows and stormwater runoff. Summaries of technical data. U.S. Environmental Protection Agency, Office of Water Program Operations, Washington, DC.

Vogan, C. R. 1996. Pollution abatement and control expenditures, 1972–94. Survey of current business. Vol. 76, No. 9, pp. 48–67. U.S. Department of Commerce, Bureau of Economic Analysis.

WEF. 1997. Profiles in water quality: Clear success, continued challenge. Water Environment Federation, Alexandria, VA.

CHAPTER 2

An Examination of BOD Loadings Before and After the CWA

Chapter 1 introduced the "three-legged stool" approach to assess the success of the CWA's mandate for POTW upgrades to secondary and greater than secondary wastewater treatment. The premise is that each "leg" of the approach must provide cumulative support for the stool to stand up firmly and success to be declared. Chapter 2 presents the results of the first leg. Specifically, this chapter focuses on whether there was a significant reduction in the discharge of oxygen-demanding materials from POTWs to the nation's waterways after implementation of the 1972 CWA.

To help put this analysis into perspective, Chapter 2 begins with a background discussion of the historical consequences of ignoring the wastewater treatment component of the urban water cycle on the aquatic ecosystem (Section A), and then explains how scientists and engineers eventually harnessed the power of decomposers and developed the process now known as secondary treatment (Section B). Section C traces the legislative and regulatory history of the federal role in water pollution control and how the 1972 CWA accelerated the national trend of upgrading POTWs to at least secondary treatment. Section D presents national trends in influent BOD loading (BOD entering POTWs) and effluent BOD loading (BOD discharged from POTWs into surface waters) for select years between 1940 and 1996, as well as effluent loading projections into the twenty-first century.

During the mid-1990s (ca. 1995), pollutant loading from municipal wastewater treatment facilities accounted for only about one-fifth of the estimated total national point and nonpoint source load of BOD discharged to surface waters. Section E presents comparative estimates of the remaining four-fifths of the total national load accounted for by industrial wastewater dischargers, combined sewer overflows (CSOs), and nonpoint (rural and urban) sources. Section F examines the national public and private investment costs associated with water pollution control. Section G provides a summary, conclusions, and a perspective on future trends for municipal wastewater loads.

A. HISTORY OF WATER SUPPLY AND ITS EFFECTS ON PUBLIC HEALTH

The urban water cycle can be divided into a water supply side and a wastewater disposal side (see Figure 1-1). The basic technological framework for the water supply side began as far back as 5,000 years ago when people from the Nippur of Sumeria, a region of the Middle East, built a centralized system to deliver water into populated areas (Viessman and Hammer, 1985). The Minoans at Knossos, some 1,000 years later, improved on the concept with the installation of a system of cisterns and stone aqueducts designed to provide a continuous flow of water from the surrounding hills to dwellings in the central city. Basic concepts and instructions related to purity of water, cleanliness, and public sanitation are also recorded in the books of Leviticus and Deuteronomy (23:12–13) in the Old Testament. Talmudic public sanitation laws were enacted in Palestine to protect water quality in the centuries before and after the early Christian era ca. 200 B.C. to 400 A.D. (Barzilay et al., 1999).

The ancient Athenians were some of the first people to develop the wastewater disposal side of the urban water cycle. The Greeks moved sanitary wastes away from their central city through a system of ditches to a rural collection basin. The wastewater was then channeled through brick-lined conduits for disposal onto orchards and agricultural fields. In the ancient world, though, the Roman Empire attained the highest pinnacle for developing the knowledge and technology to select the best water supplies and to construct far-reaching networks of aqueducts to bring water supplies to Rome for distribution through pipes to wealthy homes and public fountains. The Romans also built large-scale public sanitation projects for collecting and controlling sewage and stormwater drainage. The great Roman sewer *Cloaca Maxima* still drains the Forum in Rome today after 2,000 years of operation.

In expanding their empire throughout North Africa and Europe, the Romans introduced the technologies needed to develop water supplies and to construct aqueducts and urban drainage systems to promote rudimentary standards of public sanitation. With the collapse of the Roman Empire, however, the public sanitation infrastructure was neglected, and the technology was lost and forgotten for a thousand years as the "Dark Ages" descended on the western world. Filth, garbage, excrement in the streets, polluted water sources, disease, plague, and high mortality rates were common consequences of the dismal public sanitary conditions that persisted well into the nineteenth century (Barzilay et al., 1999).

Throughout history, two components of the urban water cycle were often absent: wastewater treatment and the transport of treated wastewater for discharge back to natural waterbodies. For towns situated near coastal areas, estuaries, or large rivers, short-circuiting the cycle caused no immediate consequences, because these waterbodies had some capacity to assimilate raw sewage without causing water pollution problems. For many inland communities, however, water pollution problems were more acute. As populations increased, even coastal towns were forced to reckon with the consequences of ignoring the wastewater treatment component of the urban water cycle (see Rowland and Heid, 1976).

Much of the blame for incomplete urban water cycles up until the middle of the nineteenth century can be traced to a general ignorance about the consequences of allowing untreated wastewater to flow into surface waters used for drinking water downstream. As the relationship between this practice and its effects on public health became better understood, however, a community's refusal to adopt effective wastewater treatment in its cycle was more often based in politics and economics, rather than a lack of technological knowledge (see Rowland and Heid, 1976). No matter the reason, bypassing the wastewater treatment side of the urban water cycle affected both water supply and water resource users.

Impacts on Water Supply Users and "The Great Sanitary Awakening"

The introduction of household piped water in the mid-nineteenth century was the key technological development that cemented the two sides of the urban water cycle—water supply and wastewater disposal. Unfortunately, although piped water supply systems gave urban dwellers more convenient access to water, people were also held hostage to the water supply source chosen by the local water company. For many city dwellers, drinking piped water became hazardous to one's health, as massive epidemics of waterborne diseases, such as cholera and typhoid fever, broke out in many cities in Great Britain and the United States (Table 2-1).

Dr. John Snow, physician to England's Queen Victoria, was one of the first to scientifically link waterborne diseases to contaminated source water supplies. Examining records of some 14,600 Londoners who had died in an 1854 cholera epidemic, Snow found that people who had received water from an intake downstream of London's sewage outlets in the lower Thames River had a much higher death rate (8.5 times higher) than those receiving Thames River water from an intake upstream of the sewage discharges (Snow, 1936). The threat of contaminated water sources did little, however, to quell the construction boom of new water supply systems in the second half of the nineteenth century, especially in the United States. In 1850, there were about 83 water systems in the United States. By 1870, the count had risen to 243 systems (Fuhrman, 1984).

Like Londoners, American city dwellers with piped water faced an increased risk of waterborne diseases. Beginning in 1805, the New York City Council had the authority and responsibility for sanitary conditions in the city. Despite this early recognition of governmental responsibility for public health, epidemics of typhoid fever broke out in 1819, 1822, 1823, and 1832. Cholera ravaged workers on the Erie Canal (Rowland and Heid, 1976). Between 1832 and 1896, cities in North America and Europe suffered four devastating outbreaks of cholera that were spread by polluted urban water supply systems (Garrett, 1994). Cholera epidemics in New York City in 1832 and 1849 claimed 3,500 and 5,000 lives, respectively. In 1891, typhoid fever caused the deaths of 2,000 people in Chicago (Fair et al., 1971). Hundreds more succumbed to typhoid in Atlanta and Pittsburgh in the 20-year period between 1890 and 1910 (Bulloch, 1989). The importance of an unpolluted source water for public drinking water was clearly shown in the earliest public health studies of waterborne diseases and drinking water supplies. Typhoid death rates in 61 cities of the United States during 1902–1906, for example, ranged from a high of 120 per 100,000 for a run-of-river supply for Pittsburgh, Pennsylvania, to a low of 15 per 100,000 for the upland watershed supply of New York City (Okun, 1996).

This trend of epidemics would have certainly continued for a few more decades if not for the discovery of a new purification technology: chlorination of drinking water. As a disinfecting agent, chlorine gained widespread use in the years 1908–1911, soon bringing typhoid fever and cholera outbreaks under control in virtually all com-

TABLE 2-1 Pathogens and Their Associated Diseases

Pathogen	Disease	Effects
	Bacteria	
Escherichia coli	Gastroenteritis	Vomiting, diarrhea, death in susceptible populations
Legionella pneumophila	Legionellosis	Acute respiratory illness
Leptospira sp.	Leptospirosis	Jaundice, fever (Weil's disease)
Salmonella typhi	Typhoid fever	High fever, diarrhea, ulceration of the small intestine
Salmonella sp.	Salmonellosis	Diarrhea, dehydration
Shigella sp.	Shigellosis	Bacillary dysentery
Vibrio cholerae	Cholera	Heavy diarrhea, dehydration
Yersinia enterolitica	Versinosis	Diarrhea
	Protozoa	
Balantidium coli	Balantidiasis	Diarrhea, dysentery
Cryptosporidium sp.	Cryptosporidiosis	Diarrhea
Entamoeba histolytica	Amedbiasis	Diarrhea with bleeding, abscesses on liver and small intestine
Giardia lamblia	Giardiasis	Mild to severe diarrhea, nausea, indigestion
Naegleria fowleri	Amoebic meningo-encephalitis	Fatal disease; brain inflammation
	Viruses	
Adenovirus (31 types)	Respiratory disease	
Enteroviruses (67 types)	Gastroenteritis	Heart anomalies, meningitis
Hepatitis A	Infectious hepatitis	Jaundice, fever
Norwalk agent	Gastroenteritis	Vomiting, diarrhea
Reovirus	Gastroenteritis	Vomiting, diarrhea
Rotavirus	Gastroenteritis	Vomiting, diarrhea

Source: Adapted from Metcalf and Eddy, 1991.

munities that adopted chlorination. Detailed mortality records and public water supply records compiled by the Commonwealth of Massachusetts, for example, clearly illustrate the link between the introduction of filtration and disinfection of public water supplies and the sharp reduction in typhoid fever deaths (Figure 2-1) from a peak of 125 per 100,000 in 1860 to less than 5 per 100,000 by 1920 and essentially zero from 1940 to the present time (Fair et al., 1971; Higgins, 1998; USCB, 1975).

Influenced by the Enlightenment and democratic movements of the late eighteenth century in Britain, France, and the new United States, the concept that a government had the moral and ethical responsibility to protect the general welfare of its citizens, including public health, arose in Britain and the United States during the first half of the nineteenth century. Motivated by the bleak urban conditions chronicled by Charles

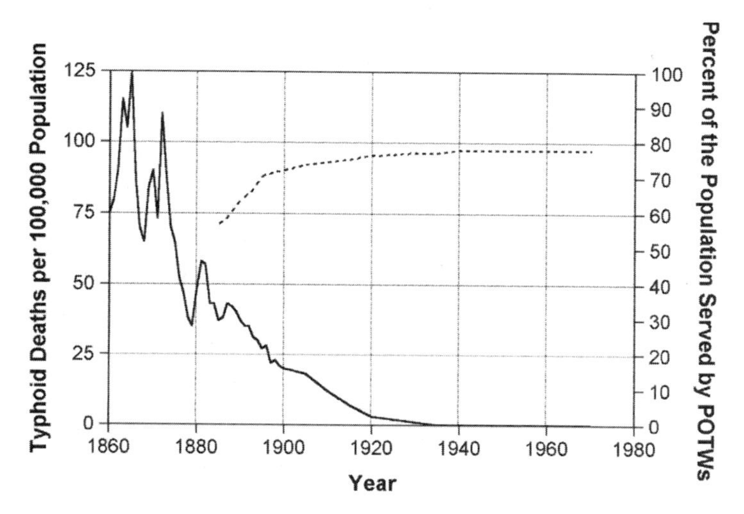

Figure 2-1 Comparison of the death rate due to typhoid and the percentage of population served by public water suppliers in Massachusetts from 1860 to 1970. *Source:* Elements of water supply and wastewater disposal, G. M. Fair, J. C. Geyer, and D. A. Okun, Copyright © 1971, Reprinted by permission of John Wiley & Sons, Inc.; USCB, 1975.

Dickens, Chadwick's (1842) Report on the Sanitary Condition of the Labouring Population of Great Britain marked the beginning of the "Great Sanitary Awakening" (Okun, 1996). Chadwick's report directly influenced passage of Great Britain's Public Health Act of 1848 and its formation of the General Board of Health, and, in the United States, the creation of the Massachusetts State Board of Health in 1869 (Okun, 1996) and the New York State Board of Health in 1880 (Rowland and Heid, 1976).

The technological impacts of the "Great Sanitary Awakening" on the origins of drinking water treatment and water pollution control systems are well documented in the records of a series of international sanitary conferences held from 1851 through 1938. The conferences addressed scientific issues related to public health, the environment, and the need to control diseases spread by contaminated food and water. The conferences highlighted serious public health and environmental issues that have since evolved as the foundation of the numerous state, local, federal, and international environmental laws and programs enacted in the latter half of the twentieth century (Howard Jones, 1975).

Impacts on Water Resources Users

Sewer is an Old English word meaning "seaward." As the name suggests, from the 1500s through mid-1800s, London's sewers were nothing more than open ditches draining wastewater seaward via the Thames River. The year 1858, also known as the year of "The Great Stink," brought matters to a head. That summer, the stench from the Thames drove people out of the city by the thousands. The windows of the Parliament building had to be draped with curtains soaked in chloride of lime. By the end of the summer session, even the most traditional members had to agree: something had to be done about wastewater.

In response, London officials abolished cesspools and made the use of water closets, drainage pipes, and centralized sewer collection systems mandatory. Over in the United States, city officials were also feeling the pressure of a populace weary of the noxious conditions associated with open sewers. In 1910, about 10% of the urban population was serviced by centralized collection systems (FWPCA, 1969). This number increased steadily in the following decades; by 1940, 70.5 million persons (53% of the population) were served by them. Unfortunately, treating drinking water with chlorine and developing efficient sewage collection systems did little to help water resources users. Raw sewage deposited into streams, lakes, and estuaries was still raw sewage, whether it was discharged through an engineered wastewater collection system or through an open ditch. Collection systems just made the dumping more efficient and complete. And though chlorine proved to be a godsend for public health, it treated only a pollution symptom, not the cause. Its success, unfortunately, tended to divert attention away from installing wastewater treatment as a means of protecting public health (Bulloch, 1989).

Several studies conducted around the turn of the twentieth century documented increasingly noxious conditions in several well-known rivers receiving untreated urban discharges. These included the Merrimack River (1908), Passaic River (1896), Chicago Ship and Sanitary Canal (1900), and Blackstone River (1890). Looking beyond the effects of water quality on supply users, scientists also began to examine the effect urban discharges were having on stream biota. Studies were conducted in places like the Sangamon River in Illinois (1929) (Eddy, 1932), the Potomac River (1913–1920), and the Shenandoah River (1947–1948) (Henderson, 1949). These and other early investigations are an invaluable starting point for assessing long-term trends in the surface water environment. At the turn of the twentieth century, public officials focused most of their attention on water supply users. The users demanded and received the two services most important to them: the delivery of clean water and the collection and removal of wastewater. Support for water resources users, on the other hand, was minimal. Generally, these users captured the attention of city leaders only when conditions reached crisis levels. Then, in most cases, the response was to deal with ways to alleviate the symptom rather than the cause of water pollution.

In Chicago, for example, officials became concerned about the increasing amount of urban water pollution flowing into their backyard water supply source, Lake Michigan. In response, they built the Chicago Drainage Canal, which diverted sewage away from the lake and directed it to the Des Plaines River, a tributary that emptied into the Mississippi River.

After the canal opened in 1900, officials in the downstream city of St. Louis fumed. They quickly initiated proceedings in the Supreme Court of the United States against the state of Illinois and the Sanitary District of Chicago. Though St. Louis eventually lost its case because the city could not prove direct harm to its water supply from its upstream neighbor, the episode underscored the fact that effective wastewater treatment was a critical component in the modern urban water cycle.

B. EVOLUTION OF WASTEWATER TREATMENT

European history as far back as 400 years ago tells of sewage being collected, dewatered, and transported as "night soil" away from population centers. In 1857, a British royal commission, in response to noxious conditions in the Thames River, directed Lord Essex to report on alternative ways to dispose of urban wastewater. Essex concluded that applying wastewater to crops would be preferable to the current practice of draining it into the river. Wastewater treatment technology has progressed tremendously since those times. Today's facilities employ a variety of sophisticated physical, chemical, and biological processes to reduce domestic and industrial wastewater to less harmful by-products.

Primary Treatment

The march toward effective wastewater treatment began in the late 1800s, when municipalities began to build facilities for the purpose of physically separating out solids and floating debris from wastewater before releasing it to a waterbody (Rowland and Heid, 1976). In many cases, this construction was promoted by city officials and entrepreneurs, who were rapidly learning that unsightly urban debris and a delightful growing phenomenon, tourists with leisure dollars to spend, did not mix. By no coincidence, one of the first of these treatment facilities was constructed in 1886 next to New York's famous Coney Island beaches. Other cities with prominent waterfront areas followed suit, and by 1909 about 10 percent of the wastewater collected by municipal sewer systems underwent some form of physical separation process, now known as primary treatment (OTA, 1987).

The practice of physically screening and settling out solids and floating debris was a critical first step in incorporating the wastewater treatment component into the urban water cycle. Even though primary treatment facilities were simple in concept, they reduced the concentrations of contaminants entering urban waterways.

Dissolved Oxygen as an Indicator of Water Quality

In 1900, the United States was primarily an agrarian society, with the majority of the population living in rural areas (Figure 2-2). In the 1920s and 1930s, a combination of population growth, the Great Depression, and the rise of urban industries with the increased employment opportunities they afforded caused the rural/urban population balance to shift in favor of cities. The increasing volumes of wastewater generated by this influx of people soon overwhelmed the primary treatment capacity of POTWs, many of which had been underdesigned from the start. Consequently, the modest water quality gains achieved in many cities by primary treatment technology were soon overwhelmed by greater volumes of sewage.

Water quality conditions grew so bad in New York Harbor that in 1906 the state legislature created the Metropolitan Sewerage Commission of New York City for the purpose of studying and dealing with the effects of municipal water pollution. Of

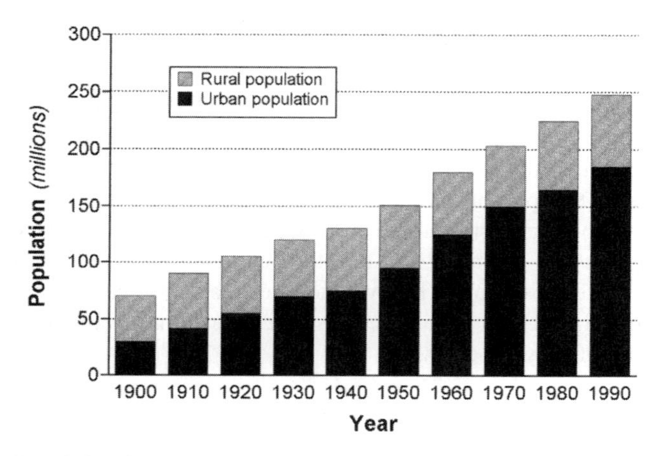

Figure 2-2 Population in the United States organized by urban and rural components from 1900 to 1990. *Source:* U.S. Census Bureau, Population Division (USCB, 2000).

immediate concern was the decline of fish and shellfish catches. The Commission concluded that a lack of oxygen in the water was the reason, and two years later, the group initiated what is now one of the longest-running water quality monitoring programs in the world. Sampling proved them right, and in 1911, the Commission set 70 percent oxygen saturation as their criterion for defining polluted waters (Cleary, 1978).

The need for adequate dissolved oxygen for aquatic respiration was well known in the late 1800s. Scientists at that time, however, were just learning about the element's critical role in the decomposition of organic matter into simple, stable end products such as carbon dioxide, water, phosphate, and nitrate (Figure 2-3). In natural waters, this process occurs when leaves, bark, dead plants and animals, and other natural carbon-based materials are eaten by bacteria, fungi, and insects. The population of these organisms rises and falls according to the amount of food available. Importantly, because the organisms are aerobic creatures, they require oxygen to breathe and carry on the task of decomposition. In addition to the carbon cycle of production and decomposition, the nitrogen cycle also influences DO through a series of sequential reactions wherein organic nitrogen compounds are hydrolyzed into ammonia and ammonia is oxidized to nitrite and nitrate (nitrification) by autotrophic nitrifying bacteria, with DO consumed as part of these sequential reactions.

The amount of oxygen water can hold at any one time is limited, however, by the saturation concentration of oxygen. The saturating amount of oxygen gas from the atmosphere that can be dissolved in water is limited by water temperature, salt content, and pressure (elevation above sea level). In a sense, then, all the aerobic aquatic life in a waterbody is in competition for that limited amount of oxygen. In natural streams, there are usually no losers because dissolved oxygen is continuously replenished from the atmosphere at about the same rate at which it is used up by aquatic organisms. A problem arises, however, when large amounts of organic material from sew-

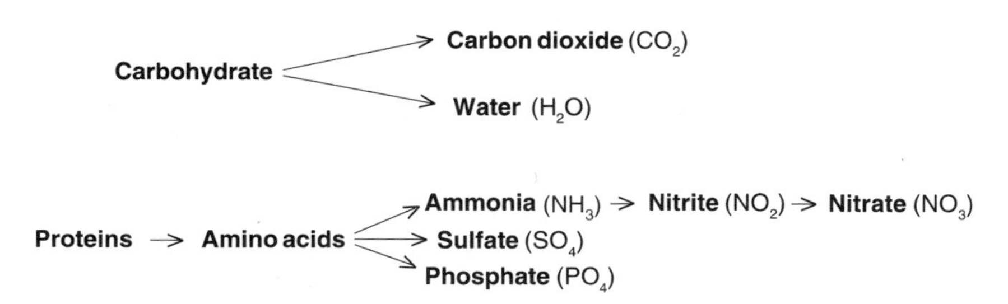

Figure 2-3 General forms of reactions involving the breakdown of carbonaceous and nitrogenous organic matter. *Source:* Dunne and Leopold, 1978.

age or other pollution sources enter the water and the decomposer population (especially bacteria) explodes in response. These organisms have the potential to lower, or even completely exhaust, oxygen in the water. When this occurs, life that depends on the presence of oxygen (aerobic) in the waterbody dies or, where possible, the biota moves on to waters with higher oxygen levels.

In the absence of oxygen in water, anaerobic bacteria further break down organic matter. These organisms obtain energy from oxygen bound into other substances such as sulfate compounds. Anaerobic processes are much slower than aerobic decomposition, however, and their end products, such as hydrogen sulfide, are usually noxious.

Secondary Treatment

Harnessing the power of decomposers to break down organic matter in wastewater is at the heart of a treatment process now known as secondary treatment. Two distinct methods of this treatment type evolved around the turn of the twentieth century. The Lawrence Experiment Station in Massachusetts pioneered the first method in 1892. Called the trickling filter method, it involves spraying wastewater onto a column of crushed stone on which a community of bacteria, fungi, protozoa, and insects resides. The organisms take in a portion of the organic matter and break it down. Some of the breakdown products, such as carbon dioxide, escape to the atmosphere. Others, like nitrate, remain in solution. Still other products are absorbed into the organisms themselves. This latter material is eventually collected in settling tanks as sludge after the organisms die or is otherwise detached from the stone.

A second method of secondary treatment was advanced around 1913 by the Lawrence Experiment Station and Edward Ardern and W. T. Lockett (1914) in England. Known as activated sludge treatment, it follows the same principles as the trickling filter, but instead of cultivating decomposers on the surface of rocks, organisms are simply suspended in a tank by a continuous flow of wastewater. Both methods of secondary treatment result in discharges with substantially less organic matter than is produced by primary treatment. City officials having problems with litigious neighbors downstream were especially eager to adopt this new technology into their urban water cycles. One of the first trickling filter facilities in the nation was constructed in the city of Gloversville, New York, in 1907. The motivation was not so much citizen demands in Gloversville for a cleaner river as it was the need to respond to a riparian rights suit filed by the downstream city of Johnstown. Chicago officials also grew tired of their ongoing battle with St. Louis, and in 1916, they constructed the first activated sludge treatment plant in the nation (Metcalf and Eddy, 1991).

Officials in most other U.S. cities, however, did not have neighbors like Johnstown or St. Louis forcing them to upgrade their wastewater treatment capabilities. Consequently, they were content to embrace a theme reflected in a leading textbook of the time, *Sewage Disposal.* In this text, Kinnicutt, Winslow, and Pratt (1913) argued that under certain circumstances, dilution of raw sewage by disposal into lakes, rivers, and tidal waters, rather than treatment of sewage, was an economical and proper method of purification.

Biochemical Oxygen Demand (BOD) as a Measure of Organic Wasteload Strength

One reason communities were slow to adopt secondary treatment into their urban water cycle was perception. There was no way to articulate the link between the organic wastes in wastewater and DO levels in natural waters. In the 1920s, these relationships became clearer with the development of an indicator called the biochemical oxygen demand (BOD). Performed in a laboratory, the BOD test measures the molecular oxygen used during a specific incubation period for the biochemical degradation of organic material, the oxidation of ammonia by nitrification, and the oxygen used to oxidize inorganic chemical compounds such as sulfides and ferrous iron.

Historically, the BOD was determined using an incubation period of 5 days at 20 degrees Celsius (C). For domestic sewage and many industrial wastes, about 70 to 80% of the total BOD is decomposed within the first five days at this temperature (Metcalf and Eddy, 1991). Because of the incubation period, BOD_5 has been adopted as the shorthand notation for this measurement in the literature. Expressed as a concentration, the BOD_5 measurement allows scientists to compare the relative pollution "strength" of different wastewaters and natural waters. The widest application of the BOD_5 test, however, is for measuring the strength and rates of wastewater loadings to and from POTWs and evaluating the BOD_5 removal efficiency of the treatment system.

Because of widespread problems with oxygen depletion in many urban rivers, several states, especially those in the more populated Northeast, Midwest, and far West, took a leadership role in the 1930s to encourage municipalities to upgrade from primary to secondary treatment. By 1950, 3,529 facilities, or about one-third of the 11,784 municipal treatment plants existing at that time, provided secondary treatment for 32 million people. At the same time, however, 35 million people were still connected to systems that discharged raw sewage, and 25 million people were provided only primary treatment (USPHS, 1951). Increasing the number of facilities that provided at least secondary treatment became a national issue as the technology was seen as a solution to the pervasive problem of low levels of DO.

C. THE FEDERAL ROLE IN IMPLEMENTING SECONDARY TREATMENT IN THE NATION'S POTWS

The story of federal involvement in water pollution control, and specifically the secondary treatment issue, is best told in two parts—before and after the passage of the Federal Water Pollution Control Act Amendments of 1972, also known as the Clean Water Act (CWA). Before 1972, regulatory authority for water pollution control rested with the states. Federal involvement was limited to cases involving interstate waters. Unfortunately, there was a great diversity among the states in terms of willingness to pay the costs of building and upgrading POTWs and to enforce pollution control laws.

At the center of the problem was the idea that water pollution could be controlled by setting ambient water quality standards and that states would go after dischargers who caused those standards to be violated. In retrospect, this approach was an enforcement nightmare for several reasons (WEF, 1997):

- The enforcing agency had to prove a particular discharger was causing a waterbody to be in violation of the ambient water quality standard. This was difficult because waste loads were allocated among all dischargers based on methods that were often open to interpretation.
- Most of the time, data with which to support the case against a discharger had to come from the discharger itself. Usually, there were no independent monitoring programs.
- Many waterbodies lacked water quality standards.
- There were few civil or criminal penalties that could be levied against dischargers who caused water quality standards to be violated.

As the state-led water quality standards approach continued to fail and water quality conditions continued to spiral downward, both water supply and water resource users looked to the federal government for leadership and relief. The CWA was designed to turn the water pollution control tables around completely, and it did. The following two subsections describe the federal role before and after passage of the CWA.

The Federal Role in Secondary Treatment Before the Clean Water Act

The public's concern about raw sewage in the nation's waterways was not entirely lost on the U.S. Congress before the turn of the twentieth century. Because of the U.S. Constitution, however, they felt powerless to act on any water resource issue unless it dealt in some way with interstate commerce. Accordingly, the first federal legislation dealing with the abatement of water pollution was tied to the fact that pollution sometimes got so bad that it impeded navigation. The Rivers and Harbors Act of 1890 specifically prohibited the discharge of any refuse or filth that would impede navigation in interstate waters. Unfortunately, this act was greatly "watered down" with the passage of the amended Rivers and Harbors Act in 1899. It conveniently exempted "refuse flowing from streets and sewers and passing therefrom in a liquid state" from the navigation impedance prohibition. After the Rivers and Harbors Act of 1899, the Public Health Service Act of 1912 authorized the federal government to investigate waterborne disease and water pollution. In 1924, the Oil Pollution Act was enacted to control discharges of oil causing damage to coastal waters.

The next few decades were lean ones in terms of federal involvement in water pollution control—but not for lack of effort. Between 1899 and 1948, more than 100 bills about water pollution were introduced. Most languished and died in the halls of Congress. One, sponsored by Senator Alben W. Barkley (later Vice President for Harry S. Truman) and Representative Fred M. Vinson (later Chief Justice of the Supreme Court), actually made it to President Roosevelt's desk. The bill, however, received a presidential veto in 1938 because of budgetary concerns. The 80th Congress finally broke the impasse and enacted the Water Pollution Control Act of 1948. This act, along with five amendments passed between 1956 and 1970, shaped the national vision and defined the federal role regarding the treatment of wastewater in the United States. It also set the stage for passage of the landmark Water Pollution Control Amendments of 1972. Figure 2-4 summarizes the key legislation enacted between 1948 and 1971.

The Water Pollution Control Act of 1948, PL 80-845 The Water Pollution Control Act of 1948 was significant on three accounts. For the first time Congress accomplished the following:

- Expressed a national interest in abating water pollution for the benefit of both water supply and water resource customers.

> The pollution of our water resources by domestic and industrial wastes has become an increasingly serious problem due to the rapid growth of our cities and industries. Large and increasing amounts of varied wastes must be disposed of from these concentrated areas. Polluted waters menace the public health through the contamination of water and food supplies, destroy fish and game life, and rob us of other benefits of our natural resources.
>
> —Senate Report No. 462 of the 80th Congress Report on the Water Pollution Control Act of 1948

- Established the view that states were primarily responsible for the control of water pollution and that the federal government's role would be to provide financial aid and technical assistance—a policy concept that has continued to the present.

> That in connection with the exercise of jurisdiction over the waterways of the Nation and in consequence of the benefits resulting to the public health and welfare by the abatement of stream pollution, it is hereby declared to be the policy of Congress to recognize, preserve, and protect the primary responsibilities and rights of the States in controlling water pollution . . . and to provide . . . financial aid to State and interstate agencies and to municipalities, in the formulation and execution of their stream pollution programs.
>
> —The Water Pollution Control Act of 1948 (PL 80-845)

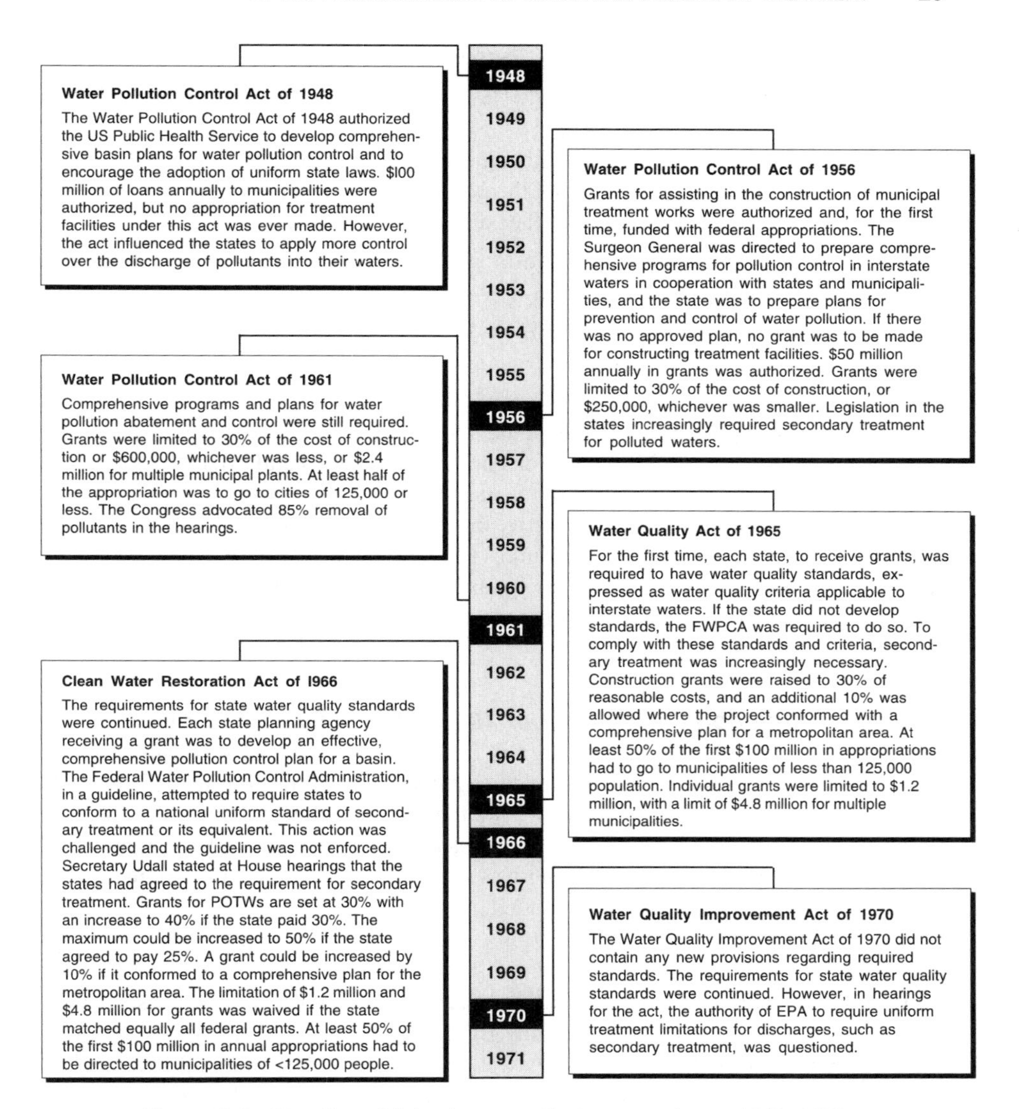

Figure 2-4 Timeline of federal water pollution control acts, 1948–1971.

- Developed activities that required states and the federal government to work as partners in solving pollution problems in interstate waters.

The act set forth a loan program designed to provide up to $100 million per year for states, municipalities, and interstate agencies to construct needed wastewater treatment works. Each loan was not to exceed $250,000 and was to bear an interest rate of 2%. Unfortunately, the loan program never saw the light of day because the program was never funded.

More successful, however, were the partnership programs developed between the states and the U.S. Public Health Service. The act required the Surgeon General to:

- Work with states and municipalities to prepare and adopt comprehensive programs for eliminating or reducing the pollution of interstate waters and improving the sanitary conditions of surface and underground waters.
- Encourage the enactment of uniform state laws relating to the prevention and control of water pollution.
- Take action against polluters of interstate waters, with the consent of the affected state.

In 1952, the Congress acknowledged that these partnership efforts were paying off and passed Public Law 82-579, which extended the activities authorized by the 1948 act for another four years. In 1955, the Senate issued a report that stated that the act caused more than half the states to improve their pollution control legislation and programs to better protect their water resources (Sen. Rep. No. 543, 84th Congress). The report also noted that some states were establishing water quality standards so stringent that they left municipalities with no choice but to implement secondary treatment at their facilities.

The Water Pollution Control Act of 1956, PL 84-660 This act was significant because it authorized a grant program for the construction of wastewater treatment facilities—and then actually funded it. A total of $150 million was earmarked over the life of the program with a provision that no more than $50 million could be spent per year. Individual grants were not to exceed 30 percent of the reasonable cost of construction, or $250,000, whichever was smaller. There was one important caveat to obtaining a grant, however: to be funded, the project must be in conformity with a plan prepared by the state water pollution control agency and approved by the Surgeon General.

Though language in the act emphasized that the law should not be "construed as impairing or in any manner affecting any right or jurisdiction of the States with respect to the waters (including boundary waters) of such States," the requirement for federal approval of a state's water pollution control plan nonetheless established a new leadership role for the federal government. If a state did not follow an approved plan, grant payments could be held up pending an appeal to a federal court.

The Federal Water Pollution Control Act of 1961, PL 87-88 Only a few changes were made in the 1961 amendment to the Federal Water Pollution Control Act. Congress's basic intent with this legislation was to extend the act through to 1967. Construction grants were authorized to the states in the total amount of $60 million for FY 1962, $90 million for FY 1963, and $100 million for each of the fiscal years 1964 through 1967.

A grant to a municipality was limited to $600,000 or 30 percent of the reasonable costs, whichever was less, with a limit of $2.4 million when the project would serve more than one municipality. At least one-half of the funds appropriated for projects were to go to cities of 125,000 population or less. The requirements for comprehensive pollution control programs and plans were carried over from the act of 1956.

Perhaps the most interesting development concerning federal involvement in water pollution control appeared not in the act itself, but in language contained in the accompanying Senate report. Here, for the first time, the Senate mentioned its desire to see secondary treatment used in municipal waste treatment plants. The same document also presented a vision for the future and an expression of hope for completion of the urban water cycle:

> There is every reason to believe that a vigorous research attack on waste treatment problems would lead to breakthroughs and new processes which will make it possible to handle ever-increasing wasteloads, and even to restore streams to a state approaching their original natural purity . . . If all waste or all water deteriorating elements could be removed by treatment, a region's water supply could be used over and over.
>
> —Senate Report No. 353, 87th Congress Report on the Water Pollution Control Act of 1961

The Water Quality Act of 1965, PL 89-234 Two important elements were established with the passage of the Water Quality Act of 1965. First, it created the Federal Water Pollution Control Administration (FWPCA) as a separate entity in the Department of Health, Education and Welfare. FWPCA did not reside there long, however. In 1966 it was transferred to the Department of the Interior. Then, in 1970, its functions were folded into the new United States Environmental Protection Agency (USEPA). Second, the act required each state desiring a grant to file a letter of intent with the FWPCA committing the state to establishing, before June 30, 1967, water quality criteria applicable to interstate waters and submitting a plan for the implementation and enforcement of water quality criteria. If the state chose not to do this, the FWPCA would do it for the state.

The state's criteria and plan were to be the water quality standards for its interstate waters and tributaries. The act mandated that these standards must protect the public health or welfare and enhance the quality of water. Consideration was also to be given to the use and value of public water supplies, propagation of fish and wildlife, recreational purposes, and agricultural, industrial, and other legitimate needs.

The Construction Grants Program was continued in this act. The federal contribution was raised to 30 percent of the reasonable costs, plus an additional 10 percent when the project conformed with the comprehensive plan for a metropolitan area. The authorized amounts for construction grants were set at $150 million for FY 1966 and FY 1967, with at least 50 percent of the first $100 million appropriated in those years used for grants for municipalities of 125,000 people or less. Grants to municipalities were limited to $1.2 million, with $4.8 million set as the limit when two or more municipalities were served by the same facility.

The Clean Water Restoration Act of 1966, PL 89-753 Basin planning was a key focus of the Clean Water Restoration Act of 1966. Each state planning agency receiving a grant had to develop an effective comprehensive pollution control and abatement plan for basins. A basin was defined as rivers and their tributaries, streams, coastal waters, sounds, estuaries, bays, lakes, and portions thereof, as well as the lands drained thereby. Congress mandated that the plan must:

- Be consistent with water quality standards.
- Recommend effective and economical treatment works.
- Recommend maintenance and improvement of water quality standards within the basin, as well as methods for financing necessary facilities.

Grants for wastewater treatment facilities were set at 30 percent of the reasonable cost, which could be increased to 40 percent if the state agreed to pay not less than 30 percent of the reasonable costs. This maximum could be increased to 50 percent if the state agreed to pay not less than 25 percent of the estimated reasonable costs of all such grants. A grant could also be increased by 10 percent of the amount of a grant if it was in conformance with a plan developed for the metropolitan area. To be eligible for any grant, a project must be included in a comprehensive water pollution program and the state water pollution control plan. Grants were again limited to $1.2 million for individual projects and $4.8 million for multimunicipality projects. This limitation was waived, however, if the state agreed to match equally all federal grants made for the project.

Authorized amounts for grants gradually increased from a total of $550 million for FY 1968 to $1.250 billion for FY 1971. The total of $2 billion was authorized for FY 1972 by the Extensions of Certain Provisions of the Federal Water Pollution Control Act of 1971, PL 92-240.

The Water Quality Improvement Act of 1970, PL 91-224 On March 18, 1968, FWPCA announced that the water quality standards of 28 states had been approved, and all of the states were expected to have approved standards by June. Soon afterwards, however, FWPCA attempted to cause states to amend their standards to include an effluent limitation of "best practicable treatment" or its equivalent for all discharges:

> No standards shall be approved which allow any waste amenable to treatment or control to be discharged into any interstate water without treatment or control regardless of the water quality criteria and water use or uses adopted.
>
> Further, no standard will be approved which does not require all wastes, prior to discharge into any interstate water, to receive the best practicable treatment or control unless it can be demonstrated that a lessor degree of treatment or control will provide for water quality and enhancement commensurate with proposed present and future water uses.
>
> —FWPCA Guideline, 1968

People questioned what authority the FWPCA thought they had to set "best practicable treatment" as the minimum level of treatment and what they meant by that term. In House hearings leading up to the Water Quality Improvement Act of 1970, Secretary Udall explained that "in practice, this guideline usually, but not always, means secondary treatment of municipal wastes . . . generally the States have agreed with us with regard to the requirement of secondary treatment." A number of officials from different states begged to differ with Secretary Udall and FWPCA's guideline. Not surprisingly, states offered up legal opinions that bluntly concluded that the FWPCA had no authority to set discharge limitations.

Against this backdrop, the Water Quality Improvement Act of 1970 was passed. The act continued the authority of the states to set standards of water quality and the authority of the FWPCA to approve such standards. Congress, however, chose not to include any new provisions regarding standards or treatment levels.

Deciding that the battle for secondary treatment in municipal wastewater plants would best be fought on another stage, the FWPCA stepped back and issued a new construction grant regulation (36 FR 13029) in July 1971 that called for primary treatment as the minimum level of treatment:

> To be eligible for a grant, a project must be designed to result in an operable treatment works, or part thereof, which will treat or stabilize sewage or industrial wastes of a liquid nature in order to abate, control, or prevent water pollution . . . such treatment or stabilization shall consist of at least primary treatment, or its equivalent, resulting in the substantially complete removal of settleable solids.
>
> —FWPCA Construction Grant Regulation, July 1971 (36 FR 13029)

After the FWPCA was reorganized out of existence, USEPA aggressively picked up the secondary treatment torch. In June 1972, prior to the passage of the Federal Water Pollution Control Act of 1972 in October, the Agency issued regulations that required grant projects to conform to secondary treatment requirements that included the removal of 85 percent of BOD_5 from POTW influent.

The Agency ruled that secondary treatment could be waived only for projects that:

- Discharged wastes to open ocean waters through an ocean outfall if such discharges would not adversely affect the open ocean waters and adjoining shores, and receive primary treatment before discharge.
- Treated or controlled combined sewer overflows if such projects were consistent with river basins or metropolitan plans to meet approved water quality standards.

The Federal Role in Secondary Treatment After the Clean Water Act

Enactment of the 1972 Amendments to the Federal Water Pollution Control Act, now popularly known as the Clean Water Act (CWA), by the 92nd U.S. Congress redirected national policy for water pollution control onto a new path. Sparked by publication of Rachel Carson's *Silent Spring* in 1962 (Carson, 1962), national publicity about environmental issues during the 1960s led to public awareness of the existence of nationwide air and water pollution problems and political demands by the "Green Movement" for governmental action to address pollution problems (Zwick and Benstock, 1971; Jobin, 1998).

On October 18, 1972, a new era for POTWs began when the 1972 Amendments to the Federal Water Pollution Control Act (PL 92-500) were unanimously passed by the U.S. Congress and, despite a veto by President Richard M. Nixon, who thought that the $24 billion investment over 5 years was "excessive and needless overspending," the act became law (Knopman and Smith, 1993). The act established a new national policy that firmly rejected the historically accepted use of rivers, lakes, and harbors as receptacles for inadequately treated wastewater. Congress's objective was clear. They wanted to "restore and maintain the chemical, physical and biological integrity of the nation's waters" and to attain "fishable and swimmable" waters throughout the nation. With PL 92-500, the federal government took control of directing and defining the nation's water pollution control programs. This commitment led to the completion of the urban water cycle in many communities across the United States.

Congress recognized that success or failure of PL 92-500's lofty objectives hinged on a combination of money, compliance, and enforcement. Consequently, the basic framework of the act included the following.

- Establishment of the National Pollutant Discharge Elimination System (NPDES), a program that requires that every point source discharger of pollutants obtain a permit and meet all the applicable requirements specified in regulations issued under sections 301 and 304 of the act. These permits are enforceable in both federal and state courts, with substantial penalties for noncompliance.
- Development of technology-based effluent limits, which serve as minimum treatment standards to be met by dischargers.
- An ability to impose more stringent water quality–based effluent limits where technology-based limits are inadequate to meet state water quality standards or objectives.
- Creation of a financial assistance program to build and upgrade POTWs. PL 92-500 authorized $5.0 billion in federal spending for fiscal year 1973, $6.0 billion for fiscal year 1974, and $7.0 billion for fiscal year 1975. In contrast, the year before the act was passed, a total of $1.25 billion (federal dollars) was spent. Under the Construction Grants program, the federal share was 75 percent of cost from fiscal years 1973 to 1983 and 55 percent thereafter. Additional funds were made available for projects using innovative and alternative treatment processes.

The story of the Clean Water Act and its evolution from 1972 to the present day is richly complicated. The purpose of this section is not to summarize all aspects of this landmark act. Rather, the objective is to focus on the role it played in implementing secondary treatment in the nation's POTWs. Other sources, such as *The Clean Water Act, 25th Anniversary Edition,* published by the Water Environment Federation (WEF, 1997), should be consulted for a complete overview of the act. Figure 2-5 summarizes the key amendments and regulations that occurred from 1972 to 1996.

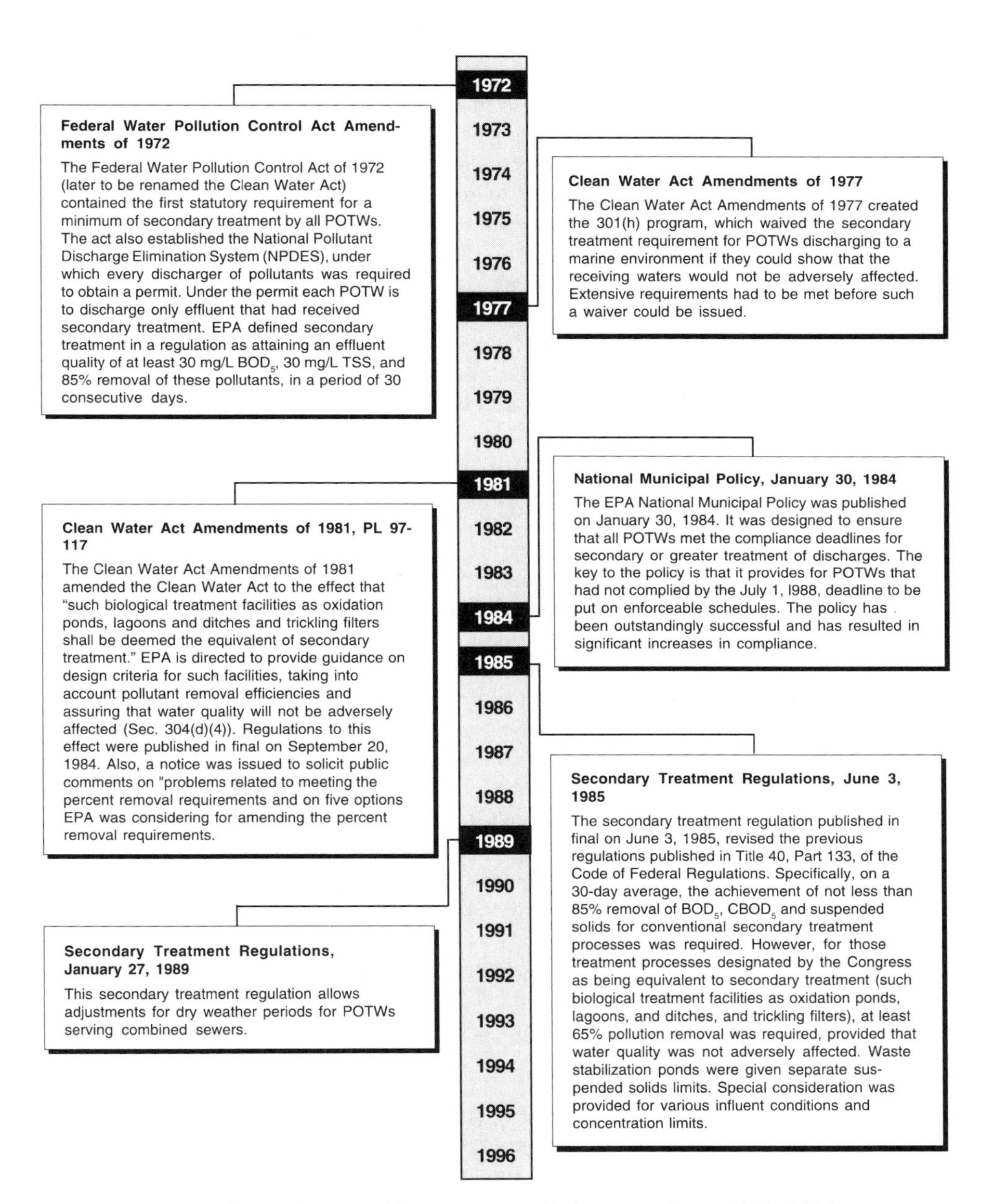

Figure 2-5 Timeline of federal water pollution control acts, 1972–1996.

The Water Pollution Control Act Amendments of 1972 (PL 92-500) and Secondary Treatment Information (38 FR 22298-22299), Published in Final on August 17, 1973 After debating the merits of secondary treatment for the better part of two decades, Congress finally put the issue to rest in the Federal Water Pollution Control Act Amendments of 1972. Section 301 required POTWs to achieve effluent limitations based on secondary treatment.

A simple, aggressive schedule was set to meet this requirement. By July 1, 1977, all existing POTWs and all facilities approved for construction before June 30, 1974, must incorporate secondary treatment. Then, by July 1, 1983, POTWs must meet an additional level of treatment described in the act as "best practicable wastewater treatment."

While developing the 1972 Amendments, Congress understood that the term "secondary treatment" needed to be carefully researched and clearly articulated before regulations could be drafted. At the time, several "working" definitions existed, including one offered by Congressman Vanik in the House debate on the amendment. He defined secondary treatment as a process that removes 80 to 90 percent of all harmful wastes from POTW influent.

Section 304(d)(1) directed USEPA to investigate and publish in the Federal Register "information, in terms of amounts of constituents and chemical, physical and biological characteristics of pollutants, on the degree of effluent reduction attainable through the application of secondary treatment." USEPA assembled a work group the next year to accomplish this task and invited outside commentators and contractors to participate.

Early on, the group decided that the effluent limitations to be used to define secondary treatment needed to include concentrations of key parameters as well as percent reduction limits. Also weighing in on the minds of the group was a congressional and public concern that, if percent removal targets were set too high, incremental environmental benefits would not be worth the cost. Consequently, economic considerations became an important part of the decision-making process. Figure 2-6 is an example of how costs were analyzed in relation to percent removal targets for BOD_5. The graph shows that costs rise rapidly beyond the 85 to 88 percent removal level. Analyses such as these helped the work group put technical capabilities in a practical (i.e., economical) context.

In April 1973, USEPA published a proposed regulation based on the group's report. After comments were addressed, the Agency issued its final regulation on August 17, 1973. It defined secondary treatment effluent concentration limits for the following parameters:

- *Five-Day Biochemical Oxygen Demand (BOD_5).* Average value for BOD_5 in effluent samples collected in a period of 30 consecutive days shall not exceed 30 milligrams per liter (mg/L). The average value for BOD_5 in effluent samples collected in a period of 7 consecutive days shall not exceed 45 mg/L.
- *Total Suspended Solids (TSS).* Average value for TSS in effluent samples collected in a period of 30 consecutive days shall not exceed 30 milligrams per liter (mg/L). The average value for TSS in effluent samples collected in 7 consecutive days shall not exceed 45 mg/L.

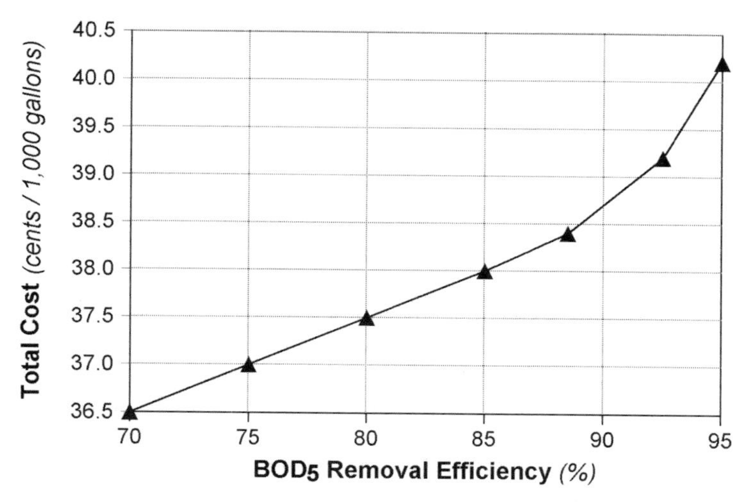

Figure 2-6 Cost versus BOD_5 removal efficiency of a new 1 million gallon per day POTW. *Source:* USEPA, 1973.

- *Fecal Coliform Bacteria.* Geometric mean of fecal coliform bacteria values for effluent samples collected in a period of 30 consecutive days shall not exceed 200 per milliliter (mL). The geometric mean of fecal coliform bacteria values for effluent samples collected in a period of 7 consecutive days shall not exceed 400 per milliliter (mL).
- *pH.* Effluent values for pH shall remain within the limits of 6.0 and 9.0.

Also included were percent removal limits for BOD_5 and TSS. Specifically, average values for BOD_5 and TSS in effluent samples collected in 30 consecutive days may not exceed 15 percent of the mean of influent samples collected at approximately the same times during the same period (85 percent removal).

The BOD and TSS limits were chosen based on an assumption that the wastewater entering a POTW (influent) contains about 200 mg/L of BOD_5 and TSS. Knowing this assumption did not hold true in every case, USEPA made a couple of allowances. Specifically, the Agency allowed a POTW to have higher BOD_5 and TSS concentrations in its effluent if the facility received more than 10 percent of its design flow from industrial facilities for which less stringent effluent limitations had been promulgated.

Special consideration was also given, on a case-by-case basis, to treatment works served by combined storm and sanitary sewer systems where increased flows during wet weather prevented the attainment of the defined minimum level of secondary treatment. Of chief concern was the 85 percent removal requirement. In stormy weather, storm water runoff dilutes the normal volume of influent, lowering BOD_5 and TSS concentrations. Expecting to reduce already reduced concentrations by 85 percent was beyond the means of many facilities.

Two subsequent amendments to the secondary treatment information were promulgated on July 26, 1976 (41 FR 30788) and October 7, 1977 (42 FR 5665). These changes provided for:

- Deletion of the fecal coliform bacteria limitations and clarification of the pH requirement.
- Special consideration for the TSS effluent limitations applicable to waste stabilization ponds with wastewater flows of less than 2 million gallons per day (mgd).

Publishing the regulation defining the minimum level of secondary treatment to be implemented by POTW facilities by 1977 was a major accomplishment for USEPA. On the horizon, however, loomed the prospect of developing a second, more stringent, level of requirements for implementation by July 1, 1983. Congress fortunately realized that this second set of requirements, or best practicable treatment, might not be needed. Section 315(b) of PL 92-500 established a national study commission to help them make this determination. Composed of five Senators, five Representatives, and five members of the public, the commission was given 3 years to accomplish this task. In the end, the group issued several general recommendations, one of which was that the secondary treatment effluent limits developed for the 1977 deadline not be changed for the 1983 deadline. Essentially, the commission determined that secondary treatment was the best practicable treatment for POTWs. Thus, the headaches associated with setting a second level of requirements were avoided.

The Clean Water Act Amendments of 1977 (PL 95-217) The tight timetable Congress established for implementing secondary treatment proved to be unrealistic for many municipalities. In fact, only about 30 percent of major POTWs (those processing 1 million—or more—gallons of wastewater per day) were in compliance when the July 1, 1977, deadline rolled around. In many cases, upgrade schedules were slowed due to delays in receiving federal funds. The Clean Water Act Amendments of 1977 (PL 95-217) responded to this situation by allowing time extensions for municipalities encountering funding problems. Time extensions aside, probably the most significant aspect of PL 95-217 in terms of secondary treatment was the fact that Congress backed off from PL 92-500's original objective of having all POTWs implement secondary treatment as a minimum technology-based standard. Municipalities discharging into ocean waters had been arguing that the benefits associated with their upgrading to secondary treatment were not worth the cost. The vastness of the marine environment, they said, effectively dilutes and incorporates wastes into the water and sea bottom without harming either the indigenous biota or users or the environment.

Congress agreed and added Section 301(h) to the Clean Water Act, allowing marine dischargers to apply for a waiver of secondary treatment requirements. USEPA would subsequently review the application and issue modified NPDES permits to POTWs that met certain environmental criteria and received state concurrence. These criteria included:

- Existence of and compliance with water quality standards.

- Protection and propagation of a balanced indigenous population of fish, shellfish, and wildlife.
- Allowance of recreational activities.
- Establishment of a monitoring program.
- Satisfactory toxics control programs, including an approved industrial pretreatment program.
- No additional treatment requirements for other sources.
- Acceptable discharge volume and pollutant limits.
- Protection of public water supplies (desalinization plants).

Municipal Wastewater Treatment Construction Grants Amendments of 1981 (PL 97-117) and Secondary Treatment Regulations (49 FR 36986-37014), Published in Final on September 20, 1984 When the decade of the 1980s dawned, the goal of implementing secondary treatment in the nation's POTWs seemed a long way off. About half of the 20,000 municipal discharges, including more than 100 larger cities, were still not in compliance with the 1977 deadline. Construction projects were bogged down with funding problems, complicated regulatory procedures, and lack of staff at state and federal agencies. To address these and other problems, Congress passed the Construction Grants Amendments of 1981. Section 301(i) recognized that funding issues were still holding up secondary treatment compliance and therefore extended the implementation deadline to July 1, 1988, on a case-by-case basis.

PL 97-117 and its companion regulations also addressed another concern involving USEPA's definition of secondary treatment effluent requirements. In theory, the requirements were not intended to favor one treatment process over another, yet they did. As it turned out, activated sludge facilities were the only ones that could consistently meet the requirement of 85 percent removal of BOD_5 and TSS limits. This situation caused an immediate problem for the many smaller communities that had invested in trickling filters, waste stabilization ponds, and other types of biological wastewater treatment. Even when their facilities performed as designed, they were in noncompliance according to USEPA's standards for secondary treatment.

Upgrading or replacing these facilities was an expensive proposition. Many questioned whether environmental benefits gained would be worth the cost. Congress agreed and PL 97-117 and its companion regulations included the following:

- Introduced the concept of "equivalent of secondary treatment" to describe facilities that use a trickling filter or waste stabilization pond as a principal treatment process and that were not meeting the secondary treatment requirements as promulgated by USEPA in 1973.
- Lowered the minimum level of effluent quality to be achieved by those facilities during a 30-day period as an average value not to exceed 45 mg/L for BOD_5 and TSS, an average 7-day value for BOD_5 and TSS of not to exceed 65 mg/L, and a percentage removal of those constituents of not less than 65 percent (30-day average).

- Required that NPDES permit adjustments for "equivalent to secondary treatment" facilities reflect the performance or design capabilities of the facility and ensure that water quality is not adversely affected.

National Municipal Policy (49 FR 3832-3833), Published on January 30, 1984 Continually pushing back the deadline for implementation of secondary treatment in POTWs created confusion. The 1972 Amendments had set the original deadline for compliance for 1977. For some municipalities, it was extended to 1983 by PL 95-217 and then to 1988 by PL 97-117. The USEPA National Municipal Policy, published in the Federal Register on January 30, 1984, was designed to eliminate this confusion and ensure that all POTWs would comply with the statutory requirements and compliance deadlines in the Clean Water Act. It also established that where there were extraordinary circumstances that precluded compliance by the July 1, 1988, deadline, POTWs would be put on enforceable schedules designed to achieve timely compliance. The policy described EPA's intentions to focus its efforts on:

- POTWs that previously received federal funding assistance and are not in compliance.
- Other POTWs.
- Minor POTWs (less than 1 mgd capacity) that are contributing significantly to impairment of water quality.

This municipal treatment policy has been outstandingly successful, with over 90 percent compliance achieved to date for major POTWs (1 mgd or over).

Secondary Treatment Regulations, Published in Final on June 3, 1985 The secondary treatment requirement of 85 percent removal of BOD_5 and TSS continued to present problems for POTWs receiving diluted influent wastewater. Whether it was a secondary treatment facility (85 percent removal) or an equivalent of secondary treatment facility (65 percent removal), to stay in compliance a facility had to install advanced technology, even if it consistently met its concentration limits. Recognizing this problem, USEPA, on November 16, 1983, published a Federal Register notice soliciting public comment on a number of options for amending the percent removal requirements.

Based on the public comments received, the Agency proposed and then finalized a revised Secondary Treatment Regulation. Published in final on June 3, 1985, it authorized USEPA to lower the percent removal requirement, or substitute a mass limit for the percent removal requirement, for certain POTWs. The Agency would make this determination on a case-by-case basis based on the removal capability of the treatment plant, the influent wastewater concentration, and the infiltration and inflow situation.

Treatment plants could apply for a permit adjustment in its percent removal limit only if:

- The treatment plant is meeting or will consistently meet its other permit effluent concentration limitations, but its percent removal requirements cannot be met due to less concentrated influent wastewater for separated sewers.
- To meet the percent removal requirement, the treatment works would have to meet significantly more stringent concentration-based limitations.
- The less concentrated influent wastewater to the treatment works was not a result of excessive infiltration and inflow.

The concentration limits in the permit would remain unchanged, and in no case was a permit to be adjusted if the permitting authority determined that adverse water quality impacts would result from a change in permit limits.

Amendment to the Secondary Treatment Regulation, Published in Final on January 27, 1989, in the Federal Register The Secondary Treatment Regulation, published in June 1985, addressed the problem POTWs with separate sewers had in meeting percent reduction standards due to the dilution of influent wastewater by wet weather conditions. The city of New York also had a problem. Its combined sewer system delivered diluted influent to city POTWs, even during dry weather. Consequently, the city petitioned to be eligible for adjustments of percent removal requirements too, arguing that nonexcessive infiltration can dilute the influent wastewater of treatment works served by combined sewers just as it does for treatment works served by separate sewers. USEPA agreed with this position and published an amendment to the regulation on January 27, 1989, to allow for percent removal adjustments during dry weather periods for POTWs with combined sewers. To obtain this adjustment, the treatment works had to satisfy three conditions:

- It must consistently meet its permit effluent concentration limitations, but the percent removal requirements cannot be met due to less concentrated influent wastewater.
- Significantly more stringent effluent concentration than those required by the concentration-based standards must be met to comply with the percent removal requirements.
- The less concentrated influent wastewater must not result from either excessive infiltration or clear water industrial discharges to the system.

Regarding the last condition, the regulation established that, if the average dry weather baseflow (i.e., the total of the wastewater flow plus infiltration) in a combined sewer system is less than 120 gallons per day per capita (gpcd) threshold value, infiltration is assumed to be nonexcessive. However, sewer systems with average dry weather flows greater than 120 gpcd might also have nonexcessive infiltration if this is demonstrated on a case-by-case basis. An applicant, therefore, has an opportunity to demonstrate that its combined sewer system is not subject to excessive infiltration even if the average total dry weather baseflow exceeds the 120 gpcd threshold value.

D. NATIONWIDE TRENDS IN BOD LOADING FROM POTWs

From 1940 to the present day, the combination of advancing wastewater treatment technology, increased public concern, various state wastewater treatment regulations, and, finally, the 1972 CWA secondary treatment mandate resulted in an increased number of POTWs with at least secondary and, in many cases, greater than secondary levels of treatment. Table 2-2 presents descriptions of the six major types of treatment found at POTWs, along with their corresponding design-based BOD_5 removal efficiency[1] (expressed as percent removal).

The total population in the United States grew rapidly in the latter half of the twentieth century, increasing from around 140 million people in 1940 to about 270 million in 1996 (see Figure 2-2). This population growth meant POTWs not only had to upgrade their treatment processes to increase pollutant removal efficiency, but they had to accomplish it while dealing with increasing influent wastewater loads. This section examines trends concerning the nation's expansion and upgrades of POTWs and analyzes how increased use of secondary and greater than secondary treatment after the 1972 CWA affected the rate of effluent BOD loading to the nation's waterways. Specifically examined are the following:

- The inventory of POTWs in the United States.
- The number of people served by those POTWs and the amount of wastewater flow they generated.
- The rate of BOD entering POTWs (influent loading).
- The rate of BOD discharged by POTWs into receiving waterways (effluent loading).
- BOD removal efficiency of POTWs.
- Projections of effluent BOD loading into the twenty-first century.

The information sources for this study include municipal wastewater inventories published by the U.S. Public Health Service from 1940 through 1968 (USPHS, 1951; NCWQ, 1976; USEPA, 1974) and USEPA's Clean Water Needs Surveys (CWNS) conducted from 1973 through 1996 (USEPA, 1976, 1978, 1980, 1982, 1984, 1986, 1989, 1993, 1997). Many of these sources categorize their information by the six types of wastewater treatment described in Table 2-2. Some sources, however, combine primary and advanced primary data and report it simply as "less than secondary" treatment data. Similarly, data for advanced secondary and advanced wastewater treatment are combined and reported as "greater than secondary" treatment data. To

[1]Designed-based BOD_5 removal efficiencies are minimum requirements typically assigned by NPDES permits according to the treatment process and treatment plant design assumptions (Metcalf and Eddy, 1991). Generally, they represent conservative estimates of BOD_5 removal efficiencies. Many modern POTWs report a higher rate of BOD_5 removal than their permitted rate. This study, however, focuses on designed-based BOD_5 removal efficiencies because it is assumed that these conservative rates would provide a more effective and consistent comparison of BOD_5 removal over the entire historical period of record used in the analysis.

keep the categories consistent, this convention was followed in the analyses presented in this section.

Types of BOD Reported in this Trends Analysis

BOD_5 is the most widely used measurement of BOD. In spite of its popularity, there are important limitations of this measurement. The subscript "5" refers to the laboratory incubation period of 5 days at 20° C. Many biochemical reactions that determine the ultimate consumption of DO in both wastewater and natural waters are not completed within the 5-day limit, however. Therefore, an estimate of "ultimate" BOD (BOD_u) of a sample requires consideration of all the biochemical processes that consume DO over a longer time scale. Figure 2-7 presents the relationships among the components of BOD_u.

Familiar to most environmental engineers is the oxygen demand associated with the bacterial decomposition of carbonaceous organic matter under aerobic conditions. Through respiration, organic matter is broken down and oxygen is consumed. Parameters in Figure 2-7 relating to carbonaceous BOD are:

- *$CBOD_5$:* BOD at 5 days that includes only the carbonaceous component of oxygen consumption.
- *CBOD:* BOD at an unspecified time that includes only the carbonaceous component of oxygen consumption.
- *$CBOD_u$:* Ultimate BOD of carbonaceous component of oxygen consumption at completion of decomposition process.

Along with the decomposition of carbonaceous matter is an additional oxygen demand associated with nitrification, the process that converts ammonia to nitrate. Nitrogen in wastewater generally appears as organic nitrogen compounds (urea, proteins, etc.) and ammonia. Over time, the nitrogen compounds are hydrolyzed and are converted to ammonia. Autotrophic bacteria of the genus *Nitrosomonas* convert the ammonia to nitrite, using oxygen in the process. Nitrite, in turn, is converted to nitrate by bacteria of the genus *Nitrobacter*, consuming additional oxygen in the process. Parameters in Figure 2-7 relating to nitrogenous BOD are:

- *NBOD:* BOD at an unspecified time that includes only the nitrogenous component of oxygen consumption from nitrification.
- *$NBOD_u$:* Ultimate BOD of the nitrogenous component of oxygen consumption at completion of nitrification process.

In Figure 2-7, carbonaceous and nitrogenous BOD components combine to yield the following parameters:

- *BOD_5:* BOD at 5 days that includes the carbonaceous and nitrogenous components of oxygen consumption.
- *Total BOD:* BOD at an unspecified time that includes the carbonaceous and nitrogenous components of oxygen consumption.

TABLE 2-2 Six Levels of Municipal Wastewater Treatment

Treatment Type	Design BOD_5 Removal Efficiency (percent)	Description
Raw	0	Wastewater is collected and discharged to surface waters without treatment, or removal, of pollutants from the influent stream.
Primary	35	Incorporates physical processes of gravitational settling to separate settleable and floatable solids material from the raw wastewater. The removal of settleable solids results in the removal of pollutants associated with solid particles such as organic matter, nutrients, toxic chemicals, heavy metals, and pathogens. Other physical processes such as fine screens and filters can also be used.
Advanced Primary	50	Enhancement of the primary clarification process using chemical coagulants such as metal salts and organic polyelectrolytes.
Secondary	85	Biological processes are added to break down organic matter in the primary effluent by oxidation and production of bacterial biomass. Biological waste treatment systems, based on bacterial decomposition of organic matter, can be classified as activated sludge, waste stabilization ponds (suspended bacterial growth), and trickling filters (attached bacterial growth). 84 to 89 percent removal of TSS and 30 mg/L effluent concentration for BOD_5 and TSS.
• Activated sludge		Involves the use of bacteria to decompose suspended solids in the sewage so that they can be settled out. Oxygen to speed the bacteriological process is generaed by mechanical aeration or by the infusion of additional oxygen. The solids produced (sludge) by the biological action are settled out and removed, except for a portion of the bacteria-rich sludge that is returned to the head of the secondary treatment process to activate the biological processes to treat sewage. This is the standard method of treatment for medium and large cities.
• Waste stabilization		Pools in which mechanical aeration is used to supply oxygen to the bacteria. In other processes, oxygen is supplied by natural surface aeration or by algal photosynthesis with no mechanical aeration.

TABLE 2-2 *Continued*

Treatment Type	Design BOD_5 Removal Efficiency (percent)	Description
• Trickling filters		Employs a bed of highly permeable media such as crushed stone or plastic to which are attached microcosms for treating sewage sprayed on the media by a mechanical arm.
Advanced Secondary	90	The conventional secondary treatment process incorporates chemically enhanced primary clarification and/or innovative biological treatment processes to increase the removal efficiency of suspended solids, BOD, and total phosphorus. Sludge production is typically increased overall as a result of the chemical enhancement of primary clarification and biological processes. Effluent concentrations of BOD_5 range from 10 to 30 mg/L and processes included to remove ammonia and phosphorus in excess of effluent levels typical for secondary treatment.
Advanced Wastewater Treatment	95	Advanced wastewater treatment (AWT), or tertiary treatment, facilities are designed to achieve high rates of removal of nutrients (nitrogen or phosphorus), BOD, and suspended solids. Nitrogen removal is achieved by enhancement of the biological processes to incorporate nitrification (ammonia removal) and denitrification (nitrate removal). Phosphorus removal is accomplished by either chemical or biological processes. Addition of high doses of metal salts removes phosphorus while biological processes are dependent on the selection of high-phosphorus microorganisms. Additional removal of nutrients and organic carbon can be accomplished using processes such as high lime, granular activated carbon, and reverse osmosis. Effluent BOD_5 is generally less than 10 mg/L, and total-N removal is more than 50 percent.

Note: Readers desiring more technical details about these processes should review standard engineering reference texts (e.g., Metcalf & Eddy, 1991) or technical reports on wastewater treatment (e.g., NRC, 1993). Effluent removal efficiency and effluent concentrations are taken from the 1978 USEPA Needs Survey (USEPA, 1978).

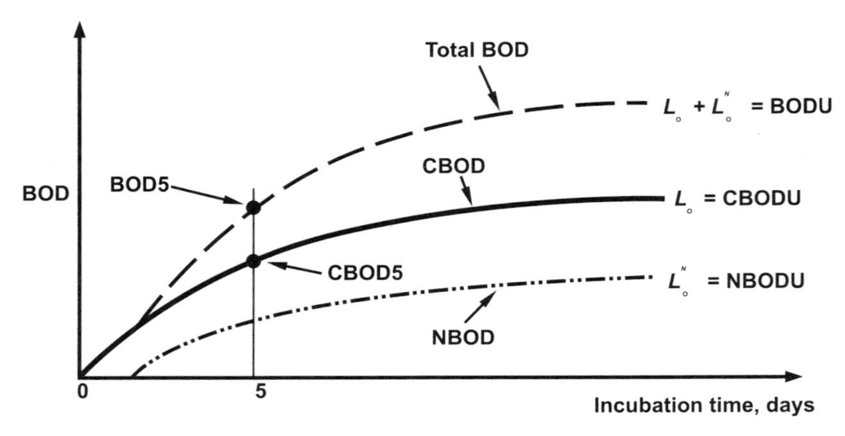

Figure 2-7 Relationship between the carbonaceous, nitrogenous, and total BOD. *Source:* Principles of surface water quality modeling and control by R. V. Thomann and J. A. Mueller, © 1987, Reprinted by permission of Pearson Education, Inc., Upper Saddle River, NJ.

- *BOD_u:* Ultimate BOD of the carbonaceous and nitrogenous components of oxygen consumption at completion of both the carbonaceous decomposition and nitrification processes.

The length of time needed to reach the "ultimate" endpoints of the carbonaceous and nitrogenous components designated in Figure 2-7 ($CBOD_u$ and $NBOD_u$) depends on several factors, including the composition of the wastewater and the corresponding rate of decomposition for its components. For predominately labile fractions of organic carbon that are easy for bacteria to decompose (e.g., mostly sugars, short chain molecules), decomposition can be completed within about 20 to 30 days. In contrast, for refractory organic matter that is strongly resistant to bacterial decomposition (e.g., mostly cellulose, long chain molecules such as pulp and paper waste), complete decomposition might require an incubation period of anywhere from 100 to 200 days. Decomposition rates for a sample of wastewater effluent from a POTW with secondary treatment, consequently, tend to be lower than rates for raw wastewater because the easily decomposed sugars have already been removed by the treatment process.

Timing of the nitrification process is also dependent on several factors. These include the ratio of organic nitrogen compounds to ammonia and the lag time necessary to hydrolyze and convert the compounds to ammonia, the presence of adequate numbers of nitrifying bacteria in the water to begin the nitrification process, alkaline pH levels, and the presence of sufficient oxygen for bacterial respiration. The net effect of these factors is to inhibit nitrification immediately downstream from POTW outfalls (Chapra, 1997). Similarly, in a laboratory sample if a "seed" population is not available for nitrification during the 5-day incubation period, the measured BOD_5 will reflect only the carbonaceous component (i.e., $CBOD_5$). If, however, factors are sufficient for nitrification to occur in the laboratory sample, the measured BOD_5 will reflect both the carbonaceous and nitrogenous components (see Hall and Foxen, 1984).

Is incorporating the nitrogenous component and using BOD_u important enough to eschew the more familiar carbonaceous $CBOD_5$ when presenting BOD information? The answer is yes. Chapra (1997) calculates that the oxygen consumed in nitrification is about 30 percent of the oxygen consumed in carbonaceous oxidation of pure organic matter. If this finding was not persuasive enough for the inclusion of nitrification in an analysis of BOD, he also presents evidence that concentrations of NBOD and CBOD are actually nearly equivalent in untreated wastewater. Chapra theorizes that the discrepancy between calculated and the actual concentrations may be attributed to the fact that not all organic matter might be decomposable under the conditions of the BOD test and that nitrogen in wastewater might not all come from organic matter. Fertilizers and other sources likely add to the nitrogen pool, increasing the significance of NBOD in the environment.

In sum, the true measure of the long-term oxygen demand of influent and effluent BOD loading and its effect on water quality in streams and rivers can be determined only if both the carbonaceous and nitrogenous components of BOD are combined and analyzed as BOD_u. Since it is impractical for most monitoring programs and laboratories to extend the incubation period beyond the traditional 5 days associated with the determination of BOD_5, other surrogate methods must be used to determine $CBOD_u$, $NBOD_u$ and BOD_u. Discussed below are the methods used in this study to determine these parameters.

Determination of $CBOD_5$ and $CBOD_u$ BOD loading data for municipal and industrial wastewater dischargers are most often reported in NPDES permit limits and Discharge Monitoring Reports (DMRs) as either BOD_5 or $CBOD_5$. Unfortunately, in analyzing historical loading trends of municipal effluent, it is impossible to determine if BOD_5 data compiled by various data sources included the suppression of possible nitrification during the laboratory analysis. In compiling long-term BOD loading trends, therefore, it is frequently assumed that BOD_5 is approximately equal to $CBOD_5$ (see Lung, 1998). This study makes the same assumption. Consequently, for the purposes of this study, all BOD_5 data reported to the USEPA are considered to be $CBOD_5$.

Leo et al. (1984) and Thomann and Mueller (1987) point out that conversion ratios for estimating $CBOD_u$ concentrations based on either BOD_5 or $CBOD_5$ concentrations are dependent on the level of wastewater treatment. The proportion of easily degraded (labile) organic matter in the effluent declines as the efficiency of wastewater treatment is improved by upgrading a facility. In an analysis of effluent data from 114 primary to advanced municipal wastewater treatment plants, Leo et al. (1984) determined mean values of 2.47 and 2.84 for the $CBOD_u$:BOD_5 and $CBOD_u$:$CBOD_5$ ratios, respectively. The differences in the two ratios reflect the oxygen demand from nitrification associated with the BOD_5 data (see Hall and Foxen, 1984).

The assumption in this book that all BOD_5 concentrations reported to USEPA are actually $CBOD_5$ concentrations reduces the focus to only $CBOD_u$:$CBOD_5$ ratios as they relate to various levels of municipal wastewater treatment. Table 2-3 presents conversion ratios for four wastewater treatment types—raw, less than secondary, secondary, and greater than secondary. The formula for this conversion is:

$$\mathrm{CBOD_u} = \mathrm{CBOD_5}\,(\,\mathrm{CBOD_u{:}CBOD_5}\ \text{ratio}\,) \tag{2.1}$$

Determination of NBOD$_u$ Recall that nitrogen in wastewater generally appears as organic nitrogen compounds and ammonia and that the organic nitrogen fraction can be remineralized to ammonia and contribute to the oxygen demand in a receiving water. NBOD, therefore, is defined as the oxygen equivalent of the sum of organic nitrogen and ammonia. Conveniently, total Kjeldahl nitrogen (TKN) is defined as the sum of organic nitrogen and ammonia-nitrogen and can be used with the stoichiometric equivalent oxygen–nitrogen ratio (O2:N). A total of 4.57 g oxygen per 1 g nitrogen consumed in the nitrification process provides the basis for estimating the $NBOD_u$ of a sample. The formula for converting TKN to $NBOD_u$ concentration is:

$$\mathrm{NBOD_u} = 4.57\,(\,\mathrm{TKN}\,) \tag{2.2}$$

Determination of BOD$_u$ The ultimate biochemical oxygen demand is determined by simply adding the carbonaceous and nitrogenous components:

$$\mathrm{BOD_u} = (\,\mathrm{CBOD_u}\,) + (\,\mathrm{NBOD_u}\,) \tag{2.3}$$

Trends in POTW Inventory

USPHS municipal wastewater inventories and the USEPA Clean Water Needs Surveys were the primary data sources used to document the inventory of POTWs in the United States before and after the CWA. Table 2-4 presents the national inventory for select years from 1940 to 1996 organized by treatment type. Figure 2-8 is a column chart displaying the POTW inventory data. The "No Discharge" category (data available beginning in 1972) refers to facilities that do not discharge their effluent to surface waters. Most facilities that fall into this category are oxidation or stabilization

TABLE 2-3 $CBOD_u$:$CBOD_5$ Conversion Ratios and Type of Municipal Wastewater Treatment

Municipal Wastewater Treatment Type	$CBOD_u$:$CBOD_5$
Raw	1.2
Less than secondary	1.6
Secondary	2.84
Greater than secondary	2.9

Sources: Thomann and Mueller (1987), Leo et al. (1984).

Notes: Less than secondary = Primary and advanced primary wastewater treatment. Greater than secondary = Advanced secondary and advanced wastewater treatment.

TABLE 2-4 Inventory of POTWs by Wastewater Type, 1940–1996

		Treatment Type				
Year	Total	Raw	Less than Secondary	Secondary	Greater than Secondary	No Discharge
1940	NA	NA	2,938	2,630	0	NA
1950	11,784	5,156	3,099	3,529	0	NA
1962	11,698	2,262	2,717	6,719	0	NA
1968	14,051	1,564	2,435	10,042	10	NA
1972	19,355	2,265	2,594	9,426[a]	461	142
1978	14,850	91	4,278	6,608	2,888	985
1982	15,662	237	3,119	7,946	2,760	1,600
1988	15,708	117	1,789	8,536	3,412	1,854
1992	15,613	0	868	9,086	3,678	1,981
1996	16,024	0	176	9,388	4,428	2,032

Sources: U.S. Public Health Service Municipal Wastewater Inventories and USEPA Clean Water Needs Surveys.

Notes: Less than secondary facilities = Primary + advanced primary; Greater than secondary facilities = Advanced secondary + advanced treatment; No discharge = Facilities that do not discharge effluent to surface waters.

[a]This total excludes 4,467 oxidation ponds and 142 land application facilities classified as secondary treatment facilities in USEPA's 1972 inventory of municipal wastewater facilities (USEPA, 1972). They were excluded because (1) USEPA did not categorize oxidation ponds as secondary treatment facilities in any other year covered in this analysis and (2) land application facilities are classified as "no discharge" facilities in subsequent years. To be consistent with data compiled after 1972, the 142 land application facilities were included in the "no discharge" category for 1972.

ponds designed for evaporation and/or infiltration of effluent. Other examples of "No Discharge" facilities include recycling, reuse, and spray irrigation systems.

Key observations from Table 2-4 and Figure 2-8 include the following:

- The total number of POTWs in the nation increased by about 36 percent between 1950 and 1996.
- POTWs providing only raw and less than secondary treatment decreased in proportion to facilities providing secondary and greater than secondary treatment during the 1950–1996 time period. In 1950, only 30 percent of POTWs nationwide (3,529 of 11,784 facilities) provided secondary treatment. In 1968, 4 years before the CWA, 72 percent of the POTWs (10,052 of 14,051 facilities) had secondary treatment or greater. By 1996, 24 years after the 1972 CWA, 99 percent of the Nation's 16,024 POTWs were providing either secondary treatment or greater or were not discharge facilities.
- In 1968, 72 percent of the nation's POTWs were providing secondary treatment and less than 1 percent were providing greater than secondary treatment. By 1996, 59 percent of POTWs were providing secondary treatment and 27 percent had greater than secondary treatment.

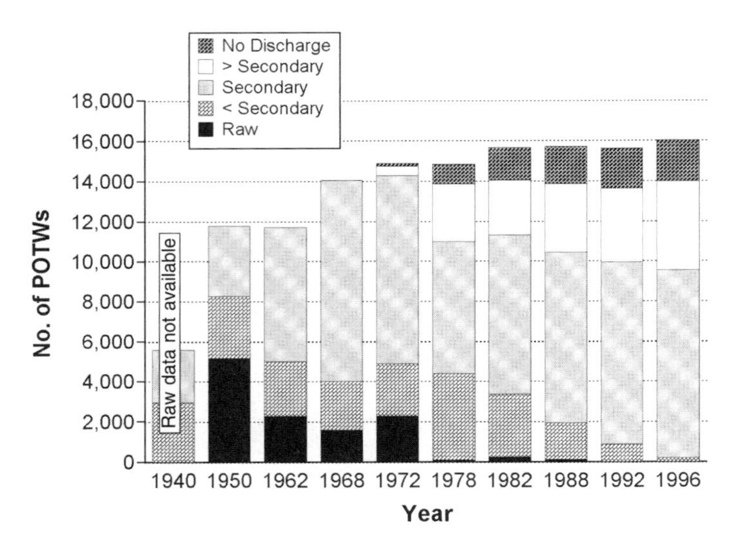

Figure 2-8 Number of POTWs nationwide for select years between 1940 and 1996 organized by wastewater treatment type. *Sources:* U.S. Public Health Service Municipal Wastewater Inventories and USEPA Clean Water Needs Surveys.

Trends in Population and Influent Wastewater Flow to POTWs

U.S. Public Health Service (USPHS) municipal wastewater inventories and the USEPA Clean Water Needs Surveys were the primary data sources used to document the population served by POTWs and the rate of influent wastewater flow to them between 1940 and 1996. Actual influent wastewater flow data were available from reports prepared for 1978, 1980, 1982, 1984, and 1986. For the years in which these data were not available, influent wastewater flow data were estimated based on the population served and an assumed constant normalized flow rate of 165 gallons per capita per day (gpcd). The influent wastewater flow rate of 165 gpcd is based on the mean of the total population served and wastewater flow data compiled in the USEPA Clean Water Needs Surveys for the five years for which actual wastewater flow data were reported (data ranged from 160 to 173 gpcd).

Influent wastewater includes residential (55 percent), commercial and industrial (20 percent), stormwater (4 percent), and infiltration and inflow (20 percent) sources of wastewater flow (AMSA, 1997). The constant per capita flow rate of 165 gpcd used in this study is identical to the typical U.S. average within the wide range (65 to 290 gpcd) of municipal water use that accounts for residential, commercial and industrial, and public water uses in the United States (Metcalf and Eddy, 1991).

Table 2-5 presents the population served by POTWs and the rate of influent wastewater flow to POTWs nationally for select years from 1940 to 1996. Figure 2-9 is a column chart displaying the population data.

TABLE 2-5 Population Served and Influent Wastewater Flow to POTWs by Wastewater Treatment Type, 1940–1996

	Population Served by POTWs (millions)					
Year	Total	Raw	Less than Secondary	Secondary	Greater than Secondary	No Discharge
1940	70.8	32.2	18.5	20.1	0.0	NA
1950	91.8	35.3	24.6	31.9	0.0	NA
1962	118.3	14.6	42.2	61.5	0.0	NA
1968	140.1	10.1	44.1	85.6	0.3	NA
1972	141.7	4.9	51.9	76.3	7.8	0.8
1978	155.2	3.6	44.1	56.3	49.1	2.2
1982	163.5	1.9	33.6	67.6	56.3	4.2
1988	177.5	1.4	26.5	78.0	65.7	6.1
1992	180.6	0.0	21.7	82.9	68.2	7.8
1996	189.7	0.0	17.2	81.9	82.9	7.7
	Influent Wastewater Flow (mgd)					
Year	Total	Raw	Less than Secondary	Secondary	Greater than Secondary	No Discharge
1940	11,682	5,313	3,053	3,317	0	NA
1950	15,141	5,819	4,059	5,263	0	NA
1962	19,520	2,409	6,963	10,148	0	NA
1968	23,117	1,667	7,277	14,124	50	NA
1972	23,384	815	8,560	12,585	1,288	136
1978	26,800	601	7,152	10,139	8,545	363
1982	27,203	310	5,301	11,010	10,092	491
1988	29,294	226	4,370	12,863	10,832	1,003
1992	29,801	0	3,583	13,680	11,258	1,281
1996	31,302	0	2,834	13,521	13,683	1,264

Sources: U.S. Public Health Service Municipal Wastewater Inventories and USEPA Clean Water Needs Surveys.

Notes: Less than secondary facilities = Primary + advanced primary; Greater than secondary facilities = Advanced secondary + advanced treatment; No discharge = Facilities that do not discharge effluent to surface waters.

Key observations from Table 2-5 and Figure 2-9 include the following:

- The population served by POTWs in the nation increased significantly, from about 91.8 million people in 1950 to about 140.1 million in 1968 (four years before the 1972 CWA). By 1996, 189.7 million people were connected to POTWs, a 35 percent increase from 1968.
- The number of people relying on POTWs with less than secondary treatment dropped rapidly after passage of the 1972 CWA. In 1968 (4 years before the

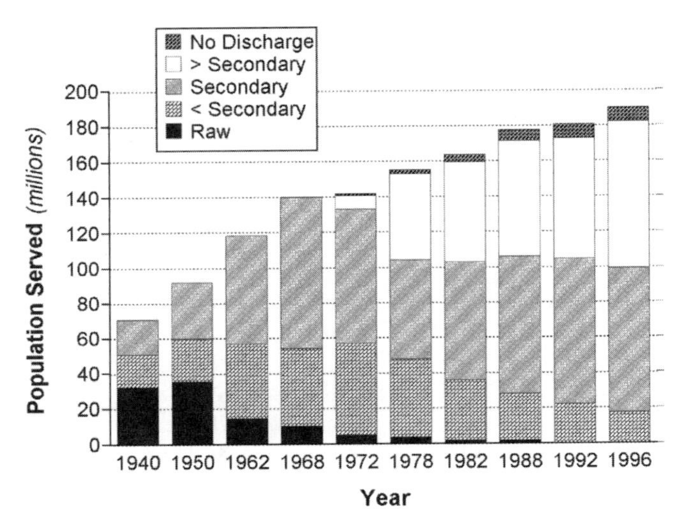

Figure 2-9 Population served by POTWs nationwide for select years between 1940 and 1996 organized by wastewater treatment type. *Sources:* U.S. Public Health Service Municipal Wastewater Inventories and USEPA Clean Water Needs Surveys.

CWA), about 39 percent of the 140.1 million people were served by POTWs providing only raw or less than secondary wastewater treatment. By 1996 (24 years after the 1972 CWA), this percentage was reduced to about 9 percent; only 17.2 million people of the 189.7 million served by POTWs received less than secondary wastewater treatment.

- Stated another way, the U.S. population served by POTWs with secondary or greater treatment almost doubled between 1968 and 1996, from 85.9 million people in 1968 to 164.8 million people in 1996! [It is noted that 5.1 million of the 17.2 million people served by less than secondary facilities in 1996 were connected to 45 POTW facilities granted CWA Section 301(h) waivers (nine pending final waiver decision as of November 1998), which allow the discharge of primary or advanced primary effluent to deep, well-mixed ocean waters.]
- Although the number of people served by POTWs with secondary treatment remained fairly constant between 1968 and 1996 (a slight decrease of 3.7 million people or about 4 percent of the population), the number of people provided with greater than secondary treatment increased significantly (from 0.3 million people in 1968 to 82.9 million people in 1996). This is consistent with the trend since 1968 in increasing numbers of POTWs providing greater than secondary treatment, as shown in Table 2-4.

Trends in Influent BOD Loading to POTWs

Table 2-6 presents nationwide influent loading of $CBOD_5$, $CBOD_u$, $NBOD_u$, and BOD_u organized by wastewater treatment type for select years from 1940 to 1996.

TABLE 2-6 Influent BOD Loading to POTWs by Wastewater Treatment Type, 1940–1996

	Influent $CBOD_5$ Loading (mt/day)					
Year	Total	Raw	Less than Secondary	Secondary	Greater than Secondary	No Discharge
1940	9,508	4,324	2,484	2,699	0	NA
1950	12,323	4,736	3,303	4,283	0	NA
1962	15,886	1,961	5,667	8,259	0	NA
1968	18,814	1,356	5,922	11,495	40	NA
1972	19,032	663	6,967	10,242	1,049	111
1978	21,253	489	5,721	8,222	6,526	295
1982	21,170	252	4,280	8,623	7,616	400
1988	23,841	184	3,557	10,468	8,816	816
1992	24,254	0	2,916	11,134	9,162	1,043
1996	25,476	0	2,307	11,004	11,136	1,029

	Influent $CBOD_u$ Loading (mt/day)					
Year	Total	Raw	Less than Secondary	Secondary	Greater than Secondary	No Discharge
1940	11,409	5,189	2,981	3,239	0	NA
1950	14,787	5,683	3,964	5,140	0	NA
1962	19,063	2,353	6,800	9,910	0	NA
1968	22,576	1,628	7,107	13,794	48	NA
1972	22,838	796	8,360	12,291	1,258	133
1978	25,503	587	6,865	9,866	7,831	354
1982	25,405	302	5,136	10,348	9,139	479
1988	28,609	220	4,268	12,562	10,579	980
1992	29,105	0	3,499	13,360	10,995	1,251
1996	30,571	0	2,768	13,205	13,363	1,235

	Influent NBOD Loading (mt/day)					
Year	Total	Raw	Less than Secondary	Secondary	Greater than Secondary	No Discharge
1940	6,123	2,785	1,600	1,738	0	NA
1950	7,936	3,050	2,128	2,758	0	NA
1962	10,232	1,263	3,650	5,319	0	NA
1968	12,117	874	3,814	7,403	26	NA
1972	12,257	427	4,487	6,597	675	71
1978	14,047	315	3,749	5,314	4,479	190
1982	14,259	162	2,778	5,771	5,290	257
1988	15,355	118	2,291	6,742	5,678	526
1992	15,621	0	1,878	7,171	5,901	672
1996	16,408	0	1,486	7,087	7,172	663

TABLE 2-6 *Continued*

	Influent BOD_u Loading (mt/day)					
Year	Total	Raw	Less than Secondary	Secondary	Greater than Secondary	No Discharge
1940	17,532	7,974	4,581	4,977	0	NA
1950	22,723	8,734	6,092	7,898	0	NA
1962	29,295	3,615	10,450	15,229	0	NA
1968	34,693	2,501	10,921	21,197	74	NA
1972	35,095	1,223	12,847	18,887	1,933	204
1978	39,551	901	10,614	15,181	12,310	544
1982	39,663	465	7,914	16,118	14,429	737
1988	43,964	339	6,558	19,304	16,257	1,506
1992	44,726	0	5,377	20,531	16,896	1,923
1996	46,979	0	4,254	20,292	20,536	1,897

Sources: U.S. Public Health Service Municipal Wastewater Inventories and USEPA Clean Water Needs Surveys.

Notes: Less than secondary facilities = Primary + advanced primary; Greater than secondary facilities = Advanced secondary + advanced treatment; No discharge = Facilities that do not discharge effluent to surface waters.

Data Sources and Calculations The USEPA Clean Water Needs Surveys were the primary data source used to estimate the nationwide rate of influent $CBOD_5$ loading to POTWs. Actual influent $CBOD_5$ loading data were reported for 1978, 1980, 1982, 1984, and 1986. For the years for which data were not available, per capita influent loading was assumed to be 0.296 lb $CBOD_5$ per person per day. This rate was based on an estimated constant normalized flow rate of 165 gpcd and an influent $CBOD_5$ concentration of 215 mg/L. The use of 215 mg/L as the influent $CBOD_5$ concentration is consistent with several other estimates of raw wastewater strength (e.g., AMSA, 1997; Tetra Tech, 1999; Metcalf and Eddy, 1991). It also is the mean nationally aggregated ratio of the total influent $CBOD_5$ loading rate normalized to total wastewater flow reported in the USEPA Clean Water Needs Surveys for the 5 years that actual wastewater flow data were reported (range from 209 to 229 mg/L). Sources of influent BOD include residential, commercial and industrial, and infiltration and inflow contributions.

Some readers might note that an influent loading rate of 0.296 lb $CBOD_5$ per person per day is almost twice the typical "textbook" value of 0.17 lb $CBOD_5$ per person per day, sometimes referred to as the "population equivalent" (PE) loading rate. Textbook values, however, usually account for only the average per capita residential load contributed by combined stormwater and domestic wastewater. The industrial and commercial components are excluded (see Fair et al., 1971). To provide a more complete characterization of influent BOD loading inclusive of all sources, the higher figure was used in this study.

$CBOD_u$ data were determined using $CBOD_5$ data and Equation 2.1 as follows:

$$CBOD_u = 1.2\,(\,CBOD_5\,) \tag{2.4}$$

where the multiplier factor of 1.2 is the $CBOD_u$:$CBOD_5$ ratio associated with raw wastewater.

The USEPA Clean Water Needs Surveys were the primary data source used to estimate the nationwide rate of influent $NBOD_u$ loading to POTWs. Actual influent TKN loading data were reported for 1978, 1980, 1982, 1984, and 1986. For the years for which wastewater flow data were not available, per capita influent loading was assumed to be 0.191 lb NBOD per person per day. This rate was based on an estimated constant normalized flow rate of 165 gpcd and an influent TKN concentration of 30.3 mg/L, a level derived from an analysis of about 100 wastewater facilities (AMSA, 1997). Influent $NBOD_u$ loading was determined using influent TKN data and Equation 2.2.

Trends in Influent $CBOD_5$ and BOD_u Loading to POTWs Figure 2-10 is a column chart that compares total influent $CBOD_5$ and BOD_u loading from 1940 to 1996. Figures 2-11 (*a*) and (*b*) display influent $CBOD_5$ and BOD_u loading data, respectively, organized by wastewater treatment type.

Key observations from Table 2-6 and Figures 2-10 and 2-11 include the following:

- Influent BOD loading to the nation's POTWs more than doubled from 1940 to 1996, reflecting population growth, increases in the number of facilities, and expanding service areas.

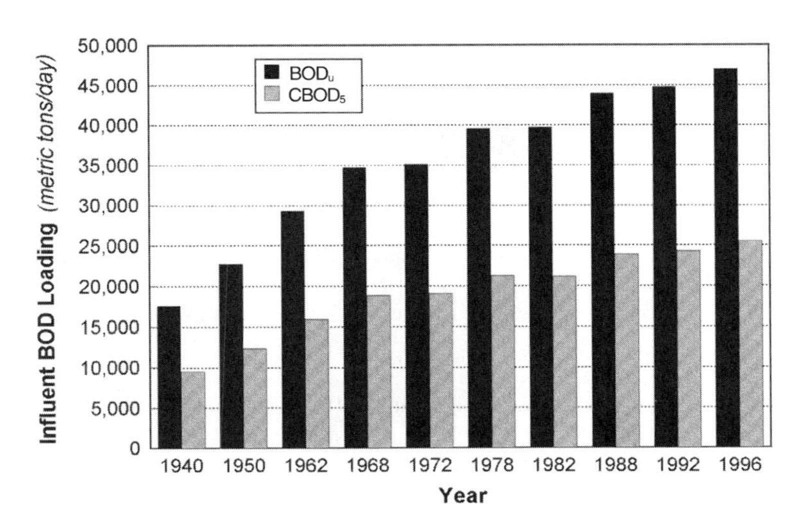

Figure 2-10 Total influent BOD_u and $CBOD_5$ loading, 1940 to 1996. *Sources:* U.S. Public Health Service Municipal Wastewater Inventories and USEPA Clean Water Needs Surveys.

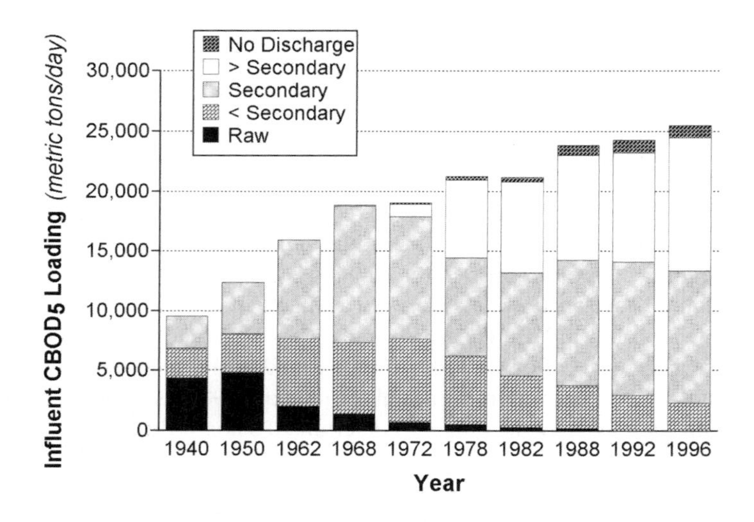

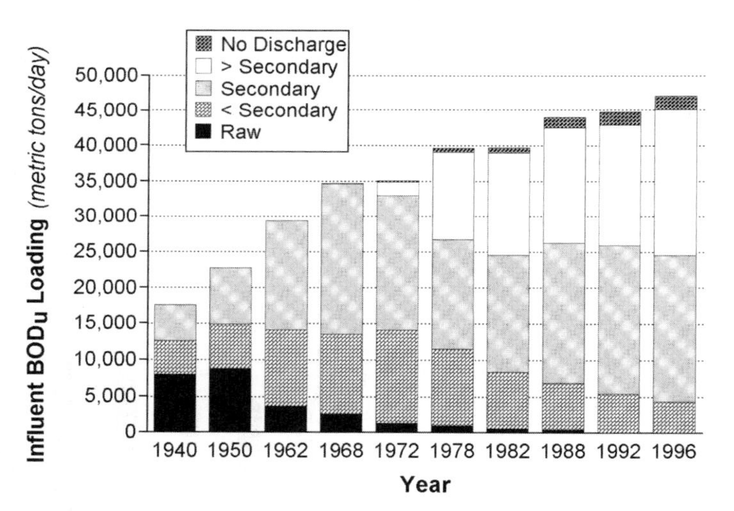

Figure 2-11 Influent loading of (*a*) $CBOD_5$ and (*b*) BOD_u to POTWs nationwide for select years between 1940 and 1996 organized by wastewater treatment type. *Sources:* U.S. Public Health Service Municipal Wastewater Inventories and USEPA Clean Water Needs Surveys.

- Influent $CBOD_5$ loading increased from 9,508 mt/day in 1940 to 18,814 mt/day in 1968. By 1996, influent $CBOD_5$ loading stood at 25,476 mt/day, a 35 percent increase from 1968.
- Influent BOD_u loading increased from 17,532 mt/day in 1940 to 34,693 mt/day in 1968. By 1996, influent BOD_u loading stood at 46,979 mt/day, a 35 percent increase from 1968.
- In 1940, 72 percent of influent BOD_u loading nationwide was being treated by

facilities with less than secondary treatment (12,555 of 17,532 mt/day of BOD_u). By 1968, 39 percent of influent BOD_u loading nationwide was being treated by facilities with less than secondary treatment (13,422 of 34,693 mt/day of BOD_u). Twenty-four years after the 1972 CWA, only 9 percent of influent BOD_u loading was being treated by facilities with less than secondary treatment (4,254 of 46,979 mt/day of BOD_u).

Trends in Effluent BOD Loading from POTWs

Table 2-7 presents nationwide effluent loading of $CBOD_5$, $CBOD_u$, $NBOD_u$, and BOD_u organized by wastewater treatment type for select years from 1940 to 1996.

Data Sources and Calculations Effluent $CBOD_5$ loading rates were estimated based on influent $CBOD_5$ loading rates and $CBOD_5$ removal efficiencies (expressed as a percentage) associated with each type of municipal wastewater treatment (see Table 2-2). In keeping with the convention of combining primary and advanced primary treatment and designating the result as "less than secondary" treatment, $CBOD_5$ removal efficiencies for these two categories were averaged to derive a "less than secondary" treatment removal efficiency of 42.5 percent. Likewise, $CBOD_5$ removal efficiencies assigned to advanced secondary treatment (90 percent) and advanced wastewater treatment (95 percent) were averaged to derive a "greater than secondary" treatment removal efficiency of 92.5 percent. Table 2-8 presents $CBOD_5$ removal efficiencies by municipal wastewater treatment type and corresponding effluent $CBOD_5$ concentrations.

Recall that the $CBOD_5$ removal efficiencies used in this study are percentages typically assigned to NPDES permits according to the treatment process and POTW design assumptions (USEPA, 1978; Metcalf and Eddy, 1991). Use of "design-based" removal efficiencies may, in some cases, result in a conservative (i.e., high) estimate of effluent $CBOD_5$ loading. USEPA's Clean Water Needs Surveys for the years 1976, 1978, and 1982, for example, report 41 and 64 percent $CBOD_5$ removal efficiency for primary and advanced primary facilities, respectively. These same reports present removal efficiencies for secondary (82 to 86 percent), advanced secondary (89 to 92 percent), and advanced wastewater treatment (87 to 94 percent) either in the range or very near to the range of design-based removal efficiencies. The design-based $CBOD_5$ removal efficiencies were chosen for use in this study over actual reported efficiencies because it was assumed that a conservative approach would provide a more effective and consistent comparison of trends for POTW BOD removal over the entire period of record analyzed.

Effluent $CBOD_u$ loading rates were estimated for each category of wastewater treatment from effluent $CBOD_5$ loading rates and the corresponding $CBOD_u$:$CBOD_5$ ratio (see Table 2-3) using Equation 2.1. $CBOD_u$ percent removal efficiencies for each treatment category were then computed from the influent (I) and effluent (E) loading rates as:

$$\text{Percent removal efficiency} = 100\,(I - E)/I \qquad (2.5)$$

TABLE 2-7 Effluent BOD Loading to POTWs by Wastewater Treatment Type, 1940–1996

	Effluent $CBOD_5$ Loading (mt/day)					
Year	Total	Raw	Less than Secondary	Secondary	Greater than Secondary	No Discharge
1940	6,344	4,324	1,615	405	0	NA
1950	7,526	4,736	2,147	642	0	NA
1962	6,883	1,961	3,684	1,239	0	NA
1968	6,932	1,356	3,849	1,724	2	NA
1972	6,768	663	4,501	1,536	68	0
1978	5,510	489	2,654	1,596	771	0
1982	4,380	252	1,975	1,539	614	0
1988	4,460	184	2,045	1,570	661	0
1992	4,034	0	1,677	1,670	687	0
1996	3,812	0	1,326	1,651	835	0
	Effluent $CBOD_u$ Loading (mt/day)					
Year	Total	Raw	Less than Secondary	Secondary	Greater than Secondary	No Discharge
1940	8,922	5,189	2,584	1,150	0	NA
1950	10,943	5,683	3,436	1,825	0	NA
1962	11,765	2,353	5,894	3,518	0	NA
1968	12,689	1,628	6,159	4,897	6	NA
1972	12,558	796	7,201	4,363	198	0
1978	11,621	587	4,246	4,533	2,255	0
1982	9,582	302	3,160	4,371	1,749	0
1988	9,869	220	3,272	4,460	1,918	0
1992	9,418	0	2,683	4,743	1,993	0
1996	9,232	0	2,122	4,688	2,422	0
	Effluent NBOD Loading (mt/day)					
Year	Total	Raw	Less than Secondary	Secondary	Greater than Secondary	No Discharge
1940	5,146	2,785	1,248	1,113	0	NA
1950	6,475	3,050	1,660	1,765	0	NA
1962	7,514	1,263	2,847	3,404	0	NA
1968	8,591	874	2,975	4,738	4	NA
1972	8,273	427	3,500	4,222	125	0
1978	7,526	315	2,924	3,401	886	0
1982	7,168	162	2,167	3,693	1,145	0
1988	7,327	118	1,787	4,315	1,107	0
1992	7,205	0	1,465	4,589	1,151	0
1996	7,093	0	1,159	4,536	1,399	0

TABLE 2-7 *Continued*

	Effluent BOD_u Loading (mt/day)					
Year	Total	Raw	Less than Secondary	Secondary	Greater than Secondary	No Discharge
1940	14,068	7,974	3,832	2,262	0	NA
1950	17,419	8,734	5,095	3,590	0	NA
1962	19,278	3,615	8,740	6,922	0	NA
1968	21,281	2,501	9,134	9,635	11	NA
1972	20,831	1,223	10,701	8,585	322	0
1978	19,147	901	7,171	7,934	3,141	0
1982	16,750	465	5,327	8,064	2,894	0
1988	17,196	339	5,059	8,774	3,025	0
1992	16,623	0	4,147	9,332	3,144	0
1996	16,325	0	3,281	9,224	3,821	0

Sources: U.S. Public Health Service Municipal Wastewater Inventories and USEPA Clean Water Needs Surveys.

Notes: Less than secondary facilities = Primary + advanced primary; Greater than secondary facilities = Advanced secondary + advanced treatment; No discharge = Facilities that do not discharge effluent to surface waters.

Table 2-9 presents the calculated $CBOD_u$ removal efficiencies by municipal wastewater treatment type and corresponding effluent $CBOD_u$ concentrations.

Effluent $NBOD_u$ loading rates were estimated based on influent $NBOD_u$ and $NBOD_u$ removal efficiencies reported for TKN (expressed as a percentage) associated with each category of wastewater treatment. Removal efficiencies for TKN were based on data compiled in Gunnerson et. al (1982) for primary facilities, AMSA (1997) for secondary facilities, and AMSA (1997) and MWCOG (1989) for advanced wastewater treatment facilities. Since $NBOD_u$ is estimated from TKN and the constant stoichiometric ratio of 4.57 g O_2 per g N, removal efficiencies for TKN and $NBOD_u$

TABLE 2-8 $CBOD_5$ Removal Efficiencies by Municipal Wastewater Treatment Type and Corresponding Effluent $CBOD_5$ Concentrations

Muncipal Wastewater Treatment Type	Removal Efficiency (%)	Effluent Concentration (mg/L)
Raw	0.0	215.0
Less than secondary	42.5	123.6
Secondary	85.0	32.3
Greater than secondary	92.5	16.1

Notes: Raw = Effluent concentration of 215 mg/L $CBOD_5$ equivalent to raw (untreated) influent concentration; Less than secondary = Primary and advanced primary wastewater treatment; Greater than secondary = Advanced secondary and advanced wastewater treatment.

TABLE 2-9 $CBOD_u$ Removal Efficiencies by Municipal Wastewater Treatment Type and Corresponding Effluent $CBOD_u$ Concentrations

Muncipal Wastewater Treatment Type	Removal Efficiency (%)	Effluent Concentration (mg/L)
Raw	0.0	258.0
Less than secondary	23.3	197.8
Secondary	64.5	91.6
Greater than secondary	81.9	46.8

Notes: Raw = Effluent concentration of 258 mg/L $CBOD_u$ equivalent to raw (untreated) influent concentration; Less than secondary = Primary and advanced primary wastewater treatment; Greater than secondary = Advanced secondary and advanced wastewater treatment.

have the same value for the various categories of wastewater treatment. Table 2-10 presents TKN removal efficiencies and effluent concentrations as TKN and $NBOD_u$.

The effluent BOD_u loading rates were determined by adding the calculated $CBOD_u$ and $NBOD_u$ loading rates. BOD_u removal efficiencies for each treatment category were then computed from the influent (I) and effluent (E) BOD_u loading rates according to Equation 2.4. Table 2-11 presents the calculated BOD_u removal efficiencies by municipal wastewater treatment type and corresponding effluent BOD_u concentrations.

Trends in Effluent $CBOD_5$ and BOD_u Loading From POTWs Figure 2-12 is a chart that compares effluent $CBOD_5$ and BOD_u loading over the same time period. Figures 2-13(a) and (b) display effluent $CBOD_5$ and BOD_u loading data organized by wastewater treatment type.

Key observations from Table 2-7 and Figures 2-12 and 2-13 include the following:

TABLE 2-10 TKN and NBOD Removal Efficiencies by Municipal Wastewater Treatment Type and Corresponding Effluent TKN and NBOD Concentrations

Muncipal Wastewater Treatment Type	Removal Efficiency (%)	Effluent Concentration (mg/L)	
		TKN	NBOD
Raw	0.0	30.3	138.5
Less than secondary	22.0	23.6	108.0
Secondary	36.0	19.4	88.6
Greater than secondary	80.5	5.9	27.0

Notes: Raw = Effluent concentration of 30.3 mg/L TKN and 138.5 NBOD equivalent to raw (untreated) influent concentrations; Less than secondary = Primary and advanced primary wastewater treatment; Greater than secondary = Advanced secondary and advanced wastewater treatment.

TABLE 2-11 BOD_u Removal Efficiencies by Municipal Wastewater Treatment Type and Corresponding Effluent BOD_u Concentration

Muncipal Wastewater Treatment Type	Removal Efficiency (%)	Effluent Concentration (mg/L)
Raw	0.0	396.5
Less than secondary	22.9	305.8
Secondary	54.5	180.2
Greater than secondary	81.4	73.8

Notes: Raw = Effluent concentration of 396.5 mg/L BOD_u equivalent to raw (untreated) influent concentration; Less than secondary = Primary and advanced primary wastewater treatment; Greater than secondary = Advanced secondary and advanced wastewater treatment.

- Effluent BOD loading by POTWs was significantly reduced between 1968 and 1996. In 1968, 4 years before the 1972 CWA, effluent $CBOD_5$ and BOD_u loadings were 6,932 and 21,281 mt/day, respectively. By 1996, $CBOD_5$ and BOD_u loadings were reduced to 3,812 and 16,325 mt/day, respectively. This represents a 45 percent decline in $CBOD_5$ and a 23 percent decline in BOD_u between 1968 and 1996. Notably, these declines were achieved even though influent $CBOD_5$ and BOD_u loading to POTWs each increased by 35 percent during the same time period!

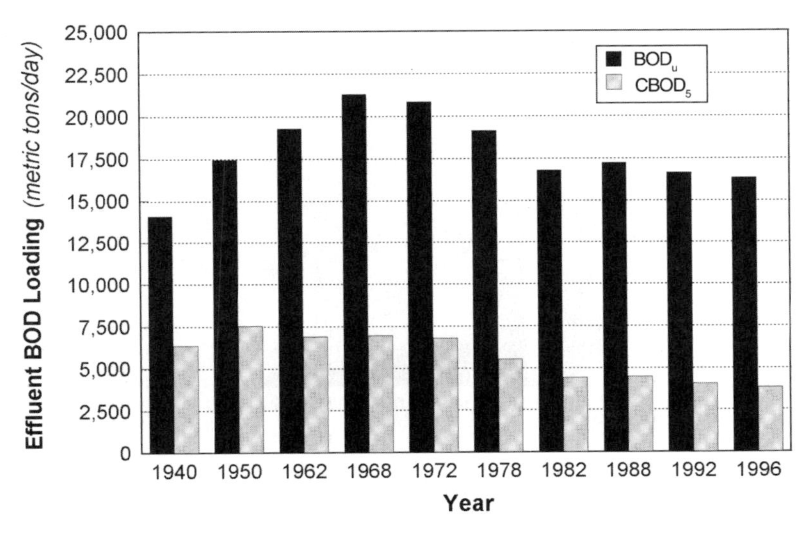

Figure 2-12 Total effluent BOD_u and $CBOD_5$ loading, 1940 to 1996. *Sources:* U.S. Public Health Service Municipal Wastewater Inventories and USEPA Clean Water Needs Surveys.

- The proportion of effluent $CBOD_5$ loading attributable to raw and less than secondary wastewater treatment was reduced from about 94 percent in 1940 to 35 percent in 1996 [see Figure 2-13(*a*)]. The proportion of effluent BOD_u loading attributable to raw and less than secondary wastewater treatment was reduced from about 84 percent in 1940 to 20 percent in 1996 [see Figure 2-13(*b*)].

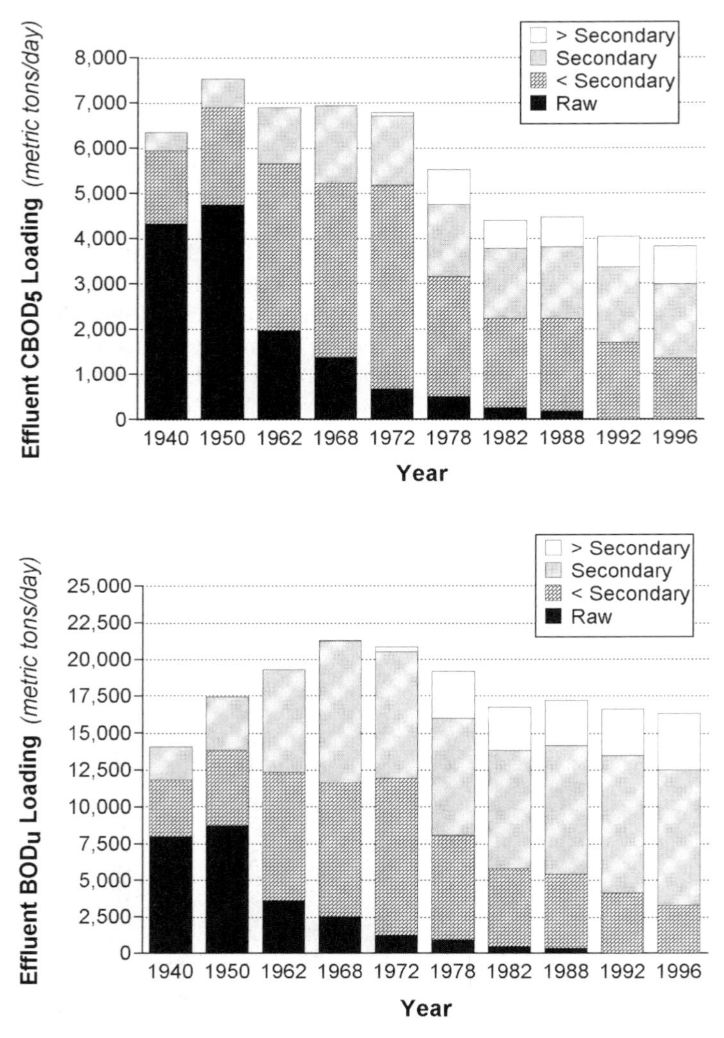

Figure 2-13 Effluent loading of (*a*) $CBOD_5$ and (*b*) BOD_u from POTWs nationwide for select years between 1940 and 1996 organized by wastewater treatment type. *Sources:* U.S. Public Health Service Municipal Wastewater Inventories and USEPA Clean Water Needs Surveys.

Trends in BOD Removal Efficiency

The rate of effluent BOD loading from a POTW is determined by two main factors: the rate of influent BOD loading and the BOD removal efficiency of the facility. Influent BOD loading, in turn, is determined by the number of people connected to the system and the rate at which they generate and export BOD in their wastewater flow. The analysis above indicates that tremendous progress was achieved between 1968 and 1996 in reducing effluent BOD loading from POTWs into the nation's waterways. Notably, this reduction occurred at the same time the number of people served by POTWs was increasing rapidly. Figures 2-14 and 2-15 present influent and effluent loadings and removal efficiencies for $CBOD_5$ and BOD_u, respectively.

Key observations from Figures 2-14 and 2-15 include the following:

- BOD removal efficiency nationwide significantly increased between 1940 and 1996. In 1940, the aggregate national removal efficiency stood at about 33 percent for $CBOD_5$ and 20 percent for BOD_u. By 1968, removal efficiencies had increased to 63 percent for $CBOD_5$ and 39 percent for BOD_u. By 1996, they had further increased to nearly 85 percent for $CBOD_5$ and 65 percent for BOD_u!
- The BOD removal efficiency increased substantially between 1972 and 1978, the 6-year period after the passage of the CWA (from 64 to 74 percent for $CBOD_5$ and from 41 to 52 percent for BOD_u). Between 1978 and 1996, removal

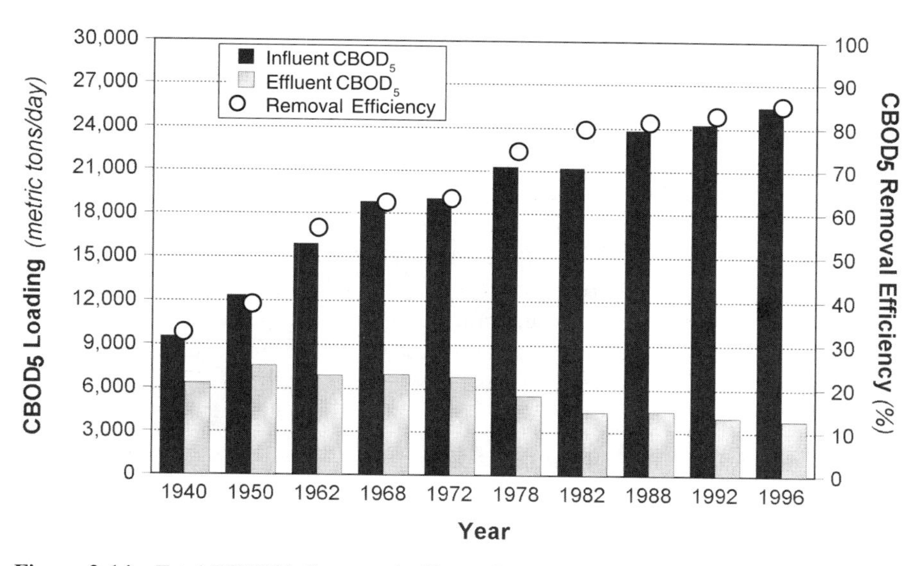

Figure 2-14 Total POTW influent and effluent $CBOD_5$ loading and corresponding $CBOD_5$ removal efficiency for select years between 1940 and 1996. *Sources:* U.S. Public Health Service Municipal Wastewater Inventories and USEPA Clean Water Needs Surveys.

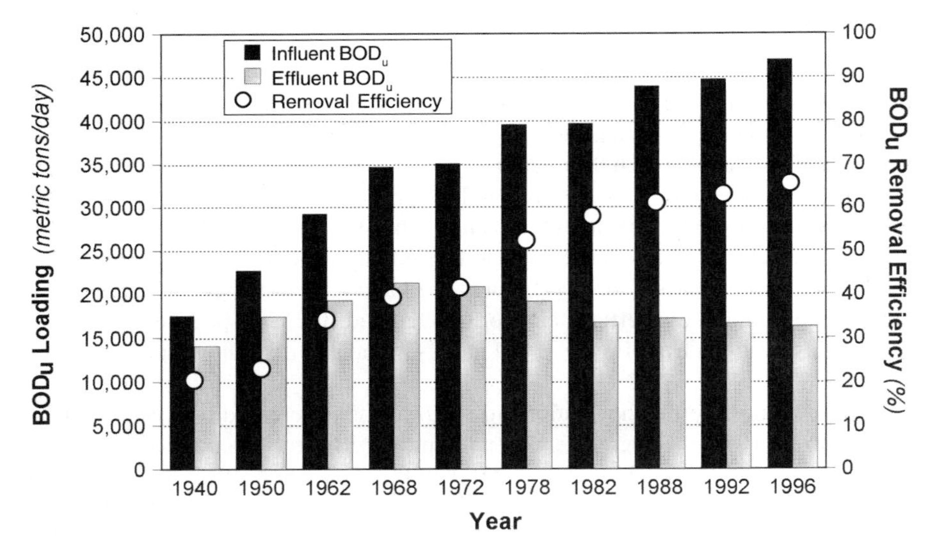

Figure 2-15 Total POTW influent and effluent BOD_u loading and corresponding BOD_u removal efficiency for select years between 1940 and 1996. *Sources:* U.S. Public Health Service Municipal Wastewater Inventories and USEPA Clean Water Needs Surveys.

efficiency increased an additional 11 percent for $CBOD_5$ and 13 percent for BOD_u. Those larger increases in BOD_u removal efficiency reflect the ever-increasing role of greater than secondary POTWs over this time period.

Figure 2-16, a three-dimensional graph of the population data presented earlier in Table 2-5 and Figure 2-9, is useful for visualizing the trends in population served by POTW treatment type. The population served by secondary treatment facilities declined sharply between 1968 (85.6 million) and 1978 (56.3 million) and then leveled off at about 82 million in the 1990s. In contrast, the number of people served by greater than secondary treatment surged between 1968 and 1978 (0.3 to 49.1 million) and then increased steadily to about 82.9 million in 1996. Unlike secondary treatment, advanced wastewater treatment enhances biological processes to incorporate nitrification (ammonia removal) and denitrification (nitrate removal), thus reducing the NBOD fraction of effluent BOD_u loading.

Future Trends in BOD Effluent Loading

The data presented in the previous sections indicate that the increase in BOD removal efficiency between 1940 and 1996 resulted in significant reductions in BOD effluent loading to the nation's waterways even though the number of people served by POTWs greatly increased. Given that the population served by POTWs is projected to continue to increase well into the twenty-first century, will the trend of effluent

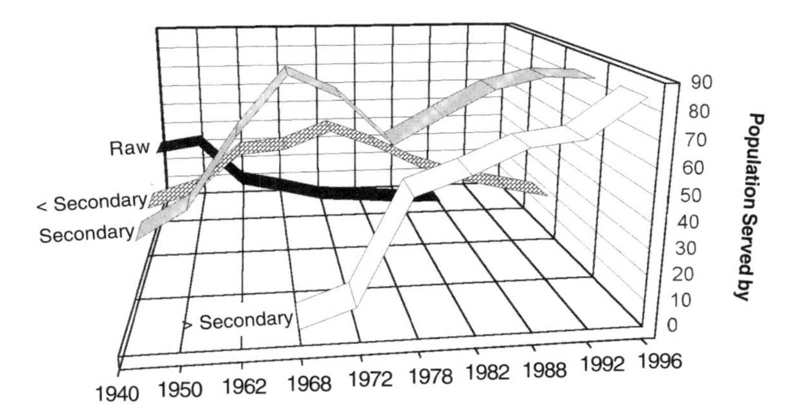

Figure 2-16 Population served by POTWs nationwide for select years between 1940 and 1996 organized by wastewater treatment type. *Sources:* U.S. Public Health Service Municipal Wastewater Inventories and USEPA Clean Water Needs Surveys.

BOD loading reductions also continue into the future? A preliminary examination of estimated influent and effluent BOD loadings based on USEPA projections of POTW facility inventories and population served for the year 2016 indicates that the answer might be "no."

Table 2-12 presents a summary of the population served, wastewater flow, influent and effluent BOD loading rates, and BOD removal efficiencies for 1996 and corresponding projections for 2016 and 2025. Figure 2-17 is a column chart that extends the influent and effluent BOD_u loading totals and POTW removal efficiencies originally presented in Figure 2-15 into the twenty-first century by adding columns for the years 2016 and 2025 to the chart. These projections are based on the following assumptions:

- USEPA Clean Water Needs Survey (USEPA, 1997) estimates that 275 million people will be served by POTWs in the year 2016. This figure is based on middle-level population projections from the Census Bureau (USBC, 1996) and the assumption that 88 percent of the population will be served by POTWs in 2016. Assuming that 88 percent of the population projected for 2025 is also served by POTWs, about 295 million people will be served by POTWs.
- Design-based BOD_u removal efficiency will increase from a nationwide average of 65 percent in 1996 to 71 percent by 2016 based on projections of population served by the different categories of POTWs. This removal efficiency is assumed to remain at that level through 2025.
- Influent wastewater flow will remain a constant 165 gpcd and influent BOD_u concentration will remain a constant 396.5 mg/L for the projection period from 1996 to 2025.

TABLE 2-12 1996 Estimates and 2016 and 2025 Projections of POTW Infrastructure and Influent and Effluent Loading by Treatment Type

	Total[a]	Raw	Less than Secondary	Secondary	Greater than Secondary	On-Site
		1996 Estimates				
Inventory of POTWs	16,024	0	176	9,388	4,428	2,032
Population of U.S. (millions)	263.4	—	—	—	—	—
Population served (millions)	189.7	0	17.2	81.9	82.9	7.7
Percent of population served	72%	—	—	—	—	—
Influent wastewater flow (mgd)	31,302	0	2,834	13,521	13,683	1,264
Unit flow (gpcd)	165	—	—	—	—	—
Influent $CBOD_5$ loading (mt/day)	25,476	0	2,307	11,004	11,136	1,029
Influent $CBOD_u$ loading (mt/day)	30,571	0	2,768	13,205	13,363	1,235
Influent $NBOD_u$ loading (mt/day)	16,408	0	1,486	7,087	7,172	663
Influent BOD_u loading (mt/day)	46,978	0	4,254	20,292	20,536	1,897
Effluent $CBOD_5$ loading (mt/day)	3,812	0	1,326	1,651	835	—
Effluent $CBOD_u$ loading (mt/day)	9,232	0	2,122	4,688	2,422	—
Effluent $NBOD_u$ loading (mt/day)	7,093	0	1,159	4,536	1,399-	—
Effluent BOD_u loading (mt/day)	16,325	0	3,281	9,224	3,821	—
$CBOD_5$ percent removal	85%	—	42%	85%	92%	—
$CBOD_u$ percent removal	70%	—	23%	64%	82%	—
$NBOD_u$ percent removal	57%	—	22%	36%	80%	—
BOD_u percent removal	65%	—	23%	54%	81%	—

	Total[a]	Raw	Less than Secondary	Secondary	Greater than Secondary	On-Site
		2016 Projections				
Inventory of POTWs	18,303	0	61	9,738	6,135	2,369
Population of U.S. (millions)	311.5	—	—	—	—	—
Population served (millions)	274.7	0	5.5	102.3	152.7	14.2
Percent of population served	88%	—	—	—	—	—
Influent wastewater flow (mgd)	45,329	0	910	16,883	25,200	2,337
Unit flow (gpcd)	165	—	—	—	—	—
Influent $CBOD_5$ loading (mt/day)	36,892	0	740	13,740	20,509	1,902
Influent $CBOD_u$ loading (mt/day)	44,270	0	888	16,489	24,611	2,282
Influent $NBOD_u$ loading (mt/day)	23,760	0	477	8,850	13,209	1,225
Influent BOD_u loading (mt/day)	68,030	0	1,365	25,338	37,819	3,507
Effluent $CBOD_5$ loading (mt/day)	4,025	0	426	2,061	1,538	—
Effluent $CBOD_u$ loading (mt/day)	10,995	0	681	5,853	4,461	—
Effluent $NBOD_u$ loading (mt/day)	8,611	0	372	5,664	2,576	—
Effluent BOD_u loading (mt/day)	19,607	0	1,053	11,517	7,036	—
$CBOD_5$ percent removal	89%	—	42%	85%	92%	—
$CBOD_u$ percent removal	75%	—	23%	64%	82%	—
$NBOD_u$ percent removal	64%	—	22%	36%	80%	—
BOD_u percent removal	71%	—	23%	54%	81%	—

TABLE 2-12 *Continued*

	Total[a]	Raw	Less than Secondary	Secondary	Greater than Secondary	On-Site
		2025 Projections				
Inventory of POTWs	—					
Population of U.S. (millions)	335.1					
Population served (millions)	295.5					
Percent of populaton served	88.2%					
Influent wastewater flow (mgd)	48,760					
Unit flow (gpcd)	165					
Influent $CBOD_5$ loading (mt/day)	39,684					
Influent $CBOD_u$ loading (mt/day)	47,620					
Influent $NBOD_u$ loading (mt/day)	25,558					
Influent BOD_u loading (mt/day)	73,179					
Effluent $CBOD_5$ loading (mt/day)	4,330					
Effluent $CBOD_u$ loading (mt/day)	11,827					
Effluent $NBOD_u$ loading (mt/day)	9,263					
Effluent BOD_u loading (mt/day)	21,090					
$CBOD_5$ percent removal	89%					
$CBOD_u$ percent removal	75%					
$NBOD_u$ percent removal	64%					
BOD_u percent removal	71%					

Sources: USEPA (1997) 1996 Clean Water Needs Survey and U.S. Census Bureau Population Projections.

Notes: Less than secondary = Primary and advanced primary wastewater treatment; Greater than secondary = Advanced secondary and advanced wastewater treatment.

[a]Totals only shown for 2025 projections.

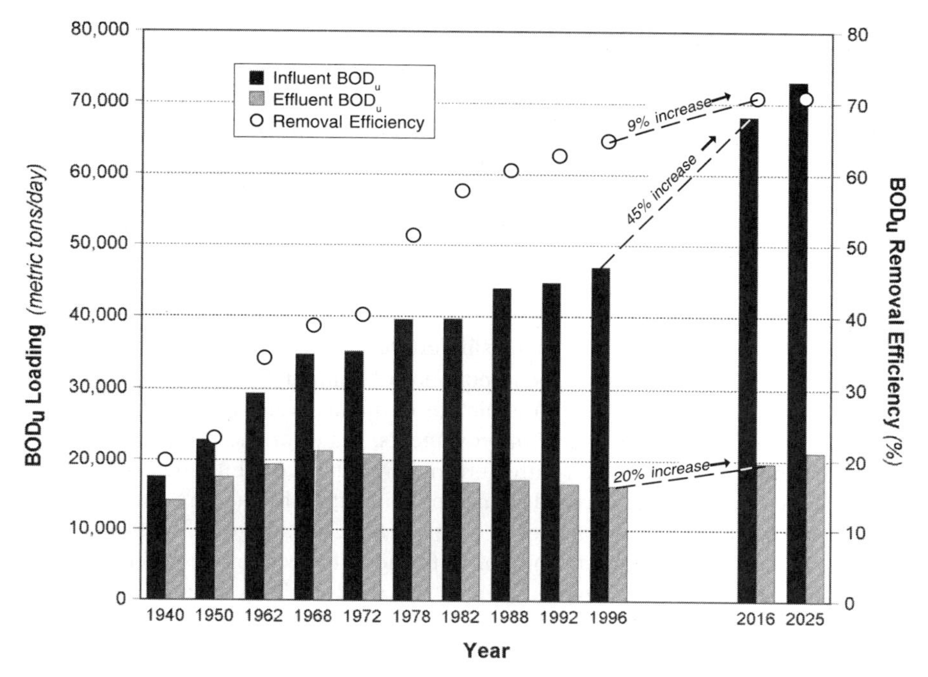

Figure 2-17 POTW influent and effluent BOD$_u$ loading and removal efficiency for select years between 1940 and 1996 and 2016 and 2025. *Sources:* U.S. Public Health Service Municipal Wastewater Inventories, USEPA Clean Water Needs Surveys, and U.S. Census Bureau Population Projections.

Key observations from Figure 2-17 include the following:

- Population growth from 1996 to 2016 will increase influent BOD$_u$ loading nationwide to 68,030 mt/day, an increase of 45 percent. By 2025, influent loading will be about 73,057 mt/day, a 56 percent increase from 1996.
- In spite of a projected national increase in BOD$_u$ removal efficiency from 65 to 71 percent by 2016 (a 9 percent increase), it is estimated that the trend of decreasing effluent BOD$_u$ loadings experienced in the 24-year period from 1968 to 1996 will be reversed. It is predicted that effluent BOD$_u$ loadings will increase from 16,325 mt/day in 1996 to 19,606 mt/day in 2016, an increase of 20 percent. The effluent BOD$_u$ loading rate estimated for 2016 is about equal to effluent loading rates that existed in the mid-1970s, only a few years after the CWA was enacted!
- By 2025, the projected effluent BOD$_u$ loading will be 21,090 mt/day, an increase of 29 percent from 1996. This rate is about equal to effluent loading rates experienced in 1968 (21,280 mt/day), the year when the discharge of oxygen-demanding material from POTWs had reached its historical peak!

- By 2016, when the projected needs for wastewater treatment are expected to be met (USEPA, 1997), the overall BOD_u removal efficiency of 71 percent and increases in population will result in a 20 percent increase of effluent loads relative to the 1996 loading rate. To maintain an effluent BOD_u loading rate comparable to 1996 conditions through 2016 (i.e., "running in place"), the national aggregate removal efficiency would have to be increased from 71 to 76 percent. This would need to be accomplished by shifting the projected population served from secondary to advanced secondary and advanced wastewater treatment facilities.

The estimated projections of increasing effluent loading rates of BOD_u over the next quarter-century underscore the importance of continually investing in improvements to wastewater treatment infrastructure to maintain and improve pollutant removal efficiencies. Without these improvements, many of the environmental successes of the water pollution control efforts over the past three decades may be overwhelmed by the future demand from population growth. The very real risk of losing the environmental gains achieved by federal (Construction Grants Program), state, and local water pollution control efforts under the technology-based and water quality–based effluent limit regulations of the 1972 CWA is also documented by Jobin (1998) and the Water Infrastructure Network (WIN, 2000).

E. BOD_5 LOADINGS FROM POINT AND NONPOINT SOURCES

The primary purpose of this chapter is to examine whether there was a significant reduction in effluent BOD loading to the nation's waterways after the technology-based and water quality–based treatment provisions of the CWA were implemented. To fully address this subject, however, it is important to recognize the following:

- Effluent BOD loading comes from several point and nonpoint sources in addition to POTWs.
- BOD is only one of several contaminants that have the potential to affect aquatic resources and the lives and livelihoods of water resource users. Table 2-13 presents some of the concerns and conditions associated with several types of water pollutants.

This section is divided into two subsections. The first subsection briefly describes non-POTW sources of BOD loading, including industrial wastewater treatment facilities, combined sewer overflows (CSOs), urban stormwater runoff, and rural nonpoint sources of pollution. For the purposes of this comparison, urban stormwater runoff includes areas both outside (termed "nonpoint source") and within [meeting the legal definition of a point source in section 502(14) of the CWA] the NPDES stormwater permit program.

The second subsection introduces the National Water Pollution Control Assess-

TABLE 2-13 Pollutant Groups and Related Water Resource Issues

Pollutant Group	Water Quality Conditions and Concerns	
Nutrients (N & P)	Eutrophication	Nuisance algal blooms
	Ammonia toxicity	Toxic algal blooms
	Anoxia/hypoxia	Fish kills
	Water clarity/transparency	Shellfish bed closure/loss
	Reduced diversity	Loss of seagrass beds/habitat
Metals and Toxics	Fish body burden	Birds body burden
	Shellfish body burden	Sediment contamination
	Mammals body burden	Drinking water supply
Organic Matter	Anoxia/hypoxia	Fish kills
	Adsorption/desorption of toxic chemicals	
Pathogens	Shellfish bed closure	Drinking water supply
	Recreational beach closure	
Sediments	Anoxic sediments	Habitat destruction/fish spawning
	Damage to benthic biota	Water clarity/transparency
Hazardous materials	Oil spills	Fish kills
	Chemical spills	Drinking water supply

ment Model (NWPCAM)(Bondelid et al., 2000), a modeling tool that can be used to simulate the water quality impact of current (ca. 1995) BOD_5 effluent loadings from point and nonpoint sources nationwide. The primary purpose of this exercise is to compare BOD_5 effluent loadings from $POTW_s$ with BOD_5 effluent loadings from other point and nonpoint sources.

Pollutant Loading from Sources Other than POTWs

Industrial Wastewater Treatment Facilities Many industrial facilities discharge treated wastewater directly to surface waters. Similar to municipal wastewater treatment, industrial wastewater treatment consists of a sequence of physical, biological, and chemical processes designed to remove pollutants that are specific to an industrial facility's manufacturing operations. USEPA's effluent guidelines, prepared for specific categories of industrial groups, define effluent limits in terms of the industry's output production rate (e.g., n kilograms of pollutant discharged per 1,000 kilograms of factory production). Table 2-14 presents median effluent concentrations for conventional and nonconventional pollutants for the industrial categories that account for the largest contributions to effluent loading rates for BOD_5.

In contrast to direct industrial dischargers, industrial facilities can also discharge wastewater to sanitary sewer systems, where it mixes with domestic sources of wastewater (indirect industrial dischargers). This wastewater often contains a variety of metals, organic chemicals, and oily wastes that are not common to domestic sources

TABLE 2-14 Effluent Characteristics for Select Major Industry Groups

	Inorganic Chemical Products	Organic Chemical Products	Feedlots	Food & Beverages	Iron & Steel	Petroleum Refining	Pulp & Paper
Parameter (mg/L)							
BOD_5	6.5	6.3	6.0	11.8	6.0	8.8	24.5
TOC	9.4	11.2	N/A	N/A	N/A	12.0	N/A
NH_3-N	1.3	1.2	0.7	0.6	1.0	2.0	1.2
Total N	1.9[a]	33.4[a]	28.5[a]	17.9[a]	2.9[a]	N/A	1.4[a]
Total P	0.4	N/A	1.4	6.7[a]	N/A	N/A	0.6
TSS	10.6	11.8	13.1	12.0	9.9	12.9	29.4
DO	N/A	N/A	7.7	N/A	6.6	N/A	5.8
Number of facilities	273	232	32	62	186	203	309
Average median design flow (mgd)							
Major facilities	2.9	2.3	N/A	0.3	3.9	3.0	5.0
Minor facilities	0.2	1.7	0.3	0.1	0.2	0.3	0.8

Sources: Tetra Tech, 1999; NOAA, 1994.

Note: Table 2-14 presents the median value of effluent data extracted from PCS by Tetra Tech (1999) for the period 1991 to 1998. Effluent values for Total N and Total P that are flagged by [a] indicate that typical pollutant concentration (TPC) effluent data compiled by NOAA (1994) are used.

of wastewater. Because of the high degree of variability, most municipal treatment systems are not designed to treat a vast array of industrial wastes. Consequently, these wastes can interfere with the operation of treatment plants, contaminate receiving waterbodies, threaten worker health and safety, and increase the cost and risks of sludge treatment and disposal. Using proven pollution control technologies and practices that promote the reuse and recycling of material, however, industrial facilities can provide "pretreatment" by removing pollutants from their wastewater before discharging to the municipal wastewater system. In addition to the categorical standards for pretreatment established as part of the industrial effluent guideline process, local pretreatment limits are enforced by various municipal facilities to protect treatment processes, worker health and safety, and equipment. USEPA's National Pretreatment Program, a cooperative effort of federal, state, and local officials, is fostering this practice nationwide.

Combined Sewer Overflows In many older cities of the United States, urban sewer systems were originally designed to convey both raw sewage and stormwater runoff collected during rainstorms. These combined sewer overflow systems were also explicitly designed to discharge (overflow) the mixture of raw sewage and stormwater into the river if a heavy rainstorm exceeded the hydraulic capacity of the combined sewer system. As a vestige of public works practices from approximately 1850 to 1900, about 880 cities, mostly in the central and northeastern states, have combined sewer systems that continue to function in this manner (USEPA, 1997). Table 2-15 presents characteristic discharge concentrations of conventional and nonconventional pollutants in combined sewer overflows (CSOs).

In addition to raw sewage, a CSO system can discharge pretreated industrial waste and street debris washed off during a storm. Although pollutant loading from CSO systems is intermittent, occurring only under heavy rainstorm conditions, the high loading rates of sewage from CSO outlets frequently result in the closure of recreational beaches and shellfish beds to protect public health. Discharges from CSOs also are associated with depressed oxygen levels in poorly flushed waterbodies, accumulation of organics in sediments, and generally noxious conditions and odors.

National assessments show that the relative significance of annual loading of BOD_5 from CSO systems is about the same as the effluent loading from secondary wastewater treatment facilities in the same urban area. In contrast to BOD_5, annual loading of suspended solids and lead is about 15 times greater from CSO systems than from secondary wastewater treatment facilities. Annual loading rates of total nitrogen and phosphorus from CSOs, however, are only about one-fourth (total N) and one-seventh (total P) of the annual loads contributed by secondary facilities (Novotny and Olem, 1994).

Urban and Rural Nonpoint Sources Organic and inorganic materials, both naturally occurring and related to human activities, are transported to waterbodies within a drainage basin by surface runoff over the land as nonpoint, or diffuse, sources of pollutants. The magnitude and the timing of nonpoint pollutant loads are

TABLE 2-15 Effluent Characteristics of Urban Runoff and CSOs

Parameter	Urban Runoff Range[a,b]	CSO Range [c,d]	CSO (event mean)
BOD_5 (mg/L)	10–13	60–200	(115)
$CBOD_u$:BOD_5	ND	ND	(1.4)[e]
TSS (mg/L)	141–224	100–1,100	(370)
TKN (mg/L)	1.68–2.12	ND	(6.5)
NH_3—N (mg N/L)	ND	ND	(1.9)
NO_2—N + NO_3—N (mg N/L)	0.76–0.96	ND	(1.0)
Total N (mg N/L)	3–10	3–24	(7.5)
Total P (mg P/L)	0.37–0.47	1–11	(10)
Total lead (mg/L)	161–204	ND	(370)
Total coliforms(MPN/100 mL)	10^3–108	105–107	(ND)

Notes: ND = No data.

[a]Range of urban runoff concentrations reflects variability of coefficient of variation of event mean concentrations for median urban sites. Data from USEPA (1983) presented in Novotny and Olem (1994, Table 1.3, p.36).

[b] Range of urban runoff concentrations for total N and total coliforms from Novotny and Olem (1994, Table 1.3, p. 36).

[c] Range of CSO concentrations for BOD_5, TSS, total N, and total coliforms from Novotny and Olem (1994, Table 1.3, p. 36).

[d] Mean CSO concentrations of BOD_5, TSS, and total lead from USEPA (1978) presented in Novotny and Olem (1994); median CSO concentrations of nitrogen constituents from Driscoll (1986); mean CSO concentration of total phosphorus from Ellis (1986).

[e] $CBOD_u$:BOD_5 ratio from Thomann and Mueller, 1987.

dependent on many complex, and interacting, processes within a drainage basin. In contrast to the relatively continuous input of pollutants from point sources, the timing of loading from diffuse sources is highly variable with intermittent loading related primarily to meteorological events (storms and snowmelt). The magnitude of pollutant loads is dependent on the area of the drainage basin, the characteristics of land uses, including ground cover, and distribution of the volume of precipitation between infiltration into shallow aquifers and surface runoff into streams and rivers.

Within a watershed undisturbed by human activities, naturally occurring biogeochemical processes account for the continual cycles of organic and inorganic materials (as uncontrollable nonpoint source loads) transported from the land to rivers, lakes, and estuaries, with eventual discharge of these materials to the coastal ocean. Since it is the uses of the land and the associated activities that occur on the land within a drainage basin that contribute anthropogenic organic and inorganic materials to surface waters, nonpoint source loading rates have been related to the type of land use (Table 2-16). The most critical factor, however, in understanding the management of nonpoint source loading is characterizing the transition from one land use to another (e.g., forest to agriculture, agriculture to suburban/urban).

Beginning with the four natural land classifications (arid lands, prairie, wetland,

TABLE 2-16 Nonpoint Source Runoff Export Coefficients for General Land Uses

Parameter	Land Use		
	Urban	Agriculture	Forest
BOD_5[a,b]	34–90	26	5
TSS[a,b]	3,360–672	1,600	256
Total N[b,c]	7.8–11.2	16.5	2.9
Total P[b,c]	1.6–3.4	1.1	0.2

Note: Units are kg/hectare-year.

[a] Export coefficients for BOD_5 and TSS for agriculture and forest categories from Thomann and Mueller (1987).

[b] Range of export coefficients for urban land use categories I, II, and III from PLUARG studies (Marsalek, 1978), presented by Novotny and Olem (1994, Table 8.2, p. 449).

[c] Mean export coefficients for total N and total P for mixed agricultural and forest land uses from Reckhow et al. (1980).

and woodland), the transformation of a watershed's land uses progresses over many years through several intermediate stages of development to a fully developed urban-industrial watershed (Novotny and Olem, 1994). With the irreversible transformation to the endpoint of urban-industrial land uses of a watershed, the emphasis in water quality management needs to incorporate strategies for control of both nonpoint sources of runoff and the point source discharges within the "urban-industrial" water cycle. In contrast to the control strategy for point sources (build a wastewater treatment facility) as the most effective technology for removal of pollutants from a point source waste discharge, the reduction of nonpoint source loading of pollutants is focused on the design and implementation of "best management practices" to control, and manage, land use activities and surface runoff. As with urban runoff control measures, the technical aspects of the numerous practices available for controlling nonpoint source runoff from forest, agricultural, and other rural land uses are presented in detail by Novotny and Olem (1994).

As part of its public works infrastructure, practically every town and city in the nation has an urban stormwater sewer system designed to collect and convey water runoff from rainstorms and snowmelt. Depending on the development characteristics of an urban area, stormwater runoff can result in significant intermittent loading of pollutants to surface waterbodies. Based on findings from the National Urban Runoff Project (NURP), conducted by USEPA from 1978 to 1983, USEPA (1983) concluded that urban runoff accounted for significant wet weather loading to the nation's surface waters of pathogens, heavy metals, toxic chemicals, and sediments. The origins of the diffuse discharges of these pollutants include contaminants contained in wet and dry atmospheric deposition, erosion of pervious lands, accumulation of debris on streets, traffic emissions, and washoff of contaminants from impervious land surfaces. Table 2-15 presents typical discharges of conventional and nonconventional pollutants in urban runoff.

BOD_5 Loads from the National Water Pollution Control Assessment Model (NWPCAM)

The National Water Pollution Control Assessment Model (NWPCAM) is a national-scale water quality model designed to link point and nonpoint source loadings and resultant calculated in-stream concentrations of $CBOD_5$, $CBOD_u$, DO, TKN, total suspended solids, and fecal coliform bacteria with a "water quality ladder" of beneficial uses (Carson and Mitchell, 1983). The framework for the model is USEPA's Reach File Version 1 (RF1) and Version 3 (RF3) national databases of streams, rivers, lakes, and estuaries. The national model uses mean summer streamflow data to characterize steady-state loads, transport, and fate of water quality constituents. Presented for comparison purposes is current (ca. 1995) BOD_5 loading information derived using available NWPCAM national data for municipal and industrial discharges, CSOs, and urban and rural nonpoint sources (Bondelid et al., 2000).

BOD_5 Loading from Municipal and Industrial Sources The input data used to estimate municipal and industrial effluent loading of BOD_5 within the NWPCAM come from USEPA's Permit Compliance System (PCS), the Clean Water Needs Survey (CWNS) databases, and default assumptions derived from the literature. The PCS database contains discharge monitoring data for major POTWs and industrial dischargers (facilities with a discharge greater than 1 mgd). The CWNS database provides a more comprehensive database of all POTWs and generally reliable population, flow, and treatment level information. Less confidence is placed on the effluent concentration data reported in the CWNS database. Therefore, when actual discharge data were available from PCS, those data were used. PCS data were also used to develop default effluent concentrations to apply when a facility's actual concentration was not available or was outside normal ranges expected for a given level of treatment.

Municipal Table 2-17 presents a compilation of characteristic effluent concentrations of conventional and nonconventional pollutants used in NWPCAM for different types of municipal POTWs. The data sets extracted from USEPA's PCS and CWNS databases are supplemented by influent and effluent data taken from the literature (e.g., AMSA, 1997; Metcalf and Eddy, 1991; Clark et al., 1977; Leo et al.; 1984; Thomann and Mueller, 1987).

A total of 1,632 of the 2,111 hydrologic catalog units in the contiguous United States are subject to municipal effluent loading. Figure 2-18 presents distributions of municipal BOD_5 loading by percentile of catalog units with nonzero municipal loads according to (1) loading rate and (2) fraction of total municipal loading. Figure 2-19 presents a map showing the magnitude of municipal effluent loading of BOD_5 aggregated for the 1,632 catalog units with nonzero municipal loads. Figure 2-20 displays the proportion of the total nonpoint and point sources load contributed by municipal waste loads.

Key observations from Figures 2-18 through 2-20 include the following:

- Less than 1 percent of the 1,632 catalog units subject to municipal loading receive effluent BOD_5 loading at a rate greater than 25 mt/day (Figure 2-18*a*). About 20 percent of the catalog units account for about 90 percent of the total municipal BOD_5 loading to the nation's waterways (Figure 2-18*b*).
- Relatively low municipal BOD_5 loading rates (less than 0.5 mt/day) characterize many of the catalog units within the western and central portions of the contiguous 48 states.
- Higher rates of municipal loading (0.5 to 5 mt/day) are characteristic of the Mississippi River valley and the Northeast, Midwest, and Southeast. The highest loading rates (> 25 mt/day) are for major urban centers, including New York, Boston, Los Angeles, San Diego, Dallas–Ft. Worth, Detroit, and San Francisco.
- The municipal wastewater component of total point and nonpoint source load of BOD_5 tracks closely with the results of the loading magnitude calculation. The municipal wastewater component is highest around major urban centers and lowest in the western and central portions of the contiguous 48 states.

TABLE 2-17 Effluent Characteristics for POTWs

Parameter (mg/L)	Raw	Primary	Advanced Primary	Secondary	Advanced Secondary	Advanced Wastewater Treatment
BOD_5						
Mean	205.0	143.5	102.5	16.4	6.2	4.1
% Removal	0	30	50	92	97	98
Reference/notes	*a,j*	*b*	*c*	*a*	*a,d*	*a,d*
$CBOD_u$:$CBOD_5$						
Mean	1.2	1.6	1.6	2.84	2.84	3.0
Reference/notes	*e*	*f*	*f*	*f*	*f*	*f*
TSS (mg/L)						
Mean	215	107.5	64.5	17.2	6.5	4.3
% Removal	0	50	70	92	97	98
Reference/notes	*a,j*	*b*	*c*	*a*	*a,d*	*a,d*
NH_3-N (mg N/L)						
Mean	18.0	14.4	14.4	12.2	3.4	2.0
% Removal	0	20	20	32	81	89
Reference/notes	*a*	*b*	*b*	*a*	*a,d*	*a,d*
TKN (mg N/L)						
Mean	30.0	23.4	23.4	16.5	12.9	3.6
% Removal	0	22	22	45	57	88
Reference/notes	*a*	*b*	*b*	*a*	*a,d*	*a,d*
Total N (mg N/L)						
Mean	30.0	23.4	23.4	18.3	18.4	14.4
% Removal	0	22	22	39	39	52
Reference/notes	*g*	*h*	*h*	*a*	*a,d*	*a,d*

TABLE 2-17 *Continued*

Parameter (mg/L)	Raw	Primary	Advanced Primary	Secondary	Advanced Secondary	Advanced Wastewater Treatment
Total P (mg P/L)						
Mean	6	5.2	5.2	2.5	0.4	0.4
% Removal	0	13	13	58	94	94
Reference/notes	*a*	*b*	*b*	*a*	*a,d*	*a,d*
DO (mg/L)						
Mean	4.1	4.3	4.3	6.6	6.6	7.1
Reference/notes	*i*	*j*	*j*	*j*	*j*	*j*
Total organic carbon (mg/L)						
Mean	148.6	107.5	76.8	21.8	8.2	5.8
% Removal	0	28	48	85	94	96
Reference/notes	*g*	*b,k*	*k*	*b,k*	*k*	*k*

[a]AMSA, 1997. Influent concentration, percent removal, and TKN TN, NH_3 TKN, and PO_4 TP ratios for secondary, advanced secondary, and advanced wastewater treatment.

[b]Gunnerson et al., 1982.

[c]NRC, 1993. Percent removal for advanced primary with "low dose chemical addition."

[d]MWCOG, 1989. Percent removal and TKN TN, NH_3 TKN, and PO_4 TP ratios for advanced secondary and advanced wastewater treatment.

[e]Thomann and Mueller, 1987.

[f]Leo et al., 1984.

[g]Metcalf and Eddy, 1991. TKN TN, NH_3 TKN, and PO_4 TP ratios of influent concentration for "medium" strength wastewater, raw total organic carbon influent concentration based on BOD_5, $CBOD_u$ BOD_5, oxygen carbon, and ratios of C DW.

[h]ICPRB, 1991. TKN TN, NH_3 TKN, and PO_4 TP ratios of effluent concentration for primary, advanced primary, and secondary treatment.

[i]Assume 50 percent saturation at 25°C and 50 mg/L chlorides at sea level.

[j]Tetra Tech, 1999. Mean effluent oxygen concentrations based on PCS database for primary, secondary, and advanced treatment. Mean influent concentrations for BOD_5 (207 mg/L) and TSS (209 mg/L) from CWNS database consistent with influent data from AMSA (1997).

[k]Effluent total organic carbon concentration computed from effluent BOD_5, $CBOD_u$ BOD_5, oxygen carbon ratio and assumption that 80 percent of organic carbon is accounted for by BOD_5 measurement. Removal efficiencies computed for primary and secondary treatment are consistent with data from Gunnerson et al. (1982).

Industrial Similar to the two municipal maps, Figure 2-21 presents the magnitude of the industrial effluent loading of BOD_5 aggregated for a total of 1,504 catalog units with nonzero industrial loads. Figure 2-22 displays the proportion of the total nonpoint and point sources load accounted for by industrial waste loads.

Key observations include the following:

- Relatively low industrial BOD_5 loading rates ($>$ 0.5 mt/day) characterize many of the catalog units in the western and central portions of the 48 states.
- Higher rates of industrial loading (0.5 to 5 mt/day) are characteristic of the Mis-

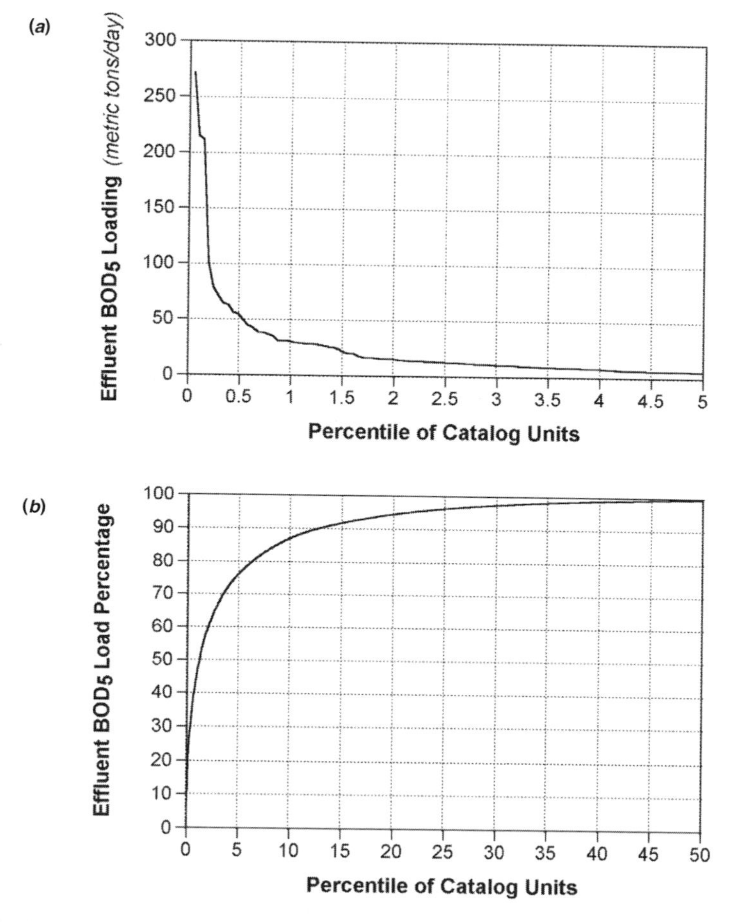

Figure 2-18 Distribution of municipal BOD_5 loading by percentile of catalog units subject to municipal loading (N = 1,632) as (*a*) metric tons/day and (*b*) fraction of total municipal loading rate. *Source:* Bondelid et al., 2000.

sissippi River valley, the Northeast, Midwest, and Southeast. The highest loading rates (> 25 mt/day) are indicated for major urban industrial watersheds including Austin-Oyster in Texas, East-Central in Louisiana, Buffalo–San Jacinto and Galveston Bay, and the Locust River, Upper Black Warrior, and Middle Coosa basins in Alabama.

- Industrial loads are the dominant component (> 75 percent) of the total point and nonpoint source load in many catalog units associated with major urban-industrial areas, particularly in the Southeast. Although not shown, the frequency distributions of industrial BOD_5 loads (as a percentile of catalog units with nonzero industrial loads) are very similar to those presented for municipal BOD_5 loads.

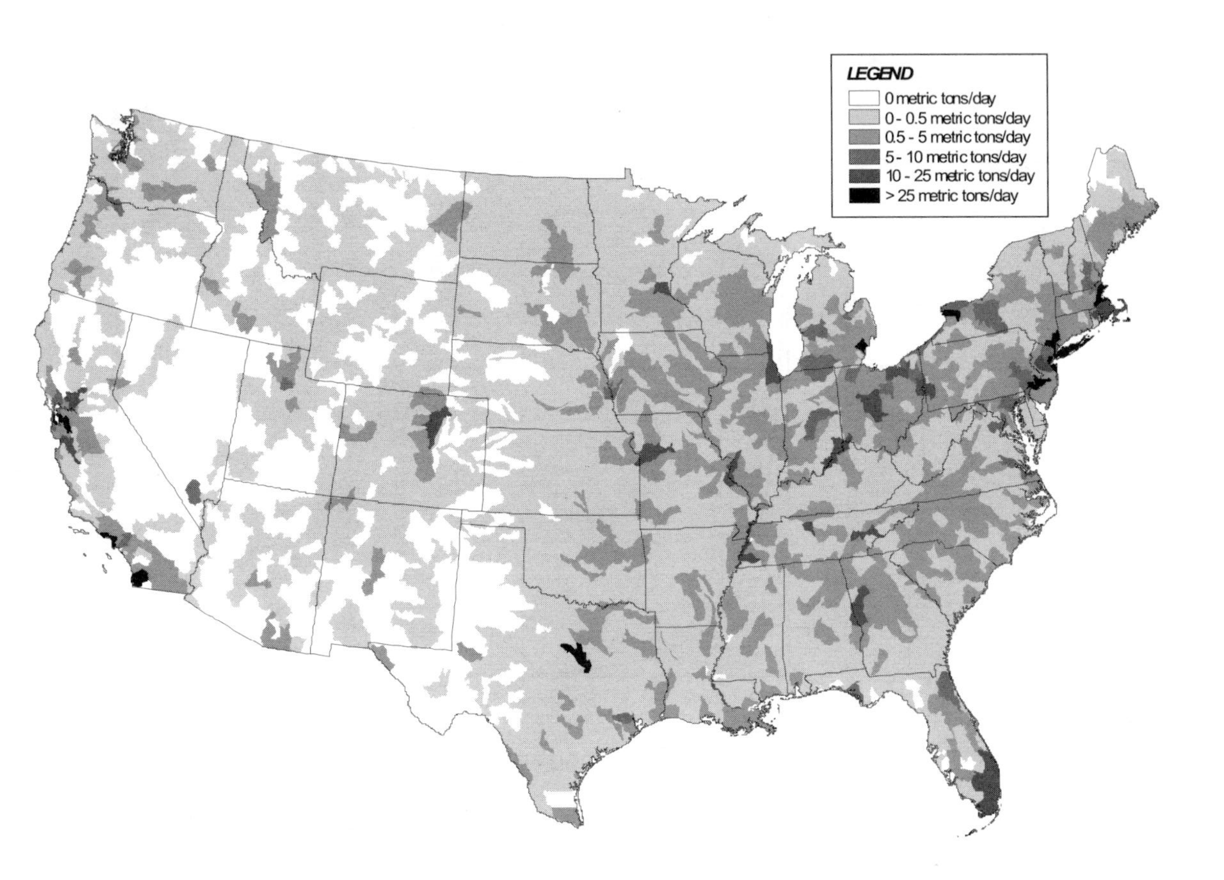

Figure 2-19 Municipal wastewater loading of BOD_5 ca. 1995 by catalog unit (metric tons per day). *Source:* Bondelid et al., 2000.

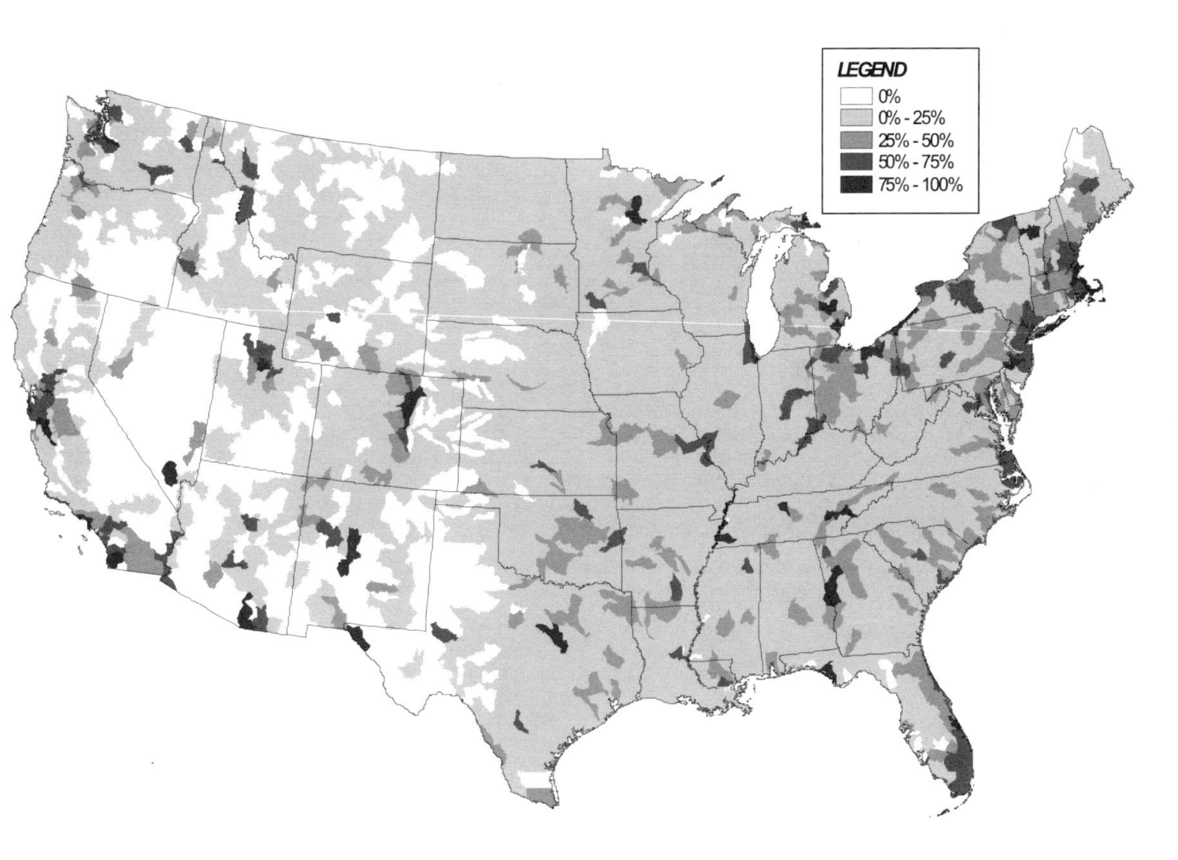

Figure 2-20 Municipal wastewater component of total point and nonpoint source loading of BOD_5 ca. 1995 by catalog unit (percent of total). *Source:* Bondelid et al., 2000.

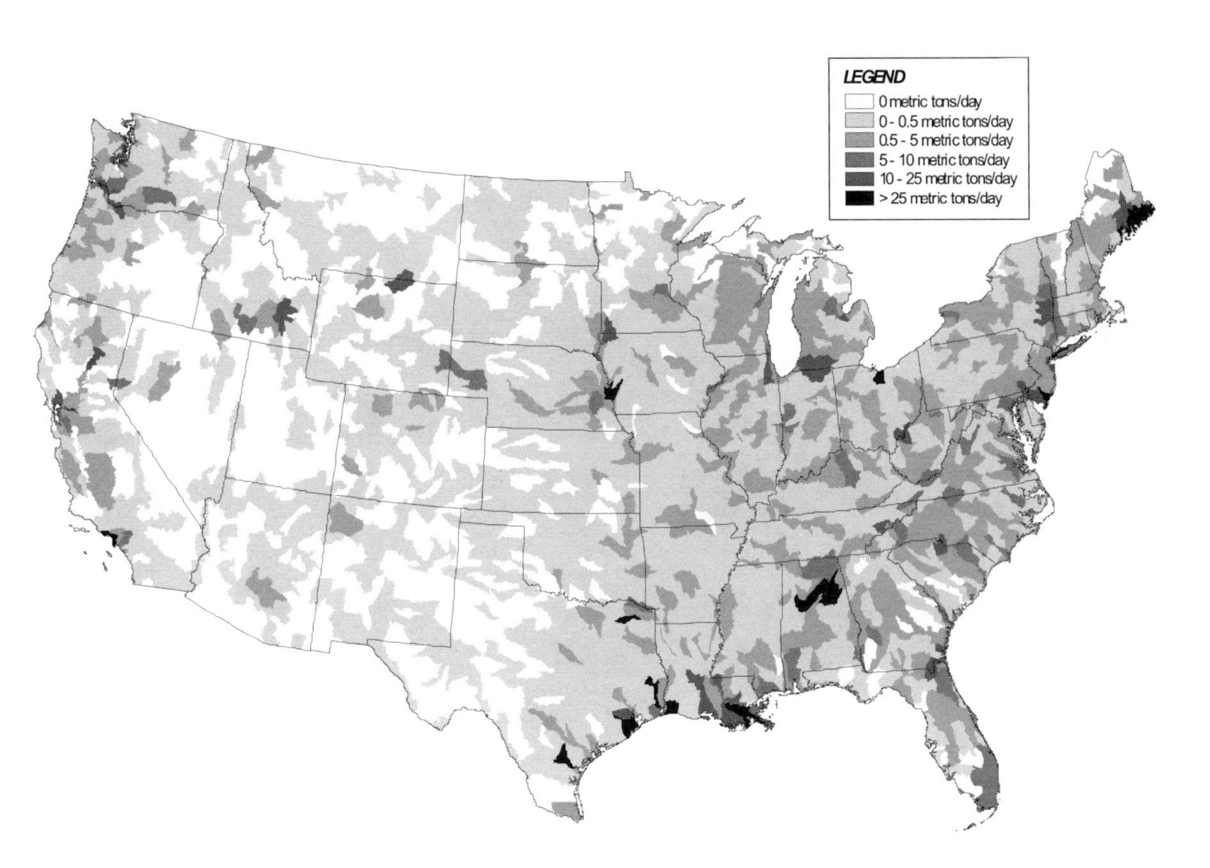

Figure 2-21 Industrial wastewater loading of BOD_5 ca. 1995 by catalog unit (metric tons per day). *Source:* Bondelid et al., 2000.

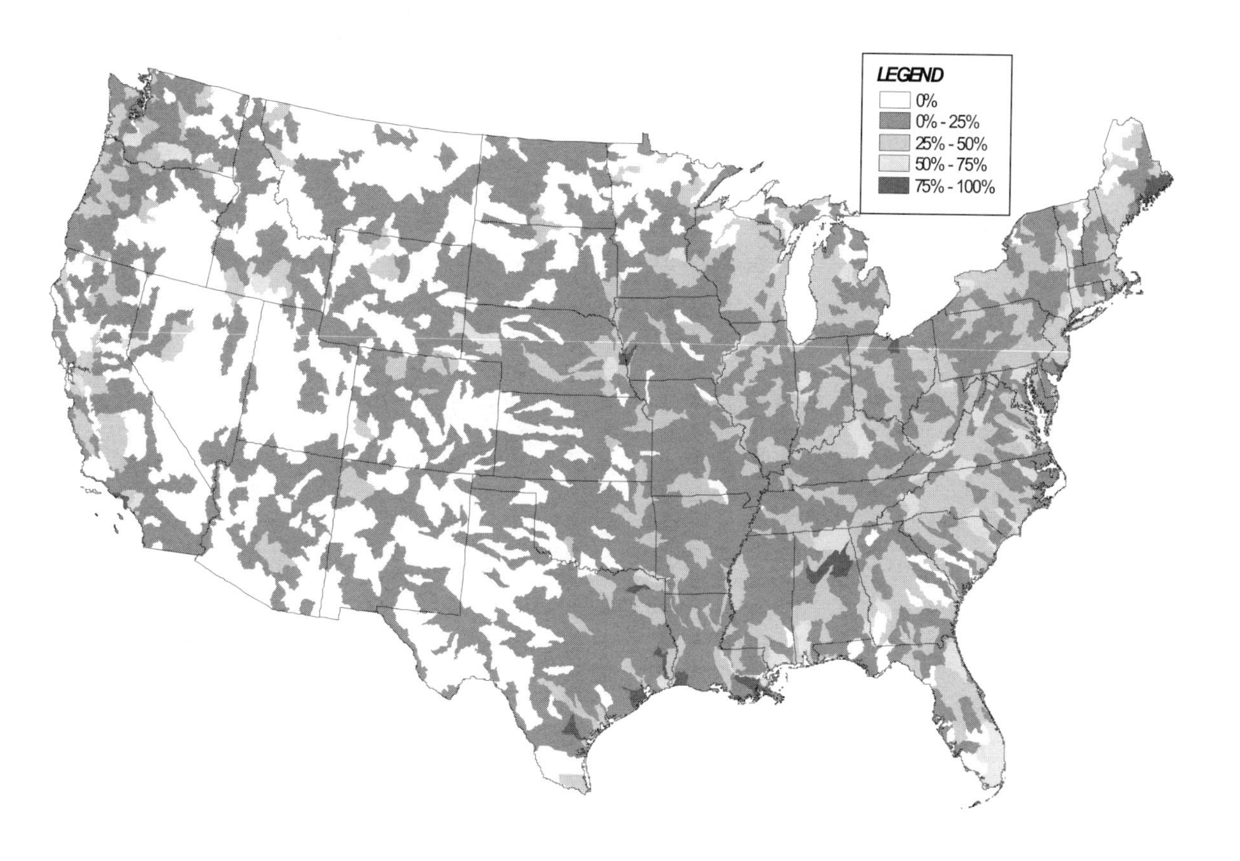

Figure 2-22 Industrial wastewater component of total point and nonpoint source loading of BOD_5 ca. 1995 by catalog unit (percent of total). *Source:* Bondelid et al., 2000.

BOD_5 Loading From CSOs Effluent loadings for CSOs were based on an analysis performed in support of the 1992 Clean Water Needs Survey (CWNS) (Tetra Tech, 1993) and subsequently adopted for the NWPCAM. During this 1992 CWNS, it was estimated that there were approximately 1,300 CSO facilities in the United States (USEPA, 1993). The number of facilities was substantially reduced to 880 during the 1996 CWNS.

The effluent loading for CSOs used in the NWPCAM is based on computing a pulse load based on the runoff volume and pollutant load associated with a 5-year, 6-hour storm event. Runoff was computed as a function of the combined sewer system's population, service area, and imperviousness. For the purposes of the NWPCAM, the pollutant loading used in the model was estimated to yield a national BOD_5 loading of 15 mt/day (Bondelid et al., 2000). As expected, most of the CSO loading is accounted for by older cities in the New England, Middle Atlantic, Great Lakes, Ohio River, and Upper Mississippi basins.

BOD_5 Loading From Urban Stormwater Runoff and Rural Nonpoint Sources Nonpoint source BOD_5 loading data were developed on a county-level basis by Lovejoy (1989) and Lovejoy and Dunkelberg (1990), with urban stormwater runoff and rural runoff loadings reported separately. These values were converted into loadings allocated to each catalog unit in the contiguous 48 states based on the proportion of a county's area in a given catalog unit. For the NWPCAM, the rural loadings were disaggregated based on stream length in a given county while urban loadings were disaggregated based on stream length and population associated with a given stream.

Using the loading data compiled for the NWPCAM, the national catalog unit-based distributions of urban stormwater and rural BOD_5 loading are presented in Figures 2-23 and 2-24 (urban) and Figures 2-25 and 2-26 (rural). The map sets present both the magnitude of the loading rate (as metric tons per day) and the percentage of the total point and nonpoint source load accounted for by the urban and rural runoff contributions, respectively.

Key observations include the following:

- With the exception of urban areas on the west coast and in the Midwest and Northeast, low loading rates ($<$ 0.5 mt/day) characterize most of the nation's watersheds for urban runoff loads.
- In urban areas, loading rates are typically less than 5 mt/day, accounting for about 25 to 75 percent of the total point and nonpoint source BOD_5 load discharged to a catalog unit.
- Rural loading rates of BOD_5 are characterized by a distinctly different geographic distribution, with the highest rates ($>$ 25 mt/day) estimated for the upper Missouri basin. Intermediate loading rates of 5 to 25 mt/day of BOD_5 characterize rural runoff in the Missouri, Upper Mississippi, and Ohio river basins. The lowest rates ($<$ 0.5 mt/day) are estimated for the coastal watersheds of the east coast and Gulf of Mexico and the arid areas of the western states.

- Rural nonpoint source loads of BOD_5 are the dominant component (> 75 percent) of total point and nonpoint source loads in vast areas of the nation, principally west of the Mississippi River and in the Ohio River Basin.
- The geographic distribution of relatively low contributions of rural runoff (< 25 percent) is consistent with the locations of large urban-industrial areas (e.g., New York, Boston, Miami, New Orleans, Chicago, Seattle, San Francisco, Los Angeles).

Comparison of Point and Nonpoint Sources of BOD_5 at the National Level

From a national perspective, BOD_5 loading from municipal facilities currently (ca. 1995) accounts for only about 38 percent of total point source loadings and only 21 percent of total loadings (point and nonpoint). Industrial facilities (major and minor) account for about 62 percent of total point source BOD_5 loadings and 34 percent of total BOD_5 loadings. Rural nonpoint source loads account for about 40 percent of the total BOD_5 loading rate. Urban stormwater runoff and CSOs, although significant in most urban waterways, account for a small share (5 percent) of the total nationwide load (Bondelid et al., 2000).

Based on this analysis of contemporary sources of loading of BOD_5, continued maintenance and improvement of water quality conditions of the nation's surface waters will clearly require an integrated, watershed-based strategy, such as that presented in the USEPA's (1998) Clean Water Action Plan, including the appropriate management of point and nonpoint sources of BOD_5 and other pollutants (e.g., nutrients, suspended solids, toxic chemicals, pathogens).

F. INVESTMENT COSTS FOR WATER POLLUTION CONTROL INFRASTRUCTURE

The analysis presented in Section D indicates that nationwide effluent BOD_u loadings from POTWs were reduced by 23 percent between 1968 and 1996. Examination of historical trends in industrial wastewater loads also suggests substantial declines in BOD loads from industrial point sources have been achieved since the early 1970s (see Luken et al., 1976). Declines can be credited to industrial pretreatment programs, upgrades of industrial wastewater treatment as required by the NPDES permit program, abandonment of obsolete manufacturing facilities in the Midwest and Northeast "rustbelt" (Kahn, 1997), and improved efficiency in industrial water use (Solley et al., 1998). The purpose of this section is to provide an overview of the costs of implementing public and private water pollution control programs.

The Construction Grants Program

The Water Pollution Control Act of 1956 was significant because it both established and funded a grant program for the construction of POTWs for the purpose of ensuring the implementation of adequate levels of municipal waste treatment as a national

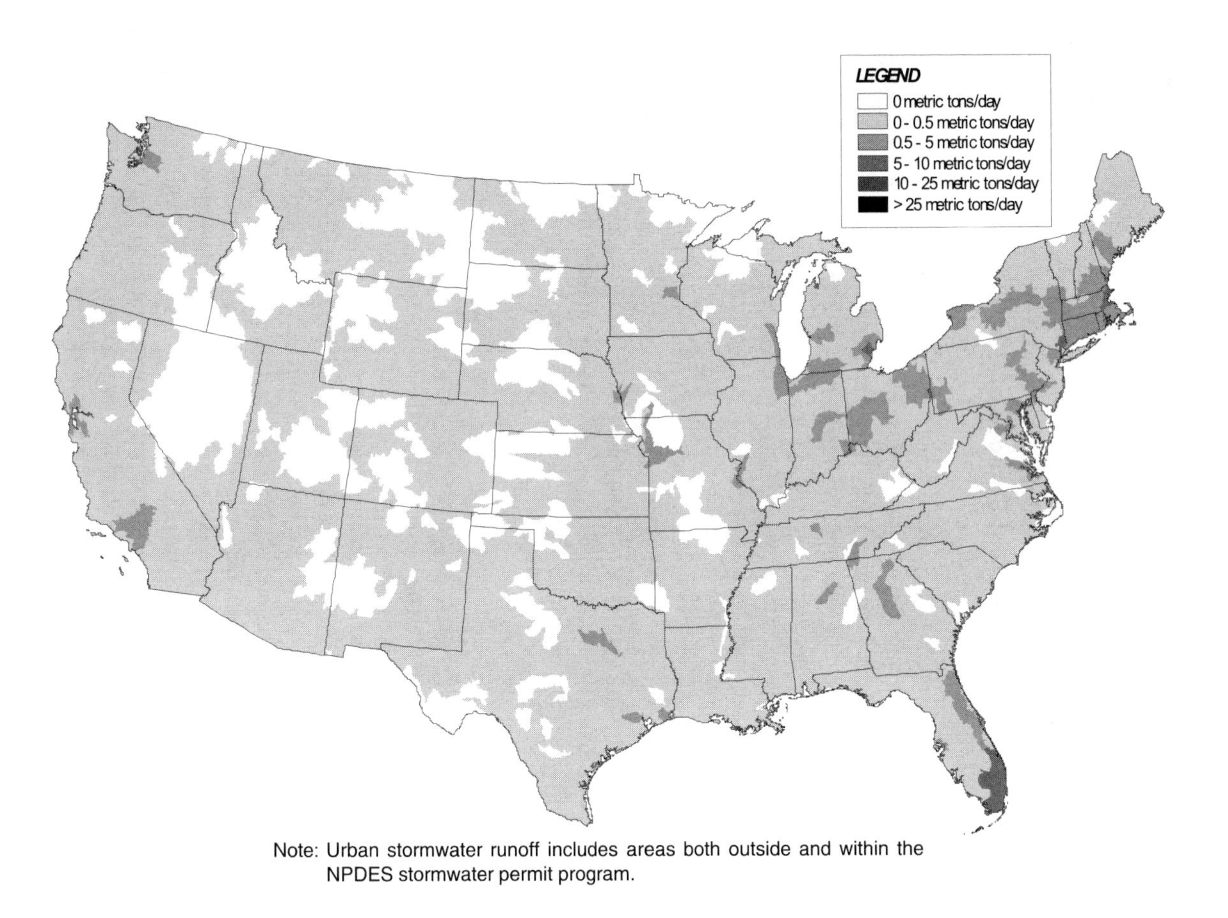

Figure 2-23 Urban nonpoint loading of BOD_5 ca. 1995 by catalog unit (metric tons per day). *Source:* Bondelid et al., 2000.

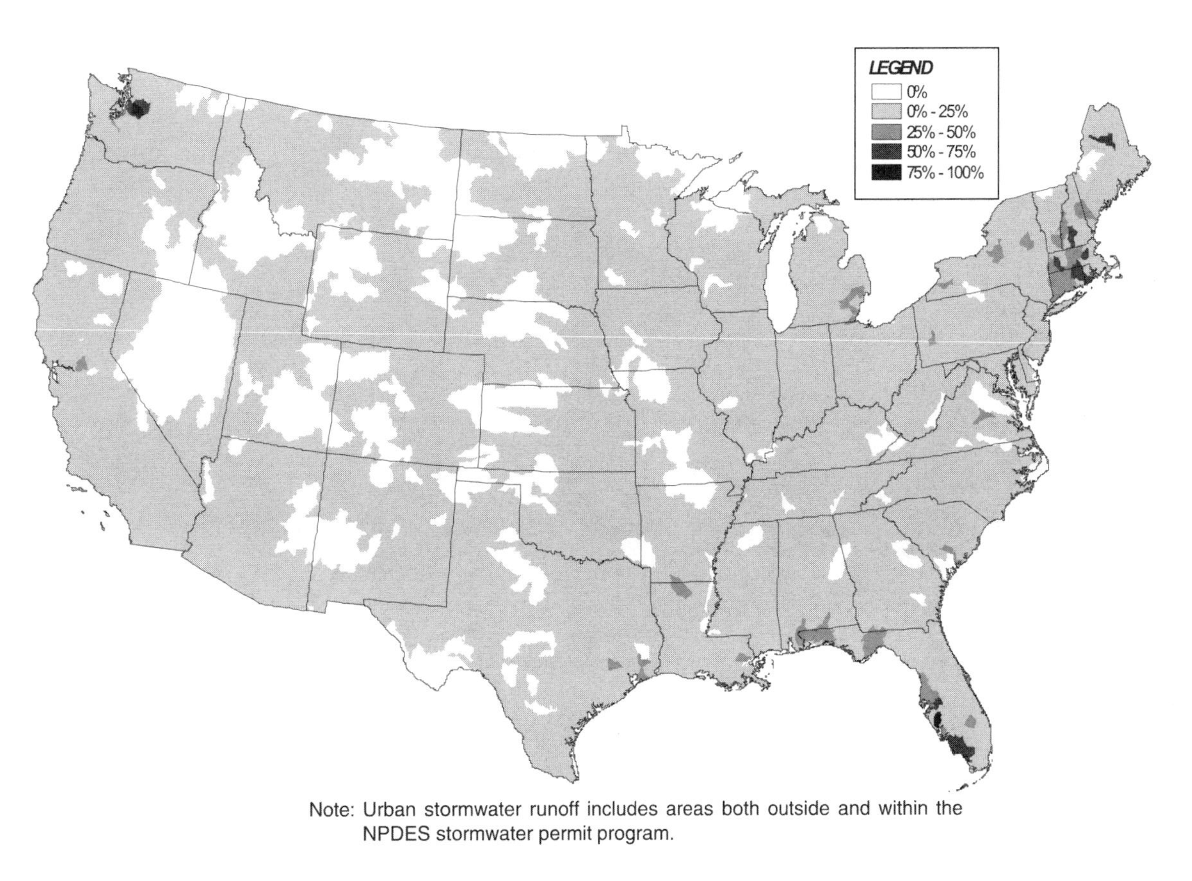

Note: Urban stormwater runoff includes areas both outside and within the NPDES stormwater permit program.

Figure 2-24 Urban nonpoint component of total point and nonpoint source loading of BOD_5 ca. 1995 by catalog unit (percent of total). *Source:* Bondelid et al., 2000.

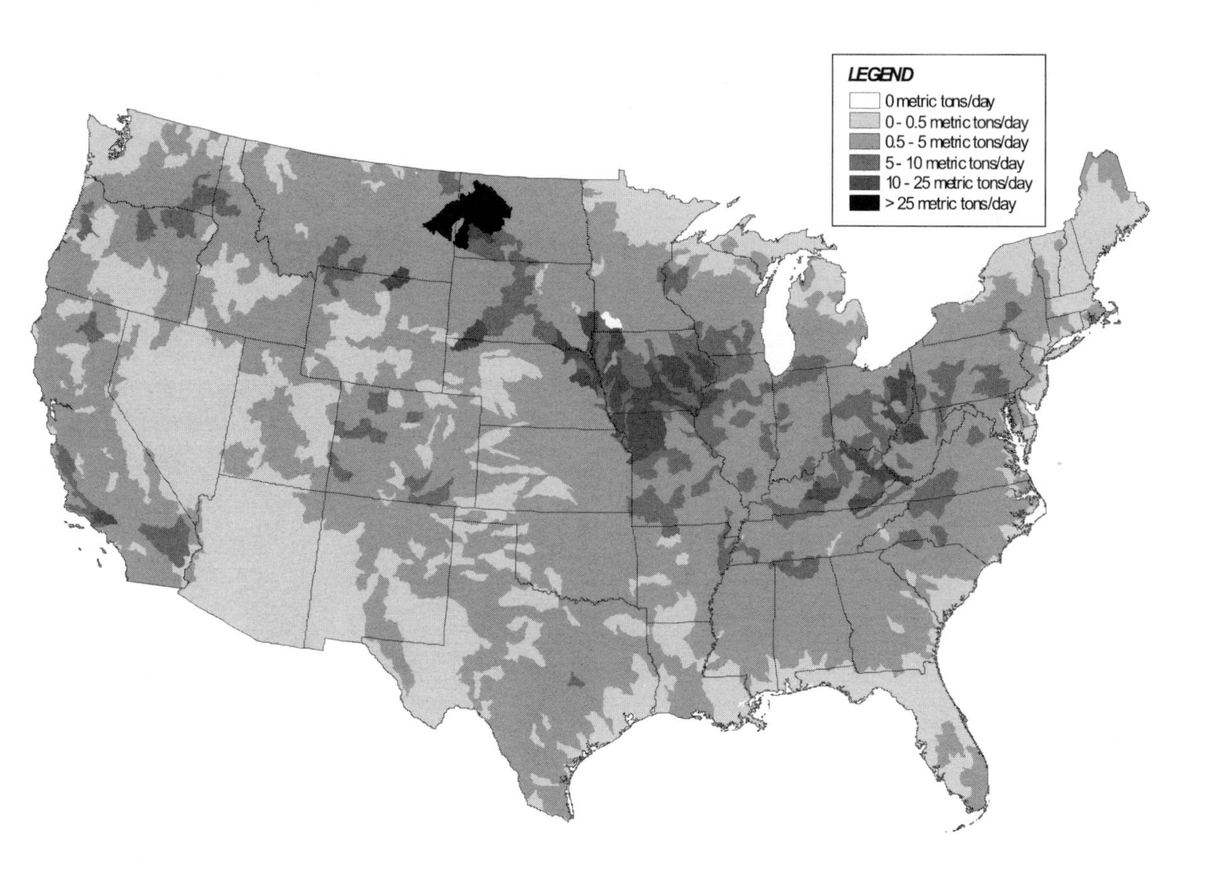

Figure 2-25 Rural nonpoint loading of BOD_5 ca. 1995 by catalog unit (metric tons per day). *Source:* Bondelid et al., 2000.

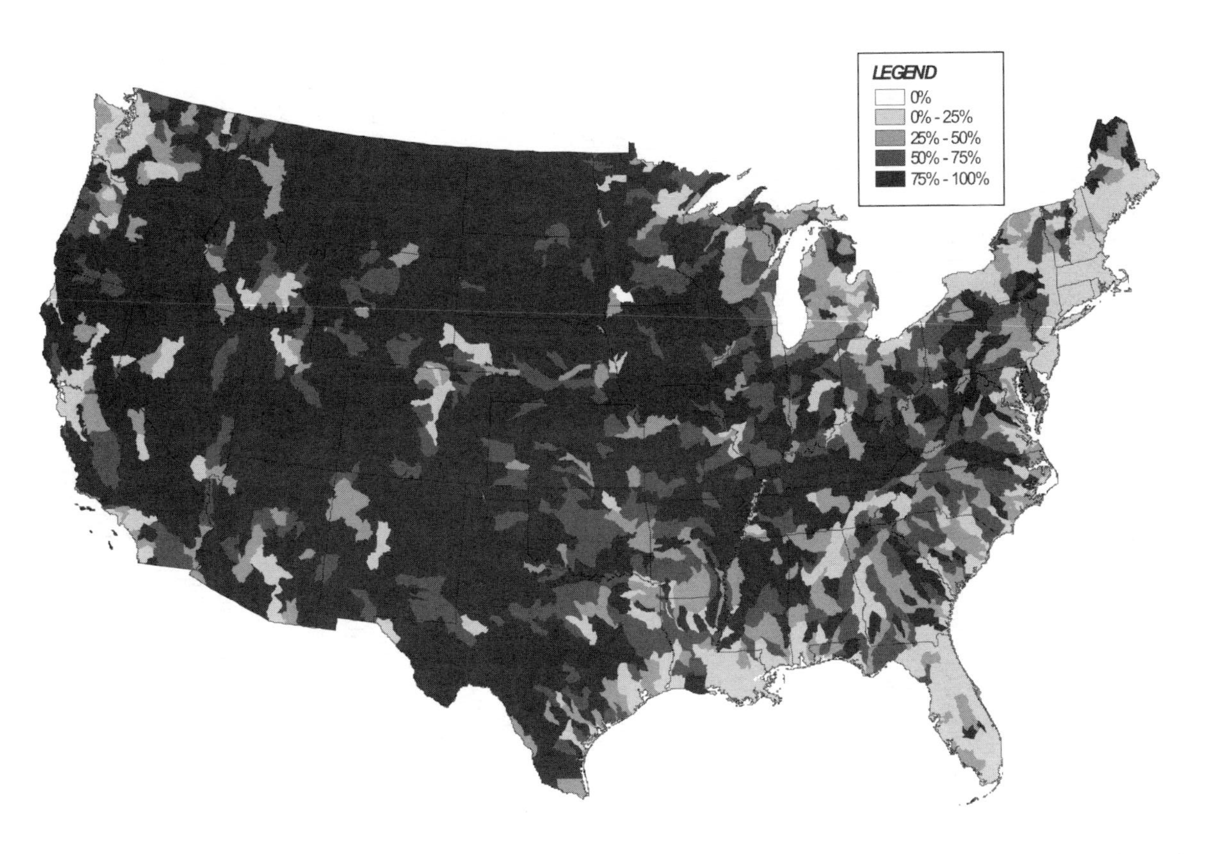

Figure 2-26 Rural nonpoint component of total point and nonpoint source loading of BOD_5 ca. 1995 by catalog unit (percent of total). *Source:* Bondelid et al., 2000.

policy for water pollution control. Following the 1956 Amendments, however, federal funding ($5.1 billion allotted from 1957 to 1972) accounted for only a small portion of the total construction costs for municipal facilities (FWPCA, 1970). The CWA made it a national policy to provide federal grants to assist in the upgrade and construction of municipal wastewater facilities. The 1972 act authorized $5.0 billion in federal spending for fiscal year 1973, $6.0 billion for fiscal year 1974, and $7.0 billion for fiscal year 1975. Under the revamped Construction Grants Program, the federal share was 75 percent of cost from fiscal years 1973 to 1983, and 55 percent thereafter.

USEPA's Grants Information and Control System (GICS) database is the central repository of Construction Grants Program data. For the following financial analysis, grant awards in the GICS database were indexed to constant 1995 dollars using the Chemical Engineering Plant Cost Index (CE, 1995) for the purpose of providing a suitable indicator of the inflation of wastewater treatment facility construction costs.

National Summary During the 29-year period from 1970 to 1999, the Construction Grants Program distributed a total of $61.1 billion in federal contributions ($96.5 billion as constant 1995 dollars) to municipalities for new construction and upgrades of POTWs to secondary and greater levels of wastewater treatment (Figure 2-27). An additional $16.1 billion (capitalization) in federal contributions was also distributed to the states through the Clean Water State Revolving Fund (CWSRF) Program from 1988 through 1999 (Figure 2-27). Additional state match, state-leveraged bonds, loan repayments, and fund earnings increased CWSRF assets by $18.4 billion. Since 1988, therefore, the CWSRF loan program assets have grown to over $30 billion, and they are funding about $3 billion in water quality projects each year.

Summaries by Catalog Unit Awards data extracted from the GICS database were assigned to each of the 2,111 catalog units of the 48 contiguous states by matching city names and counties with corresponding catalog units. Of the total amount of funding awards in the GICS database ($61.1 billion), only a small fraction (less than 1 percent) of the awards could not be assigned to a specific catalog unit. In addition, approximately 2 percent of the GICS funding was awarded to watersheds located outside the 48 contiguous states. (This accounts for the discrepancy between a total national investment of $61.1 billion and the investment of $59.2 billion that was allocated to the 48 contiguous states.)

Figure 2-28 presents the cumulative distribution of the GICS funding awards (total $59.2 billion) as a percentile of the 2,111 catalog units within the contiguous 48 states. Twenty percent of the catalog units account for about 88 percent of the funding. There is also a relationship between the municipal BOD_5 loading rate (ca. 1995) and the Construction Grants award allocated to each catalog unit. Increased municipal loading rates related to larger facilities resulted in increased grant awards from the Construction Grants Program (Figure 2-29).

Other Investment Costs for Water Pollution Control Infrastructure

In addition to the federal expenditures through the Construction Grants Program, state and local governments and private industries have made significant investments to comply with the water pollution control requirements of the CWA and other state

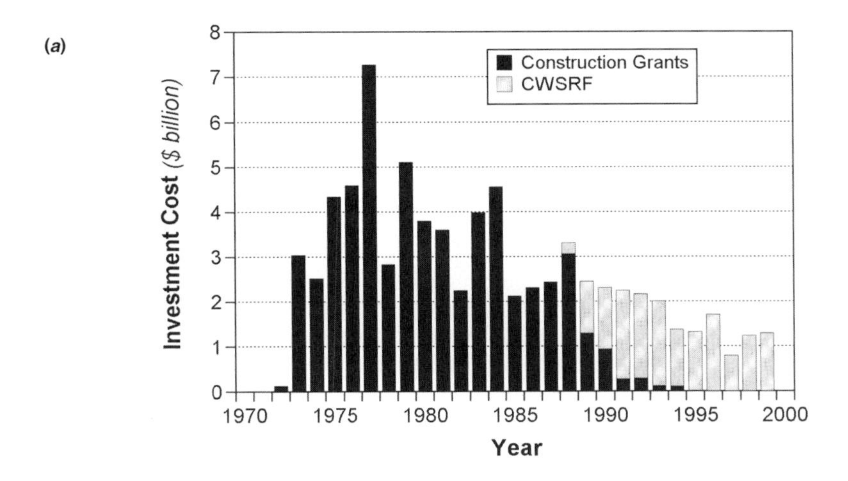

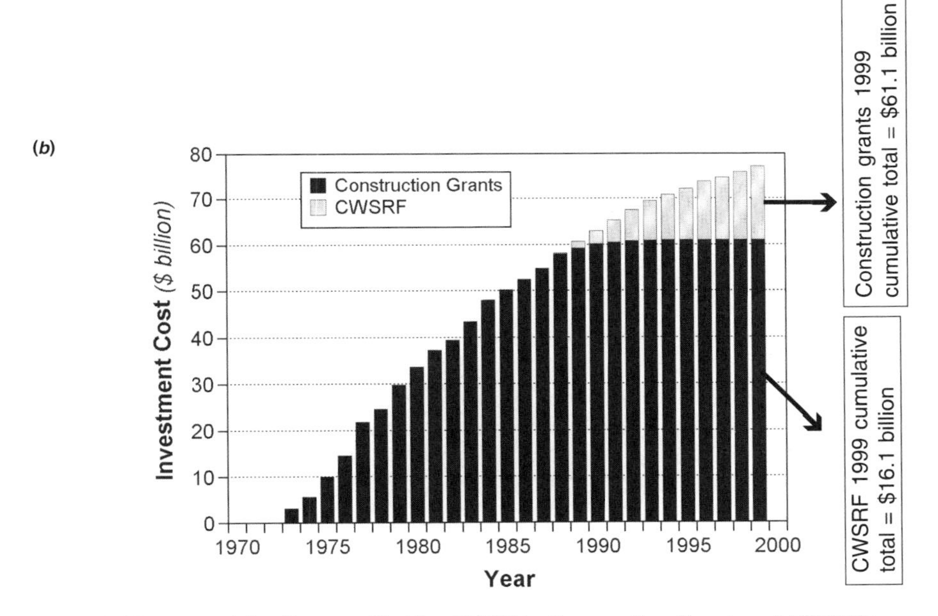

Figure 2-27 Annual funding provided by USEPA's Construction Grants and CWSRF programs to local municipalities for improvements in water pollution control infrastructure as (*a*) annual allotments for each program and (*b*) cumulative funding from both programs from 1970 to 1999. *Source:* USEPA GICS database and CWSRF Program.

and local environmental legislation. On a nationwide basis, actual expenditure data compiled by the U.S. Department of Commerce, Bureau of Economic Analysis in the annual Pollution Abatement Cost Expenditures (Vogan, 1996) document a cumulative public and private sector capital expenditure of approximately $200.6 billion and an additional $210.1 billion as operating expenditures (capital and operation and

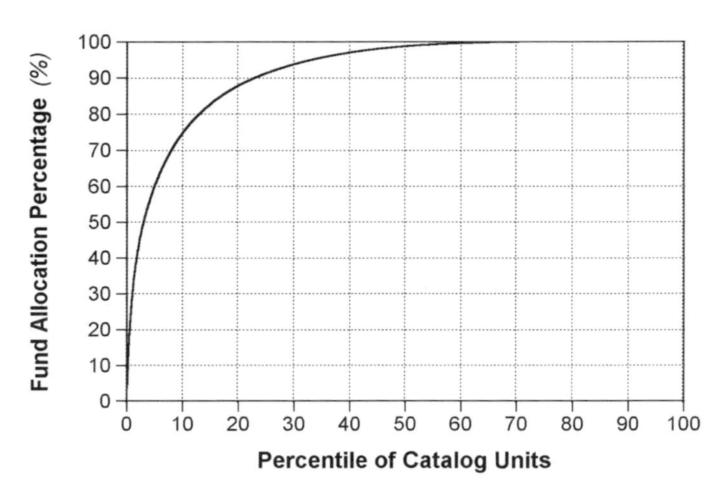

Figure 2-28 Cumulative funding of Construction Grants Program awards as a percentile of 2,111 catalog units. *Source:* USEPA GICS and Reach File Version 1 (RF1) databases.

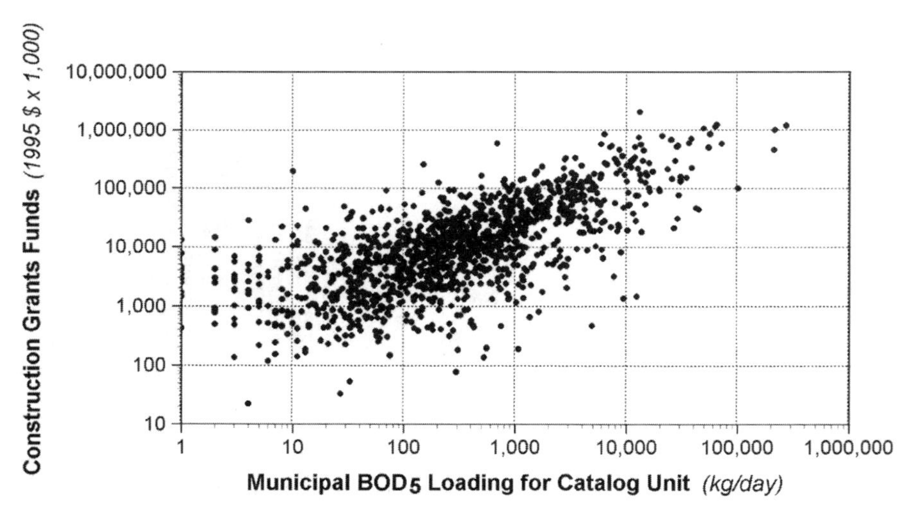

Figure 2-29 Relationship of municipal BOD_5 loading rate ca. 1995 and EPA Construction Grants Program awards by catalog unit. *Source:* USEPA GICS database and Bondelid et al., 2000.

maintenance costs as current year dollars) for water pollution control activities during the period from 1972 through 1994 (Figure 2-30).

As shown in Table 2-18, current year dollars compiled in the annual survey have been indexed to constant 1995 dollars using the Chemical Engineering Plant Cost Index for capital costs and the consumer-based Gross Domestic Product for operating costs as appropriate indices. The Construction Grants Program provided federal grant support to local municipalities that amounted to almost one-half of the public sector costs and about one-third of the total public and private sector capital investment for water pollution control.

Future Infrastructure Needs

USEPA (1997) estimates that by 2016 approximately 2,400 new facilities with secondary or greater than secondary levels of treatment will be needed to service an additional 85 million people (a 45 percent increase of total population). Further, during that time period the Agency estimates that 115 of the approximately 176 POTWs currently providing less than secondary treatment will upgrade their facilities to meet the minimum technology requirements of secondary treatment under the CWA. USEPA estimates the costs for POTW construction and upgrades to be $75.9 billion (indexed to constant 1996 dollars).

Further, USEPA plans to put more emphasis on "wet weather" sources of pollution, including CSOs and stormwater drainage from agricultural, silvicultural, city, and suburban lands. USEPA (1997) has estimated these associated federal costs to include the following:

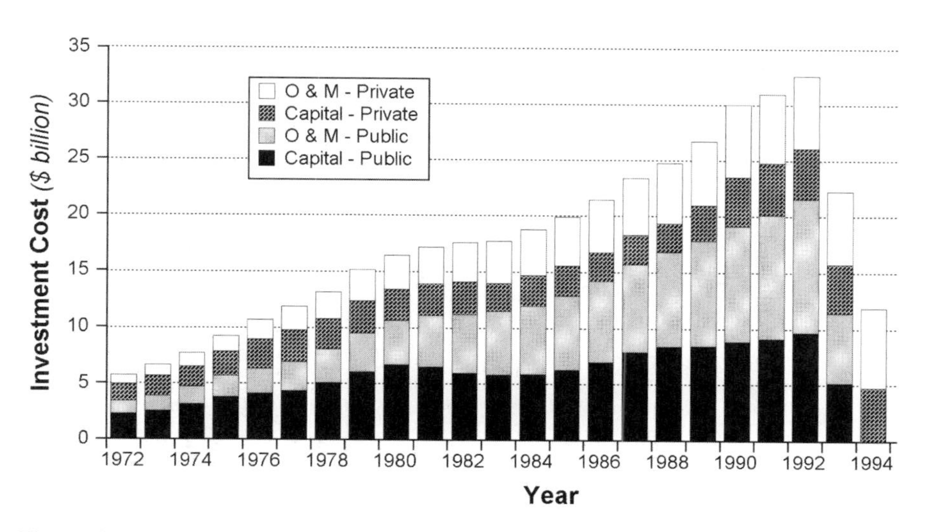

Figure 2-30 Annual water pollution control expenditures (as current year dollars) by the public and private sectors for capital and operations and maintenance costs from 1972 through 1994. *Source:* Vogan, 1996.

TABLE 2-18 National Public and Private Sector Investment in Water Pollution Control Infrastructure, 1956–1999

	EPA Construction Grants		EPA CWSRF,	Public Sector,	Private Sector,	Public + Private Sectors,
	1956–1972[a]	1970–1995[b]	1988–1999[c]	1972–1994[d]	1972–1994[d]	1972–1994[d]
			Current Year Dollars			
Capital	$5.1	$61.1	$16.2	$132.4	$68.2	$200.6
O&M	n/a	n/a	n/a	$121.2	$88.9	$210.1
Totals	$5.1	$61.1	$16.2	$253.6	$157.1	$410.7
			Equivalent as Constant 1995 Dollars[e]			
Capital	$14.3	$96.5	n/a	$178.9	$93.5	$272.4
O&M	n/a	n/a	n/a	$175.5	$128.1	$303.6
Totals	$14.3	$96.5	n/a	$354.4	$211.6	$576.0

[a]EPA Construction Grants Program (1956–1972): data obtained from EPA-OWM files compiled by R. K. Bastian, March 1992.
[b]EPA Construction Grants Program (1970–1995): data obtained from EPA GICS database, August 1995.
[c]EPA Clean Water State Revolving Fund (CWSRF) (1988–1999): data from EPA-OWM files by R. K. Bastian, April 2000.
[d]Public and private sector (1972–1994): Data from Vogan (1996). Data obtained from T. Gilliss, EPA-OPPE, 1997.
[e]Current year dollars adjusted to equivalent constant 1995 dollars. Plant Cost Index obtained from Chemical Engineering (CE, 1995) for capital expenditures. Gross Domestic Product for O&M costs obtained from Council of Economic Advisors (1997).

- $44.7 billion (indexed to constant 1996 dollars) to meet infrastructure needs associated with CSOs.
- $7.4 billion (indexed to 1996 dollars) to meet the Clean Water State Revolving Fund (CWSRF)-eligible portion of the costs that the municipal separate storm sewer systems are expected to incur for the development and implementation of a stormwater management program in response to the Phase I NPDES stormwater program regulations.
- $3.8 billion (indexed to 1996 dollars) to meet the CWSRF-eligible projects related to cropland, pastureland, and rangeland.
- $2.1 billion (indexed to 1996 dollars) to meet the CWSRF-eligible projects related to confined animal facilities with fewer than 1,000 animal units.
- $3.5 billion (indexed to 1996 dollars) to meet the CWSRF-eligible projects related to silviculture.

G. SUMMARY, CONCLUSIONS, AND FUTURE TRENDS

The purpose of this chapter is to address the first leg of the three-legged stool approach for answering the question posed in Chapter 1: Has the Clean Water Act's regulation of wastewater treatment processes at POTWs been a success? Recall that the basic goal of this first leg is to examine the extent to which the nation's investment in building and upgrading POTWs to secondary and greater than secondary wastewater treatment resulted in a decrease in effluent BOD loading to the nation's waterways. If evidence showed that these investments achieved significant reductions in the discharge of oxygen-demanding organic wasteload to the nation's waterways, the first leg of the investigation could add cumulative support for the conclusion that the CWA's mandates were successful.

This section summarizes the key points presented in Sections A through F of this chapter, discusses conclusions, and addresses future trends in wastewater infrastructure requirements.

Key Points of the Background Sections

Specifically discussed in Sections A and B is the significance of water supply and wastewater treatment in the urban water cycle, the invention of secondary treatment, and the use of biochemical oxygen demand (BOD) as a measure of the pollutional strength of organic wasteloads. Section C focuses on the roles the federal government and the CWA played in establishing, and funding, secondary and greater than secondary treatment in the nation's POTWs.

Key points made in Sections A through C include the following:

- All components of the urban water cycle must be in place and functioning properly to satisfy the needs of both water supply and water resource users.

- In the "Great Sanitary Awakening" in the late nineteenth and early twentieth centuries, public infrastructure investment was focused primarily on the water supply side of the urban water cycle and sewage collection systems for the control of waterborne diseases and protection of public health.
- Increasing urban populations in the first half of the twentieth century exacerbated water quality problems associated with the discharge of inadequately treated sewage in urban waterways.
- Secondary treatment proved to be a breakthrough discovery in treating wastewater; by 1930, several cities, especially in the Northeast, Midwest, and far West, had incorporated the technology into their wastewater treatment systems.
- Before 1972 and the passage of the CWA, municipal and industrial wastewater discharges were regulated by individual states based on state ambient water quality standards. The federal government's authority for water pollution control was restricted to interstate waterways under the Commerce clause of the U.S. Constitution.
- The passage of the CWA resulted in the federal government's assuming a greater role in directing and defining water pollution control programs in the nation. The states' water quality–based approach for regulating wastewater discharges was replaced by the CWA's two-pronged approach—a mandatory technology-based approach supplemented by a water quality–based approach on an as-needed basis—and enforced under the National Pollutant Discharge Elimination System permit program.
- Section 301 of the CWA required POTWs to achieve effluent limitations based on secondary treatment as the minimum level of technology.

Key Points of the BOD Loading Analysis Sections

Establishing a national policy requiring secondary treatment of municipal wastewater as the minimum acceptable technology supplemented by more stringent water quality–based effluent controls on a site-specific, as-needed basis was a key provision of the 1972 CWA. This mandate, coupled with an increase in funding assistance to municipalities through the Construction Grants Program, led to a dramatic increase in the number of POTWs with secondary and greater than secondary treatment capabilities.

Section D examines several national POTW trends, including the population they serve, influent and effluent BOD loadings, and BOD removal efficiencies. Key findings include the following:

- The U.S. population served by POTWs with secondary or greater treatment almost doubled between 1968 and 1996 from 85.9 million people in 1968 to 164.8 million people in 1996!
- BOD_u loading to POTWs (influent loading) increased significantly. In 1968, influent BOD_u loading was 34,693 mt/day. By 1996, influent BOD_u loading stood at 46,979 mt/day, a 35 percent increase from 1968! The same trend was seen for influent BOD_5 loading to POTWs.
- Effluent BOD_u loading discharged by POTWs was significantly reduced. In

1968, effluent BOD_u loading was 21,281 mt/day. By 1996, effluent BOD_u loading stood at 16,325 mt/day, a 23 percent decrease from 1968! Effluent BOD_5 loading was also significantly reduced (by 45 percent) over the same time period.

- BOD removal efficiency increased significantly. In 1940, the aggregate national removal efficiency stood at about 33 percent for BOD_5 and 20 percent for BOD_u. By 1968, removal efficiencies had increased to 63 percent for BOD_5 and 39 percent for BOD_u. By 1996, these had increased to nearly 85 percent for BOD_5 and 65 percent for BOD_u!
- Increasing numbers of people served by POTWs in the twenty-first century will likely reverse the trend established between 1968 and 1996 of decreasing effluent BOD loading to the nation's waterways. Assuming that national aggregate design-based BOD_u removal efficiency will increase to 71 percent, influent wastewater flow will remain a constant 165 gpcd, and influent BOD_u concentrations will remain a constant 396.5 mg/L, population projections indicate that, by 2016, effluent BOD_u loading will increase by 20 percent to 19,606 mt/day, equivalent to the rate in the mid-1970s. It is projected that, by 2025, the effluent BOD_u loading will be 21,280 mt/day, a rate approximately equal to that observed in 1968 when the discharge of oxygen-demanding material from POTWs reached its historical peak!
- By 2016, when the projected needs for wastewater treatment are expected to be met (USEPA, 1997), the overall BOD_u removal efficiency of 71 percent and increases in population will result in a 20 percent increase of effluent loads relative to the 1996 loading rate. To maintain an effluent BOD_u loading rate comparable to 1996 conditions through 2016 (i.e., "running in place"), the national aggregate removal efficiency would have to be increased from 71 to 76 percent. This would need to be accomplished by shifting the projected population served from secondary to advanced secondary and advanced wastewater treatment facilities.

Section E presents a national "snapshot" comparison of contemporary (ca. 1995) BOD_5 loadings from POTWs and other point and nonpoint sources based on available data from PCS and the Clean Water Needs Survey. Using the NWPCAM (Bondelid et al., 2000), BOD_5 loadings were estimated for municipal (POTW) and industrial point sources (major and minor), CSOs, and rural runoff and urban nonpoint sources. Loading data for each category were aggregated by catalog units and major river basins. The inclusion of other loading sources in this modeling exercise helps put the municipal loading component in perspective with total nationwide BOD_5 loading from all sources. Key findings include the following:

- Of the 2,111 catalog units in the contiguous United States, 1,632 receive municipal discharges.
- Twenty percent of catalog units account for 90 percent of the total municipal BOD_5 loading. Highest rates of municipal loading of BOD_5 occurred in the Mississippi River Valley and the Northeast and Midwest.
- Municipalities (POTWs) are the dominant source of the BOD_5 component in catalog units associated with major urban areas. Several urban areas had rates greater than 25 mt/day.

- Municipal BOD_5 loadings account for about 38 percent of total point source loadings and 21 percent of total loadings (point and nonpoint).
- Industrial (major and minor) BOD_5 loadings account for about 62 percent of total point source loadings and 34 percent of total loadings (point and nonpoint).
- Urban stormwater and CSOs account for about 5 percent of total nonpoint source loadings and 2 percent of total loadings (point and nonpoint).
- Rural nonpoint source BOD_5 loadings account for about 95 percent of total nonpoint source loadings and 43 percent of total loadings.

Clearly, continued improvement in water quality conditions of the nation's waterways will require an integrated strategy to address all pollutant sources, including both point and nonpoint sources.

Key Points of the Investment Costs Section

Section F focuses on investment costs associated with water pollution control. It includes a discussion of the Construction Grants Program and provides summaries of program spending for new construction and upgrades of POTWs. Also included in this section are summaries of public and private investment totals in point source water pollution control. Key findings include the following:

- From 1970 to 1995, the Construction Grants Program has distributed $61.1 billion (as current year dollars) to municipalities for POTW building and upgrades. The federal share was 75 percent of total costs from fiscal years 1973 to 1983, and 55 percent thereafter.
- From 1988 to 1999, an additional $16.1 billion (capitalization) in federal contributions was also distributed to the states through the Clean Water State Revolving Fund.
- From 1972 to 1994, approximately $200.6 billion in capital costs and $210.1 billion in operation and maintenance costs (as current year dollars) were spent by the public and private sectors for point source water pollution control. Based on these figures, the Construction Grants Program has contributed almost one-half of the public sector costs and about one-third of the total public and private sector capital investment for point source water pollution control.
- Excluding combined sewer systems and urban stormwater controls, USEPA estimates $75.9 billion (1996 dollars) will be required to meet traditional wastewater treatment plant (and sewer) needs through the year 2016 (USEPA, 1997).

Conclusions and Future Trends

Based on the results of the analyses presented in this chapter, the study authors propose the following conclusion regarding the first leg of the three-legged stool approach concerning the nation's investment in building and upgrading POTWs to achieve at least secondary treatment: The CWA's mandated POTW upgrades to at least secondary treatment, combined with financial assistance from the Construction

Grants Program and Clean Water State Revolving Fund, resulted in a dramatic decrease in effluent BOD loading from POTWs to the nation's waterways. This decrease was realized in spite of significant increases in influent BOD loading that occurred due to increases in the population served by POTWs.

Based on needs data submitted by the states, USEPA projects that by the year 2016, 18,303 POTWs in the United States will be serving a population of 274.7 million (USEPA, 1997). Excluding combined sewer systems and stormwater controls, the Agency estimates that $75.9 billion (1996 dollars) will be required to meet traditional wastewater treatment plant and sewer needs at this projected level of service. Based on these projections, influent BOD_u loading in 2016 is estimated to be about 68,030 mt/day, a 45 percent increase in influent BOD_u loading from 1996 (see Section D). Assuming a BOD_u removal efficiency of 71 percent based on the effluent loads contributed by different categories of POTWs (USEPA, 1997), effluent BOD_u loading in 2016 would be about 19,606 mt/day.

The projected effluent BOD_u loading of 19,606 mt/day in 2016 is a concern. Directly and indirectly due to the implementation of the CWA, there was a downward trend of effluent BOD_u loading rates beginning in the early 1970s through at least 1996 (the endpoint year of this study). The highest effluent BOD_u loading rate, 21,281 mt/day, was estimated to have occurred in 1968, four years before the passage of the CWA, and the lowest, 16,325 mt/day, in 1996. The 2016 effluent BOD_u loading estimate reverses the downward trend, with a 20 percent increase in effluent loading over the 20-year period from 1996 to 2016. This level of loading is equivalent to the effluent BOD_u loading rates in the mid-1970s. Further, effluent loading rates projected to 2025 reveal that the nation may experience loading rates similar to those occurring in 1968, a time when the symptoms of water pollution were especially acute.

These findings underscore the importance of incorporating pollutant loading estimates and corresponding water quality improvements into POTW needs surveys. Projected large increases in service population have the potential to overwhelm the gains made to date in effluent BOD loading reductions due to the CWA. To continue the downward trend in effluent BOD loading to the nation's waterways, further improvements need to be made in technologies and actions that decrease influent BOD loading to POTWs (through conservation methods) and increase BOD removal efficiency in the nation's POTWs (through more advanced wastewater treatment methods).

In the 30 years since the passage of the CWA in 1972, a majority of the national water pollution control efforts have focused on controlling pollutants from POTWs and other point sources. National standards ensure that every discharger meets or beats the performance of the best technology available. Continuing the success achieved to date in reducing BOD and other pollutants, however, will require additional investment as older facilities wear out and increasing population pressures demand that existing facilities expand and new facilities be constructed. If these investments are not made and treatment services do not keep pace with growth, many of the gains achieved by the effluent loading reductions that have occurred in the years after the CWA will be lost (WIN, 2000). If this occurs, the wastewater treatment component of the urban water cycle will again assume "weak link" status, with corresponding detrimental consequences to water resource users.

REFERENCES

AMSA. 1997. The AMSA financial survey. American Metropolitan Sewerage Association, Washington, DC.

Arden, E. and W. T. Lockett. 1914. Experiments on the oxidation of sewage without the aid of filters. J. Soc. Chem. Ind., 33, 523, 1122 (1914).

Barzilay, J. L., W. G. Weinberg, and J. W. Eley. 1999. The water we drink: Water quality and its effects on health. Rutgers University Press, New Brunswick, NJ.

Bondelid, T., S. Unger, and A. Stoddard. 2000. National water pollution control assessment model (NWPCAM) Version 1.1. Final report prepared by Research Triangle Institute, Research Triangle Park, NC for U.S. Environmental Protection Agency, Office of Policy, Economics and Innovation, Washington, DC, November, RTI Project Number 92U-7640-031.

Bulloch, D. K. 1989. The wasted ocean. Lyons & Burford Publishers, American Littoral Society, Highlands, NJ.

Carson, R. 1962. Silent spring. Houghton Mifflin Co., Boston.

Carson, R. T., and R. C. Mitchell. 1983. The Value of Clean Water. The Public's Willingness to Pay for Boatable, Fishable, and Swimmable Quality Water. *Water Resources Research,* Vol. 29, No. 7, pp. 2445–2454. July.

CE. 1995. Economic indicators: Chemical engineering plant cost index. Chem. Eng. (McGraw-Hill Companies) 102(7):192.

Chadwick, E. 1842. Report on the sanitary condition of the labouring population of Great Britain. Edinburgh University Press, Edinburgh, Scotland.

Chapra, S. C. 1997. Surface water quality modeling. McGraw-Hill, New York.

Clark, J. W., W. Viessman, and M. J. Hammer. 1977. Water supply and pollution control, 3rd ed. John Wiley & Sons, New York.

Cleary, E. J. 1978. Perspective on river-quality diagnosis. J. WPCF 50(5):825–831.

Council of Economic Advisors. 1997. Table B-3, Chain type price indexes for gross domestic product, 1959–96, Appendix B, Statistical tables relating to income, employment and production. In: Economic report to the president, Transmitted to the Congress, February, 1997, together with the Annual Report of the Council of Economic Advisors. U.S. Government Printing Office, Washington, DC.

Driscoll, E. D. 1986. Lognormality of point and nonpoint source pollution concentrations. In: Urban runoff quality: Impact and quality enhancement technology. Proceedings, Engineering Foundation Conference, American Society of Civil Engineers.

Dunne, T. and L. B. Leopold. 1978. Water in environmental planning. W. H. Freeman and Company, San Francisco, CA.

Eddy, S. 1932. The plankton of the Sangamon River in the summer of 1929. Illinois Natural History Survey Bulletin 19:469–486. Cited in: Biology of water pollution: A collection of selected papers on stream pollution, waste water and water treatment. U.S. Department of the Interior, Federal Water Pollution Control Administration, Cincinnati, OH.

Ellis, J. B. 1986. Pollutional aspects of urban runoff. In: Urban runoff pollution, ed. H. C. Torno, J. Marsalek, and M. Desborder. Springer Verlag, New York, NY.

Fair, G. M., J. C. Geyer, and D. A. Okun. 1971. Elements of water supply and wastewater disposal, 2d ed. John Wiley & Sons, New York.

Fuhrman, R. E. 1984. History of water pollution control. J. WPCF 56(4):306–313.

FWPCA. 1969. The cost of clean water and its economic impact: Volume I, The report. U.S.

Department of the Interior, Federal Water Pollution Control Administration, Washington, DC.

FWPCA. 1970. The economics of clean water: Volume I, Detailed analysis. U.S. Department of the Interior, Federal Water Pollution Control Administration, Washington DC.

Garrett, L. 1994. The coming plague: Newly emerging diseases in a world out of balance. Penguin Books, New York.

Gunnerson, C. G., et al. 1982. Management of domestic waste. In: Ecological stress and the New York Bight: Science and management, ed. G. F. Mayer, pp. 91–112. Estuarine Research Foundation, Columbia, SC.

Hall, J. C. and R. J. Foxen. 1984. Nitrification in BOD_5 test increases POTW noncompliance. J. WPCF 55(12): 1461–1469.

Henderson, C. 1949. Value of the bottom sampler in demonstrating the effects of pollution on fish-food organisms in the Shenandoah River. Progressive Fish Culturist (11):217–230. Cited in: Biology of water pollution: A collection of selected papers on stream pollution, waste water and water treatment. U.S. Department of Interior, Federal Water Pollution Control Administration, Cincinnati, OH.

Higgins, J. 1998. Massachusetts Department of Environmental Protection, Boston, MA. Personal communication. September.

Howard Jones, N. 1975. The scientific background of the international sanitary conferences, 1851–1938. World Health Organization (WHO), Geneva, Switzerland.

ICPRB. 1991. The Potomac River model data report. Interstate Commission on Potomac River Basin, Rockville, MD.

Jobin, W. 1998. Sustainable management for dams and waters. Lewis Publishers, Boca Raton, FL.

Kahn, M. 1997. The silver lining of rust belt manufacturing decline: Killing off pollution externalities. CES 97–7. U.S. Department of Commerce, Washington, DC.

Kinnicutt, L., C. E. A. Winslow, and R. W. Pratt. 1913. Sewage Disposal. John Wiley & Sons, New York, NY.

Knopman, D. S. and R. A. Smith. 1993. Twenty years of the Clean Water Act: Has U.S. water quality improved? Environment 35(1):17–41.

Leo, W. M., R. V. Thomann, and T. W. Gallagher. 1984. Before and after case studies: Comparisons of water quality following municipal treatment plant improvements. EPA 430/9-007. Office of Water, Program Operations, U.S. Environmental Protection Agency, Washington, DC.

Lovejoy, S. B. 1989. Changes in cropland loadings to surface waters: Interim report no. 1 for the development of the SCS National Water Quality Model. Purdue University, West Lafayette, IN.

Lovejoy, S. B. and B. Dunkelberg. 1990. Water quality and agricultural policies in the 1990s: Interim report no. 3 for the development of the SCS National Water Quality Model. Purdue University, West Lafayette, IN.

Luken, R. A., D. J. Basta, and E. H. Pechan. 1976. The national residuals discharge inventory: An analysis of the generation, discharge, cost of control and regional distribution of liquid wastes to be expected in achieving the requirements of Public Law 92-500. Report No. NCWQ 75/104, NTIS Accession No. PB0252-288. Prepared for the Study Committee on Water Quality Policy, Environmental Studies Board, Commission on Natural Resources, National Research Council. National Commission on Water Quality, Washington, DC.

Lung, W. 1998. Trends in BOD/DO modeling for waste load allocations. J. Environ. Eng., ASCE 124(10): 1004–1007.

Marsalek, J. 1978. Pollution due to urban runoff: Unit loads and abatement measures. Pollution from Land Use Activities Reference Group, International Joint Commission, Windsor, Ontario.

Metcalf and Eddy, Inc. 1991. Wastewater engineering: Treatment, disposal and reuse. McGraw-Hill Series in Water Resources and Environmental Engineering, McGraw-Hill, New York.

MWCOG. 1989. Potomac River water quality, 1982 to 1986: Trends and issues in the metropolitan Washington area. Metropolitan Washington Council of Governments, Department of Environmental Programs, Washington, DC.

NCWQ. 1976. Staff report to the National Commission on Water Quality. U.S. Government Printing Office, Washington, DC.

NOAA. 1994. Gulf of Maine point source inventory; A summary by watershed for 1991. National Coastal Pollutant Discharge Inventory. National Oceanic Atmospheric Administration, Strategic Environmental Assessments Division, Pollution Sources Characterization Branch, Silver Spring, MD.

Novotny, V. and H. Olem. 1994. Water quality: Prevention, identification, and management of diffuse pollution. Van Nostrand Reinhold, New York.

NRC. 1993. Managing wastewater in coastal urban areas. Committee on Wastewater Management for Coastal Urban Areas, Water Science and Technology Board, Commission on Engineering and Technical Systems, National Research Council. National Academy Press, Washington, DC.

Okun, D. A. 1996. From cholera to cancer to *Cryptosporidium*. J. Environ. Eng., ASCE 122 (6):453–458.

OTA. 1987. Wastes in marine environments. OTA-O-334. U.S. Congress, Office of Technology Assessment. U.S. Government Printing Office, Washington, DC. April.

Reckhow, K. H., M. N. Beaulac, and J. T. Simpson. 1980. Modeling phosphorus loading and lake response under uncertainty: A manual and compilation of export coefficients. EPA-440/5-80-011. U.S. Environmental Protection Agency, Washington, DC.

Rowland, W. G. and A. S. Heid. 1976. Water and the growth of the Nation. J. WPCF 48(7): 1682–1689.

Solley, W. B., R. R. Pierce, and H. A. Perlman. 1998. Estimated use of water in the United States, 1995. USGS Circular 1200. U.S. Geological Survey, Reston, VA.

Snow, J. 1936. Snow on cholera, Being a reprint of two papers by John Snow, M. D. [1849 and 1855]. Oxford University Press, London, England.

Tetra Tech. 1993. Support to the 1992 Needs Survey CSO Cost Assessment: CSO water quality modeling. EPA Contract 68-C9-0013, Work Assignment 3-157. Draft report Tetra Tech, Inc., Fairfax, VA.

Tetra Tech. 1999. Improving point source loadings data for reporting national water quality indicators. Final technical report prepared for U.S. Environmental Protection Agency, Office of Wastewater Management, Washington, DC, by Tetra Tech, Inc., Fairfax, VA.

Thomann, R. V. and J. A. Mueller. 1987. Principles of surface water quality modeling and control. Harper & Row, Inc., New York.

USCB. 1975. Historical statistics of the United States: Colonial times to 1970. Series B 193-200. U.S. Census Bureau, Washington, DC.

USCB. 1996. Population projections of the United States by age, sex, race and Hispanic origin: 1995–2050. Current Population Reports, Series pp. 25–1130. Population Division, U.S. Census Bureau, Washington, DC.

USCB. 2000. Urban and rural population: 1900 to 1990. In: 1990 Census of Population and Housing, Population and Housing Unit Counts, CPH-2-1. Population Division, U.S. Census Bureau, Washington, DC.

USEPA. 1972. 1972 NEEDS survey, conveyance and treatment of municipal wastewater: Summaries of technical data. U.S. Environmental Protection Agency, Office of Water Program Operations, Washington, DC.

USEPA. 1973. Secondary treatment parameters. U.S. Environmental Protection Agency. Fed. Regist., April 30, 1973, 38:12973.

USEPA. 1974. National water quality inventory, 1974. EPA-440/9-74-001. U.S. Environmental Protection Agency, Office of Water Planning and Standards, Washington, DC.

USEPA. 1976. 1976 NEEDS survey, conveyance and treatment of municipal wastewater: Summaries of technical data. U.S. Environmental Protection Agency, Office of Water Program Operations, Washington, DC.

USEPA. 1978. 1978 NEEDS survey, conveyance and treatment of municipal wastewater: Summaries of technical data. U.S. Environmental Protection Agency, Office of Water Program Operations, Washington, DC.

USEPA. 1980. 1980 NEEDS survey, conveyance and treatment of municipal wastewater: Summaries of technical data. U.S. Environmental Protection Agency, Office of Water Program Operations, Washington, DC.

USEPA. 1982. 1982 NEEDS survey, conveyance, treatment, and control of municipal wastewater, combined sewer overflows and stormwater runoff: Summaries of technical data. U.S. Environmental Protection Agency, Office of Water Program Operations, Washington, DC.

USEPA. 1983. Results of the nationwide urban runoff program, 1, final report, December, 1983. NTIS PB84-185552.

USEPA. 1984. 1984 NEEDS survey, conveyance, treatment, and control of municipal wastewater, combined sewer overflows and stormwater runoff: Summaries of technical data. U.S. Environmental Protection Agency, Office of Water Program Operations, Washington, DC.

USEPA. 1986. 1986 NEEDS survey, conveyance, treatment, and control of municipal wastewater, combined sewer overflows and stormwater runoff: Summaries of technical data. U.S. Environmental Protection Agency, Office of Water Program Operations, Washington, DC.

USEPA. 1989. 1988 NEEDS survey, conveyance, treatment, and control of municipal wastewater, combined sewer overflows and stormwater runoff: Summaries of technical data. U.S. Environmental Protection Agency, Office of Water Program Operations, Washington, DC.

USEPA. 1993. 1992 Clean Water Needs Survey (CWNS), conveyance, treatment, and control of municipal wastewater, combined sewer overflows and stormwater runoff: Summaries of technical data. EPA-832-R-93-002. U.S. Environmental Protection Agency, Office of Water Program Operations, Washington, DC.

USEPA. 1997. 1996 Clean Water Needs Survey (CWNS), conveyance, treatment, and control of municipal wastewater, combined sewer overflows and stormwater runoff: Summaries of technical data. EPA-832-R-97-003. U.S. Environmental Protection Agency, Office of Water Program Operations, Washington, DC.

USEPA. 1998. Clean water action plan: Restoring and protecting America's waters. U.S. Environmental Protection Agency, Office of Water, Washington, DC.

USPHS. 1951. Water pollution in the United States. A report on the polluted conditions of our waters and what is needed to restore their quality. U.S. Federal Security Agency, Public Health Service, Washington, DC.

Viessman, W. and M. J. Hammer. 1985. Water supply and pollution control, 4th ed. Harper & Row, New York.

Vogan, C. R. 1996. Pollution abatement and control expenditures, 1972–94. Survey of current business. Vol. 76, No. 9, pp. 48–67. U.S. Department of Commerce, Bureau of Economic Analysis, Washington, DC.

WEF. 1997. The Clean Water Act. 25th anniversary ed. Water Environment Federation, Alexandria, VA.

WIN. 2000. Clean and safe water for the 21st century: A renewed national commitment to water and wastewater infrastructure. Water Infrastructure Network, Washington, DC.

Zwick, D. and M. Benstock. 1971. Ralph Nader's study group report on water pollution: Water wasteland. Center for Study of Responsive Law, Grossman Publishers and Bantam Books, New York.

CHAPTER 3

An Examination of "Worst-Case" DO in Waterways Below Point Sources Before and After the CWA

Chapter 2 discussed the evolution of the BOD measurement, the impact of BOD loadings on DO levels in natural waters, and the massive amount of public and private money invested in municipal wastewater treatment to meet the mandates of the CWA. Key conclusions from the first leg of the three-legged stool approach are:

- The nation's investment in building and upgrading POTWs significantly reduced BOD effluent loading to the nation's waterways.
- This reduction occurred in spite of a significant increase in influent BOD loading caused by an increase in population served by POTWs.

The second leg follows up on the first leg with another question: Has the CWA's push to reduce BOD loading resulted in improved water quality in the nation's waterways? And, if so, to what extent? The key phrase in the question is "to what extent?" Earlier studies by Smith et al. (1987a, 1987b) and Knopman and Smith (1993) concluded that any improvements in DO conditions in the nation's waterways are detectable only within relatively local spatial scales downstream of wastewater discharges.

> Perhaps the most noteworthy finding from national-level monitoring is that heavy investment in point-source pollution control has produced no statistically discernible pattern of increases in water's dissolved oxygen content during the last 15 years [1972–87]. . . . The absence of a statistically discernible pattern of increases suggests that the extent of improvement in dissolved oxygen is limited to a small percentage of the nation's total stream miles. This is notable because the major focus of pollution control expenditures under the act [CWA] has been on more complete removal of oxygen-demanding wastes from plant effluents.
>
> —Knopman and Smith, 1993

The purpose of the second leg of this investigation is to examine evidence that may show that the CWA's municipal wastewater treatment mandates benefited water quality on a broad scale, as well as in reaches immediately downstream from POTW discharges. The systematic, peer-reviewed approach used in this investigation includes the following steps:

- Developing before- and after-CWA data sets composed of DO summary statistics derived from monitoring stations that were screened for worst-case conditions. The purpose of the screening exercise is to mine data that inherently contain a response "signal" linking point source discharges with downstream water quality.
- Calculating a worst-case DO summary statistic for each station for each before- and after-CWA time period and then aggregating station data at sequentially larger spatial scales (reaches, catalog units, and major river basins).
- Conducting an analysis of spatial units that have before- and after-CWA worst-case DO summary statistics and then documenting the direction (improvement or degradation) and magnitude of the changes in worst-case DO concentration.
- Assessing how the point source discharge/downstream DO signal changes over progressively larger spatial scales.

Section A of this chapter provides background on the relationship between BOD loading and stream water quality and discusses the two key physical conditions (high temperature and low flow) that create "worst-case" conditions for DO. Section B describes the development and application of a set of screening rules to select, aggregate, and spatially assess before- and after-CWA DO data drawn from USEPA's STORET database. Section C presents the results of the comparison analysis of worst-case DO from before and after the CWA for reach, catalog unit, and major river basin scales. The chapter concludes with Section D, which provides the summary and conclusions for the second leg of this investigation.

A. BACKGROUND

In both terrestrial and aquatic ecosystems, the continuous cycles of production and decomposition of organic matter are the principal processes that determine the balance of organic carbon, nutrients, carbon dioxide, and DO in the biosphere. Plants (autotrophs) use solar energy, carbon dioxide, and inorganic nutrients to produce new organic matter and, in the process, produce DO by photosynthesis. Bacteria and animals (heterotrophs) use the organic matter as an energy source (food) for respiration and decomposition, and in these processes, consume DO, liberate carbon dioxide, and recycle organic matter back into the ecosystem as simpler inorganic nutrients. Water quality problems, such as depleted levels of DO, nutrient enrichment, and eutrophication (overproduction of aquatic plants), occur when the aquatic cycle of production and decomposition of organic matter becomes unbalanced from excessive amounts of anthropogenic inputs of organic carbon and inorganic nutrients from wastewater discharges and land use–influenced watershed runoff.

DO is the most meaningful and direct signal relating municipal and industrial discharges of organic matter to downstream water quality responses over a wide range of temporal and spatial scales. In addition to DO's significance as a measure of aquatic ecosystem health, there are several other practical reasons for choosing DO as the signal for assessing changes in water quality, including the following:

- Historical records go as far back as the early twentieth century for many major waterbodies. New York City, for example, began monitoring DO in New York Harbor in 1909 and records exist for the Upper Mississippi River beginning in 1926, for the Potomac estuary in 1938, and for the Willamette River in 1929 (see Wolman, 1971).
- Basic testing procedures for measuring DO have introduced few biases over the past 90 years, thereby providing the analytical consistency needed for comparing historical and modern data (Wolman, 1971).

This section provides background on sources of DO data, the relationship between BOD loading, downstream DO levels, and the two key physical conditions (high temperature and low flow) that create "worst-case" DO conditions. As will be explained, DO data collected under worst-case conditions inherently contain the sharpest signal of the point source discharge/downstream DO relationship.

Sources of DO Data

Key to this analysis is the existence of DO data with which a before- and after-CWA comparison can be made. Fortunately, systematic water pollution surveillance of many of the nation's waterways began in 1957 in response to the 1956 Amendments to the Federal Water Pollution Control Act. Figure 3-1 is a map, developed by Gunnerson (1966), displaying minimum DO concentrations throughout the United States using data collected from 1957 through 1965. It illustrates both the spatial extent of historical data and the poor DO conditions found in many of the nation's waterways in the late 1950s and early 1960s.

These and more recent water quality data collected by state, federal, and local agencies are in USEPA's STORET database and are available for a before- and after-CWA comparison (Gunnerson, 1966; Ackerman et al., 1970; Wolman, 1971; USEPA, 1974). Currently, the system holds over 150 million testing results from more than 735,000 sampling stations, about 4.6 million of which are DO observations recorded from 1941 to 1995 (Figure 3-2). The challenge was to figure out how to mine STORET's mountain of DO data and create before- and after-CWA data sets that inherently contain the best response "signal" linking point source discharges with downstream DO. This task is not unlike panning for gold. What was needed was a series of screens to divert away all the "rubble and debris" (noisy data), leaving a clean set of "nuggets" (signal data). Using a systematic comparison of before- and after-CWA water quality data sets, the national policy for technology- and water quality–based effluent controls can be considered a success if downstream waterways with poor water quality before the CWA can be shown to have improved significantly after the CWA.

"Worst-Case" Conditions as a Screening Tool

The first step in developing the before- and after-CWA data sets was to analyze the relationship between point source BOD loading and downstream DO levels. As the reader will see in Section B, the rules subsequently adopted and applied to screen out

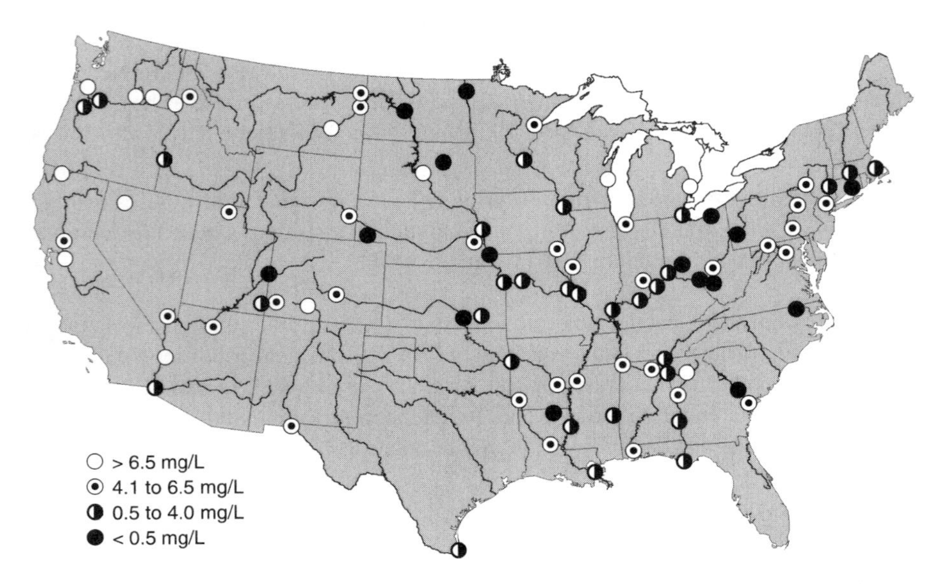

Figure 3-1 Location of sample stations and minimum DO concentrations in the contiguous 48 states from 1957 to 1965. *Source:* Gunnerson, 1966.

noisy data were based on eliminating hydrologic and other physical factors that interfered with, or confounded, the point source discharge/downstream DO signal (see Hines et al., 1976). As it turned out, the DO data that contained the strongest signal were the data collected under conditions that yielded the lowest DO levels (high water temperature and low flow). The purpose of this subsection is to explain the physical processes and spatial characteristics that make worst-case conditions the appropriate screening tool for developing the before- and after-CWA data sets.

Worst-Case Conditions from a Temporal Perspective In an unpolluted stream, DO concentrations in most of the water column are typically at or near saturation. Saturation, however, varies inversely with water temperature and elevation. At typical winter water temperatures of about 10° C, the solubility of oxygen is about 11.3 mg/L at sea level. At a higher summer temperature of 25° C, the solubility is only about 8.2 mg/L. This high water temperature–low solubility relationship makes hot weather an especially critical period for aquatic organism survival. Higher water temperatures mean a lower reserve of oxygen is available to buffer against any additional oxygen demands made by wastewater effluent discharges.

Wastewater effluent typically has an oxygen deficit (a DO concentration below saturation). Therefore, its initial entry into a waterway causes an immediate drop in stream DO near the outfall. The effluent becomes diluted as it mixes with the stream water and flows down the channel. The BOD of the stream water thus becomes the discharge-weighted average BOD of the effluent and the stream above the discharge. The volume of streamflow (the dilution factor), therefore, is a critical variable in de-

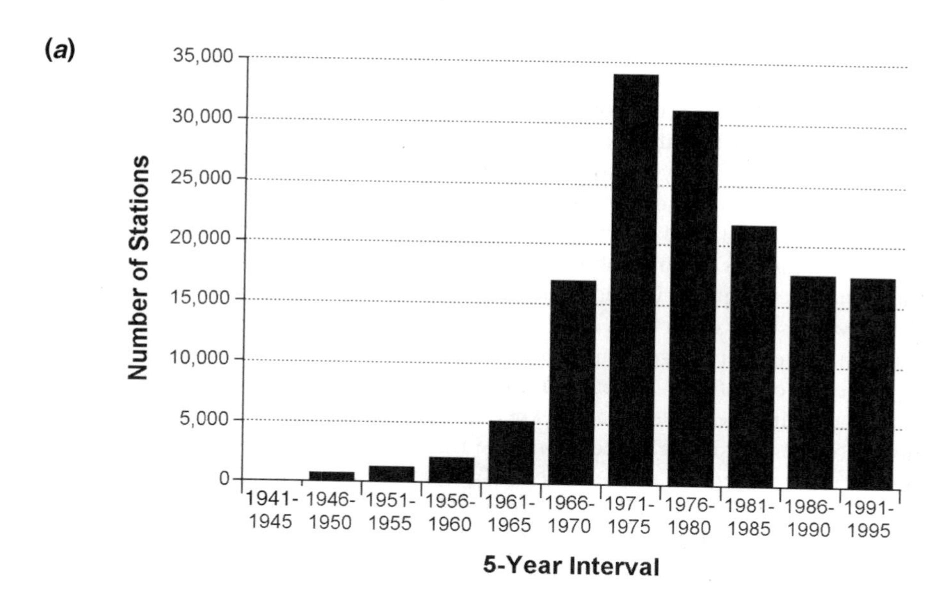

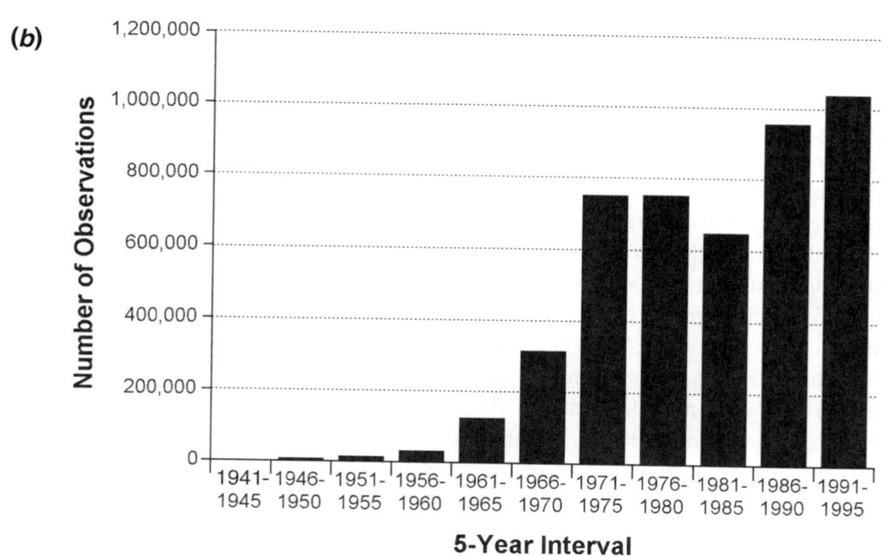

Figure 3-2 National inventory of (*a*) stations collecting DO data and (*b*) the number of DO observations made by those stations aggregated by 5-year intervals from 1941 through 1995. *Source:* USEPA STORET.

termining the concentration of oxygen-demanding waste. Consequently, periods of low flow in the stream channel yield the highest concentration of BOD.

The combination of unnaturally high levels of BOD inputs, high water temperature, and low stream flow creates worst-case DO levels in streams and, in turn, the most critical conditions for the survival of aquatic organisms, that is, conditions of increased oxygen demand, low oxygen solubility, and low dilution potential. Fortunately, worst-case conditions do not occur all the time. Although the BOD loading component tends to remain relatively stable over the course of a year, there are usually distinct seasonal variations in temperature and rainfall (directly related to flow). On an annual basis in the contiguous United States, the highest water temperatures and minimal flow levels usually occur from early summer to late fall. Therefore, the months of July through September are generally considered "worst-case" months for DO.

Observations of year-to-year variations in climate reveal that many areas on the earth, including the United States, experience runs of wet and dry years, a phenomenon known as persistence. The short time frame of historical record-keeping makes it difficult for scientists to predict exactly when these wet and dry year cycles will occur; however, more than 100 years of rainfall data have proven that they are not uncommon. Importantly, persistence tends to have a cumulative effect on stream conditions. Therefore, the worst-case scenario for DO in waterways from a temporal perspective can be further refined to include the months of July through September (worst-case months) during a run of dry years (worst-case persistence).

Defining the periods of years before and after the CWA to represent worst-case persistence was accomplished in three steps. In the first step, USGS flow data taken from approximately 5,000 gages with over 20 years of record during the period from 1951 to 1980 was used to calculate long-term mean summer flow from July through September. The mean flow for each gage was then normalized as runoff (cfs per square mile) over the drainage area contributing flow at each gage. The normalized runoff data for each gage was spatially interpolated to determine the long-term mean summer runoff for each catalog unit (see Figure F-1 in Appendix F). Summer mean flows for each gage were computed for each year from 1961 through 1995 and classified as "dry," "normal," and "wet" years by calculating the relative ratio of flow for each summer to the long-term (1951–1980) summer mean flow computed for each gage. Years with ratios less than 0.75 were considered dry; normal years had ratios from 0.75 to 1.5, and wet years were defined as having ratios greater than 1.5.

Figure 3-3 illustrates how widely mean summer flow can vary over time. The figure displays USGS gage data from the Upper Mississippi River at St. Paul, Minnesota, for the years 1960 through 1995. The scale on the left vertical axis is streamflow measurements as thousands of cubic feet per second (cfs). The scale on the right vertical axis is the interannual-to-long-term mean (10,658 cfs) streamflow ratio. Note that the benchmark ratio of 0.75 (which distinguishes dry from normal years) is represented by the dashed horizontal line. This graph shows that dry summers with low flow occurred in St. Paul in the years 1961, 1970, 1976, 1980, and 1987–1989. The data from this gage also show the enormous wet conditions that occurred primarily in response to the "Great Flood of 1993." That year, the mean summer flow was about 4.5 times greater than the normal mean summer flow.

For the second step, a sliding window methodology was used as an algorithm to

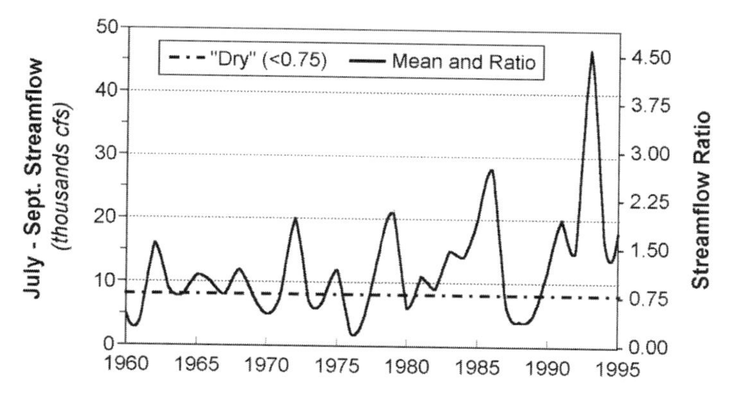

Figure 3-3 Time series of mean summer (July–September 1960–1995) streamflow and ratio of interannual to long-term (1951–1980) summer mean. *Source:* Data from USGS Gage 05331000 on the Upper Mississippi River near Minneapolis-St. Paul, Minnesota.

weight and interpolate normalized streamflow ratios for multiple gages within a catalog unit. The outcome was a weighted streamflow ratio assigned to each catalog unit for each year from 1961 through 1995. Similar to the gage-scale streamflow ratio, the catalog unit–scale streamflow ratio was used to classify catalog units into dry (< 0.75), normal (0.75–1.5), and wet (> 1.5) years.

The third and final step used to define the periods of worst-case dry persistence before and after the CWA involved grouping the 35-year period from 1961 to 1995 into consecutive 5-year "time-blocks." Then for each catalog unit, the number of years within each time-block during which the catalog unit scale streamflow ratio was below 0.75 (i.e., dry) was determined. Rather than using the seemingly obvious 5-year time-block of 1966–1970 to characterize water quality conditions "before" the 1972 CWA, 1961–1965 was selected instead to represent conditions "before" the CWA, while 1986–1990 was used to characterize conditions "after" the CWA.

Widespread drought conditions, a critical factor for "worst-case" water quality conditions, occurred in 1961, 1962, 1963, 1964, 1965, 1966, 1987, and 1988 in the Northeast, Middle Atlantic, Midwest, and Central states. Drought conditions thus occurred during both of these "before and after" 5-year time-blocks of record (i.e., 1961–1965 and 1986–1990). The widespread extent of drought conditions during the "before and after" time-blocks is shown in Figure 3-4 with maps of normalized streamflow ratios computed for each catalog unit for 1963 and 1988.

For the 5-year time-block of 1961–1965, selected to represent before-CWA conditions, 1,923 (91 percent) of the 2,111 catalog units of the 48 contiguous states were characterized by at least one year of "dry" streamflow conditions. Similarly, for the 5-year time-block of 1986–1990 "after" the CWA, 1,776 (84 percent) of the 2,111 catalog units of the 48 contiguous states were characterized by at least one year of "dry" streamflow conditions. For the catalog units characterized as "dry," low-flow conditions occurred for a mean period of 2.5 years during 1961–1965 and 2.7 years during 1986–1990 (Figure 3-5). Hydrologic conditions for the summers of 1963 and

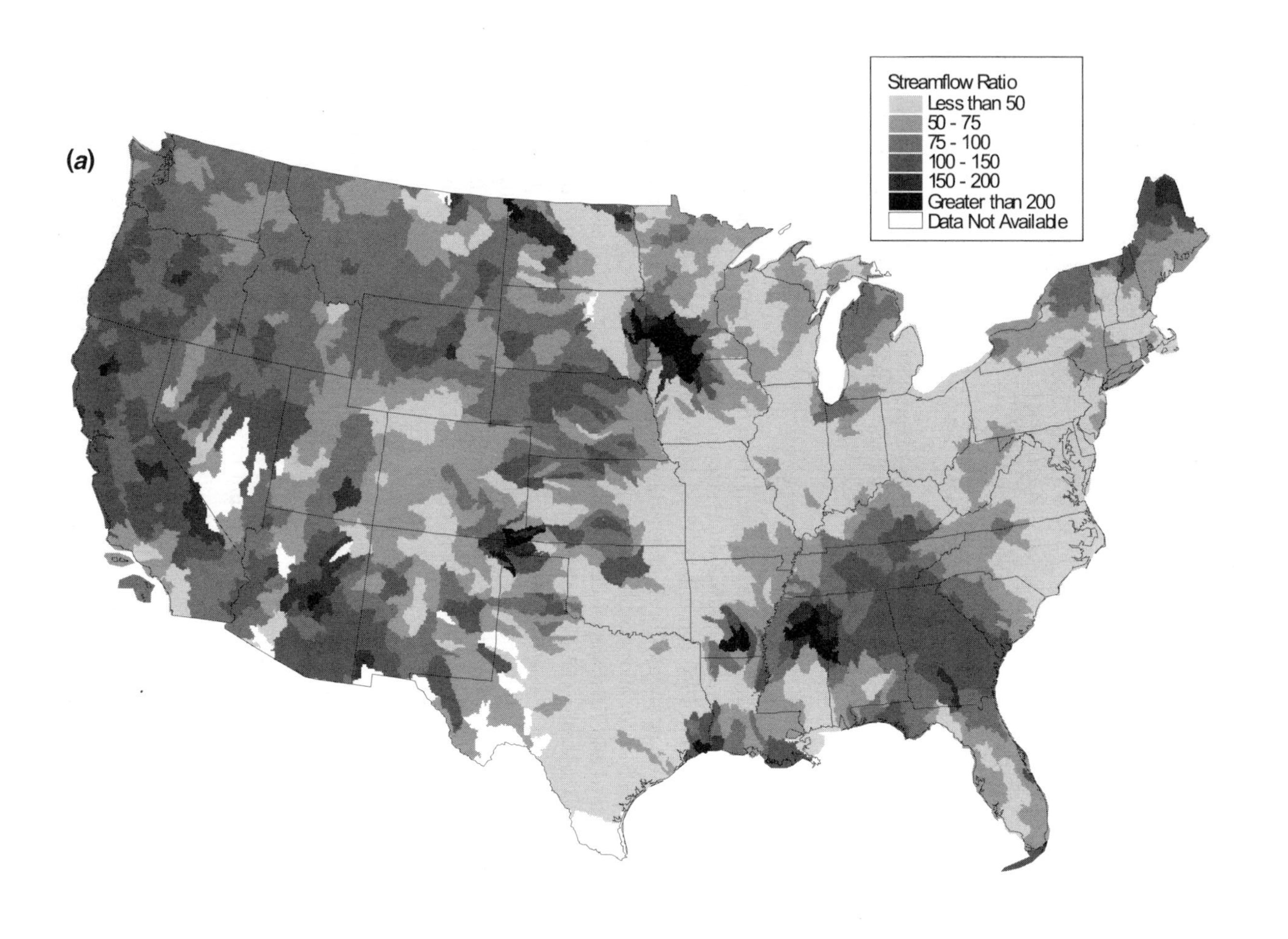

(a)
Streamflow Ratio
Less than 50
50 - 75
75 - 100
100 - 150
150 - 200
Greater than 200
Data Not Available

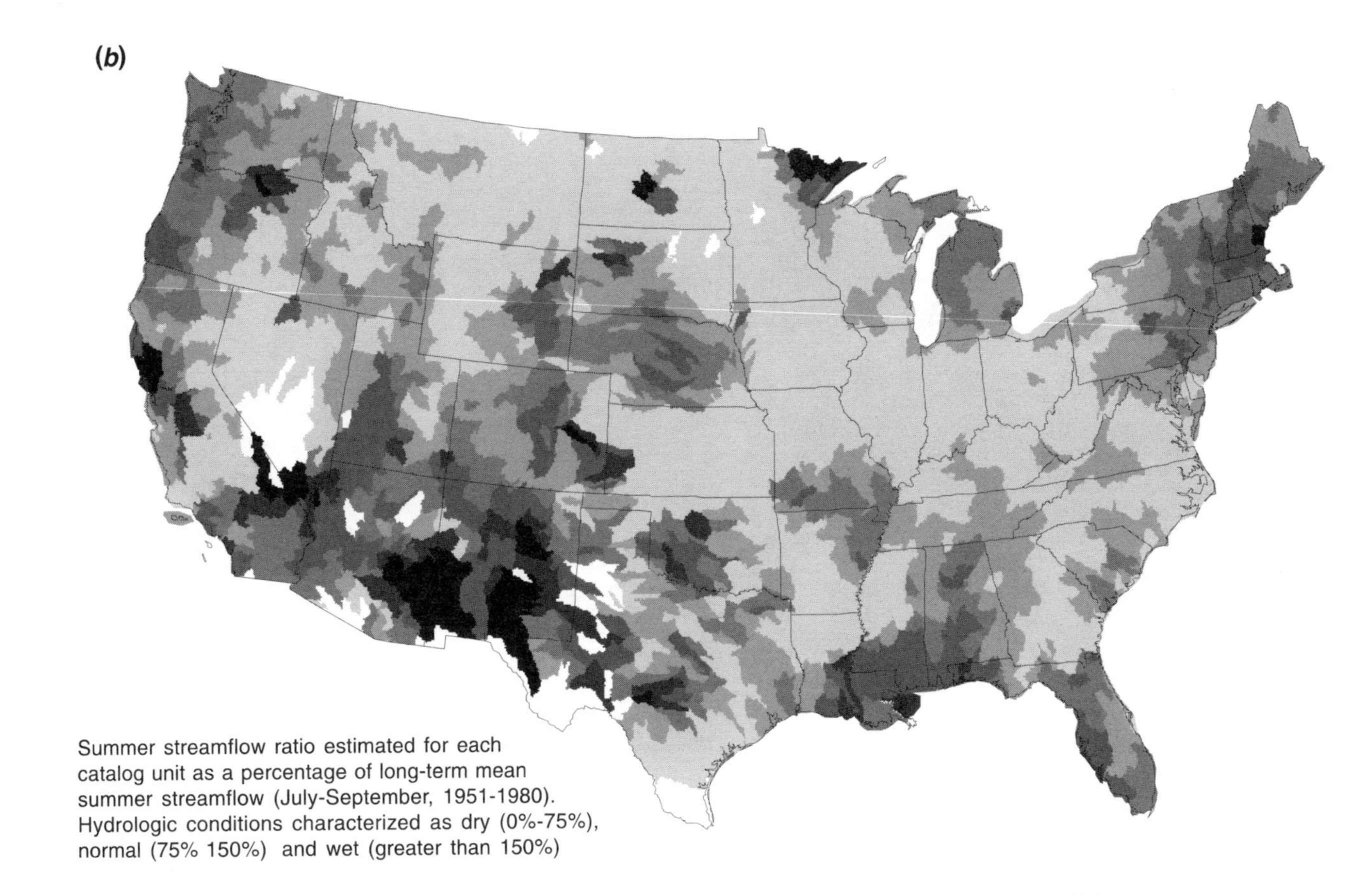

Figure 3-4 Hydrologic conditions during July–September of (*a*) 1963 and (*b*) 1988.

(*a*)

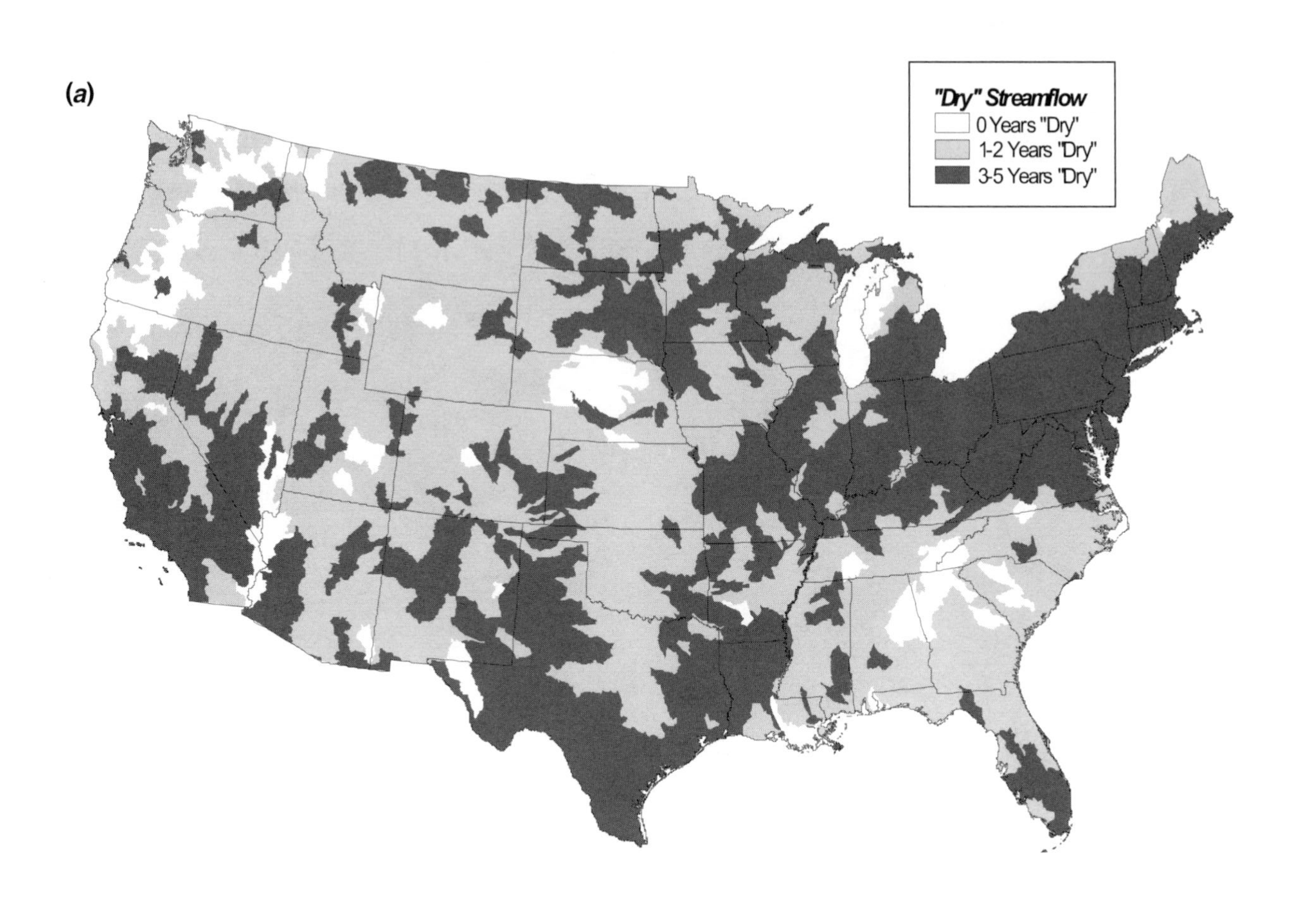

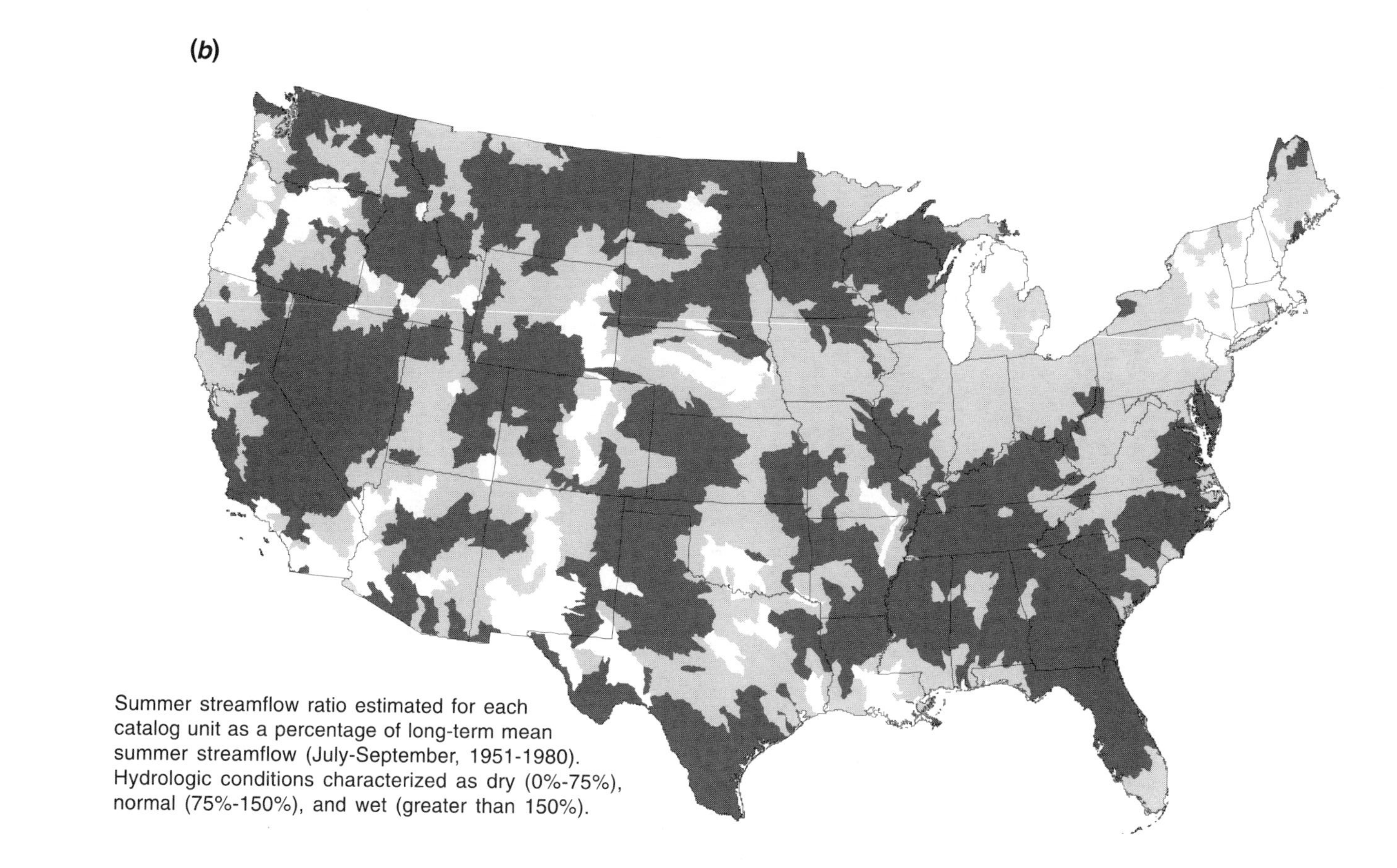

Figure 3-5 Persistence of dry hydrologic conditions during July–September of (*a*) 1961–1965 and (*b*) 1986–1990.

1988 are shown to illustrate the similarity of the spatial extent of drought conditions within the 48 contiguous states during the before- and after-CWA time-blocks. Using this station selection approach based on summer streamflow ratios, trends identified for "before versus after" changes in DO can then be correctly attributed to changes in pollutant loadings (under comparable "dry" streamflow conditions) rather than to differences in hydrologic conditions.

Worst-Case Conditions from a Spatial Perspective In a clean river, upstream of any wastewater inputs, DO levels are typically near saturation. Downstream of an effluent discharge, however, measurements of DO lower than saturation exhibit a characteristic spatial pattern influenced by the loss of oxygen from degradation of organic matter and nitrification and the replenishment of oxygen transferred from the atmosphere into the river (see Thomann and Mueller, 1987; Chapra, 1997). An understanding of the spatial pattern of DO in rivers was critical for the design of the screening methodology used to detect "worst-case" conditions from a spatial perspective. Using river miles from a downstream confluence as a measure of distance along the river, Figure 3-6 illustrates spatial patterns of carbon (CBOD), nitrogen (organic N, NH_3-N, and $NO_2-N + NO_3-N$), and DO in zones identified as "clean water," "degradation," "active decomposition," and "recovery" that are upstream and downstream of a POTW discharge. The distributions were computed using data adapted from Thomann and Mueller (1987) to describe upstream inputs for a flow of 100 cfs and wastewater loads from a 7.5 mgd primary treatment plant (see Table 2-17) discharging into a river 1 m deep and 30 m wide, with a water temperature of 25°C.

In streams and rivers, DO levels are maintained near saturation by the continuous transfer of atmospheric oxygen into solution in a thin surface layer of the river. The rate of transfer of atmospheric oxygen into the river (i.e., mixing of oxygen as a gas from the air into solution in the water) depends on how fast the river is running, how deep the water is, how "bubbly" the river appears to be, the water temperature, and how much oxygen is already in solution in the river. The less oxygen that is in solution in the river, the faster more oxygen can be transferred from the air into the water. In the "degradation" zone, more oxygen is being consumed by decomposition than can be replenished from the atmospheric supply of oxygen, and DO levels quickly drop. In the "active decomposition" zone, more oxygen is gained by the mixing of oxygen from the air into the water than is lost by the continued decomposition of a diminishing amount of carbon (CBOD) and nitrogen (NBOD), and oxygen gradually increases. In the "recovery" zone, the rate of atmospheric replenishment of oxygen greatly exceeds the oxygen lost due to small levels of CBOD and NBOD remaining in the river, and oxygen returns to the saturation level.

Immediately downstream from the POTW, the carbon concentration (CBOD) jumps from the low upstream level to a much higher flow-weighted CBOD concentration as the effluent load is diluted with the ambient upstream load (Figure 3-6*a*). Bacterial decomposition of the carbon results in a steady decrease of in-stream CBOD and a steep drop in oxygen in the "degradation" zone, followed by a continued decline of CBOD with a gradual increase in oxygen in the "active decomposition" zone.

As shown in Figure 3-6*b* for the spatial patterns of nitrogen, organic nitrogen (organic N) and ammonia nitrogen (NH_3-N), both jump from a low upstream level fol-

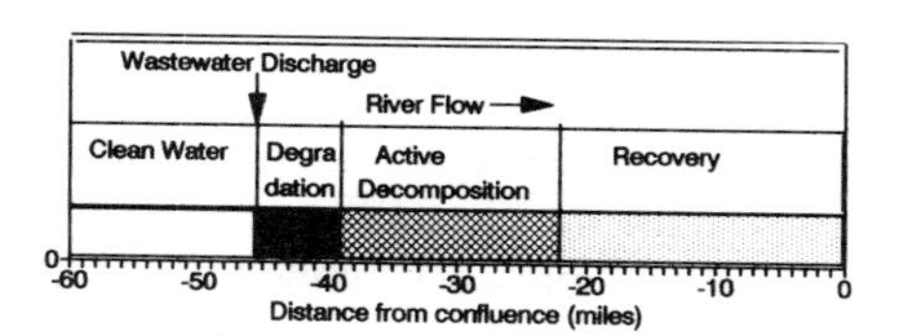

(a)

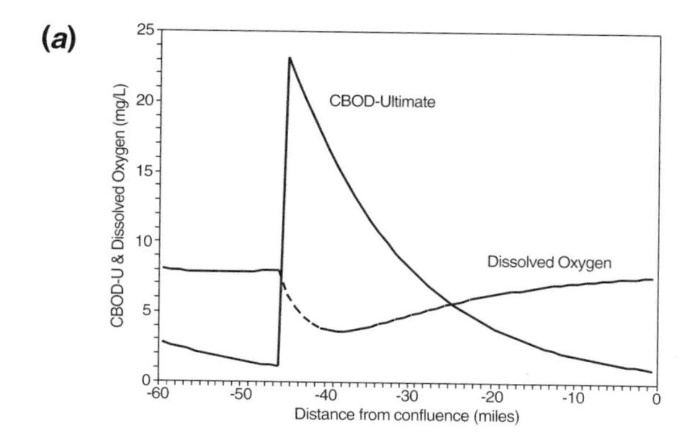

(b)

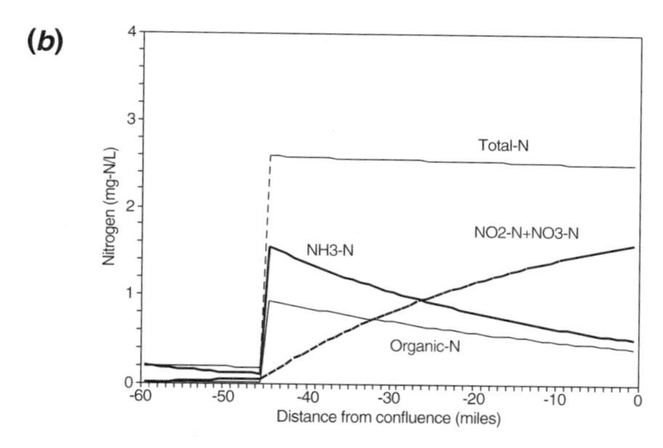

(c)

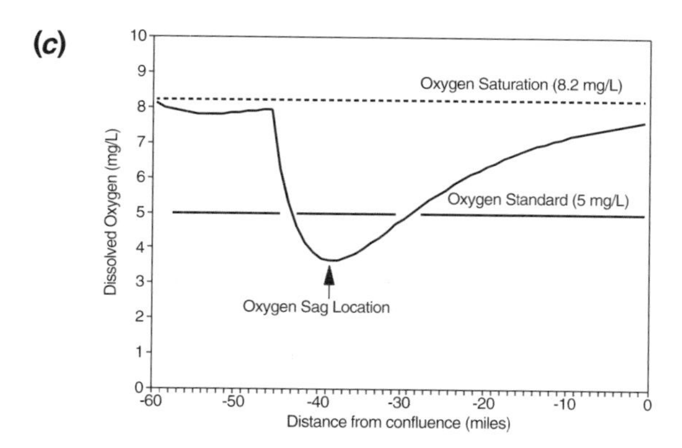

Figure 3-6 Spatial distribution of (*a*) organic carbon (CBOD), (*b*) nitrogen (organic nitrogen, ammonia, nitrite + nitrate), and (*c*) DO, downstream of a wastewater discharge into a river. *Source:* Adapted from Chapra, 1997, and Thomann and Mueller, 1987.

lowing mixing of the wastewater load with the ambient upstream load. As organic nitrogen declines by hydrolysis, the nitrification process begins (if a sufficient "seed" population of nitrifying bacteria is present), ammonia is oxidized to nitrite, and nitrite is quickly oxidized to nitrate. In the figure, nitrite and nitrate are shown combined as the sum (NO_2—N + NO_3—N) of these two inorganic forms of the nitrogen cycle. As the sequential reactions of the nitrogen cycle proceed downstream, the concentration of total nitrogen (total N) remains unchanged to maintain the mass balance of the reactions between the organic and inorganic forms of nitrogen. In these sequential oxidation reactions of nitrification, the nitrogenous oxygen demand (NBOD) consumes oxygen faster than it can be replenished by atmospheric reaeration, and thus, oxygen drops.

The combined effect of the carbon and nitrogen reactions causes a characteristic critical low DO zone identified by a "sag" in the spatial distribution of oxygen (Figure 3-6*c*). Two key features of the "oxygen sag" curve are especially important for the purposes of this study:

- The magnitude of the minimum DO concentration.
- The distance downstream from a waste discharge affected by "degradation" and "active decomposition."

In designing the screening methodology to detect the "worst case" for oxygen from a spatial perspective, it is important to recognize that water quality monitoring stations located immediately downstream of wastewater inputs will most likely be within the zones of "degradation" or "active decomposition" but not necessarily at the minimum, or critical, location of the sag. For monitoring stations located considerably farther downstream from a wastewater discharge location, it is less likely that the station will be within the "degradation" or "active decomposition" zones of the river. It is more likely, rather, that the station(s) will be located in the "recovery" zone. For any stream or river, the actual locations that mark the beginning and end of these zones are highly variable. The spatial pattern of oxygen shown in Figure 3-6*c* is dependent on a number of factors, including streamflow and river velocity (travel time), depth, water temperature, the type and makeup of effluent discharged, the magnitude of the wastewater discharge load, and the degree of turbulent mixing. Rather than attempting to select stations that are located in the exact sag zone, which would undoubtedly show the sharpest downstream DO signal but in the smallest area of the waterbody, the opposite approach was taken. That is, location of the station relative to the sag zone is purposely not controlled or selected, thereby allowing representation of far larger spatial areas but at the possible sacrifice of the downstream DO signal strength.

The question originally posed in Chapter 1 is broad-based: How have the nation's water quality conditions changed since implementation of the 1972 CWA's mandate for secondary treatment as the minimum acceptable technology for POTWs? The focus of the analysis is on detecting improvements in water quality conditions downstream of POTWs in the nation as a whole, not just areas immediately below outfalls. Consequently, when the term "worst-case DO data" is used in this document, it should be taken to refer to data collected primarily during times of high water temperature and low-flow conditions (i.e., "worst-case" from a temporal perspective). Spatially, no screens were developed for selecting monitoring stations located at the deepest part of the sag curve, nor even for stations in the sag curve itself. The only screening rule applied was that the water quality station had to be downstream from a point source. Thus, a station might be anywhere from within a few yards to hundreds of miles below any particular outfall. As a result, the data sets developed for the comparative before- and after-CWA analysis contain a mix of DO data from within and outside DO sag curves.

The Role of Spatial Scale in this Analysis

Recall that the objectives for this portion of the study are as follows:

- Develop before- and after-CWA data sets made up of DO summary statistics derived from monitoring stations that inherently contain a response "signal" linking point source discharges with downstream water quality.
- Calculate a DO summary statistic (tenth percentile) for each station for each before- and after-CWA time period, and then aggregate station data at sequentially larger spatial scales (reaches, catalog units, and major river basins).
- Conduct an analysis of all spatial units having both a before- and an after-CWA summary statistic, and then document the direction and magnitude of the changes in worst-case (summer, mean tenth percentile) DO concentration.
- Assess the change in the point source discharge/downstream DO signal over progressively larger spatial scales.

The use of spatial scale is a key attribute of this analysis. Detection of positive change in signal at large (river basin) as well as small (stream reach) scales would provide evidence that the CWA's technology and water quality-based controls yielded broad as well as localized benefits (i.e., reaches both within and beyond the immediate sag curve have benefitted from the CWA). If true, therefore, the second leg of the three-legged stool approach would provide further support for the claim that the effluent control regulations of the CWA were a broad success.

B. DATA MINING

As discussed in the previous section, the key objective in the data mining process was to screen out data collected under conditions or factors that might interfere with, or confound, the point source discharge/downstream DO signal. This section presents the six-step, peer-reviewed data mining process designed and implemented to develop the before- and after-CWA data sets to be used in the comparison analysis.

Step 1—Data Selection Rules

The data selection step incorporated three screening rules:

- DO, expressed as a concentration (mg/L), will function as the signal relating municipal and industrial discharges to downstream water quality responses.

- DO data will be extracted only from the July–September (summer season) time period.
- Only surface DO data (DO data collected within 2 meters of the water surface) will be used.

DO Concentration (mg/L) as the Water Quality Indicator The rationale for selecting DO as the water quality indicator for this study was discussed earlier in this chapter and in Chapter 2. The only question remaining was how this parameter should be expressed in the analysis—by concentration or by percent of DO saturation. The latter measurement has some advantages because it would reduce the noise introduced by changes in temperature. However, DO expressed as mg/L concentration was ultimately selected because it is more intuitive to a broader audience. For example, USEPA has established a DO concentration of 5.0 mg/L as the minimum concentration to be achieved at all times for early life stages of warm-water biota (see Table 1-1). For this reason, this level of DO is used as a benchmark for assessing acceptable versus nonacceptable conditions. In contrast, it is somewhat more difficult, perhaps, to comprehend whether a DO saturation of 50, 60, or 70 percent is protective.

DO from the Time Period of July to September Summer and early fall (July through September) is usually the best time for evaluating worst-case impacts of wastewater loading on water quality in general and DO in particular in most areas of the continental United States. Typically, this is when water temperatures are highest and flow is the lowest (i.e., lowest oxygen solubility and lowest dilution potential). Selecting DO data from only this time period screens out noise introduced by seasonal variations in temperature, precipitation, and flow. In addition, BOD loadings from nonpoint sources of pollution are reduced during low precipitation periods, thus minimizing this contribution to DO signals.

DO from Surface Waters In lakes, reservoirs, estuaries, coastal waters, and deep rivers, scientists typically measure DO at several depths in the water column. Often these measurements reveal significant differences between surface and bottom DO concentrations because of thermal stratification and the lack of reaeration of the bottom layer. By limiting DO data selection to the top 2 meters of a waterway, one can screen out much of the noise associated with the physical, chemical, and biological processes that occur in the lower layers and maintain some level of comparability between shallow streams and deeper waters.

Step 2—Data Aggregation Rules from a Temporal Perspective

The data aggregation from a temporal perspective step incorporated the following rules:

- 1961–1965 will serve as the time-block to represent conditions before the CWA, and 1986–1990 will serve as the time-block to represent conditions after the CWA.

- To remain eligible for the before- and after-CWA comparison, DO data must come from a station residing in a catalog unit that had at least 1 year classified as dry (streamflow ratio < 0.75) out of the 5 years in each before- and after CWA time-block.

An analysis of catalog units revealed that 1,923 (91 percent) of the 2,111 catalog units in the contiguous United States experienced at least one dry summer in the 1961–1965 time-block. Further, a total of 1,475 catalog units (70 percent) experienced at least two dry summers, and 886 catalog units (42 percent) experienced at least three dry summers in the before-CWA time-block. Of the catalog units that remained eligible for the comparison analysis (note that only 188 were screened out), low flow conditions persisted for an average of 2.5 years. In the 1986–1990 time-block, 1,776 (84 percent) of the 2,111 catalog units in the contiguous United States experienced at least one dry summer. A total of 1,420 catalog units (64 percent) experienced at least two dry summers and 1,073 catalog units (51 percent) experienced at least three dry summers in the after-CWA time-block. Of the catalog units that remained eligible for the comparison analysis (335 were screened out), low flow conditions persisted for an average of 2.7 years.

Step 3—Calculation of the Worst-Case DO Summary Statistic Rules

The calculation of the worst-case DO summary statistic step incorporated the following rules:

- For each water quality station, the tenth percentile of the DO data distribution from the before-CWA time-block (July through September, 1961–1965) and the tenth percentile of the DO data distribution from the after-CWA time-block (July through September, 1986–1990) will be used as the DO worst-case statistic for the comparison analysis.
- To remain eligible for the before- and after-CWA comparison, a station must have a minimum of eight DO measurements within each of the 5-year time-blocks.

Typically, the mean or median statistic is used to summarize a distribution of data because it describes the central tendency of the distribution. In this study, however, the emphasis is on worst-case (low) DO. Consequently, a summary statistic describing the lowest DO measurements of the data distribution was needed because these data would inherently carry a sharper point source discharge/downstream water quality signal. In other words, the objective was to characterize the worst of the DO data collected under the worst-case physical conditions (high temperature and low flow).

Because simply choosing the minimum measurement might introduce anomalous results, the tenth percentile, a more robust statistic (i.e., one that conveys information under a variety of conditions and is not overly influenced by data values at the extremes of the data distribution) was selected as the appropriate summary statistic to characterize the worst DO of a station's range of DO measurements within a time-block. An example of how one might interpret a tenth percentile value for a station is to say that 90 percent of the values collected at that station were higher than the tenth percentile value. To minimize statistical errors associated with calculating extreme percentiles, the requirement was added that a station must have a minimum of eight observations within each 5-year time-block to remain eligible for the before- and after-CWA comparison.

Step 4—Spatial Assessment Rules

The spatial assessment step incorporated one screening rule:

- Only water quality stations on portions of streams and rivers affected by point sources will be included in the before- and after-CWA comparison analysis; stations influenced only by nonpoint sources are excluded from the analysis.

The objective was to develop before- and after-CWA data sets that contain data that inherently contain a response signal linking point source discharges with downstream water quality. Consequently, a screening rule reflecting the need to ensure that DO data came from stations located downstream, rather than upstream, from point sources was required. As noted in Section A of this chapter, the distance downstream was not relevant for the screening rule; the only requirement was that the station was somewhere in the downstream network.

Although the focus of this book is on effluent loading from POTWs, changes in DO are tied to industrial discharges as well. Estimates of current (ca. 1995) BOD_5 loading using the NWPCAM indicate that industrial loads are the dominant component of total point and nonpoint source loading in many catalog units associated with major urban-industrial areas (see Section E in Chapter 2). For this reason, and because of the fact that it is not always possible to satisfactorily distinguish between industrial and POTW outfalls because of their close proximity in many areas, this leg of the study defines "point source discharges" to include both industrial and municipal dischargers.

The upstream/downstream relationship between point source discharges and water quality monitoring stations was established using USEPA's Reach File, Version 1 (RF1). RF1 is a computerized network of 64,902 river reaches in the 48 contiguous states, covering 632,552 miles of streams (see Figure 1-2). Using this system, one can traverse stream networks and establish relative positions along the river basin network of both free-flowing and tidally influenced rivers.

A list of point source dischargers was developed from USEPA's Permit Compliance System (PCS), Clean Water Needs Survey (CWNS), and Industrial Facilities Discharge File (IFD). Spatially integrating the dischargers with RF1 resulted in identifying 12,476 reaches that are downstream of point source dischargers (Figure 3-7) (Bondelid et al., 2000). These reaches, in turn, reside in 1,666 out of a total of 2,111 catalog units in the contiguous United States.

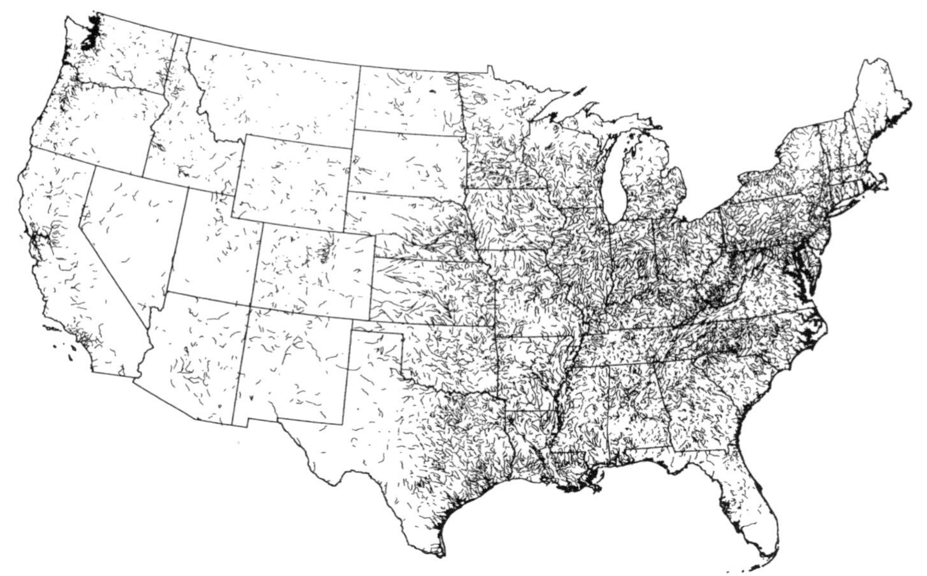

Figure 3-7 Reach File version 1 stream reach network of the 48 contiguous states with point source inputs discharging to a reach.

Example Application of the Screening Rules on DO Data from a Single Water Quality Monitoring Station Figure 3-8 illustrates how the above screening rules were applied to monitoring station data to obtain worst-case DO data for the before- and after-CWA comparison analysis. A station located on the Upper Mississippi River at Lock and Dam No. 2 at Hastings, Minnesota, is used in this example. Figure 3-8*a* displays a time series of the entire historical record (225 observations) of raw ambient DO measurements for the station from 1957 to 1997. Note that DO concentrations fluctuate from close to zero to slightly over 15 mg/L. The apparent noise (rapid up and down movement of the DO line) is due to many factors, including annual seasonal changes in streamflow and water temperature. Long-term interannual changes, on the other hand, might be due to persistent dry or wet weather or to changes in pollutant loading from the St. Paul METRO wastewater facility.

In the data selection step, the study authors extracted from the raw data set surface measurements collected at the station during the summer season (52 observations). Then, in the data aggregation step, they grouped the data in 5-year time-blocks and focused in on the data from the before- and after-CWA persistent dry weather time-blocks of 1961–1965 (10 observations) and 1986–1990 (15 observations). Because (1) the catalog unit in which the station resides had at least one dry year (streamflow ratio < 0.75) in each of the before- and after-CWA time-blocks [streamflow ratios: 1961 (0.31); 1964 (0.65); 1987 (0.59); 1988 (0.22); and 1989 (0.40)] and (2) the number of observations for each grouping was confirmed to be greater than eight, the groupings remained eligible for the next phase.

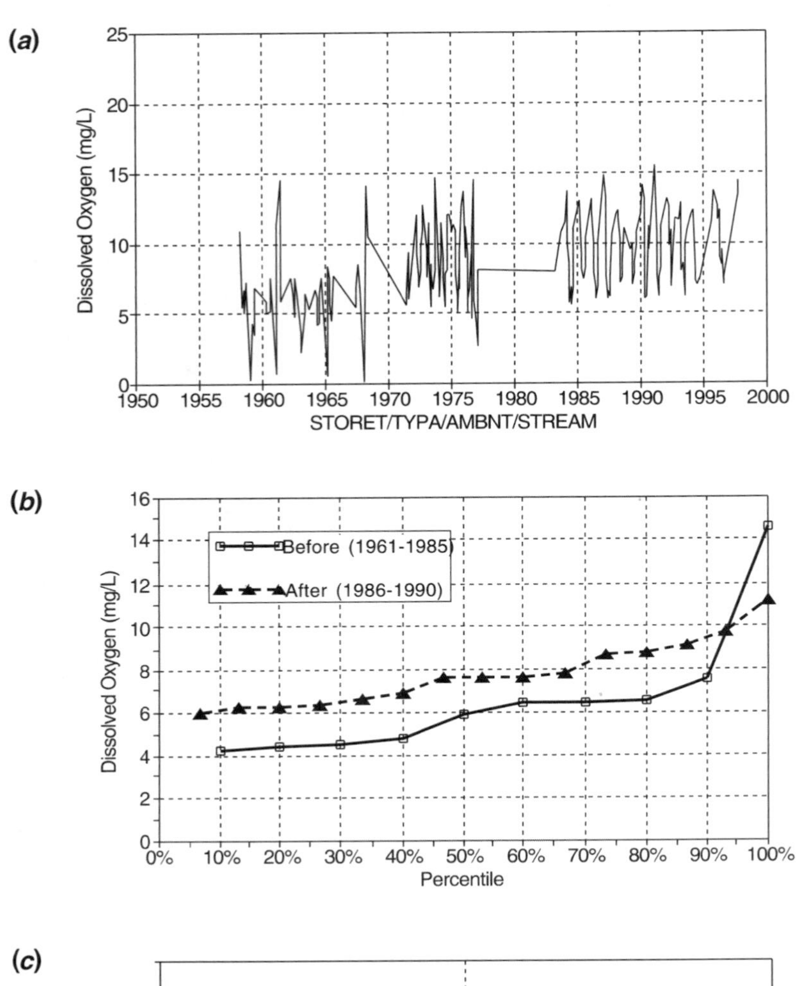

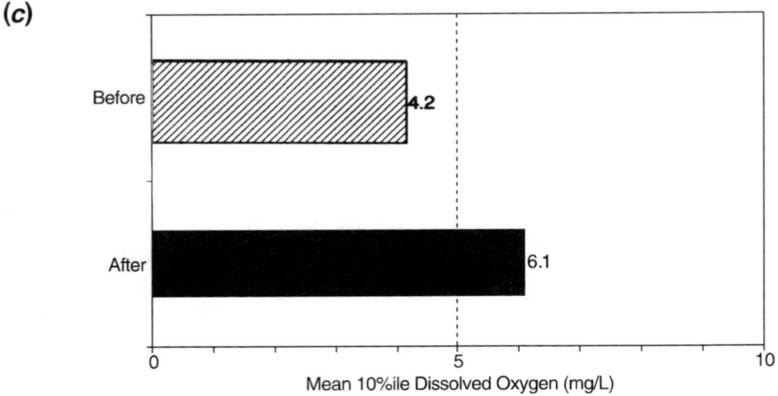

Figure 3-8 Application of the screening rules for station 21 MINN MSU-815-BB15E58 located in the Upper Mississippi River: (*a*) time series of historical DO observations from 1957–1997; (*b*) before- and after-CWA frequency distribution; and (*c*) tenth percentile values. *Source:* USEPA STORET.

Distributions were made for each group and the tenth percentile determined. Figure 3-8*b* displays the before- and after-CWA DO frequency distribution. Figure 3-8*c* is a bar chart comparing the tenth percentile DO values of the before- and after-CWA time-blocks. Note that the tenth percentile statistic associated with the before-CWA period is below the USEPA's minimum concentration of 5 mg/L, the level required to be achieved at all times for early life stages of warm-water biota.

Finally, the spatial assessment phase confirmed that the monitoring station where the DO data were collected was on the Upper Mississippi River downstream of the St. Paul METRO water pollution control plant. Therefore, the station remained eligible for the comparison analysis.

Step 5—Data Aggregation Rules from a Spatial Perspective

The data mining steps described above were used to develop before- and after-CWA sets of monitoring station data. Recall that:

- The before- and after-CWA data sets are collections of DO summary statistics that characterize worst-case DO at individual water quality monitoring stations across the United States for the 1961–1965 and 1986–1990 time-blocks, respectively (one DO summary statistic per station per time-block).
- The summary statistic used to characterize worst-case DO at a station is the tenth percentile value of a data distribution of actual DO measurements taken at the station during the specified time-block and recorded in STORET. For the station to be eligible for inclusion in the data set, at least eight measurements had to have been taken during the 5-year time-block.

The purpose of the data aggregation from a spatial perspective step was to assign a worst-case DO summary statistic to every eligible spatial unit defined at the reach and hydrologic unit scales for the before- and after-CWA time-blocks. This task was accomplished in two steps. First, for each data set and time-block, the mean tenth percentile value from each eligible station was computed within the spatial unit. (Since the scales are hierarchical, a station's summary statistic was effectively assigned to a reach and a catalog unit.) Second, the mean tenth percentile summary statistic was calculated and assigned to the spatial unit for the purpose of characterizing its worst-case DO.

If a spatial unit had only one monitoring station within its borders meeting the screening criteria, the tenth percentile DO value from that station simply served as the unit's worst-case summary statistic. If, however, there were two or more stations within a spatial unit's borders, the tenth percentile values for all the eligible stations were averaged, and this mean value was used to characterize worst-case DO for the unit. This averaging process reduced the correlation between stations that were located near each other. (Increased correlation reduces the effective sample size and complicates statistical comparisons. Averaging across larger spatial scales tends to reduce the correlation the most. As demonstrated later in this section, the results from the different spatial scales are generally consistent, and the impact of spatial correlation is believed to be minimal.)

Step 6—Development of the Paired Data Sets (at Each Spatial Scale)

The purpose of the sixth and final step was to prepare the before- and after-CWA data sets for the comparison analysis to be conducted at each of the three sequentially larger hydrologic spatial scales (RF1 reach, catalog unit, and major river basin). The screening rule associated with this step was as follows:

- To be eligible for the paired (i.e., before vs. after) comparison analysis, a hydrologic unit must have both a before-CWA and an after-CWA summary statistic assigned to it.

After each eligible reach and catalog unit was assigned a worst-case DO summary statistic for the appropriate before- and after-CWA time-blocks, a check was made to see which spatial units had both a before- and an after-CWA summary statistic. For many reaches and catalog units, factors such as the absence of dry flow conditions, station removal or change in station location, or change in water quality sampling technique over time (see Figure 3-2) caused a summary statistic to be available for one time-block but not the other. In this case, the spatial unit was removed from the analysis since a comparison could not be evaluated.

Implementation of this final step of the data mining process yielded the following results:

- Of the 12,476 reaches identified as being downstream from point sources, 311 reaches had both before- and after-CWA worst-case DO summary statistics.
- Of the 1,666 catalog units identified as being impacted by point sources, 246 catalog units had both before- and after-CWA worst-case DO summary statistics.
- The 311 reach-aggregated DO summary statistics were pooled by the 18 major river basins in the contiguous United States. Using the statistical requirements to conduct a paired t-test as a criterion, 11 of the 18 major river basins had sufficient reach-aggregated worst-case DO data to conduct the comparison analysis at the river basin level. The number of reaches and catalog units with both before- and after-CWA DO data was constrained by the limited availability of station records for the 1961–1965 before-CWA period. Figure 3-2 shows that a much larger database of station records was available for the 1966–1970 and 1971–1975 time blocks. Although these periods could have been selected, the known occurrence of severe drought conditions during the early 1960s was the major reason for selecting 1961–1965 as the before-CWA time block.

C. COMPARISON OF WORST-CASE DO IN WATERWAYS BELOW POINT SOURCE DISCHARGES BEFORE AND AFTER THE CWA AT THREE SPATIAL SCALES

This section presents the comparative before- and after-CWA analysis of worst-case DO data derived using the screening criteria described in Section B and then aggregated by spatial units defined by three scales (reach, catalog unit, and major river basin). In the discussion that follows, the term "worst-case DO" should be interpreted to mean the average tenth percentile DO statistic computed for the corresponding spatial level unless specifically noted otherwise. Also, the reader should note that a worst-case DO concentration of 5 mg/L has been adopted in this report as a general benchmark threshold for defining "desirable" versus "undesirable" levels of worst-case DO. This benchmark value was chosen primarily because USEPA has established it as the minimum concentration to be achieved at all times for early life stages of warm-water biota (see Table 1-1).

Before and After DO at Reach Scale

A total of 311 river reaches had monitoring stations with both before- and after-CWA data, and thus were eligible for comparison. Notably, these 311 evaluated reaches represent a disproportionately high amount of urban/industrial population centers, with approximately 13.7 million people represented (7.2 percent of the total population served by POTWs in 1996). Of this total, 215 reaches (69 percent) showed improvements in worst-case DO after the CWA. Figure 3-9 presents a frequency distribution of the before- and after-CWA data.

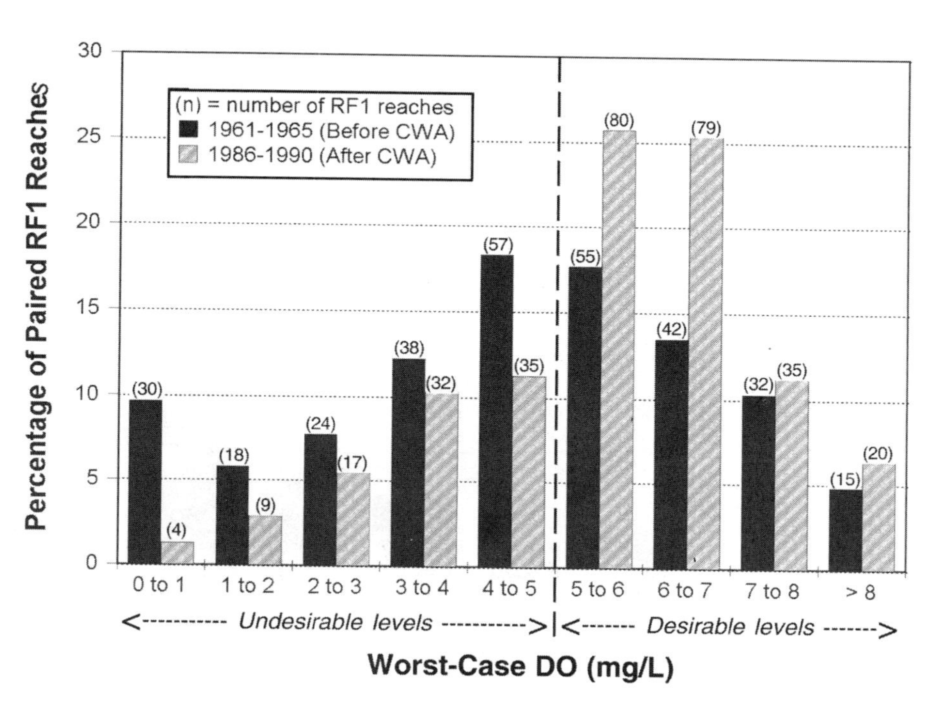

Figure 3-9 Frequency distribution comparing worst-case DO concentration of evaluated reaches before and after the CWA. *Source:* USEPA STORET.

Key observations from Figure 3-9 include the following:

- The percentage of evaluated RF1 reaches characterized by "very low" worst-case DO (< 2 mg/L) was reduced from 15 to 4 percent. Before the CWA, 48 reaches had very low worst-case DO. After the CWA, only 13 reaches had very low worst-case DO.
- The percentage of evaluated reaches characterized by undesirable worst-case DO (below the 5 mg/L threshold) was reduced from 54 to 31 percent. Before the CWA, 167 reaches had undesirable levels of worst-case DO. After the CWA, 97 reaches had undesirable levels of worst-case DO.
- The percentage of evaluated reaches characterized by desirable worst-case DO (above the 5 mg/L threshold) increased from 46 to 69 percent. Before the CWA, 144 reaches had desirable levels of worst-case DO. After the CWA, 214 reaches had desirable levels of worst-case DO.

By tracking individual reaches, it was revealed that 85 of the 167 reaches characterized by undesirable worst-case DO before the CWA improved to greater than 5 mg/L after the act. On the flip side, only 15 of the 144 reaches characterized by desirable worst-case DO before the CWA dropped below the 5 mg/L benchmark after the act. Thus, the net change was 70 reaches moving from undesirable levels of worst-case DO to desirable levels of worst-case DO.

Figure 3-10 is a column graph that breaks down the 85 reaches that had undesirable DO levels before the CWA and then improved past the benchmark threshold of 5 mg/L after the act according to their before-CWA worst-case DO concentration.

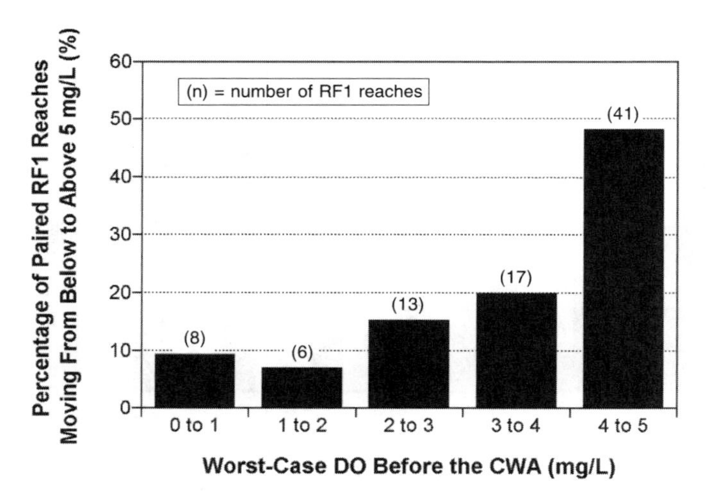

Figure 3-10 Frequency distribution of worst-case DO levels before the CWA for the 85 evaluated reaches that were < 5 mg/L before the CWA and > 5 mg/L after the CWA. *Source:* USEPA STORET.

Key observations from Figure 3-10 include the following:

- Approximately 48 percent of the evaluated reaches (41 out of 85) that had undesirable worst-case DO levels before the CWA and then improved past the benchmark threshold of 5 mg/L after the act had before-CWA worst-case DO in the 4–5 mg/L range.
- Approximately 16 percent of the evaluated reaches (14 out of 85) that had undesirable worst-case DO levels before the CWA and then improved past the benchmark threshold of 5 mg/L after the act had very low worst-case DO (< 2 mg/L) before the CWA.

Of the 311 evaluated reaches with paired before- and after-CWA data, 215 reaches (69 percent) had increased worst-case DO and 96 (31 percent) had decreased worst-case DO after the CWA. Figure 3-11*a* and *b* display the magnitude of degradation and improvement, respectively. Key observations from Figure 3-11 include the following:

- Approximately 36 percent of the evaluated reaches that had increases in worst-case DO (78 of the 215 improving reaches) increased by 2 mg/L or more.
- Approximately 15 percent of the evaluated reaches that had decreases in worst-case DO (14 of 96 degrading reaches) decreased by 2 mg/L or more.
- Approximately 41 percent of all evaluated reaches either stayed the same or improved or degraded by 1 mg/L or less (129 of the 311 reaches).

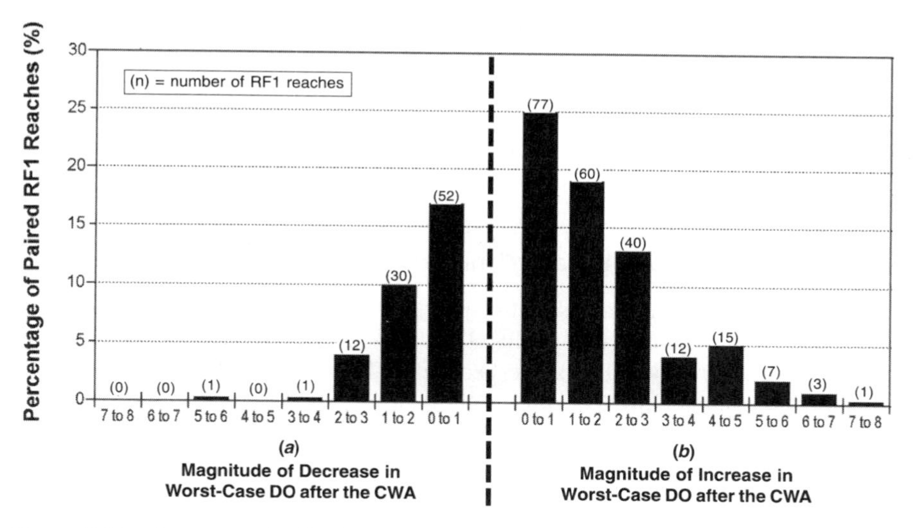

Figure 3-11 Frequency distribution of worst-case DO for evaluated RF1 reaches that (*a*) decreased in concentration ($n = 96$); and (*b*) increased in concentration ($n = 215$) after the CWA. *Source:* USEPA STORET.

Reaches with Greatest Improvements Table 3-1 lists the 25 river reaches with the greatest before- and after-CWA improvements in worst-case DO. Figure 3-12 presents a location map of these reaches along with a stacked column graph that shows their before- and after-CWA worst-case DO data. Key observations from Table 3-1 and Figure 3-12 include the following:

- All but one of the top 25 river reaches with the greatest before- and after-CWA improvements had before-CWA worst-case DO levels below the benchmark threshold of 5 mg/L. Five reaches had a before-CWA worst-case DO concentration of 0 mg/L.
- For 20 of the 24 reaches with before-CWA worst-case DO levels below the threshold value of 5 mg/L, after-CWA worst-case DO improved to levels greater than 5 mg/L.
- The four reaches that did not break the threshold value of 5 mg/L after the CWA all had a before-CWA worst-case DO concentration of 0 mg/L.
- Worst-case DO in the top 10 improving river reaches typically improved by about 4 to 7 mg/L (from about 0–3 mg/L in the 1961–1965 time-block to about 6–8 mg/L in the 1986–1990 time-block).

Long-Term Trend of DO at Reach Scale Time series for dissolved oxygen, averaged over 5-year periods from 1961–1995 and filtered for "dry" hydrologic conditions, are presented in Figures 3–13 through 3–18 for selected RF1 reaches identified in Appendix E with significant improvements in DO concentrations. As documented in the time-series data extracted for the following reaches: Delaware Bay, New Jersey and Delaware (Figure 3-13), Fox River, Ilinois (Figure 3-14), Root River, Illinois (Figure 3-15), Milwaukee River, Wisconsin (Figure 3-16), Maumee River, Ohio (Figure 3-17), River Raisin, Michigan (Figure 3-18), there have been gradual improvements in tenth percentile dissolved oxygen within numerous reaches since enactment of the CWA in 1972. "Before and after" changes in DO for these, and other RF1 reaches, are presented in Table 3-1 with a complete ranking of all 311 RF1 reaches evaluated in this study presented in Appendix E.

Before and After DO at Catalog Unit Scale

Figures 3-19 and 3-20 are maps that display the locations and worst-case DO concentrations of catalog units potentially eligible for the paired analysis for the 1961–1965 and 1986–1990 time-blocks, respectively. The before-CWA data set contained a total of 333 catalog units. The after-CWA data set had 905 catalog units.

In the before-CWA map (Figure 3-19),

- 45 of the 333 catalog units (14 percent) have worst-case DO less than 2.5 mg/L.
- 102 of the catalog units (31 percent) have levels from 2.5 to 5 mg/L.
- 186 of the catalog units (56 percent) are characterized by worst-case DO greater than 5 mg/L.

TABLE 3-1 Twenty-five RF1 River Reaches with Greatest Improvements in Worst-Case (Mean Tenth Percentile) DO After the CWA

Rank	RF1 Reach ID	RF1 Name	Catalog Unit Name	1961–1965 (mg/L)	1986–1990 (mg/L)	DO Change (mg/L)	Number of Stations 1961	Number of Stations 1986
1	10170203037	Big Sioux River	Lower Big Sioux	0.00	7.22	7.22	1	1
2	04100002001	River Raisin	Raisin	1.60	8.34	6.74	2	2
3	04110002001	Cuyahoga River	Cuyahoga	0.30	6.50	6.21	2	24
4	05030103007	Mahoning River	Mahoning	1.09	7.16	6.07	1	1
5	07070002034	Wisconsin River	Lake Dubay	0.88	6.84	5.96	1	1
6	05120201004	White River	Upper White	0.69	6.42	5.73	5	1
7	05080002008	Great Miami River	Lower Great Miami	0.20	5.86	5.66	1	1
8	07120004018	DuPage River, East Branch	Des Plaines,IL	0.58	5.92	5.35	4	3
9	07090001004	Rock River	Upper Rock	2.76	8.05	5.29	1	1
10	05020006031	Casselman River	Youghiogheny	2.96	8.00	5.04	1	1
11	04040002005	Root River	Pike-Root	0.94	5.94	5.00	1	1
12	02040201011	Neshaminy River	Crosswicks-Neshaminy	2.60	7.56	4.96	1	1
13	04030101012	Manitowoc River	Manitowoc-Sheboygan	5.95	10.90	4.95	1	1
14	03170006007	Pascagoula River	Pascagoula	0.00	4.92	4.92	1	7
15	06010102004	Holston River, South Fork	South Fork Holston	1.60	6.48	4.88	1	2
16	08030203006	Enid Lake	Yocona	0.00	4.87	4.87	1	3
17	04040003001	Milwaukee River	Milwaukee	2.18	6.96	4.78	2	3
18	04030104002	Oconto River	Oconto	0.50	5.20	4.70	1	1
19	08030205018	Grenada Lake	Yalobusha	0.00	4.62	4.62	1	4
20	05050008006	Kanawha River	Lower Kanawha	0.00	4.57	4.57	2	3
21	04120102002	Cattaraugus Creek	Cattaraugus	3.30	7.60	4.30	1	2
22	03050109053	Reedy River	Saluda	1.95	6.23	4.28	4	10
23	07120004002	Des Plains River	Des Plaines	1.76	6.00	4.24	2	1
24	05120201013	White River	Upper White	2.23	6.38	4.15	3	2
25	03050103037	Catawba River	Lower Catawba	1.68	5.80	4.12	5	1

Source: USEPA STORET.

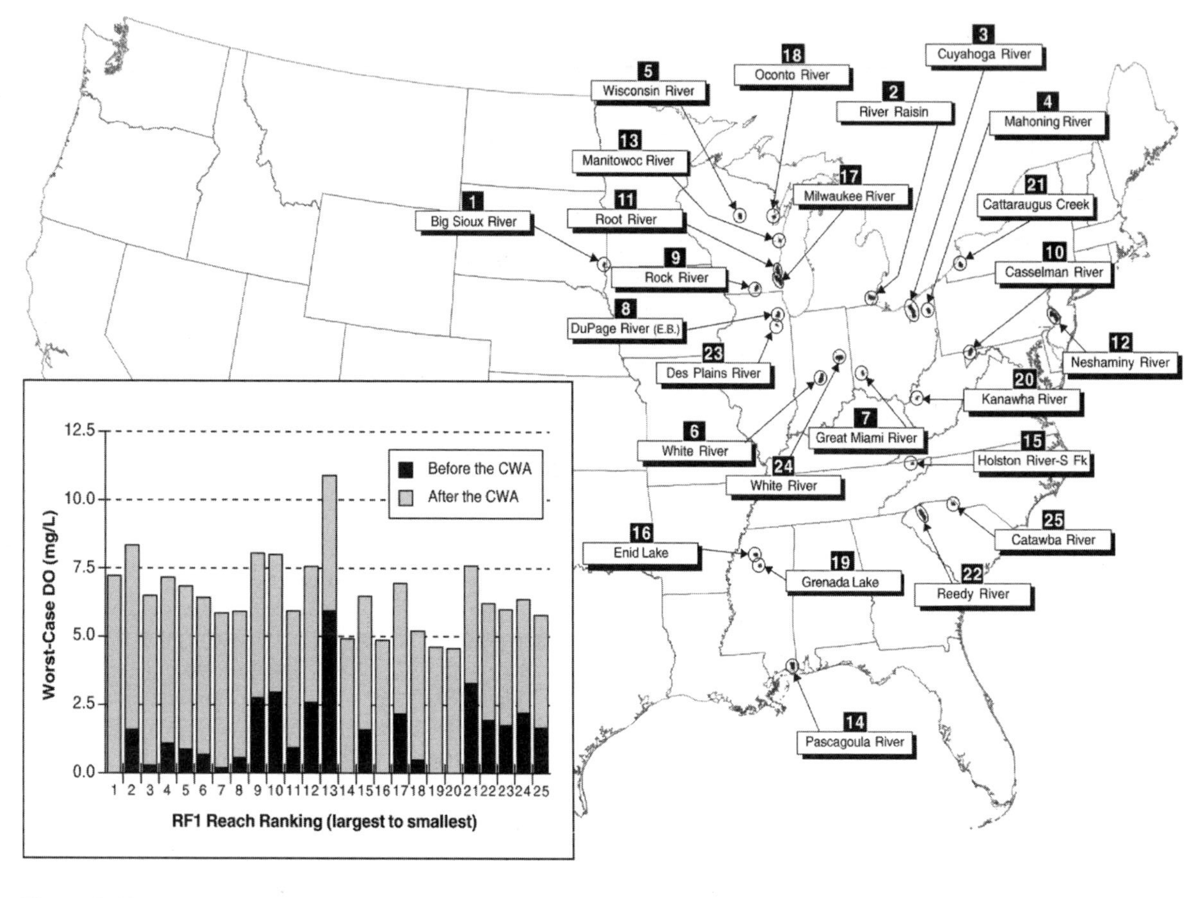

Figure 3-12 Location map and distribution chart of the 25 RF1 reaches with the greatest after-CWA improvements in worst-case DO. *Source:* USEPA STORET.

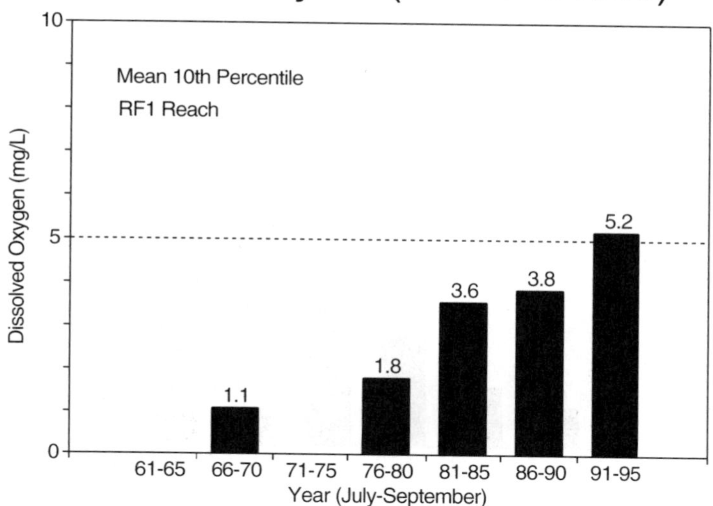

Figure 3-13 Time series from 1961–1995 of tenth percentile dissolved oxygen for RF1 reach: Delaware Bay (02040204035). *Source:* USEPA STORET.

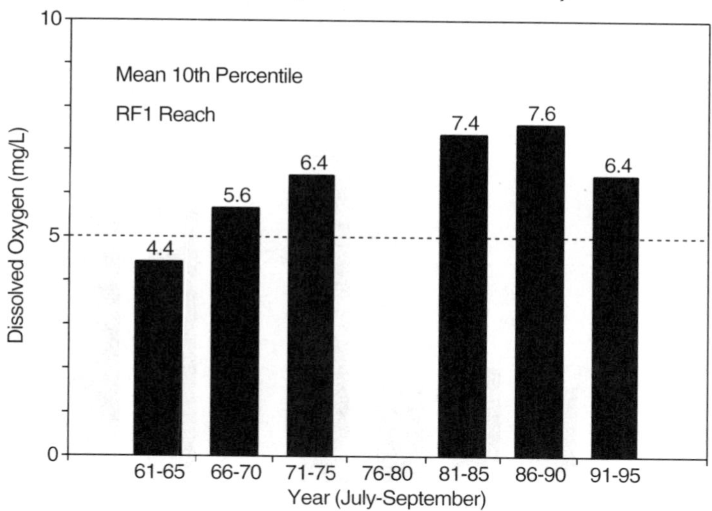

Figure 3-14 Time series from 1961–1995 of tenth percentile dissolved oxygen for RF1 reach: Fox River, Illinois (07120007006). *Source:* USEPA STORET.

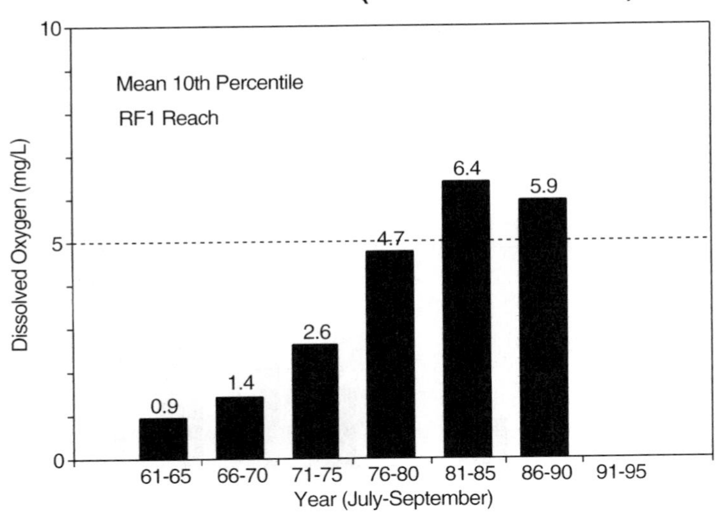

Figure 3-15 Time series from 1961–1995 of tenth percentile dissolved oxygen for RF1 reach: Root River, Illinois (04040002005). *Source:* USEPA STORET.

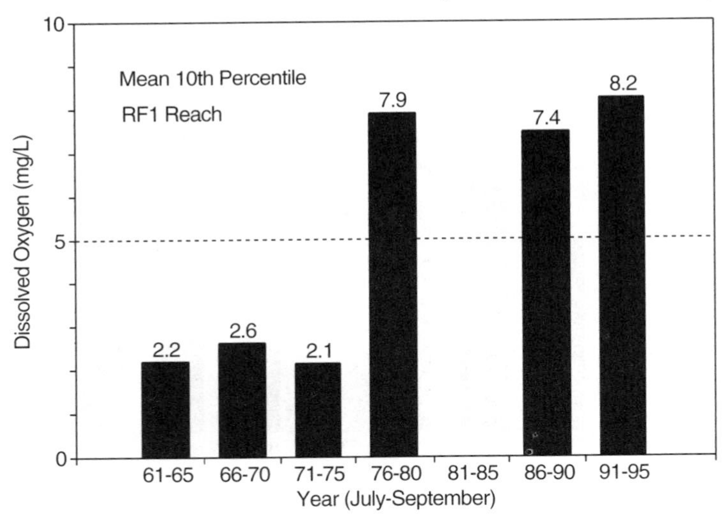

Figure 3-16 Time series from 1961–1995 of tenth percentile dissolved oxygen for RF1 reach: Milwaukee River, Wisconsin (04040003001). *Source:* USEPA STORET.

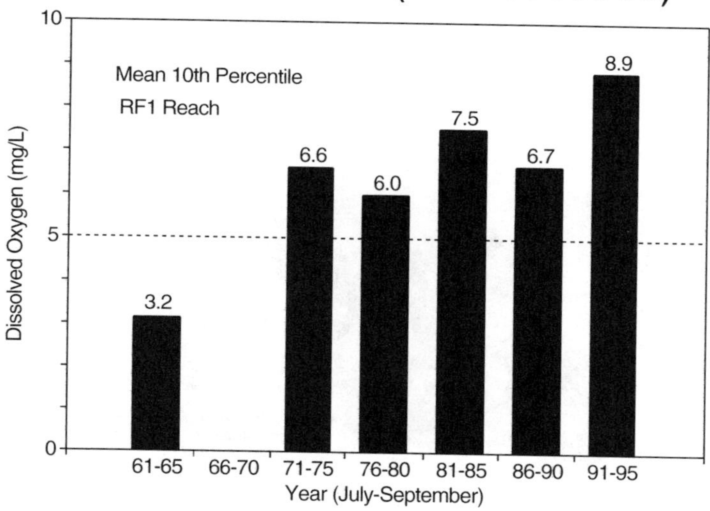

Figure 3-17 Time series from 1961–1995 of tenth percentile dissolved oxygen for RF1 reach: Maumee River, Ohio (04100009005). *Source:* USEPA STORET.

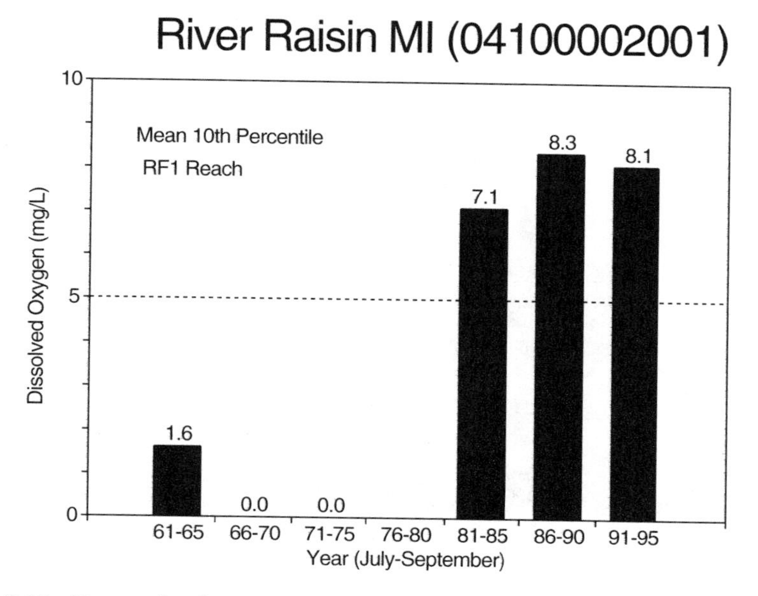

Figure 3-18 Time series from 1961–1995 of tenth percentile dissolved oxygen for RF1 reach: River Raisin, Michigan (04100002001). *Source:* USEPA STORET.

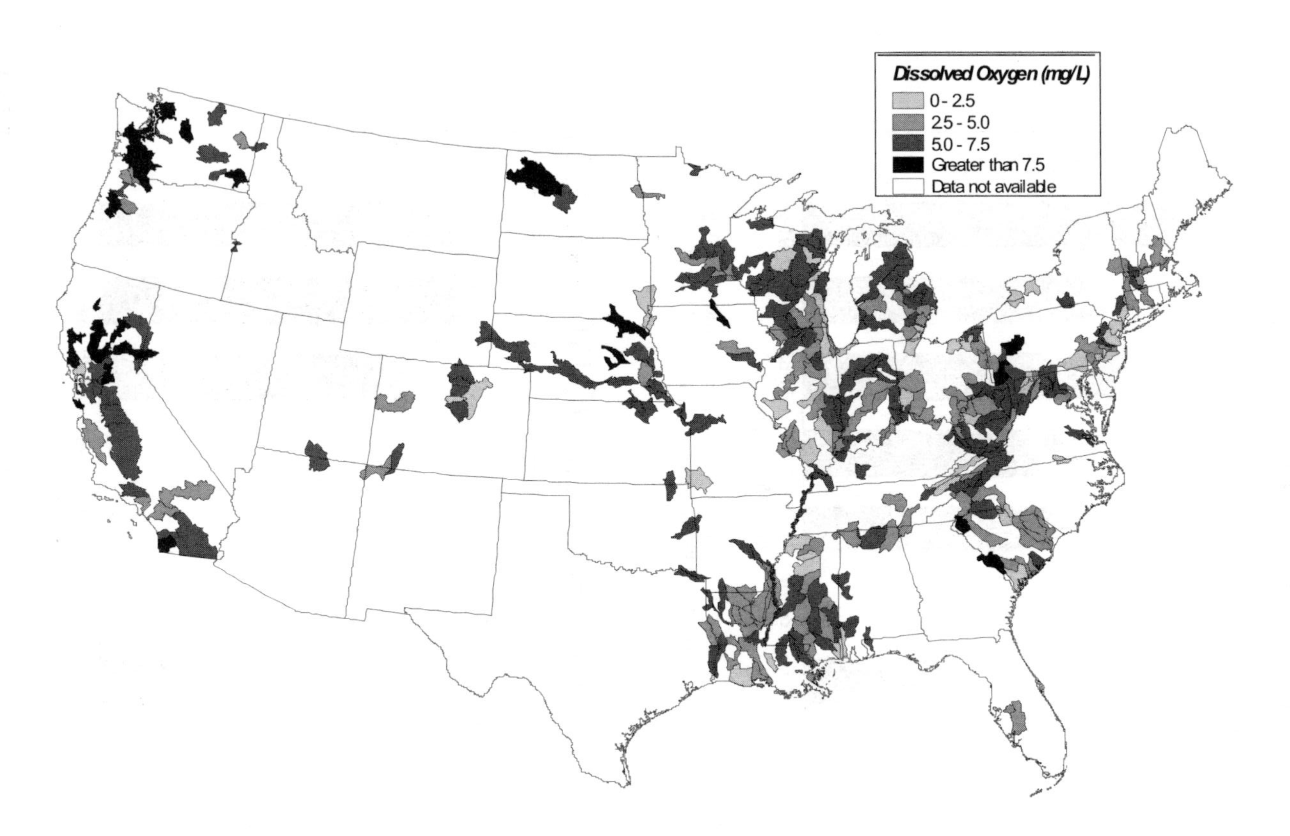

Figure 3-19 Locations and worst-case DO concentrations of catalog units potentially eligible for the paired analysis for the 1961–1965 time-block (before CWA). N = 333 catalog units. *Source:* USEPA STORET.

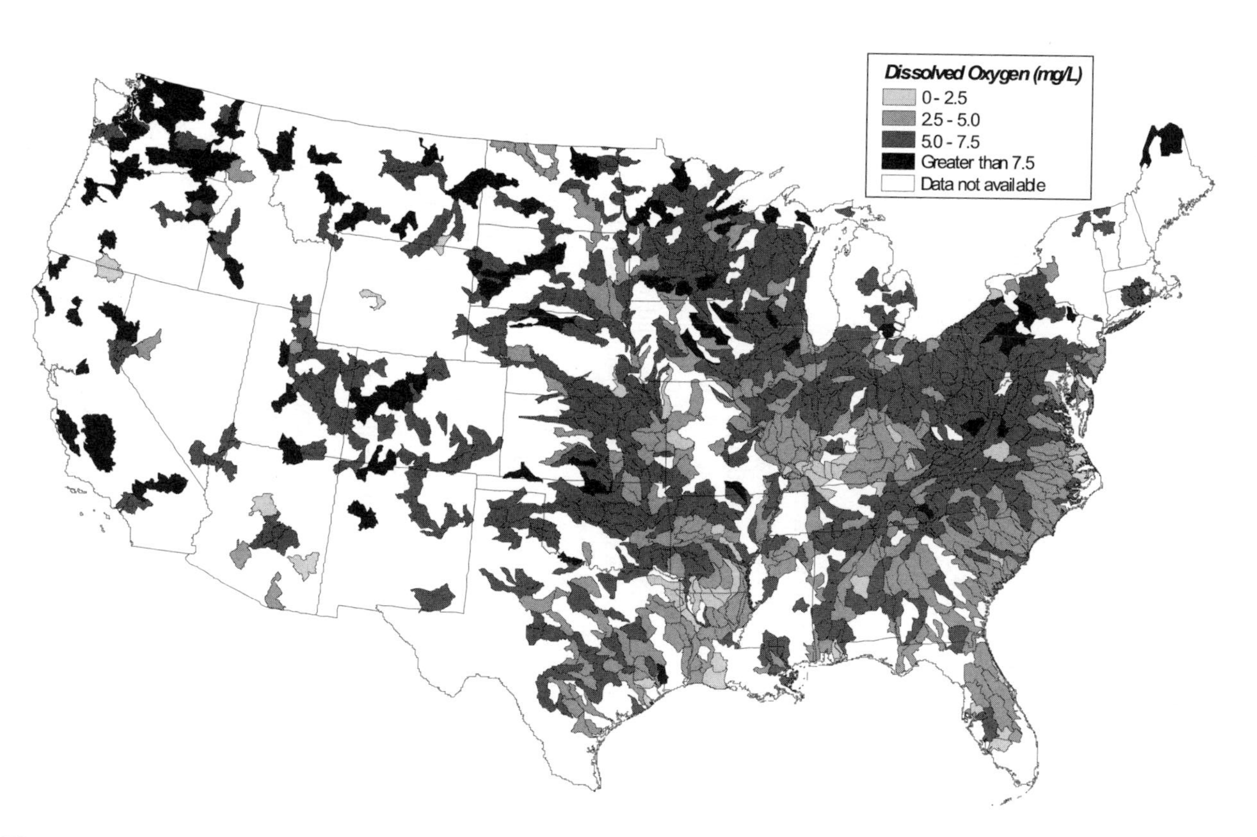

Figure 3-20 Locations and worst-case DO concentrations of catalog units potentially eligible for the paired analysis for the 1986–1990 time-block (after CWA). N = 905 catalog units. *Source:* USEPA STORET.

In comparing these results with the historical data from the FWPCA surveillance network (see Figure 3-1 in Section A of this chapter), many of the catalog units characterized by poor DO conditions (DO less than 5 mg/L) in 1961–1965 correspond to the areas represented by many of the stations compiled by Gunnerson (1966) with minimum DO less than 0.5 and minimum DO between 0.5 and 4 mg/L in the 1957–1965 data set.

In the after-CWA map (Figure 3-20),

- 49 of the 905 catalog units (5 percent) have worst-case DO less than 2.5 mg/L.
- 252 of the catalog units (28 percent) have levels from 2.5 to 5 mg/L.
- 604 of the catalog units (67 percent) are characterized by worst-case DO greater than 5 mg/L.

Undesirable levels of worst-case DO (less than 5 mg/L) are still quite prevalent after the CWA in some midwestern and southeastern watersheds, with a pattern of moderately low worst-case DO (2.5 to 5 mg/L) that appears to be characteristic of the Atlantic coastal plain from Florida to New Jersey. Higher worst-case DO (5 to 7.5 mg/L) characterizes the Piedmont region and the watersheds of the Appalachian Mountains and is likely due to cooler water temperatures. The coastal plain pattern of moderately low worst-case DO most likely reflects natural factors such as warmer summer temperatures, higher decomposition rates, and relatively long residence times within sluggish rivers and tidal waters rather than municipal or industrial point source loading within these watersheds.

Overlaying the 333 eligible catalog units in the before-CWA data set with the 905 eligible units in the after-CWA data set yielded a total of 246 intersecting catalog units that had both before- and after-CWA data. Notably, these 246 evaluated catalog units represent a disproportionately high amount of urban/industrial population centers, with approximately 61.6 million people represented (32.5 percent of the total population served by POTWs in 1996). Figure 3-21 presents maps that display the locations and worst-case DO concentrations of the evaluated catalog units. Figure 3-21*a* displays the catalog units that had improvement in worst-case DO after the CWA. Figure 3-21*b* displays the catalog units that had degradation in worst-case DO after the CWA. Figure 3-22 presents a frequency distribution of the before- and after-CWA DO data.

Key observations from Figures 3-21 and 3-22 include the following:

- 167 (68 percent) of the 246 evaluated catalog units had increases in worst-case DO after the CWA; 79 (32 percent) of the catalog units had decreases in worst-case DO after the CWA.
- The percentage of evaluated catalog units characterized by "very low" worst-case DO ($<$ 2 mg/L) was reduced from 11 to 2 percent. Before the CWA, 26 catalog units had very low worst-case DO; after the CWA, only 6 catalog units had very low worst-case DO.
- The percentage of evaluated catalog units characterized by undesirable worst-case DO (below the 5 mg/L threshold) was reduced from 47 to 26 percent. Before the CWA, 115 catalog units had undesirable levels of worst-case DO; after the CWA, 65 catalog units had undesirable levels of worst-case DO.

- The percentage of evaluated catalog units characterized by desirable worst-case DO (above the 5 mg/L threshold) increased from 53 to 74 percent. Before the CWA, 131 catalog units had desirable levels of worst-case DO; after the CWA, 181 catalog units had desirable levels of worst-case DO.

Figure 3-23 is a column graph that describes the changes in worst-case DO that occurred after the CWA for the 246 evaluated catalog units in relation to the 5 mg/L threshold. Key observations from this figure include the following:

- 67 percent of the evaluated catalog units (166 out of 246 units) remained either above (47 percent) or below (20 percent) the 5 mg/L worst-case DO threshold.
- Of the 115 catalog units that had worst-case DO concentrations below the threshold of 5 mg/L before the CWA, 57 percent (65 catalog units) increased to above the threshold after the CWA.
- Of the 131 catalog units that had worst-case DO concentrations above the benchmark threshold of 5 mg/L before the CWA, only 11 percent (15 catalog units) fell below the threshold after the CWA.

Of the 246 evaluated catalog units with paired before- and after-CWA data, 167 catalog units (68 percent) had increased worst-case DO and 79 (32 percent) had decreased worst-case DO after the CWA. Figure 3-24*a* and *b* display the magnitude of degradation and improvement, respectively. Key observations from Figure 3-24 include the following:

- Approximately 32 percent of the evaluated catalog units that had increases in worst-case DO (53 of the 167 improving catalog units) increased by 2 mg/L or more.
- Approximately 13 percent of the evaluated catalog units that had decreases in worst-case DO (10 of 76 degrading catalog units) decreased by 2 mg/L or more.
- Approximately 44 percent of all evaluated catalog units either stayed the same or improved or degraded by 1 mg/L or less (108 of the 246 catalog units).

Catalog Units with Greatest Improvements Table 3-2 lists the 25 catalog units with the greatest before- and after-CWA improvements in worst-case DO. Appendix D presents a complete listing of all 246 catalog units with before- and after-CWA changes in DO. Figure 3-25 presents a location map of the top 10 of these units along with a stacked column graph that shows their before- and after-CWA worst-case DO concentration. Key observations from Table 3-2 and Figure 3-25 include the following:

- All of the top 25 catalog units with the greatest before- and after-CWA improvements had before-CWA worst-case DO levels below the benchmark threshold of 5 mg/L. Four catalog units had a before-CWA worst-case DO concentration of 0.0 mg/L.

(*a*)

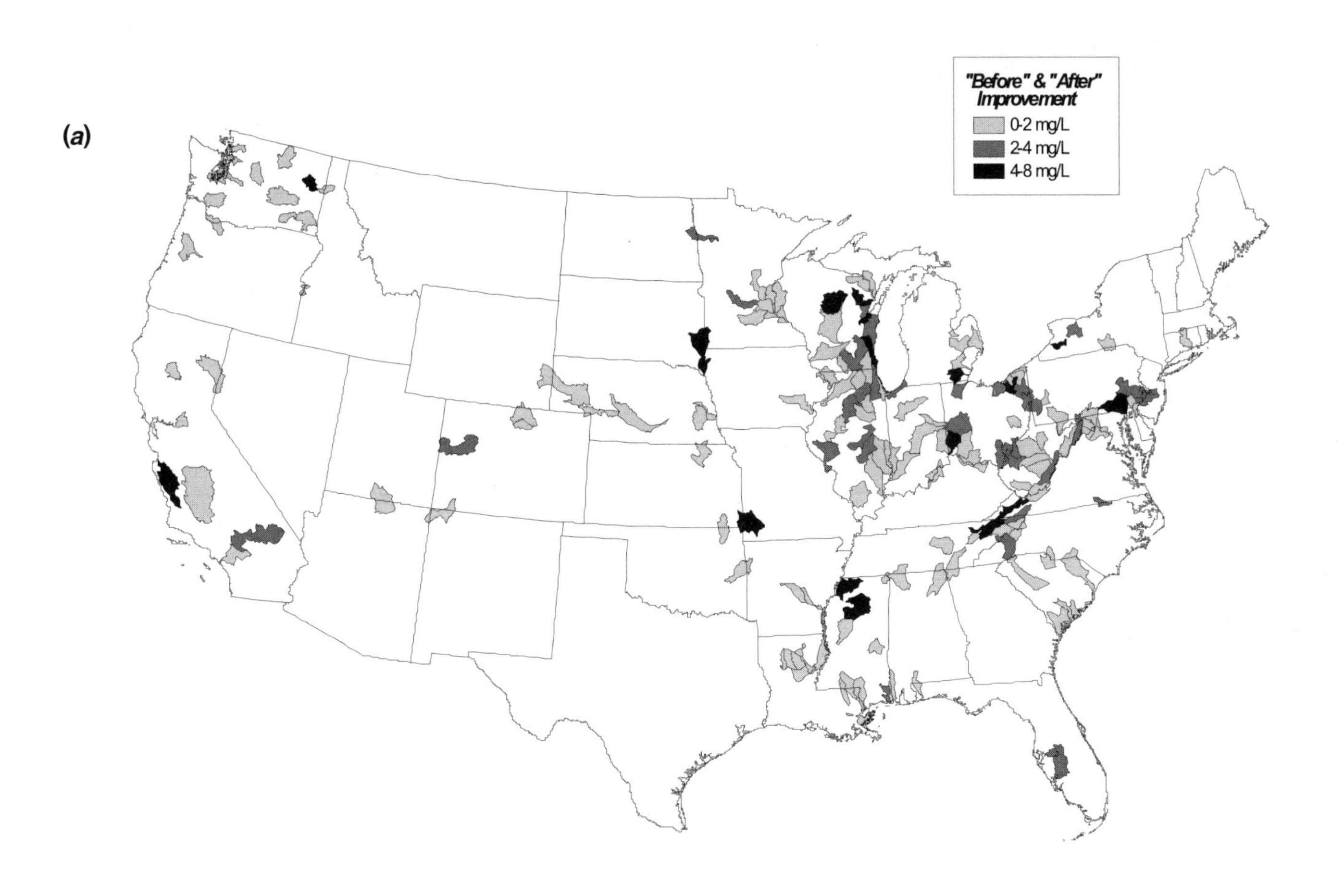

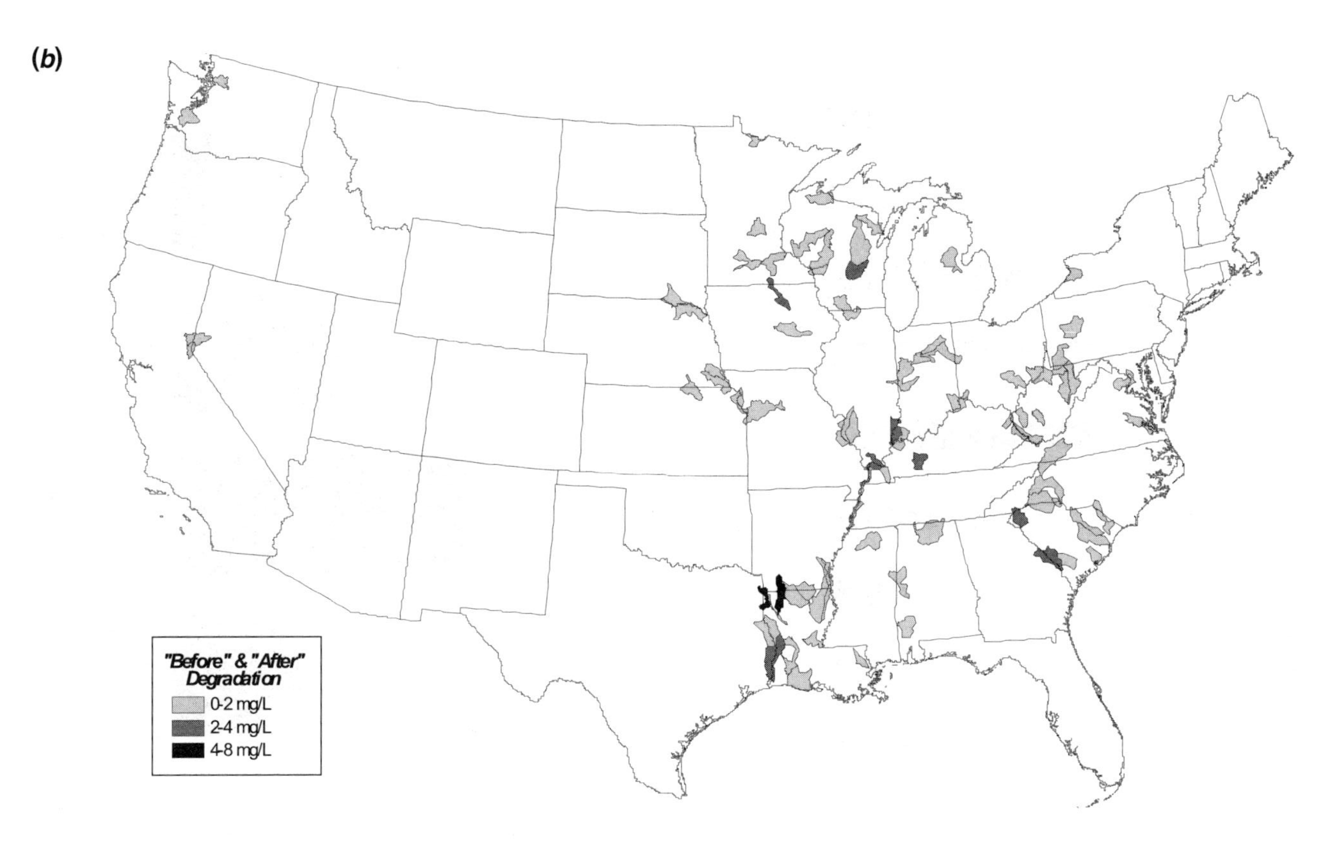

Figure 3-21 Locations and change in worst-case DO concentrations of evaluated catalog units where (*a*) shows improving units (N = 167) and (*b*) shows degrading units (N = 79) before (1961–1965) versus after (1986–1990) the CWA. *Source:* USEPA STORET.

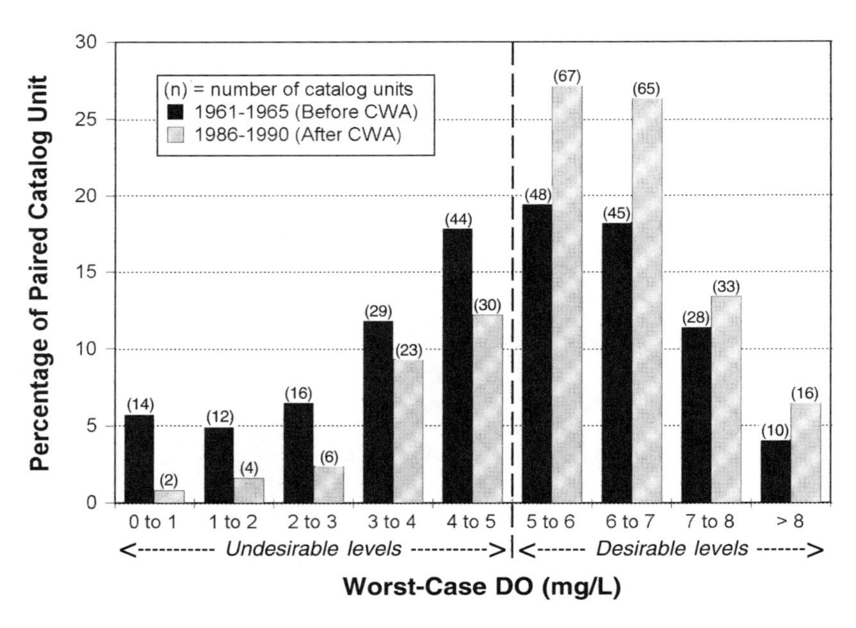

Figure 3-22 Frequency distribution comparing worst-case DO concentration of evaluated catalog units before and after the CWA. $N = 246$ catalog units. *Source:* USEPA STORET.

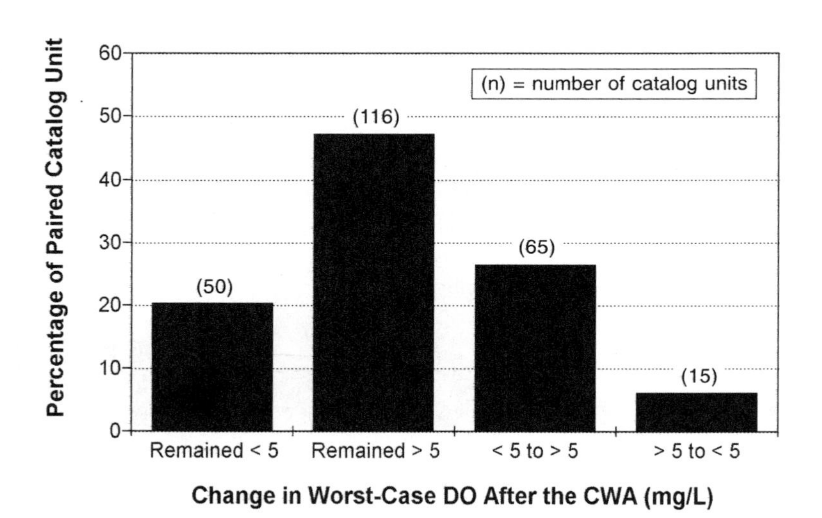

Figure 3-23 Frequency distribution of changes in worst-case DO levels after the CWA using 5 mg/L as the threshold value. $N = 246$ catalog units. *Source:* USEPA STORET.

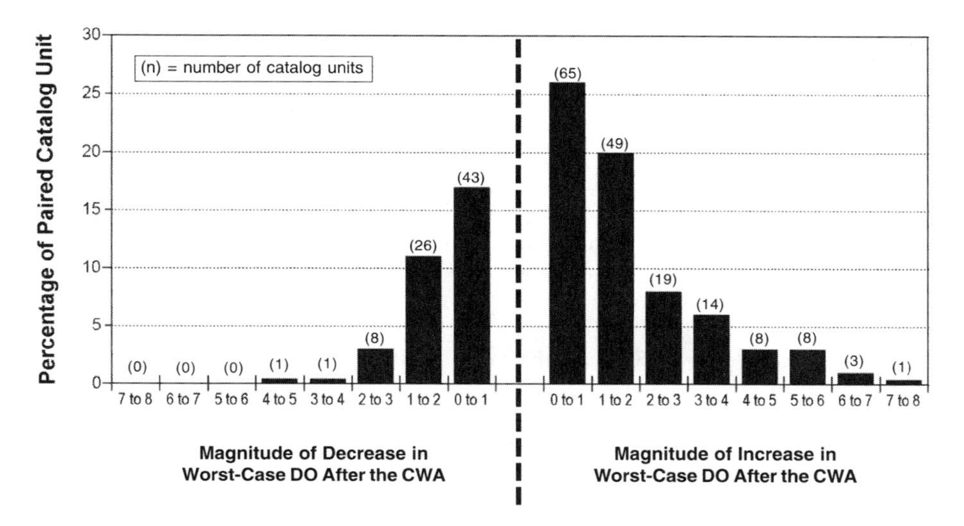

Figure 3-24 Frequency distribution of change in worst-case DO for evaluated catalog units that (*a*) decreased in concentration ($n = 79$) and (*b*) increased in concentration ($n = 167$) before and after the CWA. *Source:* USEPA STORET.

- For 20 of the 25 catalog units, after-CWA worst-case DO improved to levels greater than 5 mg/L.
- The five catalog units that did not break the threshold value of 5 mg/L after the CWA all had concentrations of 0.6 mg/L or less in the before-CWA time-block.

Long-Term Trends of DO and BOD_5 at Catalog Unit Scale Time series for dissolved oxygen and BOD_5, averaged over 5-year periods from 1961–1995 and filtered for "dry" hydrologic conditions, are presented in Figures 3-26 through 3-33 for some of the catalog units identified with the greatest improvements in dissolved oxygen concentrations (Table 3-2). Note that time series data for BOD_5 were not available for the Lower Spokane River, Washington, and the Salinas River, California. As can be seen for the Lower Fox River, Wisconsin (Figure 3-26), Cattaraugus Creek, New York (Figure 3-27), Lake Dubay (Wisconsin River), Wisconsin (Figure 3-29), Oconto River, Wisconsin (Figure 3-31), Lower Great Miami River, Ohio (Figure 3-32), and the Lower Big Sioux River, Iowa (Figure 3-33), the decreasing "before and after" trend in mean ninetieth percentile BOD_5 levels is consistent with the progressive improvements in dissolved oxygen over the 35-year period. As can be seen in these time series data sets, many of the catalog units are characterized by mean ninetieth percentile BOD_5 concentrations during the 1960s ranging from ~3 to 15 mg/L. Very high BOD_5 concentrations, however, were recorded for the Lower Big Sioux River, Iowa (62 mg/L in 1961–1965) and Cattaraugus Creek, New York (54 mg/L and 108 mg/L in 1961–1965 and 1966–1970). Appendix D presents a complete listing of before- and after-CWA BOD_5 data for the 97 catalog units with paired data sets.

TABLE 3-2 Twenty-five Catalog Units with Greatest Improvements in Worst-Case (Mean Tenth Percentile) DO after the CWA

Rank	Catalog Unit ID	Catalog Unit Name	DO (mg/L) (1961–1965)	DO (mg/L) (1986–1990)	DO (mg/L) Change
1	04030204	Lower Fox, WI	0.1600	7.2050	7.0450
2	04120102	Cattaraugus, NY	1.3230	7.6000	6.2770
3	04110002	Cuyahoga, OH	0.2950	6.5008	6.2058
4	17010307	Lower Spokane, WA	3.5000	9.7000	6.2000
5	07070002	Lake Dubay, WI	0.8800	6.6833	5.8033
6	18060005	Salinas, CA	3.1800	8.7500	5.5700
7	02050306	Lower Susquehanna, MD	0.8800	6.1960	5.3160
8	04030104	Oconto, WI	0.5000	5.8000	5.3000
9	05080002	Lower Great Miami, IN	1.1850	6.4675	5.2825
10	08030204	Coldwater, MS	0.0000	5.2082	5.2082
11	10170203	Lower Big Sioux, IA	0.0000	5.1433	5.1433
12	04040002	Pike-Root, IL	0.9400	5.9400	5.0000
13	08030203	Yocona, MS	0.0000	4.8543	4.8543
14	04040003	Milwaukee, WI	2.1800	6.9567	4.7767
15	06010104	Holston, TN	0.1570	4.8686	4.7116
16	08030205	Yalobusha, MS	0.0000	4.6295	4.6295
17	06010205	Upper Clinch, TN	1.6140	6.0819	4.4679
18	02040204	Delaware Bay, NJ	0.5300	4.9100	4.3800
19	04100002	Raisin, MI/OH	4.0588	8.3400	4.2812
20	11070207	Spring, KS/MO	1.6000	5.6250	4.0250
21	04040001	Little Calumet, IL/IN	0.5700	4.5553	3.9853
22	18090208	Mojave, CA	4.0200	7.9767	3.9567
23	07120007	Lower Fox, IL	3.7800	7.5764	3.7964
24	07130011	Lower Illinois, IL	1.9400	5.7225	3.7825
25	04100009	Lower Maumee, OH	2.0676	5.8471	3.7795

Source: USEPA STORET.

The "signal" of DO improvement has been clearly identified at both the catalog unit (watershed) and RF1 reach level spatial scales. In contrast to the results obtained by aggregation of the data at the catalog unit scale, the "before" and "after" "signals" identified at the smaller scale of an RF1 river reach appear to provide a better indicator of the long-term improvements in "worst-case" dissolved oxygen.

On the Upper Mississippi River in the vicinity of Minneapolis–St. Paul, Minnesota, for example, the mean tenth percentile dissolved oxygen levels for station records averaged over the catalog unit 07010206 showed an improvement from 3.7 mg/L "before" to 5.4 mg/L "after" the CWA (see Figure 1-5). This trend is clearly an improvement in water quality that reflects the substantial reductions in effluent loading from the St. Paul METRO facility, the "worst case" summer conditions (~0 to 2 mg/L) observed in the oxygen sag zone about 10 miles downstream of the Twin Cities, and the much higher DO conditions observed in other reaches of the Upper Mississippi River upstream of the Twin Cities, including tributaries to the Upper Mississippi River.

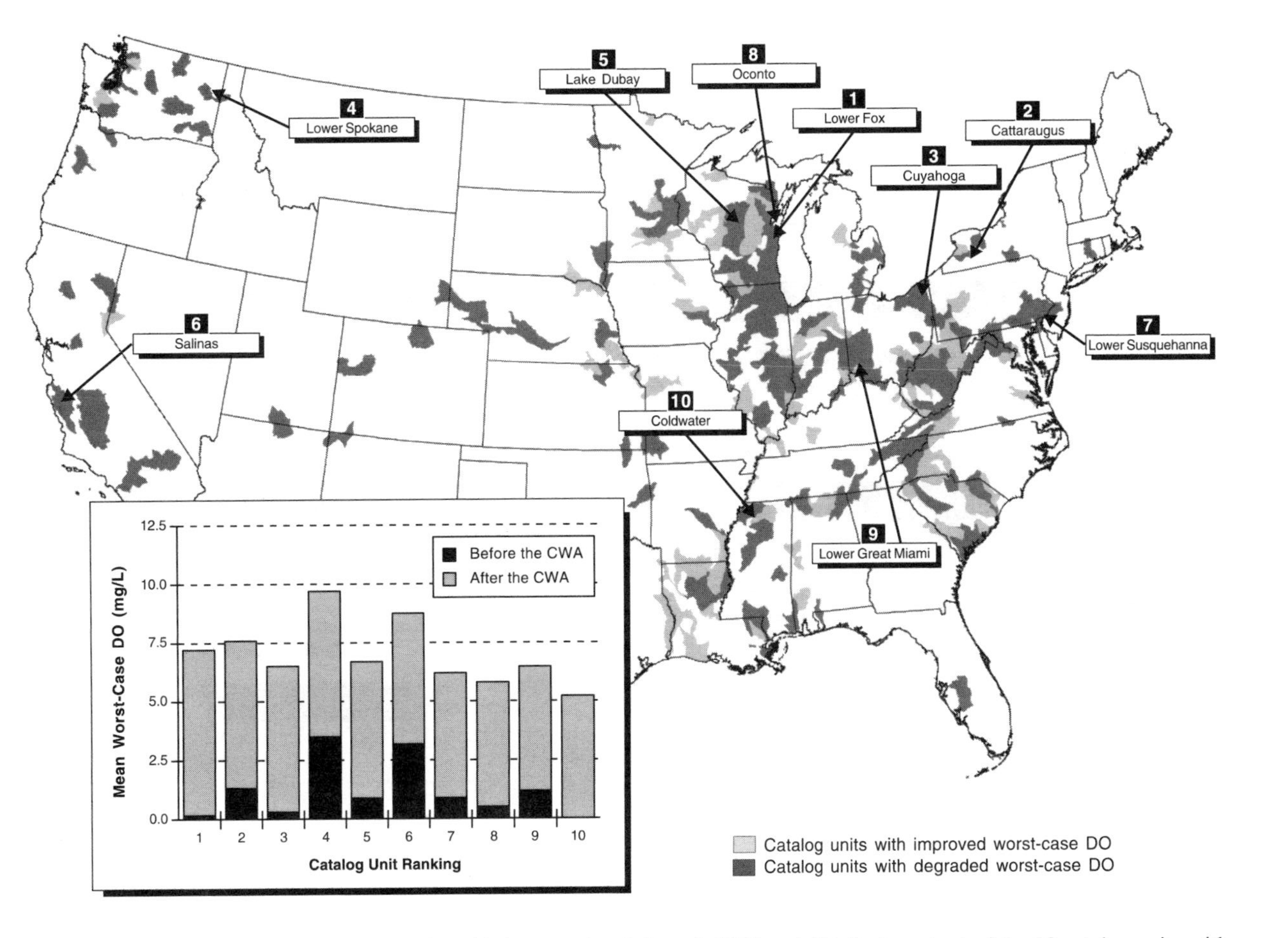

Figure 3-25 Location map of catalog units with improved and degraded DO and distribution chart of the 10 catalog units with the greatest before- versus after-CWA improvements in worst-case DO. *Source:* USEPA STORET.

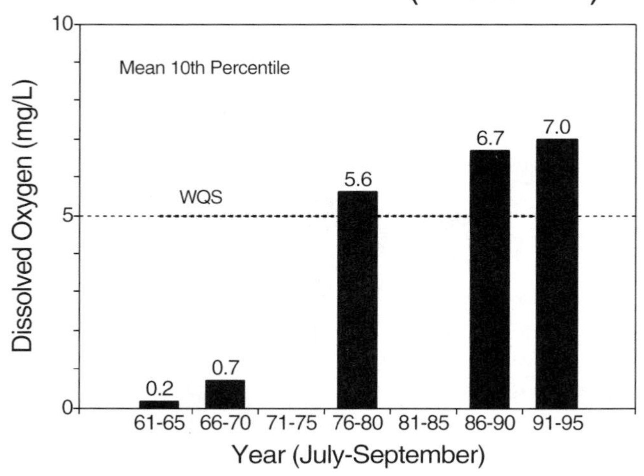

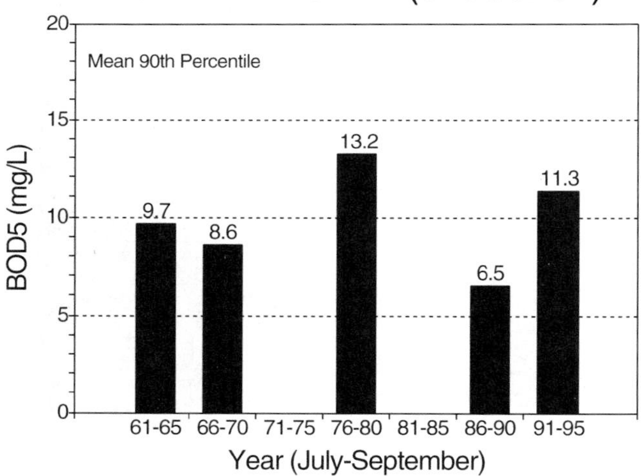

Figure 3-26 Time series from 1961–1995: (*a*) of tenth percentile dissolved oxygen and (*b*) of ninetieth percentile BOD_5 for catalog unit: Lower Fox River, Wisconsin (04030204). *Source:* USEPA STORET.

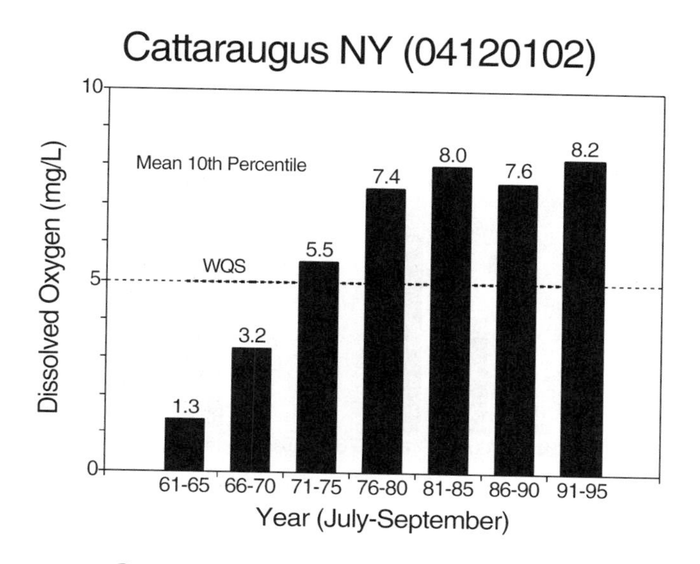

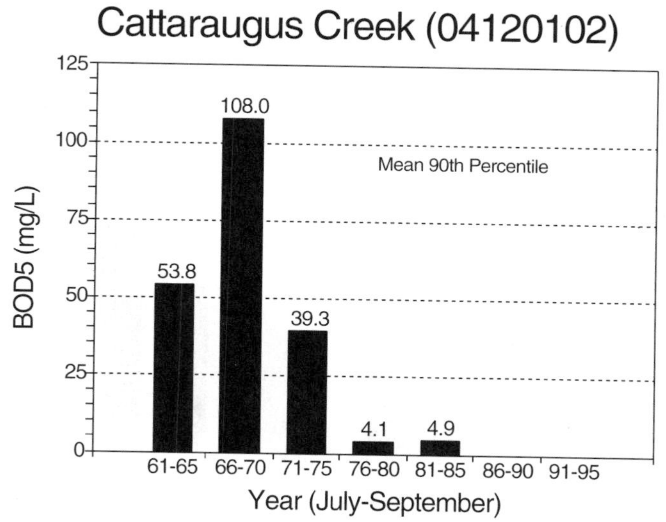

Figure 3-27 Time series from 1961–1995: (*a*) of tenth percentile dissolved oxygen and (*b*) of ninetieth percentile BOD_5 for catalog unit: Cattaruagus Creek, New York (04120102). *Source:* USEPA STORET.

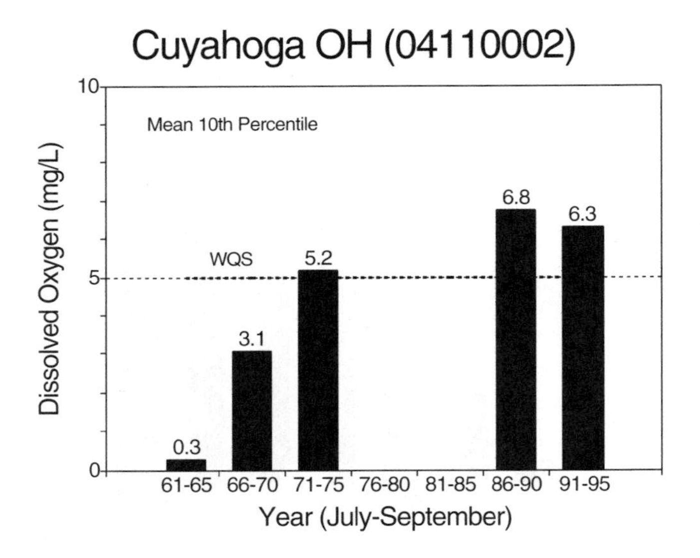

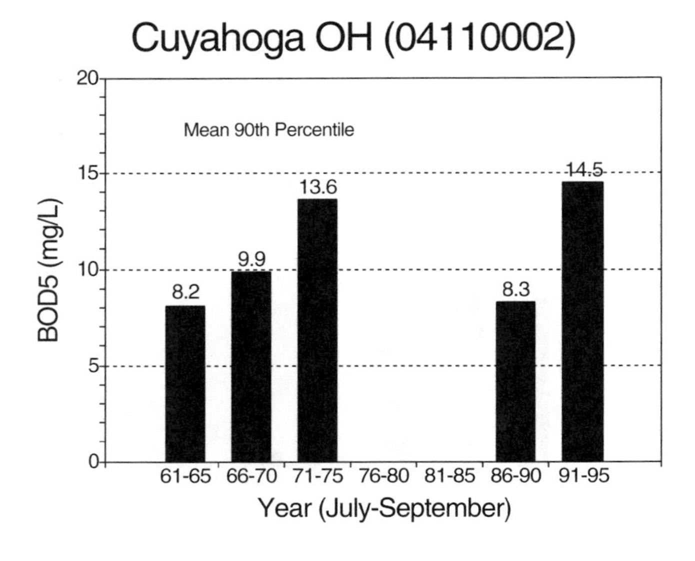

Figure 3-28 Time series from 1961–1995: (*a*) of tenth percentile dissolved oxygen and (*b*) of ninetieth percentile BOD_5 for catalog unit: Cuyahoga River, Ohio (04110002). *Source:* USEPA STORET.

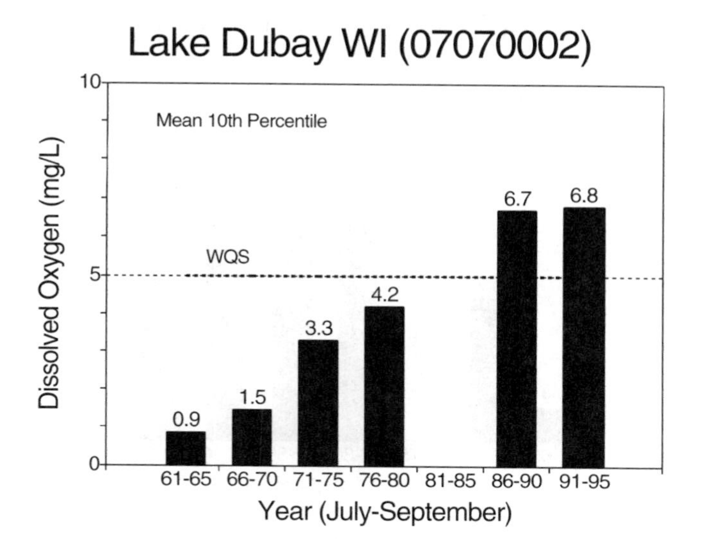

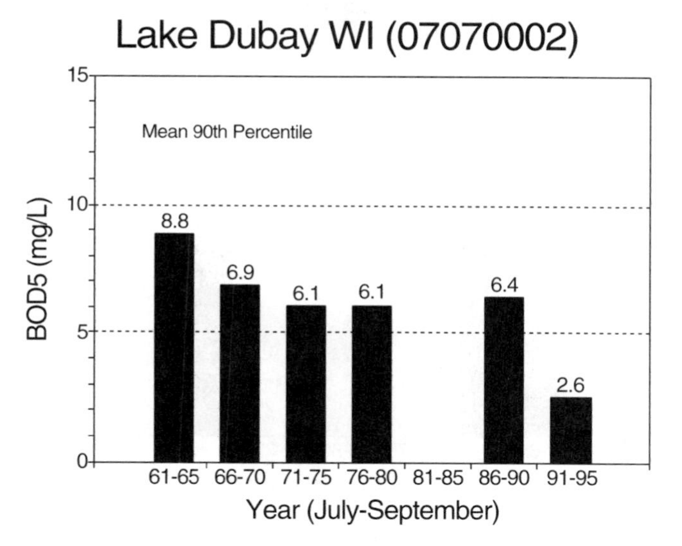

Figure 3-29 Time series from 1961–1995: (*a*) of tenth percentile dissolved oxygen and (*b*) of ninetieth percentile BOD_5 for catalog unit: Lake Dubay (Wisconsin River), Wisconsin (07070002). *Source:* USEPA STORET.

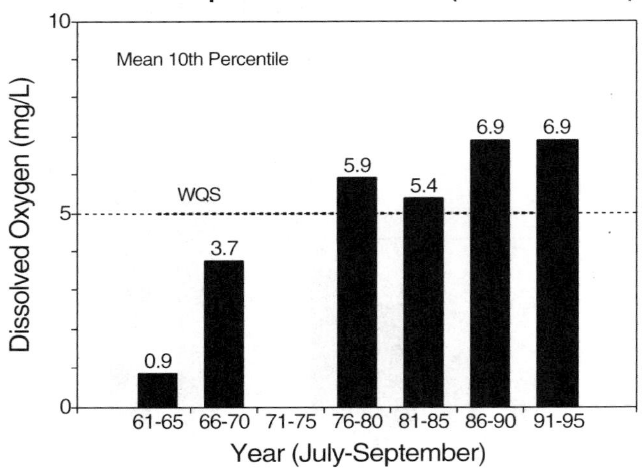

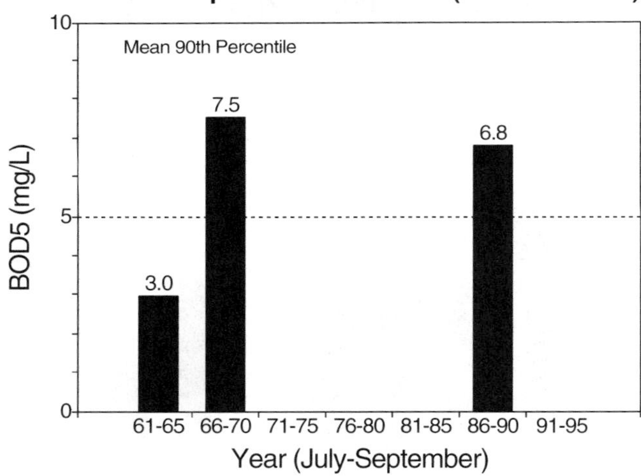

Figure 3-30 Time series from 1961–1995: (*a*) of tenth percentile dissolved oxygen and (*b*) of ninetieth percentile BOD_5 for catalog unit: Lower Susquehanna River, Maryland (02050306). *Source:* USEPA STORET.

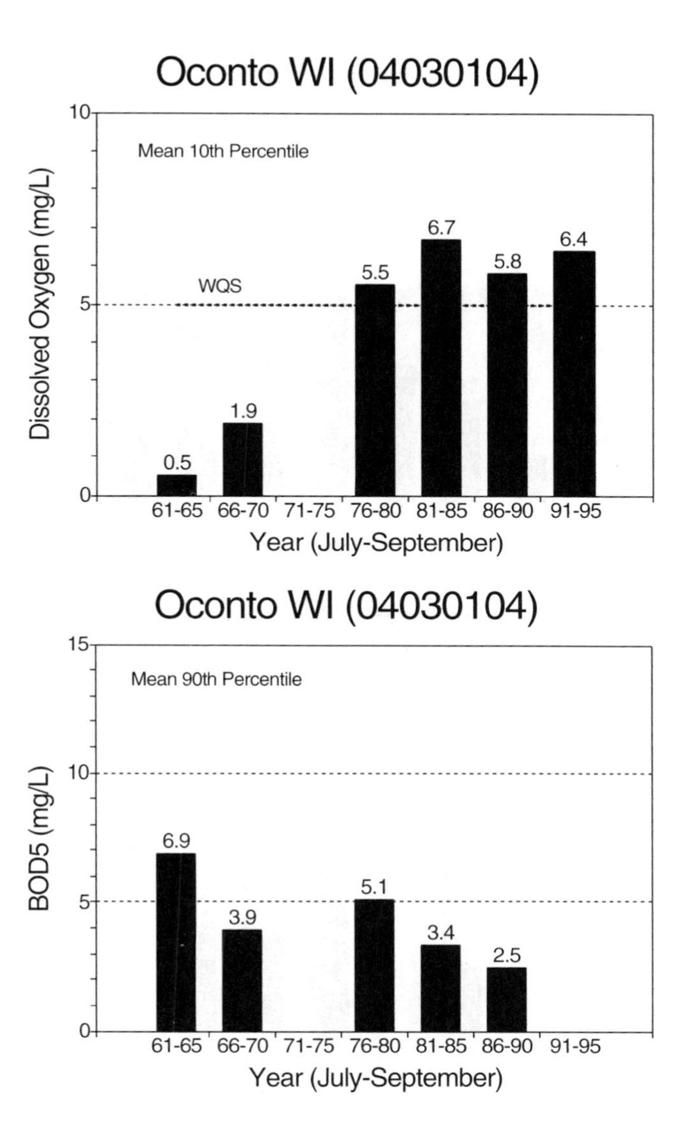

Figure 3-31 Time series from 1961–1995: (*a*) of tenth percentile dissolved oxygen and (*b*) of ninetieth percentile BOD_5 for catalog unit: Oconto River, Wisconsin (04030104). *Source:* USEPA STORET.

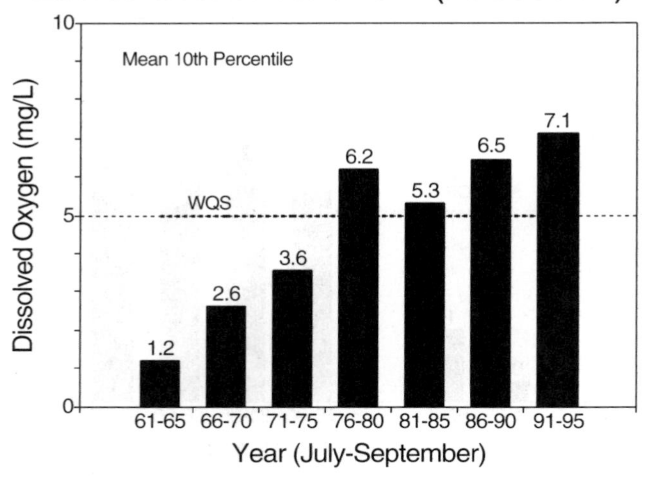

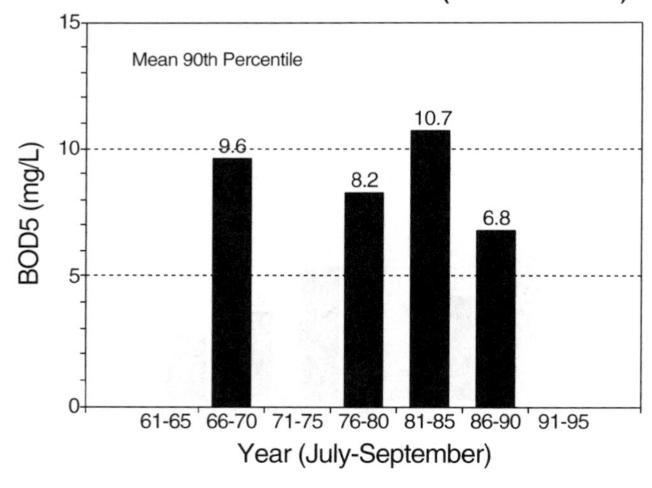

Figure 3-32 Time series from 1961–1995: (*a*) of tenth percentile dissolved oxygen and (*b*) of ninetieth percentile BOD_5 for catalog unit: Lower Great Miami River, Ohio (05080002). *Source:* USEPA STORET.

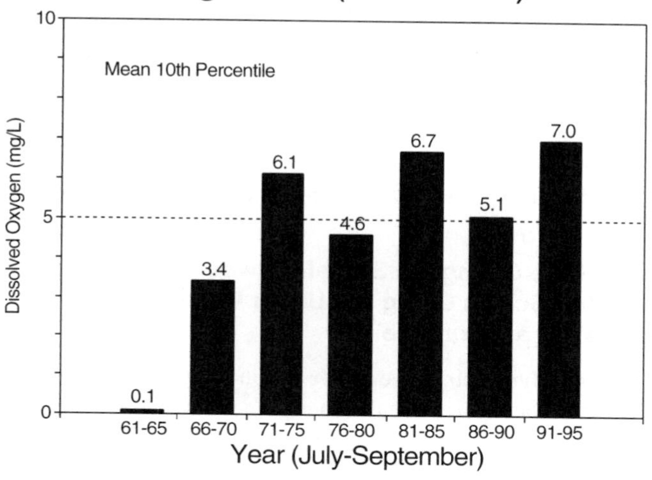

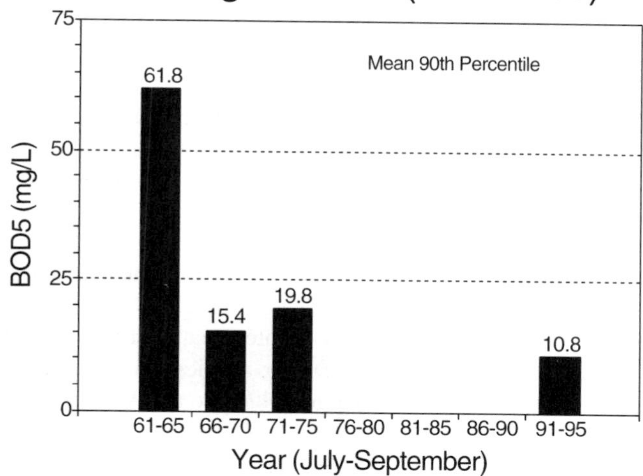

Figure 3-33 Time series from 1961–1995: (*a*) of tenth percentile dissolved oxygen and (*b*) of ninetieth percentile BOD_5 for catalog unit: Lower Big Sioux River, Iowa (10170203). *Source:* USEPA STORET.

For stations extracted from the 33-mile-long RF1 reach (07010206001) from the Minnesota River (UM 844.7) to the St. Croix River (UM 811.6) where the oxygen sag zone is located, however, the "signal" of worst-case DO improvement is much sharper, with mean tenth percentile dissolved oxygen concentrations as low as 2.5 mg/L "before" and 6.0 mg/L "after" the CWA (see Figure 1-5). The RF1 reach scale results reflect the considerably stronger influence of the wastewater discharge from the St. Paul METRO facility for stations located in closer proximity to the point source input. A detailed time series of summer DO data extracted from stations located only within the oxygen sag zone of the Upper Mississippi River (UM 820–830) during "dry" years documents an even stronger "signal" of improvements in DO from < 1 mg/L in 1961 to > 6 mg/L in 1987–1989 (see Figure 12-15 in Chapter 12).

Comparison of the Change in Signal Between the Reach and Catalog Unit Scales Using the Upper White River Basin (Indiana) as an Example

Recall that the underlying objective of the second leg of the three-legged stool approach of this study was to measure the change in the response "signal" linking point source discharges with downstream water quality before and after the CWA at sequentially larger aggregations of spatial scales (reach, catalog unit, and major river basin). The theory is that, if a signal change can be detected at sequentially larger scales, this would provide evidence that the CWA's technology-based and water quality–based effluent control requirements yielded broad as well as localized benefits (that is, stream reaches both within and beyond the immediate sag curve have benefited from the CWA).

The purpose of this subsection is to provide a practical comparison of reach scale and catalog unit scale signals using worst-case DO from monitoring stations in the Upper White River Basin, the catalog unit (05120201) in which the city of Indianapolis, Indiana, and several smaller municipalities are located.

Background In the 1960s, the citizens of the city of Indianapolis depended on primary treatment. Secondary treatment was added in the 1970s, and in 1983, the city further upgraded its POTWs to advanced wastewater treatment (AWT) to achieve compliance with water quality standards for DO. Two municipal facilities, designed to treat up to 379 cfs (245 mgd), currently discharge effluent to the White River. The baseflow of the river is low; the 10-year, 7-day minimum (7Q10) flow is about 50 cfs in the channel upstream of the two POTWs. Consequently, under these low-flow conditions, Indianapolis's wastewater effluent accounts for about 88 percent of the downstream flow.

In addition to Indianapolis, the 2,655-square-mile drainage area of the Upper White River Basin contains several smaller municipalities that also discharge municipal wastewater into the White River network. Population centers upstream from Indianapolis include Muncie, Anderson, and Noblesville. Waverly, Centerton, and Martinsville are towns located downstream of the city. Land use in the basin includes agricultural uses (65 percent) and urban-industrial uses (25 percent), with other uses accounting for the remaining 10 percent (Crawford and Wangness, 1991).

Using point and nonpoint source loading estimates of BOD_5 for contemporary conditions (16.3 mt/day ca. 1995) compiled for the NWPCAM (Bondelid et al., 2000), municipal loads in the basin are estimated to account for 50 percent of the total loading to basin waterways. The remaining one-half of the total BOD_5 load is contributed by major and minor industrial sources (11 percent), rural runoff (24 percent), urban runoff (13 percent), and CSOs (2 percent).

In a pre-AWT (1978–1980) and post-AWT (1983–1986) study of changes in water quality of the White River following completion of the upgrade to AWT from secondary activated sludge facilities for the city of Indianapolis, Crawford and Wangness (1991) concluded that there were statistically significant improvements in ambient levels of DO, BOD_5, and ammonia-nitrogen downstream of the upgraded municipal wastewater facilities. DO, in particular, improved by about 3 mg/L as a result of reductions in carbonaceous (BOD_5) and nitrogenous (ammonia) oxygen demands. For this study, Crawford and Wangness (1991) selected monitoring stations located about 10 and 15 miles downstream of Indianapolis's outfalls to collect data within the critical oxygen sag location of "degradation" and "active decomposition" (Waverly) and the "recovery" zone (Centerton) (see Figure 3-6).

During the before-CWA period from 1961 to 1965, streamflow conditions in the Upper White River Basin were characterized as dry, with persistent drought conditions for three consecutive summers from 1963 through 1965. During these three summers, streamflow ratios ranged from 40 to 63 percent of the long-term summer mean flow (see Figure 3-4*a* for 1963). Similarly, during the after-CWA period of 1986–1990, the Upper White River Basin was affected by the severe drought conditions of 1988 (streamflow ratio of only 34 percent of mean summer flow) that extended over large areas of the Midwest, Northeast, and upper Midwest (see Figure 3-4*b*). The hydrologic conditions of the White River are particularly critical in assessing before and after changes in DO because the municipal effluent flow of the upgraded AWT facilities (after 1983) accounted for about 88 percent of the river flow downstream of Indianapolis under low-flow conditions of the White River.

Upper White River Catalog Unit Level Signal The analysis of before- and after-CWA worst-case DO data for the Upper White River catalog unit revealed that this catalog unit improved by 1.75 mg/L, from 3.80 mg/L (mean value of worst-case DO from 37 stations) before the CWA to 5.55 mg/L (mean value of worst-case DO from 14 stations) after the CWA. This level of improvement ranked it 64th out of the 246 catalog units with before and after data sets (see Appendix D). A companion examination of BOD_5 revealed that worst-case (ninetieth percentile) loading in the catalog unit was reduced from 34.8 mg/L before the CWA (1961–1965) to 6.9 mg/L after the CWA (1986–1990).

The signal change detected provides evidence that:

- The signal linking point source discharges with downstream water quality inherently resides in the before- and after-CWA worst-case DO data collected at stations throughout the Upper White River catalog unit.
- The signal is strong enough to be detected using a catalog unit scale summary statistic (mean of tenth percentile worst-case DO measurements for stations within the catalog unit).

- Improved wastewater treatment by the city of Indianapolis, as well as upgrades of wastewater treatment from other small municipal facilities throughout the basin, resulted in broad water quality improvements in the Upper White River after the CWA.

Upper White River RF1 Reach Level Signals The POTW discharge/downstream water quality signal detected at the catalog unit scale is, in reality, a statistical aggregation of signals associated with all the monitored point source–influenced reaches in the Upper White River watershed. If one breaks the catalog unit down and examines the before- and after-CWA summary statistics for individual reaches, one would expect to find that the reaches in the "degradation" and "active decomposition" zones have more pronounced DO changes than reaches located outside those zones. An examination of reaches in the Upper White River catalog unit revealed this hypothesis to be true. Figure 3-34 includes the locations and before- and after-CWA bar charts for each of the seven reaches in the Upper White River that have paired worst-case DO data. Figure 3-35 presents data showing changes in worst-case (ninetieth percentile) BOD_5 concentrations for the same reaches.

Key observations include the following:

- The reach with the greatest reduction of BOD_5 and greatest improvement in DO was the reach located immediately downstream of Indianapolis (05120201004) in the vicinity of Waverly. DO in this reach, which ranked sixth out of 311 reaches with before- and after-CWA DO data nationwide (see Table 3-1), moved from 0.7 to 6.4 mg/L, an increase of 5.7 mg/L. In this same reach, the ninetieth percentile BOD_5 concentration declined from 58.1 mg/L to 4.3 mg/L.
- Reaches located immediately upstream of Indianapolis showed little change in before- and after-CWA DO conditions (Eagle Creek 05120201032; White River 05120201007, 05120201009; and Fall Creek 05120201006). BOD_5, however, decreased from 20.6 to 7.0 mg/L in reach 05120201007 and from 12.4 to 3.0 mg/L in reach 05120201009. The decline in BOD_5 levels most likely reflects upgrades in municipal facilities for the small towns upstream of Indianapolis.
- Farther upstream, in the vicinity of Muncie and Anderson, greater improvements in DO were detected (along with decreasing trends in ninetieth percentile BOD_5 concentrations). In reach 05120201013 (Muncie), DO in the White River improved by 4.2 mg/L, from 2.2 mg/L before the CWA to 6.4 mg/L after the act. In the compilation of 311 reaches with the greatest before and after improvements in DO, this reach ranked 24th. For the reach in the vicinity of Anderson (05120201011), located downstream of Muncie, DO improved by 2.8 mg/L, from 3.4 mg/L to 6.2 mg/L. This reach ranked 44th in the nationwide ranking of stream reaches with DO improvements (see Appendix E).
- The Lower White River catalog unit is located downstream from the Upper White River unit. Before and after station records from the most upstream reach of the basin reflect the impact of the wastewater discharges from the small towns of Centerton and Martinsville, as well as the recovery zone of the sag curve associated with the Indianapolis point source inputs. In this recovery reach of the White River (05120202031), DO improved by 1.9 mg/L, from 3.4 mg/L to 5.3 mg/L.

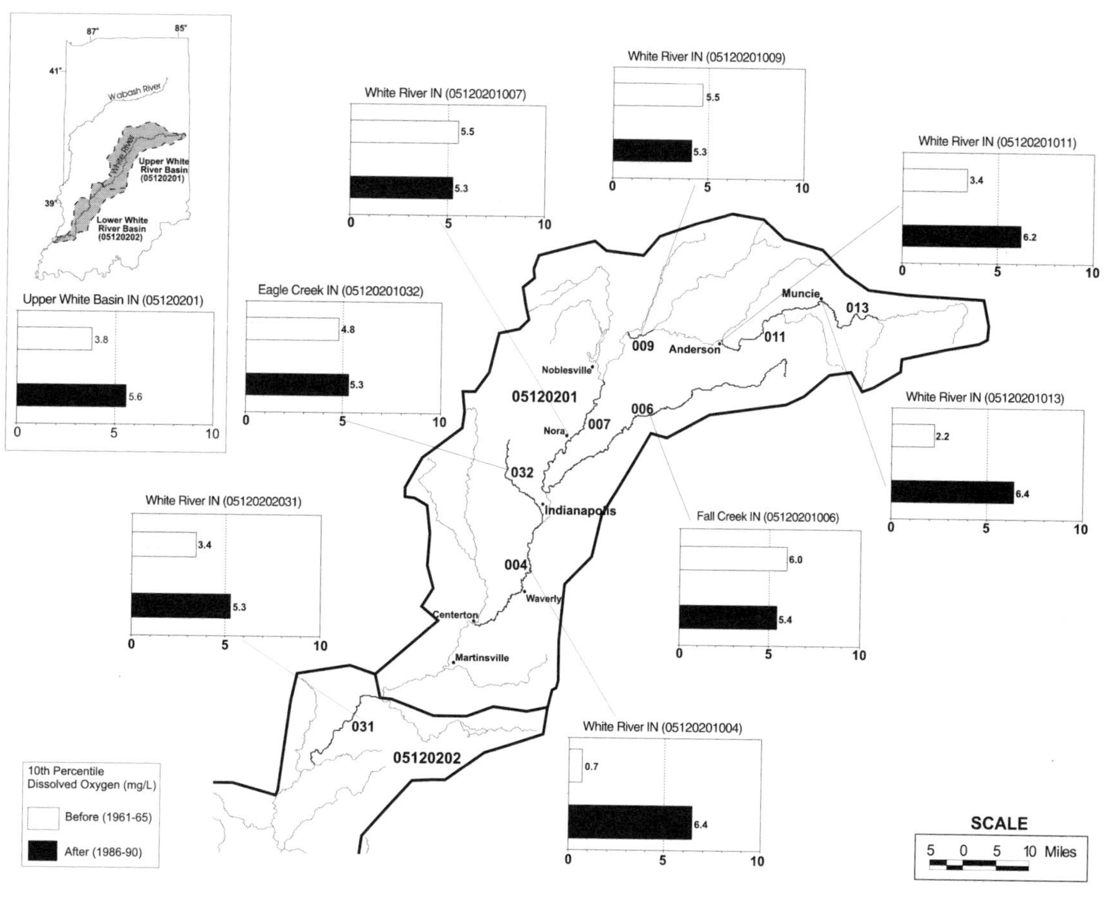

Figure 3-34 Before and after changes in worst-case DO (mg/L) for RF1 reaches of the Upper White River Basin (05120201) in Indiana.

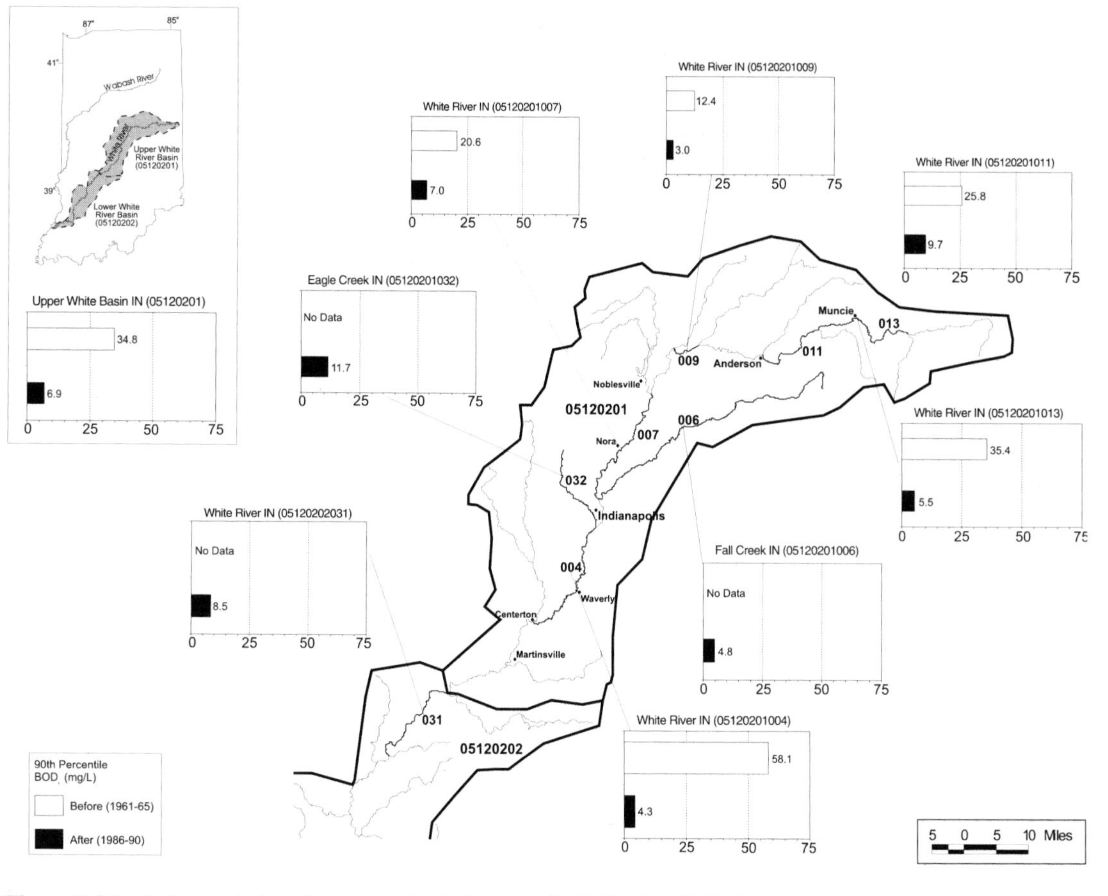

Figure 3-35 Before and after changes in ninetieth percentile BOD_5 (mg/L) for RF1 reaches of the Upper White River Basin (05120201) in Indiana.

The aggregation of worst-case before- and after-CWA station records at the reach scale produced a variety of signals. As expected, the signal linking point source discharges with downstream water quality is most pronounced in reaches located immediately below point source discharges (in the critical portion of the oxygen sag zone). The signal became weaker farther downstream; however, in most reaches it was detectable, especially in the recovery zone of the oxygen sag curve associated with the Indianapolis discharges.

Comparisons of the catalog unit and reach-scale results for both the Upper Mississippi and Upper White River Basins tend to support the conclusions reached by Smith et al. (1987a, 1987b) and Knopman and Smith (1993), where they suggest that any improvements in dissolved oxygen conditions related to upgrading wastewater treatment facilities are detectable only within relatively local spatial scales downstream of wastewater discharge locations.

> The weak association between observed trends in dissolved oxygen deficit and recorded changes in municipal biochemical oxygen demand loads suggests that detectable improvements in dissolved oxygen conditions do not generally extend to the locations of the [USGS] stations studied here. Given the bias toward high biochemical oxygen demand loads at these stations in comparison to river reaches in general, one is forced to conclude that detectable effects of improved municipal treatment on dissolved oxygen are limited to a small fraction of total river miles.
>
> —Smith et al., 1987a

The "signal" of dramatic improvements in dissolved oxygen has been clearly detected at both the catalog unit and the smaller RF1 reach scales. In contrast to the results obtained by aggregation of station records at the catalog unit scale, the "before and after" "signal" detected at the smaller RF1 reach scale appears to provide a better indicator of the long-term trends in improvements of "worst-case" dissolved oxygen. In addition to the DO and BOD_5 data sets presented for the Upper White River, time series data sets of DO from the Lower Big Sioux River from Sioux Falls, South Dakota, to Sioux City, Iowa, provide another example of the difference in "signal" strength obtained at the catalog unit and RF1 reach scales.

Comparison of the time series of dissolved oxygen compiled for the Lower Big Sioux catalog unit (10170203) and a single reach in the Big Sioux River (10170203037) (Figure 3-36) shows that the trend at the RF1 reach scale is consistently increasing from 1961–1965 through 1981–1985, whereas the trend detected for the larger catalog unit scale indicates an apparent erratic "signal" of improvements from 1961–1965 through 1991–1995 that is then marked by declines in DO during 1976–1980 and 1986–1990.

Evaluation of time series DO data sets (Figure 3-37) for two other reaches (10170203039 and 10170203042) in the Lower Big Sioux catalog unit shows that the apparently inconsistent catalog unit trend in Figure 3-36 resulted from averaging the station tenth percentile values over each 5-year time period for stations located in reaches that were characterized by both continuously increasing (10170203037 and 10170203042) and increasing followed by decreasing (10170203039) trends. In contrast to a "before and after" improvement from 0.0 to 5.1 mg/L identified for dissolved oxygen of the Lower Big Sioux catalog unit, ranked #11 of the 246 paired catalog

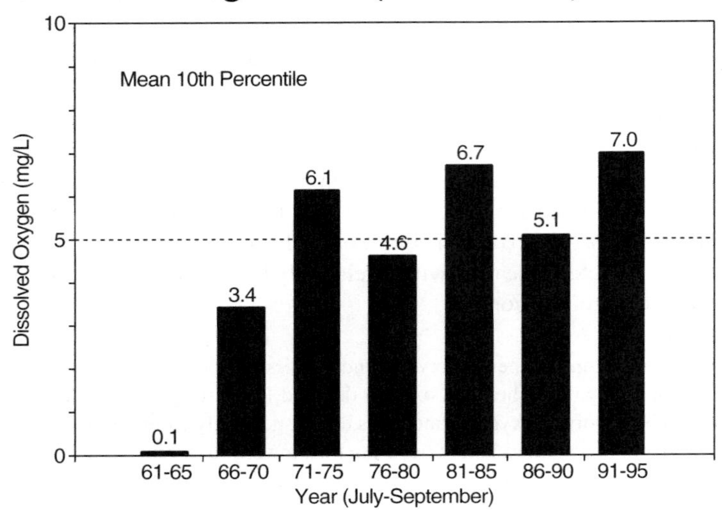

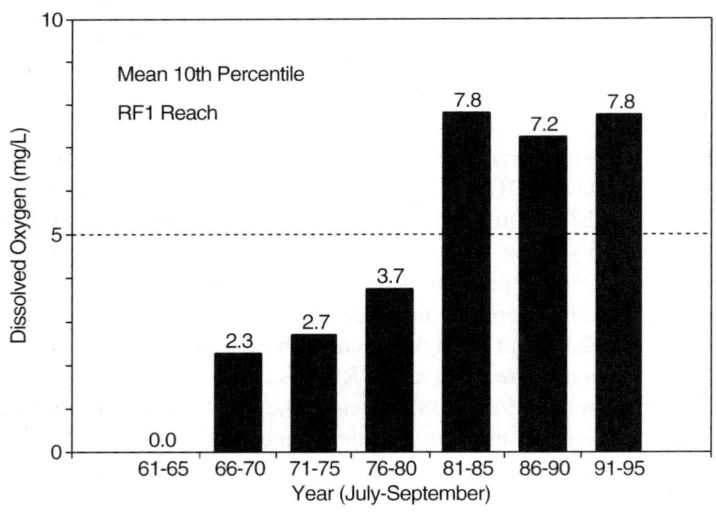

Figure 3-36 Time series from 1961–1995 of tenth percentile dissolved oxygen: (*a*) for Lower Big Sioux catalog unit (10170203), and (*b*) for Big Sioux River RF1 reach (10170203037). *Source:* USEPA STORET.

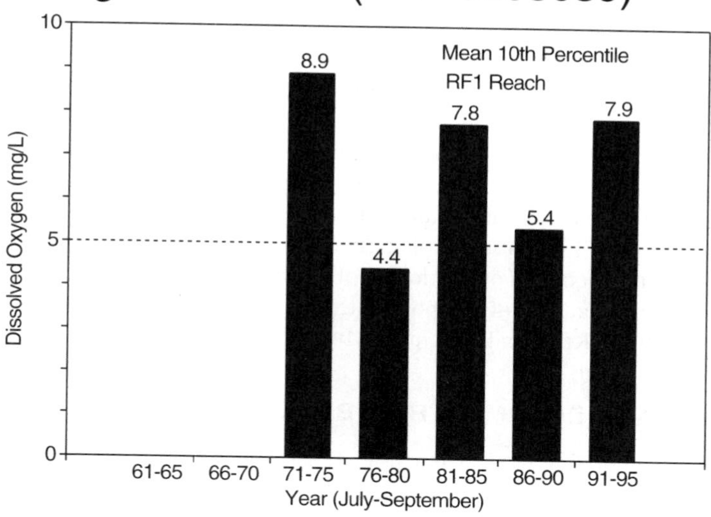

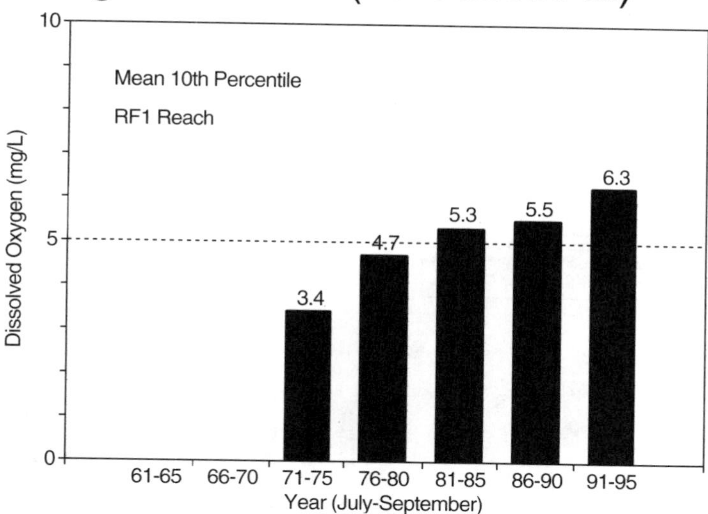

Figure 3-37 Time series from 1961–1995 of tenth percentile dissolved oxygen for Big Sioux River: (*a*) RF1 reach 10170203039, and (*b*) RF1 reach 10170203042. *Source:* USEPA STORET.

units (see Table 3-2), the strongest "signal," ranked #1 for all 311 paired RF1 reaches nationwide, is the "before and after" improvement from 0.0 to 7.2 mg/L identified for a reach of the Big Sioux River near Sioux Falls, South Dakota (see Table 3-1).

With an "after" improvement of 7.2 mg/L for the RF1 reach that is considerably greater than the catalog unit–based "after" improvement of 5.1 mg/L and less detected for other reaches of the Lower Big Sioux catalog unit, the highest ranked RF1 reach "signal" demonstrates the ability of the "data mining" methodology to accurately detect the influence of upgrades to municipal wastewater discharges from the city of Sioux Falls within the critical oxygen sag location in a reach downstream of the wastewater inputs. Similar comparisons of the strengths of "signals" detected for catalog units and RF1 reach scales of station aggregation have also been evaluated to identify the critical location of the impact of upgrades in wastewater discharges from Indianapolis, Indiana, and Minneapolis–St. Paul, Minnesota, on water quality trends in the Upper White River and the Upper Mississippi River.

Before and After DO at Major River Basin Scale

The stations comprising the 311 reach-aggregated worst-case DO data were pooled by the 18 major river basins of the contiguous United States for statistical analyses of the significance of changes in DO concentration before and after the CWA. These analyses were limited to the 311 evaluated reaches to improve the assurance that the data were collected from the same sample population.

Table 3-3 presents the number of observations, the results of the paired t-test (95 percent confidence level), and the mean of the pooled before and after worst-case DO data. The null hypothesis assumes that there is not a significant difference between the mean concentrations for the before and after periods. The means of the pooled worst-case DO data are presented as column graphs in Figure 3-38.

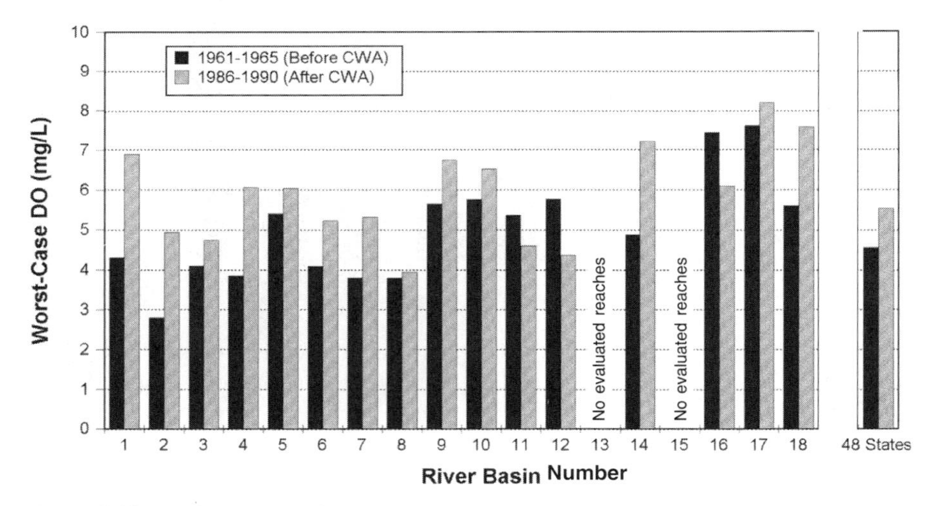

Figure 3-38 Before versus after trends in worst-case DO for major river basins: 1961–1965 versus 1986–1990. *Source:* USEPA STORET.

TABLE 3-3 Statistical Significance of Trends in Mean Tenth Percentile (Worst-Case) DO by Major River Basin: Before Versus After CWA (1961–1965 vs. 1986–1990)

					DO (mg/L)	
	Basin Name	Number of Reaches	Paired t-Test	Kolmogorov Smirnov Test	1961–1965	1986–1990
	All USA (01–18)	311	Yes	Yes	4.56	5.53
01	New England Basin	1	*	*	4.30	6.90
02	Middle Atlantic Basin	17	Yes	Yes	2.80	4.94
03	South Atlantic-Gulf	61	Yes	Yes	4.10	4.73
04	Great Lakes Basin	26	Yes	Yes	3.85	6.06
05	Ohio River Basin	66	Yes	Yes	5.40	6.04
06	Tennessee River Basin	19	Yes	No	4.08	5.23
07	Upper Mississippi Basin	48	Yes	Yes	3.80	5.31
08	Lower Mississippi Basin	25	No	No	3.79	3.94
09	Souris-Red Rainy Basin	2	*	*	5.65	6.75
10	Missouri River Basin	10	No	No	5.76	6.53
11	Arkansas-Red–White Basin	7	No	No	5.36	4.60
12	Texas-Gulf Basin	2	*	*	5.77	4.37
13	Rio Grande Basin	0	*	*	—	—
14	Upper Colorado River Basin	1	*	*	4.88	7.22
15	Lower Colorado River Basin	0	*	*	—	—
16	Great Basin	2	*	*	7.45	6.10
17	Pacific Northwest Basin	17	Yes	No	7.61	8.21
18	California Basin	7	Yes	Yes	5.61	7.58

Source: USEPA STORET.

Notes: Paired t-test: 95 percent confidence—2-sided test. Kolmogorov Smirnov test: 90 percent confidence, 2-sided test.

*Insufficient data for analysis.

Figure 3-39 maps the results of the paired t-test. The darker shaded (yes) river basins indicate that there is a statistically significant difference at the 95 percent confidence level; the river basins marked with lighter shading (no) indicate that there is not a statistically significant difference between the means. Discounting the river basins, mostly in arid western states, with insufficient data for the paired t-test (river basins 01, 09, 12, 13, 14, 15, and 16), 8 of the 11 river basins in the Midwest, Southeast, West Coast, and Middle Atlantic states showed a statistically significant improvement in DO using the paired t-test. The visual decreases in DO in the Texas-Gulf (12), Arkansas-Red-White (11), and Great Basins (16) are not statistically significant.

Recalling that the planning and design of wastewater treatment plant upgrades are often targeted at improving worst-case (low) DO conditions, it is expected that incremental improvements for waters with higher DO conditions (e.g., approaching saturation levels of about 8 to 10 mg/L) are less likely to accrue. As a result, it was suspected that most of the gains would be for the river basins with the lowest DO concentrations before the upgrades, with fewer gains identified for basins that had not been characterized by low DO concentrations. Therefore, frequency distributions are compared in addition to the comparison of means described above.

Figures 3-40 through 3-50 present the "before and after" DO frequency distributions based on the mean tenth percentile DO concentration computed for paired RF1 reaches in each major river basin with sufficient data for the analysis. "Before and after" DO frequency distributions are presented for the Mid-Atlantic (Figure 3-40), South Atlantic (Figure 3-41), Great Lakes (Figure 3-42), Ohio River (Figure 3-43), Tennessee River (Figure 3-44), Upper Mississippi (Figure 3-45), Lower Mississippi (Figure 3-46), Missouri River (Figure 3-47), Arkansas-Red-White (Figure 3-48), Pacific Northwest (Figure 3-49), and California (Figure 3-50) major river basins. It is important to note that not only has the mean changed, but the distribution from low to high DO has also changed significantly. The frequency distributions shown for the major river basins suggest that there have been improvements at the lower percentile levels of DO (i.e., tenth and twentieth percentiles) for these river basins. Before the CWA in the 1961–1965 time-block, worst-case tenth percentile DO was typically at 1 mg/L or lower for all basins except for the Arkansas-Red-White, the Pacific Northwest, and California. During the 1986–1990 period after the CWA, worst-case conditions had improved to levels of about 2 to 5 mg/L for most of the river basins.

The Kolmogorov–Smirnov test was used to statistically compare whether the before and after distributions are significantly different. The Kolmogorov–Smirnov test is a goodness of fit test that compares the empirical distributions from the two time periods. Figures 3-40 to 3-50, showing the empirical cumulative distribution functions of DO from the before and after periods, can be used to visualize what the Kolmogorov–Smirnov test is comparing on a statistical basis. The vertical axis presents the DO concentration corresponding to a given percentile on the horizontal axis. Referring to the Mid-Atlantic basin (Figure 3-40), for example, it can be seen that about 70 percent of the observations from the before period were less than 4 mg/L, whereas in the after period only 30 percent of the observations were less than 4 mg/L. The Kolmogorov–Smirnov test is a statistical comparison of the maximum distance between these curves. The results from the Kolmogorov–Smirnov test are provided in Table 3-3.

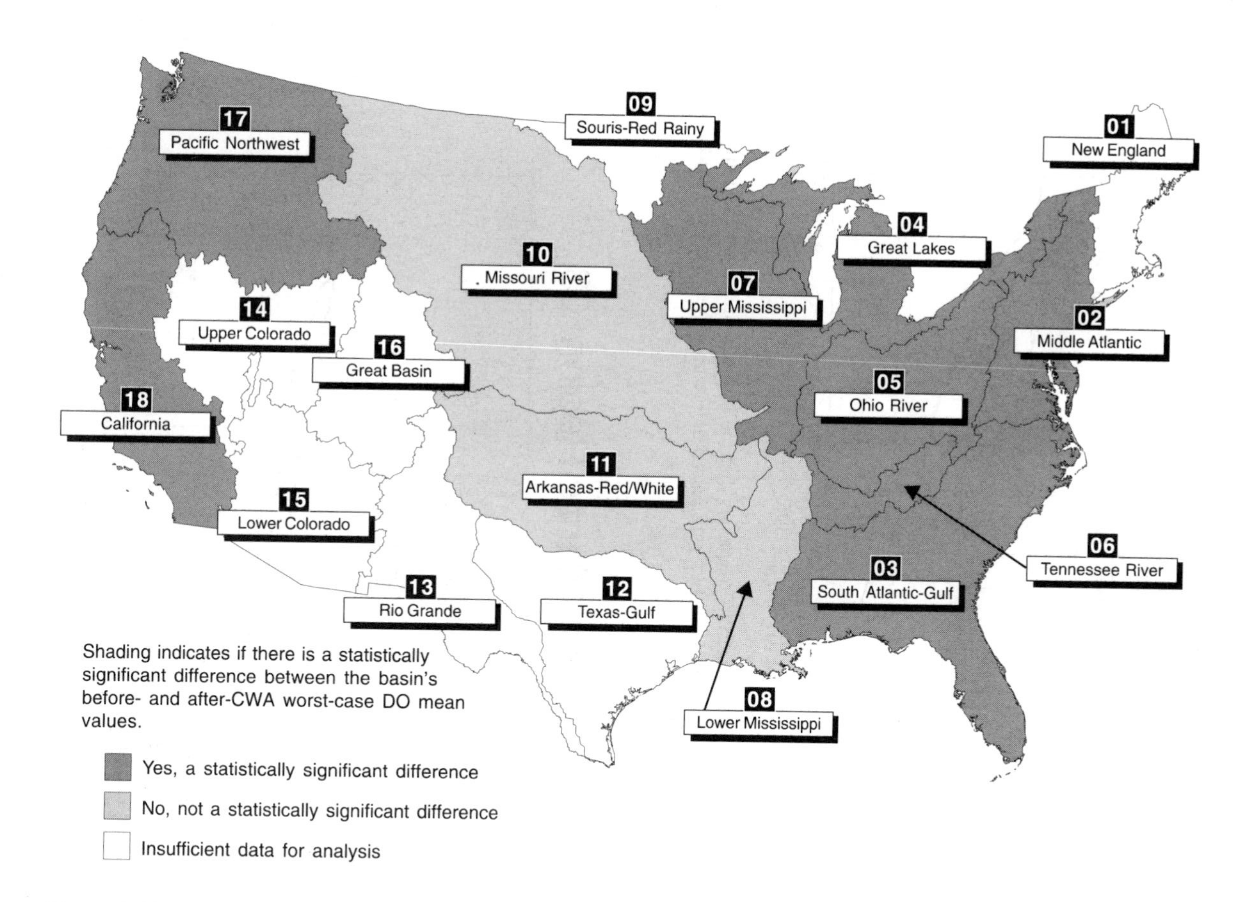

Figure 3-39 Statistical significance of the difference between before- and after-CWA worst-case DO mean values for the 18 major river basins in the 48 contiguous states. *Source:* USEPA STORET.

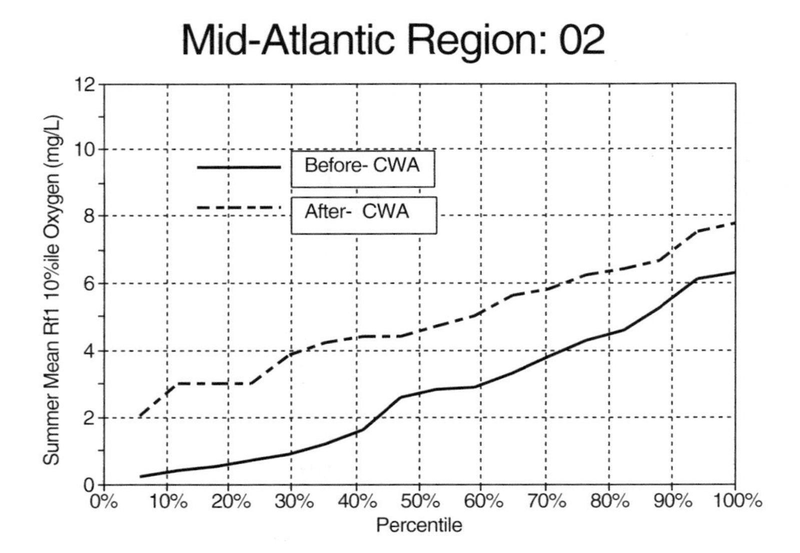

Figure 3-40 Before- and after-CWA frequency distributions of worst-case DO aggregated by major river basin for n = 17 RF1 reaches with paired before and after data sets: Mid-Atlantic (02). *Source:* USEPA STORET.

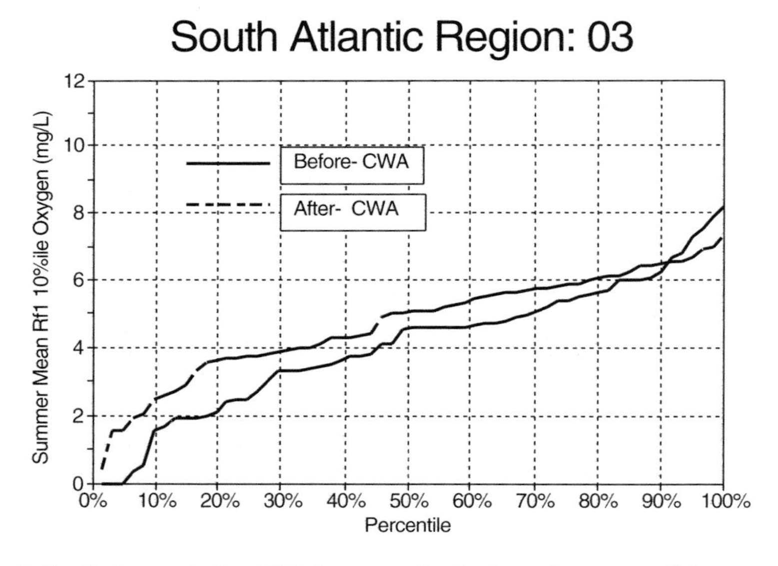

Figure 3-41 Before- and after-CWA frequency distributions of worst-case DO aggregated by major river basin for n = 61 RF1 reaches with paired before and after data sets: South Atlantic (03). *Source:* USEPA STORET.

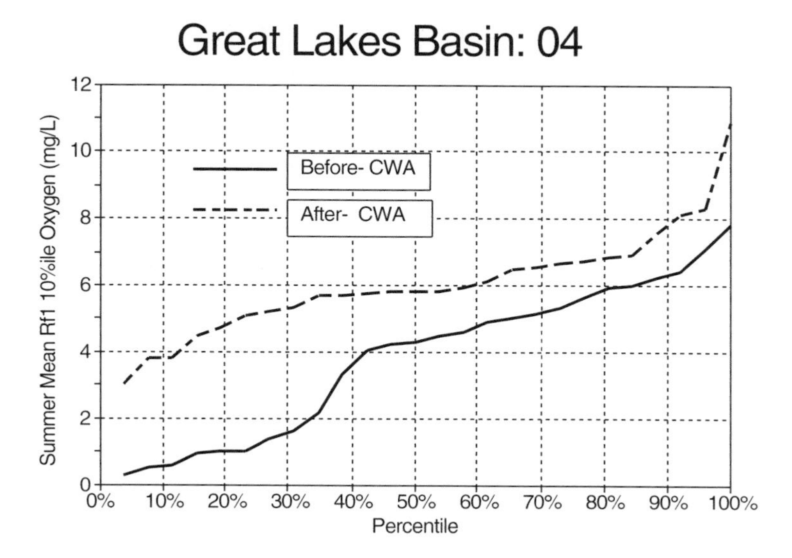

Figure 3-42 Before- and after-CWA frequency distributions of worst-case DO aggregated by major river basin for $n = 26$ RF1 reaches with paired before and after data sets: Great Lakes (04). *Source:* USEPA STORET.

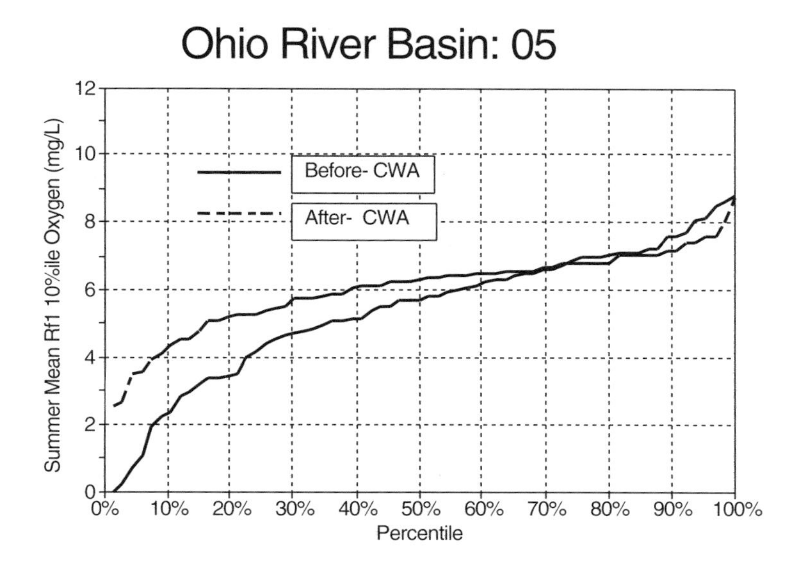

Figure 3-43 Before- and after-CWA frequency distributions of worst-case DO aggregated by major river basin for $n = 66$ RF1 reaches with paired before and after data sets: Ohio River (05). *Source:* USEPA STORET.

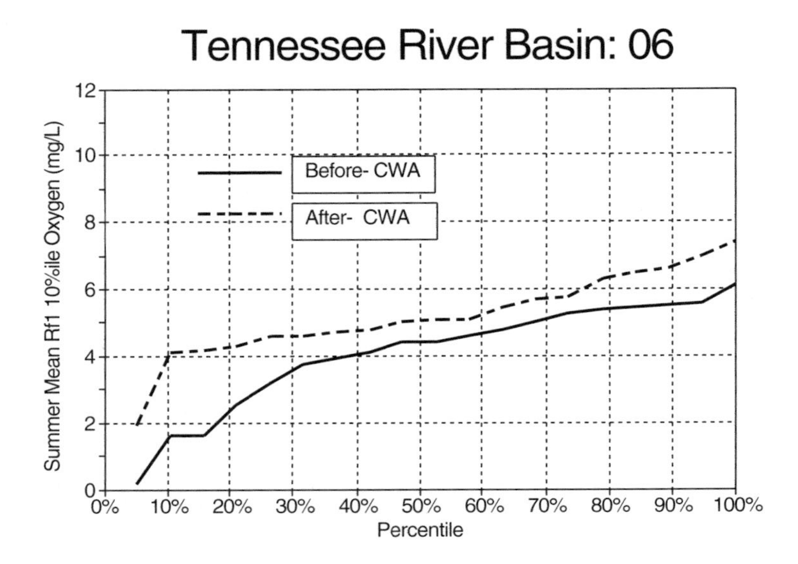

Figure 3-44 Before- and after-CWA frequency distributions of worst-case DO aggregated by major river basin for $n = 19$ RF1 reaches with paired before and after data sets: Tennessee River (06). *Source:* USEPA STORET.

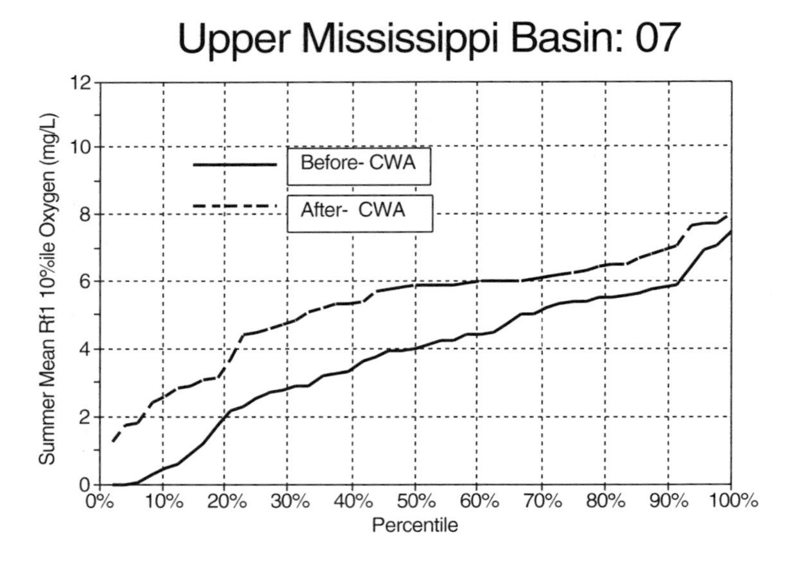

Figure 3-45 Before- and after-CWA frequency distributions of worst-case DO aggregated by major river basin for $n = 48$ RF1 reaches with paired before and after data sets: Upper Mississippi River (07). *Source:* USEPA STORET.

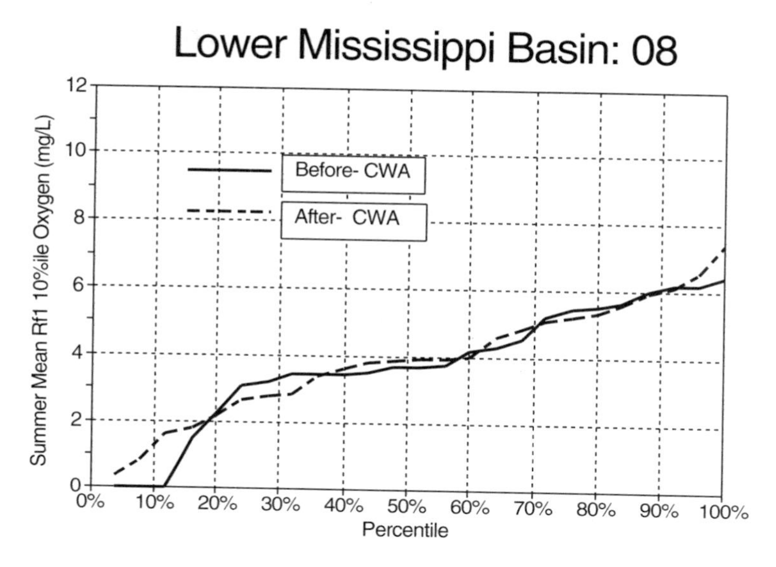

Figure 3-46 Before- and after-CWA frequency distributions of worst-case DO aggregated by major river basin for $n = 25$ RF1 reaches with paired before and after data sets: Lower Mississippi River (08). *Source:* USEPA STORET.

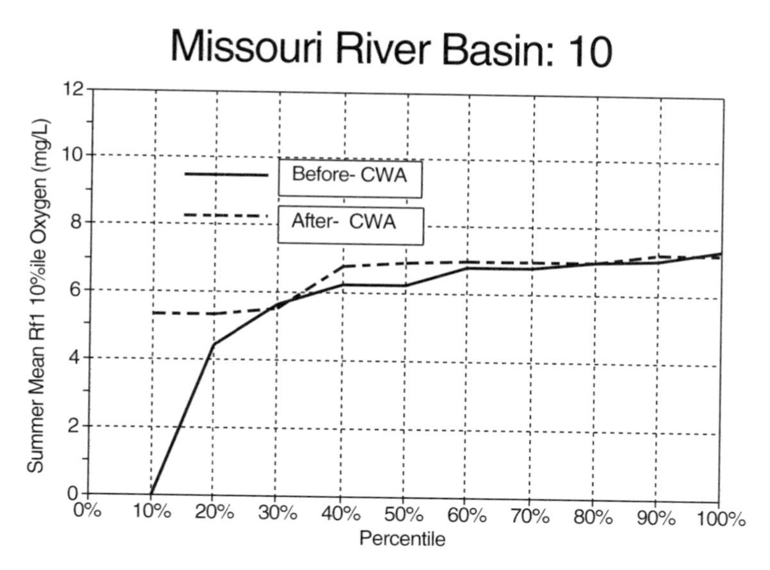

Figure 3-47 Before- and after-CWA frequency distributions of worst-case DO aggregated by major river basin for $n = 10$ RF1 reaches with paired before and after data sets: Missouri River (10). *Source:* USEPA STORET.

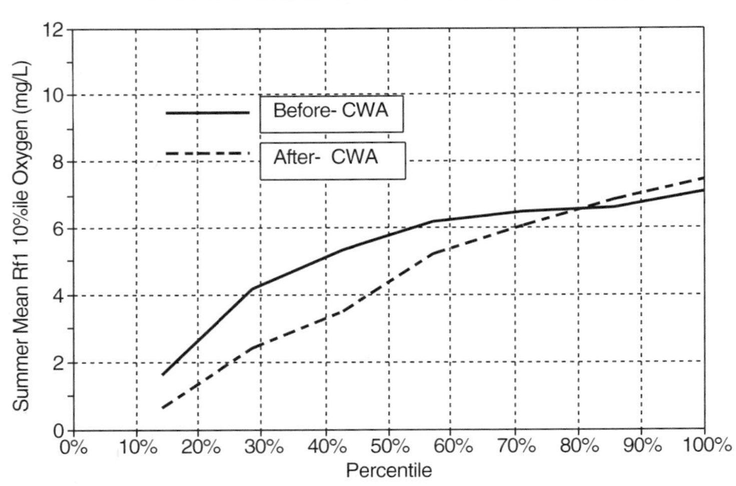

Figure 3-48 Before- and after-CWA frequency distributions of worst-case DO aggregated by major river basin for $n = 7$ RF1 reaches with paired before and after data sets: Arkansas-Red-White River (11). *Source:* USEPA STORET.

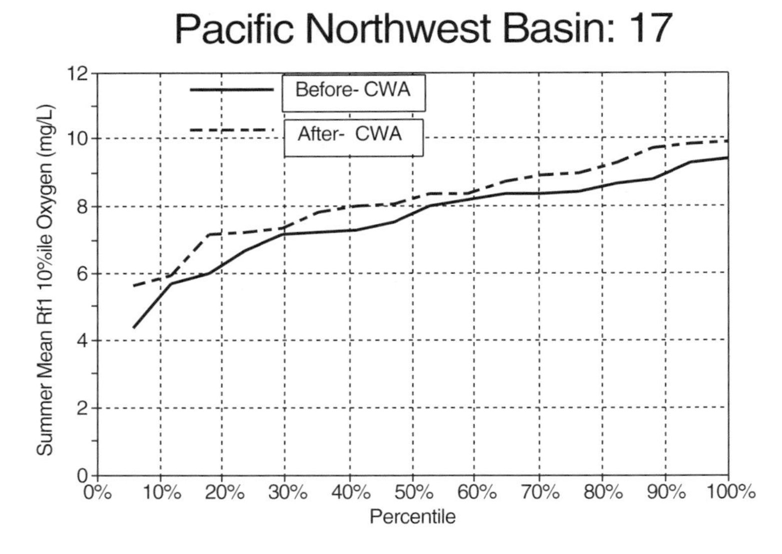

Figure 3-49 Before- and after-CWA frequency distributions of worst-case DO aggregated by major river basin for $n = 17$ RF1 reaches with paired before and after data sets: Pacific Northwest (17). *Source:* USEPA STORET.

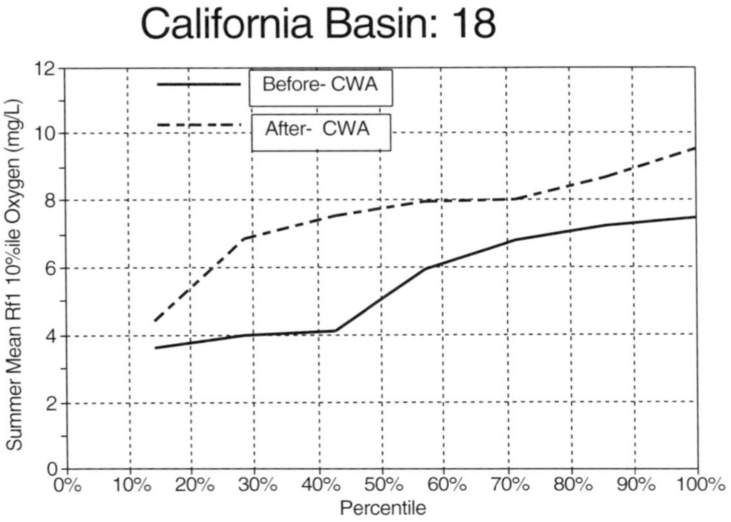

Figure 3-50 Before- and after-CWA frequency distributions of worst-case DO aggregated by major river basin for $n = 7$ RF1 reaches with paired before and after data sets: California (18). *Source:* USEPA STORET.

Based on the two different statistical tests, and discounting the 7 river basins with limited data, 8 of the 11 remaining river basins can be characterized by a statistically significant improvement in worst-case DO using at least one of the two tests. Mixed results (yes and no) were obtained for two basins with the Kolmogorov–Smirnov test indicating no significant improvement for the Tennessee (6) and the Pacific Northwest (17) basins, whereas the paired t-test indicated significant improvements (yes) in these basins. Overall, there is a statistically significant improvement in worst-case DO trends using both statistical tests at 6 out of 11 river basins with sufficient data. Of the five basins with at least one "nonsignificant" change, three basins (Missouri River, Arkansas-Red-White, and Pacific Northwest) had a mean worst-case pooled DO level greater than 5 mg/L in the before time period and were less likely to be targeted for improved point source pollution control. It is also noteworthy that in the 25-year interval between the before- and after-CWA periods, there were no statistically significant conditions of degradation of worst-case DO for any of the major river basins. It is also noteworthy that when all 311 paired reaches are analyzed together, both tests indicate significant national-scale increases in worst-case DO [see Figure 3-51 and top row (All USA) of Table 3-3].

D. SUMMARY AND CONCLUSIONS

The purpose of this chapter is to address the second leg of the three-legged stool approach for answering the question posed in Chapter 1: How has the nation's water quality changed since implementation of the 1972 CWA's mandate for secondary

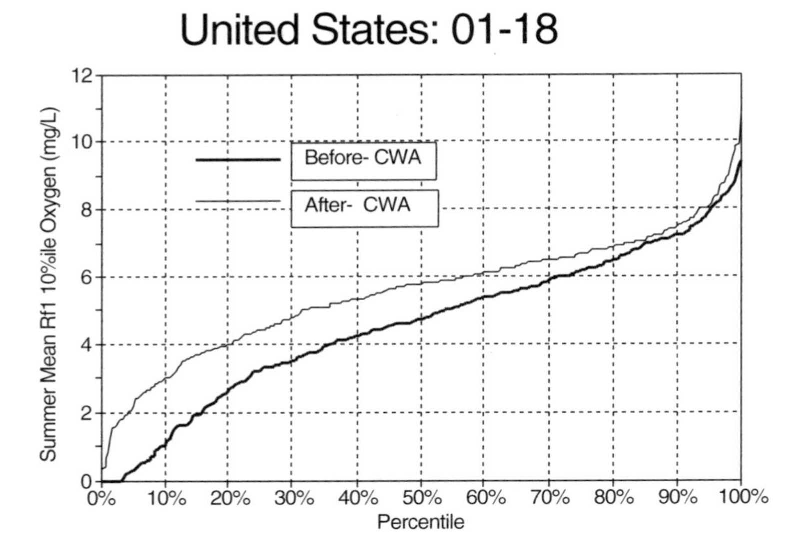

Figure 3-51 Before- and after-CWA frequency distributions of worst-case DO aggregated by all major river basins or $n = 311$ RF1 reaches with paired before and after data sets: 48 States. *Source:* USEPA STORET.

treatment as the minimum acceptable technology for POTWs? Recall that the basic goal of the second leg was to determine the extent to which water quality improvements could be linked to the CWA's push for secondary and greater levels of treatment in the nation's POTWs. If evidence showed that worst-case DO concentrations improved at broad, as well as localized, spatial scales, the second leg of the investigation could add cumulative support for the conclusion that the CWA's mandates were successful. The following objectives were established to guide this part of the study:

- Develop before- and after-CWA data sets composed of DO summary statistics derived from monitoring stations screened for worst-case conditions.
- Develop a worst-case DO summary statistic for each station for each before- and after-CWA time period and then aggregate these data by sequentially larger spatial scales (reaches, catalog units, and major river basins).
- Conduct an analysis of the spatial units having both a before- and after-CWA summary statistic and assess the magnitude of worst-case DO change between the two time periods.
- Assess the before- and after-CWA change in the point source discharge/downstream DO signal over the progressively larger spatial scales.

Key Points of the Background Section

Section A provided background concerning the source of DO data used in this study, why worst-case conditions are an appropriate screening tool for developing the before- and after-CWA data sets, and the role spatial scale played in the second leg of this study. Key points include the following:

- The sharpest signal linking point source loading and downstream DO inherently resides in data collected in worst-case (high temperature and low flow) conditions. These worst-case conditions typically occur in the summer months (July through September) during consecutive runs of dry years (persistent drought).
- Widespread persistent drought was most pronounced in the summers in 1961–1965 (before the CWA) and 1986–1990 (after the CWA). These time-blocks were used to define the before- and after-CWA time periods for the comparison analysis.
- From a spatial perspective, worst-case critical, or minimum, DO below a point source occurs in the "degradation" or "active decomposition" zone of the oxygen sag curve. However, screening rules were not developed to select monitoring stations located within these zones because the goal of this second leg is to examine changes in the point source discharge/downstream DO at broad scales as well as localized scales. Consequently, the only screening rule regarding location of stations eligible for the before- and after-CWA analysis is that the station must be somewhere downstream from, and therefore potentially influenced by, a municipal and/or industrial point source discharge.

Key Points of the Data Mining Section

Section B presented the six-step data mining process used to create the before- and after-CWA data sets to be used in the comparison analysis. The screening rules associated with each step are listed below:

Step 1—Data Selection Rules

- DO, expressed as a concentration (mg/L), will function as the signal relating municipal and industrial point source discharges of oxidizeable organic matter to downstream water quality responses.
- DO data are extracted only from the July–September (summer season) time period.
- Only surface DO data (DO data collected within 2 meters of the water surface) are used.

Step 2—Data Aggregation Rules from a Temporal Perspective

- 1961–1965 serves as the time-block to evaluate persistent drought before the CWA and 1986–1990 serves as the time-block to evaluate persistent drought after the CWA.
- To remain eligible for the before- and after-CWA comparison, DO data must come from a station residing in a catalog unit that had at least one year classified as dry (streamflow ratio 75 percent of long-term summer mean) out of the 5 years in each before- and after-CWA time-block.

Step 3—Calculation of the Worst-Case DO Summary Statistic Rules

- For each water quality station, the tenth percentile of the DO data distribution from the before-CWA time period (July–September, 1961–1965) and the tenth percentile of the DO data distribution from the after-CWA time period (July–September, 1985–1990) are used as the station's DO worst-case statistics for the comparison analysis.
- To remain eligible for the before- and after-CWA statistical comparison, a station must have a minimum of eight DO measurements within each of the 5-year time-blocks.

Step 4—Spatial Assessment Rules

- Only water quality stations located on streams and rivers affected by point sources are included in the before- and after-CWA comparison analysis.

Step 5—Data Aggregation Rules from a Spatial Perspective

- The before- and after-CWA data sets are collections of DO summary statistics that characterize worst-case DO at individual water quality monitoring stations across the United States for the 1961–1965 time-block and the 1986–1990 time-block, respectively (one DO summary statistic per station per time-block).
- For each data set and time-block, the tenth percentile value from each eligible station is aggregated within the spatial hydrologic unit. (Since the scales are hierarchical, a station's summary statistic is effectively assigned to both a reach and a catalog unit.) A summary statistic is then calculated and assigned to the spatial unit for the purpose of characterizing its worst-case DO. If a spatial unit has only one monitoring station within its borders that meets the screening criteria, the tenth percentile DO value from that station simply serves as the unit's worst-case summary statistic. If, however, there are two or more stations within a spatial unit's borders, the tenth percentile values for all the eligible stations are averaged, and this value is used to characterize worst-case DO for the unit.
- The mean tenth percentile value is computed from the eligible station's tenth percentile values for the before- and after-CWA periods.

Step 6—Development of the Paired Data Sets (at Each Spatial Scale)

- To be eligible for the paired comparison analysis, a hydrologic unit must have both a before-CWA and an after-CWA summary statistic assigned to it.

Key Points of the Comparison Analysis Section

Section C presented the results of the comparative before- and after-CWA analysis of worst-case DO data derived using the screening criteria described in Section B and aggregated by spatial units defined by three scales (RF1 reach, catalog unit, and major river basin). Listed below are key observations for each spatial scale:

RF1 Reach Scale

- Sixty-nine percent of the reaches evaluated showed improvements in worst-case DO after the CWA. [Three hundred eleven reaches (out of a possible 12,476 downstream of point sources) survived the data screening process with comparable before- and after-CWA DO summary statistics. The number of reaches available for the paired analysis was limited by the historical data for the 1961–1965 period].
- These 311 evaluated reaches represent a disproportionately high amount of urban/industrial population centers, with approximately 13.7 million people represented (7.2 percent of the total population served by POTWs in 1996). The top 25 improving reaches saw their worst-case DO increase by anywhere from 4.1 to 7.2 mg/L!
- The number of evaluated reaches characterized by worst-case DO below 5 mg/L was reduced from 167 to 97 (from 54 to 31 percent).
- The number of evaluated reaches characterized by worst-case DO above 5 mg/L increased from 144 to 214 (from 46 to 69 percent).
- The long-term trends of worst-case DO presented for selected RF1 reaches identified with the greatest before and after improvement document progressive improvements in DO as sewage treatment plants were upgraded to comply with the CWA requirements for a minimum level of secondary treatment.

Catalog Unit Scale

- Sixty-eight percent of the catalog units evaluated showed improvements in worst-case DO after the CWA. [Two hundred forty-six catalog units (out of a possible 1,666 downstream of point sources) survived the data screening process with comparable before- and after-CWA DO summary statistics].
- The number of evaluated catalog units characterized by worst-case DO below 5 mg/L was reduced from 115 to 65 (from 47 to 26 percent). The number of evaluated catalog units characterized by worst-case DO above 5 mg/L increased from 131 to 181 (from 53 to 74 percent).
- Fifty-three of the 167 improving catalog units (32 percent) improved by 2 mg/L or more, while only 10 of 79 degrading catalog units (13 percent) degraded by 2 mg/L or more.
- These 246 evaluated catalog units represent a disproportionately high amount of urban/industrial population centers, with approximately 61.6 million people represented (32.5 percent of the total population served by POTWs in 1996).
- The long-term trends of worst-case DO and BOD_5 presented for selected catalog units identified with the greatest before and after improvement document progressive increases in DO with corresponding decreases in BOD_5 as sewage treatment plants were upgraded to comply with the CWA requirements for a minimum level of secondary treatment.

Major River Basin Scale

- A total of 11 out of 18 major river basins had sufficient reach-aggregated worst-case DO data for a before- and after-CWA comparison analysis.
- Based on two statistical tests, 8 of the 11 major river basins can be characterized as having statistically significant improvement in worst-case DO levels after the CWA. The three basins that did not statistically improve under either test also did not have statistically significant degradation.
- When all the 311 paired (i.e., before vs. after) reaches were aggregated and the statistical tests run on all 18 of the major river basins of the contiguous states as a whole, worst-case DO also showed significant improvement.

Conclusions

The statistical analyses developed for this study are not ideal. One major concern is the potential bias introduced in the ambient monitoring programs used to collect the data archived in STORET. It is believed that the analysis of data sets with data in the before and after time periods alleviates some of these concerns and that results are generally comparable for the two different statistical tests. Based on the systematic, peer-reviewed approach designed to identify and evaluate the national-scale distribution of water quality changes that have occurred since the 1960s, this study has compiled strong evidence that the technology-based and water quality–based policies of the CWA for point source effluent controls have been effective in significantly reducing loads and improving DO. In this retrospective analysis, DO was used as the key indicator because the reduction of organic carbon and nitrogen (BOD_u) loading from municipal and industrial point sources was one of the major goals of the CWA's technology-based policy, which required industrial effluent limits and a minimum level of secondary treatment for municipal facilities. Based on ambient DO records, significant before and after improvements in many rivers and streams have been identified over national, major river basin, catalog unit, and RF1 reach-level spatial scales.

The "signal" of downstream water quality responses to upstream wastewater loading and the changes in this signal since the 1960s have been successfully decoded from the "noise" of millions of archived water quality records. Given the very large spatial scale of the major river basins, it is remarkable to observe statistically significant before and after DO improvements as detected using the systematic methodology described in this book. Previous evaluations of the effectiveness of the CWA (e.g., Smith et al., 1987a, 1987b; Knopman and Smith, 1993) were not able to report conclusively significant improvements in DO. In these earlier studies, however, the methodologies used were not specifically designed to separate the signal of downstream water quality response from the noise within large national databases. Using appropriate data screening rules and spatial aggregations, it has been demonstrated that improvements in water quality, as measured by improvements in worst-case DO, have been achieved since the 1960s.

The findings of this national-scale water quality assessment demonstrate three important points:

- As new monitoring data are collected, it is crucial for the success of future performance measure evaluations of pollution control policies that the data be submitted, with appropriate QA/QC safeguards, to accessible databases (e.g., Alexander et al., 1998). If the millions of records archived in STORET had not been readily accessible, it would have been impossible to conduct this analysis to identify the signals of water quality improvements that have been achieved since the early 1960s.
- Significant after-CWA improvements in worst-case summer DO conditions have been quantitatively documented with credible statistical techniques in this study over different levels of spatial data aggregation from the small sub-watersheds of RF1 river reaches (mean drainage area of approximately 115 square miles) to the very large watersheds of major river basins (mean drainage area of 434,759 square miles).
- The data mining and statistical methodologies designed for this study can potentially be used to detect long-term trends in signals for water quality parameters other than DO (e.g., suspended solids, nutrients, toxic chemicals, pathogens) to develop new performance measures to track the effectiveness of watershed-based point source and nonpoint source controls. The key element needed to apply the data mining methodology to other water quality parameters is the careful specification of rules for data extraction that reflect a thorough understanding of the hydrologic processes that influence the spatial and temporal distributions of a water quality constituent, as well as the relevant sources of associated pollutants (see Hines et al., 1976).

Population Affected by Reaches with Improved DO To quantify, in financial terms, the environmental benefits derived from various environmental policy decisions, USEPA developed the National Water Pollution Control Assessment Model (NWPCAM) (Bondelid et al., 2000), which includes a link between 1990 population and RF1 river reaches. As discussed in Section E, this model does not include all estuarine and coastal waters, and as a result, does not account for the entire U.S. population. It is estimated that about one-third of the U.S. population is not accounted for in the model. At the same time, if a person is located near two rivers, that person is counted twice, since he or she can derive a benefit from environmental improvements in either river.

Recognizing these limitations of this accounting procedure, the model accounts for 197.7 million people in 23,821 reaches. In the 311 reaches analyzed here (1.3 percent of reaches in the model), the model accounts for 13.7 million people (6.9 percent of the population in the model). The ratio of the percent population to percent reaches in the model demonstrates that the screening process developed for this analysis is reasonably successful in finding reaches with data near urban centers, although 57 of the 311 reaches did not have population associated with them. Of the 13.7 million people represented by the 311 reaches, 11.8 million of them (86 percent) are associated with reaches that have an increased worst-case DO from before to after the CWA. Almost one-half (45 percent) of the selected population are associated with

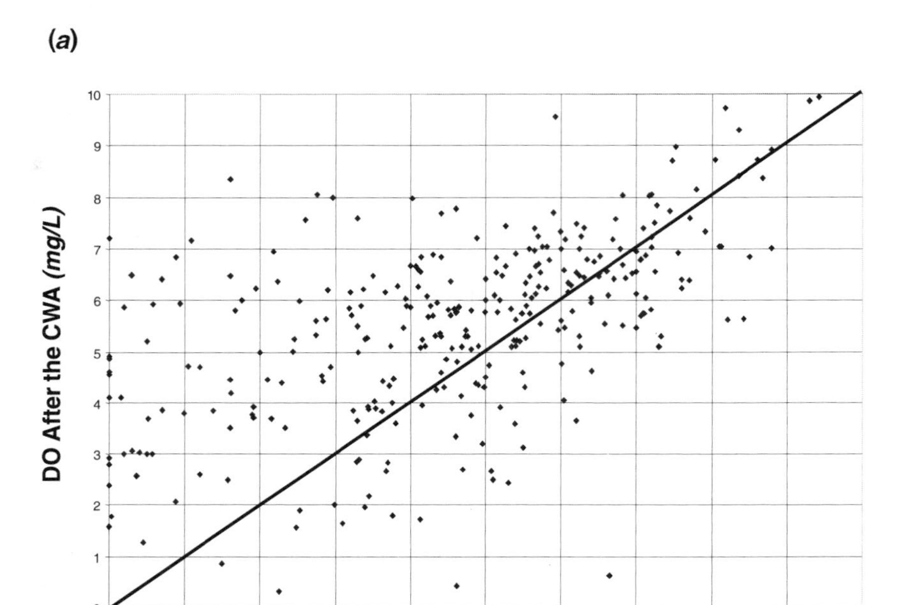

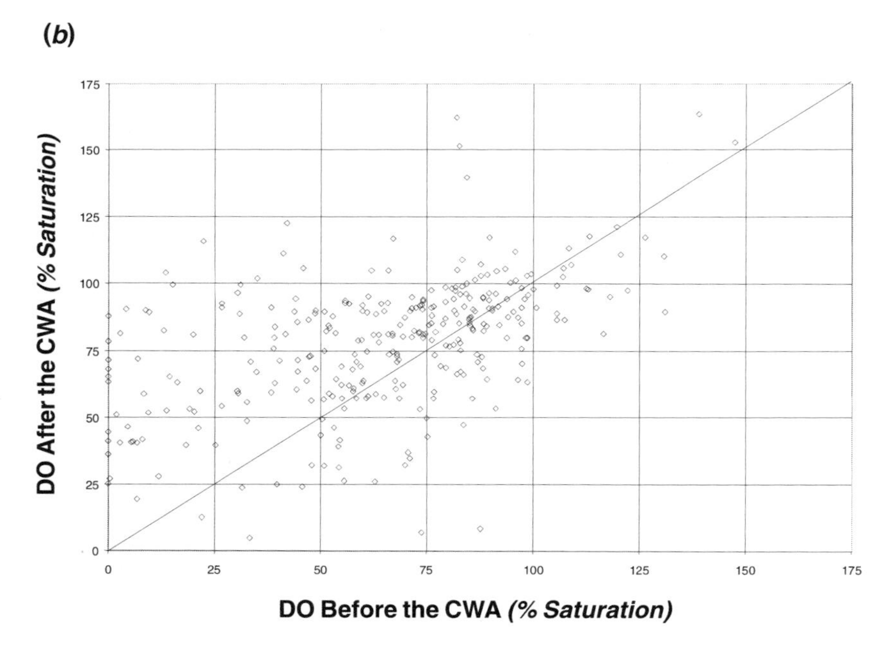

Figure 3-52 Comparison of the tenth percentile DO before and after the CWA: (*a*) as a function of DO concentration as mg/L; and (*b*) as a function of DO saturation as percent saturation. *Source:* USEPA STORET.

reaches that went from worst-case DO below 5 mg/L before the CWA to greater than 5 mg/L after the CWA. Although it is unfortunate that more reaches are not considered in the current analysis (mainly because of limitations in available monitoring data for the before-CWA periods as shown in Figure 3-2), it is helpful to consider that the corresponding 246 catalog units include 61.6 million (31.2 percent) of the 197.7 million people accounted for in the model. And three-fourths (46.5 million) of the 61.6 million people are in catalog units that had an increase in worst-case DO between the before and after time periods.

Sensitivity of Evaluation to Using DO Concentration Versus Percent Saturation The beginning of this chapter describes the physical processes associated with atmospheric reaeration, oxygen demand, and dilution, as well as the impact of water temperature and elevation. During the initial development of the screening methodology, considerable effort was spent evaluating various indicators for water quality. Ultimately, DO concentration was selected. Another strong candidate was DO expressed as percent saturation. Use of percent saturation would effectively normalize the DO data to account for geographic differences in elevation, chlorides, and water temperature. Saturation levels of DO decrease with higher elevations, increasing chloride content, and warmer water temperatures (Chapra, 1997). Correcting for elevation would improve spatial comparisons such as those in Figures 3–19 through 3–21, and correcting for chlorides and water temperature would account for some of the unexplained variability that might exist between the before and after time periods.

To evaluate the impact that selection of DO concentration over DO as percent saturation might have on the analysis, two scatter plots with data aggregated to the reach level were compared. Figure 3-52*a* presents the DO (concentration) after the CWA as a function of DO (concentration) before the CWA. Figure 3-52*b* presents the DO (percent saturation) after the CWA as a function of DO (percent saturation) before the CWA aggregated to the reach level. The values for DO (percent saturation) were computed using the same procedure used for DO (concentration). Points above the diagonal line in either figure indicate that the DO (concentration) or DO (percent saturation) increased. Although the two figures use different scales, a visual comparison suggests that there would be little difference if DO (percent saturation) were adopted over DO (concentration). Given that the public presumably has a more intuitive understanding of DO measured as concentration, the analysis in this chapter uses DO concentration rather than percent saturation as an appropriate indicator.

REFERENCES

Ackerman, W. C., R. H. Harmeson, and R. A. Sinclair. 1970. Some long-term trends in water quality of rivers and lakes. Trans. Amer. Geophys. Union 51:516–522.

Alexander, R. B., J. R. Slack, A. S. Ludtke, K. K. Fitzgerald, and T. L. Schertz, 1998. Data from selected U.S. Geological Survey national stream water quality monitoring networks. Water Resources Research, 34(9):2401–2405.

Bondelid, T., S. Unger, and A. Stoddard. 2000. National water pollution control assessment model (NWPCAM) Version 1.1. Final report prepared by Research Triangle Institute, Research Triangle Park, NC for U.S. Environmental Protection Agency, Office of Policy, Economics and Innovation, Washington, DC, November, RTI Project Number 92U-7640-031.

Chapra, S. C. 1997. Surface water quality modeling. McGraw-Hill, New York. 844 pp.

Crawford, C. G. and D. J. Wangness. 1991. Effects of advanced wastewater treatment on the quality of White River, Indiana. Water Resources Bull. 27(5): 769–779.

Gunnerson, C. G. 1966. An atlas of water pollution surveillance in the United States, October 1, 1957 to September 30, 1965. Water Quality Activities, Division of Pollution Surveillance, Federal Water Pollution Control Adminstration, U. S. Department of the Interior, Cincinnati, OH.

Hines, W. G., D. A. Rickert, and S. W. McKenzie, 1976. Hydrologic Analysis and River-Quality Data Programs, River-quality assessment of the Willamette River Basin, Oregon. U. S. Geological Survey Circular 715-D, U. S. Department of the Interior.

Knopman, D. S. and R. A. Smith. 1993. Twenty years of the Clean Water Act: Has U. S. water quality improved? Environment 35(1):17–41.

Smith, R. A., R. B. Alexander, and M. G. Wolman. 1987a. Analysis and interpretation of water quality trends in major U. S. rivers, 1974–81. Water-Supply Paper 2307. U. S. Geological Survey, Reston, VA.

Smith, R. A., R. B. Alexander, and M. G. Wolman. 1987b. Water quality trends in the nation's rivers. Science 235 (27 March): 1607–1615.

Thomann, R. V. and J. A. Mueller. 1987. Principles of surface water quality modeling and control. Harper & Row, New York. 644 pp.

USEPA. STOrage and RETrieval Water Quality Information System. U. S. Environmental Protection Agency, Office of Wetlands, Oceans, and Watersheds, Washington, DC.

USEPA. 1974. National water quality inventory, 1974. EPA-440/9-74-001. U. S. Environmental Protection Agency, Office of Water Planning and Standards, Washington, DC.

Wolman, A. 1971. The nation's rivers. Science 174:905–918.

CHAPTER 4

Case Study Assessments of Water Quality

In the previous chapter, the national-scale evaluation of long-term trends in water quality conditions identified numerous waterways that were characterized by substantial improvements in worst-case DO after the CWA (from 1961–1965 to 1986–1990). The signals of worst-case DO improvements that have been detected from the "noise" of the STORET database document the tremendous progress that has been achieved as a result of implementation of the effluent control regulations of the CWA in 1972. Having identified numerous watersheds and RF1 reaches, however, the inquisitive reader could easily list a number of questions to fill in the information needed to tell a more complete history about environmental management and water pollution control decisions in these watersheds.

Typical questions might include the following: What are the population trends? Are point or nonpoint sources the largest component of pollutant loading? What have been the long-term trends in effluent loading from municipal and industrial sources over the past 25–50 years? Has industrial wastewater loading declined because obsolete manufacturing facilities have been abandoned? What have been the long-term trends in key water quality parameters over the past 25–50 years? Have reductions in wastewater loads had any impact on biological resources or recreational activities?

This third leg of the three-legged stool approach focuses on answering these types of questions. The uniqueness of each watershed requires an investigator to go beyond STORET and other centralized databases to identify, obtain, and compile sufficient historical data to answer these questions and others. By necessity, the selection of specific waterways based on case studies has often been used as an appropriate technique for policy evaluations of the environmental effectiveness of water pollution control decisions. That technique is used in Chapters 5 through 13 of this study.

A. BACKGROUND

Less than a decade after enactment of the 1972 CWA, Congress and the public began to raise policy questions about the national-scale effectiveness of the technology-based controls of the CWA. In attempting to provide some answers to these questions, case studies of water pollution control and water quality management were compiled for a number of streams, rivers, lakes, and estuarine waterbodies. To meet a variety of objectives, both anecdotal and quantitative data and information have been collected for case studies evaluating water quality conditions.

Anecdotal accounts of historical water pollution problems and changes in the water quality of streams, rivers, estuaries, and coastal waters that had been achieved by the early 1980s were reported by state agencies and compiled by USEPA (1980) and the Association of State and Interstate Water Pollution Control Administrators (ASIWPCA, 1984). Twenty-five years after enactment of the 1972 CWA, USEPA (1997) and the Water Environment Federation (WEF, 1997) reported on the substantial water quality improvements that had been achieved in rivers, lakes, estuaries, and coastal waters. Based on anecdotal evidence, these reports concluded that the CWA had produced substantial gains in water quality. No quantitative data were presented, however, in either of these reports to support the conclusion that the goals of the CWA were being achieved.

In a 1988 quantitative synthesis of before-and-after studies, USEPA (1988) compiled the results of 27 case studies to document water quality changes that had resulted from upgrades to municipal wastewater treatment facilities (primary to secondary, or secondary to advanced treatment). With the exception of only a few cases (e.g., Potomac estuary near Washington, DC, and Hudson River near Albany, New York), most of the 27 cases accounted for both minor and major facilities (< 0.1 to 30 mgd) discharging to small receiving waters with 7Q10 low flows ranging from < 1 cfs to 100 cfs. Based on pollutant loading and water quality data sets, 23 of the 27 case studies were characterized by at least moderate improvements in water quality conditions after upgrades of the POTWs. Included in USEPA's 1988 synthesis were the well-documented before-and-after findings of Leo et al. (1984), based on 13 case studies of water quality changes that were linked to upgrades from secondary to advanced treatment. Also included in USEPA's synthesis were four case studies prepared by GAO (1986a) of municipal wastewater treatment plant upgrades for rivers in Pennsylvania: Lehigh River, Allentown (30 mgd); Neshaminy Creek, Lansdale (2.36 mgd); Little Schuykill River, Tamaqua (1.09 mgd); and Schuykill River, Hamburg (0.46 mgd).

A number of case studies other than those presented in this book have documented trends in improvements in water quality conditions and biological resources following site-specific upgrades. Estuarine case studies of pollutant loading, water quality trends, fisheries, and other biological resources have been prepared for Narragansett Bay (Desbonnet and Lee, 1991), Galveston Bay (Stanley, 1992a), the Houston Ship Channel (EESI, 1995), and Pamlico-Albemarle Sound (Stanley, 1992b).

For Lake Washington in Seattle, Edmondson (1991) documented the long-term ecological impact of the diversion during the mid-1960s of municipal wastewater on cultural eutrophication and recovery of a large urban lake. The rejuvenation of Lake Erie, declared "dead" during the 1960s, is positive evidence that the regulatory controls of the 1972 Clean Water Act and the 1972 Great Lakes Water Quality Agreement between Canada and the United States, designed to mitigate bottom water hypoxia and cultural eutrophication by reducing pollutant loads of organic matter and phosphorus, have been successful in greatly improving water quality (Burns, 1985; Charlton et al., 1995; Sweeney, 1995) and ecological conditions (Krieger et al., 1996; Koonce et al., 1996; Makarewicz and Bertram, 1991) in this once ecologically devastated lake. The Cuyahoga River, a major tributary to Lake Erie at Cleveland, Ohio, sparked national attention when the river caught fire in 1969, helping to push the U.S. Congress to pass the Clean Water Act in 1972 (NGS, 1994). Three decades after the infamous fire, although some water quality problems remain to be solved (e.g., urban runoff and CSOs), water quality is greatly improved. Tourist-related businesses and recreational uses along the riverfront are thriving, as are populations of herons, salmon, walleye, and smallmouth bass (Hun, 1999; Brown and Olive, 1995).

In freshwater river systems, Isaac (1991) presented long-term trends (1969–1980) of DO in the Blackstone, Connecticut, Hoosic, and Quinebaug rivers in Massachusetts to document water quality improvements after upgrades of municipal facilities to secondary treatment. Using a wealth of historical data compiled for New England, Jobin (1998) presents a number of case studies documenting long-term trends in pollutant loading and water quality for freshwater rivers (e.g., Neponset, Charles, Taunton, Blackstone) and estuarine systems (e.g., Boston Harbor, Narragansett Bay). In the Midwest, Zogorski et al. (1990) prepared a case study of the Upper Illinois River basin to evaluate the availability and suitability of water quality and effluent loading data as a demonstration of the methodology for use in national assessments of water quality trends. Zogorski et al. concluded that, although a large amount of the required data is available from national and state databases, "the suitability of the existing data to accomplish the objectives of a national water-quality assessment is limited."

In another midwestern river, a statistical before-and-after analysis of water quality in the White River near Indianapolis, Indiana, clearly showed improvements in DO, ammonia, and BOD_5 after an upgrade from secondary to advanced treatment (Crawford and Wangness, 1991). (See discussion in section C of Chapter 3.) Similar water quality improvements have also been documented for the Flint River in Georgia and the Neches River in Texas (Patrick et al., 1992). Becker and Neitzel (1992) have compiled case studies of the impacts from water pollution and other human activities on water quality, fisheries, and biological resources for a number of major North American rivers. Another success story in the Pacific Northwest has documented both water quality and economic benefits achieved by water pollution control in the Boise River in Idaho (Hayden et al., 1994; Noah, 1994).

B. SELECTION OF CASE STUDY WATERWAYS

Following the precedent established by these earlier before-and-after assessments of changes in water quality that can be attributed, in part, to the effluent control regulations of the CWA, a number of freshwater and estuarine waterbodies were selected as case studies for this report. Criteria for the selection of case study sites included the following:

- The major river or estuarine system was identified in the 1960s as having gross water pollution problems.
- The major river or estuarine system lies in a major urban-industrial region.
- Municipal wastewater is a significant component of the point source pollutant load to the system.
- Water quality models were available to evaluate the water quality impact of simulated primary, secondary, and actual effluent scenarios for municipal dischargers.
- Historical data were readily available.

Table 4-1 provides the 1996 population for the Metropolitan Statistical Areas (MSAs) and counties included in the case study, and the types of data and information compiled for each river or estuarine waterbody selected as a case study. The population of the case study MSAs (43.2 million) accounted for 16 percent of the nation's total population in 1996 (265.2 million) (USDOC, 1998). Figure 4-1 shows the location of the case study watersheds. In contrast to some of the other case study assessments discussed previously, the case studies in this report were specifically selected because they represent large cities located on major waterways known to have been plagued by serious water pollution problems during the 1950s and 1960s (Table 4-2). Many of the case study waterways either were the sites of interstate enforcement conferences from 1957 to 1972 or were listed by the federal government as being potential waterways to convene state-federal enforcement conferences in 1963 (Zwick and Benstock, 1971) (see Appendix A). Two of the case studies, the Ohio River and tributaries to New York Harbor (Passaic River and Arthur Kill), were identified by the federal government in 1970 in a list of the top ten most polluted rivers (Zwick and Benstock, 1971). The Department of the Interior identified all the estuarine case study sites as waterways suffering from either low oxygen levels or bacterial contamination in a national study of estuarine water quality (USDOI, 1970). All but two of the case study areas were the subject of water quality evaluation reports prepared for the National Commission on Water Quality (NCWQ) to provide baseline data to track the effectiveness of the technology-based effluent controls required under the newly enacted 1972 CWA (see Mitchell, 1976). With the exception of the James River, enforcement actions were initiated by the Federal government for all the case study rivers (see Appendix A-4).

For all the case studies, data have been compiled to characterize long-term trends (more than 50 years) beginning in 1940 for population, upgrades to municipal wastewater facilities, effluent loading, water quality, environmental resources, and recreational uses. Additional data have been obtained from validated water quality models for the Upper Mississippi River, Potomac estuary, Delaware estuary, and James estuary to quantify improvements in water quality achieved by municipal upgrades from primary to secondary or advanced treatment levels. Data sources include published scientific and technical literature, USEPA's STORET database, and unpublished technical reports ("grey" literature) prepared by consultants and state, local, and federal agencies.

C. BEFORE AND AFTER CWA

Using water quality data extracted from USEPA's STORET database (as described in Chapter 3), before-and-after conditions for summer (July–September), tenth percentile DO levels in RF1 reaches selected from the case study watersheds (Figure 4-1) clearly demonstrate dramatic improvements during the period after the CWA from 1986–1995 for all the case study sites (Figure 4-2). Before the CWA, during the 10-year period from 1961 to 1970, "worst-case" DO levels were in the range of 1 to 4 mg/L for most of the case study sites. After the CWA, worst-case DO levels had improved substantially to levels of about 5 to 8 mg/L during 1986–1995, with the worst-case oxygen levels of less than 2 mg/L before the CWA improving to 5 mg/L or higher after the

TABLE 4-1 Case Study Assessments of Trends in Water Quality and Environmental Resources

Case Study	Study Area Population, 1996	Information Presented					
		Population	Pollutant Loads	Water Quality	Environmental Resources	Recreational Uses	Water Quality Model
Connecticut River	1,109,000	X	X	X	X	X	
Hudson-Raritan estuary	16,991,000	X	X	X	X	X	
Delaware estuary	5,973,000	X	X	X	X	X	X
Potomac estuary	4,635,000	X	X	X	X	X	X
James estuary	2,237,000	X	X	X	X	X	X
Chattahoochee River	3,528,000	X	X	X	X	X	
Ohio River	3,779,000	X	X	X	X	X	
Upper Mississippi River	2,760,000	X	X	X	X	X	X
Willamette River	2,149,000	X	X	X	X	X	

Source: Study area population data from USDOC, 1998.

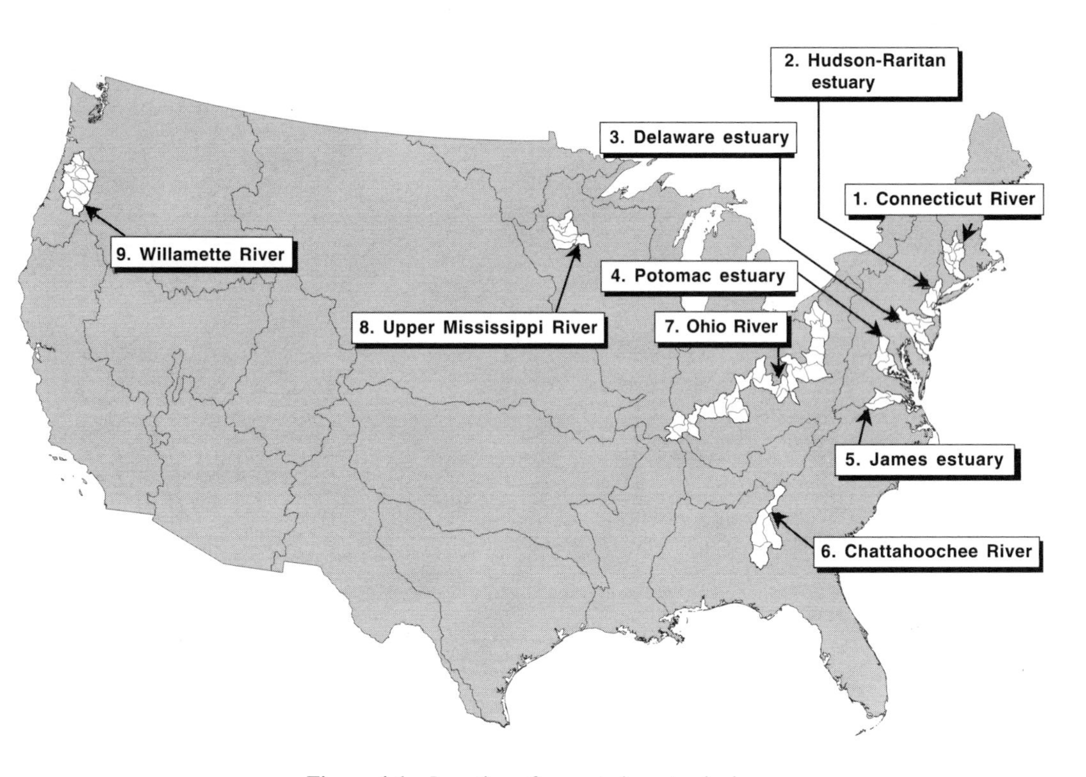

Figure 4-1 Location of case study watersheds.

TABLE 4-2 Identification of Water Pollution Problems for Case Study Waterways

Case Study	Potential Enforcement Conference 1963	Enforcement Conference 1957–1972	Top Ten Polluted Waterways 1970	NCWQ Case Studies 1976	National Estuarine Pollution Study 1970	Enforcement Actions 1957–1972
Connecticut River	X	X		X	X	X
Hudson-Raritan estuary	X	X	X	X	X	X
Delaware estuary				X	X	X
Potomac estuary	X	X		X	X	X
James estuary					X	
Chattahoochee River	X	X		X		X
Ohio River	X	X	X	X		X
Upper Mississippi River	X	X		X		X
Willamette River						X

Sources: Zwick and Benstock, 1971; USDOI, 1970; Mitchell, 1976.

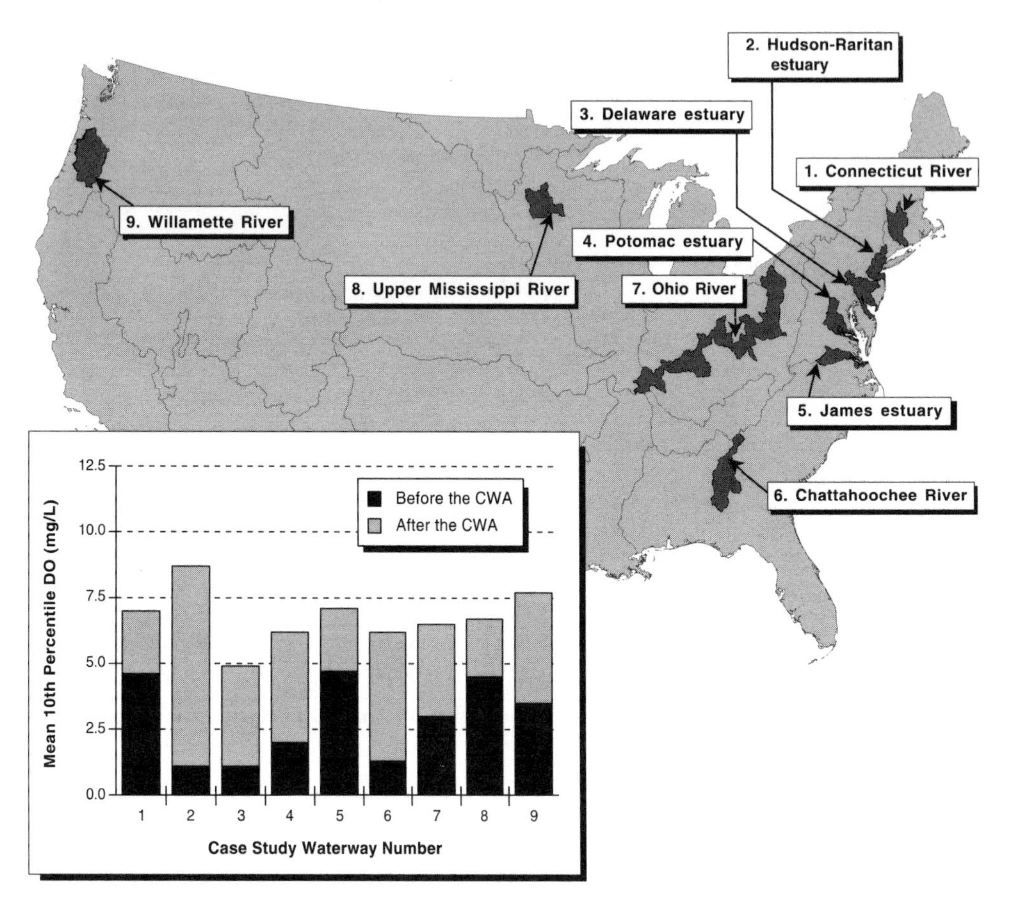

Figure 4-2 Location map of case study waterways and distribution chart of their before- and after-CWA mean tenth percentile DO for case study RF1 reaches: 1961–1970 versus 1986–1995. *Source:* USEPA STORET.

CWA. Great progress has been achieved in improving DO conditions in New York Harbor, the Chattahoochee River, the Delaware River, and the Potomac River.

Water quality improvements in other constituents, including BOD_5, suspended solids, coliform bacteria, heavy metals, nutrients, and algal biomass, have also been linked to reductions in municipal and industrial point source loads for many of the case studies. Figure 4-3 correlates long-term trends in the reduction of effluent loading of BOD_5 with improvements in summer DO in the Hudson-Raritan estuary (New York City), the Upper Potomac estuary (Washington, DC), the Upper Mississippi River (Minneapolis–St. Paul, MN), and the Willamette River (Portland, OR). Finally, improvements in water quality have also been linked to the post-CWA restoration of important biological resources (e.g., fisheries and submersed aquatic vegetation in the Potomac estuary) and increased recreational demand and aesthetic values of waterways once considered extremely unsightly (e.g., Upper Mississippi River).

D. POLICY SCENARIOS FOR MUNICIPAL EFFLUENT DISCHARGES

Before the 1972 CWA, state officials made waterbody-dependent decisions about the required level of municipal wastewater treatment needed to attain compliance with ambient water quality criteria or standards. After the 1972 CWA, the USEPA implemented a technology-based policy to regulate pollutant loading from municipal and industrial point sources. Under the 1972 CWA, municipalities were required to achieve at least a minimum level of secondary treatment to remove approximately 85 percent of the oxygen-demanding material from wastewater. In cases where the minimum level of secondary treatment was not sufficient to meet water quality criteria or standards, ambient criteria were used to determine a water quality–based level of wastewater treatment greater than secondary treatment. From a policy and planning perspective, the key question for water quality management decision makers is: What level of municipal wastewater treatment is needed to ensure compliance with water quality criteria or standards under critical conditions?

For the Delaware, Potomac, James, and Upper Mississippi case studies, validated water quality models have been used to provide quantitative answers to evaluate the changes in water quality conditions achieved as a result of either actual or hypothetical upgrades to municipal wastewater treatment facilities. Effluent loading rates for the primary and secondary loading scenarios were based on existing population served and effluent flow data with typical effluent concentrations characteristic of primary and secondary treatment facilities; existing loading rates were used to define the better-than-secondary (actual) scenario. Receiving water streamflow was based on the existing "dry" summer streamflow measurements used to validate the models. The water quality models were used to simulate the impact of the primary, secondary, and actual better than secondary loading scenarios on the spatial distributions of DO, BOD_5, nitrogen, phosphorus, and algal biomass.

Figure 4-4 summarizes the key results for the model simulations for dissolved oxygen simulated at the worst-case critical oxygen sag location along the length of each river. As shown in these results, the primary effluent scenario results in ex-

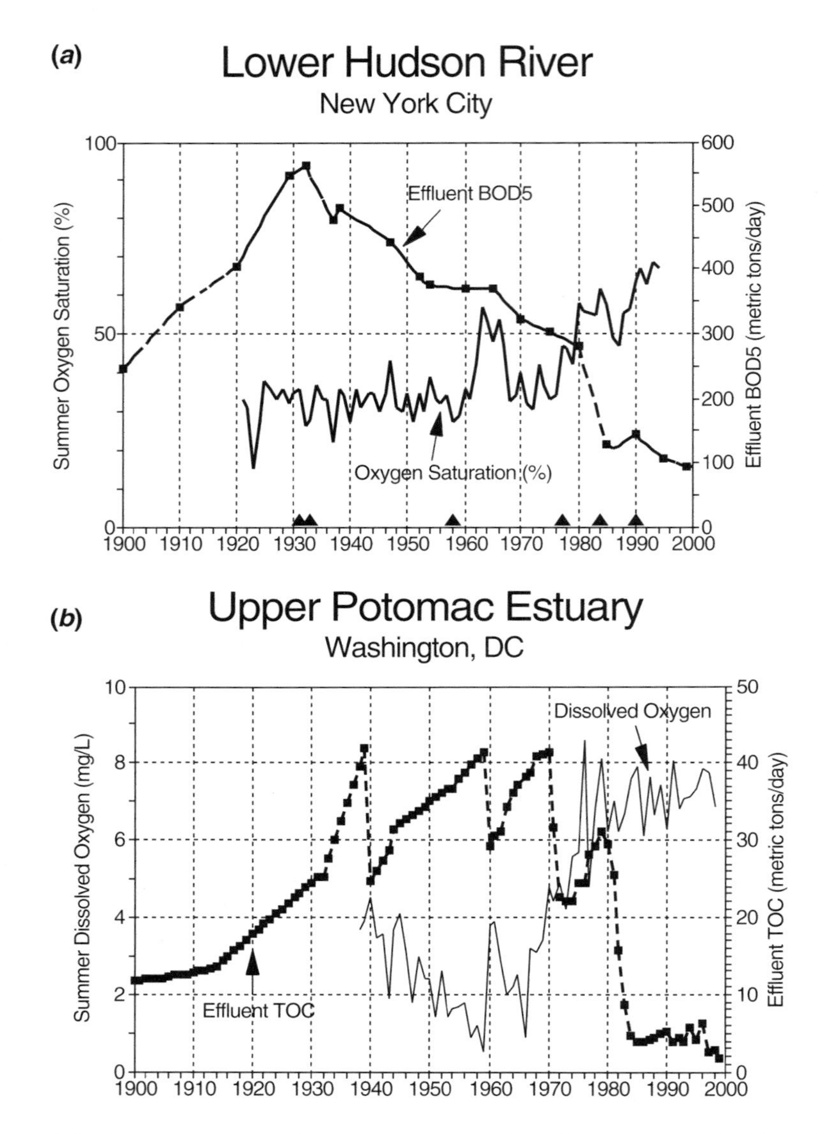

Figure 4-3 Long-term trends of improvements in ambient DO and declines in effluent BOD_5 and TOC loading: (*a*) for Hudson-Raritan estuary; (*b*) for Potomac estuary.

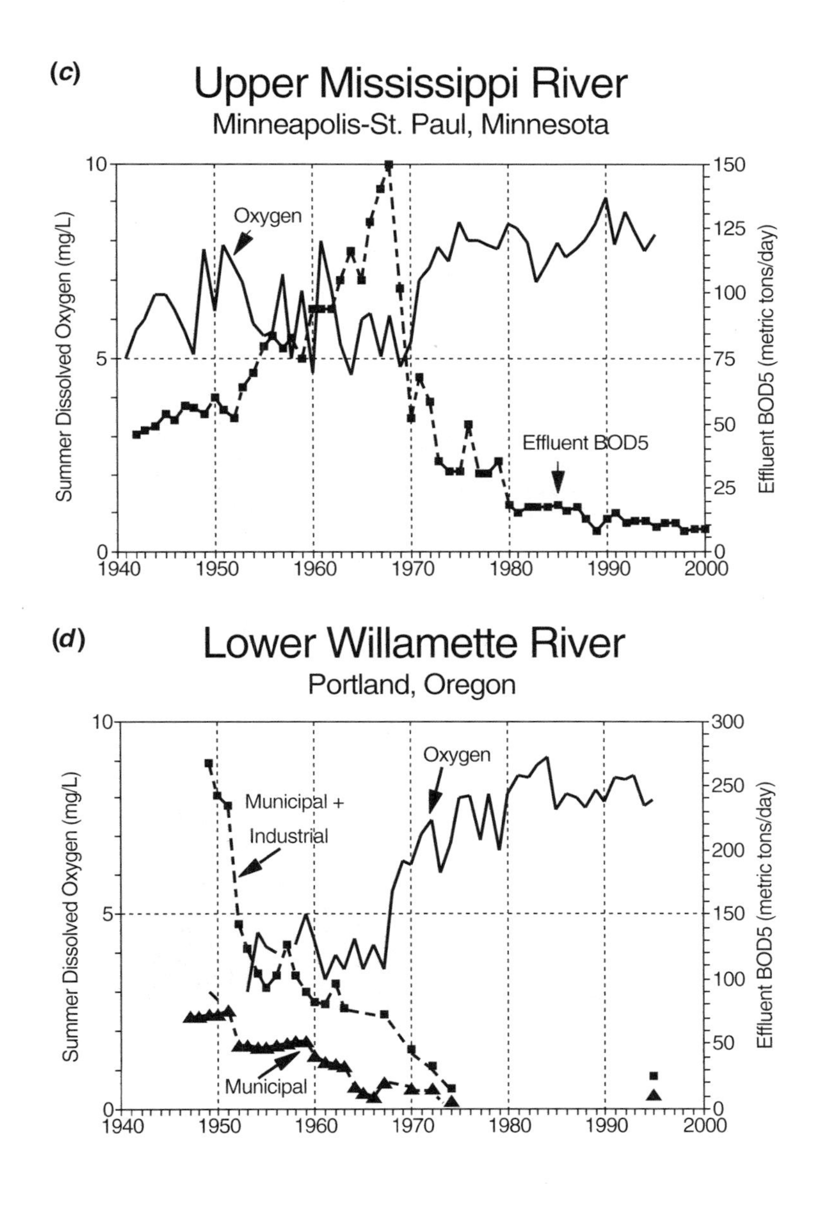

Figure 4-3 ***(continued)*** (*c*) for Upper Mississippi River; and (*d*) for Willamette River. *Sources:* Brosnan and O'Shea, 1996; Brosnan et al., 2002; Hetling et al., 2001; Johnson and Aasen, 1989; Larson, 1999, 2001; Gleeson, 1972; Jaworksi, 1990, 2001; MWCOG, 1989; ODEQ, 1970; USEPA STORET; Bondelid et. al., 2000.

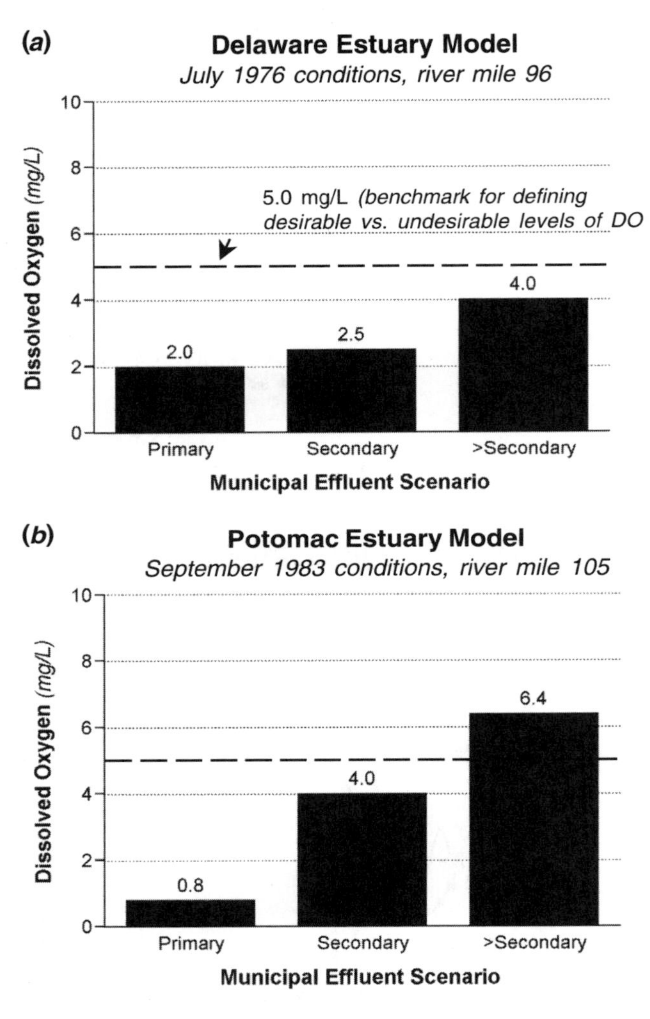

Figure 4-4 Model simulation of DO under summer "dry" streamflow conditions at the critical oxygen sag location for primary, secondary, and better-than-secondary effluent scenarios for case studies of: (*a*) Delaware estuary; (*b*) Potomac estuary.

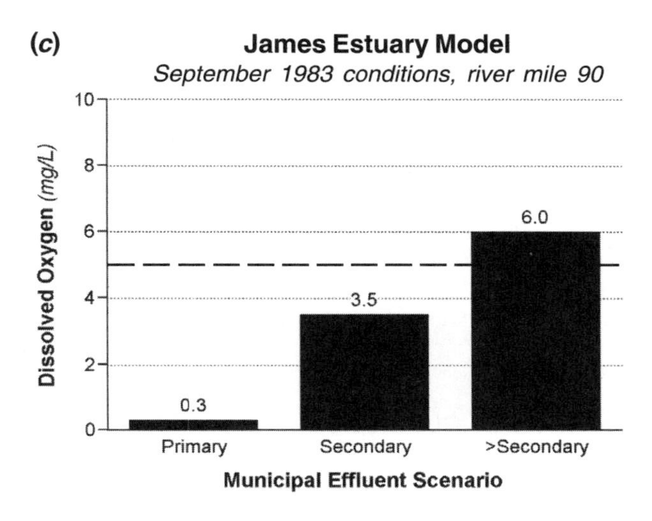

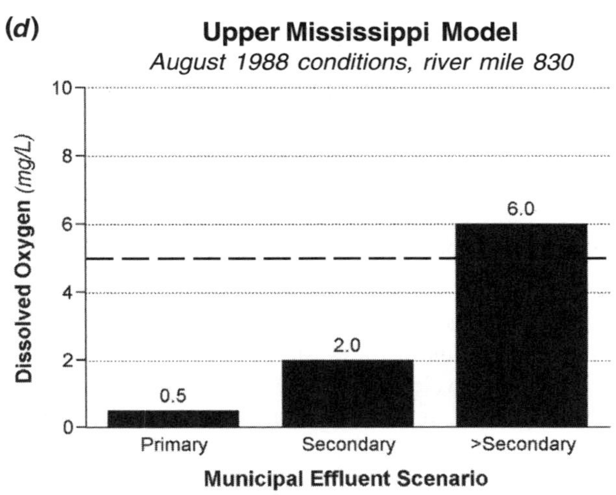

Figure 4-4 *(continued)* (*c*) James estuary; and (*d*) Upper Mississippi River. *Sources:* Lung, 1991a; Fitzpatrick and Di Toro, 1991; Fitzpatrick, 1991; HydroQual, 1986; Lung, 1998; Lung, 1991b.

tremely poor conditions with DO levels of less than 1 mg/L for the Potomac, James, and Upper Mississippi cases, and 2 mg/L for the Delaware.

The model results for the primary scenario of severe oxygen depletion are, in fact, consistent with historical oxygen data recorded for these rivers during the 1960s. Simulating an upgrade to secondary treatment, as mandated by the 1972 CWA for municipal facilities, DO conditions are improved but are still less than the benchmark concentration of 5 mg/L often used to describe compliance with water quality standards. As demonstrated with the models, and actually achieved, better than secondary levels of municipal treatment are needed to exceed a benchmark of 5 mg/L for DO. In contrast to the poor water quality conditions common in these rivers during the 1960s, the occurrence of low DO levels has been effectively eliminated, even under severe drought conditions, as a result of upgrades beyond primary treatment to better than secondary levels of wastewater treatment.

E. DISCUSSION AND CONCLUSIONS

In developing a methodology to evaluate the effectiveness of USEPA's Construction Grants Program, GAO (1986b) posed four questions to evaluate the water quality benefits obtained from upgrading municipal wastewater treatment facilities:

1. Did upgrading the POTW decrease the amount of pollutants discharged?
2. Did water quality improve downstream from the POTW?
3. Is there a relationship between changes in a plant's effluent and changes in stream water quality indicators?
4. Can other reasonable explanations of a stream's water quality be excluded?

Although many of the case studies in this report (Chapters 5 through 13) include a mix of multiple municipal and industrial wastewater discharges and might not be applicable to the methodology developed by GAO (1986b), the dramatic improvements that have been documented for effluent loading, water quality, environmental resources, and recreational uses clearly suggest that the answer to the questions raised by GAO (1986b) for all nine case studies is an overwhelming "yes."

In addition to the case study questions posed by GAO, the national policy questions raised by Congress and the public can be modified slightly to use for evaluations of the case study waterways: Has water quality improved as a result of public and private capital improvement expenditures for water pollution control? Has the waterbody achieved the "fishable and swimmable" goals set forth in the CWA? Has the CWA worked?

For all the case study waterways, tremendous progress has been made in improving water quality, restoring valuable biological resources, and creating thriving water-based recreational uses of the waterways that contribute to the local economies. Although significant progress has been achieved in eliminating noxious water pollution conditions, nutrient enrichment, and sediment contamination, heavy metals and toxic organic chemicals continue to pose threats to human health and aquatic organisms. Serious ecological problems remain to be solved for many of the nation's waterways, including the case study sites. The evidence is overwhelming, however, that the national water pollution control policy decisions of the 1972 CWA have achieved significant successes in many waterways. With the new watershed-based strategies for managing pollutant loading from point and nonpoint sources detailed in USEPA's Clean Water Action Plan (USEPA, 1998), the nation's state-local-private partnerships will continue to work to attain the original "fishable and swimmable" goals of the 1972 CWA for all surface waters of the United States.

REFERENCES

ASIWPCA. 1984. America's clean water: The states evaluation of progress 1972–1982. Association of State and Interstate Water Pollution Control Administrators, Washington, DC.

Becker, C. D. and D. A. Neitzel, eds. 1992. Water quality in North American river systems. Battelle Press, Columbus, OH.

Bondelid, T., S. Unger, and A. Stoddard. 2000. National water pollution control assessment model (NWCPCAM) Version 1.1. Final report prepared by Research Triangle Institute, Research Triangle Park, NC for U.S. Environmental Protection Agency, Office of Policy, Economics and Innovation, Washington, DC, November, RTI Project Number 92U-7640-031.

Brosnan, T. M. and M. L. O'Shea. 1996. Long-term improvements in water quality due to sewage abatement in the Lower Hudson River. Estuaries 19(4): 890–900.

Brosnan, T., A. Stoddard, and L. Hetling. 2002. Sewage inputs and impacts: Past and present. In: The Hudson River Ecosystem, ed. J. Levinton, Section 4: Human impacts on the Hudson River. Book in preparation for publication by Hudson River Foundation, New York.

Brown, B. J. and J. H. Olive. 1995. Diatom communities in the Cuyahoga River (USA): Changes in species composition between 1974 and 1992 following renovations in wastewater management. Ohio J. Sci. 95(3): 254–260.

Burns, N. M. 1985. Erie, the lake that survived. Rowman & Allanheld Publishers, Division of Littlefield, Adams & Co., Totowa, NJ.

Charlton, M. N., S. L'Italien, E. S. Millard, R. W. Bachmann, J. R. Jones, R. H. Peters, and D. M. Soballe, eds., 1995. Recent changes in Lake Erie water quality. Annual International Symposium of the North American Lake Management Society, Toronto, Ontario, Canada, November 6–11. Lake Reservoir Management 11(2): 125.

Clark, L. J., R. B. Ambrose, and R. C. Crain. 1978. A water quality modeling study of the Delaware Estuary. Technical report 62. U.S. Environmental Protection Agency, Region 3, Annapolis Field Office, Annapolis, MD.

Crawford, C. G. and D. J. Wangness. 1991. Effects of advanced wastewater treatment on the quality of White River, Indiana. Water Resources Bull. 27(5): 769–779.

Desbonnet, A. and V. Lee. 1991. Historical trends: Water quality and fisheries in Narragansett Bay. Rhode Island Sea Grant, University of Rhode Island, Narragansett, RI, and U. S. Department of Commerce, National Oceanic and Atmospheric Administration, Coastal Ocean Office, National Ocean Pollution Program, Rockville, MD.

Edmondson, W. T. 1991. The uses of ecology, Lake Washington, and beyond. University of Washington Press, Seattle, WA.

EESI. 1995. The Houston Ship Channel: An environmental success story? Proceedings of workshop, May 18–19. Publ. No. EESI-01. Energy & Environmental Systems Institute, Rice University, Houston, TX.

Fitzpatrick, J. J. and D. M. DiToro. 1991. Development and calibration of a two functional algal group model of the Potomac Estuary. Final Report to the Metropolitan Washington Council of Governments, Washington, DC. Prepared by HydroQual, Inc., Mahwah, NJ.

Fitzpatrick, J. P. 1991. Trends in BOD/DO modeling for wasteload allocations of the Potomac estuary. Technical memorandum prepared under sub-contract for Tetra Tech, Inc., Fairfax, VA, by HydroQual, Inc., Mahwah, NJ, September 18.

Forstall, R. L. 1995. Population by counties by decennial census: 1900 to 1990. U. S. Bureau of the Census, Population Division, Washington, DC. www.census.gov/population/www/censusdata/cencounts.html

GAO. 1986a. The nation's water: Key unanswered questions about the quality of rivers and streams. GAO/PMED-86-6. U. S. General Accounting Office, Program, Evaluation and Methodology Division, Washington, DC.

GAO. 1986b. Water quality: An evaluation method for the Construction Grants Program—methodology. Report to the Administrator, U. S. Environmental Protection Agency. Vol. 1, GAO/PMED-87-4A. U. S. General Accounting Office, Program, Evaluation and Methodology Division, Washington, DC.

Gleeson, G. W. 1972. The return of a river, the Willamette River, Oregon. Oregon State University, Corvallis, OR.

Hayden, D. F. G., J. L. Peter, W. C. Peterson, and D. A. Ball. 1994. Valuation of non-market goods: A methodology for public policy; making a social fabric matrix cultural and environmental interaction model. Draft working paper. Boise State University, College of Business, Boise, Idaho.

Hetling, L. J., A. Stoddard, D. Hammerman, and T. Brosnan. 2001. Changes in wastewater loadings to the Hudson River over the past century. Paper presented at New York Water Environment Association (NYWEA) Meeting, New York, February 4.

Hun, T. 1999. Cuyahoga no longer burns. Water Environ. Technol. 11(6): 31–33.

HydroQual, Inc. 1986. Water quality analysis of the James and Appomattox Rivers. Report prepared for Richmond Regional Planning District Commission, Richmond, VA, by HydroQual, Inc., Mahwah, New Jersey. June.

Isaac, R. A. 1991. POTW improvements raise water quality. Water Environ. Technol. June: 69–72.

Jaworski, N. A. 1990. Retrospective study of the water quality issues of the Upper Potomac estuary. Rev. Aquat. Sci. 3(2): 11–40, CRC Press, Boca Raton, FL.

Jaworski, N. 2001. Wakefield, RI. Personal Communication, September 5.

Jobin, W. 1998. Sustainable management for dams and waters. Lewis Publishers, Washington, DC.

Johnson, D. K., and P. W. Aasen. 1989. The metropolitan wastewater treatment plant and the Mississippi River: 50 years of improving water quality. Jour. Minnesota Academy of Science, 55(1): 134–138.

Koonce, J. F., W. D. N. Busch, and T. Czapla. 1996. Restoration of Lake Erie: Contribution of water quality and natural resource management. Can. J. Fish. Aquat. Sci. 53 (Supplement 1): 105–112.

Krieger, K. A., D. W. Schloesser, B. A. Manny, C. E. Trisler, S. E. Heady, J. J. H. Ciborowski, and K. M. Muth. 1996. Recovery of burrowing mayflies (*Ephemeroptera: Ephemeridae: Hexagenia*) in western Lake Erie. J. Great Lakes Res. 22(2): 254–263.

Larson, C. E. 1999. Metropolitan Council Environmental Services, St. Paul, Minnesota. Personal communication, March 12.

Larson, C. E. 2001. Metropolitan Council Environmental Services, St. Paul, Minnesota. Personal communication, August 30.

Leo, W. M, R. V. Thomann, and T. W. Gallagher. 1984. Before and after case studies: Comparisons of water quality following municipal treatment plant improvements. EPA 430/9-007. U. S. Environmental Protection Agency, Office of Water, Program Operations, Washington, DC.

Lung, W. S., 1991a. Trends in BOD/DO modeling for waste load allocations of the Delaware River estuary. Technical memorandum prepared for Tetra Tech, Inc., Fairfax, VA, by Enviro-Tech, Inc., Charlottesville, VA.

Lung, W. S., 1991b. Trends in BOD/DO modeling for waste load allocations of the James River estuary. Technical memorandum prepared under sub-contract for Tetra Tech, Inc., Fairfax, VA, by Enviro-Tech, Inc., Charlottesville, VA.

Lung, W. S. 1998. Trends in BOD/DO modeling for waste load allocations. J. Environ. Eng., ASCE 124(10): 1004–1007.

Lung, W. S. and C. E. Larson. 1995. Water quality modeling of Upper Mississippi River and Lake Pepin. J. Environ. Eng., ASCE 121(10): 691–699.

Lung, W. S. and N. Testerman. 1989. Modeling fate and transport of nutrients in the James estuary. J. Environ. Eng., ASCE 115(5): 978–991.

Makarewicz, J. C. and P. Bertram. 1991. Evidence for the restoration of the Lake Erie ecosystem. Bioscience 41(4): 216–223.

Mitchell, D. 1976. NCWQ reports available. Jour. WPCF 48(10): 2427–2433.

MWCOG. 1989. Potomac River water quality, 1982 to 1986: Trends and issues in the Washington metropolitan area. Annual report. Metropolitan Washington Council of Governments, Department of Environmental Programs, Washington, DC.

NCWQ. 1976. Public Law 92-500. Water quality analysis and environmental impact assessment: Technical report. NTIS No. PB252298. National Commission on Water Quality, Washington, DC.

NGS. 1994. A cleaner Cuyahoga is flammable no more. Earth Almanac. National Geographic Society, Washington, DC. February.

Noah, T. 1994, The river that runs through Boise runs clear once again. The Wall Street Journal, New York, April 22.

ODEQ. 1970. Water quality control in Oregon. Oregon Department of Environmental Quality, Portland, OR. December.

OMB. 1999. OMB Bulletin No. 99-04. U. S. Census Bureau, Office of Management and Budget, Washington, DC. www.census.gov/population/www/estimates/metrodef.html

Patrick, R., F. Douglass, D. M. Palarage, and P. M. Stewart. 1992. Surface water quality: Have the laws been successful? Princeton University Press, Princeton, NJ.

Stanley, D. W. 1992a. Historical trends: Water quality and fisheries: Galveston Bay. Publication No. UNC-SG-92-03. University of North Carolina Sea Grant College Program. Institute for Coastal and Marine Resources, East Carolina University, Greenville, NC.

Stanley, D. W. 1992b. Historical trends: Water quality and fisheries: Albemarle-Pamlico Sound, with emphasis on the Pamlico River estuary. Publication No. UNC-SG-92-04. University of North Carolina Sea Grant College Program. Institute for Coastal and Marine Resources, East Carolina University, Greenville, NC.

Sweeney, R. A. 1995. Rejuvenation of Lake Erie. GeoJournal 35(1): 65–66.

USDOC. 1998. Census of population and housing. Prepared by U. S. Department of Commerce, Economics and Statistics Administration, U. S. Census Bureau, Population Division, Washington, DC.

USDOI. 1970. The national estuarine pollution study. Report of the Secretary of Interior to the U. S. Congress pursuant to Public Law 89-753, the Clean Water Restoration Act of 1966. Document No. 91-58. Washington, DC, March.

USEPA (STORET). STOrage and RETrieval Water Quality Information System. U. S. Environmental Protection Agency, Office of Wetlands, Oceans, and Watersheds, Washington, DC.

USEPA. 1980. National accomplishments in pollution control: 1970–1980. Some case histories. U. S. Environmental Protection Agency. Office of Planning and Management, Program Evaluation Division, Washington, DC.

USEPA. 1988. POTW's and water quality: In search of the big picture. A status report on EPA's ability to address several questions of ongoing importance to the nation's municipal pollution control program. U. S. Environmental Protection Agency, Office of Water, Office of Municipal Pollution Control, Washington, DC. April.

USEPA. 1997. The Clean Water Act: A snapshot of progress in protecting America's waters. Vice President Al Gore's Clean Water Act remarks on the 25th Anniversary of the CWA. U.S. Environmental Protection Agency, Washington, DC.

USEPA. 1998. Clean water action plan: Restoring and protecting America's waters. U. S. Environmental Protection Agency, Office of Water, Washington, DC.

USGS. 1999. Streamflow data downloaded from the United States Geological Survey's National Water Information System (NWIS)-W. Data retrieval for historical streamflow daily values. http://waterdata.usgs.gov/nwis-w

WEF. 1997. Profiles in water quality: Clear success, continued challenge. Water Environment Federation, Alexandria, VA.

Zogorski, J. S., S. F. Blanchard, R. D. Randal, and F. A. Fitzpatrick. 1990. Availability and suitability of municipal wastewater information for use in a national water-quality assessment: A case study of the Upper Illinois River Basin in Illinois, Indiana and Wisconsin. Open File Report 90–375. U. S. Geological Survey, Urbana, IL.

Zwick, D. and M. Benstock. 1971. Water wasteland. Ralph Nader's Study Group report on water pollution. The Center for Study of Responsive Law, Grossman Publishers, New York.

CHAPTER 5

Connecticut River Case Study

The New England Basin (Hydrologic Region 1), covering a drainage area of 64,071 square miles from Maine to southwestern Connecticut, includes some of the major rivers in the continental United States. The Connecticut River, the largest river in New England, originates from a series of small lakes just south of the Canadian border and flows 400 miles south over a drainage area of 11,250 square miles through Vermont, New Hampshire, Massachusetts, and Connecticut to Long Island Sound (Figure 5-1). An estimated 1.1 million people lived in the Lower Connecticut River basin in 1996. Densely populated urban centers border the river from Springfield, Massachusetts, downstream to Middletown, Connecticut. The major urban centers along the river are Holyoke-Chicopee-Springfield, Massachusetts, and Hartford, Connecticut. A diverse mix of manufacturing, trade, finance, agriculture, recreation, and tourism forms the economic base of the basin.

Figure 5-2 highlights the location of the Lower Connecticut River case study watersheds (catalog units) identified in this major river basin as a major urban-industrial area affected by severe water pollution problems during the 1950s and 1960s (see Table 4-2). In this chapter, information is presented to characterize long-term trends in population, municipal wastewater infrastructure and effluent loading of pollutants, ambient water quality, environmental resources, and uses of the Lower Connecticut River. Data sources include USEPA's national water quality database (STORET), published technical literature, and unpublished technical reports ("grey" literature) obtained from local agency sources.

BACKGROUND

Although the Connecticut River has been characterized as one of the nation's most scenic rivers, the river was so grossly polluted in the 1960s that it was classified as suitable only for transportation of sewage and industrial wastes (Conniff, 1990). The deplorable condition of the river discouraged development along the waterfront and adjacent shorelands over long reaches of the lower river. In recent years, improvements in the river's water quality have resulted in the Lower Connecticut River's becoming a popular place for boating and recreation. Perhaps most telling of all, the shorelines of the Connecticut River are now under the new threat of suburban development. The historic turnaround in the quality of the river can be correlated with the

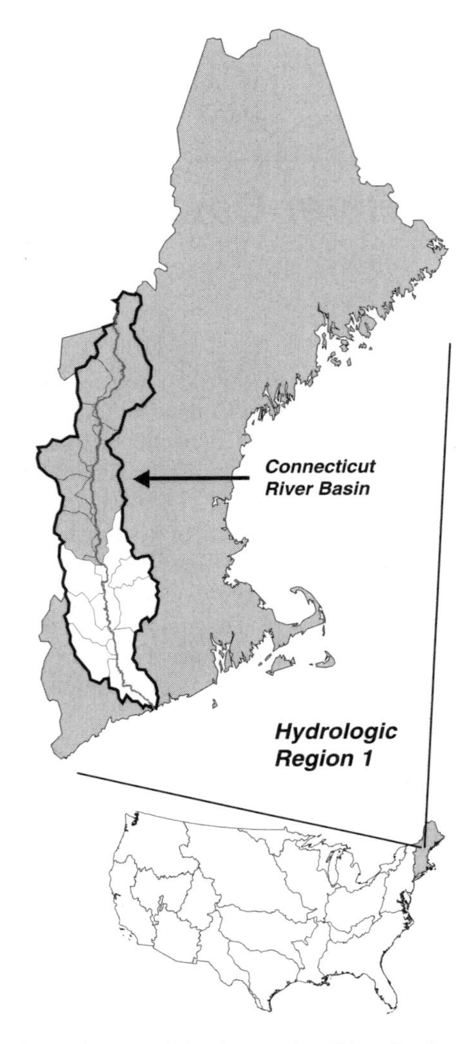

Figure 5-1 Hydrologic Region 1 and the Connecticut River Basin.

enactment of the 1972 CWA, which resulted in the construction and upgrading of wastewater treatment plants along the length of the river, including three major treatment plants serving the Hartford area.

PHYSICAL SETTING AND HYDROLOGY

The Connecticut River forms the border between Vermont and New Hampshire and bisects west-central Massachusetts and central Connecticut. The topography of the Connecticut River's 11,250-square-mile watershed varies from the rugged terrain of

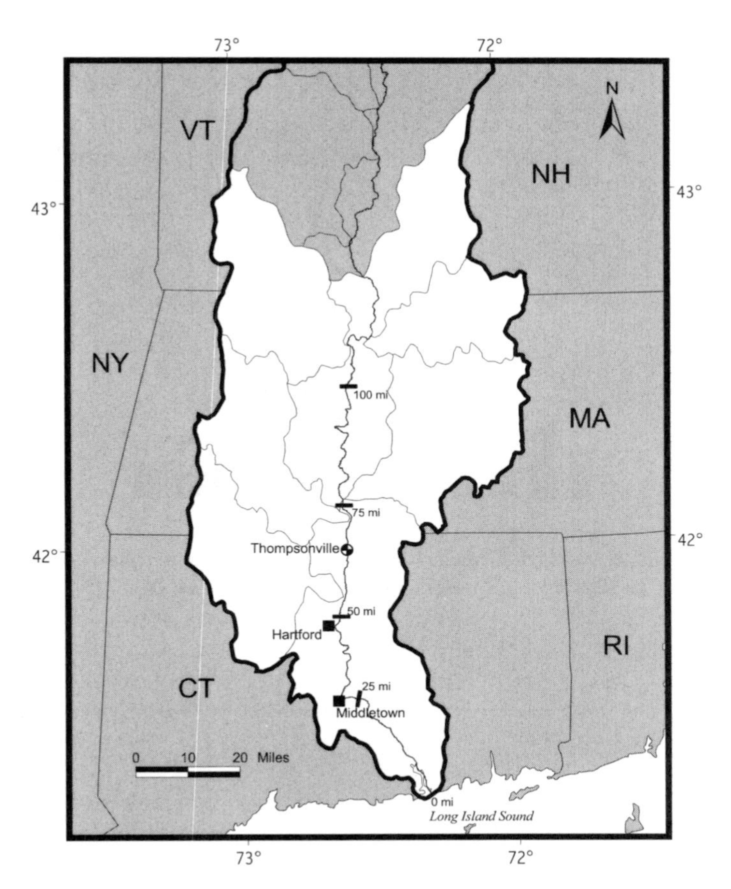

Figure 5-2 Location map for Lower Connecticut River Basin. (River miles shown are distances from Long Island Sound.)

the White Mountains in New Hampshire and the rounded hills and mountains in Vermont and Massachusetts to the lowlands of the floodplains along the river's banks in Massachusetts and Connecticut. Rising in the semi-mountainous area of northern New Hampshire, the Connecticut River drops more than half of its 2,650 feet in elevation in the first 30 miles of its course. The river is tidally influenced from Hartford to Long Island Sound (Figure 5-2).

Long-term trends in summer streamflow from the USGS gage at Thompsonville, Connecticut, shown in Figure 5-3, illustrate the interannual variability of discharge during the critical summer months. Seasonal flow conditions reflect the long, cold winters and the relatively short summers characteristic of New England. High flows are generally experienced in the spring (March–May), corresponding to large snowmelt events (Figure 5-4). Low flows occur during the summer months. In the past, flow regulation for hydropower production at Holyoke Dam (Massachusetts) periodically reduced flows in the Connecticut River to a minimum of near zero, but minimum re-

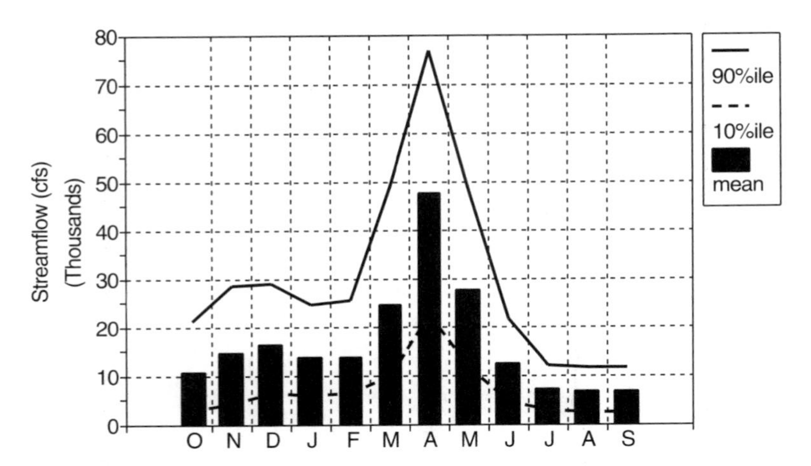

Figure 5-3 Monthly trends of mean, tenth, and ninetieth percentile streamflow for the Connecticut River at Thompsonville, Connecticut (USGS Gage 01184000), 1951–1980. *Source:* USGS, 1999.

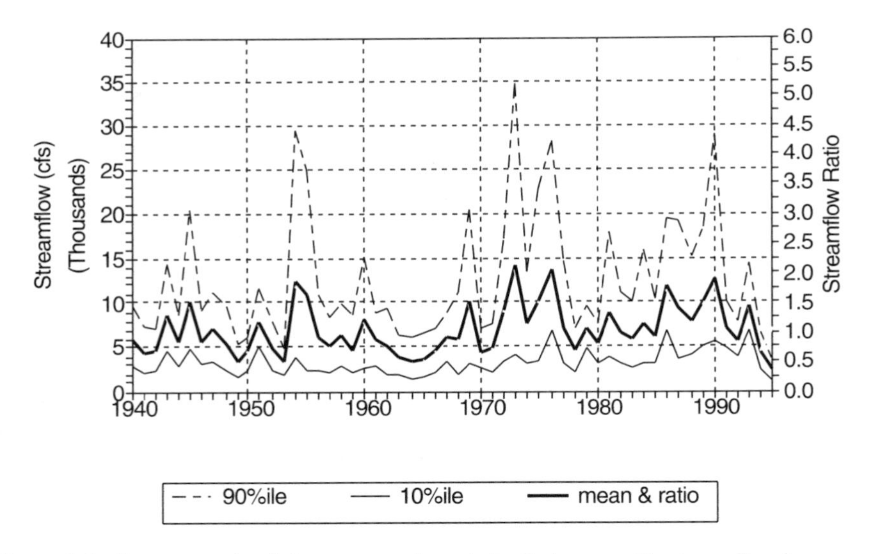

Figure 5-4 Long-term trends in mean, tenth, and ninetieth percentile streamflow in summer (July–September) for the Connecticut River at Thompsonville, Connecticut (USGS Gage 01184000). *Source:* USGS, 1999.

lease requirements have been established to maintain the summer low flow at a higher level. Currently, the flow is regulated by a number of headwater lakes and reservoirs, as well as power plants, with a combined usable capacity of 107 billion cubic feet (USGS, 1989) at Thompsonville, Connecticut. The 7-day, 10-year low flow (7Q10) discharge at Thompsonville is 2,200 cubic feet per second (cfs). The minimum recorded daily discharge was 519 cfs on September 30, 1984, below the Holyoke Dam and 968 cfs on October 30, 1963, at Thompsonville, Connecticut (USGS, 1989).

POPULATION, WATER, AND LAND USE TRENDS

The population density in the Connecticut River Basin generally increases from the north to the south. Approximately 85 percent of the river basin's residents live in Massachusetts and Connecticut. Approximately 1.1 million people live in Connecticut municipalities adjacent to the river; the largest city, Hartford, had a 1990 estimated population of 139,739 (CSDC, 1991). The Connecticut River case study area includes a number of counties identified by the Office of Management and Budget (OMB, 1999) as Metropolitan Statistical Areas (MSAs) or Primary Metropolitan Statistical Areas (PMSAs). The Hartford, Connecticut, MSA and three Connecticut counties, Fairfield, Middlesex, and Tolland, are included in this case study. Figure 5-5 presents long-term population trends (1940–1996) for the three counties. From 1940 to 1996, the population in the Connecticut River case study area about doubled (Forstall, 1995; USDOC, 1998).

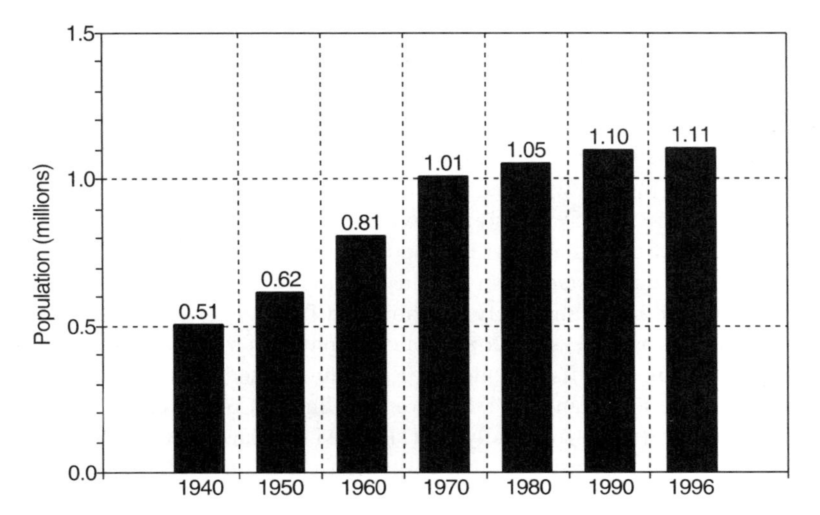

Figure 5-5 Long-term trends in population in the Lower Connecticut River Basin. *Sources:* Forstall, 1995; USDOC, 1998.

The first European settlements in the Connecticut River Basin were centered around Hartford in the 1630s. During the initial 100 years of development, the water and lands of the Connecticut River Valley provided a transportation route to the interior, as well as food and vast quantities of timber for shelter and fuel. Timber exploitation from 1700 to 1850 removed about three-fourths of the basin's forest cover. Following the timber-cutting era, cleared land was used for raising sheep and goats. The farm economy dwindled by the 1850s, and the land began to revert back to its forested condition.

The upper basin in New Hampshire and Vermont has retained a more rural character, although suburbanization is replacing traditional farm areas in some locations as the small northern towns expand. The 52-mile-long tidal section of the river in Connecticut between Long Island Sound and Hartford has traditionally supported shipbuilding and has been used as a major route for waterborne commerce, mostly petroleum products. Land use in this lower basin includes large-scale industrial and commercial development in Hartford. In the past, major industries in the Hartford area included woolen mills, paper mills, and machine tool factories. In recent decades, the economy of the lower basin has shifted from manufacturing toward a service economy. Beginning with the Hartford Fire Insurance Company in 1794, insurance has become a multibillion-dollar industry. Hartford has been deemed "the insurance capital of the world." The Connecticut River is not currently used as a public water supply in the state of Connecticut. Most of the Connecticut River water used by agriculture in the Connecticut River Valley is used to irrigate tobacco, vegetable crops, fruits, and nursery stock. In 1980, approximately 11,500 acres of the 33,922 acres of harvested cropland in Hartford County were irrigated with water from the Connecticut River or the Farmington River (a major tributary just north of Hartford) (USACE, 1981).

HISTORICAL WATER QUALITY ISSUES

Water quality problems in the Hartford area of the Connecticut River date back to the late 1800s. In July 1914, the level of DO in the Connecticut River near Hartford (~4–6 mg/L) was 2 to 3 mg/L lower than levels during the late 1980s (7.4 to 7.9 mg/L in 1988) (CTDEP, 1982, 1988). Early in the history of European settlement along the river, the construction of dams for hydropower had significantly exacerbated water quality problems, due to stagnation and the creation of faunal barriers. By 1872, Atlantic salmon had been completely exterminated from the river system because of poor water quality as well as the construction of physical barriers that prohibited the migration of anadromous fish (Center for Environment and Man, 1975).

In 1955, the New England Interstate Water Pollution Control Commission classified the Connecticut River from Holyoke Dam in Massachusetts to Middletown, Connecticut, as a Class D waterway suitable for "transportation of sewage and industrial wastes without nuisance and for power, navigation, and certain industrial uses" (Kittrell, 1963). Severe water pollution problems in this reach of the Connecticut River have resulted from two sources: industrial effluent and municipal sewage disposal.

One of the major industries responsible for degradation of water quality has been paper mills. Before the late 1970s, paper mills in the Massachusetts segment of the river discharged effluent with high concentrations of BOD_5 and suspended solids into the river (Center for Environment and Man, 1975). Downstream of the paper mills in Holyoke, Massachusetts, it was reported that the river flowed different colors depending on the dye lot used at the paper mill that day.

In 1963, it was reported that in the stretch of river from central Massachusetts to south of Hartford, Connecticut, 9 of the 22 jurisdictions responsible for discharge of sewage provided no wastewater treatment. Twelve of the 22 provided only primary treatment, and 1 provided secondary treatment (Kittrell, 1963). Large discharges of municipal and industrial wastes caused a steady depletion of DO downstream of the Holyoke Dam in Massachusetts. Minimum DO levels reached nearly zero during a low-flow survey in 1966, and DO levels of less than 2 mg/L were recorded in 1971. Connecticut River data collected in the summer of 1971 documented other forms of pollution, with a minimum density of coliform bacteria of 75,000 MPN/100 mL and a maximum of over 1 million MPN/100 mL (Center for Environment and Man, 1975).

LEGISLATIVE AND REGULATORY HISTORY

On the basis of reports indicating that pollution in this reach of the Connecticut River was endangering the health and welfare of persons in Connecticut, the Secretary of Health, Education and Welfare convened a conference under Section 8 of the Federal Water Pollution Control Act (33 U.S. C. 466g *et seq.*) in 1963 to investigate the pollution of the Connecticut River in Massachusetts and Connecticut (See Appendix A-2). This conference documented the appalling water quality of the Connecticut River and initiated strategies to begin to clean up the river (Kittrell, 1963). By the early 1960s, the steadily increasing public concern regarding water pollution issues resulted in organized planning for implementation of primary and secondary wastewater treatment in several municipalities, including Hartford, Connecticut.

Since 1963, USEPA's Construction Grants Program has been responsible for elimination of vast amounts of untreated or partially treated wastewater entering the Connecticut River. The process of reducing the loadings and substantially improving the quality of the Connecticut River was significantly influenced by the 1972 CWA and the available federal funding. Subsequent to the enactment of this legislation, 125 new or upgraded treatment plants were constructed along the Connecticut River at a cost of nearly $900 million (Conniff, 1990). From 1972 through 1984, eligible projects were funded 75 percent by federal grants, 15 percent by state grants, and 10 percent by local financing; prior to 1972, the federal share was 55 percent (CTDEP, 1982). Three secondary wastewater treatment plants in the Hartford area (Hartford, East Hartford, and Rocky Hill) were completed by the mid-1970s (Gilbert, 1991).

One of the major problems still facing this important New England waterway, however, is combined sewer overflows (CSOs). Overflows during storm events can still cause discharge of untreated sewage into the Connecticut River between Springfield,

Massachusetts, and Middletown, Connecticut. CSO problems are the principal reason the Connecticut River does not consistently meet the Class B fishable/swimmable standard for fecal coliform in northern Connecticut (above Middletown) (Mauger, 1991).

IMPACTS OF WASTEWATER TREATMENT

Pollutant Loading and Water Quality Trends

As a result of implementation of municipal and industrial wastewater treatment in the Connecticut River Basin, total pollutant loading has decreased substantially since the 1960s. The approximate total population served by the 22 sewer systems in the Connecticut and Massachusetts portions of the river basin in 1963 was 734,265 people; of these, 282,590 (38 percent) resided in East Hartford and Hartford, Connecticut (Kittrell, 1963). In 1990 the sewered population of the greater Hartford metropolitan area was 366,574, served by the Hartford, East Hartford, and Rocky Hill facilities. The largest of these, the Hartford water pollution control plant, currently has secondary treatment upgraded from 60 mgd to 80 mgd in 1993.

Since implementation of the 1972 CWA, substantial reductions in point source loads of oxidizable materials have been achieved as a result of technology-based and water quality–based effluent controls on municipal and industrial dischargers in the Connecticut River watershed. Nonpoint source runoff, driven by the land uses and hydrologic characteristics of the watershed, also contributes a pollutant load to the Connecticut River that must be considered in a complete evaluation of the impact of regulatory policy and controls on long-term water quality trends. To evaluate the relative significance of point and nonpoint source pollutant loads, inventories of NPDES point source dischargers, land uses, and land use–dependent export coefficients (Bondelid et al., 2000) have been used to estimate catalog unit–based point (municipal, industrial, CSOs) source and nonpoint (rural, urban[1]) source loads of BOD_5 for contemporary (ca. 1995) conditions in the case study area (Figure 5-6). Municipal facilities contribute 42 percent (10.5 metric tons per day, mt/day) of the total estimated BOD_5 load, while industrial dischargers account for 10 percent (2.4 mt/day) of the total BOD_5 load. Nonpoint sources of BOD_5 account for a total of 47 percent, with rural runoff contributing 13 percent (3.3 mt/day) and urban land uses accounting for 34 percent (8.5 mt/day) of the total load (Figure 5-6).

Oxygen depletion and high BOD_5 levels historically have been documented downstream from the major wastewater discharges in the Massachusetts and Connecticut segments of the river. Prior to upgrading publicly owned treatment works (POTWs) in the southern Massachusetts sections of the river, water quality monitoring data near the Connecticut/Massachusetts border documented that DO concentrations in the river violated the Massachusetts state standard (5 mg/L for non-low-flow periods) 22 percent of the days recorded in the early 1970s (June–October) (Isaac, 1991).

[1]For the purposes of this comparison, urban stormwater runoff includes both areas outside (termed "nonpoint sources") and within [meeting the legal definition of a "point source" in section 502(14) of the CWA] the NPDES stormwater permit program.

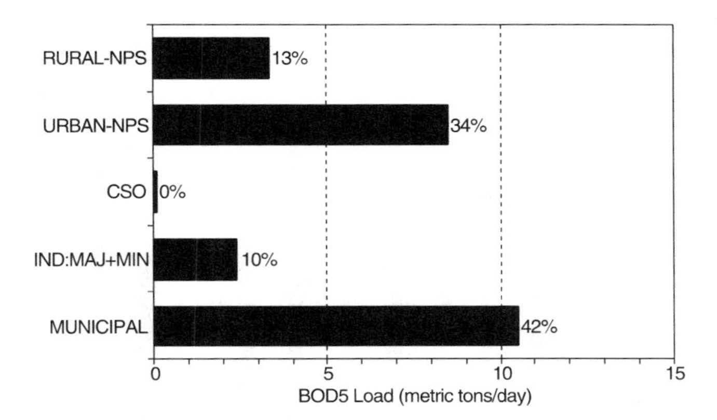

Figure 5-6 Comparison of point and nonpoint source loads of BOD_5 (ca. 1995) for the Lower Connecticut River Basin. *Source:* Bondelid et al., 2000.

Minimum recorded DO levels reached nearly 0 mg/L in a 1966 survey and less than 2.0 mg/L in 1971 (NCWQ, 1975) in Massachusetts. After POTW upgrades, by 1974 violations had dropped to only 6 percent of the days of record with DO less than 5 mg/L (Isaac, 1991) (Figure 5-7).

The average summer DO concentrations in the Lower Connecticut River in northern Connecticut (Hartford to Windsor) have also improved steadily since the mid-1960s (Figure 5-8). Corresponding to the increase in DO shown has been a progressive decline in ambient BOD_5 that reflects upgrades to Hartford area wastewater treatment facilities (Figure 5-9). Since the early 1970s, the average summer (July to September) DO levels in the Lower Connecticut River from Haddam to Middletown have remained above 7 mg/L (Figure 5-10). In a September 1988 intensive survey of water quality in the Lower Connecticut River, the DO concentrations ranged from 7.3 to 7.9 mg/L for all 10 stations sampled from the Massachusetts/Connecticut border to near the mouth of the river (CTDEP, 1988). The improvement in water quality in the Lower Connecticut River as a result of the significant reductions in oxidizable pollutant loading over the past 30 years has been substantial.

Recreational and Living Resources Trends

Information on biotic populations in the Connecticut River is scarce for most of the period previous to 1975 (Center for the Environment and Man, 1975). The precolonial salmon population was very large and supplied Native Americans and, later, early colonists with an abundant food supply. A long absence of Atlantic salmon in the river was noted between 1874 and the late 1970s. An Atlantic salmon caught in 1977 was the first documented occurrence of the fish in the river since 1874 (USEPA, 1980).

The absence of salmon can be attributed partially to dam construction, which prevented the fish from migrating upstream to spawn, and partially to water pollution. The first dam across the river was constructed in 1798 at Turners Falls, Massachusetts (Jobin, 1998). Fish ladders were built around dams when people began to understand

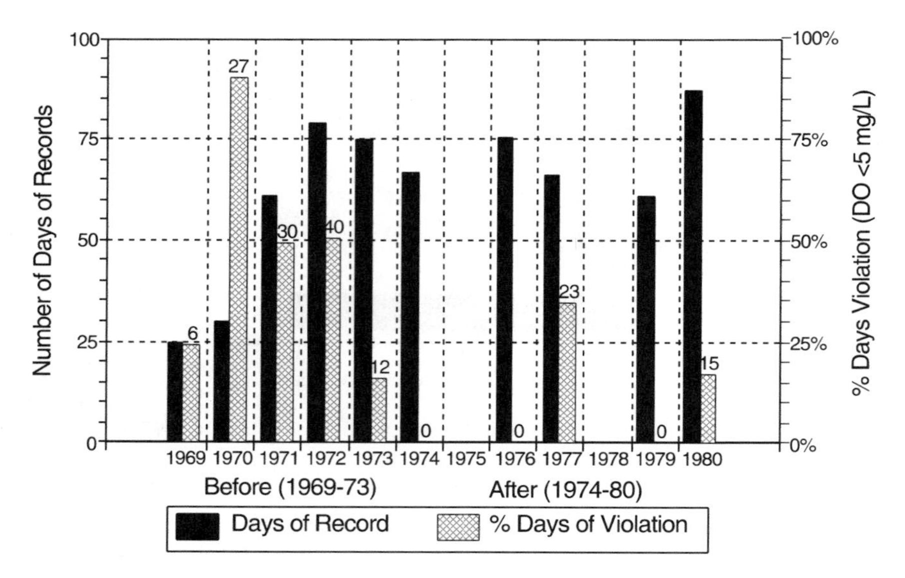

Figure 5-7 Trends in violations of DO standard (DO < 5 mg/L) in summer (July–September) for the Connecticut River before (1969–1973) and after (1974–1980) construction and upgrade of municipal wastewater treatment facilities at Agwam, Massachusetts. *Source:* Isaac, 1991.

that the dams prevented migration, yet 200,000 hatchery salmon placed in the river between 1968 and the early 1970s failed to return to the river to spawn, presumably because of the poor water quality (USEPA, 1980). Efforts to clean up the river began after passage of the 1972 CWA, and the return of the salmon in the late 1970s can be attributed to improved water quality.

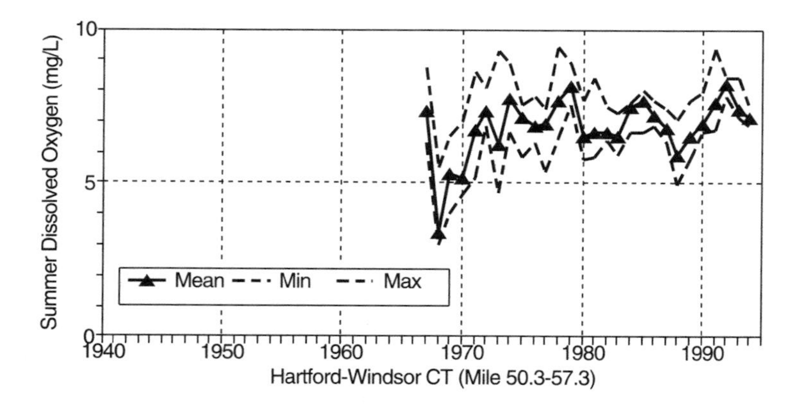

Figure 5-8 Long-term trends in DO for the Lower Connecticut River from Hartford to Windsor, Connecticut (RF1-01080205029, River Miles 50.3–57.3). *Source:* USEPA STORET.

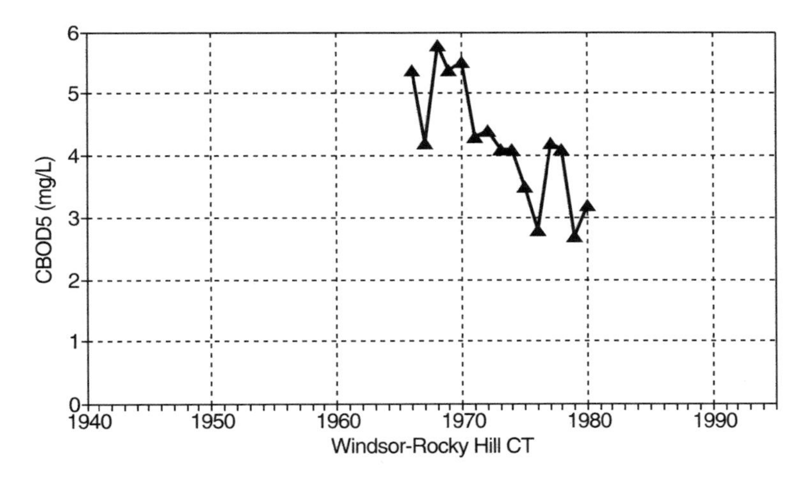

Figure 5-9 Long-term trends in BOD_5 in the Lower Connecticut River from Windsor to Rocky Hill, Connecticut. *Source:* Reimold, 1991.

Another anadromous fish species historically important to commercial and recreational fishing on the Connecticut River is the American shad (*Alosa sapidissima*). Shad had a precarious existence in the river before 1975 (Center for Environment and Man, 1975), but their population increased afterward (Figure 5-11). The estimated mean population for the years 1975–1989 was 841,265 (Savoy, 1991). The 1990 estimated population was 654,885, lower than the previous 14-year mean but considered by the Connecticut Department of Environmental Protection (CTDEP) to be stable.

Other indices lead to the conclusion that the shad population is faring well in the Connecticut River. The 1990 commercial catch of shad in the river (x = 9,687, adjusted for angler effort) was nearly twice the 1989 catch (x = 5,243) and reversed a

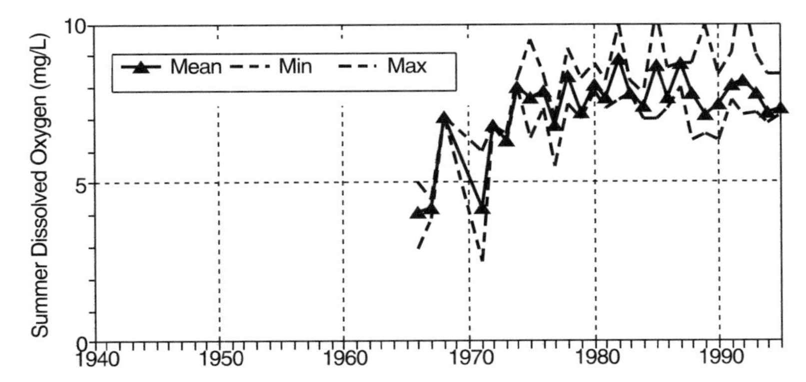

Figure 5-10 Long-term trends in DO for the Lower Connecticut River from Haddam to Middletown, Connecticut. (RF1-01080205021, River Miles 16.3–21.6). *Source:* USEPA STORET.

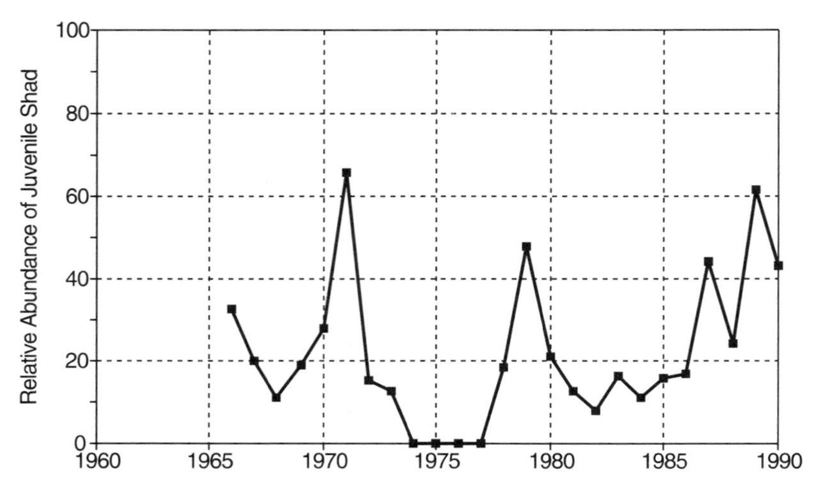

Figure 5-11 Long-term trends of shad relative abundance for the Lower Connecticut River. *Source:* Savoy and Shake, 1991.

general declining trend that lasted from 1986 to 1989 (Savoy, 1991). Similarly, juvenile shad had strong relative abundances from 1987 to 1990 (Figure 5-11), indicating good reproductive success (Savoy, 1991). Juvenile fish are generally less tolerant than adults of low DO concentrations, so an improvement in reproductive success is a good indicator of improving water quality.

SUMMARY AND CONCLUSIONS

The federal, state, and local funding for construction of municipal wastewater treatment facilities in the Connecticut River Basin has led to significant improvement in water quality since the 1960s. A river basin that during the early 1970s was considered a flowing sewer is now a popular recreational area. One measure of the improvement in the fishable/swimmable quality of the river is documented by the U.S. Fish and Wildlife Service. Dramatic improvements in water quality, along with the installation of fish ladders to eliminate physical barriers to migration, have resulted in the successful return of Atlantic salmon to the Connecticut River.

Concentrations of total nitrogen and total phosphorus in the Connecticut River case study area since the CWA have followed the national trends—phosphorus and ammonia-N have decreased with associated increases in nitrate-N and total nitrogen, indicating that improved wastewater treatment has improved water quality (Garabedian et al., 1998). In its report Water Quality in the Connecticut, Housatonic, and Thames River Basins, Connecticut, Massachusetts, New Hampshire, New York, and Vermont, 1992–1995 (Garabedian et al., 1998), the U.S. Geological Survey concluded that increasing nitrate concentrations may contribute to eutrophication in Long Island Sound.

REFERENCES

Bondelid, T., S. Unger, and A. Stoddard. 2000. National water pollution control assessment model (NWPCAM) Version 1.1. Final report prepared by Research Triangle Institute, Research Triangle Park, NC for U.S. Environmental Protection Agency, Office of Policy, Economics and Innovation, Washington, DC, November, RTI Project Number 92U-7640-031.

Center for Environment and Man. 1975. Environmental impact assessment. Water quality analysis. Connecticut River. Report No. NCWQ 75/51. Prepared for the National Commission on Water Quality, Washington, DC, by the Center for Environment and Man., Hartford, CT, National Technical Information Service No. PB-250 924. April.

Conniff, R. 1990. The transformation of a river from sewer to suburbs in 20 years. Smithsonian 21(1): 71–84.

CSDC. 1991. 1990 Census population by municipality. State of Connecticut Office of Policy and Management, Connecticut State Data Center, Hartford, CT.

CTDEP. 1982. The Connecticut River—Worth the cost! Connecticut Department of Environmental Protection, Water Compliance Division, Hartford, CT.

CTDEP. 1988. Connecticut River intensive survey, September 7–8, 1988. Connecticut Department of Environmental Protection, Water Compliance Division, Hartford, CT.

Forstall, R. L. 1995. Population by counties by decennial census: 1900 to 1990. U.S. Bureau of the Census, Population Division, Washington, DC. http://www.census.gov/population/www/censusdata/cencounts.html

Garabedian, S. P., J. F. Coles, S. J. Grady, E. C. T. Trench, and M. J. Zimmerman. 1998. Water quality in the Connecticut, Housatonic, and Thames River Basins, Connecticut, Massachusetts, New Hampshire, New York, and Vermont, 1992–1995. U.S. Geological Survey Circular 1155. U.S. Geological Survey, Reston, VA.

Gilbert, P. 1991. Metropolitan District Commission, Hartford, CT. Personal communication.

Isaac, R. A. 1991. POTW improvements raise water quality. Water Environ. Technol. 3(6): 69–72.

Jobin, W. R. 1998. Sustainable management for dams and waters. CRC Press, Boca Raton, FL.

Kittrell, F. W. 1963. Report to the conference in the matter of pollution of the interstate waters of the Connecticut River; Massachusetts-Connecticut. U.S. Department of Health, Education and Welfare, Washington, DC. December.

Mauger, A. 1991. Water Compliance Division, State of Connecticut Department of Environmental Protection, Hartford, CT. Personal communication.

OMB. 1999. OMB Bulletin No.99-04. Revised statistical definitions of Metropolitan Areas (MAs) and Guidance on uses of MA definitions. U.S. Census Bureau, Office of Management and Budget, Washington, DC. http://www.census.gov/population/www/estimates/metrodef.html

Reimold, R. 1991. Metcalf & Eddy, Inc., Boston, MA. Personal communication.

Savoy, T. 1991. Sturgeon status in Connecticut waters. Completion Report. Project No. AFC-18. State of Connecticut, Department of Environmental Protection, Division of Marine Fisheries, Waterford, CT. June.

Savoy, T. and D. Shake. 1991. Population dynamics studies of American shad, *Alosa sapidissima,* in the Connecticut River. U.S. Department of Commerce, National Oceanic and Atmospheric Administration, National Marine Fisheries Service. June 30.

USACE. 1981. Water resources development in Connecticut 1981. U.S. Army Corps of Engineers, New England Division, Waltham, MA.

USDOC. 1998. Census of Population and Housing. Prepared by U.S.

Department of Commerce, Economics and Statistics Administration, Bureau of the Census—Population Division, Washington, DC.

USEPA. 1980. National accomplishments in pollution control: 1970–1980. Some case histories. The Connecticut River: Salmon are caught again. U.S. Environmental Protection Agency, Office of Planning and Evaluation, Program Evaluation Division, Washington, DC. December.

USEPA. STOrage and RETrieval Water Quality Information System. U.S. Environmental Protection Agency, Office of Wetlands, Oceans, and Watersheds, Washington, DC.

USGS. 1989. Water resources data, Connecticut water year 1988. USGS Water-Data Report CT-88-1. U.S. Geological Survey, Hartford, CT.

USGS. 1999. Streamflow data downloaded from the U.S. Geological Survey's National Water Information System (NWIS)-W. Data retrieval for historical streamflow daily values. http://waterdata.usgs.gov/nwis-w

CHAPTER 6

Hudson-Raritan Estuary Case Study

The Mid-Atlantic Basin (Hydrologic Region 2), covering a drainage area of 111,417 square miles, includes some of the major rivers in the continental United States. Figure 6-1 highlights the location of the basin and the Hudson-Raritan estuary, the case study watershed profiled in this chapter.

With a length of 306 miles and a drainage area of 13,370 square miles, the Hudson River ranks seventy-first among the 135 U.S. rivers that are more than 100 miles in length. On the basis of mean annual discharge (1941–1970), the Hudson ranks twenty-sixth (19,500 cfs) of large rivers in the United States (Iseri and Langbein, 1974). Figure 6-2 highlights the location of the Hudson-Raritan estuary case study catalog units identified by major urban-industrial areas affected by severe water pollution problems during the 1950s and 1960s (see Table 4-2). This chapter presents long-term trends in population, municipal wastewater infrastructure and effluent loading of pollutants, ambient water quality, environmental resources, and uses of the Hudson-Raritan estuary. Data sources include USEPA's national water quality database (STORET), published technical literature, and unpublished technical reports ("grey" literature) obtained from local agency sources.

BACKGROUND

The Hudson-Raritan estuary, with its rich and diverse populations of birds, fish, and shellfish, is unmatched in terms of the historical abundance of its natural resources. New York City, in fact, owes its existence as a major urban center to the bounty of the estuary (Trust for Public Lands, 1990). The estuarine and coastal waters around New York City currently support significant fish and wildlife resources (Sullivan, 1991). For example, the extensive wetland systems along the Arthur Kill on northwest Staten Island, adjacent to one of the most industrialized corridors in the northeastern United States, has been colonized by several species of herons, egrets, and ibises (Trust for Public Lands, 1990). Current heron populations represent up to 25 percent of all nesting wading birds along the coast from Cape May, New Jersey, to the Rhode Island line (HEP, 1996). Today, despite mounting pressures for industrial and residential development, there is a growing awareness of the estuary's unique ecological function

Figure 6-1 Hydrologic Region 2 and Hudson-Raritan estuary watershed.

and a new appreciation of its almost limitless potential as a recreational, cultural, and aesthetic resource (Trust for Public Lands, 1990).

For more than 300 years, New York Harbor and the New York metropolitan region have been a focal point of urban development, transportation, manufacturing, and commerce. New York City has been characterized by tremendous population increases and economic growth and has traditionally been a major harbor. As a large estuary with vast wetlands and marsh areas, New York Harbor offered an abundance of natural resources that supported a commercially important shellfish industry until its decline in the early 1900s. With a relatively deep protected estuary that was ideal for navigation, the harbor developed as a key shipping and transportation link for commerce and passenger traffic between the inland states and Europe.

PHYSICAL SETTING AND HYDROLOGY

New York Harbor is formed by a network of interconnected tidal waterways along the shores of New York and northern New Jersey; bounded by the Hudson River to the north, Long Island Sound to the east, and the Atlantic Ocean to the south (Figure 6-2). Freshwater tributaries discharging into the estuary drain an area of 16,290 square miles and contribute approximately 81 percent of the total freshwater inflow to the harbor. The remainder of the freshwater input is contributed by wastewater (15 percent), urban runoff (4 percent), CSOs (1 percent), and industrial discharges, landfill leachate, and precipitation (0.5 percent) (Brosnan and O'Shea, 1996a). Fresh water is also imported into the New York City water supply system from the combined watershed areas of the Delaware and Catskills mountains, with eventual discharge via the wastewater drainage system into the harbor.

Seasonal and interannual variation of streamflow of the Hudson River recorded at Green Island, New York, near Troy (USGS gage 01358000) is characterized by high flow during March through May, with the monthly mean peak flow of 32,719 cfs observed in April (Figures 6-3 and 6-4). High spring flows result from spring snowmelt and runoff over the mountainous drainage basin. Low-flow conditions occur during July through September, with the mean monthly minimum of 5,797 cfs observed during August. In dramatic contrast to the long-term (1951–1980) summer (July–September) mean of 6,396 cfs, during the extreme drought conditions of 1962–1966, mean summer flow was only 49 to 70 percent of the long-term mean summer flow. The driest conditions occurred during the summer of 1964, with a mean flow of 3,104 cfs and a minimum flow of only 1,010 cfs (Bowman and Wunderlich, 1977; O'Connor et al., 1977). Inspection of the long-term trend data (1947–1995) for summer streamflow clearly shows the persistent drought conditions of the 1960s, as well as the high-flow conditions recorded a decade later (Figure 6-4).

Population, Water, and Land Use Trends

In 1628, New York City was a small village of 270 settlers; today it is an urban metropolis of 16 million (Figure 6-5). The physical environment of the New York region has contributed greatly to its enormous growth and economic development. The natural port of the harbor has made commerce and shipping a major component of the economy since the colonial era. The Watchung and Ramapo mountains, west and northwest of the city, also focused growth around the harbor by constraining transportation routes and land development patterns. In 1810, New York emerged as the largest city in the new nation, surpassing Boston and Philadelphia. New transportation routes—the Erie Canal in 1825 and railroad connections between New York and Phila-

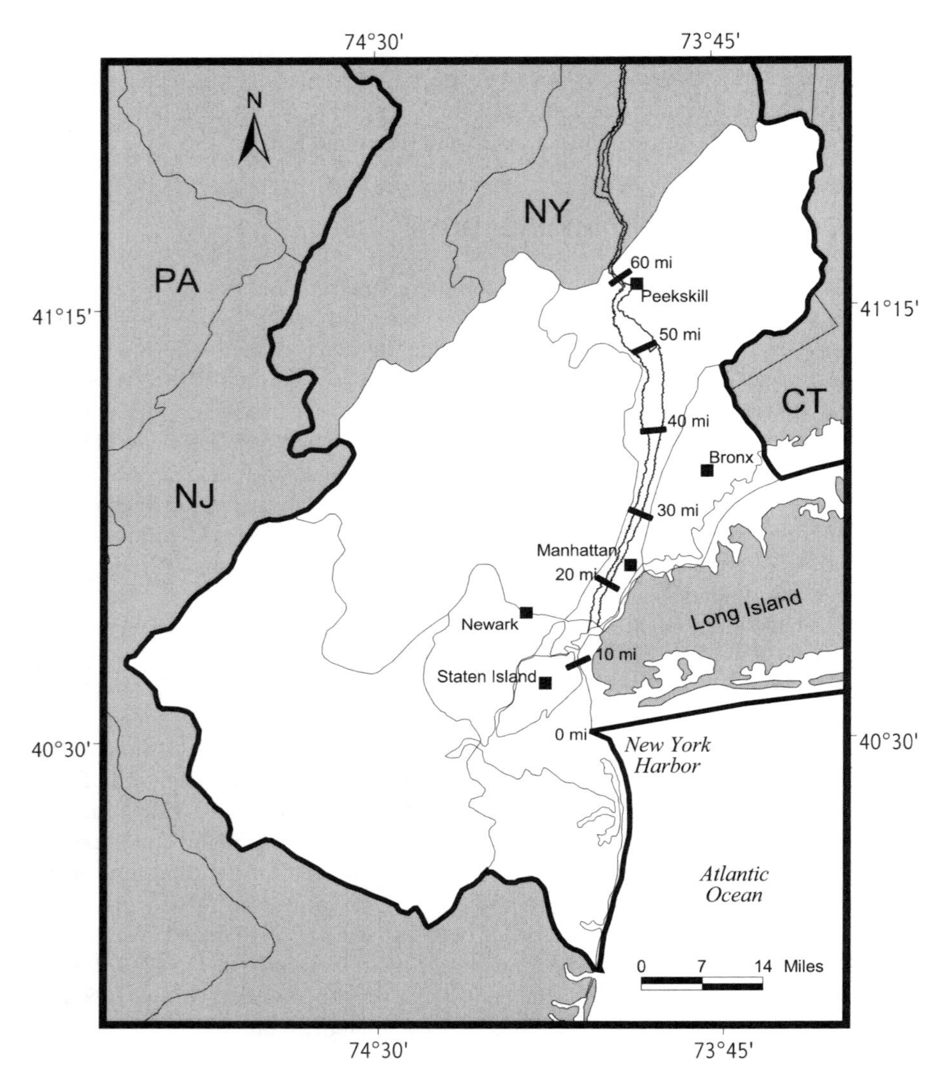

Figure 6-2 Location map of the Hudson-Raritan estuary. (River miles shown are distances from Sandy Hook–Rockaway transect of Atlantic Ocean.)

delphia in 1839—strengthened the city's links to Europe and to the nation's interior. During the massive European immigration period of the mid-1800s to the early 1900s, immigrants to the United States passed through Ellis Island in New York Harbor, a main port of entry. Many chose to remain and contribute to the growth of the city.

The Hudson-Raritan estuary case study area includes a number of counties identified by the Office of Management and Budget (OMB) that define Consolidated Metropolitan Statistical Areas (CMSAs) or Primary Metropolitan Statistical Areas (PMSAs).

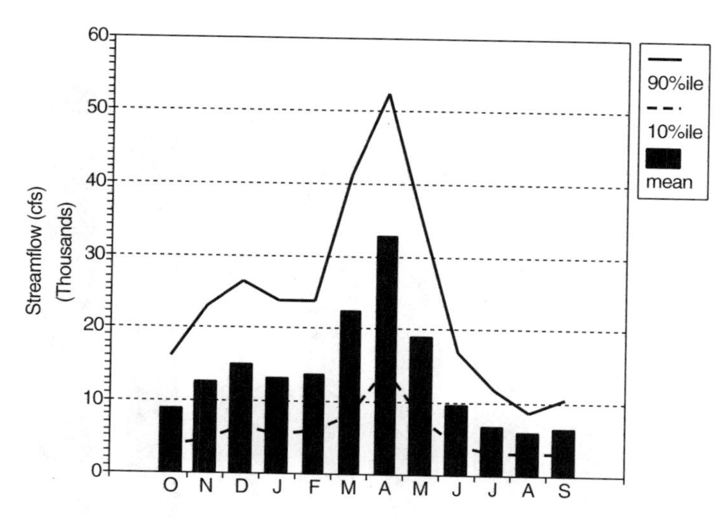

Figure 6-3 Monthly trends of mean, tenth, and ninetieth percentile streamflow for the Hudson River at Green Island, New York (USGS Gage 01358000), 1951–1980. *Source:* USGS, 1999.

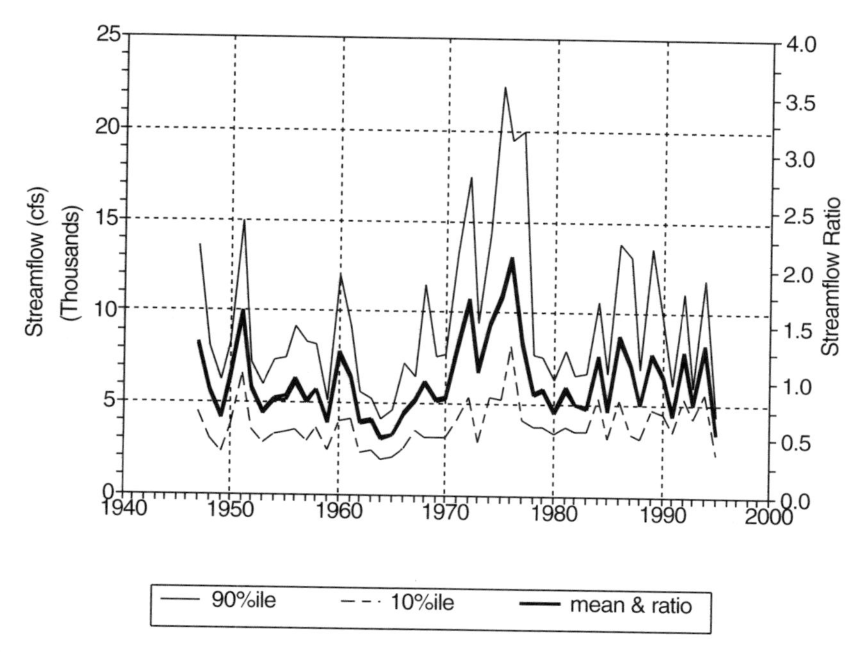

Figure 6-4 Long-term trends in mean, tenth, and ninetieth percentile streamflow in summer (July–September) for the Hudson River at Green Island, New York (USGS Gage 01358000). *Source:* USGS, 1999.

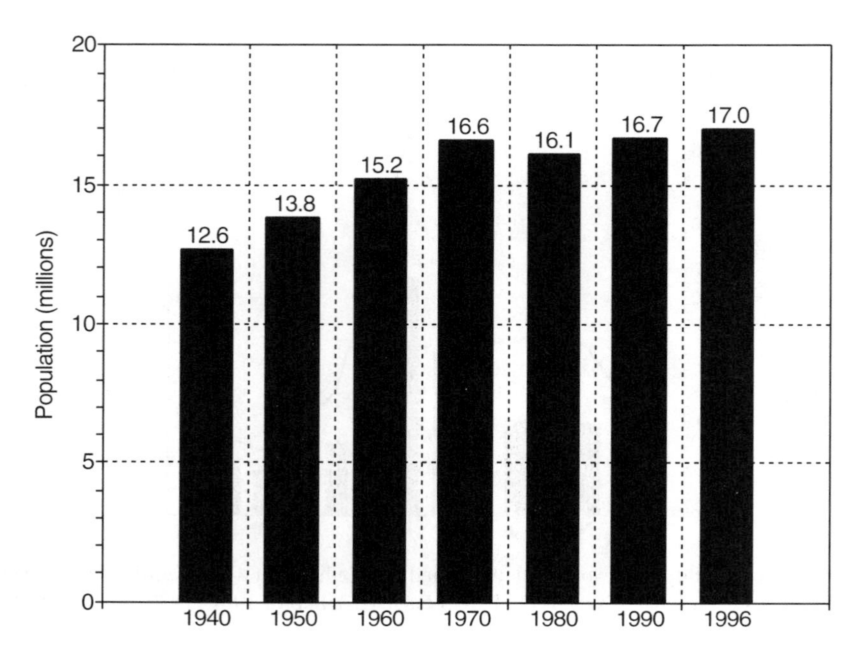

Figure 6-5 Long-term trends in population for the New York–Northern New Jersey–Long Island CMSA counties for the Hudson-Raritan estuary metropolitan region. *Source:* Forstall, 1995; USDOC, 1998.

Table 6-1 lists the CMSA counties included in this case study. Figure 6-5 presents long-term population trends (1940–1996) for the counties listed in Table 6-1. From 1940 to 1996, the population in the Hudson-Raritan estuary case study area increased by 34 percent, from 12.6 million in 1940 to 17.0 million in 1996 (Forstall, 1995; USDOC, 1998).

Because of the proximity to shipping and other transportation routes, manufacturing and industrial development evolved as a major component of the region's industrial economy and a major contributor to the environmental decline of the area's once bountiful wetlands. New Jersey, the most densely populated state in the nation, is second only to California in industrialization, and most of the industrial activity of New Jersey is centered around New York Harbor. Within New York City, economic growth has depended on manufacturing, services, world trade, and the city's position as a national and international center for banks, finance, culture, and the arts. Since the turn of the twentieth century, and particularly since the development of the automobile and highways, progressive suburban development radiating from the city has transformed the once agricultural region into a densely populated metropolitan area. At a distance of about 60 miles from New York City, however, farmland, rural lands, and low-density suburban towns still characterize the outer fringes of the metropolitan region.

TABLE 6-1 Consolidated Metropolitan Statistical Area (CMSA) Counties in the Hudson-Raritan Estuary Case Study, New York-Northern New Jersey-Long Island NY-NJ-CT-PA CSMA

Fairfield County, CT
Litchfield County, CT
New Haven County, CT
Bergen County, NJ
Essex County, NJ
Hudson County, NJ
Hunterdon County, NJ
Middlesex County, NJ
Monmouth County, NJ
Morris County, NJ
Ocean County, NJ
Passaic County, NJ
Somerset County, NJ
Sussex County, NJ
Union County, NJ
Warren County, NJ
Bronx County, NY
Dutchess County, NY
Kings County, NY
New York County, NY
Orange County, NY
Putnam County, NY
Queens County, NY
Richmond County, NY
Rockland County, NY
Westchester County, NY
Pike County, PA

Source: OMB, 1999.

Despite intense development and the loss of wildlife habitat due to wetland conversion, the New York–New Jersey Harbor and the New York Bight do contain significant fish and wildlife resources. Water uses of the Hudson River and New York Harbor include public water supply of the freshwater river upstream of Poughkeepsie, New York, municipal and industrial wastewater disposal, commercial shipping and navigation, recreational boating, swimming, and commercial and recreational fishing. Although commercial fishing was once a significant component of the New York–New Jersey regional economy, the abundance of commercially important fish and shellfish has declined considerably during the twentieth century. The loss of once abundant fishery resources has been attributed to disease, overfishing, loss of habitat, and poor water quality conditions. Despite the significant reductions in fishery resources, commercial fishing of more than 60 species of seafood contributed approximately $500 million to the regional economies of New York and New Jersey during the mid-1990s (Schwartz and Porter, 1994). Recreational fishing in New York Harbor, Long Island Sound, and the New York Bight is also quite important accounting for more than $1 billion annually in economic activity for New York and New Jersey during the mid-1990s (Schwartz and Porter, 1994).

HISTORICAL WATER QUALITY ISSUES

Waste disposal issues in New York did not emerge only since the 1950s. Contemporary residents of the New York metropolitan area would be surprised to learn that public policy debates related to waste disposal and water pollution issues began only a few decades after colonial settlers arrived in the New World. The early settlers' practice of simply dumping pails of sewage and other refuse into the harbor became such a problem that, in 1680, the Governor ordered that a common sewer be constructed in Lower Manhattan. In 1683, the Common Council decreed "that none doe cast any dung, drought, dyrte or any other thing to fill up or annoy the mould or Dock or the neighborhood near the same, under the penalty of twenty shill" (Gross, 1976). Construction of a sewer and wastewater collection system in New York City began in 1696, with many sewers in lower and central Manhattan constructed two centuries later between 1830 and 1870 (O'Connor, 1990). Pollution problems existed, however, in both New York City and Newark, New Jersey, since the harbor received untreated wastewater from the sewers.

In 1868, unsanitary conditions were described as "poisoning the water and contaminating the air" (Suszkowski, 1990). During the 1920s, the overpowering stench of hydrogen sulfide from polluted water in the Passaic River near Newark, New Jersey, forced excursion boat passengers to seek refuge in the cabins (Cleary, 1978). During that period, all the regional New York and New Jersey communities discharged raw sewage into the harbor "to conduct by the cheapest route to the nearest waterway, giving no thought whatever to its effect on the waterway and on adjacent waters" (Franz, 1982). In the 1920s, New York City discharged approximately 600 mgd of raw sewage into the harbor (Brosnan and O'Shea, 1996a).

The earliest water pollution surveys of New York Harbor began with the formation of the Metropolitan Sewerage Commission of New York in 1906. In a 1910 report on conditions of the harbor, the Commission stated that "Bathing in New York Harbor above the Narrows is dangerous to health, and the oyster industry must soon be entirely given up." The Commission further noted that a number of tributaries and tidal channels in the harbor "have become little else than open sewers. Innumerable local nuisances exist along the waterfront of New York and New Jersey where the sewage of the cities located about the harbor is discharged . . . ." Finally, the Commission concluded that "the water which flows in the main channels of the harbor . . . is more polluted than considerations of public health and welfare should allow" (Suszkowski, 1990). As with many other urban areas around the turn of the twentieth century, development of a combined drainage network for stormwater and sewage collection evolved to address public health problems resulting from inadequate methods of waste disposal that created a nuisance in the streets and contaminated groundwater supplies (Fuhrman, 1984).

With vast marshlands, embayments, and interconnecting tidal channels, New York Harbor once supported abundant populations of fish, shorebirds, and shellfish that were important local food resources and essential to certain commercial activities. The progressive decline of the once thriving oyster industry provides an important ecological indicator of the trends in environmental quality of New York Harbor. Commercially important oyster beds were harvested during the 1800s in Raritan Bay, the Kill Van Kull, Jamaica Bay, and Newark Bay, and in the Shrewsbury River. By the turn of the twentieth century, waste disposal from industries and towns began to seriously affect the survival of seed beds. In addition to industrial waste and sewage discharges, dredging and disposal of dredge spoils, illegal dumping of cellar dirt, street sweepings, and refuse all contributed to the demise of this once valuable estuarine resource.

Although a century of pollution, disruption of habitat, and mismanagement of seed beds all contributed to the decline of oyster abundance, bacterial contamination from raw sewage disposal was the catalyst for the death of the commercial oyster industry. As early as 1904, typhoid cases were linked to consumption of contaminated oysters. By 1915, 80 percent of the city's 150 typhoid cases were attributed to contaminated oysters harvested from the harbor. In 1924 and 1925, another major outbreak ocurred, even though many of the beds had been closed in 1921 because of public health concerns (Franz, 1982). More than three decades later, consumption of sewage-contaminated hard clams from Raritan Bay again resulted in serious public health problems with an outbreak of infectious hepatitis in 1961.

Oysters, however, were not the only natural resource to suffer serious depletion of once-abundant stocks. In the closing decades of the nineteenth century, pollution and habitat destruction had begun to seriously degrade water quality and affect the abundance of marine resources. A century-long record of commercial fishery landings for New York and New Jersey clearly documents the adverse impact of water pollution and habitat destruction on the rich natural resources of the estuary (Esser, 1982). Combined landings of important estuarine and anadromous species, such as shad, alewives, striped bass, sturgeon, American oysters, hard clams, and bay scallops, have declined 90 percent over the past century, from 58 million pounds in 1887 to 6.6 million pounds in 1996 (McHugh et al., 1990; Wiseman, 1997) (Figure 6-6). In interpreting this long-term trend, it is important to realize that, even a century ago, resource abundance was already considered depleted in comparison to reports of abundance recorded through 1850. Contemporary degradation of the resources of the estuary, marked by successive anthropogenic assaults and incremental improvements in wastewater treatment, is believed to have begun as early as 1870 (Carriker et al., 1982).

The connection between raw sewage disposal and the decline of the oyster beds in the lower Hudson River eventually led to the creation of the New York Bay Pollution Commission in 1903 and the Metropolitan Sewerage Commission in 1906 (Franz, 1982). Routine water quality surveys have been conducted in New York Harbor since 1909. This unique data set represents the longest historical record of water quality in the nation and one of the longest historical records in the world (O'Connor, 1990; Brosnan and O'Shea, 1996a). Historically, water quality problems in the harbor have

Figure 6-6 Long-term trend of commercial landings of major anadromous and estuarine species in New York Harbor. *Sources:* McHugh et al., 1990; Wiseman, 1997.

included severe oxygen depletion and closure of shellfish beds and recreational beaches due to bacterial contamination. More recently, nutrient enrichment, algal blooms, heavy metals, sediment contamination, and bioaccumulation of toxics, such as polychlorinated biphenyls (PCBs) in striped bass (Faber, 1992; Thomann et al., 1991) and bald eagles (Revkin, 1997), have also become issues of concern.

By the 1920s, summer oxygen within much of the harbor had deteriorated to critical levels of less than 20 percent saturation (Brosnan and O'Shea, 1996a). Along with oyster industry records, long-term DO records document a progressive decline in the environmental quality of the harbor from 1910 through about 1930, as a result of increased population growth and raw sewage loading to the harbor (Brosnan and O'Shea, 1996a; Wolman, 1971). Following a period of very low oxygen saturation from about 1920 through 1950, the subsequent increasing trend generally corresponds chronologically to incremental improvements in construction and upgrades of sewage treatment plants, beginning in 1938 (Brosnan and O'Shea, 1996a).

With the completion of New York City's last two sewage treatment plants in 1986–1987, one of the major remaining water pollution problems in the harbor results from combined sewer overflows that discharge raw sewage and street debris. Following storm overflows, high bacteria levels require the closing of shellfish beds and bathing beaches. Although an aggressive industrial pretreatment program reduced the total industrial metal contribution from New York City plants from 3,000 lb/day in 1974 to 227 lb/day in 1991 (Brosnan et al., 1994), early ambient data still suggested violations of state water quality standards for metals in many locations of the harbor. More recent investigations conducted under the auspices of the NY/NJ Harbor Estuary Program (HEP), indicated significantly lower metal concentrations, with harborwide exceedances found only for mercury. Current monitoring and modeling efforts have greatly reduced the extent of waters suspected to be in violation of standards for nickel, lead, and copper (Stubin, 1997).

Additional toxic chemical problems in the harbor are associated with PCB contamination of sediments and striped bass and other marine organisms, resulting from PCB discharges from two General Electric plants upstream of Albany from the 1940s through the mid-1970s (Thomann et al., 1991). With a commercial fishing ban imposed because of PCB contamination (Faber, 1992), the striped bass population is thriving to the extent that the abundance of contaminated bass caught in nets and then returned to the estuary is actually creating an economic hardship for the commercial shad fishery (Suszkowski, 1990). More recent state-of-the-art monitoring and analysis technologies have detected trace level concentrations of PCBs in regional sewage treatment plant effluents. Current track-down programs, again initiated under the auspices of HEP, seek to determine the sources of these PCB contributions to the municipal waste stream.

LEGISLATIVE AND REGULATORY HISTORY

Responding to the increasingly polluted conditions of the estuary, the New York State legislature directed the city of New York to form the Metropolitan Sewerage Commission of New York in 1906. This commission was charged with the dual tasks of investigating the extent of water pollution in the harbor and formulating a plan to improve city sanitary conditions. In addition to recommendations for upgrades of waste treatment, which eventually were implemented beginning in the 1930s, the Commission also recommended that outfalls be relocated from nearshore areas to a central diffuser in the Lower Bay. A central diffuser system, however, was never adopted (Suszkowski, 1990).

Construction of primary wastewater treatment plants in the Hudson-Raritan estuary began with Passaic Valley, New Jersey, coming on line in 1924, followed by Yonkers, New York, in 1933. During the construction of the first treatment plants in the 1930s and 1940s in New York City, the New York City Department of Public Works maintained an active role in research and development of waste treatment processes, particularly in the area of biological waste treatment. Although the federal government's primary role was to provide technical advice through the Public Health Service, the Roosevelt Administration did provide federal public works funding for sewage treatment plant construction as a relief program during the Great Depression (O'Connor, 1990). Because of the regional nature of water pollution problems, New York, New Jersey, and Connecticut established the Interstate Sanitation Commission (ISC) to develop water quality standards and to report on progress in water pollution control in the harbor.

Following the passage of the Federal Water Pollution Control Act and amendments in 1948 and 1956, the state and federal governments gradually began to assume a larger role in providing funds for water pollution control. Following World War II, the State of New York initiated a water pollution control program in 1949 that was primarily directed toward classification of streams and rivers and inventories of municipal and industrial wastewater sources (Hetling and Jaworksi, 1995). Beginning in 1956, and continuing on a much larger scale with the 1972 CWA, the Construction Grants Program has provided federal funding for the construction of municipal wastewater treatment plants (see Chapter 2). Following the 1965 amendments to the Federal Water Pollution Control Act, federal funding through the Public Health Service and the Federal Water Pollution Control Administration was also available to provide technical assistance in monitoring and analysis to investigate water quality management issues (FWPCA, 1965, 1969). Stimulated by severe water quality problems exacerbated by the persistent drought conditions of the early 1960s, the State of New York passed a $1.7 billion bond issue in 1965 to provide funds under the "Pure Waters Program" for the construction of municipal wastewater treatment plants (Hetling and Jaworksi, 1995).

Under the 1972 CWA, 208 studies were conducted areawide to evaluate regional water quality management solutions related to waste treatment facility needs (Hazen and Sawyer, 1978; O'Connor and Mueller, 1984). Authorization for New York City Department of Environmental Protection (NYCDEP) to oversee its own industrial pretreatment program for corrosion control in 1987 has led to significant reductions in heavy metal loadings (Brosnan et al., 1994). A citywide CSO Abatement Program is currently under way to comply with USEPA's national CSO strategy. New York City has allocated $1.5 billion for construction of CSO abatement facilities over the next 10 years and is proceeding with water quality studies and facility planning. In the meantime, the city implemented the "Nine Minimum Controls" issued by USEPA as part of the 1994 National CSO Control Policy, with significant improvements in water quality conditions (Brosnan and Heckler, 1996; Heckler et al., 1998). Since enactment of the 1965 amendments to the Federal Water Pollution Control Act, $7.5 billion has been invested by federal, state, and local governments to upgrade 11 of 12 water pollution control plants and to construct and upgrade the North River and Red Hook plants (Adamski and Deur, 1996).

With limited open land area, sludge disposal has always been a major problem for the New York–New Jersey region. In 1924, New York City began routine ocean disposal of sewage sludge at a dump site 12 miles south of Rockaway Inlet off Long Island. Over the following five decades, New Jersey and Westchester County also used ocean dumping to dispose of sewage sludge. By 1979, 5.4 million metric tons (mt) of sewage sludge solids (5 percent) had been dumped into the shallow (30-meter) site (Mueller et al., 1982). Because of the ecological effects, and the resulting political and public controversy (NACOA, 1981), ocean dumping at the 12-mile site was abandoned in 1985. Sludge disposal was then moved to a deepwater site 106 miles offshore until this practice was ended in 1992. New York City has subsequently constructed eight sludge dewatering facilities. Various private contractors then further process approximately 1,200 tons per day of dewatered sludge, known as biosolids for beneficial reuse. New York City's biosolids program is designed to be diverse, employing direct land application, thermal drying, composting and lime stabilization (Ryan, 2001).

IMPACTS OF WASTEWATER TREATMENT

Pollutant Loading and Water Quality Trends

Beginning with decisions by local authorities to construct an organized sewerage collection system in Lower Manhattan as early as 1696, a complex network of stormwater and sewage collection systems and wastewater treatment plants has evolved over the past 300 years, initially to minimize nuisances and protect public health, and most recently to restore and protect the estuarine environment. In 1886, the first wastewater treatment plant was constructed to protect bathing beaches at Coney Island. Following recommendations of a 1910 master plan for sewage treatment by the Sanitary Commission, New York City, Passaic Valley, New Jersey, and Yonkers, New York, initiated construction programs, beginning in the mid-1920s at Passaic Valley, for wastewater plants (O'Connor, 1990).

Following the master plan from the Sanitary Commission, the City of New York began construction of the first modern wastewater treatment facility at Coney Island in 1935 and three plants discharging to the East River in 1938. Other locations also constructed municipal wastewater treatment plants at this time. Modern treatment plants went on-line in 1938 at North and South Yonkers, New York, designed for a combined discharge of 130 mgd into the Hudson River; Passaic Valley, New Jersey, first constructed a plant in 1924 and upgraded it in 1937 to 250 mgd capacity. By 1952, a total of 11 water pollution control facilities were operational in New York City, with 7 of these 11 facilities providing primary treatment. Upgrades to the seven primary facilities during the 1950s and 1960s gradually resulted in improvements in water quality within the harbor. By 1967, the largest New York City plant, Newton Creek, came online discharging 310 mgd into the East River, with New York City's wastewater treatment facilities accounting for a total effluent discharge of approximately 1,000 mgd.

Driven by the regulatory controls of the 1972 Clean Water Act, public works officials in New York City, New Jersey, Connecticut, and Westchester County embarked upon programs to upgrade all municipal treatment facilities to full secondary treatment during the 1970s and 1980s. In 1986, completion of the North River water pollution control plant ended the discharge of 170 mgd of raw sewage into the Hudson River from Manhattan, with secondary treatment attained in 1991. In 1987, completion of the Red Hook water pollution control plant abated the discharge of 40 mgd of raw sewage into the Lower East River from Brooklyn, with secondary treatment attained in 1988. An additional 0.7 mgd of previously unsewered discharge was captured beginning in 1993 when wastewater from Tottenville, Staten Island, was connected to the 40-mgd Oakwood Beach water pollution control plant. Since the completion of the North River plant in 1986 and Red Hook plant in 1987 as advanced primary plants, all wastewater collected in the total sewered area of about 2,000 square miles (Figure 6-7) in the New York metropolitan region has been treated before discharge into the Hudson-Raritan estuary.

From 1979 to 1994, 13 of the 14 municipal water pollution control plants operated by the city of New York were upgraded to full secondary treatment, as defined by the 1972 Clean Water Act (Schwartz and Porter, 1994). The North River (170 mgd) and

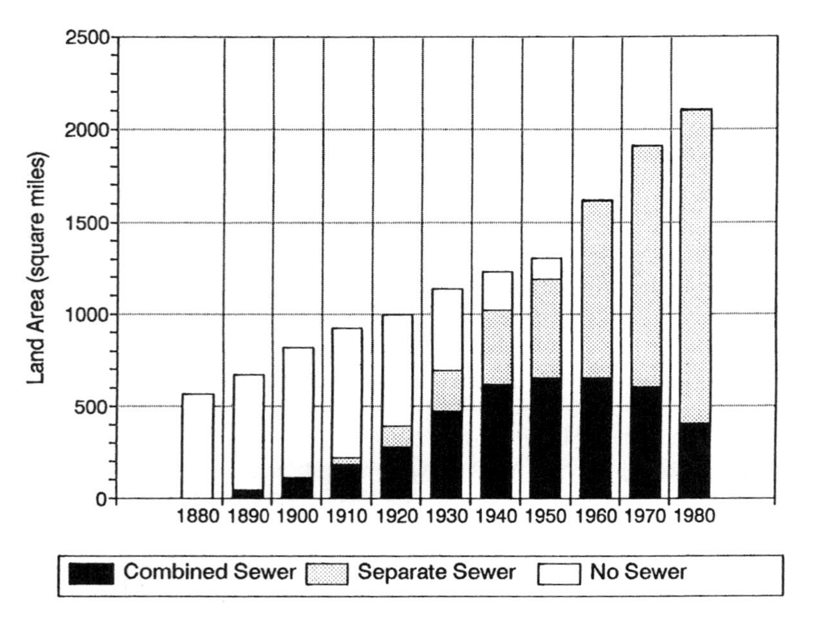

Figure 6-7 Long-term trend of sewerage collection in the Hudson-Raritan estuary, metropolitan region, 1880–1980. *Source:* Suskowski, 1990.

Red Hook (45 mgd) plants, originally on line in 1986–1987 as advanced primary facilities, were upgraded to full secondary plants in 1991 and 1989, respectively (Brosnan and O'Shea, 1996a). The Newton Creek water pollution control plant is expected to be upgraded to full secondary treatment by 2007 (Schwartz and Porter, 1994). As a result of upgrades to existing plants and construction of the North River and Red Hook plants, the discharge of raw sewage has been reduced from 1,070 mgd in 1936 to less than 1 mgd by 1993. Intermittent raw discharges, caused by malfunctions or construction bypasses, have been reduced from 3.8 mgd in 1989 to 0.85 mgd by 1995 (O'Shea and Brosnan, 1997; Brosnan and O'Shea, 1996b). The locations of municipal water pollution control plants (WPCP) discharging flows > 10 mgd into the Hudson-Raritan estuary are shown in Figure 6-8.

Historical data have been compiled from 1900–2000 to show long-term trends for untreated, primary, and secondary municipal wastewater treatment plants. Population served, effluent flow and effluent loading rates of biochemical oxygen demand (BOD_5), total suspended solids (TSS), total nitrogen (TN), and total phosphorus (TP) are compiled for the Middle and Lower Hudson River basins (Hetling et al., 2001). A budget of the point and nonpoint source contributions of these pollutant loads is also presented as a contemporary "snapshot" of loads for the 1990s. The data sources, methodology, and assumptions used to estimate population served, wastewater flow, and pollutant loads are documented in Johnson (1994), Johnson and Hetling (1995), Hetling and Jaworski (1995), and Hetling et al. (2001).

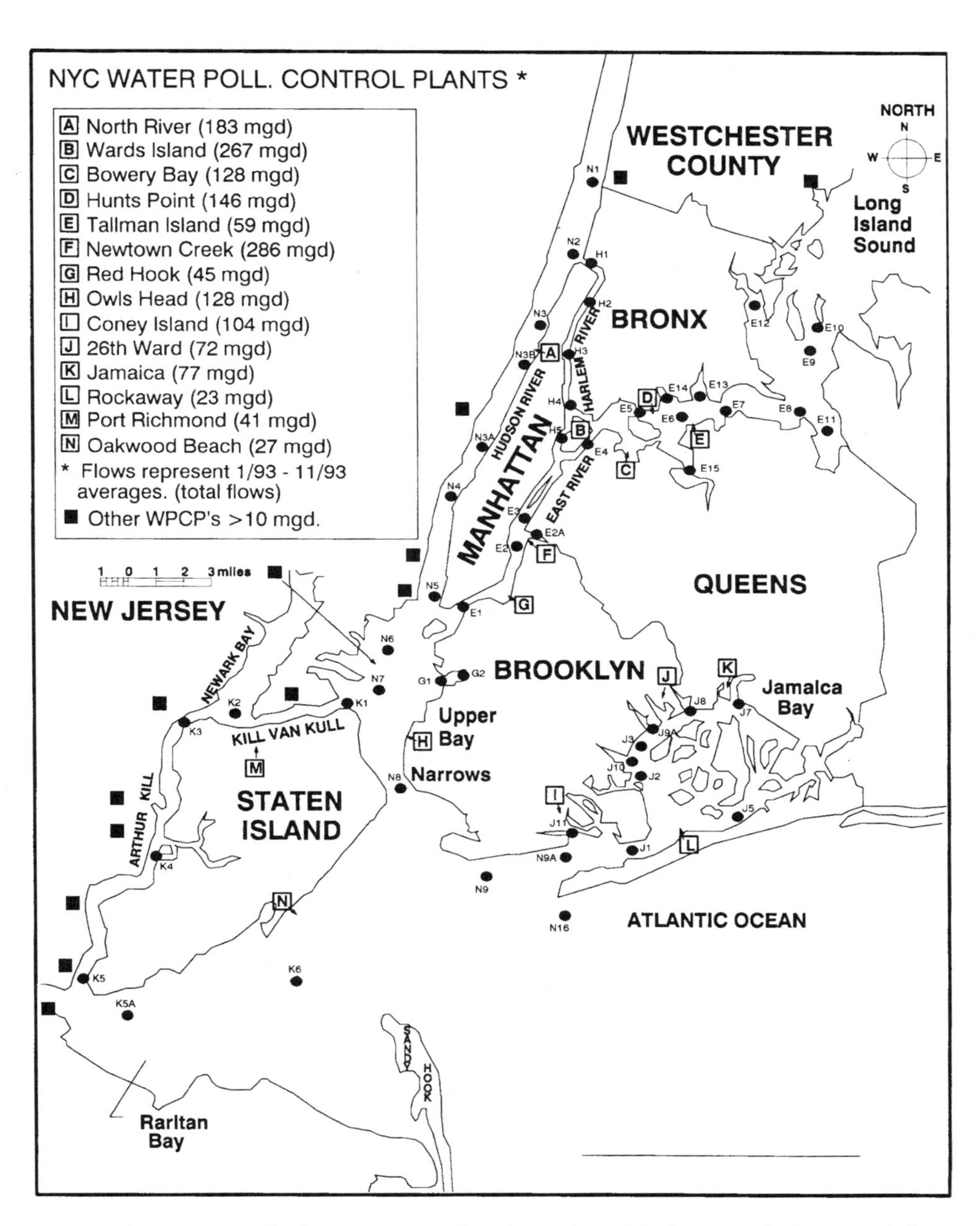

Figure 6-8 Location of harbor survey sampling sites and municipal water pollution control plants in New York Harbor. *Source:* Brosnan and O'Shea, 1996b. Copyright © Water Environment Federation, Alexandria, VA. Reprinted with permission.

The Middle Hudson watershed, from the Federal Lock and Dam at Troy, New York, to the New York City boundary, includes most of the area of Albany, Columbia, Duchess, Greene, Orange, Putnam, Rensselaer, Rockland, Ulster, and Westchester counties. The Lower Hudson watershed, from the New York City boundary to the Atlantic Ocean, includes the five boroughs of New York City (Queens, Bronx, Richmond, Kings, and New York counties) and Bergen, Passaic, Essex, Morris, Hudson, and Union counties in New Jersey. For the effluent flow and pollutant load trend estimates presented in this chapter, the Lower Hudson basin includes discharges to the Hudson River from the New York City line to the Verrazano-Narrows Bridge, the Harlem River, the East River to Throgs Neck and the Kill van Kull. Wastewater discharges to Western Long Island Sound, Jamaica Bay, the Hackensack and Raritan Rivers, Raritan Bay, Newark Bay, and the Arthur Kill are not included in the long-term trend estimates of population served, effluent flow, and effluent loads for the Lower Hudson basin.

During the course of the twentieth century, the total population served by municipal wastewater facilities in the Middle and Lower Hudson basins more than doubled, from 3.4 million in 1900 to 8.5 million by 2000 (Figure 6-9). This increase in the population served by wastewater facilities reflects the growth of the population of the New York and New Jersey metropolitan region (see Figure 6-5) and, beginning in about 1880, the increasing proportion of the population in the metropolitan drainage

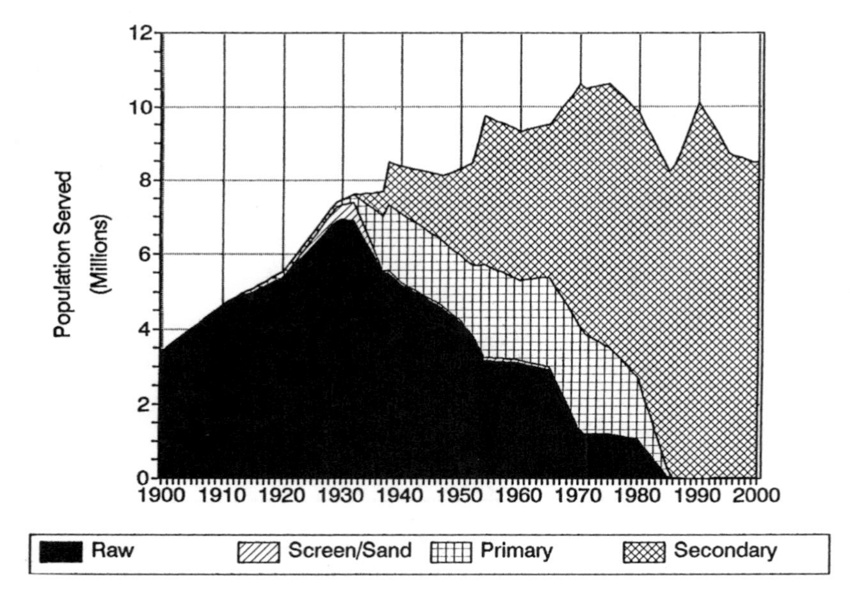

Figure 6-9 Long-term trends (1900–2000) of population served by raw, fine screen/sand filter, primary, and secondary municipal wastewater facilities in the Middle and Lower Hudson basins. *Source:* Hetling et al., 2001. Presented at the NYWEA conference February, 2001.

basin that was connected to urban sewerage collection systems (Suskowski, 1990) (see Figure 6-7). Reflecting the migration of people from the suburbs following World War II, the sewered population served by wastewater facilities in the Middle Hudson basin rose from 10 percent of the total population served in the mid-1950s to 18 percent by 2000.

The population served by facilities discharging untreated (raw) sewage steadily increased during the period from 1900 to the 1930 (Figure 6-9). In the Lower Hudson basin, raw sewage was discharged by 3.0 million people in 1900 increasing to a peak of 6.4 million by 1932. From the mid-1930s to the late-1980s, the population served by untreated sewage facilities steadily declined to 2.9 million in 1960 and 1.07 million in 1980 as raw sewage discharge facilities were upgraded to primary or secondary treatment. The population served by primary facilities increased from 1.05 million in 1937 to a peak of 1.92 million in 1954 (Figure 6-9). Completion of Manhattan's North River plant in 1986 and Brooklyn's Red Hook facility in 1987 as advanced primary facilities finally eliminated the discharge of 121 mgd of raw sewage into the East River and the Lower Hudson River off the west side of Manhattan (Brosnan and O'Shea, 1996a).

As a result of the regulatory requirements of the 1972 CWA and the availability of federal and state construction grants, the population served by full secondary treatment facilities has increased from 1.17 million in 1938 to 8.58 million by 1990 in the Lower Hudson basin. A similar trend is recorded for the Middle Hudson basin, where the population served by secondary increased tenfold during the 1950s and 1960s to 0.25 million by 1970. After the 1972 CWA and an establishment of a state construction grants program, the population served by secondary treatment plants in the Middle Hudson increased to 1.5 million by 1999. Based on an analysis of all the wastewater discharges to the Hudson River and all other adjacent waterways, municipal sewage treatment plants in the New York metropolitan region served approximately 16 million people and discharged about 2,500 mgd during the mid-1990s (Brosnan and O'Shea, 1996a).

The erratic trends in total population served presented in Figure 6-9, particularly the noticeable decline between 1980 and 1990, result from year to year inconsistencies in the population served data obtained from Interstate Sanitation Commission (ISC) annual reports on municipal wastewater facilities discharging to the Lower Hudson basin and Long Island Sound. The time series estimates of effluent flow and effluent loads of BOD_5, TSS, TN, and TP are dependent on estimates of population served. Any inconsistencies in the trends for population served also result in inconsistencies for the characterization of effluent trends, particularly between 1980 and 1990, for effluent flow (Figure 6-10) and BOD_5 (Figure 6-11), TSS (Figure 6-12), and TN loads (Figure 6-13).

Following the long-term trend in population served by sewers, effluent flow from municipal facilities in the Lower Hudson basin also increased steadily over the course of the twentieth century. Figure 6-10 shows the trends in effluent flow for the Middle and Lower Hudson basins estimated from the population served and per capita water use rates. Based on a per capita water use rate of 125 gpcd at the turn of the twentieth century, a total of 410 mgd of untreated sewage was discharged to the Middle and

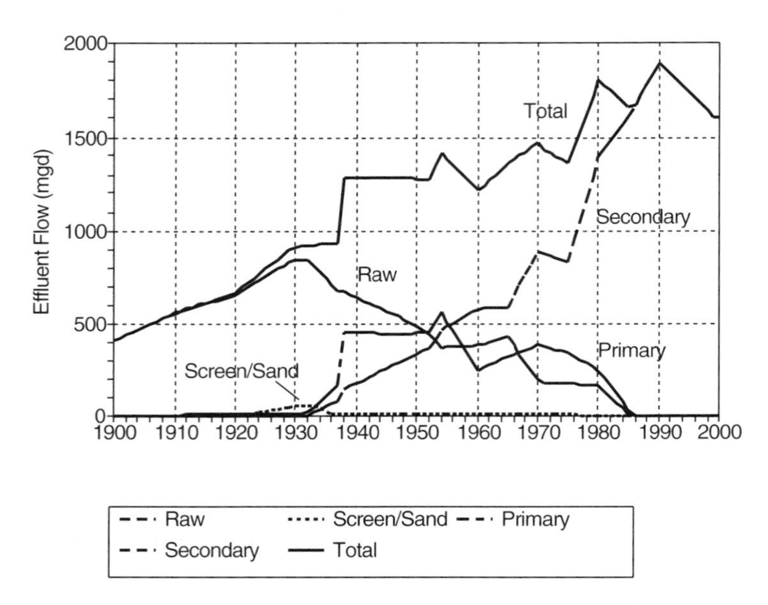

Figure 6-10 Long-term trends (1900–2000) of effluent flow by raw, fine screen/sand filter, primary, and secondary municipal wastewater facilities in the Middle and Lower Hudson basins. *Source:* Hetling et al., 2001. Presented at the NYWEA conference, February, 2001.

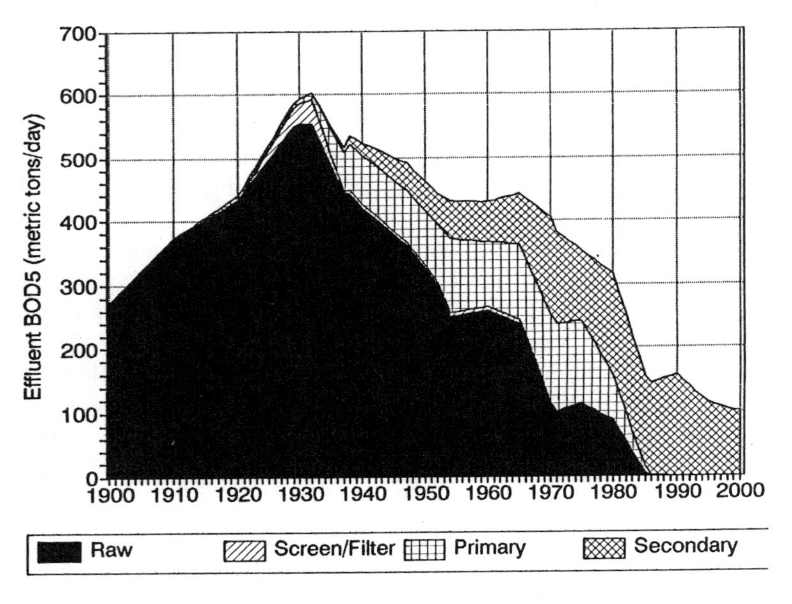

Figure 6-11 Long-term trends (1900–2000) of effluent BOD_5 loads by raw, fine screen/sand filter, primary, and secondary municipal wastewater facilities in the Middle and Lower Hudson basins. *Source:* Hetling et al., 2001. Presented at the NYWEA conference, February, 2001.

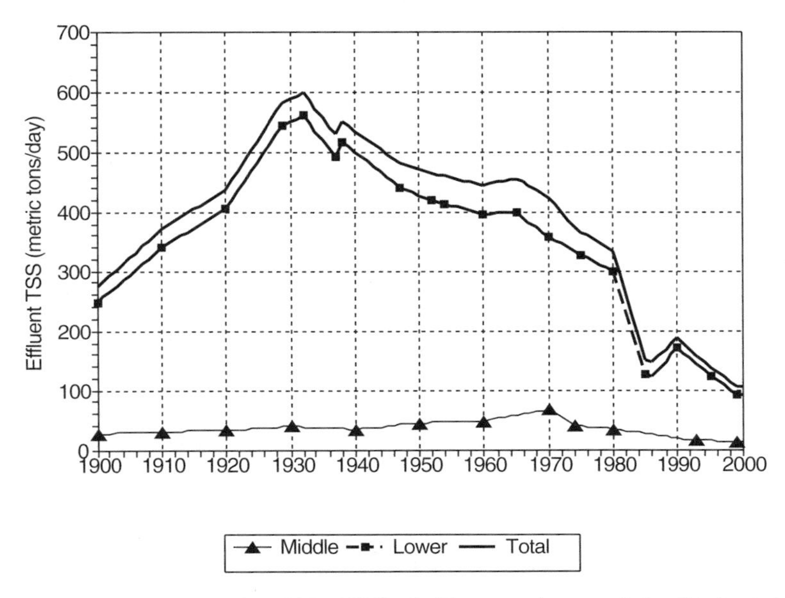

Figure 6-12 Long-term trends (1900–2000) of effluent total suspended solids loads by municipal wastewater facilities in the Middle and Lower Hudson basins. *Source:* Hetling et al., 2001. Presented at the NYWEA conference, February, 2001.

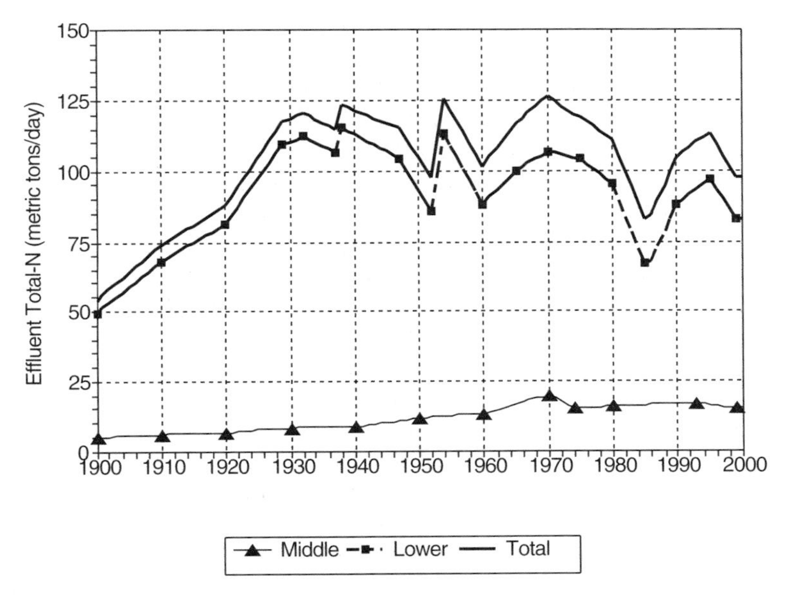

Figure 6-13 Long-term trends (1900–2000) of effluent total nitrogen loads by municipal wastewater facilities in the Middle and Lower Hudson basins. *Source:* Hetling et al., 2001. Presented at the NYWEA conference, February, 2001.

Lower Hudson River basins. During the 1940s and 1950s, effluent flow ranged from 1,250 to 1,344 mgd with about 30 percent from raw, 24 to 38 percent from primary plants, and 30 to 45 percent from secondary facilities. Increasing per capita consumption of water (150 gpcd in 1965 to 170 gpcd in 1985) and a small increase in population served resulted in a steady increase in effluent flow from 1,250 mgd in the early 1960s to 1,900 mgd by 1990. With the upgrade of the Coney Island plant to full secondary in 1994, effluent flow from less than secondary plants has been almost completely abated in the Hudson-Raritan estuary; the Newton Creek plant is expected to be upgraded to secondary by 2007 (Schwartz and Porter, 1994). Declines in population served from 1990 to 2000 and a New York City water conservation program then resulted in a decline of total effluent flow to 1,600 mgd by 2000. Effluent flow from municipal wastewater facilities in the Middle Hudson basin accounted for about 8 percent of the total effluent flow in 1900, 5 percent in the 1940s and 1950s, and 15 percent by 1990–2000.

With the same per capita loading rate (0.18 lb per capita per day) used to estimate effluent loads for BOD_5 and TSS (Hetling et al., 2001), the long-term trends for BOD_5 (Figure 6-11) and TSS (Figure 6-12) loads for the Middle and Lower Hudson basins are quite similar. The small differences in the estimated effluent loads are dependent on the BOD_5 and TSS removal efficiency assigned to primary (15 percent BOD_5 and 50 percent TSS removal) and secondary (85 percent BOD_5 and 85 percent TSS removal) treatment plants (Hetling et al., 2001). Following a steadily increasing trend similar to that shown for effluent flow and population served, BOD_5 and TSS loads from raw discharges to the Middle and Lower Hudson basins increased from 273 metric tons per day (mt/day) in 1900 to a peak load of 554 mt/day by 1932. With the construction of primary treatment plants in the late 1920s and 1930s and upgrades to secondary during the 1940s, 1950s, and 1960s, untreated BOD_5 and TSS effluent loads declined by 80 percent from the peak of 554 mt/day in 1932 to 112 mt/day in 1970. The combined raw, primary, and secondary loads declined from 600 mt/day in 1932 to 400 mt/day by 1970. The steady increase in BOD_5 loading from 1900 to 1932 is attributed to population growth (Figure 6-5) and an expanding urban sewage collection system (Figure 6-7), while the reduction in loads from the 1930s to the 1950s is attributed to the construction of three primary treatment plants during the 1930s. After the mid-1960s, the decline in total TSS and BOD_5 effluent loading to 103 mt/day by 2000 was driven by upgrades to full secondary treatment and the elimination of raw sewage discharges from the west side of Manhattan (North River plant) and Brooklyn (Red Hook plant) with the construction of advanced primary plants in 1986–1987 and eventual upgrades to full secondary in 1988–1991. Effluent loads of BOD_5 and TSS from municipal wastewater facilities in the Middle Hudson basin accounted for about 10 percent of the total BOD_5 and TSS effluent load in 1900, 10 to 25 percent in the 1950s and 1960s, 15 percent during the 1980s, and 10 to 14 percent by 1990–2000.

Following a steadily increasing trend similar to that shown for effluent flow and population served, TN loads (Figure 6-13) from raw sewage discharges to the Middle and Lower Hudson basins increased from 54 mt/day in 1900 to a peak loading rate of 124 mt/day by 1938. With the construction of primary treatment plants in the late 1920s and 1930s and subsequent upgrades to secondary treatment during the 1940s,

1950s, and 1960s, effluent TN loads declined by only 15 percent from the peak of 124 mt/day in 1938 to 105 mt/day during the early 1960s. After enactment of the CWA in 1972, and the required upgrades of water pollution control plants in the Middle Hudson and the Lower Hudson metropolitan region to full secondary treatment, effluent loads of TN declined from 125 mt/day in 1970 by only ~22 percent to 97 mt/day by 1999. Full secondary plants, although not specifically designed for the removal of nitrogen, typically can achieve about 40 percent removal of TN (see Table 2–17). Note, however, that New York City wastewater treatment plant removals for TN are only about 20 percent or less, primarily due to weak influent (O'Shea and Brosnan, 2000). Effluent loads of TN from municipal wastewater facilities discharging to the Middle Hudson accounted for about 9 percent of the total TN effluent load in 1900, 7 to 14 percent in the 1940s and 1950s, 18 to 22 percent during the 1970s and 1980s, and 18 percent by the 1990s.

Following a steadily increasing trend similar to that shown for effluent flow and population served, TP loads (Figure 6-14) from raw sewage discharges to the Middle and Lower Hudson basins increased by 150 percent from 6 mt/day in 1900 to a loading rate of 15 mt/day by 1938. Even with the construction of primary treatment plants in the late 1920s and 1930s, and subsequent upgrades to secondary treatment from the 1940s through the 1960s, effluent TP loads continued to increase to 14.5 mt/day by 1950, with a rapid increase over the next two decades to a peak of 36 mt/day by 1970. Effluent loads of TP increased from 1938 to 1970 even as raw sewage discharges

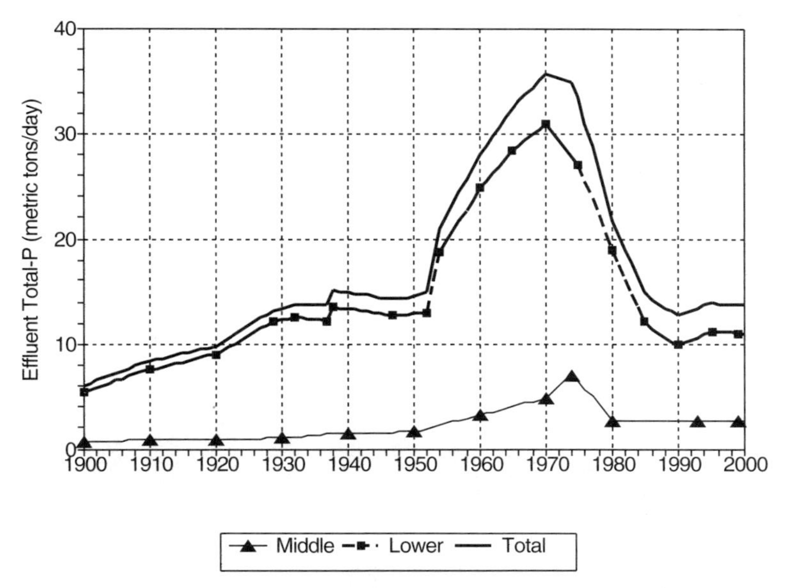

Figure 6-14 Long-term trends (1900–2000) of effluent total phosphorus loads by municipal wastewater facilities in the Middle and Lower Hudson basins. *Source:* Hetling et al., 2001. Presented at the NYWEA conference, February, 2001.

were eliminated and water pollution control plants were upgraded to primary and secondary treatment for three reasons: (1) population served and influent wastewater flow increased; (2) removal efficiency of TP for both primary and secondary plants is only ~30 percent; and (3) influent concentration of TP steadily increased from ~2 mg P/L to a peak of ~10 mg P/L in 1973 after the introduction of phosphorus-based detergents in 1945 (Hetling and Jaworski, 1995). After state legislative bans of phosphorus-based detergents in 1973 reduced influent levels of TP to ~3 mg P/L by the mid-1990s (Hetling and Jaworski, 1995), and after water pollution control plants in the Middle Hudson and the Lower Hudson metropolitan region were upgraded to full secondary treatment, effluent loads of TP declined sharply by 63 percent from the peak of 36 mt/day in 1971 to only 13.7 mt/day by 2000. Since the removal efficiency of 30 percent for phosphorus is similar for both primary and secondary treatment, the decline in effluent loading of TP has resulted from the ban on phosphorus-based detergents (Clark et al., 1992; Hetling and Jaworski, 1995).

The long-term trend (1880–1980) of historical loading of copper and lead to New York Harbor (Figure 6-15) reflects increasing urbanization and uncontrolled wastewater discharges from industrial activity in the metropolitan region from 1880 through 1970 (Rod et al., 1989). The reduction in loading of these metals after 1970, resulting from the industrial pretreatment program, corrosion controls, and effluent controls on industrial discharges, corresponds to a decrease in sediment levels of copper and lead in the Hudson estuary (Valette-Silver, 1993) (Figure 6-16). Studies con-

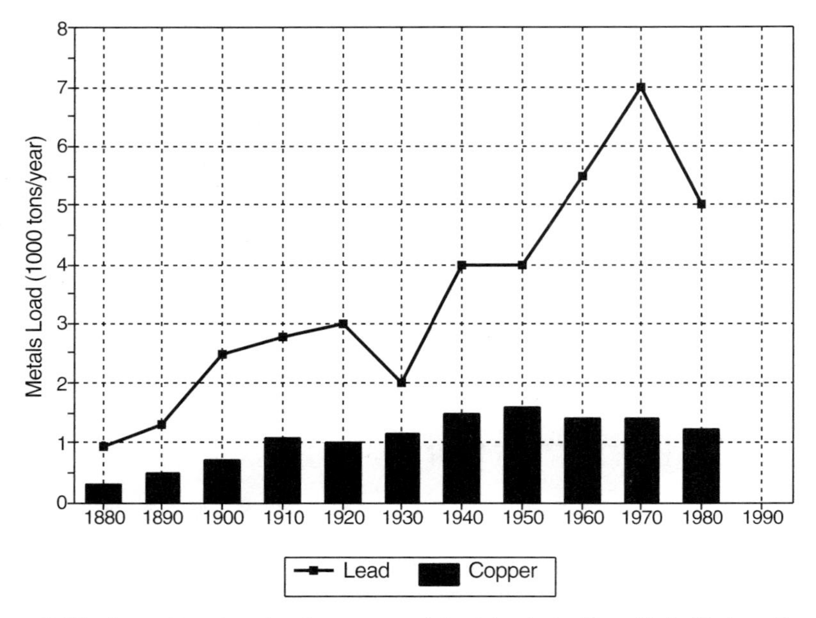

Figure 6-15 Long-term trends of copper and lead loads to New York Harbor. *Sources:* Suskowski, 1990; Rod et al., 1989.

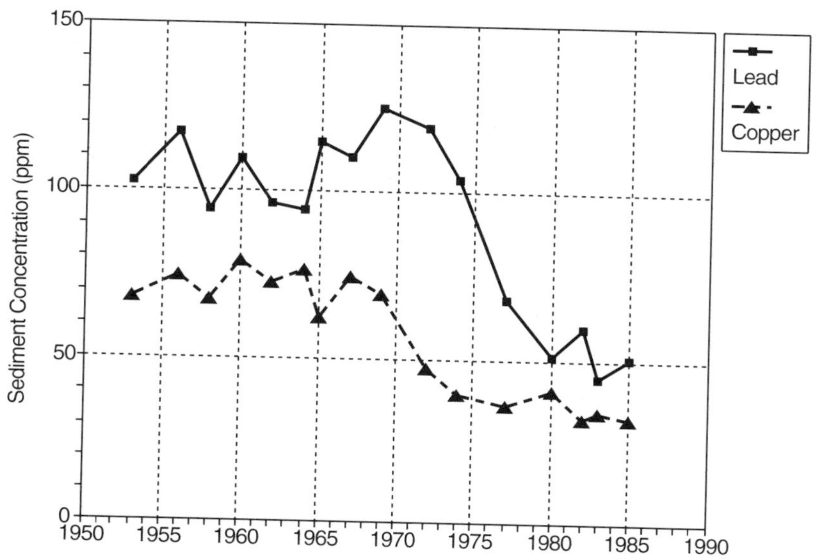

Figure 6-16 Time history of copper and lead in a sediment core in the low-salinity reach of the Hudson estuary. Sediment core collected in 1986, 55 miles (88.6 km) from the mouth of the estuary. *Source:* Valette-Silver, 1993.

ducted at Columbia University have documented a 50 to 90 percent reduction from the 1960s and 1970s in most trace metals and chlorinated organic compounds in fine-grained sediments of the Hudson River (Chillrud, 1996). Sediment toxicity, however, has been identified for the Upper East River, Arthur Kill, Newark Bay, and Sandy Hook Bay. The observed distribution of sediment toxicity appears to be most strongly related to polynuclear aromatic hydrocarbons (PAHs) rather than trace metals (Wolfe et al., 1996). Historical point and nonpoint source loading estimates for the Hudson-Raritan estuary are presented elsewhere for other trace metals, PCBs, total suspended solids (HydroQual, 1991), total organic carbon (Swaney et al., 1996; Howarth et al., 1996) and nutrients (HydroQual, 1991, 1999; Carpenter, 1987). Using a steady-state toxics model, the New York–New Jersey Harbor Estuary Program has also developed mass balance analyses for copper, nickel, and lead and a preliminary mass balance for mercury and PCBs (HydroQual, 1995b).

Municipal wastewater discharges account for only one point source component of the total loading of pollutants to the Hudson-Raritan estuary from point and nonpoint sources. In order to properly place the magnitude of municipal wastewater loads in the context of the total point and nonpoint source pollutant loads discharged to the estuary, estimates of the contributions from tributaries, municipal and industrial point sources, CSOs and urban stormwater of the Middle and Lower Hudson basins have been compiled as a budget based on data compiled for conditions during the 1990s. Table 6-2 illustrates that the relative significance of different sources is dependent on

TABLE 6-2 Pollutant Loadings to the Hudson-Raritan Estuary (in Percent)

Parameter	Tributary (%)	Municipal Effluents (%)	Combined Sewer Overflow (%)	Stormwater (%)	Other[a] (%)	Total Load (Units)
Flow	81	15	1.0	4	0.5	765 m^3/sec
Fecal coliform	2	0.1	89.0	9	0.1	2.1E + 16 mpn/day
BOD	16	58	19	5	2	5.7E + 05 kg/day
TSS	80	11	5	3	1	2.4E + 06 kg/day
Nitrogen	29	63	2	2	4	2.8E + 05 kg/day
Phosphorus	16	75	4	4	0.5	2.3E + 05 kg/day

Sources: Data in table taken from Brosnan and O'Shea, 1996a is modified from HydroQual (1991), based on data from the late 1980s.

Note: Values in a row may not equal 100% because of rounding.

[a]Other = combined loading from industrial discharges, landfill leachate, and direct atmospheric deposition.

the pollutant considered. Combined sewer overflows, for example, account for only 1 percent of the total freshwater input to the harbor but contribute 89 percent of the total loading of fecal coliform bacteria (Brosnan and O'Shea, 1996a). Effluent from water pollution control plants contributes about one-half to three-quarters of the total load of BOD_5 and nutrients, while watershed runoff via tributaries accounts for 80 percent of the total suspended solids (TSS) load. Table 6-3 presents a summary of the distribution of effluent flows from municipal water pollution control plants (WPCPs) and industrial point source discharges to New York Harbor waterways (HydroQual, 1991; O'Shea and Brosnan, 1997). As presented in Table 6-3, approximately 2,500 mgd of treated wastewater was discharged in 1995 from 81 water pollution control facilities located in New York City, six New Jersey coastal counties, two coastal Connecticut counties, and Westchester and Rockland counties in New York (O'Shea and Brosnan, 1997). Of the total 2,500 mgd, facilities operated by the City of New York accounted for 1,490 mgd in 1995.

Long-term water quality records for most locations within the estuary clearly illustrate degradation from population growth and inadequate sewage treatment through the mid-1960s and gradual improvement following construction of wastewater treatment plants and implementation of secondary treatment. Using historical data collected at 40 stations in the harbor from 1968 to 1993, an analysis of harborwide long-term trends clearly documents more than an order-of-magnitude improvement in total coliform and fecal coliform concentrations (Figure 6-17). The dramatic decline in bacterial levels is attributed to water pollution control infrastructure improvements that eliminated raw sewage discharges and upgraded all water pollution control plants to include disinfection by chlorination (O'Shea and Brosnan, 1997). Other improvements, reductions of approximately 50 percent in bacterial levels for most areas of the

TABLE 6-3 Distribution of Wastewater Flows into New York Harbor Waterways

	WPCPs (2,500 mgd)		Direct Industrial Discharges (52 mgd)	
Waterway	Flow (mgd)	% Total	Flow (mgd)	% Total
Hudson River	375	15	3.1	6
East River	1,050	42	0.0	0
Upper New York Bay	375	15	1.0	2
Jamaica Bay	300	12	0.0	0
Lower New York Bay	125	5	0.0	0
Arthur Kill	100	4	40.0	78
Kill van Kull	50	2	0.0	0
Raritan River	25	1	2.1	4
Hackensack River	100	4	4.2	8
Passaic River	0	0	1.0	2
Total	2,500	100	51.4	100

Sources: HydroQual, 1991; O'Shea and Brosnan, 1997.

Note: Some municipal dischargers (WPCPs) include industrial discharges.

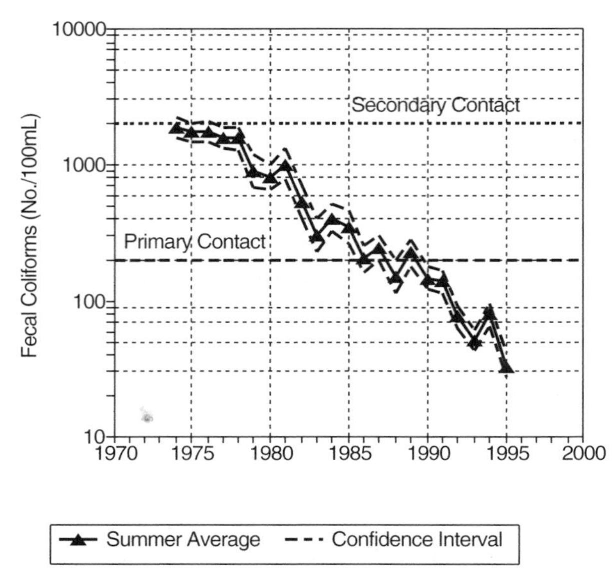

Figure 6-17 Long-term trends in summer geometric mean fecal coliform bacteria. Data represents harborwide composite of 40 stations monitored since at least 1970. *Source:* O'Shea and Brosnan, 1997.

harbor, are attributed to increased surveillance and maintenance of the entire sewage distribution system, including the capture of combined sewage during rain events (Brosnan and O'Shea, 1996b).

Long-term summer DO saturation records, collected almost continuously since 1909, at a station in the Hudson River near 42nd Street on the west side of Manhattan (Figure 6-18) and stations at Baretto Point and 23rd Street in the Upper and Lower East River (Figure 6-19) clearly document the beneficial impact of upgrading water pollution control facilities to full secondary treatment. Over a 40-year period from the 1920s through the 1960s, summer oxygen saturation levels were only about 35 percent to 50 percent at the surface and 25 percent to 40 percent in bottom waters. As a result of significant reductions in biochemical oxygen demand loading (see Figure 6-11), DO saturation levels increased to about 90 percent in the surface and greater than 60 percent in the bottom (Brosnan and O'Shea, 1996a). DO concentrations have increased significantly since the 1980s harborwide (Brosnan and O'Shea, 1996a; Parker and O'Reilly, 1991). In many waterways, the greatest oxygen and BOD_5 improvements were recorded between 1968 and 1984, coinciding with the greatest water pollution control plant construction and upgrading activity (O'Shea and Brosnan, 1997). Analysis of data for stations from 1968 to 1995 document reductions in ammonia-nitrogen (Figure 6-20) and decreases in BOD_5 (Figure 6-21) throughout New York Harbor; exceptions to these decreasing trends include stations in Jamaica Bay and scattered stations in Lower New York Bay and the Upper East River (O'Shea and Brosnan, 1997).

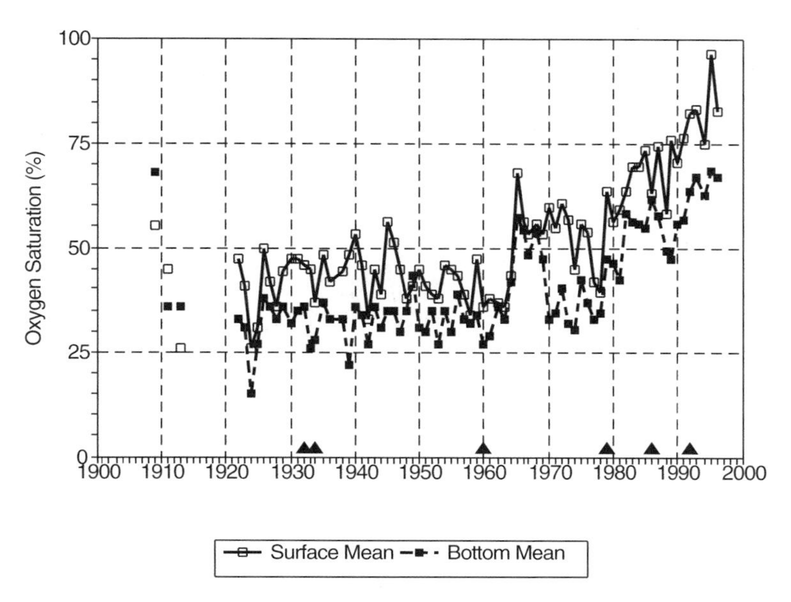

Figure 6-18 Long-term trends of DO saturation (summer average) at 42nd Street in the Hudson River. Triangle markers identify years of upgrades for Yonkers WPCP (1932, 1934, 1960, 1979) and North River WPCP (1986, 1993). *Source:* Brosnan and O'Shea, 1996a.

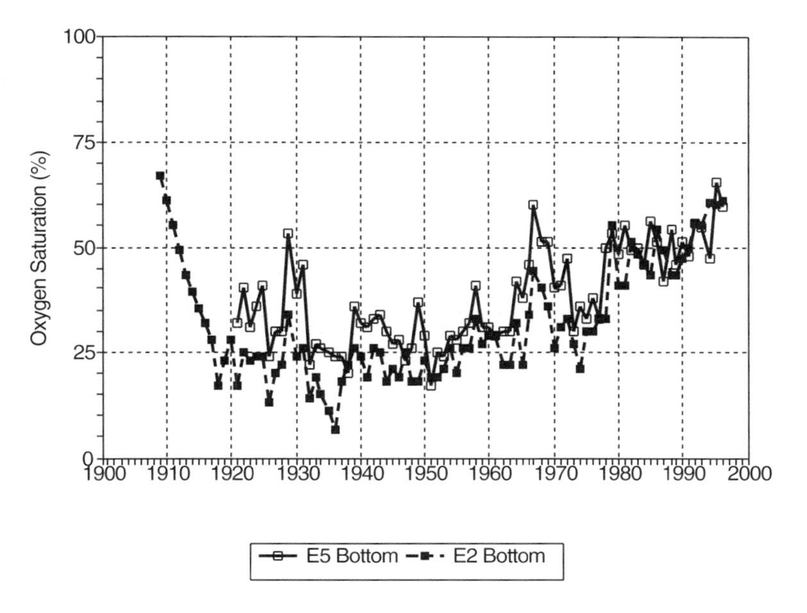

Figure 6-19 Long-term trends of DO saturation (summer average) at Baretto Point (Station E5) in the Upper East River and at 23rd Street (Station E2) in the Lower East River. *Source:* O'Shea and Brosnan, 1997.

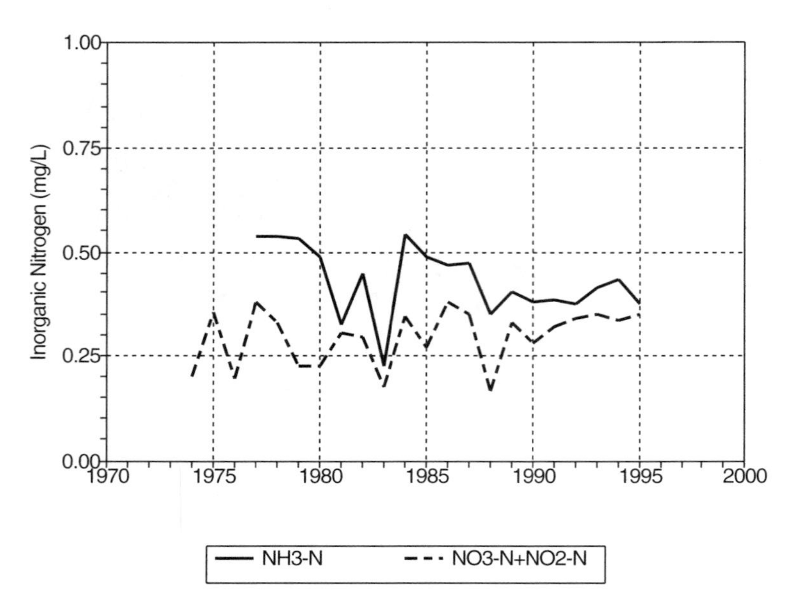

Figure 6-20 Long-term trends in summer mean inorganic nitrogen. Data represent harbor-wide composite of 40 stations monitored since at least 1970. *Source:* NYCDEP, 1999.

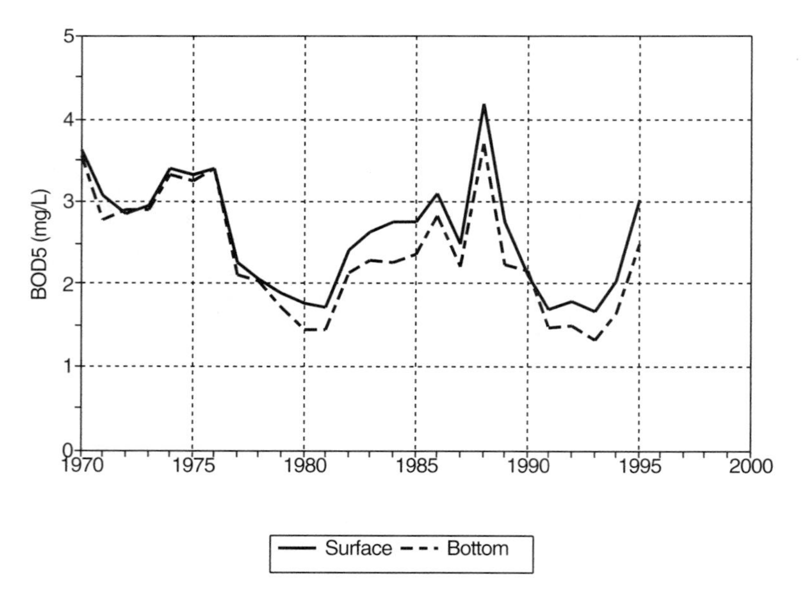

Figure 6-21 Long-term trends in summer mean BOD_5. Data represent harbor-wide composite of 40 stations monitored since at least 1970. *Source:* O'Shea and Brosnan, 1997.

Although not generally appreciated, the poor water quality conditions, particularly low DO levels, that characterized New York Harbor for most of the twentieth century actually had a beneficial effect for shipping activities, since wooden pilings and other submerged wooden structures were *not* destroyed by pollution-intolerant marine borers. During the nineteenth century, before water quality had deteriorated in the harbor, abundant populations of shipworms (*teredos*) and gribbles (*limnoria*) quickly devoured driftwood (naturally occurring) and wooden pilings (man-made). This natural ecological activity probably kept the harbor clear of driftwood, but created severe problems for commercial shipping interests, since untreated wooden pilings needed to be replaced after only about 7 to 10 years (Port Authority of New York, 1988). As water pollution problems increased in the harbor, populations of marine borers declined to such a level that, ironically, wooden pilings and other submerged wooden structures were preserved for many years while submerged in the noxious, oxygen-depleted waters laden with oil, bilge waste, and toxic chemicals. The dramatic improvements in water quality conditions in New York Harbor, as well as other East and West Coast harbors, have resulted in a resurgence of thriving populations of marine borers since the mid-1980s (Gruson, 1993) (Figure 6-22). The population boom of marine borers has resulted in severe infestation and rapid deterioration and collapse of wooden pilings and other submerged wooden structures in New York Harbor from JFK International Airport to New Jersey, including Brooklyn, Staten Island, and the east and west sides of Manhattan (Randolph, 1998; Abood et al., 1995; Metzger and Abood, 1998; Schwartz and Porter, 1994).

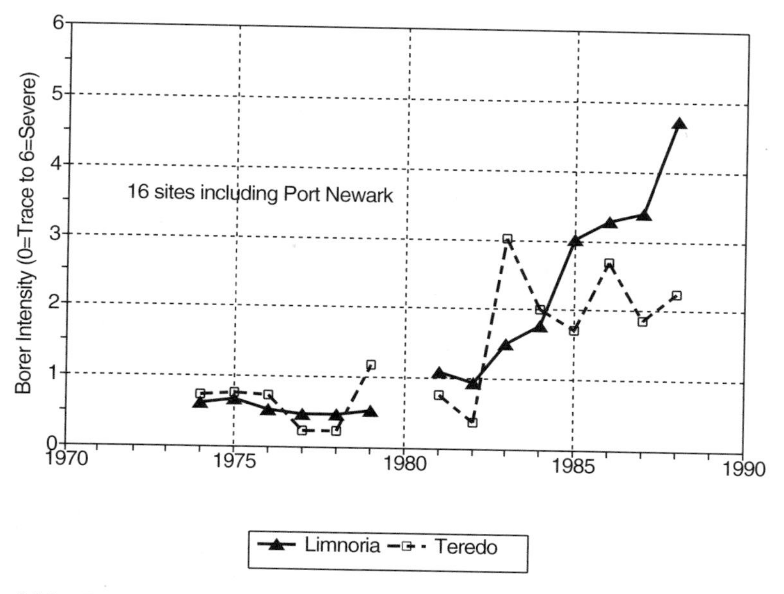

Figure 6-22 Trends in marine borer activity in Upper New York Harbor. *Source:* Port Authority of New York, 1988.

Over the past several years, advanced hydrodynamic and water quality models have been developed for water quality management studies of the harbor, including New York City's Harbor-Wide Eutrophication Model and, most recently, the System-Wide Eutrophication Model (SWEM) (HydroQual, 1995a, 1996, 1999). Earlier models, developed for USEPA's 208 Study of the harbor (Hazen and Sawyer, 1978; Higgins et al., 1978; Leo et al., 1978; O'Connor and Mueller, 1984), have been used to assess the impact of secondary treatment requirements on DO in the harbor.

The more recent New York City models, employing improved loading estimates and modern hydrodynamics (Blumberg et al., 1997), are being used to determine the feasibility and effectiveness of management alternatives for New York City point sources of nitrogen. For example, SWEM will enable New York City to evaluate options, as part of the facility planning for the Newton Creek water pollution control plant, the last remaining plant operated by the City of New York to be upgraded to secondary treatment (HydroQual, 1999). This model is further assisting the New York–New Jersey Harbor Estuary Program in understanding the complex relationships between physical transport processes, nitrogen loading, algal biomass, and DO in New York Harbor, the New York Bight, and Long Island Sound (HEP, 1996).

Recreational and Living Resources Trends

Since 1968, the New York City Council has required the New York City Department of Public Health to notify the Department of Parks of beaches that pose a potential health risk to the public. Such beaches were traditionally posted with wet weather advisories, following occurrences of heavy or prolonged rainstorms. These postings have long been replaced with seasonal wet weather advisory postings. The advisories are based upon the occurrence of high coliform concentrations, which may indicate the presence of raw or partially treated sewage and the likely presence of waterborne pathogens. Diseases associated with recreational swimming waters include typhoid fever, gastroenteritis, swimmer's itch, swimmer's ear, and some viral infections such as infectious hepatitis (NJDEP, 1990).

The most important source of pollution, contributing about 89 percent of the total fecal coliform bacteria load to the harbor, is wet weather CSOs (Brosnan and O'Shea, 1996a). Large volumes of water generated during rainstorm events, when combined with the regular volume of sewage, overwhelm the capacity of the collection system and discharge the mixture of storm runoff and raw wastewater directly into the harbor. During wet weather events, water quality may be seriously degraded.

Before 1900, untreated wastewater caused severe outbreaks of disease associated with exposure pathways by shellfish consumption and recreational swimming. Conditions improved somewhat as sewage treatment plants adopted primary treatment as a practice to settle out the solids in wastewater before discharge to the harbor. Pathogen reduction was further enhanced by upgrading water pollution control facilities to secondary treatment with chlorination of the effluent for disinfection. Improvements in municipal wastewater treatment practices have significantly reduced the incidence of waterborne disease outbreaks. Typhoid fever, once a serious swimming-related disease, for example, has not been reported in the last 30 to 40 years (NJDEP, 1990).

During the 1970s and 1980s, significant efforts were made to construct and upgrade water pollution control plants in the Hudson-Raritan Estuary to attain secondary levels of wastewater treatment as mandated by the 1972 Clean Water Act. With upgrades and chlorination of effluent, the discharge of raw wastewater to the entire Hudson-Raritan estuary has been reduced from 450 mgd in 1970 to less than 5 mgd by 1988 and essentially zero by 1993. The most dramatic improvement in bacterial conditions in the harbor occurred in 1986 with the completion of the North River water pollution control plant in Manhattan. Before construction of the primary facility, 170 mgd of raw sewage was discharged into the Hudson River from 50 outfalls on the west side of Manhattan (Brosnan and O'Shea, 1996a). Treatment of the raw sewage and year-round disinfection resulted in a dramatic decline in fecal coliform concentrations. The 1986 data revealed a 78 percent decrease in the fecal coliform concentrations in the Hudson River compared to values measured in 1985 before the primary plant came on-line. When the 45-mgd Red Hook water pollution control plant went on-line in 1987 in Brooklyn, abating the raw sewage discharge from 33 outfalls, fecal coliform concentrations in the East and Lower Hudson Rivers declined by 69 percent within 2 years. Continued improvements in water quality and decreases in fecal coliform bacterial concentrations on the order of 50 percent from 1989 to 1995 are attributed to improved maintenance and surveillance of the sewage treatment system. Management actions that have contributed to these improvements in water quality include abatement of illegal connections, reduced raw sewage bypasses from dry-weather malfunctions of pumping stations and sewage regulators, and increased capture of combined sewage during rain events (Brosnan and O'Shea, 1996b).

"Snapshots" of the distribution of total coliform bacteria from 1972 to 1995 in surface waters of the harbor (Figure 6-23) clearly document the significant reductions in bacterial concentrations that resulted from implementation of controls to reduce water pollution in the harbor. Following completion of the North River and Red Hook water pollution controls plants in 1986 and 1987, respectively, total coliform distributions in 1988 demonstrated significant improvements compared to 1985 before these two plants came on-line. The improvements attributed to CSO controls and reduction of the volume of raw sewage bypasses are also quite apparent with total coliform levels in the harbor declining by more than 50 percent at 45 of 52 stations in 1995 and compliance with water quality standards improving from 87 percent in 1989 to 98 percent in 1995 (Brosnan and Heckler, 1996).

Historically, many public bathing beaches in Lower New York Harbor have been closed to swimming to protect public health because of high bacterial levels that consistently violated water quality standards of 2,400 MPN/100 mL (total coliforms) and 200 MPN/100 mL (fecal coliforms) for primary contact. As a result of the construction and upgrade of water pollution control plants in the harbor, the capture of combined sewage and the reduction of raw sewage bypasses, the significant harborwide reductions in coliform bacteria levels (see Figure 6-17) allowed the reopening of public beaches that had been closed for decades. In 1988, Seagate Beach on Coney Island was opened to swimming for the first time in 40 years. South Beach and Midland Beach on Staten Island, closed since the early 1970s, were opened for swimming in 1992.

In July 1997, the New York City Department of Environmental Protection initiated

TOTAL COLIFORM TRENDS IN SURFACE WATERS
Summer Geometric Means

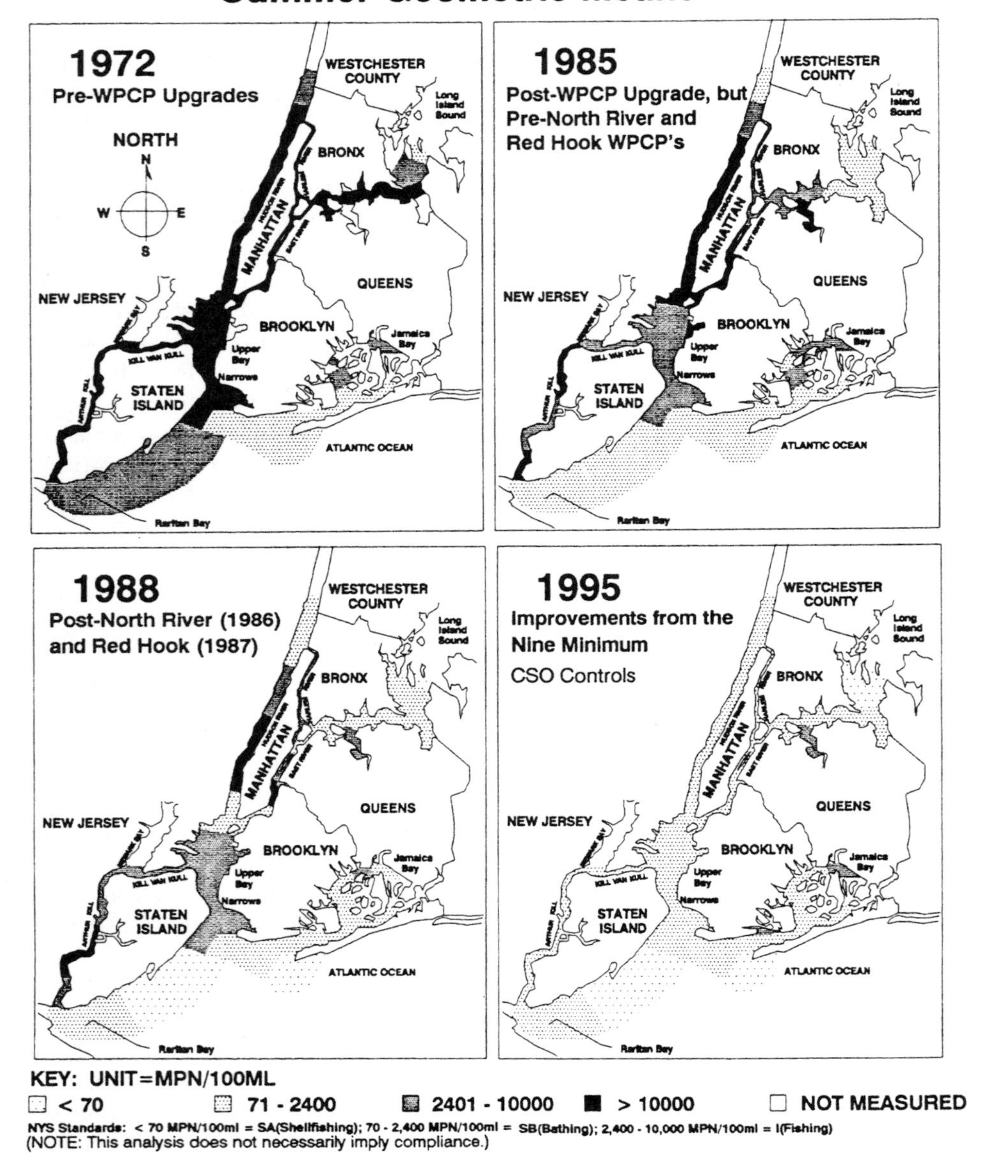

Figure 6-23 Total coliform trends in surface waters of New York Harbor. Summer geometric means for 1972, 1985, 1988, and 1995. *Source:* Brosnan and Heckler, 1996. Copyright © Water Environment Federation, Alexandria, VA. Reprinted with permission.

the Enhanced Beach Protection Program as a response to beach closings caused by a major failure of a pump station and the discharge of raw sewage into western Long Island Sound. Since 1997, the Enhanced Beach Protection Program has succeeded in significantly reducing the occurrence and duration of raw sewage bypasses resulting from operational failures of the conveyance system of 90 pump stations and 490 sewage overflow regulators (Oliveri et al., 2001); no beaches have been closed because of collection facilities bypasses (Loncar, 2002). As a result of management actions, other than the Enhanced Beach Protection Program, implemented by the City of New York to improve the surveillance, maintenance and operation of pump stations and sewage overflow regulators, the volume of raw sewage bypasses from operational failures has been reduced by over 96% from 1,845 million gallons per year as of 1989 to 61 million gallons per year as of 1998 (NYCDEP, 1999).

In addition to beach closings because of high bacterial levels, recreational beaches are also closed because of strandings of floatable debris on the beaches. Floatable debris includes ordinary trash washed off from city streets and illegally disposed medical waste such as hypodermic syringes and blood bags. As a result of increased abatement and control of discharges of floatable debris, beach closings in New York and New Jersey were greatly reduced during the 1990s. Except for 1998 when Rockaway Beach was closed for a day because of medical waste, no beaches in New York City have been closed because of floatables since 1989 (Brosnan and Heckler, 1996).

In addition to closing bathing beaches, the presence of pathogens and pathogenic indicator organisms directly impacts shellfish resources. Because pathogen levels were significantly reduced by improved wastewater treatment and year-round chlorination, 67,864 acres of shellfish beds in the estuary have been upgraded since 1985, including removal of seasonal restrictions for 16,000 acres in the New York Bight in 1988, and 13,000 acres in Raritan Bay in 1989 (Gottholm et al., 1993; NJDEP, 1990). Additionally, 1,000 acres of shellfish waters in the Navesink River are being considered for upgrading to a "seasonally approved" classification (HEP, 1996). Shellfish resources to a greater extent than finfish populations are directly related to improvements in wastewater treatment (Sullivan, 1991).

Although the long-term trends in the abundance of fish such as American shad and striped bass in coastal waters may be the result of degraded water quality (Summers and Rose, 1987), the National Marine Fisheries Service has indicated that overfishing, rather than changes in water quality, is probably the most significant cause of present changes in resource abundance for many species (Sullivan, 1991; Summers and Rose, 1987). The principal commercial fishery of the lower Hudson River estuary is for American shad. Shad landings from 1979 to 1989 were maintained, whereas landings of whiting, red hake, scup, and weakfish decreased during the last decade (Woodhead, 1991). Improved water quality has expanded the spawning area available for American shad (Sullivan, 1991). The prevalence of fin rot in winter flounder declined tenfold in the New York Bight region between 1973 and 1978 for reasons that are not clear (Swanson et al., 1990). Although the causes of fin rot are not well understood, it tends to be more prevalent in shallow inshore waters receiving municipal effluents, and therefore the decline in the incidence of fin rot lesions might reflect improvements at wastewater treatment plants (Sullivan, 1991).

Populations of some birds in the Hudson-Raritan estuary have historically been influenced by many aspects of this complex urban ecosystem other than water quality. Notable among these factors is the decimation of local bird populations in the latter half of the nineteenth century by the hunting and milliner's trade (Sullivan, 1991). Before the passage of federal protective legislation, such as the 1913 Migratory Bird Treaty, annual catches for food and feathers totaled more than a million birds per year. Even small songbirds, such as robins, were sought for sale in commercial markets. By 1884, the once abundant populations of common terns, least terns, and piping plovers, formerly present between Coney Island and Fire Island, had been reduced to but a few individuals. Populations of common and roseate terns, herons, snowy egrets, and many other species were similarly affected by hunting.

Populations gradually recovered over the next several decades until development and associated draining and spraying of wetlands for mosquito control encroached on, and degraded, waterfowl habitat. Between the late 1940s and its ban in 1972, DDT was heavily applied to the salt marshes of Long Island and New Jersey, with New Jersey marshes receiving the heaviest applications for the longest period of time. The DDT was transferred up the food chain to fish and shellfish, which are an important food source for many coastal birds in the harbor. DDT accumulated in bird tissues and contributed to the decline in reproductive success by affecting eggshell thickness (e.g., Hickey and Anderson, 1968). The osprey was probably the species most affected in the Hudson-Raritan Estuary area, although bald eagles and herons were also affected.

Recent concerns for shorebirds include the high concentrations of industrial chemicals, such as PCBs, measured in mallards, black ducks, scaup, and osprey. Due to the many factors contributing to the abundance of shore birds and the fact that they can be exposed to more than one geographic area through migration, there is only a tenuous linkage between improved water pollution control efforts and bird populations. Overwintering populations of waterfowl, however, have generally remained stable since the 1980s (Sullivan, 1991). For example, Canada goose populations of New Jersey increased from about 6,000 in 1975 to 23,200 in 1981 to 124,000 in 1990, a record high for the state. This increase is most likely the result of displacement of geese from other states, particularly Maryland.

Most remarkable among bird population recoveries is the return of herons to the heavily industrialized northwestern portion of Staten Island along the Arthur Kill and Kill Van Kull waterways. In these urban wetlands, undaunted by nearby oil refineries and chemical manufacturing plants, herons and other wading birds are making a comeback. The Harbor Herons Complex, first documented in the industrial Arthur Kill waterway in the 1970s, has become a regionally significant heron and egret nesting rookery (HEP, 1996). In 1974, snowy egrets, cattle egrets, and black-crowned night-herons began nesting on Shooters Island; in 1978, nesting snowy egrets and cattle egrets were found on Prall's Island, a 88-acre high marsh that in the past has served as a disposal site for dredged spoils. By 1981, these birds were joined by glossy ibises, great egrets, and black-crowned night-herons. In 1989, snowy egrets, glossy ibises, cattle egrets, black-crowned night-herons, yellow-crowned night-herons, little blue herons, and great egrets were found on the nearby Isle of Meadows. Ospreys,

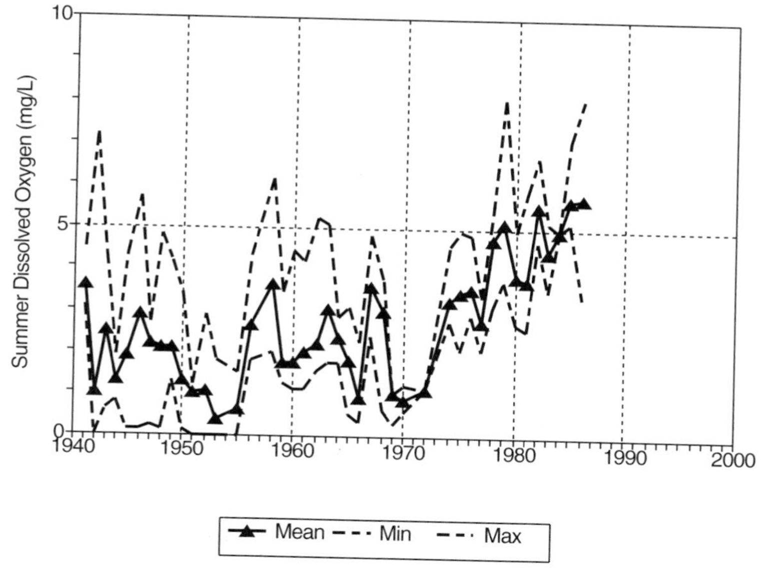

Figure 6-24 Long-term trend in summer (July–September) mean, minimum, and maximum DO in the Arthur Kill (RF1-02030104003). *Source:* USEPA STORET.

now nesting in portions of the harbor core area where they had been absent for decades (primarily because of bioaccumulation of DDT), have rapidly become so numerous as to be considered a nuisance by boaters and fishermen. Ten percent of the East Coast population of the federally endangered peregrine falcon is located in the Hudson-Raritan estuary metropolitan area (HEP, 1996).

Fish-eating bird populations have thrived despite the fact that sluggish circulation and urban runoff and municipal and industrial wasteloads characterize these waterways (HydroQual, 1991). The Arthur Kill waterway is possibly one of the sites of poorest water quality in New York Harbor. Summer mean DO concentrations in the Arthur Kill, ranging from less than 1 mg/L to about 3 mg/L from 1940 to the mid-1970s, however, steadily improved during the 1970s to concentrations above 5 mg/L by the mid-1980s (Figure 6-24) (Keller et al., 1991; O'Shea and Brosnan, 1997). Average summer concentrations of DO at Shooters Island in the Kill Van Kull further reflect this trend of improvements, increasing from 30 percent in 1974 to near 60 percent saturation in 1995 (O'Shea and Brosnan, 1997). Improvements in DO concentrations, as well as habitat protection efforts by the New York City Audubon Society, may have contributed to the success of populations of herons that feed on pollution-intolerant young fish in the Arthur Kill and its associated tidal creeks and wetlands. A 1988–1989 census of wading bird breeding populations indicated approximately 900 to 1,200 pairs of breeding herons, egrets, and ibises that constitute possibly the largest colonial waterbird rookery complex in New York State (Trust for Public Lands, 1990).

SUMMARY AND CONCLUSIONS

In the three centuries since the Governor of New York ordered a sewer system to be constructed in Lower Manhattan, New York City has made considerable progress in protecting public health and improving the water quality of the harbor. Since the early 1900s, when the city of New York instituted one of the nation's first long-term water quality monitoring programs in New York Harbor, the city's efforts to improve the waters of New York Harbor have included constructing, maintaining, and upgrading the infrastructure for wastewater collection and treatment, pollution prevention and remediation, water quality monitoring, and programs to protect the natural resources of the estuary and restore disrupted natural drainage patterns to mitigate urban runoff problems.

Although construction and upgrades of municipal wastewater treatment facilities resulted in some water quality improvements beginning in the 1950s, the greatest strides in improving ecological conditions in the harbor can be attributed to new construction and upgrades of municipal wastewater plants in the Hudson-Raritan metropolitan region during the 1970s, largely stimulated by the effluent control requirements of the 1972 CWA. Based on assessments of long-term water quality monitoring data and other environmental indicators, the ecological and water quality conditions of New York Harbor are the best they have been since the early 1900s (NYCDEP, 1999).

Biological indicators of environmental improvement in New York Harbor include the reestablishment of breeding populations of waterfowl (e.g., peregrine falcons, ospreys, herons) in many areas of the estuary, the recovery of Hudson River shortnose sturgeon to record populations, the decline of PCBs in striped bass, and a relaxation of New York State advisories for human consumption of striped bass in parts of the Hudson River. Marine organisms that were long absent from the waters of the harbor because of poor water quality conditions are now thriving as a result of the cleanup of the harbor. The resurgence of pollution-intolerant benthic organisms in Lower New York Bay and the heavy reinfestation of submerged wooden pilings by marine borers throughout the Hudson-Raritan estuary are strong evidence of the improvement in the ecological condition of the harbor.

Water quality indicators of environmental improvement in the harbor that can be attributed to upgrades of wastewater treatment facilities include significant declines in total and fecal coliform bacteria, dramatic improvements in dissolved oxygen levels, and declines in ammonia-nitrogen and BOD_5 in most areas of the Hudson-Raritan estuary. Controls on releases of heavy metals and toxic chemicals have resulted in a 50 to 90 percent reduction relative to peak levels of trace metals and chlorinated organic compounds associated with fine-grained sediments in the Hudson River. The 1972 federal ban on lead in gasoline has resulted in declines in lead in the sediments in New York Harbor and many other waterways (O'Shea and Brosnan, 1997).

Resource use indicators of environmental improvements in the harbor include the bacteria-related upgrading of the status of 68,000 acres of shellfish beds, including the lifting of restrictions on harvesting shellfish in 30,000 acres in Raritan Bay and off the Rockaways in the late 1980s. As a result of the dramatic declines in coliform bacterial levels, all New York City beaches, historically closed to swimming since the

1950s, have been open since 1992 and wet-weather swimming advisories have been lifted for all but three beaches (NYCDEP, 1999). These bacteria-related improvements in public and commercial uses of the harbor can be attributed to the continued construction and upgrading of the city's municipal water pollution control plants, the elimination of raw and illegal waste discharges, and the increased efficiency of the combined sewer system (Brosnan and O'Shea, 1996b; Brosnan and Heckler, 1996).

As a result of the clean-up efforts to date in the harbor, the public has enjoyed greatly increased opportunities for recreational uses such as swimming, boating, and fishing. The improvements in water quality also provide substantial benefits to the local economy through commercial fishing and other water-based revenue-generating activities. Although tremendous ecological improvements have resulted from water pollution control efforts implemented since the 1970s, a number of environmental problems remain to be solved for the Hudson-Raritan estuary. Some contemporary concerns and issues include, for example, contamination of sediments and restrictions on dredge spoil disposal, remaining fish advisories for human consumption, episodic low dissolved oxygen, the occurrence of nuisance algal blooms and effluent controls on nitrogen discharged to the estuary; and increasing nonpoint source runoff from overdevelopment within the drainage basin of the estuary (NYCDEP, 1999). The success of continued water pollution control efforts to remedy these concerns in the Hudson-Raritan estuary will require financial support from all levels of state, local, and federal government; enhanced public awareness about the resource value of the estuary; and strong public stakeholder support for regional coordination of environmental control programs throughout the entire Hudson-Raritan watershed.

REFERENCES

Abood, K. A., M. J. Ganas, and A. Matlin. 1995. The *Teredos* are coming! The *Teredos* are coming! In: Ports '95, Vol. 1, ed. M. A. Knott, pp. 677–690. Proceedings of Conference, March 13–15. Tampa, FL. American Society of Civil Engineers, New York.

Adamski, R. E. and A. A. Deur. 1996. History of New York City's wastewater treatment. In: Proceedings of Japan Sewage Works Association, International Symposium on Sewage Works, July 19, 1996. Tokyo Big Site, Bureau of Sewerage, Tokyo Metropolitan Goverment, Tokyo, Japan.

Blumberg, A. F., L. A. Khan, and J. P. St. John. 1997. A three-dimensional hydrodynamic model of New York Harbor, Long Island Sound and New York Bight. Paper presented at 5th International Conference on Estuarine and Coastal Modeling, October 22–24. Alexandria, VA.

Bowman, M. J. and L. D. Wunderlich. 1977. Hydrographic properties. MESA New York Bight Atlas Monograph 1. New York Sea Grant Institute, Albany, NY.

Brosnan, T. M. and P. C. Heckler. 1996. The benefits of CSO control: New York City implements nine minimum controls in the harbor. Water Environ. Technol. 8(8): 75–79.

Brosnan, T. M. and M. L. O'Shea. 1996a. Long-term improvements in water quality due to sewage abatement in the Lower Hudson River. Estuaries 19(4): 890–900.

Brosnan, T. M. and M. L. O'Shea. 1996b. Sewage abatement and coliform bacteria trends in the lower Hudson-Raritan Estuary since passage of the Clean Water Act. Water Environ. Res. 68(1): 25–35.

Brosnan, T. M., A. I. Stubin, V. Sapienza, and Y. G. Ren. 1994. Recent changes in metals loadings to New York Harbor from New York City water pollution control plants. In: Proceedings of the 26th Mid-Atlantic Industrial and Hazardous Waste Conference, August 7–10. University of Delaware, Newark, DE.

Carpenter, E. J. 1987. Nutrients and oxygen: Too much of one and too little of the other. In: Long Island Sound: Issues, resources, status and management, Seminar proceedings, NOAA Estuary-of-the-Month Seminar Series No. 3, pp. 23–45. Department of Commerce, National Oceanic and Atmospheric Administration, Estuarine Programs Office, Washington, DC.

Carriker, M. R., J. W. Anderso, W. P. Davis, D. R. Franz, G. F. Mayer, J. B. Pearce, T. K. Sawyer, J. H. Tietjen, J. F. Timoney, and D. R. Young. 1982. Effects of pollutants on benthos. In: Ecological stress and the New York Bight: Science and management, ed. G. F. Mayer, pp. 3–22. Estuarine Research Foundation, Columbia, SC.

Chillrud, S. N. 1996. Transport and fate of particle associated contaminants in the Hudson River Basin. Ph. D. thesis, Columbia University, New York.

Clark, J. F., H. J. Simpson, R. F. Bopp, and B. L. Deck. 1992. Geochemistry and Loading History of Phosphate and Silicate in the Hudson Estuary. Estuarine Coastal and Shelf Science 34:213–233.

Cleary, E. J. 1978. Perspective on river-quality diagnosis. J. WPCF 50(5): 825–831.

Esser, S. C. 1982. Long term changes in some finfishes of the Hudson-Raritan estuary. In: Ecological stress and the New York Bight: Science and management, ed. G. F. Mayer, pp. 299–314. Estuarine Research Foundation, Columbia, SC.

Faber, H. 1992. Striped bass running in the Hudson, but they're off limits to fishermen. The New York Times, April 12, p. 24.

Franz, D. R. 1982. An historical perspective on molluscs in Lower New York Harbor, with emphasis on oysters. In: Ecological stress and the New York Bight: Science and management, ed. G. F. Mayer, pp. 181–198. Estuarine Research Foundation, Columbia, SC.

Forstall, R. L. 1995. Population by counties by decennial census: 1900 to 1990. U.S. Bureau of the Census, Population Division, Washington, DC.
www.census.gov/population/www/censusdata/cencounts.html

Fuhrman, R. E. 1984. History of water pollution control. J. WPCF 56(4): 306–313.

FWPCA. 1965. Conference in the matter of pollution of the interstate waters of the Hudson River and its tributaries. Proceedings. 2 vols. Federal Water Pollution Control Administration, U.S. Department of Health, Education and Welfare. U.S. GPO, Washington, DC.

FWPCA. 1969. Conference in the matter of pollution of the interstate waters of the Hudson River and its tributaries—New York and New Jersey. Proceedings, third session, 18–19 June 1969. Federal Water Pollution Control Administration, U.S. Department of the Interior. U.S. GPO, Washington, DC.

Gottholm, B. W., M. R. Harmon, and D. D. Turgeon. 1993. Toxic contaminants in the Hudson-Raritan estuary and coastal New Jersey area. Draft Report. National Status and Trends Program for Marine Environmental Quality, National Oceanic and Atmospheric Administration, Rockville, MD. June.

Gross, M. G. 1976. Waste disposal. MESA New York Bight Atlas Monograph 26. New York Sea Grant Institute, Albany, NY.

Gruson. L. 1993. In a cleaner harbor, creatures eat the waterfront. The New York Times, June 27.

Hazen and Sawyer. 1978. Section 208 areawide waste treatment management planning program. Draft final report to City of New York, by Hazen and Sawyer, Engineers, New York.

Heckler, P. C., T. M. Brosnan, and A. I. Stubin. 1998. New York City adopts USEPA's technology requirements for CSO's. In: Proceedings of the International Symposium Paris-Quebec, Waterway Rehabilitation in Urban Environments. September.

HEP. 1996. Comprehensive conservation and management plan. New York/New Jersey Harbor Estuary Program, U.S. Environmental Protection Agency, Region 2, New York. March.

Hetling, L. J. and N. Jaworski. 1995. The evolution of federal and state roles in water quality management, In: Environmental protection, The role of the state, ed. M. Murphy. Sherkin Island Marine Station, Cork, Ireland.

Hetling, L., A. Stoddard, D. Hammerman, and T. Brosnan, 2001. Changes in wastewater loadings to the Hudson River over the past century. Paper presented at 73rd Annual Winter Meeting of the New York Water Environment Association (NYWEA). February 4–7. New York.

Hickey, J. J., and D. W. Anderson, 1968. Chlorinated hydrocarbons and eggshell changes in raptorial and fish-eating birds. Science, 162: 271–273.

Higgins, J. J., J. A. Mueller, and J. P. St. John. 1978. Baseline and alternatives: Modeling. NYC 208 Task Report 512/522. Prepared for Hazen and Sawyer, Engineers, New York, by Hydroscience, Inc., Westwood, NJ. March.

Howarth, R. W., R. Schneider, and D. P. Swaney. 1996. Metabolism and organic carbon fluxes in the tidal freshwater Hudson River. Estuaries 19(4): 848–865.

HydroQual, Inc. 1991. Assessment of pollutant loadings to New York–New Jersey Harbor. Draft final report to U.S. Environmental Protection Agency, Region 2, New York, for Task 7.1, New York/New Jersey Harbor Estuary Program, by HydroQual, Inc., Mahwah, NJ.

HydroQual, Inc. 1995a. Analysis of factors affecting historical dissolved oxygen trends in Western Long Island Sound. Job No. NENG0040. Prepared for the Management Committee of the Long Island Sound Study and New England Interstate Water Pollution Control Commission, Wilmington, MA by HydroQual, Inc., Mahwah, NJ.

HydroQual, Inc. 1995b. Development of total maximum daily loads and wasteload allocations (TMDLs/WLAs) procedure for toxic metals in NY/NJ Harbor. Job No. TETR0103. Prepared for the U.S. Environmental Protection Agency, Region 2, New York, by HydroQual, Inc., Mahwah, NJ.

HydroQual, Inc. 1996. Water quality modeling analysis of hypoxia in Long Island Sound using LIS3.0. Job No. NENG0035. Prepared for the Management Committee of the Long Island Sound Study and New England Interstate Water Pollution Control Commission, Wilmington, MA by HydroQual, Inc., Mahwah, NJ.

HydroQual, Inc. 1999. Newtown creek water pollution control project: East River water quality plan. Prepared for New York City Department of Environmental Protection, New York, by HydroQual, Inc., Mahwah, NJ.

Iseri, K. T. and W. B. Langbein. 1974. Large rivers of the United States. Circular No. 686. U.S. Department of the Interior, U.S. Geological Survey, Reston, VA.

Johnson, C. and L. Hetling. 1995. A historical review of pollution loadings to the Lower Hudson River. Paper presented at 68th Annual Winter Meeting of New York Water Environment Association (NYWEA) Meeting, January 31. New York.

Johnson, C. 1994. Evaluation of BOD, SS, N and P loadings into the Lower Hudson River Basin from point and nonpoint sources. Master of Science thesis, Rensselaer Polytechnic Institute, Troy, NY. November.

Keller, A. A., K. R. Hinga, and C. A. Oviatt. 1991. New York–New Jersey Harbor Estuary Program module 4: Nutrients and organic enrichment. Appendices to final report. Prepared for the U.S. Environmental Protection Agency, New York, by the Marine Ecosystems Research Laboratory, University of Rhode Island, Kingston, RI.

Leo, W. M., J. P. St. John, and D. J. O'Connor. 1978. Seasonal steady state model. NYC 208 Task Report 314. Prepared for Hazen and Sawyer, Engineers, New York, by Hydroscience, Inc., Westwood, NJ. March.

Loncar, F. 2002. Letter to Geoffrey Ryan, Acting Director, Bureau of Public Affairs. New York City Department of Environmental Protection Collection Facilities, Bureau of Wastewater Treatment, Corona, January 4.

McHugh, J. L., R. R. Young, and W. M. Wise. 1990. Historical trends in the abundance and distribution of living marine resources in the system. In: Cleaning up our coastal waters: An unfinished agenda, ed. K. Bricke and R. V. Thomann (co-chairmen). Regional conference cosponsored by Manhattan College and the Management Conferences for the Long Island Sound Study (LISS), the New York–New Jersey Harbor Estuary Program (HEP), and the New York Bight Restoration Plan (NYBRP), March 12–14. New York.

Metzger, S. G. and K. A. Abood. 1998. The *Limnoria* has landed! In: ed. M. A. Kraman. Ports '98. Proceedings of Conference, Long Beach, CA. American Society of Civil Engineers, New York, March 8–11.

Mueller, J. A., T. A. Gerrish, and M. C. Casey. 1982. Contaminant inputs to the Hudson-Raritan Estuary. Technical memorandum OMPA-21. National Oceanic and Atmospheric Administration, Boulder, CO.

NACOA. 1981. The role of the ocean in a waste management strategy. Special Report to the President and Congress. National Advisory Committee on Oceans and Atmospheres, Washington, DC. January.

NJDEP. 1990. Characterization of pathogen contamination in the NY-NJ Harbor estuary. Prepared by the New Jersey Department of Environmental Protection Pathogen Workshop, Trenton, NJ. (Also included was a February 6, 1992, addendum.)

NYCDEP. 1999. 1998 New York Harbor water quality regional summary. New York City Department of Environmental Protection, New York. http://www.ci.nyc.ny.us/dep

O'Connor, D. J. 1990. A historical perspective: Engineering and scientific. In: Cleaning up our coastal waters: An unfinished agenda, ed. K. Bricke and R. V. Thomann (co-chairmen), pp. 49–67. Regional conference cosponsored by Manhattan College and the Management Conferences for the Long Island Sound Study (LISS), the New York–New Jersey Harbor Estuary Program (HEP), and the New York Bight Restoration Plan (NYBRP), New York, March 12–14.

O'Connor, D. J. and J. A. Mueller. 1984. Water quality analysis of the New York Harbor complex. J. Environ. Eng., ASCE 110(6): 1027–1047.

O'Connor, D. J., R. V. Thomann, and H. J. Salas. 1977. Water quality. MESA New York Bight Atlas Monograph 27. New York Sea Grant Institute, Albany, NY.

Oliveri, F. J., F. Loncar, and M. Ellis. 2001. Enhanced Beach Protection-2000. Clearwater, 31(2), New York Water Environment Association (NYWEA), Syracuse, NY.

OMB. 1999. OMB Bulletin No. 99-04. U.S. Census Bureau, Office of Management and Budget, Washington, DC. www.census.gov/population/www/estimates/metrodef.html

O'Shea, M. L. and T. M. Brosnan. 1997. New York Harbor water quality survey. Main report and appendices 1995. New York Department of Environmental Protection, Bureau of Wastewater Pollution Control, Division of Scientific Services, Marine Sciences Section, Wards Island, NY.

O'Shea, M. L., and T. M. Brosnan. 2000. Trends in indicators of eutrophication in Western Long Island Sound and the Hudson-Raritan Estuary. Estuaries 23:6.

Parker, C. A. and J. E. O'Reilly. 1991. Oxygen depletion in Long Island Sound: A historical perspective. Estuaries 14(3): 248–264.

Port Authority of New York. 1988. Meeting of the Marine Borer Research Committee of the New York Harbor/East Coast Commission on Marine Barriers, November 10. World Trade Center, New York.

Randolph, E. 1998. Terror of the Hudson River is back. The Los Angeles Times, January 26, p. A-5.

Revkin, A. C. 1997. High PCB level is found in a Hudson bald eagle. The New York Times, September 17, p. B-4.

Rod, S. R., R. U. Ayres, and M. Small. 1989. Reconstruction of historical loadings of heavy metals and chlorinated hydrocarbon pesticides in the Hudson-Raritan basin, 1880–1980. Final report to the Hudson River Foundation, New York, Foundation Grant No. 001-86A-3.

Ryan, G. 2001. Personal communication. New York City Department of Environmental Protection (NYCDEP), Bureau of Public Affairs, Corona, New York, October 29.

Schwartz, J. J. and K. S. Porter. 1994. The state of the city's waters: 1994. The New York Harbor estuary. Prepared by the New York State Water Resources Institute, Center for the Environment, Cornell University, Ithaca, NY.

Stubin, A. 1997. New York City Department of Environmental Protection, Marine Sciences Section, Wards Island, NY. Personal communication, October 1.

Sullivan, J. K. 1991. Fish and wildlife populations and habitat status and trends in the New York Bight. A report to the Habitat Workgroup for the U.S. Environmental Protection Agency, New York Bight Restoration Plan, New York.

Summers, J. K. and K. A. Rose. 1987. The role of interactions among environmental conditions in controlling historical fisheries variability. Estuaries 10(3): 255–266.

Suszkowski, D. J. 1990. Conditions in the New York/New Jersey harbor estuary. In: Cleaning up our coastal waters: An unfinished agenda, ed. K. Bricke and R. V. Thomann (co-chairmen), pp. 105–131. Regional conference cosponsored by Manhattan College and the Management Conferences for the Long Island Sound Study (LISS), the New York–New Jersey Harbor Estuary Program (HEP), and the New York Bight Restoration Plan (NYBRP), March 12–14.

Swaney, D. P., D. Sherman, and R. W. Howarth. 1996. Modeling water, sediment and organic carbon discharges in the Hudson-Mohawk Basin: Coupling of terrestrial sources. Estuaries 19(4): 833–847.

Swanson, R. L., T. M. Bell, J. Kahn, and J. Olha. 1990. Use impairments and ecosystem impacts of the New York Bight. In: Cleaning up our coastal waters: An unfinished agenda, ed. K. Bricke and R. V. Thomann (co-chairmen). Regional conference cosponsored by Manhattan College and the Management Conferences for the Long Island Sound Study (LISS), the New York–New Jersey Harbor Estuary Program (HEP), and the New York Bight Restoration Plan (NYBRP), New York, March 12–14.

Thomann, R. V., J. A. Mueller, R. P. Winfield, and C. Huang. 1991. Model of fate and accumulation of PCB homologues in Hudson Estuary. J. Environ. Eng. 117(2): 161–178.

Trust for Public Lands. 1990. The Harbor Herons report: A strategy for preserving a unique urban wildlife habitat and wetland resource in northwestern Staten Island. Published by the Trust for Public Land in conjunction with the New York City Audubon Society, New York.

USDOC. 1998. Census of population and housing. Prepared by U.S. Department of Commerce, Economics and Statistics Administration, Bureau of the Census—Population Division, Washington, DC.

USEPA (STORET). STOrage and RETrieval Water Quality Information System. U.S. Environmental Protection Agency, Office of Wetlands, Oceans, and Watersheds, Washington, DC.

USGS. 1999. Streamflow data downloaded from U.S. Geological Survey, United States National Water Information System (NWIS)-W. Data retrieval for historical streamflow daily values. http://waterdata.usgs.gov/nwis-w/

Valette-Silver, N. J. 1993. The use of sediment cores to reconstruct historical trends in contamination of estuarine and coastal sediments. Estuaries 16(3B): 577–588.

Wiseman, W. 1997. State University of New York at Stony Brook, Marine Sciences Research Center, Stony Brook, NY. Personal communication, December 23.

Wolfe, D. A., E. R. Long, and G. B. Thursby. 1996. Sediment toxicity in the Hudson-Raritan estuary: Distribution and correlations with chemical contamination. Estuaries 19(4): 901–912.

Wolman, M. G. 1971. The nation's rivers. Science 174: 905–917.

Woodhead, P. M. J. 1991. Inventory and characterization of habitat and fish resources, and assessment of information on toxic effects on the New York–New Jersey Harbor Estuary. A report to the New York–New Jersey Harbor Estuary Program, New York.

CHAPTER 7

Delaware Estuary Case Study

The Mid-Atlantic Basin (Hydrologic Region 2), covering a drainage area of 111,417 square miles, includes some of the major rivers in the continental United States. Figure 7-1 highlights the location of the basin and the Delaware estuary, the case study watershed profiled in this chapter.

With a length of 390 miles and a drainage area of 11,440 square miles, the Delaware River ranks seventeenth among the 135 U.S. rivers that are more than 100 miles in length. On the basis of mean annual discharge (1941–1970), the Delaware ranks twenty-eighth (17,200 cfs) of large rivers in the United States (Iseri and Langbein, 1974). Figure 7-2 highlights the location of the Delaware estuary case study catalog units identified by major urban-industrial areas affected by severe water pollution problems during the 1950s and 1960s (see Table 4-2). This chapter presents long-term trends in population, municipal wastewater infrastructure and effluent loading of pollutants, ambient water quality, environmental resources, and uses of the Delaware estuary. Data sources include USEPA's national water quality database (STORET), published technical literature, and unpublished technical reports ("grey" literature) obtained from local agency sources.

The Delaware River, formed by the confluence of its east and west branches in the Catskill Mountains near Hancock, New York, on the Pennsylvania–New York state line, becomes tidal at Trenton, New Jersey (Figure 7-2). The first 86 miles of the tidal river are the Delaware River estuary, which flows by Trenton, New Jersey; Philadelphia, Pennsylvania; Camden, New Jersey; and Wilmington, Delaware. This major urban-industrial area has a tremendous impact on the water quality of the river. In this area, the Delaware River estuary flows along the boundary between the Piedmont Plateau and the Atlantic Coastal Plain. A large number of municipal and industrial wastewater facilities discharge to the Delaware River, with municipal water pollution control plants accounting for the largest component of BOD_5 loading. In general, water quality is good at the head of the tide at Trenton (RM 134.3), but it begins to deteriorate downstream.

From the 1930s through the 1970s, water quality conditions were very poor. Depleted DO levels were recorded in the region from Torresdale (RM 110.7) to Eddystone (RM 84.0) as a result of wastewater loading from Philadelphia (RM 110–93). Since the mid-1980s, water quality conditions in the estuary have improved significantly.

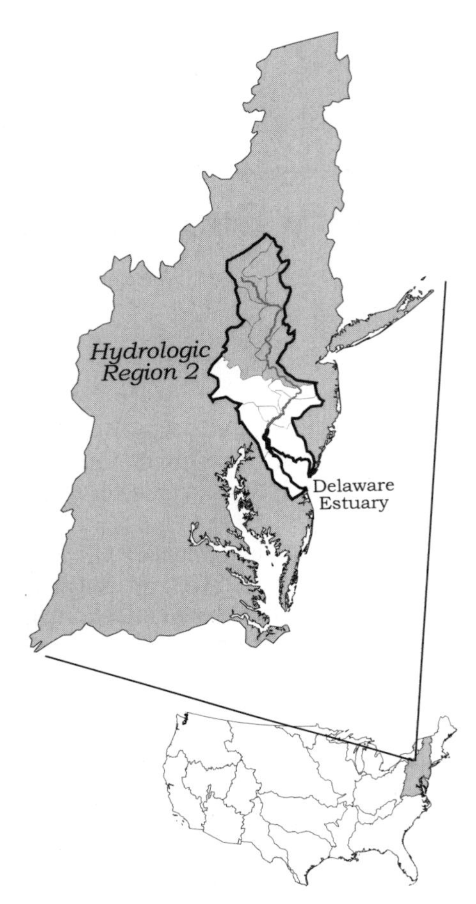

Figure 7-1 Hydrologic Region 2 and Delaware estuary watershed.

PHYSICAL SETTING AND HYDROLOGY

The Delaware River originates in the south-central area of New York State and flows 390 miles in a southerly direction to the Atlantic Ocean, separating New Jersey on its eastern bank from Pennsylvania and Delaware on its west. The total drainage area at the mouth of the river at Liston Point on Delaware Bay is 11,440 square miles, of which 6,780 square miles lie upstream of the gaging station at Trenton, New Jersey (Iseri and Langbein, 1974). The major tributary to the Delaware estuary is the Schuylkill River, which joins the main river in the vicinity of Philadelphia, Pennsylvania. The Schuylkill has a drainage area of 1,890 square miles at the Fairmount Dam, 8 miles above the mouth. In addition to the Schuykill River, the other major tributaries to the Delaware estuary include Assunpink Creek, Crosswicks Creek, Rancocas Creek,

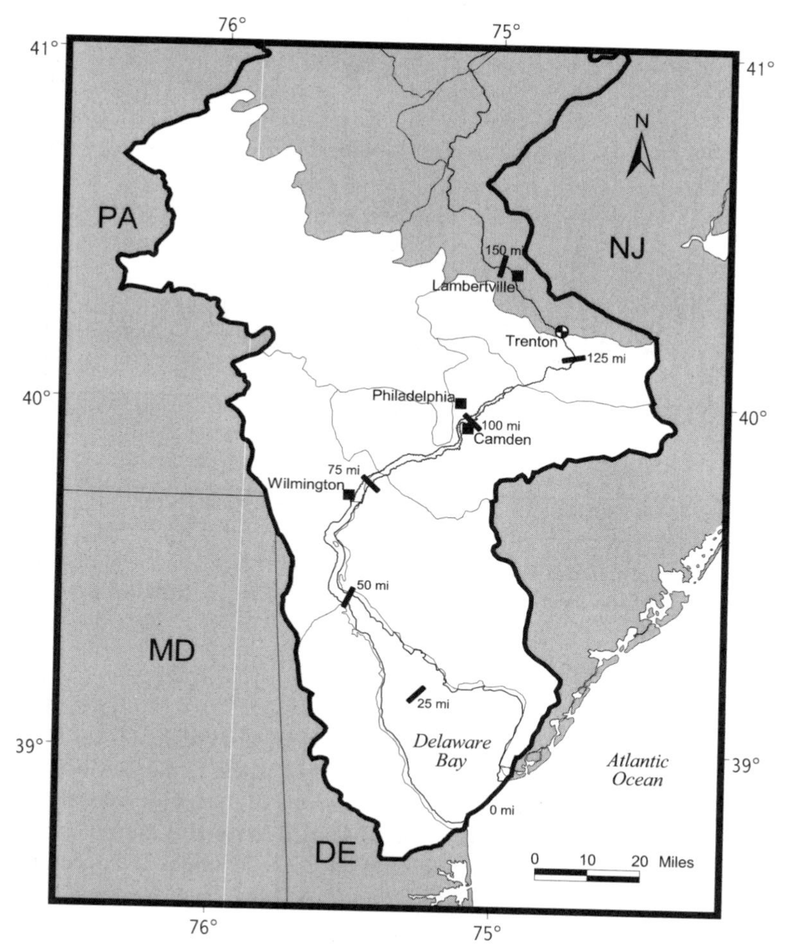

Figure 7-2 Location map of the Lower Delaware River–Delaware Bay. (River miles shown are distances from the Cape May, New Jersey, to Cape Henlopen, Delaware, transect.)

Neshaminy Creek, Cooper River, Chester Creek, the Christina River, and the Salem River.

Figure 7-3 presents long-term statistics of summer streamflow from the USGS gaging station at Trenton, New Jersey, from 1940 to 1995. The extreme drought conditions of the mid-1960s are quite apparent in the long-term record (1962–1966). Seasonal variation of freshwater flow of the Delaware River is characterized by high flow from March through May, with a peak flow of 21,423 cfs in April. Low-flow conditions typically occur from July through October, with the monthly minimum flow of 5,830 cfs recorded during September (Figure 7-4). From July through October, low-flow in the river is typically augmented by releases from reservoirs. During dry conditions,

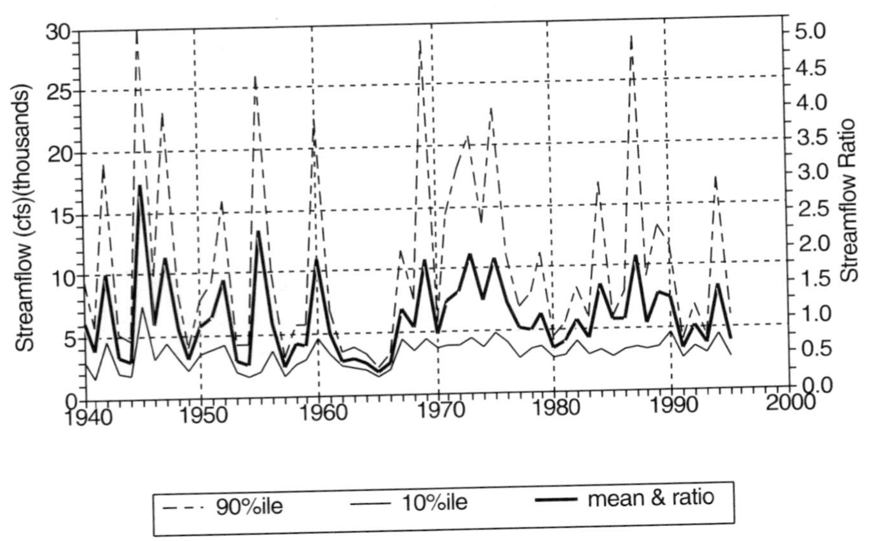

Figure 7-3 Long-term trends in mean, tenth, and ninetieth percentile streamflow in summer (July–September) for the Delaware River at Trenton, New Jersey (USGS Gage 01463500). *Source:* USGS, 1999.

reservoir releases, regulated to maintain a minimum flow of 2,500 to 3,000 cfs at Trenton, can be greater than 60 percent of the inflow to the estuary (Albert, 1997).

The Delaware River–Delaware Bay system is one of the major coastal plain estuaries of the East Coast of the United States. The tidal river and estuary extend a distance of 134 miles from the fall line at Trenton, New Jersey, to the ocean mouth of Delaware Bay along an 11-mile section from Cape May, New Jersey, to Cape Henlopen, Delaware (Figure 7-2). Because Philadelphia is a major East Coast port, a navigation channel is maintained to a depth of 12 meters (39 feet) from the entrance to the bay upstream to Philadelphia. From Philadelphia to Trenton, the channel is maintained at a depth of 8 meters (26 feet) (Galperin and Mellor, 1990). The semidiurnal tide has a mean range of 1.5 meters (4.9 feet) at the mouth of the bay and propagates upstream on the incoming tide to Trenton in approximately 7 hours; typical tidal currents are approximately 1.5 meters/second (Galperin and Mellor, 1990). Approximately 25 miles downstream from Philadelphia, tidal currents near the Tacony-Palmyra Bridge (RM 107) are characterized by vigorous vertical mixing and a marked current reversal with currents of approximately 1.0 meter/second (Thomann and Mueller, 1987).

The Delaware estuary can be characterized as three distinct hydrographic regimes based on distributions of salinity, turbidity, and biological productivity: (1) tidal freshwater, (2) transition zone, and (3) Delaware Bay zone. The tidal fresh river extends about 55 miles from the head of tide at Trenton, New Jersey (RM 134) to Marcus Hook, Pennsylvania (RM 79). Under mean freshwater flow conditions, salinity

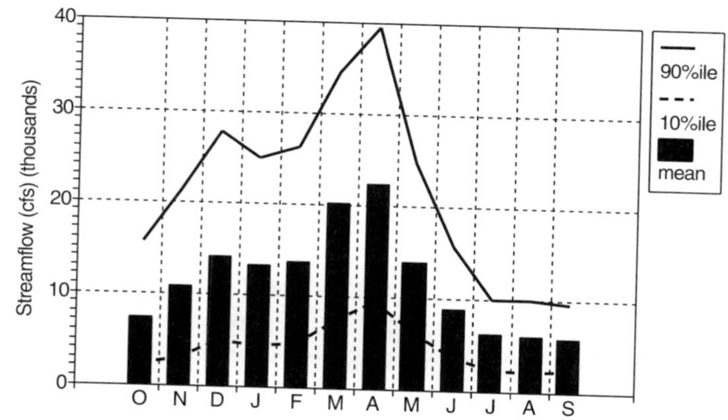

Figure 7-4 Monthly trends of mean, tenth, and ninetieth percentile streamflow for the Delaware River at Trenton, New Jersey (USGS Gage 01463500), 1951–1980. *Source:* USGS, 1999.

intrusion in the tidal fresh section of the river generally extends upstream to the reach between the Delaware Memorial Bridge at Wilmington (RM 68.7) and Marcus Hook (RM 79.1). During drought periods (e.g., 1962–1966), salinity intrusion is a concern because industrial water withdrawals are less desirable and recharge areas of the South Jersey aquifers serving the Camden metropolitan area are potentially threatened (DRBC, 1992). During drought conditions, the Delaware River Basin Commission requires releases from the upper basin reservoirs to prevent critical salinity concentrations from intruding farther upstream than Philadelphia at RM 98 (DRBC, 1992).

The transition zone, extending about 26 miles from Marcus Hook (RM 79) to Artificial Island, New Jersey (RM 53), is characterized by low salinity levels, high turbidity, and relatively low biological production (Marino et al., 1991). The estuarine region extends downstream of Artificial Island about 53 miles to the mouth of Lower Delaware Bay; salinity in this region varies from approximately 8 ppt upstream to approximately 28 ppt at the mouth of the bay (Marino et al., 1991).

POPULATION, WATER, AND LAND USE TRENDS

Four densely populated metropolitan areas have developed along a 50-mile industrialized section of the Delaware River from Philadelphia, Pennsylvania, and Trenton and Camden, New Jersey, to Wilmington, Delaware. From 1880 to 1980, urban growth accounted for most of the 236 percent increase in total population of the region. The urban proportion of the population increased from approximately 64 percent at the turn of the twentieth century to approximately 80 percent in 1980 (Marino et al., 1991). During the period after World War II, from 1950 through 1980, development in the region was characterized by urban and suburban sprawl: urban land use area increased from 460 square miles in 1950 to 3,682 square miles by 1980, while population density declined from 8,000 to 3,682 persons per square mile (Marino et al., 1991). Much of this development occurred by converting agricultural lands in close proximity to the major metropolitan areas to suburban land uses.

The Delaware River case study area includes 14 counties identified in 1995 by the Office of Management and Budget (OMB, 1999) as the Philadelphia–Wilmington–Atlantic City, PA–NJ–DE–MD Consolidated Metropolitan Statistical Area (CMSA) (Table 7-1). Long-term population trends from 1940 through 1996 for these counties are presented in Figure 7-5. Population in these counties has increased by 162 percent, from 3.67 million in 1940 to 5.97 million by 1996 (Forstall, 1995; USDOC, 1998).

The city of Philadelphia withdraws water for domestic water supply at the Torresdale intake upstream of the salt front. The city of Trenton also withdraws water for public water supply from the Delaware River. In addition to these cities, Camden, the Delaware County Sewer Authority, and Wilmington are among more than 80 dischargers of municipal wastewater directly to the estuary or the tidal portions of its tributaries. Historical water use data are not readily available at the county level of aggregation to assess the contribution of the Delaware estuary region to municipal and industrial water withdrawals compiled by the USGS for the entire Middle-Atlantic Basin from 1950 to 1995 (Solley et al., 1998). The natural resources (plankton, fisheries, marshes, and shorebirds), human uses (waste disposal, transportation and dredging, beach development), and management issues of the Delaware estuary are presented in Bryant and Pennock (1988).

TABLE 7-1 Consolidated Metropolitan Statistical Area (CMSA) Counties in the Delaware Estuary Case Study, Philadelphia–Wilmington–Atlantic City, PA–NJ–DE–MD CMSA

County
New Castle County, DE
Cecil County, MD
Atlantic County, NJ
Burlington County, NJ
Camden County, NJ
Cape May County, NJ
Cumberland County, NJ
Gloucester County, NJ
Salem County, NJ
Bucks County, PA
Chester County, PA
Delaware County, PA
Montgomery County, PA
Philadelphia County, PA

Source: OMB, 1999.

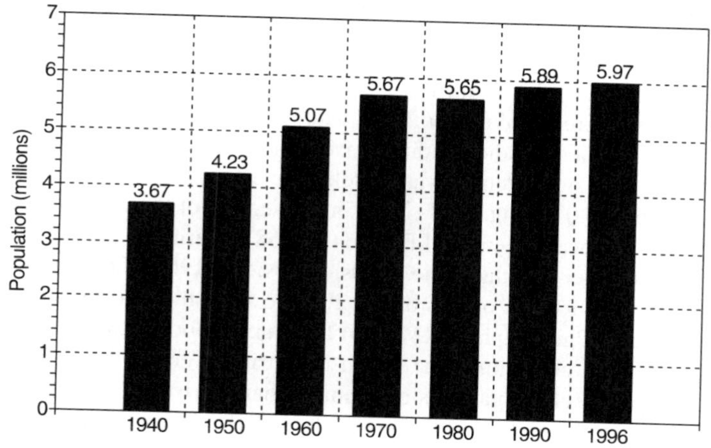

Figure 7-5 Long-term trends in population for the Philadelphia–Wilmington–Atlantic City CMSA counties of the Lower Delaware River–Delaware Bay. *Sources:* Forstall, 1995; USDOC, 1998.

HISTORICAL WATER QUALITY ISSUES

In reports sent back to Europe, Captain Thomas Young, one of the early European explorers of the Delaware estuary, noted that "the river aboundeth with beavers, otters and other meaner furrs . . . I think few rivers of America have more . . . the quantity of fowle is so great as hardly can be believed. Of fish heere is plentie, but especially sturgeon." Early colonial advertising copy like this, circulated widely in Europe, presumably inspired Old World colonists to emigrate to the Delaware Valley (Sage and Pilling, 1988). The estuary was abundant with striped bass, sturgeon, shad, oysters, and waterfowl.

Beginning with the Industrial Revolution and the development of the Delaware Valley as a major industrial and manufacturing center in the nineteenth century, waste disposal from increasing population and industrial activities resulted in progressive degradation of water quality and loss of the once-abundant natural resources of the estuary. By the turn of the twentieth century, the American shad population had collapsed. By 1912–1914, low DO conditions were all too common in the Philadelphia and Camden area of the river (Albert, 1997). Sanitary surveys conducted in 1929 and 1937 documented poor water quality conditions in the nontidal reaches of the Delaware from Port Jervis, New York, to Easton, Pennsylvania. During high-flow conditions, black water from the Lehigh River–Easton area would result in closing of the water supply intakes at Trenton, New Jersey (Albert, 1982).

In the tidal river between Trenton and Philadelphia, the discharge of raw sewage from Philadelphia, Trenton, Camden, Wilmington, and other communities, along with untreated industrial wastewater discharges, resulted in gross water pollution of the estuary. Peak bacterial densities of ~6,000 to 8,000 most probable number (MPN)/100 mL were recorded during the late 1960s and early 1970s in the vicinity of the Philadelphia Navy Yard (RM 90) (Patrick et al., 1992; Marino et al., 1991). Fecal coliform bacteria levels were high as a result of raw or inadequately treated wastewater discharges from the large municipalities. Acidic conditions from industrial waste discharges were observed in the river near the Pennsylvania-Delaware border; pH levels ranged from approximately 6.5 to 7.0 during 1968–1970 in the section of the river from Paulsboro (RM 89) to the Delaware Memorial Bridge (RM 68) (Marino et al., 1991).

During the summer months in the 1940s, 1950s, and 1960s, DO levels were typically approximately 1 mg/L or less over a 20-mile section of the river from the Ben Franklin Bridge in Philadelphia (RM 100) to Marcus Hook (RM 79). Under these anoxic and hypoxic conditions, the urban-industrial river ran black, and the foul stench of hydrogen sulfide gas was a common characteristic (Patrick, 1988). Dock workers and sailors were often overcome by the stench of the river near Philadelphia, and ships suffered corrosion damage to their hulls from the polluted waters. Aircraft pilots landing in Philadelphia reported smelling the Delaware estuary at an altitude of 5,000 feet. Water quality conditions were so bad that President Roosevelt ordered a study in 1941 to determine whether water pollution in the Delaware River was affecting the U.S. defense buildup (Albert, 1982; CEQ, 1982).

LEGISLATIVE AND REGULATORY HISTORY

Water pollution in the Delaware estuary reached its peak in the 1940s. The source of the pollution was raw sewage (350 mgd from Philadelphia alone), along with untreated industrial wastewater of all kinds. In response to steadily increasing pollution, the Interstate Commission on the Delaware River Basin (INCODEL) launched a basinwide water pollution control program in the late 1930s. Following a delay due to the war, the abatement program was finally completed by the end of the 1950s. During that time, the number of communities with adequate sewage collection and treatment facilities rose from 63 (approximately 20 percent) to 236 (75 percent) (Albert, 1982). Concurrent success was not achieved in abating industrial pollution.

The first generation of water pollution control efforts, largely completed by 1960, resulted in secondary treatment levels at most treatment plants above Philadelphia. Primary treatment was considered adequate in the estuary downstream of Philadelphia. Although most areas built the required facilities, some treatment facilities from the first-generation effort were not completed until the 1960s or 1970s.

In 1961 INCODEL became incorporated into a more powerful interstate regulatory agency, the Delaware River Basin Commission (DRBC). The DRBC, created as a result of federal and state legislation, has broad water resources responsibilities, including water pollution control. The Commission developed a clean-up program based on a 6-year $1.2 million Delaware Estuary Comprehensive Study (DECS), conducted by the U.S. Public Health Service. Nearly 100 municipalities and industries were found to be discharging harmful amounts of waste into the river. Using a water quality model (DECS), the DRBC calculated the river's natural ability to assimilate oxidizable wastewater loads and established allocations for each city and industry (Thomann, 1963; Thomann and Mueller, 1987). The objective of the DRBC wasteload allocation program and the corollary programs of Pennsylvania, New Jersey, Delaware, and the federal government, was to upgrade the somewhat improved water quality of 1960 to more acceptable levels.

For the purposes of water quality management, the Delaware estuary has been divided into six water quality zones. Zone 1 is upstream of the fall line at Trenton, New Jersey. Zones 2 through 6 are in the tidal Delaware, which is water quality–limited. Here, more stringent effluent limits are required, based on allocations of assimilative capacity, to achieve water quality standards. Based on the DECS water quality model, the DRBC, in 1967, adopted new, higher water quality standards, and then in 1968, issued wasteload allocations to approximately 90 dischargers to the estuary. These required treatment levels were more stringent than secondary treatment as defined by USEPA in the 1972 Clean Water Act.

IMPACTS OF WASTEWATER TREATMENT

Pollutant Loading and Water Quality Trends

The Delaware River from Trenton, New Jersey, to Liston Point is one of the most heavily industrialized sections of a waterway in the United States. Four major cities and a large number of oil refineries and chemical manufacturing plants are located along the river. The effect of the DRBC wasteload allocation program and the related water pollution control programs of Pennsylvania, New Jersey, Delaware, and the federal government on the Delaware Estuary is best demonstrated by the substantial reduction of ultimate CBOD loading from municipal and industrial dischargers that has been achieved since the late 1950s (Figure 7-6). Ultimate CBOD loadings to the estuary have been reduced by 89 percent, from 515.4 metric tons/day (mt/day) in 1958 (Patrick et al., 1992) to 58.2 mt/day by 1995 (HydroQual, 1998). Major wastewater treatment facilities that upgraded to secondary treatment and better in order to meet the wasteload allocations include Philadelphia NE (1985), Philadelphia SE (1986), Philadelphia SW (1980), CCMUA (1989), Trenton (1982), Bordentown MUA (1991), and Lower Bucks MUA (1980). A complete listing of the 34 municipal and 26 industrial point sources discharging to the Delaware estuary between Trenton and Liston Point is presented in HydroQual (1998). In addition to reductions of pollutant loading from direct dischargers to the estuary, the cleanup of major tributaries to the Delaware has also contributed to water quality improvements in the Delaware estuary (Albert, 1982).

Since implementation of the 1972 CWA, reductions in point source loads of oxidizable materials have been achieved as a result of technology- and water quality–based effluent controls on municipal and industrial dischargers in the Delaware River watershed. Nonpoint source runoff, driven by the land uses and hydrologic characteristics of the watershed, also contributes a pollutant load that must be considered in a complete evaluation of the impact of regulatory policy and controls on long-term

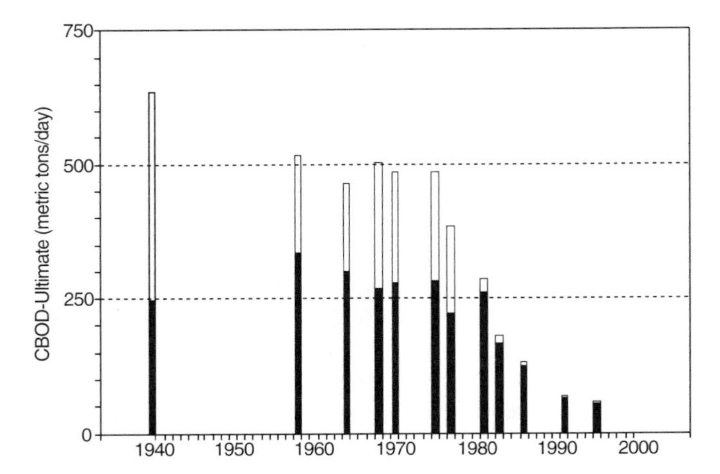

Figure 7-6 Long-term trends in ultimate CBOD loading from municipal and industrial wastewater dischargers to the Delaware estuary. *Sources:* HydroQual, 1998; Patrick, Ruth, Faith Douglass, and Drew Palavage, Surface water quality: Have the laws been successful? Copyright © 1992 by Princeton University Press, Reprinted by permission of Princeton University Press.

water quality trends. To evaluate the relative significance of point and nonpoint source pollutant loads, inventories of NPDES point source dischargers, land uses, and land use–dependent export coefficients (Bondelid et al., 2000) have been used to estimate catalog unit–based point source (municipal, industrial, and CSOs) and nonpoint source (rural and urban) loads of BOD_5 for mid-1990s conditions in the catalog units of the Delaware River case study area (see Figure 7-2). The point source load of

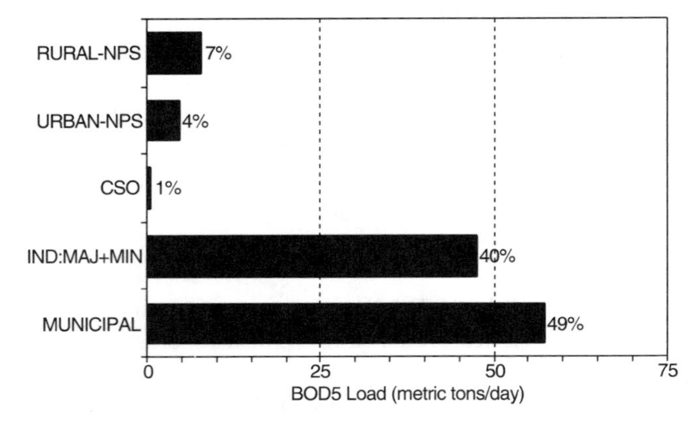

Figure 7-7 Point and nonpoint source loads of BOD_5 (ca. 1995) for the Lower Delaware River–Delaware Bay case study catalog units. *Source:* Bondelid et al., 2000.

105.4 mt/day accounts for 89 percent of the total estimated BOD_5 load of 117.1 mt/day from point and nonpoint sources. Municipal facilities contribute 57.3 mt/day (49 percent), while industrial dischargers account for 47.5 mt/day (40 percent) of the total point and nonpoint source BOD_5 load (Figure 7-7). Nonpoint sources of BOD_5 account for 12.6 mt/day; rural runoff contributes approximately 8 percent, and urban land uses account for approximately 5 percent of the total point and nonpoint load of 117.1 mt/day (Figure 7-7).

One of the major trends indicative of water quality improvement in the estuary has been that for DO. A comparison of mean summer DO levels between 1968–1972, 1975–1979, 1981–1985, and 1988–1994 (Figure 7-8) clearly shows the water quality improvements achieved as a result of the point source loading reductions of ultimate CBOD (Figure 7-6). Mean summer DO concentrations have increased by approximately 1 mg/L between River Mile 110 and River Mile 55 (DRBC Zones 3, 4, and 5) between 1957–1961 and 1981–1985 (Brezina, 1988). DO concentrations have increased from less than 2 mg/L to 5 mg/L at the critical DO sag point at the mouth of the Schuylkill River in Philadelphia (RM 92) during the period from 1968–1972 to 1988–1994. DO concentrations increased steadily farther downstream of Philadelphia (RM 83), reaching a level of approximately 5.5 mg/L during 1988–1994.

The historical summer DO spatial transects data (Figure 7-8) show that wastewater discharges from the Philadelphia area result in minimum DO conditions between the Ben Franklin Bridge (RM 100), the Philadelphia Navy Yard (RM 93), and

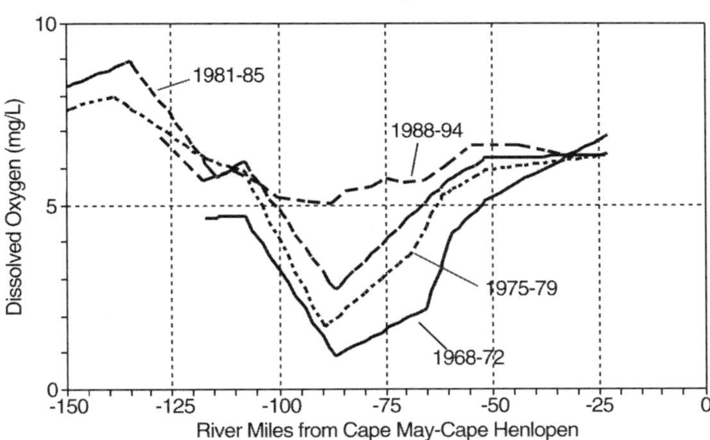

Figure 7-8 Long-term trends of the spatial distribution of summer DO in the Delaware estuary. *Sources:* Scally, 1997; Patrick, Ruth, Faith Douglass, and Drew Palavage, Surface water quality: Have the laws been successful? Copyright © 1992 by Princeton University Press, Reprinted by permission of Princeton University Press.

Marcus Hook (RM 78). Figure 7-9 shows long-term summer (July–September) trends in DO measured at stations within the RF1 reach from the Ben Franklin Bridge to the Philadelphia Navy Yard. The long-term trend documents improvements in oxygen during the 1980s and 1990s from the water pollution control efforts initiated during the 1970s. Most dramatic, however, is the progressive improvement in the minimum oxygen levels during the 1980s and early 1990s. Summer minimum values increased from approximately 1 mg/L or less in the 1960s and 1970s to approximately 4–5 mg/L during 1990–1995. Although oxygen conditions improved tremendously between the 1960s and the early 1990s, a continued trend of further improvements through the mid-1990s has not been recorded. Minimum oxygen concentrations still can approach 4 mg/L near Chester, Pennsylvania (River Mile 84) and can drop lower than 4 mg/L in the 10-mile oxygen sag reach between River Mile 95 and River Mile 85 (HydroQual, 1998).

Spatial water quality trends recorded during the late 1960s, 1970s, 1980s, and 1990s include documentation of temporal declines in BOD_5 (Figure 7-10), ammonia-N (Figure 7-11), total nitrogen (Figure 7-12), and total phosphorus (Figure 7-13). Effluent reductions of oxygen-demanding loads from industrial and municipal sources have resulted in significant declines in ambient levels of BOD_5 and ammonia-N. An interannual temporal trend for ambient ammonia-N (Figure 7-14) at a station near Marcus Hook (RM 78) shows a considerable improvement in water quality, with a steep decline from approximately 1.4 mg N/L during the late 1960s to approximately 0.5 mg N/L by the late 1970s, followed by relatively unchanging ambient concentrations (approximately 0.15 mg N/L) recorded during the mid-1980s through the mid-1990s (Santoro, 1998). The decline in ambient ammonia-N during this 30-year

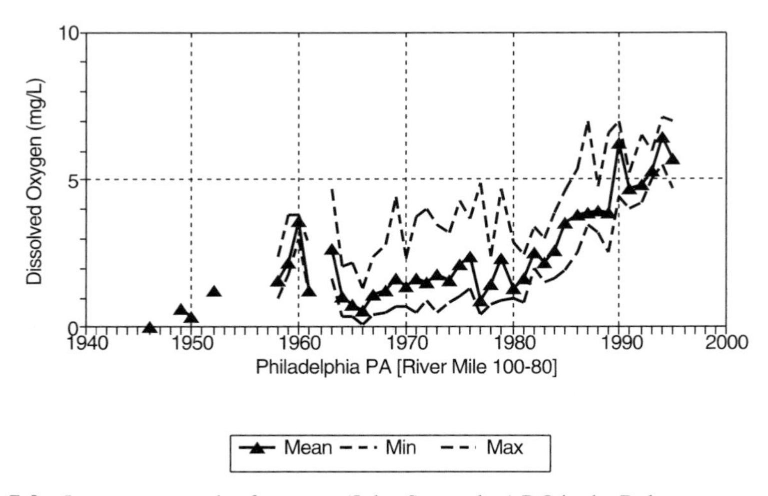

Figure 7-9 Long-term trends of summer (July–September) DO in the Delaware estuary near Philadelphia, PA (RF1–02040202030, River Mile 100–80). *Source:* USEPA STORET.

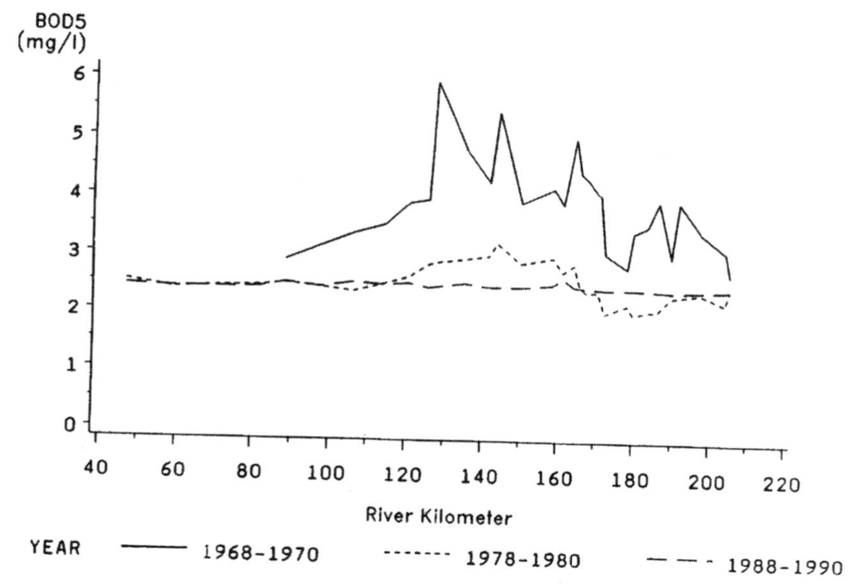

Figure 7-10 Long-term trends of the spatial distribution of BOD_5 in the Delaware estuary (mean of data from 1968–1970, 1978–1980, and 1988–1990). *Source:* Marino et al., 1991.

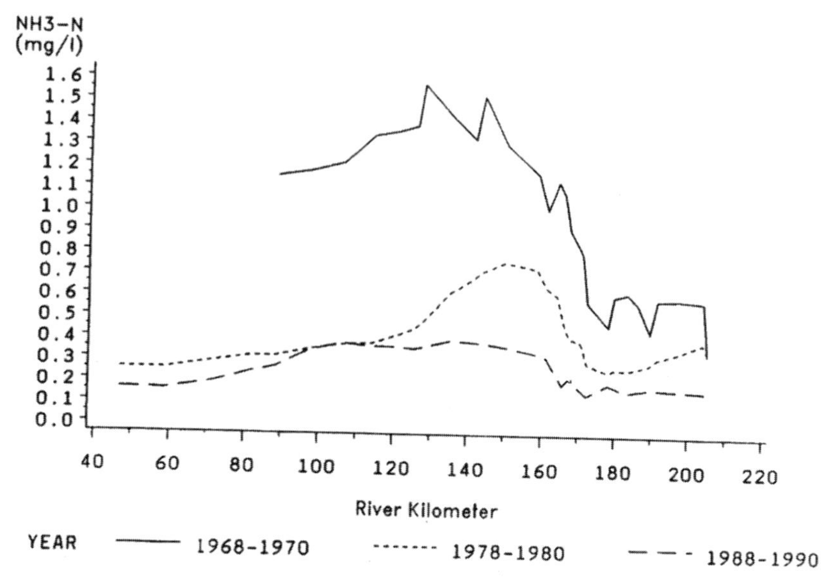

Figure 7-11 Long-term trends of the spatial distribution of ammonia-N in the Delaware estuary (mean of data from 1968–1970, 1978–1980, and 1988–1990). *Source:* Marino et al., 1991.

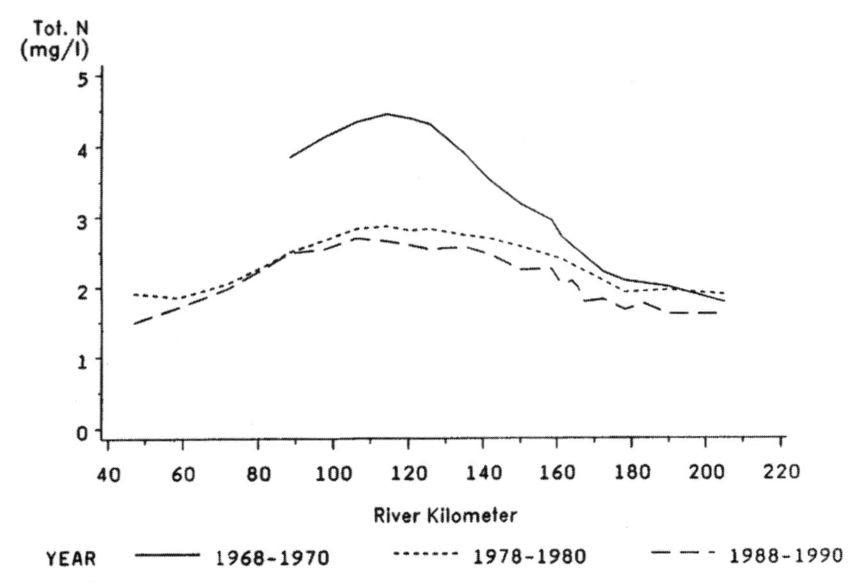

Figure 7-12 Long-term trends of the spatial distribution of total nitrogen in the Delaware estuary (mean of data from 1968–1970, 1978–1980, and 1988–1990). *Source:* Marino et al., 1991.

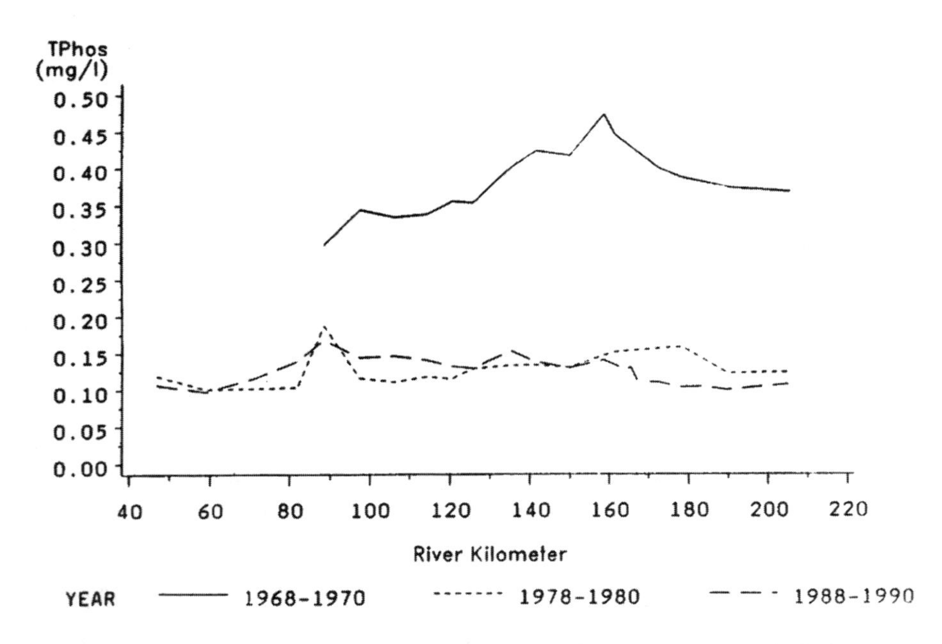

Figure 7-13 Long-term trends of the spatial distribution of total phosphorus in the Delaware estuary (mean of data from 1968–1970, 1978–1980, and 1988–1990). *Source:* Marino et al., 1991.

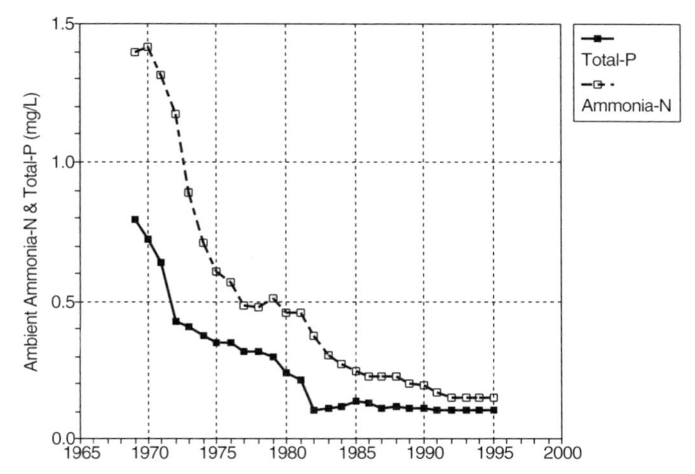

Figure 7-14 Long-term trends of summer ammonia-N and total phosphorus at a station in the Delaware estuary near Marcus Hook (mile 78) (computed as 4-year moving averages for July from DRBC boat run records). *Source:* Santoro, 1998.

period has been shown to correspond to a concurrent increase in nitrate-N from the mid-1970s through the mid-1980s as a result of nitrification (Santoro, 1998; Marino et al., 1991).

Reflecting deforestation, agricultural practices, fossil fuel combustion, and the increase in human population of an increasingly urbanized drainage basin over the much longer time scale of a century, ambient nitrate and chloride levels (Figure 7-15) have steadily increased by approximately 400 percent to 500 percent since measurements were first recorded in 1905 at a water supply intake near Philadelphia (Jaworksi and Hetling, 1996). Similar patterns of long-term increasing trends in ambient nitrate and chlorides have also been recorded at other East Coast water supply intakes for the Merrimack, Connecticut, Hudson, Schuykill, and Potomac Rivers (Jaworski and Hetling, 1996).

Total phosphorus has also declined from peak levels of approximately 0.45 mg P/L during 1968–1970 to much lower levels of approximately 0.15 mg P/L by 1988–1990 near River Mile 100 (Figure 7-13). An interannual time series of total phosphorus (Figure 7-14) for a station near Marcus Hook (River Mile 78) exhibits a trend similar to that of ammonia-N, with a sharp decline from approximately 0.8 mg P/L in the late 1960s to approximately 0.3 mg P/L by the late 1970s, followed by relatively unchanging concentrations (approximately 0.1 mg P/L) from the mid-1980s through the mid-1990s. The decline of ambient levels of total phosphorus has been attributed to the detergent phosphate ban of the early 1970s (Roman et al., 2000), reductions of effluent loads from wastewater facility upgrades (Sharp, 1988), and changes in partitioning of dissolved and soluble phases of phosphorus and changes in solubility of phosphate (Lebo and Sharp, 1993).

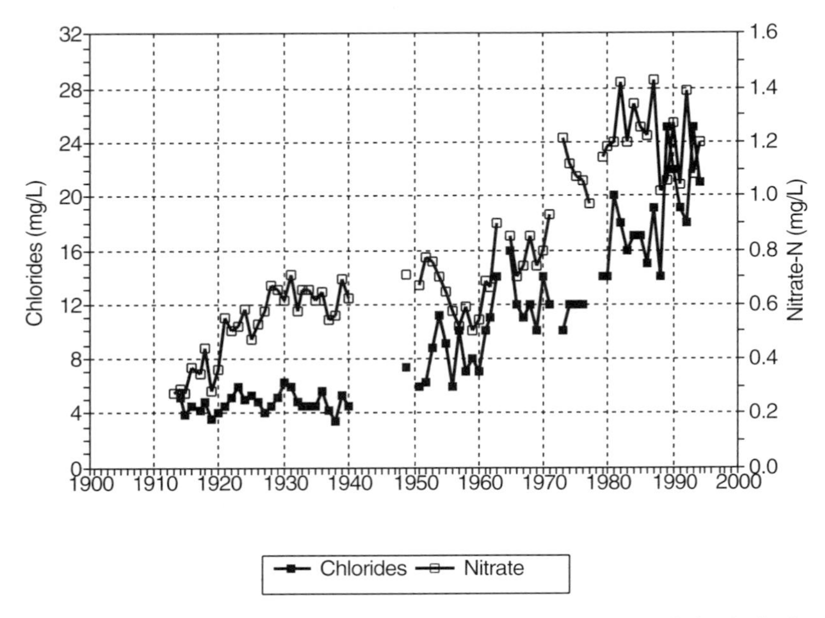

Figure 7-15 Long-term trends of chlorides and nitrate-N at a water supply intake in the tidal Delaware River near Philadelphia. *Source:* Jaworski and Hetling, 1996. Copyright © Water Environment Federation, Alexandria, VA. Reprinted with permission.

Evaluation of Water Quality Benefits Following Treatment Plant Upgrade

From a policy and planning perspective, the central question related to the effectiveness of the secondary treatment requirement of the 1972 CWA is simply: Would water quality standards for DO be attained if primary treatment levels were considered acceptable? In addition to the qualitative assessment of historical data, water quality models can provide a quantitative approach to evaluate improvements in DO and other water quality parameters achieved as a result of upgrades in wastewater treatment. Since the early 1960s four classes of water quality models, developed from the 1960s through the 1990s, have been applied to determine wasteload allocations for municipal and industrial dischargers to meet the needs for water quality management decisions for the Delaware estuary (Mooney et al., 1998).

During the 1960s, one-dimensional estuarine water quality models of DO and carbonaceous and nitrogenous BOD were developed by Thomann (1963), O'Connor et al., (1968), Pence et al. (1968), Jeglic and Pence (1968), and Feigner and Harris (1970). DRBC used a 1960s era model, known as the Delaware Estuary Comprehensive Study (DECS) model, to establish wasteload allocations for ultimate CBOD and nitrogenous BOD for the six zones of the Delaware estuary.

With funding available from the CWA Section 208 program during the 1970s, Clark et al. (1978) upgraded the kinetics of the DECS water quality model to incorporate nitrification and denitrification in a nitrogen cycle represented by organic nitrogen, ammonia, and nitrate + nitrite as state variables. The oxygen contribution by algal production and respiration was included as an empirical input term dependent on chlorophyll observations. Transport was provided to the water quality model with one-dimensional link-node hydrodynamics, and the 1970s-era model was identified as the Dynamic Estuary Model (DEM) (Mooney et al., 1998).

As a result of industrial and municipal waste treatment plant upgrades from primary to secondary levels of treatment during the late 1970s and early 1980s, the water quality model used for wasteload allocations was once again upgraded to reflect the reduced wasteloads and improvements in water quality conditions (Mooney et al., 1998). The model was upgraded from a one-dimensional (longitudinal) to a two-dimensional (longitudinal and lateral) representation of water quality and transport in the Delaware estuary. A two-dimensional hydrodynamic model was coupled with a water quality model that retained the kinetic framework of the one-dimensional model with kinetic coefficients adjusted to reflect changes in pollutant loading (LTI, 1985). The upgraded 1980s-era two-dimensional model (DEM-2D) was used to conduct a toxics analysis (Ambrose, 1987) and to reevaluate the wasteload allocations developed with the earlier models (DRBC, 1987).

Following the completion of the Delaware Estuary Use Attainability (DEL USA) Project (DRBC, 1989), a technical review of the two-dimensional DEM model recommended that a new time-variable model be developed to incorporate state-of-the-art advances, with a three-dimensional hydrodynamic model coupled to an advanced eutrophication model framework (HydroQual, 1994). Using revised kinetic coefficients to reflect reductions in waste loads and improvements in water quality, the kinetics of the water quality framework were expanded to include a eutrophication submodel, nitrogen and phosphorus cycles, labile and refractory organic carbon, and particulate and dissolved fractions of organic carbon and nutrients (HydroQual, 1998; Mooney et al., 1999). Unlike the advanced eutrophication model developed for the Chesapeake Bay (Cerco and Cole, 1993), internal coupling of particulate organic matter deposition with sediment oxygen demand and benthic nutrient fluxes was not included in the upgraded framework; benthic fluxes were assigned as model input on the basis of monitoring data (HydroQual, 1998).

To evaluate the incremental improvements in water quality conditions that can be achieved by upgrading municipal wastewater facilities from primary to secondary and better-than-secondary levels of waste treatment, Lung (1991) used the 1970s-era one-dimensional DEM model (Clark et al., 1978) to demonstrate the water quality benefits attained by the secondary treatment requirements of the 1972 CWA. With this model, Lung used existing population and municipal and industrial wastewater flow and effluent loading data (ca. 1976) to compare water quality for summer flow conditions simulated with three management scenarios for municipal facilities: (1) primary effluent, (2) secondary effluent, and (3) existing wastewater loading. Water quality conditions for these alternatives were calibrated (Figure 7-16) using data for 1976, a

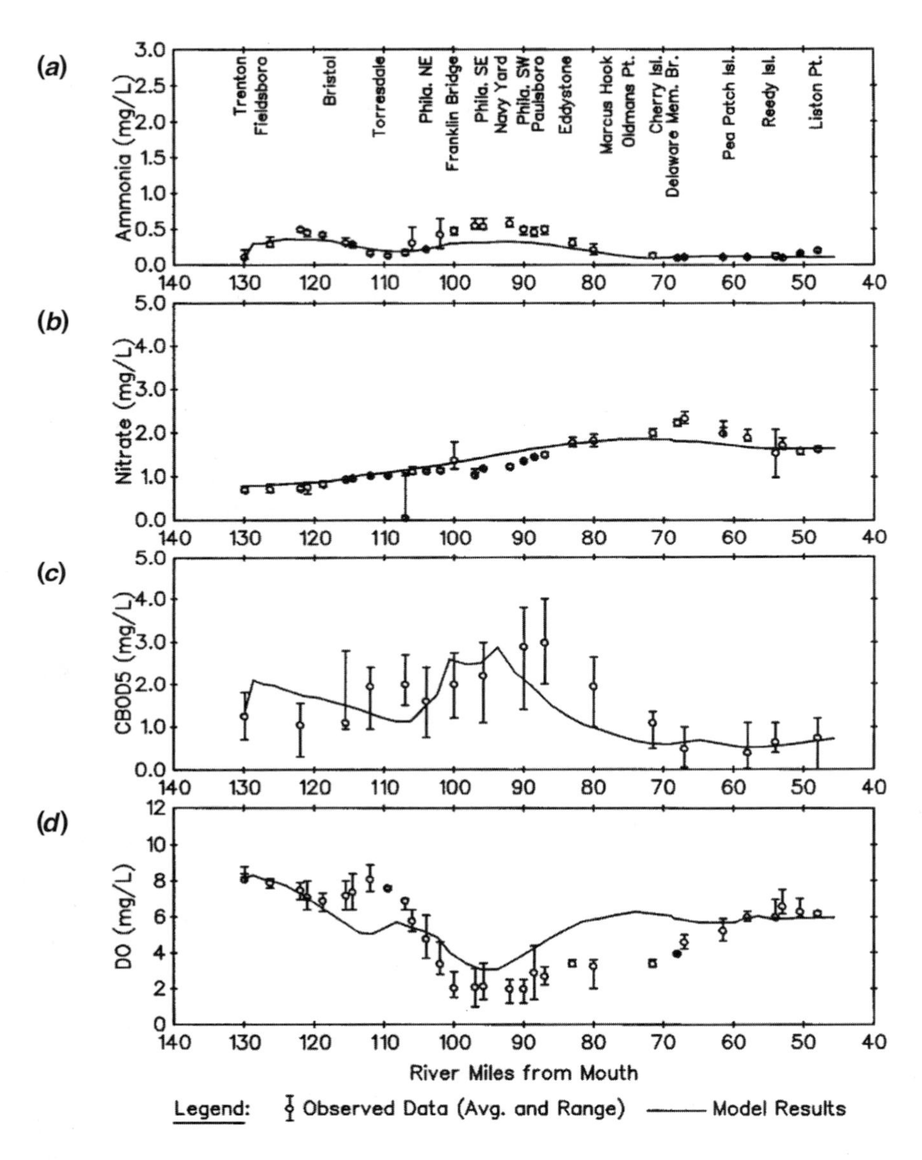

Figure 7-16 Model versus data comparison for calibration of the one-dimensional Dynamic Estuary Model (DEM) for the Delaware estuary to July 1976 conditions for: (*a*) ammonia, (*b*) nitrate, (*c*) $CBOD_5$, and (*d*) DO. *Source:* Lung, 1991.

year characterized by average summer flow of the Delaware River (see Figure 7-3). Freshwater flow at Trenton, New Jersey, was 7,700 cfs; flow in the Schuylkill River, a major tributary to the Delaware estuary, was 1,350 cfs for the 1976 calibration. Flow conditions during the summer of 1976 were 120 percent higher than the long-term (1951–1980) summer (July–September) mean streamflow of 5,986 cfs recorded at Trenton. Upstream of Trenton, flow releases from several impoundments along the free-flowing Delaware River are regulated to maintain the guideline for a minimum summer streamflow of 2,500 to 3,000 cfs at Trenton (Mooney et al., 1998).

Under the primary effluent assumption, water quality is noticeably deteriorated in comparison to the 1976 calibration results. DO concentrations are at a minimum about 35 miles downstream of Trenton, the traditional region of minimum DO levels. Under the primary scenario, an oxygen sag of 2 mg/L is computed by the model under summer (28°C), low-flow 7Q10 conditions (2,500 cfs for the Delaware at Trenton and 285 cfs for the Schuylkill River at Philadelphia) (Figure 7-17).

Using the secondary effluent assumption, the reduction in ultimate CBOD loading significantly improves DO downstream of Philadelphia at the critical oxygen sag location (RM 96). In comparison to the primary scenario, minimum oxygen levels increased to almost 4 mg/L from approximately 2 mg/L under the secondary effluent scenario (Figure 7-18). To achieve compliance with a water quality standard of 5 mg/L, advanced waste treatment is required (Albert, 1997). As shown with the historical water quality data sets, the implementation of secondary and better than secondary levels of wastewater treatment has resulted in major improvements in DO, BOD_5, ammonia, and total phosphorus conditions of the estuary (Figures 7-8 through 7-14). As demonstrated with the model, better than secondary treatment is required to achieve compliance with the water quality standard of 5 mg/L for DO downstream of Philadelphia. In contrast to the 1950s and 1960s, the historical occurrence of persistent and extreme low DO conditions has essentially been eliminated from the upper Delaware estuary. Improvements in suspended solids, heavy metals, and fecal coliform bacteria levels have also been achieved as a result of upgrades in municipal and industrial wastewater treatment.

Recreational and Living Resources Trends

With vast tidal marshes and freshwater tributaries providing spawning and nursery grounds for abundant fishery resources, the coastal plain of the Delaware estuary provided a cornucopia of fishery and waterfowl resources important for sustenance to both Native American villages and colonial settlements. Historically, the estuary produced an enormous quantity of seafood from the early colonial era (ca. 1700s) through the early twentieth century. Colonial-era reports suggest schools of herring and sheepshead thick enough to walk on in a stream (Price et al., 1988). Abundant harvests of American shad and shortnose sturgeon provided important sustenance to the growing population of the Delaware valley for about 200 years.

Since the mid-1900s, however, the abundance of these, and other, species has declined dramatically as a result of urbanization and industrialization of the drainage basin. Deterioration in water quality (e.g., severe oxygen depletion), overfishing,

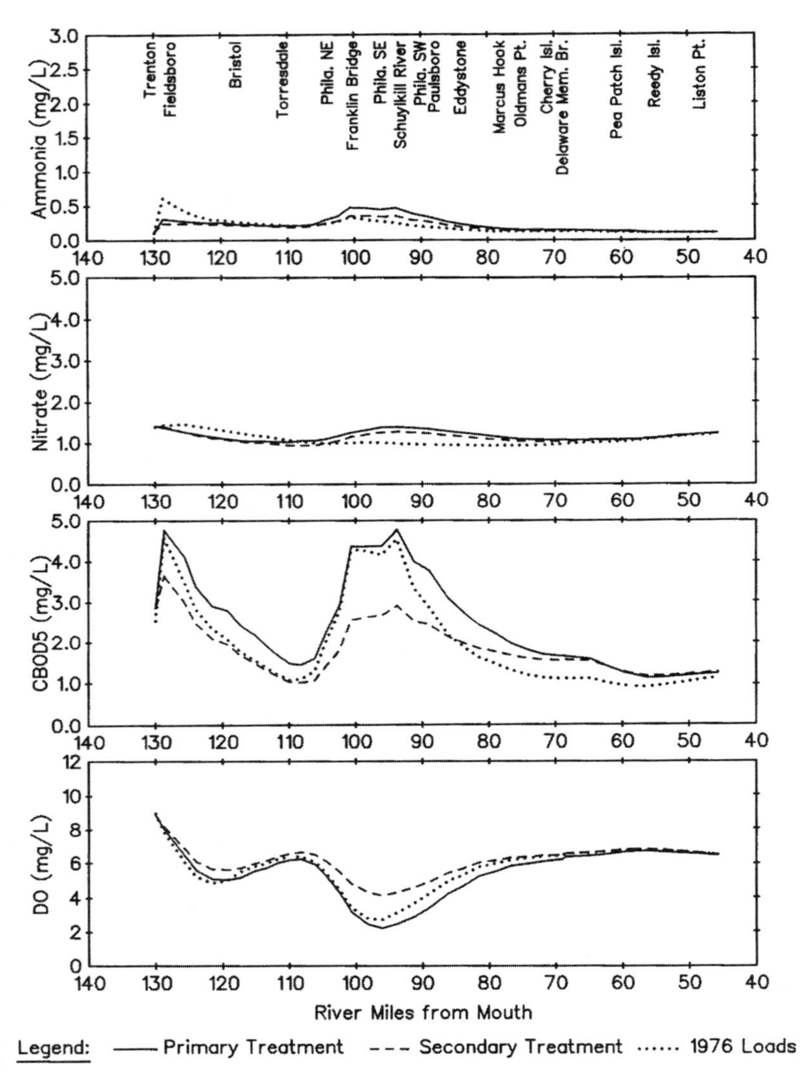

Figure 7-17 Comparison of simulated water quality impact of primary, secondary, and existing (1976) wastewater loading conditions on: (*a*) ammonia, (*b*) nitrate, (*c*) $CBOD_5$, and (*d*) DO in the Delaware Estuary, July 1976 conditions. *Source:* Lung, 1991.

construction of dams, and habitat destruction have all contributed to the decline of the river's fisheries resources, beginning around the turn of the twentieth century (Majumdar et al., 1988). Massive fish kills were a frequent occurrence along the river from about 1900 through 1970 (Albert, 1988). Former wetlands and tributaries, crit-

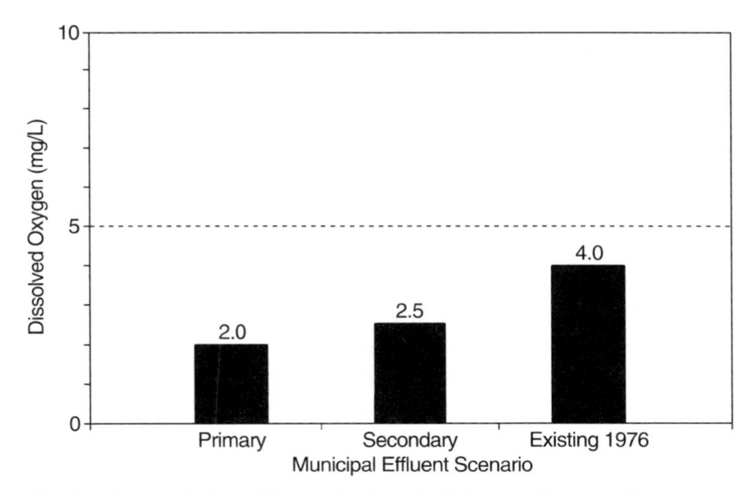

Figure 7-18 Model simulation of DO under July 1976 "normal" streamflow conditions at the critical oxygen sag location (River Mile 96) in the Delaware estuary for primary, secondary, and existing (1976) effluent loading scenarios. *Source:* Lung, 1991.

ical to the spawning success of anadromous species, have been converted into docks, wharves, industrial sites, and oil refineries (Stutz, 1992).

Decades of discharge of untreated municipal and industrial waste resulted in severe declines in the once-abundant fishery resources of the Delaware estuary. In 1836, commercial landings of the American shad (*Alosa sapidissima*), an important anadromous fish that spawns in the Upper Delaware River, were estimated at 10.5 million pounds. By the turn of the twentieth century, the average annual harvest of shad was 12 to 14 million pounds (Frithsen et al., 1991). Historically, the commercial shad harvest from the Delaware River fishery was the largest of any river system along the Atlantic coast (Frithsen et al., 1991). Primarily as a consequence of overfishing, water pollution, and low levels of DO that created a "dead zone," construction of dams, and other obstructions in the river, shad populations declined drastically in the early 1900s (Frithsen et al., 1991).

In a pattern similar to that for shad, annual commercial landings of striped bass (*Morone saxatilis*) have also dropped from hundreds of thousands of pounds per year in the early 1900s to only thousands of pounds per year by 1960. In 1969, a fishery survey showed a complete absence of striped bass larvae and eggs along the Philadelphia-Camden waterfront, which had been an important spawning and nursery area for striped bass; by 1980 there was no commercial catch of striped bass (Himchak, 1984).

The historical abundance of shortnose sturgeon (*Acipenser brevirostrum*), once prized for caviar that rivaled imported Russian caviar, also followed the same precipitous decline as shad as overfishing and water pollution took their toll on this once-thriving fishery. Historically, the range of the shortnose sturgeon was from the lower

Delaware Bay as far upstream as New Hope, Pennsylvania (RM 149) (Frithsen et al., 1991). Historical records from 1811 to 1913 document 1,949 sturgeon captured, primarily as a bycatch of the shad gill net fishery. During the period from 1913 through 1954, no documented catches of sturgeon were reported. From 1954 through 1979, 37 sturgeon were reported in fishery and ecological surveys. From 1981 to 1984, 1,371 sturgeon were collected between Philadelphia and Trenton (Frithsen et al., 1991). Using data collected from the early 1980s surveys, Hastings et al. (1987) have estimated populations of approximately 6,000 to 14,000 adult shortnose sturgeon in the upper tidal river near Trenton, with a smaller population estimated for the section of the river near Philadelphia (Frithsen et al., 1991).

Although it is difficult to assess the relative importance to these species of each of the major industrialization factors that contributed to the declines, Summers and Rose (1987) identified a connection between water quality, especially DO concentrations, and wastewater loading and shad population levels. Using records collected during the twentieth century from the Delaware, Potomac, and Hudson estuaries, historical fluctuations in American shad populations have been strongly correlated with wastewater discharges that increased biochemical oxygen demand levels and depleted oxygen resources (Summers and Rose, 1987). Albert (1988) and Sharp and Kraeuter (1989) also noted the importance of adequate oxygen concentrations to successful shad migrations. Little correlation between water quality and striped bass populations was found, but Summers and Rose (1987) noted that larval survival for both shad and striped bass is tied to DO and other water quality factors.

Beginning in the mid-1970s, however, the water pollution control efforts of the 1970s and 1980s have paid off with a dramatic recovery of once moribund fishery resources. Estimates of the American shad population fluctuated from a low of 106,202 in 1977 to a high of 882,600 in 1992 (Santoro, 1998) (Figure 7-19). As a result of im-

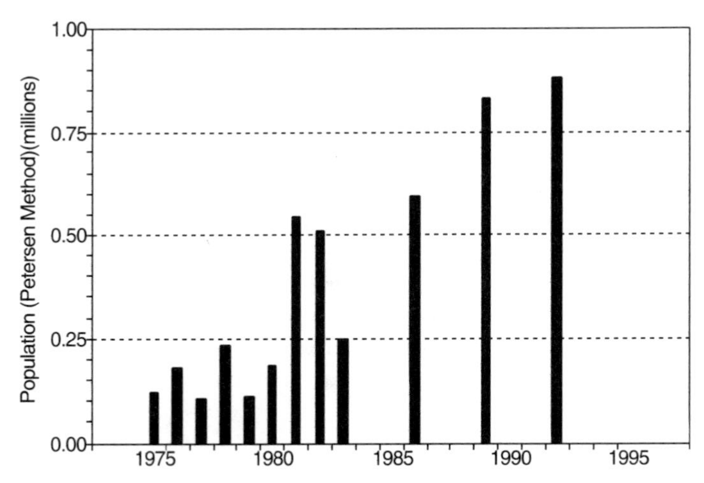

Figure 7-19 Long-term trends in population estimates of adult American shad in the Delaware estuary. *Source:* Santoro, 1998.

provements in water quality conditions, the spawning area used by shad has increased by 100 miles in the estuary (Albert, 1997). Annual shad festivals are now celebrated in the spring along the Delaware River, and the recreational shad fishery is considered to be a multimillion-dollar industry (Frithsen et al., 1991). As a result of water pollution control efforts and a well-regulated fishery, populations of striped bass in the Delaware River are also showing evidence of a resurgence of once-depleted populations (Santoro, 1998). Assessments of commercial harvest statistics for American shad (Figure 7-20), striped bass (Figure 7-21), and white perch (Figure 7-22) clearly document significant increases in the catch-per-unit effort of these species from 1985 to 1993, correlated with improvements in water quality (Weisberg et al., 1996). Trends in catch efficiency are also reported by Weisberg et al. (1996) for blueback herring and alewives. Studies of the distribution and abundance of the shortnose sturgeon, listed as an endangered species (Price et al., 1988), suggest that populations may be recovering from the historical decimation of this species during the twentieth century, a decimation resulting from water pollution and overfishing (Frithsen et al., 1991).

In addition to pelagic fishery resources, the Delaware estuary has historically provided important harvests of American oysters, blue crabs, horseshoe crabs, hard clams, and American lobsters. Following a pattern identified in New York Harbor, a sharp decline in the harvest of oysters during the 1950s has been attributed to overfishing, sediment runoff, and industrialization of the watershed, industrial and municipal wastewater discharges, oil spills, and spraying of marshes with DDT for mosquito control (Frithsen et al., 1991). In 1957, a parasitic organism (MSX) infected the oyster beds, drastically reducing abundance for decades. With the decline of the oyster harvest, the blue crab catch has accounted for most of the shellfish catch of the Delaware estuary. During the late 1800s through the 1930s, few blue crabs were harvested commercially. Since the 1930s, commercial landings have increased substan-

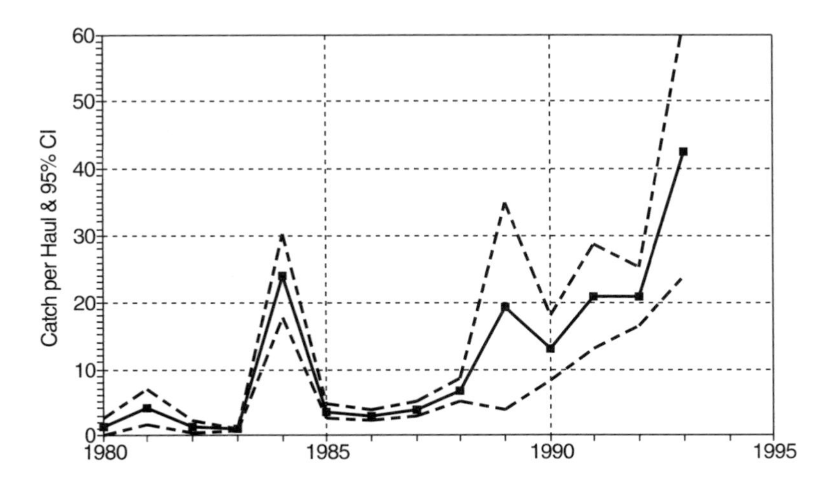

Figure 7-20 Trends in catch efficiency (mean ± 95% confidence interval) for American shad in the Delaware estuary. *Source:* Weisberg et al., 1996.

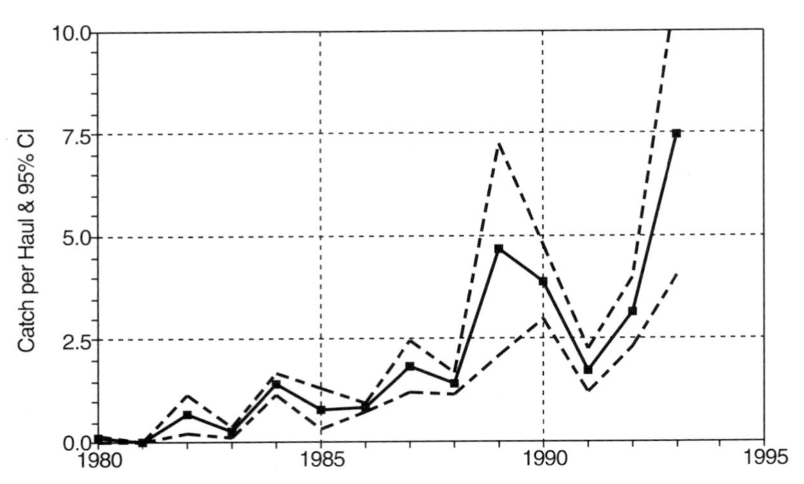

Figure 7-21 Trends in catch efficiency (mean ± 95% confidence interval) for striped bass in the Delaware estuary. *Source:* Weisberg et al., 1996.

tially, particularly during the 1980s (Figure 7-23), although large interannual variability in the Delaware estuary is characteristic of this species sensitivity to water temperature (Frithsen et al., 1991). More detailed reviews of historical trends for the shellfish and fishery resources of the Delaware estuary are given by Price et al. (1988), Patrick (1988), Patrick et al. (1992), and Frithsen et al. (1991).

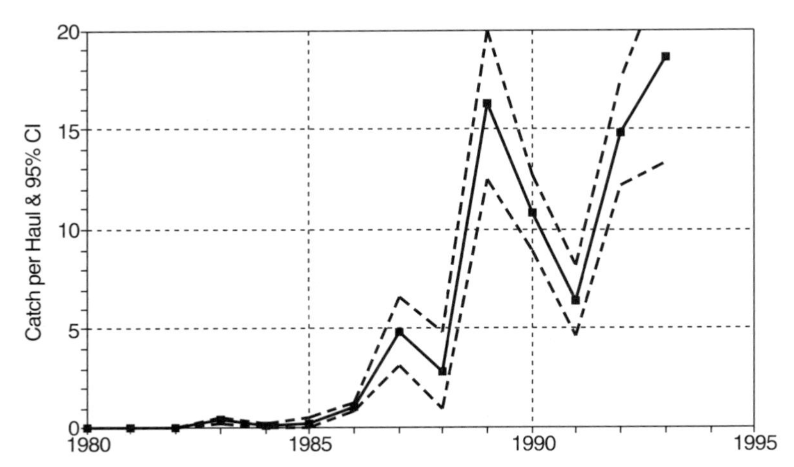

Figure 7-22 Trends in catch efficiency (mean ± 95% confidence interval) for white perch in the Delaware estuary. *Source:* Weisberg et al., 1996.

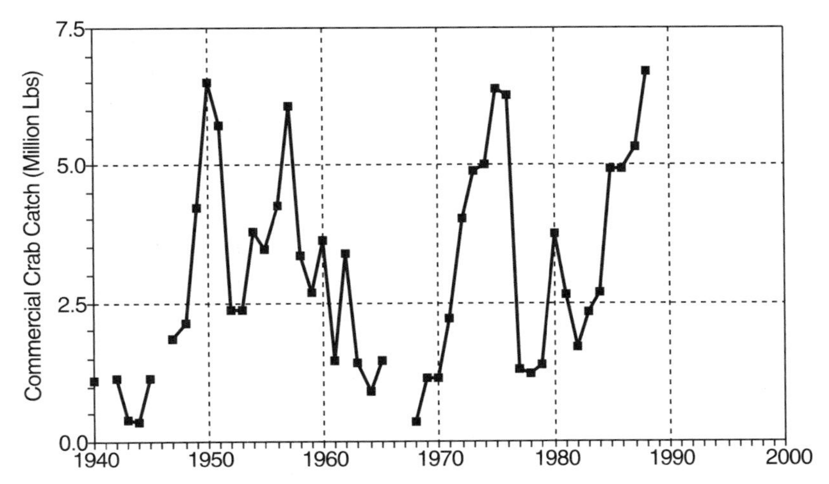

Figure 7-23 Long-term trends of commercial blue crab catch in the Delaware estuary (New Jersey and Delaware totals). *Source:* Patrick, Ruth, Faith Douglass, and Drew Palavage, Surface water quality: Have the laws been successful? Copyright © 1992 by Princeton University Press, Reprinted by permission of Princeton University Press.

SUMMARY AND CONCLUSIONS

During the 1940s, 1950s, and 1960s, the Delaware estuary was characterized by severe water quality problems, including the foul stench of hydrogen sulfide gas caused by anoxic conditions in sections of the river near Philadelphia. Uncontrolled wastewater discharges and destruction of habitat from urban and industrial growth in the Delaware watershed were responsible, along with overfishing, for the collapse of many historically important fisheries in the Delaware estuary such as American shad, striped bass, shortnose sturgeon, and American oysters. Desirable amenities such as parks, walking trails, or cafes along the riverfront were not considered for urban development because of the noxious conditions of the Delaware River.

As a result of water pollution control efforts implemented since the late 1960s in the Delaware estuary, dramatic reductions in municipal and industrial effluent discharges of ultimate CBOD, ammonia-N, total phosphorus, and fecal coliform bacteria have been achieved by upgrading wastewater treatment facilities to secondary and better-than-secondary levels of treatment. Municipal and industrial loading of ultimate CBOD to the river, for example, was reduced by 89 percent during the period from 1958 to 1995.

New construction and upgrades of municipal and industrial water pollution control facilities have resulted in significant improvements in water quality, the resurgence of important commercial and recreational fishery resources, and a renewal of economic vitality to once abandoned urban waterfronts along the Delaware River.

Assessment of long-term trends of historical water quality data at critical locations clearly documents great improvements in DO, ammonia-N, total phosphorus, and fecal coliform bacteria. DO, for example, has improved from typical summer minimum levels of less than 1 mg/L during the 1960s and 1970s along a 10-mile section of the river downstream from Philadelphia to minimum levels of 4 mg/L and higher during the 1990s. Ambient ammonia-N concentrations near Marcus Hook have declined by an order of magnitude from late 1960s levels of approximately 1.4 mg N/L to mid-1990s levels of approximately 0.15 mg N/L. Total phosphorus has exhibited a trend similar to that of ammonia-N with late 1960s levels of approximately 0.8 mg P/L dropping almost an order of magnitude to approximately 0.1 mg P/L during the mid-1990s.

A number of indicators of environmental resources of the Delaware estuary have also demonstrated tremendous improvements that can be attributed to the water pollution control efforts and associated public awareness of the importance of environmental quality initiated by the 1972 CWA. The recovery of the American shad population during the mid-1980s, for example, is a remarkable achievement. The restoration of this important fishery resource to populations that can support an extensive recreational and commercial fishery is a success story. Highly popular annual shad festivals now celebrate the seasonal migration of this fish from the ocean into the estuary as a rite of spring.

Although the restoration of valuable fishery resources is important from an economic and ecological perspective, the recreational benefits achieved by the cleanup of the Delaware River far exceed the benefits attributed to fishery improvements. Riverfront development for commercial uses and public parks, increases in sailing and boating, and numerous other economic benefits have occurred along the Delaware River. Most remarkable is that the city centers of Philadelphia, Wilmington, and Trenton, after decades of urban development activity retreating inland, are now moving back toward the riverfront. Investments in urban development along the river would simply not be feasible without the aesthetic qualities of clean water in the Delaware estuary (Albert, 1997). Urban waterfront and riverfront development activity has also been booming in many other cities (e.g., New York Harbor; Cleveland, Ohio; Boise, Idaho; Portland, Oregon; Atlanta, Georgia; Richmond, Virginia) that have successfully cleaned up polluted rivers, lakes, and harbors, making their urban waterways assets and sources of civic pride rather than disgraceful liabilities.

Despite the remarkable environmental improvements achieved by investments in water pollution control infrastructure since initiation of the 1972 CWA, challenges remain for the next generation. Water quality and resource management problems recognized only since the mid-1980s must be addressed. Contamination of the water column and sediments by heavy metals such as mercury, chromium, lead, copper, and zinc has been identified in urban-industrial areas of the river. Probable sources of heavy metals include natural geochemical processes, industrial and municipal dischargers, stormwater runoff, and atmospheric deposition (Santoro, 1998). Toxic chemicals such as PCBs, polynuclear aromatic hydrocarbons (PAHs), and pesticides have also contaminated the water column and sediments of the estuary, resulting in bioaccumulation of those chemicals in pelagic fish and benthic organisms.

Fish consumption advisories were issued in 1989 by New Jersey and Pennsylvania and in 1996 by Delaware (Santoro, 1998). Acute sediment toxicity appears to be more widespread in the estuary than previously documented, with the highest areas of sediment toxicity identified in the heavily urbanized and industrialized region between Torresdale and Marcus Hook. Chronic toxicity was also identified in the water column under particular conditions of streamflow and effluent discharges (Santoro, 1998). The design and construction of facilities to control and treat combined sewer overflow discharges of raw sewage to the tidal river during heavy rainstorms is an ongoing project. Finally, the allocations of wastewater loads for ultimate CBOD from municipal and industrial dischargers that have evolved since 1968 will need to be revised to ensure that the water quality improvements achieved since the 1970s can continue to be maintained as population and industrial activity grow during the twenty-first century (Mooney et al., 1999; HydroQual, 1998).

In 1973, a USEPA study concluded that the Delaware River would never achieve designated uses defined by "fishable standards." More than 25 years after that pessimistic pronouncement, the fishery resources of the Delaware estuary are thriving. The restoration of the vitality of the estuary is a direct result of water pollution control efforts and strong public awareness of the importance of supporting federal, state, and local environmental regulations and policies.

REFERENCES

Albert, R. C. 1982. Cleaning up the Delaware River. Status and progress report prepared under the auspices of Section 305(b) of the Federal Clean Water Act. Delaware River Basin Commission, West Trenton, NJ.

Albert, R. C. 1988. The historical context of water quality management for the Delaware estuary. Estuaries 11(2): 99–107.

Albert, R. C. 1997. Delaware River Basin Commission, West Trenton, NJ. Personal communication, July 31.

Ambrose, R. B. 1987. Modeling volatile organics in the Delaware estuary. J. Environ. Eng. ASCE 113(4): 703–721.

Bondelid, T., S. Unger, and A. Stoddard. 2000. National water pollution control assessment model (NWPCAM) Version 1.1. Final report prepared by Research Triangle Institute, Research Triangle Park, NC for U.S. Environmental Protection Agency, Office of Policy, Economics and Innovation, Washington, DC, November, RTI Project Number 92U-7640-031.

Brezina, E. R. 1988. Water quality issues in the Delaware River basin. In: Ecology and restoration of the Delaware River basin, ed. S. K. Majumdar, E. W. Miller, and L. E. Sage, pp. 30–38. Pennsylvania Academy of Science, Philadelphia, PA.

Bryant, T. L. and J. R. Pennock. 1988. The Delaware estuary: Rediscovering a forgotten resource. University of Delaware Sea Grant College Program, Newark, DE. Philadelphia Press, Burlington, NJ.

CEQ. 1982. Environmental quality 1982. 13th annual report of the Council on Environmental Quality. Executive Office of the President, Council on Environmental Quality, Washington, DC.

Cerco, C. and T. Cole. 1993. Three-dimensional eutrophication model of Chesapeake Bay. J. Environ. Eng., ASCE 119(6): 1006–1025.

Clark, L. J., R. B. Ambrose, and R. C. Crain. 1978. A water quality modelling study of the Delaware estuary. Technical Report 62. U.S. Environmental Protection Agency, Region 3, Annapolis Field Office, Annapolis, MD.

DRBC. 1987. Recalibration/verification of the Dynamic Estuary Model for current conditions in the Delaware estuary. DEL USA Project Element 19. Delaware River Basin Commission, West Trenton, NJ.

DRBC. 1989. Delaware estuary use attainability project: Final report. DEL USA Project. Delaware River Basin Commission, West Trenton, NJ.

DRBC. 1992. Delaware River and Bay water quality assessment 1990–1991 305(b) report. Delaware River Basin Commission, West Trenton, NJ.

Feigner, K. and H. S. Harris. 1970. Documentation report on FWQA Dynamic Estuary Model. U.S. Department of the Interior, Annapolis, MD.

Forstall, R. L. 1995. Population by counties by decennial census: 1900 to 1990. U.S. Bureau of the Census, Population Division, Washington, DC. http://www.census.gov/population/www/censusdata/cencounts.html.

Frithsen, J. B., K. Killam, and M. Young. 1991. An assessment of key biological resources in the Delaware River estuary. Prepared for the Delaware Estuary Program, U.S. Environmental Protection Agency, Region 2, New York, by Versar, Inc., Columbia, MD.

Galperin, B. and G. L. Mellor. 1990. A time-dependent, three-dimensional model of the Delaware Bay and River system. Part 1: Description of the model and tidal analysis. Est. Coast. Shelf Sci. 31: 231–253.

Hastings, R. W., F. C. O'Herron, II, K. Schick, and M. A. Lazzari. 1987. Occurrence and distribution of shortnose sturgeon, *Acipenser brevirostrum*, in the upper tidal Delaware River. Estuaries 10(4): 337–341.

Himchak, P. J. 1984. Monitoring of the striped bass population in New Jersey. Final report covering the period from December 1, 1983, through September 30, 1984. New Jersey Department of Environmental Protection, Division of Fish, Game and Wildlife, Bureau of Marine Fisheries, Nacote Creek Research Station, Port Republic, NJ.

HydroQual. 1994. Review of the Delaware River DEM Models. Technical report prepared for Delaware River Basin Commission, West Trenton, NJ, by HydroQual, Inc., Mahwah, NJ.

HydroQual. 1998. Development of a hydrodynamic and water quality model for the Delaware River. Technical report prepared for Delaware River Basin Commission, West Trenton, NJ. by HydroQual, Inc., Mahwah, NJ. May 29.

Iseri, K. T. and W. B. Langbein. 1974. Large rivers of the United States. Circular No. 686. U.S. Department of the Interior, U.S. Geological Survey, Reston, VA.

Jaworski, N. and L. Hetling. 1996. Water quality trends of the Mid-Atlantic and Northeast watersheds over the past 100 years. In Proceedings Watershed '96: A national conference on watershed management, pp. 980–983. Baltimore, MD, June 8–12. Water Environment Federation, Alexandria, VA.

Jeglic, J. M. and G. D. Pence. 1968. Mathematical simulation of the estuarine behavior and its application. Socio-Econ. Plan. Sci. 1: 363–389.

Lebo, M. E. and J. H. Sharp. 1993. Distribution of phosphorus along the Delaware, an urbanized coastal plain estuary. Estuaries 16(2): 290–301.

LTI. 1985. Two-dimensional DEM model of the Delaware estuary. Technical report prepared for Delaware River Basin Commission, West Trenton, NJ, by LimnoTech, Inc., Ann Arbor, MI.

Lung, W. 1991. Trends in BOD/DO modeling for waste load allocations of the Delaware River estuary. Technical memorandum prepared for Tetra Tech, Inc., Fairfax, VA, by Enviro-Tech, Inc., Charlottesville, VA.

Majumdar, S. K., E. W. Miller, and L. E. Sage (eds.). 1988. Ecology and restoration of the Delaware River basin. Pennsylvania Academy of Science, Dept. Biology, Lafayette College, Easton, PA.

Marino, G. R., J. L. Di Lorenzo, H. S. Litwack, T. O. Najarian, and M. L. Thatcher. 1991. General water quality assessment and trend analysis of the Delaware Estuary, Part One: Status and trends. Prepared for Delaware Estuary Program and Delaware River Basin Commission, West Trenton, NJ, by Najarian & Associates, Eatontown, NJ.

Mooney, K. G., T. W. Gallagher, and H. J. Salas. 1998. A review of thirty years of water quality modeling of the Delaware estuary. Paper presented at XXVI Congress InterAmericano de Ingenieria Sanitaria Y Ambiental-Associacion InterAmericano de Ingenieria Sanitaria Y Ambiental (AIDIS), Lima, Peru, November 1–5.

Mooney, K. G., P. J. Webber, and T. W. Gallagher. 1999. A new total maximum daily load (TMDL) model for the Delaware River. Paper presented at WEFTEC'99, 1999 annual conference of the Water Environment Federation, New Orleans, LA. Water Environment Federation, Alexandria, VA, October 9–13.

O'Connor, D. J., J. P. St. John, and D. M. Di Toro. 1968. Water quality analysis of the Delaware River Estuary. J. Sanit. Eng. Div., ASCE 94(SA6): 1225–1252.

OMB. 1999. OMB Bulletin No. 99-04. Revised statistical definitions of Metropolitan Areas (MAs) and guidance on uses of MA definitions. U.S. Census Bureau, Office of Management and Budget, Washington, DC.
http://www.census.gov/population/www/estimates/metrodef.html

Patrick, R. 1988. Changes in the chemical and biological characteristics of the Upper Delaware River estuary in response to environmental laws. In: Ecology and restoration of the Delaware River Basin, ed. S. K. Majumdar, E. W. Miller, and L. E. Sage, pp. 332–359. Pennsylvania Academy of Science, Dept. Biology, Lafayette College, Easton, PA.

Patrick, R., F. Douglass, D. M. Palarage, and P. M. Stewart. 1992. Surface water quality: Have the laws been successful? Princeton University Press, Princeton, NJ.

Pence, G. D., J. M. Jeglic, and R. V. Thomann. 1968. Time-varying DO model. J. Sanit. Eng. Div., ASCE 94(SA4): 381–402.

Price, K. S., R. A. Beck, S. M. Tweed, and C. E. Epifanio. 1988. Fisheries. In: The Delaware Estuary: Rediscovering a forgotten resource, ed. T. L. Bryant and J. R. Pennock, pp. 71–93. University of Delaware Sea Grant College Program, Newark, DE. Philadelphia Press, Burlington, NJ.

Roman, C. T., N. Jaworski, F. T. Short, S. Findlay, and R. Scott Warren. 2000. Estuaries of the Northeastern United States: Habitat and Land Use Signatures. Estuaries 23(6):743–764.

Sage, L. E. and F. B. Pilling. 1988. The development of a nation: The Delaware River. In: Ecology and restoration of the Delaware River Basin, ed. S. K. Majumdar, E. W. Miller, and L. E. Sage, pp. 217–233, chapter 14. Pennsylvania Academy of Science, Department Biology, Lafayette College, Easton, PA.

Santoro, E. 1998. Delaware estuary monitoring report. Report prepared for Delaware Estuary Program, U.S. Environmental Protection Agency, Region 2, New York, by Delaware River Basin Monitoring Coordinator in cooperation with the Monitoring Implementation Team of the Delaware Estuary Program, West Trenton, NJ.

Scally, P. 1997. Delaware River Basin Commission, West Trenton, NJ. Personal communication, August.

Sharp, J. H. 1988. Trends in nutrient concentrations in the Delaware Estuary. In Ecology and restoration of the Delaware River Basin, ed. S. K. Majumdar, E. W. Miller, and L. E. Sage, pp. 77–92. Pennsylvania Academy of Science, Department Biology, Lafayette College, Easton, PA.

Sharp, J. H. and J. N. Kraeuter. 1989. The state of the Delaware Estuary. Summary report of a workshop held on October 19. Scientific and Technical Advisory Committee, Delaware Estuary Program, West Trenton, NJ.

Solley, W. B., R. R. Pierce, and H. A. Perlman. 1998. Estimated use of water in the United States, 1995. USGS Circular 1200. U.S. Geological Survey, Reston, VA.

Stutz, B. 1992. The Delaware: Portrait of a river. Nature Conservancy, May/June, Arlington, VA.

Summers, J. K. and K. A. Rose. 1987. The role of interactions among environmental conditions in controlling historical fisheries variability. Estuaries 10(3): 255–266.

Thomann, R. V. 1963. Mathematical model for DO. J. Sanit. Eng. Div., ASCE 89(SA5): 1–30.

Thomann, R. V. and J. A. Mueller. 1987. Principles of surface water quality modeling and control. Harper & Row, New York.

USDOC. 1998. Census of population and housing. U.S. Department of Commerce, Economics and Statistics Administration, Bureau of the Census—Population Division, Washington, DC.

USEPA (STORET). STOrage and RETrieval Water Quality Information System. U.S. Environmental Protection Agency, Office of Wetlands, Oceans, and Watersheds, Washington, DC.

USGS. 1999. Streamflow data downloaded from the United States Geological Survey's National Water Information System (NWIS)-W. Data retrieval for historical streamflow daily values. http://waterdata.usgs.gov/nwis-w

Weisberg, S., P. Himchak, T. Baum, H. T. Wilson, and R. Allen. 1996. Temporal trends in abundance of fish in the tidal Delaware River. Estuaries 19(3): 723–729.

CHAPTER 8

Potomac Estuary Case Study

The Mid-Atlantic Basin (Hydrologic Region 2), covering a drainage area of 111,417 square miles, includes some of the major rivers in the continental United States. Figure 8-1 highlights the location of the basin and the Potomac estuary, the case study watershed profiled in this chapter.

With a length of 340 miles and a drainage area of 14,670 square miles, the Potomac River ranks forty-eighth among the 135 U.S. rivers that are more than 100 miles in length (Iseri and Langbein, 1974). Figure 8-2 highlights the location of the Potomac estuary case study catalog units identified by major urban-industrial areas affected by severe water pollution problems during the 1950s and 1960s (see Table 4-2). This chapter presents long-term trends in population, municipal wastewater infrastructure and effluent loading of pollutants, ambient water quality, environmental resources, and uses of the Potomac estuary. Data sources include USEPA's national water quality database (STORET), published technical literature, and unpublished technical reports ("grey" literature) obtained from local agency sources.

With a combined drainage area of 14,670 square miles, the freshwater and estuarine Potomac River basin is the second largest watershed in the Middle Atlantic region. The freshwater Upper Potomac River flows more than 220 miles from the headwaters of the North Branch in the eastern Appalachian Mountains to the fall line at Little Falls, Virginia, near Washington, DC. Tidal influences in the Potomac extend 117 miles from the fall line at Little Falls to the confluence with Chesapeake Bay at Point Lookout, Maryland (Figure 8-2).

In this 117-mile reach, the Potomac River is classified into three distinct hydrographic regions—tidal river, transition zone, and estuary. The tidal river, extending 38 miles from the fall line to Quantico, Virginia, is characterized as freshwater (salinity < 0.5 ppt) with net seaward flow from surface to bottom. This section of the Potomac River receives the effluent discharge from the major municipal wastewater treatment facilities in the Washington, DC, metropolitan area. The transition zone, extending 29 miles from Quantico, Virginia, to the Route 301 bridge in Maryland, is characterized by variable salinity (0.5 to 10 ppt) and significant mixing of freshwater and saltwater from Chesapeake Bay. In the mesohaline estuary region, extending 50 miles from the Route 301 bridge to Chesapeake Bay at Point Lookout, Maryland, salinity varies from 5 to 18 ppt, with estuarine circulation described as partially mixed (Haramis and Carter, 1983).

During much of the past century, the Potomac estuary has been characterized by severe water pollution problems—bacterial contamination, oxygen depletion, and nui-

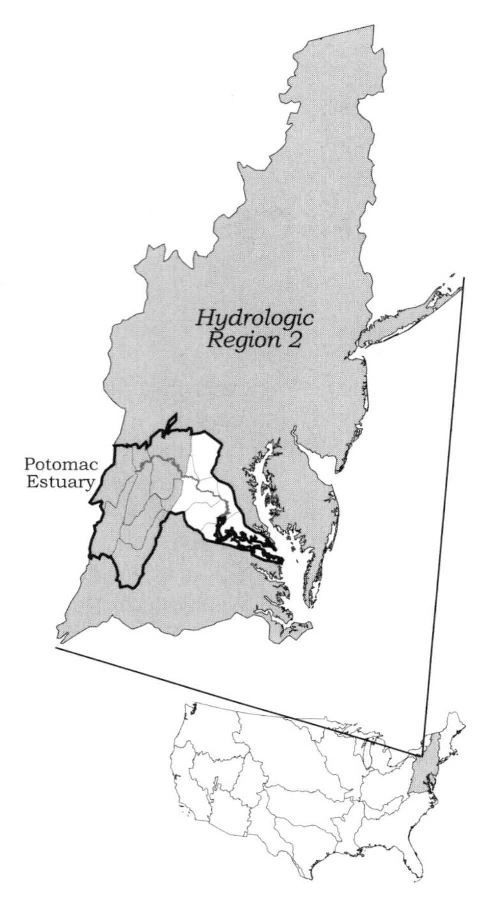

Figure 8-1 Hydrologic Region 2 and the Potomac estuary watershed.

sance algal blooms—resulting from population growth in the Washington, DC, area and inadequate levels of waste treatment. Historical DO data provide an excellent indicator to characterize long-term trends in the ecological status of the Potomac estuary. The water quality benefits attributed to implementation of secondary and advanced waste treatment by Washington, DC, area municipal wastewater dischargers to the Potomac estuary represent a major national environmental success story.

PHYSICAL SETTING AND HYDROLOGY

The Upper Potomac River, which has a drainage area of 11,560 square miles, is the major freshwater inflow to the estuary. Based on long-term (1931–1981) USGS data at Little Falls near the fall line, the mean annual daily flow is 11,406 cfs, with extreme

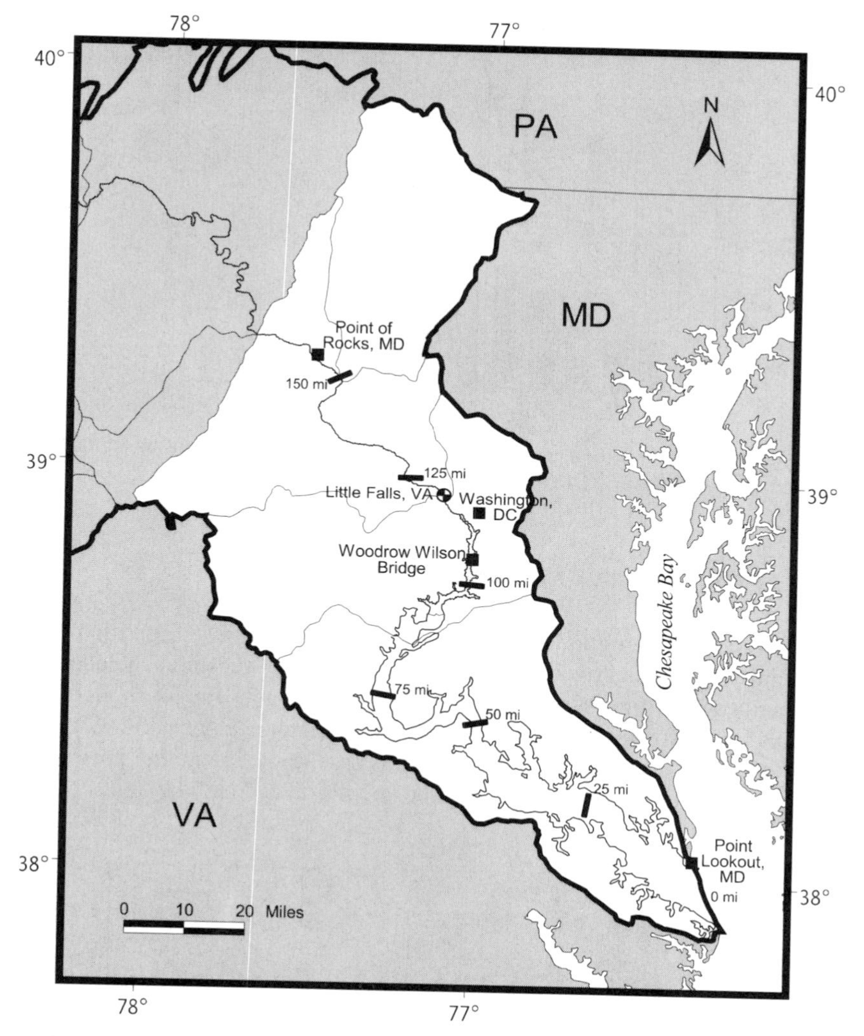

Figure 8-2 Location map of Middle and Lower Potomac River. (River miles shown are distances from the confluence of the Lower Potomac River with the Chesapeake Bay.)

discharge conditions of 374 cfs recorded during the drought of 1966 and 483,802 cfs recorded during the flood of 1936 (MWCOG, 1989). The long-term (1931–1988) mean 7-day, 10-year low flow (7Q10) at Little Falls is 628 cfs. Low-flow conditions typically occur from July through September, with the minimum monthly flow of 4,126 cfs recorded during September (Figure 8-3). The long-term (1951–1980) mean summer (July–September) flow for the Potomac River at Little Falls was 4,428 cfs (Figure 8-4).

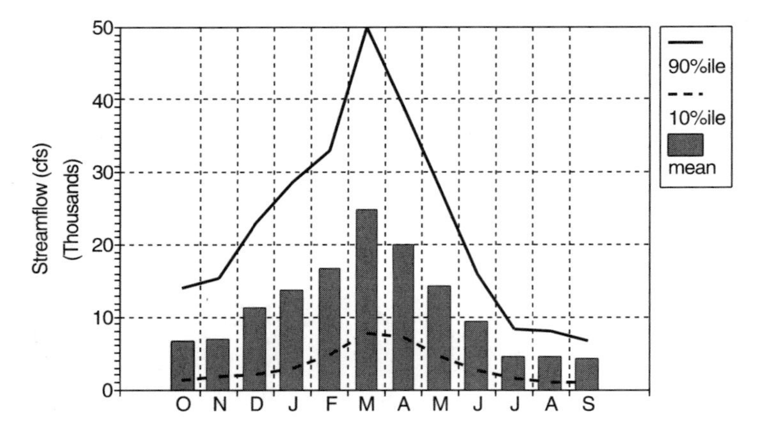

Figure 8-3 Monthly trends of mean, tenth, and ninetieth percentile streamflow for the Potomac River at Little Falls, Virginia (USGS Gage 01646500), 1951–1980. *Source:* USGS, 1999.

POPULATION, WATER, AND LAND USE TRENDS

In 1996, more than 4.5 million people lived in the Washington, DC, metropolitan area in the vicinity of the tidal river. The Potomac estuary case study area includes a number of counties identified by the Office of Management and Budget as Metropolitan Statistical Areas (MSAs) or Primary Metropolitan Statistical Areas (PMSAs). Table 8-1 lists the MSAs and counties included in this case study. Figure 8-5 presents long-term population trends (1940–1996) for the counties listed in Table 8-1. From 1940

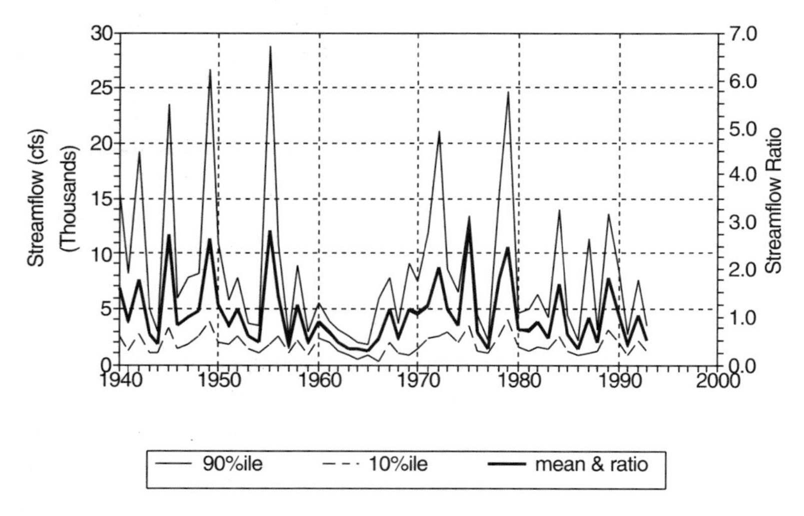

Figure 8-4 Long-term trends in mean, tenth, and ninetieth percentile streamflow in summer (July–September) for the Potomac River at Little Falls, Virginia (USGS Gage 01646500). *Source:* USGS, 1999.

TABLE 8-1 Metropolitan Statistical Area (MSAs) and Counties and Incorporated Cities in the Potomac Estuary Case Study

Calvert County, MD
Charles County, MD
Frederick County, MD
Montgomery County, MD
Prince George's County, MD
Arlington County, VA
Clarke County, VA
Culpepper County, VA
Fairfax County, VA
Fauquier County, VA
King George County, VA
Loudoun County, VA
Prince William County, VA
Spotsylvania County, VA
Stafford County, VA
Warren County, VA
Alexandria City, VA
Fairfax City, VA
Falls Church City, VA
Fredericksburg City, VA
Manassas City, VA
Manassas Park City, VA
Berkeley County, WV
Jefferson County, WV

Source: OMB, 1999.

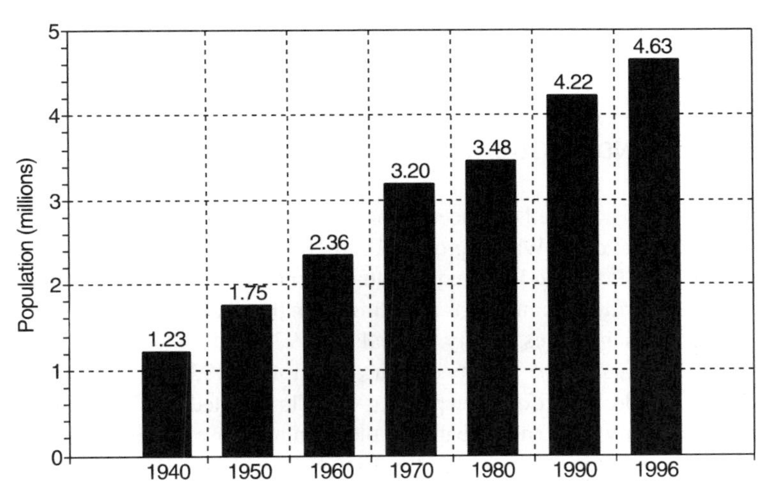

Figure 8-5 Long-term trends in population of Washington, DC, metropolitan area. *Sources:* Forstall, 1995; USDOC, 1998.

to 1996, the population in the Potomac estuary study area nearly quadrupled (Forstall, 1995; USDOC, 1998).

Within the Potomac basin, land use is characterized as forested (55 percent), agricultural (40 percent), and urban (5 percent) (Jaworski, 1990). A rapid transition from agricultural land use to suburban land use has occurred since the 1960s in the Washington, DC, metropolitan area. In contrast to other major metropolitan areas, industrial activities are a negligible component of the regional economy (and wastewater loading). Upstream of the fall line, the free-flowing Potomac is used for five municipal water supply diversions with a total mean withdrawal (ca. 1986) of 386 mgd (MWCOG, 1989). As a result of major improvements in water quality over the past decade, boating and recreational and commercial fishing have become important resource uses of the Potomac estuary.

HISTORICAL WATER QUALITY ISSUES

As in many other urban areas centered around rivers and harbors, water pollution problems have been documented in the tidal river since the turn of the century (e.g., Newell, 1897). In the late 1950s USPHS officials described the Potomac near Washington, DC, as "malodorous . . . with gas bubbles from sewage sludge over wide expanses of the river . . . and coliform content estimated as equivalent to dilution of 1 part raw sewage to as little as 10 parts clean water." Dissolved oxygen levels near Washington, DC, were typically less than 1 mg/L during summer low-flow conditions, and nuisance algal blooms and fish kills were commonplace during the 1940s, 1950s, and 1960s. Between 1955 and 1960, the stock abundance of American shad in the Potomac River dropped precipitously despite favorable hydrographic conditions for spawning and development. American shad in northeastern estuaries such as the Potomac River, although influenced by spawning success, may be influenced to a larger extent by mortality suffered by young fish as they pass seaward through regions of poor water quality (Summers and Rose, 1987).

LEGISLATIVE AND REGULATORY HISTORY

Following generally accepted engineering practices a century ago, a sewage collection system was constructed in Washington, DC, in 1870, with wastewater collected and discharged into the Potomac River without treatment. By 1913, USPHS surveys documented severely polluted conditions resulting from the discharge of raw sewage. Following the recommendations of city officials in 1920 and a study conducted in the early 1930s, the Blue Plains facility began operation in 1938 as a primary plant to serve 650,000 people. An unforeseen population influx related to World War II quickly exceeded the capacity of the new treatment plant. In response to the continuing degradation of water quality in the Potomac, the 1956 Federal Water Pollution Control Act, and subsequent amendments, served as the mechanism for establishing cooperative federal, state, and local remedial action plans for wastewater treatment. For more than two decades, federal, state, and local officials have cooperated in developing regional water quality management plans and implementing recommended effluent limits.

TABLE 8-2 Effluent Flow in 2000 from Major Tidal Potomac River Municipal Wastewater Treatment Plants

Plant	Effluent Flow (mgd)
Alexandria	36.8
Arlington	27.5
Blue Plains	317.5
Dale City Section 1	2.5
Dale City Section 4	2.4
Lower Potomac	42.8
Mattawoman	7.1
H. L. Mooney	9.6
Piscataway	21.1
Total	467.3

Source: Howard, 2001.

IMPACTS OF WASTEWATER TREATMENT

Pollutant Loading and Water Quality Trends

In the Washington, DC, region, 13 wastewater treatment plants currently discharge effluent into the Potomac estuary. As of 2000, nine major plants (Table 8-2) served about 4 million people and discharged a total of about 467 mgd (Howard, 2001). The 318-mgd discharge of the Blue Plains plant, which serves the population of Washington, DC, accounts for about two-thirds of the total effluent discharge to the Potomac estuary during low-flow conditions (Figure 8-6).

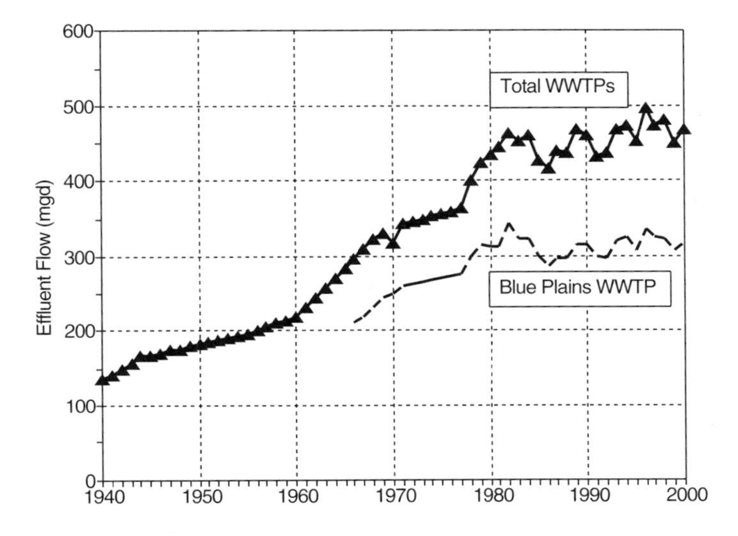

Figure 8-6 Long-term trends in effluent flow rate of municipal wastewater facilities. *Sources:* MWCOG, 1989; Howard, 2001; Jaworski, 2001.

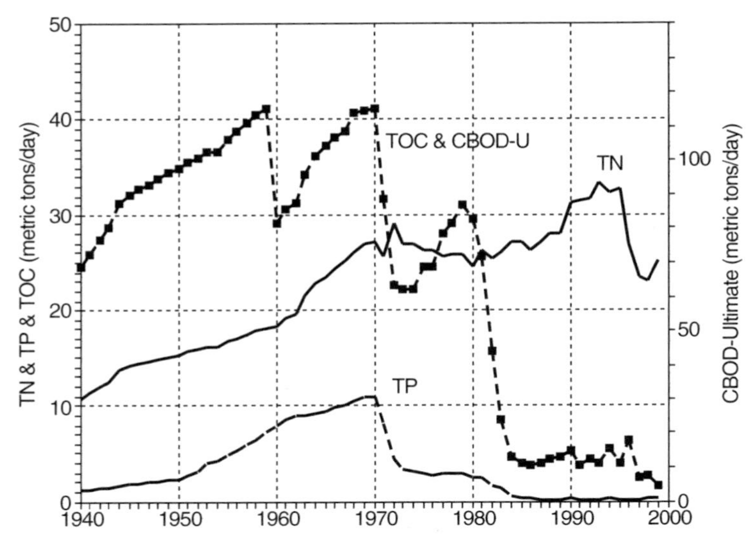

Figure 8-7 Long-term trends of TOC, total nitrogen, and total phosphorus effluent loads from municipal wastewater. *Sources:* Jaworski, 2001; Nemura, 1992; Howard, 2001.

Over the past 60 years, trends in TOC and nutrient loading have reflected a growing population and increasing levels of wastewater treatment. The reduction in BOD_5 loading resulted from the implementation of secondary treatment at Blue Plains in 1959 and at the other facilities from 1960 to 1980. The dramatic drop in phosphorus loading by 1986 resulted from phosphorus controls implemented, beginning in the early 1970s, at all the major wastewater treatment facilities to minimize eutrophication in the Potomac estuary. Nitrogen loading, however, has increased with population in the absence of controls on effluent nitrogen levels (Figure 8-7).

DO, influenced by temperature, wastewater loading, and freshwater flow, is characterized by seasonal and spatial variations, with minimum levels observed during the high-temperature, low-flow conditions of summer. Historical DO data are available to characterize long-term changes in the spatial distribution of oxygen during the summer (July–September) over a distance of approximately 55 miles from Chain Bridge to Mathias Point. These historical data sets clearly show the significant problem with oxygen depletion during the 1960s (1960–1964, 1966) with recorded concentrations of less than 2 mg/L downstream of the Blue Plains wastewater treatment plant [located at about River Mile (RM) 106] (Figure 8-8).

Mean summer (June–September) oxygen records obtained from stations directly influenced by the wastewater discharge from Blue Plains near the Woodrow Wilson Bridge (RM 95) clearly show long-term trends in the ecological condition of the estuary from 1940 to 2000 (Figure 8-9). The decline from 1945 to 1960 reflects substantial increases in population and related raw and primary effluent loading from the Washington, DC, region. Low oxygen levels recorded in the mid-1960s (1 to 2 mg/L)

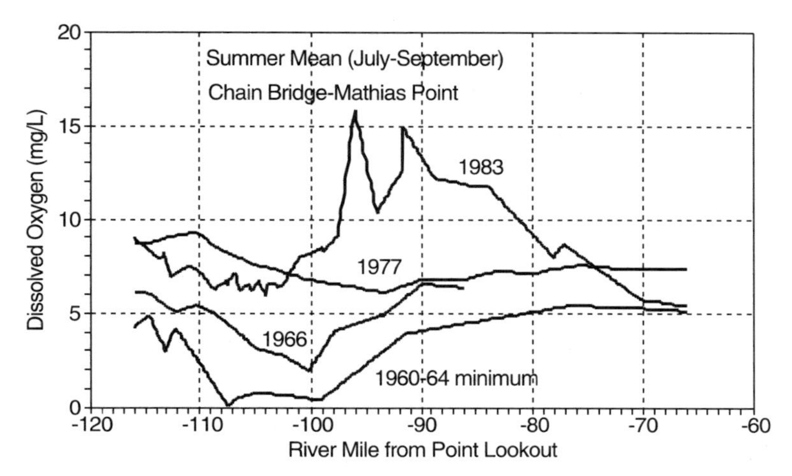

Figure 8-8 Long-term trends in summer DO levels at Chain Bridge–Mathias Point. *Sources:* Davis, 1968; Thomann and Fitzpatrick, 1982; Fitzpatrick and DiToro, 1991.

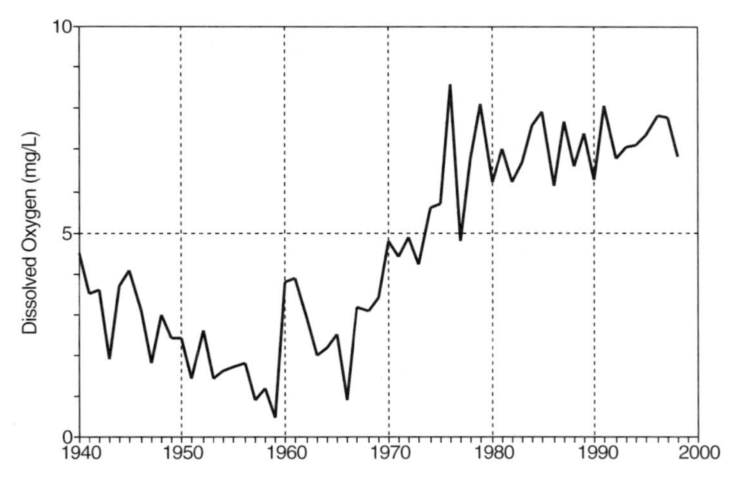

Figure 8-9 Long-term trends in summer DO levels on the Potomac River near the Wilson Bridge (River Mile 95). (Data for 1940–1986 from MWCOG averaged from June–September, data for 1987–1995 from Jaworski (2001) averaged from July–September.) *Sources:* MWCOG, 1989; Jaworski, 2001.

reflect the reduction of freshwater available for dilution because of drought conditions (see Figure 8-4) rather than any increase of pollutant load. The water quality standard for DO of 5 mg/L was typically violated during the 1960s and 1970s. Compliance was attained for the summer average condition only after all the regional wastewater treatment plants achieved secondary treatment by 1980; minimum summer levels, however, continued to periodically be less than 5 mg/L (MWCOG, 1989).

Evaluation of Water Quality Benefits Following Treatment Plant Upgrades

From a policy and planning perspective, the central question related to the effectiveness of the secondary treatment requirement of the 1972 CWA is simply: Would water quality standards for dissolved oxygen be attained if primary treatment levels were considered acceptable?

In addition to the qualitative assessment of historical data, water quality models can provide a quantitative approach to evaluate improvements in water quality achieved as result of upgrades in wastewater treatment. The Potomac Eutrophication Model (PEM), developed by Thomann and Fitzpatrick (1982), was enhanced by Fitzpatrick and Di Toro (1991) to represent two functional groups of algae. Fitzpatrick and Di Toro's (1991) version of PEM has been used in this study to demonstrate the water quality benefits attained by the technology- and water-quality–based regulations of the 1972 CWA for municipal wastewater facilities.

The Potomac Eutrophication Model (PEM) was calibrated using observed data sets collected from 1983 through 1985. In the summer of 1983, an anomalous bloom of the blue-green alga *Microcystis aeruginosa* formed a dense, brilliant green scum-like mat on the surface that extended over a distance of about 20 miles in the central estuary and embayments. Peak chlorophyll levels in the main river were ~300 μg/L, and dense concentrations as high as ~800 μg/L were recorded in the embayments (Thomann and Mueller, 1987). During the peak of the bloom in September 1983, dissolved oxygen levels computed with PEM were in reasonable agreement with the observed monthly mean data (Figure 8-10); concentrations of ~6 mg/L were observed and simulated in the vicinity of the Blue Plains wastewater treatment plant (RM 105) (Fitzpatrick and DiToro, 1991). The rapid increase from ~6 mg/L to observed (~10 to 16 mg/L) and computed (~12 mg/L) levels of dissolved oxygen ~5 to 10 miles downstream from Blue Plains is caused by high rates of phytoplankton primary productivity associated with peak algal biomass levels of ~150 to 250 μg/L (as chlorophyll *a*) in the vicinity of the Woodrow Wilson Bridge (RM 95). Further downstream in the transition zone of the Potomac estuary, observed and computed dissolved oxygen levels decline to ~5 to 7 mg/L in the vicinity of Indian Head (RM 85) to Maryland Point (RM 65) as a result of the attenuation of the *Microcystis* bloom by nitrogen limitation and the effects of salinity toxicity (Fitzpatrick and DiToro, 1991). As documented by Fitzpatrick and DiToro (1991), oxygen, algal biomass, nutrients, BOD_5, inorganic carbon, pH, and salinity simulated with the Potomac Eutrophication Model is considered well calibrated to the observed data for the summer periods during September 1983 (Figure 8-11), September 1984, and September 1985.

Using data to describe effluent flow, pollutant loading, and hydrologic conditions

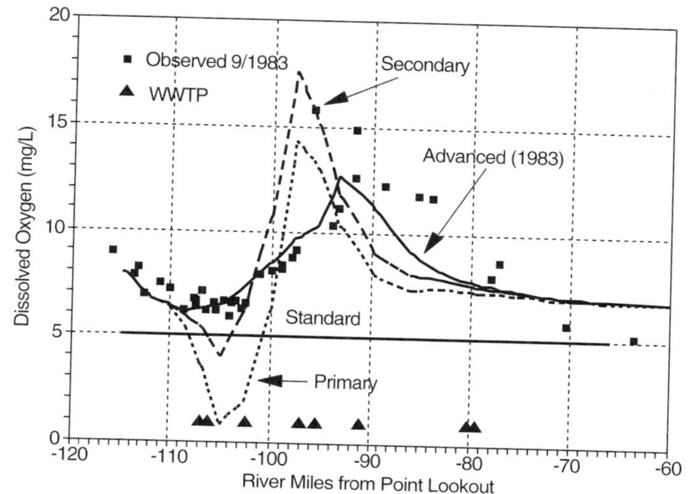

Figure 8-10 Potomac Eutrophication Model comparison of simulated DO for primary, secondary, and greater than secondary effluent loading scenarios. Observed data are for September 1983 calibration. *Sources:* Fitzpatrick, 1991; Fitzpatrick and DiToro, 1991.

during the lower-than-average flow conditions of September 1983, the calibrated model was used to evaluate the water quality impact of two regulatory control scenarios based on an assumption of (1) primary treatment and (2) secondary treatment compared to the existing (*ca.* 1983) effluent loading for all the municipal facilities in the Washington, DC, region (Fitzpatrick, 1991). Water quality conditions for these scenarios were simulated using freshwater flow data for 1983, a year characterized by summer flow (2,333 cfs) that was about 53 percent of the long-term summer mean flow of the Potomac River (see Figure 8-4). Other than the effluent characteristics, the ratio of ultimate-to-5-day BOD and the oxidation rate for CBOD (K_d) were the only parameters changed in the simulations to reflect differences in the proportion of refractory and labile organic carbon for the different levels of wastewater treatment (Fitzpatrick, 1991; Lung, 1998; Thomann and Mueller, 1987).

For the primary simulation, a value of K_d = 0.21/day, obtained from the original PEM (Thomann and Fitzpatrick, 1982), is typical of wastewater effluent characterized by the high CBOD concentrations typical of primary treatment conditions observed during the 1960s. For the secondary simulation, a value of K_d = 0.16/day, obtained from the original PEM (Thomann and Fitzpatrick, 1982), is typical of wastewater effluent characterized by the intermediate CBOD concentrations typical of secondary treatment conditions observed during the late 1970s. For the advanced (actual 1983) loading scenario, a value of K_d = 0.10/day, obtained from calibration of the updated PEM (Fitzpatrick and DiToro, 1991), is typical of wastewater effluent characterized by the low CBOD concentrations of advanced secondary and tertiary treatment conditions observed during the mid-1980s.

Under the primary effluent assumption, water quality is noticeably deteriorated in comparison to the 1983 calibration results (Figure 8-12). As a result of the effluent

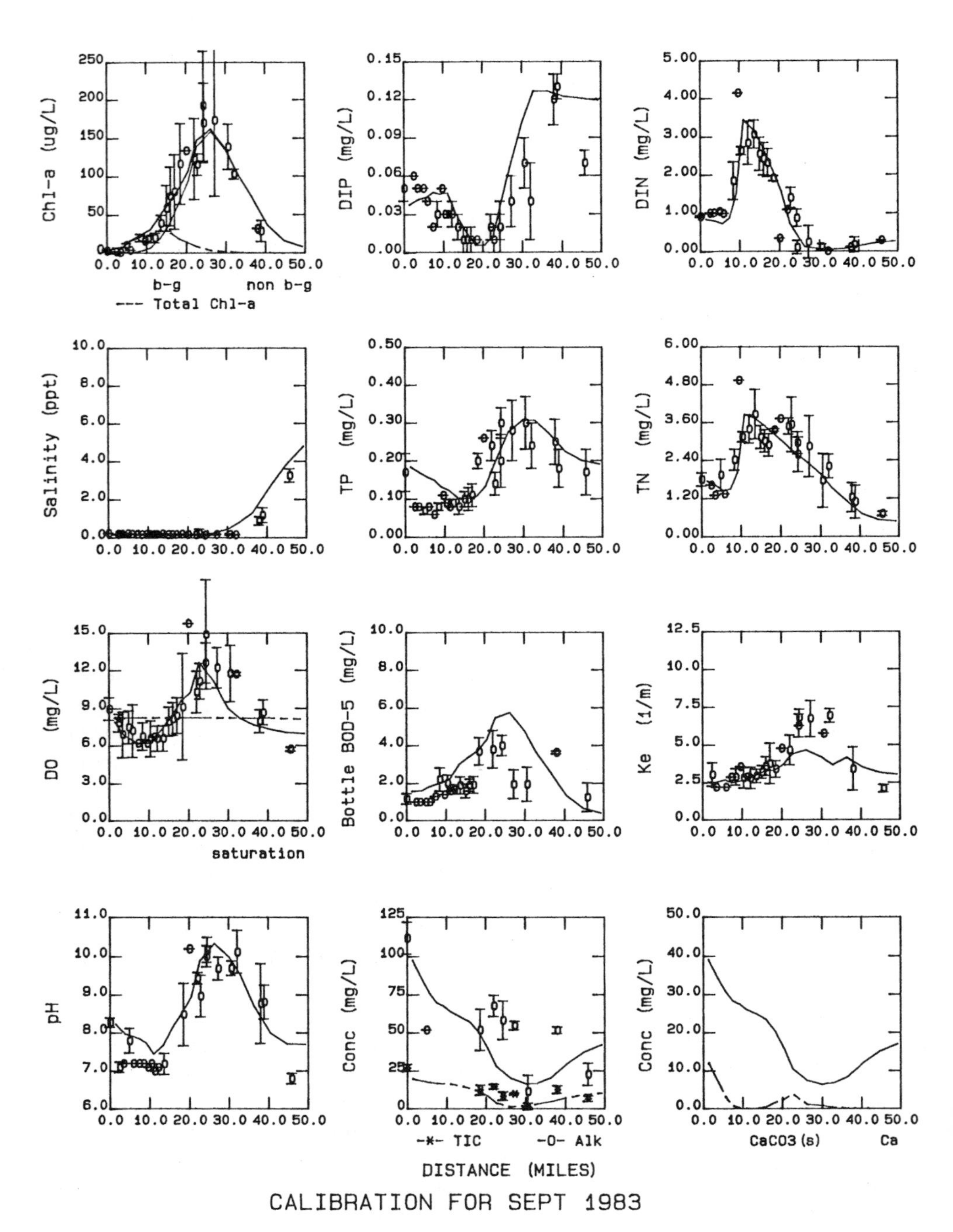

Figure 8-11 Potomac Eutrophication Model comparison of observed and simulated water quality data for actual greater than secondary effluent loading scenario. Observed data are for September 1983 calibration. *Sources:* Fitzpatrick, 1991; Fitzpatrick and DiToro, 1991.

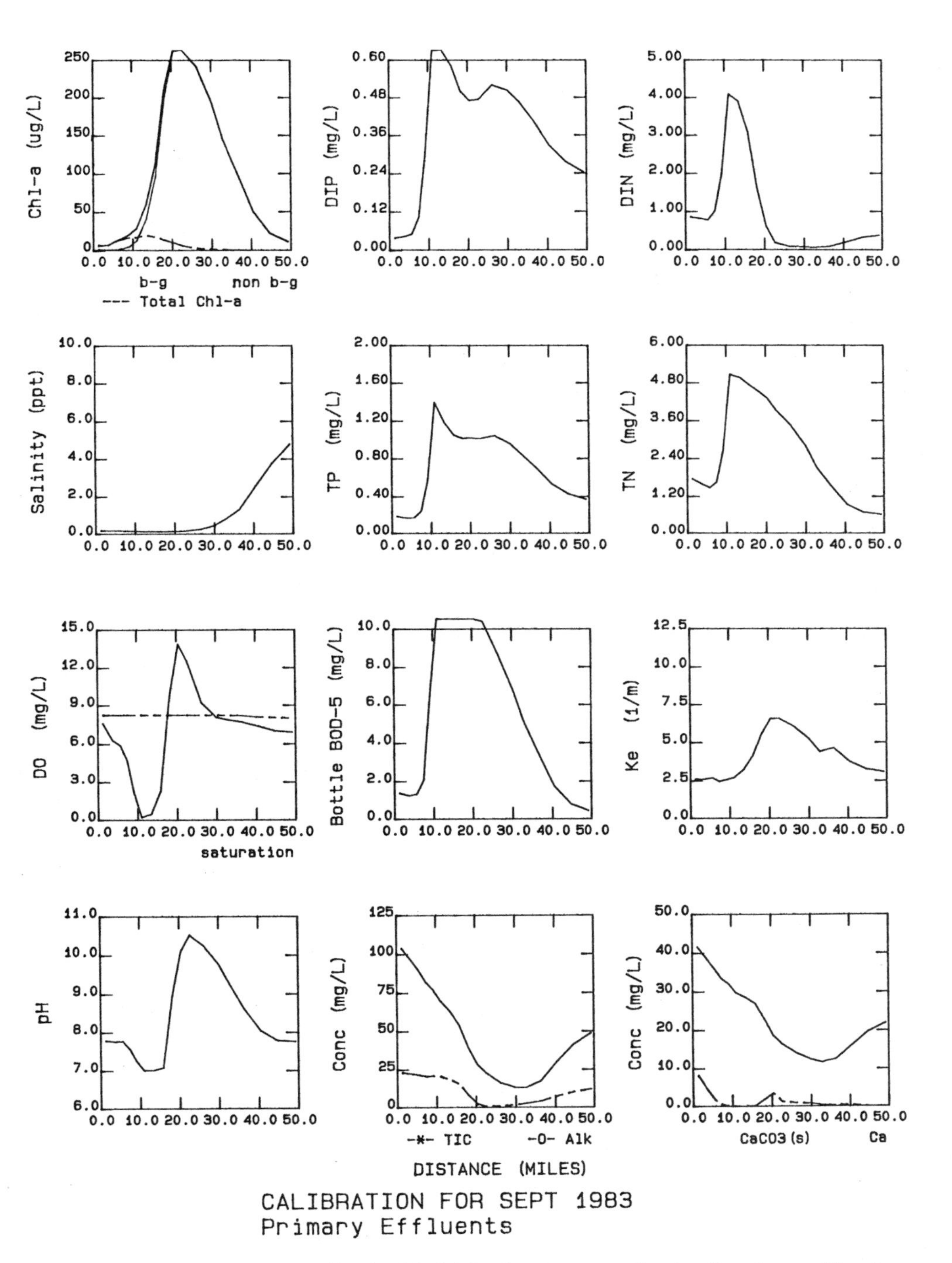

Figure 8-12 Potomac Eutrophication Model simulated water quality data for primary effluent loading scenario. Flow regime is based on September 1983 calibration. *Sources:* Fitzpatrick, 1991; Fitzpatrick and DiToro, 1991.

loading from the Blue Plains treatment plant, DO concentrations in the vicinity of Washington, DC are computed to be ~1 mg/L under the primary effluent scenario. With minimum levels less than 1 mg/L in the vicinity of the Blue Plains discharge (RM 106), the simulated results for the primary effluent scenario are remarkably similar to the historical data recorded for 1960–1964 and 1966 during the drought conditions of the 1960s (see Figure 8-8). The spatial extent of the depression for dissolved oxygen is relatively limited because of the oxygen production by photosynthesis of the algal bloom. Compared to current data and model results, the magnitude of the bloom would have intensified under conditions of primary effluent loads. Consistent with the known occurrence of dense algal blooms during the 1960s (Jaworski, 1990), peak chlorophyll-*a* concentrations greater than 250 μg/L were simulated under the primary effluent scenario in contrast to the present range of ~160 μg/L under the loading conditions of 1983. The increase in primary productivity and algal biomass resulted from the additional nutrient (TN, TP) loading delivered to the river.

Under the secondary effluent assumption (Figure 8-13), the reduction in CBOD loading significantly improved dissolved oxygen near Washington, DC. In comparison to the primary scenario, minimum monthly-averaged oxygen levels increased to almost 3.5 mg/L from approximately 0.2 mg/L under the secondary effluent scenario. When compared to the model results for the existing 1983 conditions, the results of the secondary effluent simulation show somewhat poorer water quality conditions for dissolved oxygen and algal biomass. The reason for the failure to achieve compliance with the 5 mg/L water quality standard for dissolved oxygen over only a few miles (RM 104–106) is that, under the existing loading scenario for 1983, the Blue Plains facility (the largest wastewater discharger to the Potomac River) has instituted advanced secondary treatment with greater removal of CBOD, ammonia, and phosphorus than is represented in the secondary effluent scenario (Fitzpatrick, 1991). The magnitude of the algal bloom has also been slightly attenuated as a result of reduced effluent nitrogen levels used to represent secondary effluent discharges to the Potomac estuary.

As shown with both observed data and model simulations, the implementation of secondary and better treatment has resulted in significant improvements in the DO status of the estuary. As demonstrated with the model (and actually attained), better-than-secondary treatment is required to achieve compliance with the water quality standard of 5 mg/L for DO at the critical location downstream of Blue Plains (see Figure 8-10). In contrast to the 1950s and 1960s, the occurrence of low oxygen conditions has been virtually eliminated in the Upper Potomac estuary (see Figure 8-8). Additional improvements in Potomac water quality, in terms of reduced algal biomass, increased water clarity, and still greater improvements in dissolved oxygen levels, have been achieved as a result of implementation of advanced secondary and tertiary levels of municipal wastewater treatment for the Upper Potomac estuary.

Recreational and Living Resources Trends

In addition to public water supply withdrawals (from the free-flowing river) and wastewater disposal from a number of municipalities, the uses of the Upper Potomac estuary include recreational and commercial fishing, boating and navigation, bird-watching, and secondary contact water-based recreation (e.g., wind-surfing). Although

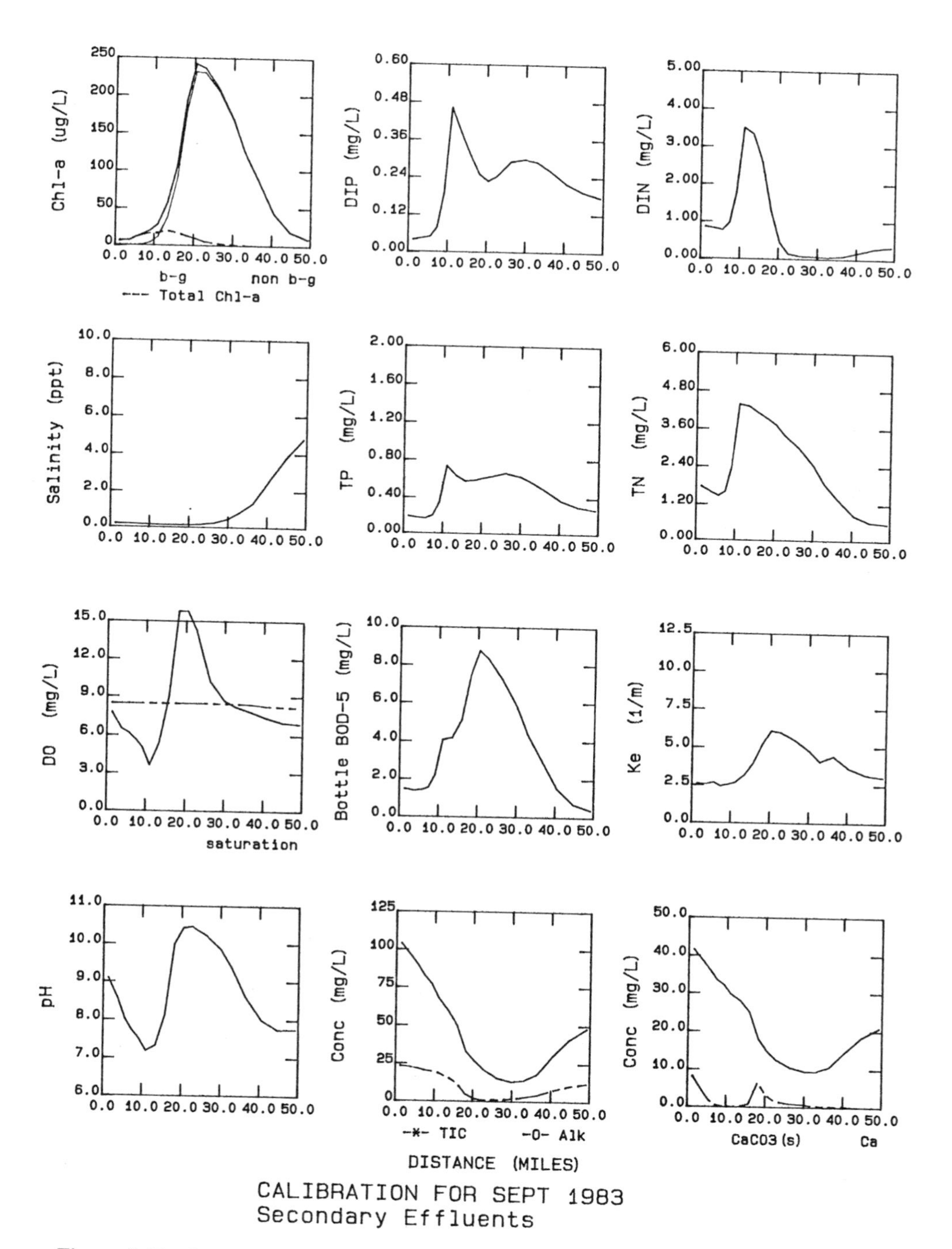

Figure 8-13 Potomac Eutrophication Model simulated water quality data for secondary effluent loading scenario. Flow regime is based on September 1983 calibration. *Sources:* Fitzpatrick, 1991; Fitzpatrick and DiToro, 1991.

recreational opportunities were severely limited during the 1940s, 1950s, and 1960s because of water pollution, the improvements in water quality during the 1980s have resulted in a significant increase in a variety of recreational uses of the river by the urban population of Washington, DC. Boating, canoeing, kayaking, windsurfing, walking, running, and bicycling on trails along the riverbanks, and recreational fishing are now extremely popular activities in the tidal river in the vicinity of Washington, DC.

Designated Uses and Bacterial Trends Unlike the uses of many other major urban waterways, swimming, because of limited access from the shoreline and a lack of public bathing beaches, is not considered a major use of the Upper Potomac estuary. Most of the Potomac River from the upper freshwater reaches near Point of Rocks, Maryland, to the estuarine waters near Point Lookout, Maryland, is designated for primary contact recreational uses (swimming). In the vicinity of Washington, DC, however, the waters of the tidal Potomac are designated for secondary contact recreational uses such as boating or windsurfing. The estuarine portions of the Potomac downstream of Smith Point, Maryland, have been designated for shellfish harvest and must comply with more stringent bacteria level standards than those set for primary or secondary contact recreational uses. To protect public health from risks resulting from direct contact with the waters of the Potomac or ingestion of shellfish from the estuary, water quality standards have been established by the state of Maryland, the District of Columbia, and the state of Virginia for the maximum log mean fecal coliform bacteria levels [as most probable number (MPN) per 100 mL] as follows:

- Primary contact $<$ 200 MPN/100 mL
- Secondary contact $<$ 1,000 MPN/100 mL
- Shellfish harvest $<$ 14 MPN/100 mL

Based on long-term historical water quality data from measurements taken downstream of the Blue Plains discharge, it is apparent that the introduction of effluent chlorination in 1968 resulted in dramatic improvements in bacterial contamination of the tidal Potomac (Figure 8-14). Prior to chlorination of wastewater effluent, summer coliform levels, typically on the order of 10^5 to 10^6 MPN/100 mL from 1940 to the mid-1960s, consistently were in violation of the secondary contact standard of 1,000 MPN/100 mL. Even with the dramatic reductions, summer bacteria levels still exceeded water quality standards during the 1970s. As bacteria loadings from the Washington area municipal wastewater plants continued to decrease during the 1980s, summer bacteria densities began to be in compliance with the water quality standard for both primary and secondary contact. Since the 1980s, periodic violations of bacteria level standards in the tidal Potomac have usually been related to storm event discharges from combined sewer overflows in the District of Columbia and Alexandria, Virginia (MWCOG, 1989).

Since the passage of the 1965 Water Quality Act, well-planned and coordinated water pollution control programs in the Washington metropolitan region have succeeded in achieving substantial reductions in pollutant discharges to the Potomac estuary. Despite the remarkable improvements in the bacteria levels of the tidal Po-

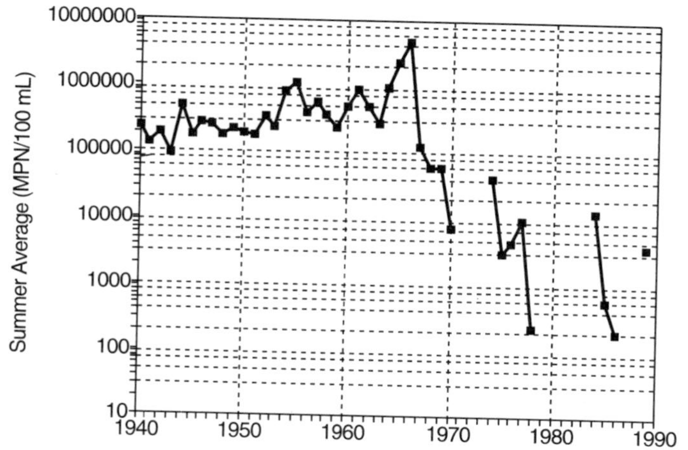

Figure 8-14 Long-term trends in total coliform densities in the Potomac River downstream of the Blue Plains POTW near Gunston Cove, Virginia. *Source:* USEPA STORET.

tomac, it is unlikely that President Johnson's 1965 pledge to "reopen the Potomac for swimming" will be fulfilled because of the lack of beaches along the shoreline and access for swimming.

Submersed Aquatic Vegetation, Fishery, and Waterfowl Resources In numerous accounts of the early colonists, the natural abundance of waterfowl and fishery resources of the Potomac basin was considered an important factor in attracting new colonists to the region. Like many freshwater and marine environments, the shallow littoral areas of the tidal Potomac River near Washington, DC, were characterized by extensive beds of a variety of species of aquatic macrophytes, or submersed aquatic vegetation (SAV), during the late 1800s and early 1900s (Carter et al., 1985). Detailed maps in 1904 and 1916, for example, showed extensive "grass" beds in Gunston Cove and shallow areas of the Maryland and Virginia sides of the river. In addition to the direct effect on the survival and condition of fish populations due to low DO concentrations caused by high organic loadings, fish populations are indirectly influenced by SAV abundance, necessary to provide nursery habitat for juvenile fish (Fewlass, 1991).

Increased municipal wastewater loading from the Washington area and the resulting poor water quality, were most likely responsible for the disappearance of SAV from the tidal Potomac River (Carter et al., 1985; Carter and Paschal, 1981), first noticed in 1939. During the 1940s and 1950s, widespread losses of SAV were common, not only in the Potomac, but also throughout the Chesapeake Bay basin (Carter et al., 1985). Although the SAV beds were severely diminished by the late 1930s, periodic nuisance "invasions" of submersed aquatic vegetation were recorded in the tidal Potomac during the 1930s (water chestnut) and from 1958 through 1965 (Eurasian watermilfoil) (Jaworski, 1990).

Trends in Suspended Solids Load and Water Clarity By the late 1970s, SAV in the Washington, DC, area had effectively disappeared from the tidal Potomac (Carter et al., 1985). The loss of SAV in the Potomac, and elsewhere in Chesapeake Bay region, has been attributed to the decreased availability of light in the littoral zone resulting from increased turbidity from the discharge of suspended solids and nutrients to the estuary (Carter et al., 1985). High levels of algae reduced light penetration and inhibited the growth of SAV. The natural abundance of fish and waterfowl of the Potomac, documented by the early colonists, was in fact directly related to the abundance and distribution of SAV in the shallow areas of the river. Redhead ducks, canvasbacks, and migrating widgeons and gadwalls feed on SAV, and other ducks such as mergansers feed on juvenile fish that depend on SAV for spawning and development (Forsell, 1992). The absence of SAV during the 1940s through 1970s resulted in a loss of habitat and food resources for fish and waterfowl dependent on the presence of the SAV beds.

The long-term ecological effects of the dramatic reductions of municipal wastewater loading of phosphorus (Figure 8-7) and suspended solids (Figure 8-15) to the estuary that began during the 1970s became apparent in the early 1980s with the surprising reappearance of SAV beds in the tidal Potomac (Carter and Rybicki, 1990). The return of the SAV beds was directly related to improvements in the clarity of the water (Figure 8-16), resulting from reductions in suspended solids and phosphorus loading from municipal wastewater discharges to the estuary and subsequent reductions in ambient phosphorus and algal biomass (Figure 8-17) (Carter and Rybicki, 1990, 1994; Jaworski, 1990). The presence of the SAV beds, in turn, has further en-

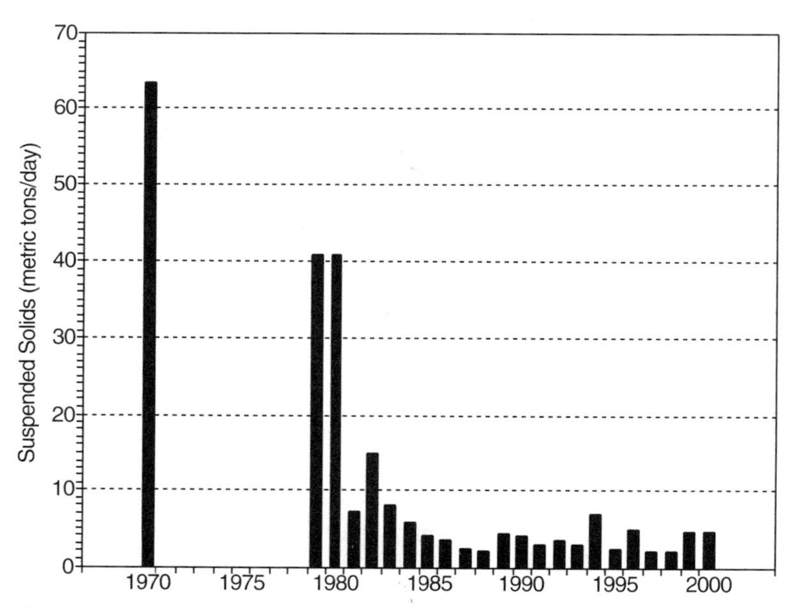

Figure 8-15 Long-term trends in municipal wastewater loading of suspended solids in the tidal Potomac River. *Sources:* Nemura, 1992; Howard, 2001.

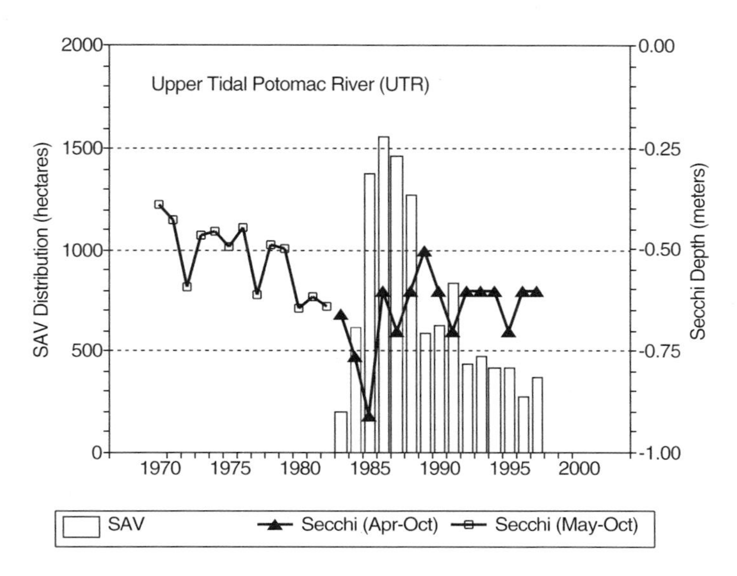

Figure 8-16 Long-term trends in SAV and water transparency in the Upper Tidal Potomac River. Secchi depth data from 1970–1982 (May–October) from USEPA STORET; SAV and secchi depth data from 1983–1997 (April–October) from Landwehr et al. (1999). *Sources:* Landwehr et al., 1999, USEPA STORET.

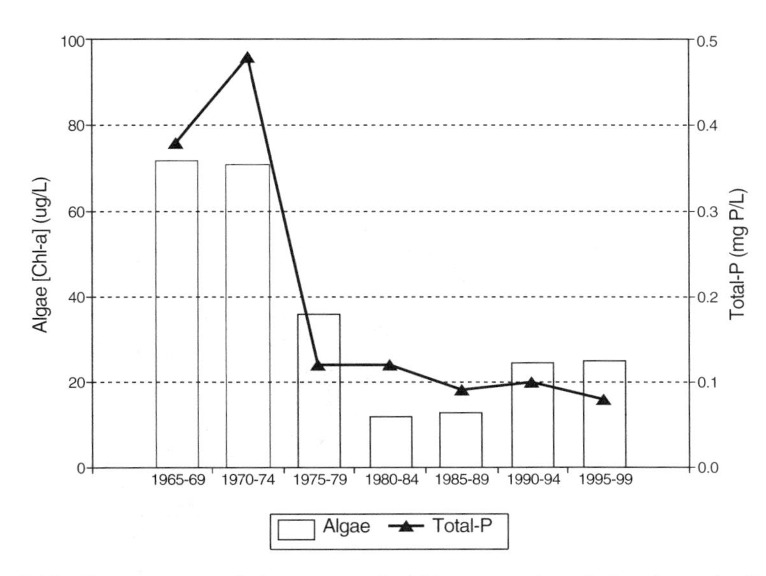

Figure 8-17 Long-term trends in summer algal biomass and total phosphorus in the Upper Potomac estuary. Data from Jaworski (2001) for Piscataway station. *Source:* Jaworski, 2001.

hanced water quality by physical settling of particulate solids, filtering of nutrients by plant uptake, and reduction of algal production in the water column (Figure 8-18). The reemergence of SAV beds in the tidal Potomac has resulted in dramatic increases in the diversity, abundance, and distribution of waterfowl (Figures 8-19 and 8-20).

Fish surveys documented significant increases in species diversity and abundance from 1984 to 1986 (MWCOG, 1989) that are consistent with the SAV and fisheries abundance data reported for the Choptank River on the eastern shore of Maryland (Kemp et al., 1984) (Figure 8-21).

Before the disappearance of the SAV beds, waterfowl populations (*ca.* 1929–1930) were about an order of magnitude greater than after the disappearance of SAV during the 1950s, when the annual average waterfowl census was 6,547 birds. Historical data from the Upper Chesapeake Bay (1958–1975) are useful to illustrate the relationship between the availability of SAV, fisheries, and waterfowl populations (Figures 8-20 and 8-21). The importance of SAV in the overall biological health of the tidal Potomac is clearly demonstrated with recent observations of a doubling of waterfowl abundance and an increase in the diversity of species (MWCOG, 1989). In 1972, only nine species of ducks wintered in the Potomac tidal river and transition zone (represented by more than one individual observed in winter transect counts); by 1992 the number of species had increased to 17 (Forsell, 1992). Fall-migrating, SAV-eating widgeons and gadwalls, absent from the estuary in winter for 15 years, have lengthened their stay in the Potomac, possibly encouraged by recent warmer winters and more plentiful food supplies. Populations of fish-eating mergansers, increasing since the 1970s, may be responding to increasing fish habitat available since the reemergence of SAV beds. Populations of Canada geese, tundra swans, and mallards, although not directly linked to SAV, are also increasing in the tidal Potomac River; this trend has also been observed in other areas of the northeast.

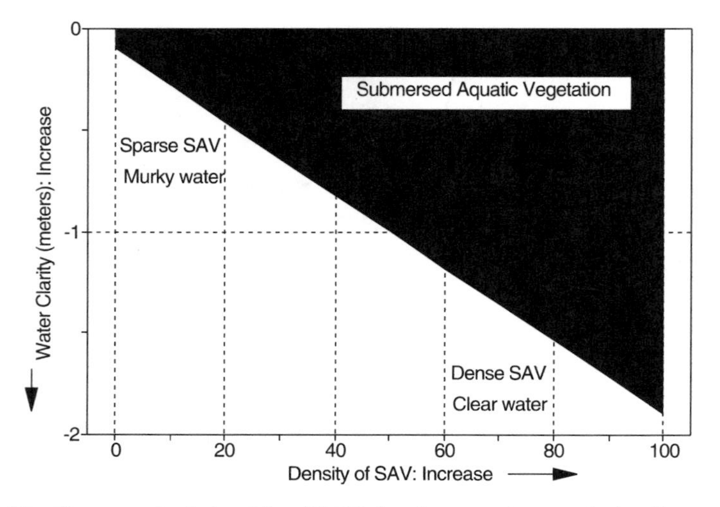

Figure 8-18 Conceptual relationship of SAV abundance and water clarity. *Sources:* Carter and Rybicki, 1990, 1994; Kemp et al., 1984.

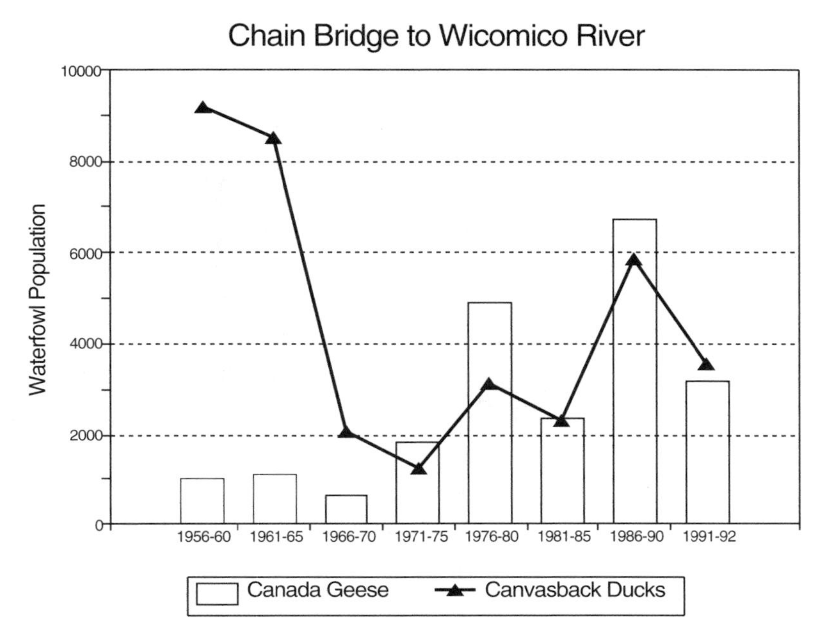

Figure 8-19 Long-term trends of waterfowl in the Upper Potomac River. Observations from Washington, DC, to Route 301 bridge. *Source:* Forsell, 1992.

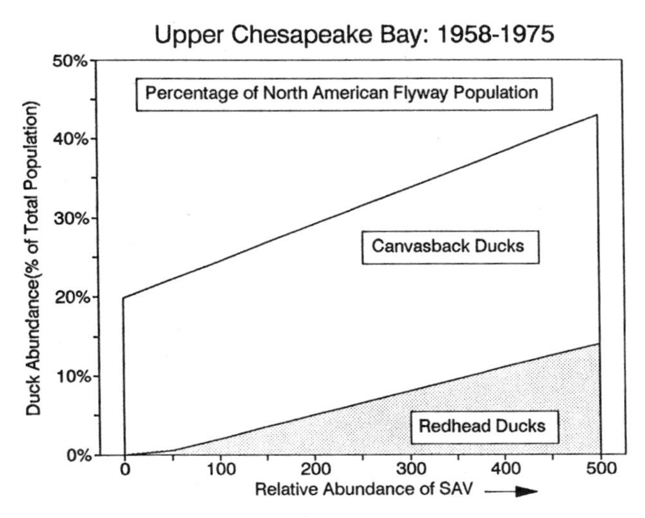

Figure 8-20 SAV and waterfowl abundance in Upper Chesapeake Bay (1958–1975). *Source:* W. M. Kemp, W. Boynton, and R. Twilley, The estuary as filter, ed. V. S. Kennedy, © 1984, Reprinted by permission of Harcourt/Academic Press, Inc.

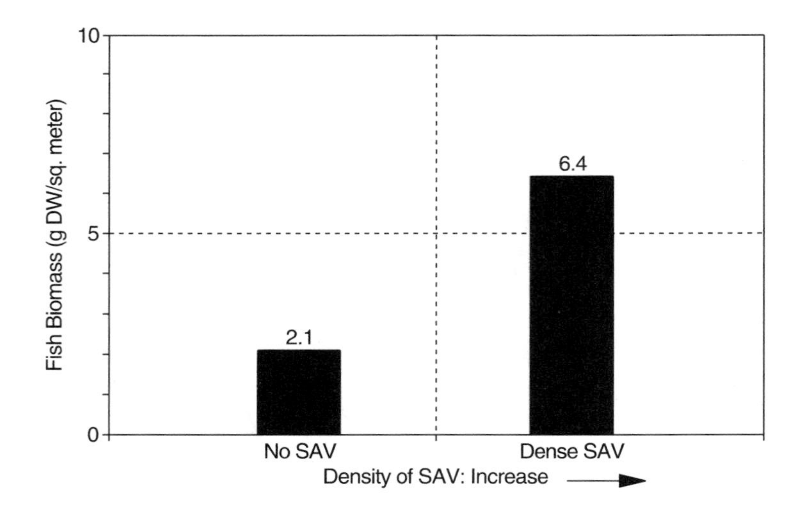

Figure 8-21 SAV abundance and fishery resources in the Choptank River. *Source:* W. M. Kemp, W. Boynton, and R. Twilley, The estuary as filter, ed. V. S. Kennedy, © 1984, Reprinted by permission of Harcourt/Academic Press, Inc.

Fishery surveys in the tidal Potomac, and elsewhere in the Chesapeake Bay, clearly document an increase in abundance and diversity of fish species. Juvenile fish survey data, collected between 1965 and 1987 at Indianhead and Fenwick in the tidal river, were analyzed using the Index of Biotic Integrity (IBI) (Figure 8-22). The IBI, developed by Karr (1981) for use in midwestern streams, has been adapted for use in other areas. This index is a composite of 12 ecological attributes of fish communities, including species richness, indicator taxa (both intolerant and tolerant), trophic guilds, fish abundance, and incidence of hybridization, disease, and abnormalities (Karr et al., 1986). IBI scores range from a low of 12 to a high of 60. A score of 12 is assigned to conditions where no fish are present even after repeated sampling; a score of 60 is assigned to conditions comparable to the best habitats without human disturbance (Karr et al., 1986). The trend in the IBI at Indianhead (Jordan, 1992) shows that the river quality for fish increased from poor, indicating an impaired or restricted habitat (IBI scores in the 20 to 30 range), to fair, indicating slightly impaired habitat (IBI scores in the 40 to 50 range).

These data indicate that, in the late 1960s and early 1970s, the fish community at Indianhead was dominated by a few tolerant species, with few fish present at all in some years. In the last 20 years, a general upward trend in river quality for fish has been observed, evidenced by increasing numbers of pollution-intolerant species and a species mix suitable to provide for a reasonably balanced trophic structure. Indicator variables currently measured at Indianhead are at about two-thirds of their expected level in undisturbed habitats. The rise in the IBI at Indianhead, where a wastewater treatment plant discharge is located, is in contrast to stable or declining trends observed at other locations that lack wastewater treatment plant outfalls (Jordan, 1992).

In addition to the direct effect on the survival and condition of fish populations due to low DO concentrations due to high organic loadings (Tsai, 1991), fish populations

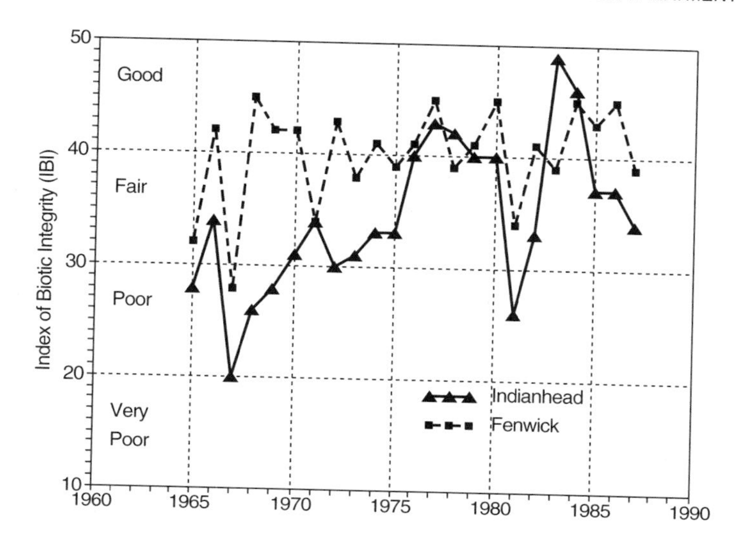

Figure 8-22 Long-term trends in fishery resources of the Potomac estuary based on Index of Biotic Integrity (IBI). *Source:* Jordan, 1992.

are indirectly influenced by SAV abundance, necessary to provide nursery habitat for juvenile fish (Fewlass, 1991). The quality of river habitat for fish has increased with the resurgence of SAV habitat in comparison to areas characterized by the absence of SAV beds. During 1984–1986, years characterized by a rapid increase in the distribution of SAV beds (primarily *Hydrilla*) (see Figure 8-16), fishery surveys near Washington, DC, clearly showed an increase in species diversity; abundance increased from 79 to 196 fish per net haul over the same 2-year period (MWCOG, 1989). The relationship of SAV and fishery data from the tidal Potomac is consistent with data reported by Kemp et al. (1984) for the Choptank River on the eastern shore of Maryland (see Figure 8-21).

SAV and Ecological Resources The evidence is clear from observations in Chesapeake Bay and the tidal Potomac River that the presence of SAV beds is critical for a healthy and diverse aquatic ecosystem. The presence of SAV beds has the following positive ecological impacts:

- Increases habitat and food resource availability
- Increases species diversity and abundance
- Increases fishery resources
- Increases waterfowl populations
- Increases recreational opportunities (fishing, hunting, bird-watching)
- Enhances water quality
- Removes nutrients

- Allows particulate material to settle out
- Reoxygenates water column by photosynthesis

SUMMARY AND CONCLUSIONS

Water quality and biological resources data clearly illustrate the cause-effect relationship of reductions in wastewater loading of BOD_5, nutrients, and suspended solids and improvements in water quality and the ecological resources of the tidal Potomac. As a result of the improvements in water quality and SAV habitat, the Upper Potomac estuary emerged during the 1990s as one of the best largemouth bass fisheries in the nation. The thriving fishery resources now support popular recreational fishing activities, including professional fishing guide services and annual Bassmasters fishing tournaments. In answering a question about why people don't consider fishing in the Potomac in the Washington, DC area, Bill Kramer, owner of Potomac Guide Service, explained that "...in the 1950s and 1960s the Potomac was a flowing cesspool... It was a disgrace, it was so polluted. If you fell in the river it was recommended you go to the hospital for examination. Now the river's much better. Pollution controls are higher and the fish population's are mostly solid. But the people still think of it as the old river. So people don't understand how good the fishing is here." (Tidwell, 1998). One of the earliest guides, Ken Penrod of Life Outdoors Unlimited, now has one of the largest freshwater fishing guide services in the nation. Guides like Ken Penrod have reported that since 1982 every year has been better than the previous year in terms of the quantity and quality of fish that have returned to the waters of the tidal Potomac (Soltis, 1992). In 1999, for example, some 500 people took part in the 1999 Bassmasters Fishing Tournament on the Potomac (Shields, 1999). In recognition of the ecological improvements of the river, the Potomac River was selected as an American Heritage River in 1998.

A River Reborn

"Time has been good to the Potomac River— at least the last 25 years have been . . . A generation ago (1960s and 1970s), when we had gone to the Potomac for thrills, the river was ugly and almost frightening in its decay. The water was an opaque red-brown sludge, smelly and foaming with unknown chemical pollutants. The shore was littered with the rotting carcasses of carp and with slime-covered tires, cans, glass and other filth. But as we sat talking this time, the water was clear—really clear—as though we were in the countryside far from a big city. We could see right to the sandy bottom. The shoreline was free of debris, and the air smelled fresh.

Best of all were the birds. Mallards swooped overhead, heading toward the water. Despite the season, some songbirds still darted through the trees. And far out in the channel, close to Virginia, a huge flock of birds circled round and round a cluster of rocks. They seemed to be feeding out there (on what type of freshwater creature?), and the sun glistened off their wings Who are the stewards of the Potomac? Who is responsible for the amazing rebirth of a beautiful river? To all, I extend a hearty thanks . . ."

—Chase, 1995

REFERENCES

Carter, V. and J. E. Paschal. 1981. Biological factors affecting the distribution and abundance of submerged aquatic vegetation in the tidal Potomac River, Maryland and Virginia. Estuaries 4(3): 300.

Carter, V., J. E. Paschal, and N. Bartow. 1985. Distribution and abundance of submersed aquatic vegetation in the tidal Potomac River and estuary, Maryland and Virginia, May 1978 to November 1981. Water-Supply Paper 2234-A. U.S. Geological Survey, Reston, VA.

Carter, V. and N. B. Rybicki. 1990. Light attenuation and submersed macrophyte distribution in the tidal Potomac River and estuary. Estuaries 13(4): 441–452.

Carter, V. and N. B. Rybicki. 1994. Role of weather and water quality in population dynamics of submersed macrophytes in the tidal Potomac River. Estuaries 17(2): 417–426.

Chase, N. 1995. A river reborn. The Washington Post, Letter to the editor, page C8, Washington, DC. January 15.

Davis, R. K. 1968. The range of choice in water management: A study of dissolved oxygen in the Potomac estuary. Published for Resources for the Future, Inc., Washington, DC, by The Johns Hopkins Press, Baltimore, MD.

Fewlass, L. 1991. Statewide fisheries survey and management study V: Investigations of largemouth bass populations inhabiting Maryland's tidal waters. Maryland Department of Natural Resources, Tidewater Administration, Annapolis, MD.

Fitzpatrick, J. J. 1991. Trends in BOD/DO modeling for wasteload allocations of the Potomac estuary. Technical memorandum prepared under sub-contract for Tetra Tech, Inc., Fairfax, VA, by HydroQual, Inc., Mahwah, NJ, September 18.

Fitzpatrick, J. J. and D. M. DiToro. 1991. Development and calibration of a two functional algal group model of the Potomac Estuary. Final Report to the Metropolitan Washington Council of Governments. Washington, DC. Prepared by HydroQual, Inc., Mahwah, NJ.

Forsell, D. 1992. U.S. Fish and Wildlife Service, Chesapeake Bay Estuary Program, Annapolis, MD. Personal communication, October 22.

Forstall, R. L. 1995. Population by counties by decennial census: 1900 to 1990. U.S. Bureau of the Census, Population Division, Washington, DC. http://www.census.gov/population/www/censusdata/cencounts.html.

Haramis, G. M. and V. Carter. 1983. Distribution of submersed aquatic macrophytes in the tidal Potomac River. Aquat. Bot. 15(1983): 65–79.

Howard, C. 2001. Metropolitan Washington Council of Governments, Department of Environmental Programs, Washington, DC. Personal communication, September 25.

ICPRB. 1990. The healing of a river, The Potomac: 1940–1990. Interstate Commission on the Potomac River Basin, Rockville, MD.

Iseri, K. T. and W. B. Langbein. 1974. Large rivers of the United States. Circular No. 686, U.S. Department of the Interior, U.S. Geological Survey, Reston, VA.

Jaworski, N. A. 1990. Retrospective study of the water quality issues of the Upper Potomac estuary. Rev. Aquat. Sci. 3(2): 11–40, CRC Press, Boca Raton, FL.

Jaworski, N. 2001. Wakefield, RI. Personal communication, September 5.

Jordan, S. 1992. Unpublished data. Potomac River Fisheries Commission, Colonial Beach, VA.

Karr, J. R. 1981. Assessment of biotic integrity using fish communities. Fisheries 6(6): 21–27.

Karr, J. R., K. D. Fausch, P. L. Angermeier, P. R. Yant, and I. J. Schlosser. 1986. Assessing biological integrity in running waters: A method and its rationale. Special Publication 5. Illinois Natural History Survey, Champaign, IL.

Kemp, W. M., W. R. Boynton, R. R. Twilley, J. C. Stevenson, and L. G. Ward. 1984. Influences of submersed vascular plants on ecological processes in Upper Chesapeake Bay. In: The estuary as a filter, ed. V. S. Kennedy, pp. 367–394. Academic Press, New York.

Landwehr, J. M., J. T. Reel, N. B. Rybicki, H. A. Ruhl, and V. Carter. 1999. Chesapeake Bay Habitat Criteria Scores and the Distribution of Sumbersed Aquatic Vegetation in the Tidal Potomac River and Potomac Estuary, 1983–1997. U.S. Geological Survey Open File Report 99–219, Reston, VA.

Lung, W. 1998. Trends in BOD/DO modeling for waste load allocation. J. Environ. Eng., 124 (10):1004–1007. ASCE.

MWCOG. 1989. Potomac River water quality, 1982 to 1986: Trends and issues in the metropolitan Washington area. Metropolitan Washington Council of Governments, Department of Environmental Programs, Washington, DC.

Nemura, A. 1992. Metropolitan Washington Council of Governments, Department of Environmental Programs, Washington, DC. Personal communication, March 3.

Newell, F. H. 1897. Pollution of the Potomac River. National Geographic (December 1897): 346–351.

OMB. 1999. OMB Bulletin No. 99-04. Revised statistical definitions of Metropolitan Areas (MAs) and guidance on uses of MA definitions. U.S. Census Bureau, Office of Management and Budget, Washington, DC.
http://www.census.gov/population/www/estimates/metrodef.html

Shields, T. 1999. Fishing tourney a tourism catch for Charles County. Washington Post, Southern Maryland Extra Section, M1, Washington, DC. October 17.

Soltis, S. 1992. Rebirth of an American river. Washington Flyer Magazine (March/April): 22–25.

Summers, K. J. and K. A. Rose. 1987. The role of interactions among environmental conditions in controlling historical fisheries variability. Estuaries 10(3): 255–266.

Thomann, R. V. and J. J. Fitzpatrick. 1982. Calibration and verification of a mathematical model of the eutrophication of the Potomac estuary. Prepared for Department of Environmental Services, Government of the District of Columbia, Washington, DC, by HydroQual, Inc., Mahwah, NJ.

Thomann, R. V. and J. A. Mueller. 1987. Principles of surface water quality modeling and control. Harper & Row, New York.

Tidwell, M. 1998. Where is this man fishing? The Washington Post, Travel Section, pp. E11, Washington, DC. August 30.

Tsai, C.-F. 1991. Rise and fall of the Potomac River striped bass stock: A hypothesis of the role of sewage. Trans. Amer. Fish. Soc. 120(1): 1–22.

USDOC. 1998. Census of population and housing. U.S. Department of Commerce, Economics and Statistics Administration, Bureau of the Census—Population Division, Washington, DC.

USEPA. 1992. Water quality goals for living resources: Role in the Chesapeake Bay nutrient reevaluation, summary and preliminary results. USEPA Chesapeake Bay Program, Annapolis, MD.

USGS. 1999. Streamflow data downloaded from the U.S. Geological Survey's National Water Information System (NWIS)-W. Data retrieval for historical streamflow daily values.
http://waterdata.usgs.gov/nwis-w

CHAPTER 9

James River Estuary Case Study

Figure 9-1 highlights the location of the James estuary case study watershed (catalog units). The James River was identified as one of the urban-industrial waterways affected by severe water pollution problems during the 1950s and 1960s (see Table 4-2). The James River basin, at the southern boundary of the Mid-Atlantic Basin, is one of the most important water resources in the Commonwealth of Virginia (Figure 9-2).

As the largest river in the state, the James River extends more than 400 miles from its mouth at the Chesapeake Bay to its headwaters near the West Virginia state line. The river is a recognized asset to the surrounding residential and metropolitan areas, providing recreational opportunities such as boating and fishing.

The James River is known for its annual national Bassmasters fishing tournaments, and it has exceptional Class IV white water rapids in the drop between the riverine and estuarine portions of the river in Richmond, Virginia. The river is also an asset to commerce and industry, serving as an important water supply and, as such, a catalyst for economic growth.

PHYSICAL SETTING AND HYDROLOGY

The James River is a typical coastal plain estuary draining to the Chesapeake Bay. The variation of depth, cross-sectional area, and tidal velocity in the James River from Richmond to the Chesapeake Bay is significant. For example, the cross-sectional depths vary from about 10 feet in areas with shallow side embayments to 25 to 30 feet in the deepwater channel. The river generally widens in the downstream direction, although natural constrictions occur at several locations. Cross-sectional area varies markedly, from the deep, narrow channel in the upstream section to broad, shallower profiles downstream.

Upstream freshwater flow to the study area is monitored at the USGS gaging station near Richmond, Virginia, on the James River. The freshwater flow to the James River is contributed by runoff from 6,758 square miles of woodland and agricultural areas upstream of the city of Richmond. A relatively small additional flow enters the study area via the Kanawha Canal, bypassing the USGS gage near Richmond. The combined average annual flow in the river at the gage is 6,946 cfs (1937–1998). A

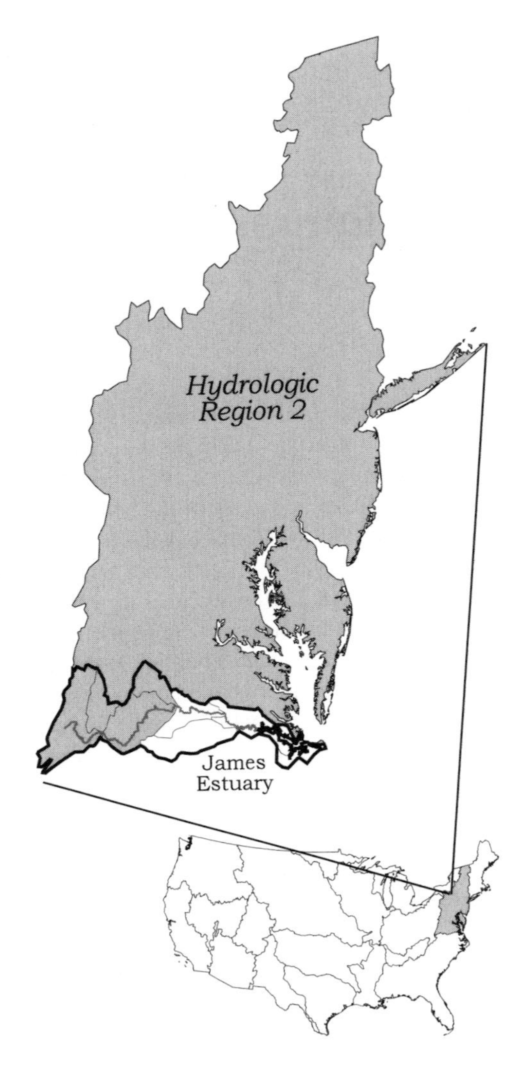

Figure 9-1 Hydrologic Region 2 and the James estuary watershed.

relatively small intervening drainage area provides a nominal increase in in-stream flow between Richmond and the confluence with the Appomattox River. Water is withdrawn from the James River for both municipal and industrial purposes and then returned to the river. Treatment is provided by all users except those who use the water solely for cooling purposes. Long-term interannual and mean monthly trends in streamflow for the James River near Richmond, Virginia, are shown in Figures 9-3 and 9-4.

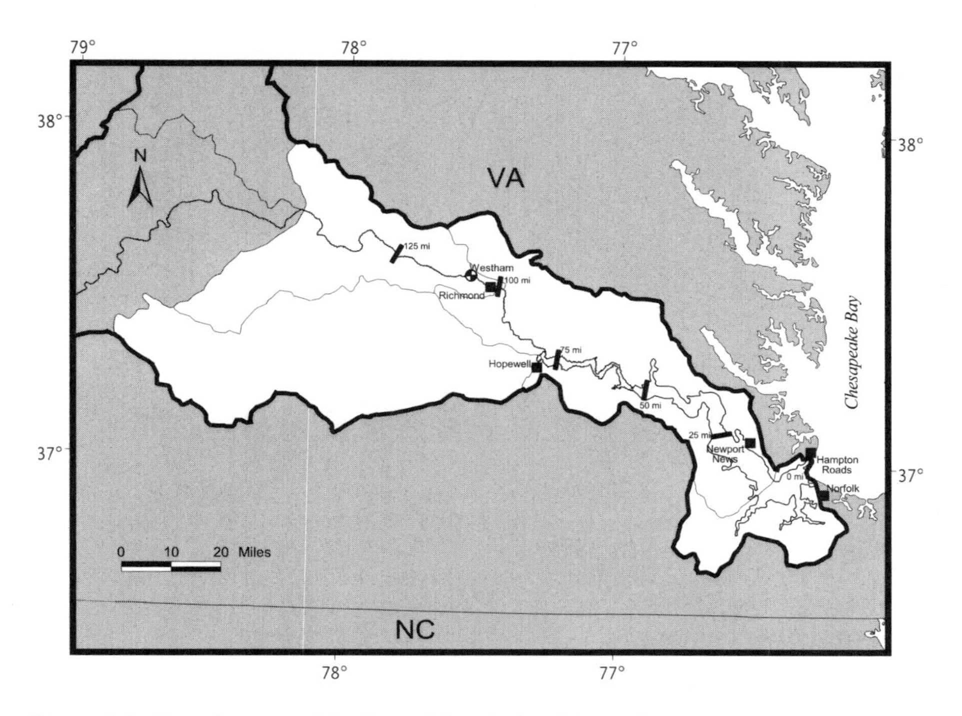

Figure 9-2 Location map of the James River basin. (River miles shown are distances from Chesapeake Bay at the mouth of the James River.)

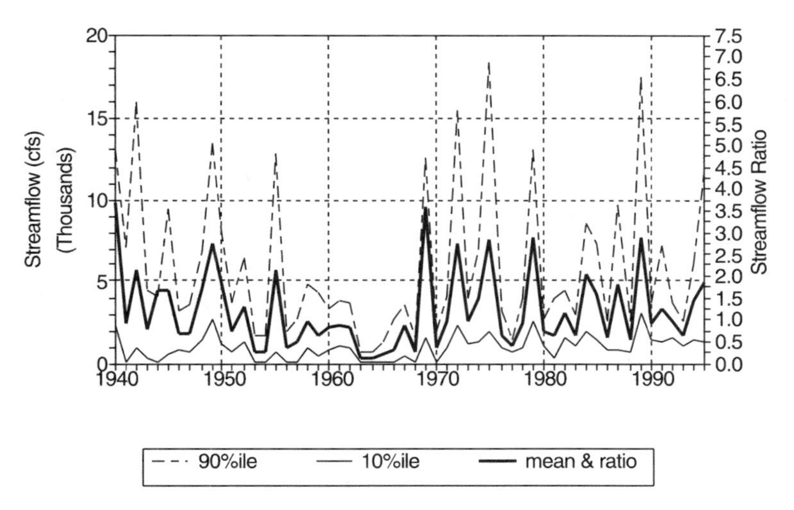

Figure 9-3 Trends of mean, tenth, and ninetieth percentile statistics computed for summer (July–September) streamflow for the James River (USGS Gage 02037500 near Richmond, Virginia). *Source:* USGS, 1999.

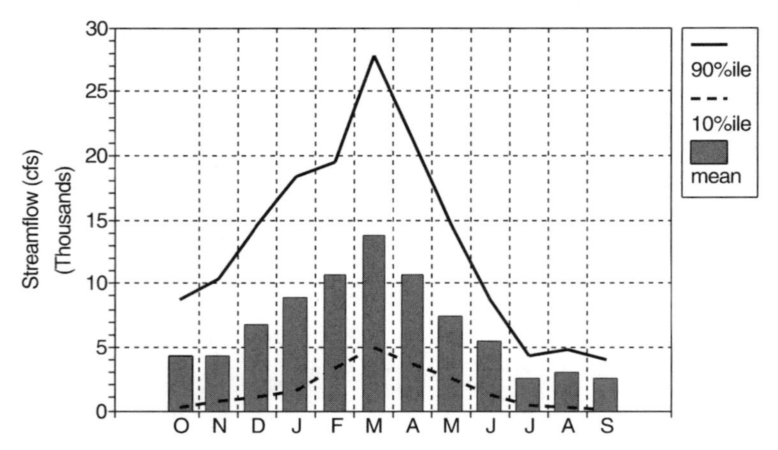

Figure 9-4 Monthly trends in streamflow for the James River. Monthly mean, tenth, and ninetieth percentile statistics computed for 1951–1980 (USGS Gage 02037500 near Richmond, Virginia). *Source:* USGS, 1999.

POPULATION TRENDS

The James estuary case study area includes a number of counties identified by the Office of Management and Budget as Metropolitan Statistical Areas (MSAs) or Primary Metropolitan Statistical Areas (PMSAs). Table 9-1 lists the MSAs and counties in-

TABLE 9-1 Metropolitan Statistical Areas (MSAs) and Counties and Incorporated Cities in the James Estuary Case Study: Norfolk–Virginia Beach–Newport News, VA–NC MSA

Currituck County, NC	*Richmond-Petersburg, VA MSA*
Gloucester County, VA	Charles City County, VA
Isle of Wight County, VA	Chesterfield County, VA
James City County, VA	Dinwiddie County, VA
Mathews County, VA	Goochland County, VA
York County, VA	Hanover County, VA
Chesapeake City, VA	New Kent County, VA
Hampton City, VA	Powhatan County, VA
Newport News City, VA	Prince George County, VA
Norfolk City, VA	Colonial Heights City, VA
Poquoson City, VA	Hopewell City, VA
Portsmouth City, VA	Petersburg City, VA
Suffolk City, VA	Richmond City, VA
Virginia Beach City, VA	
Williamsburg City, VA	

Source: OMB, 1999.

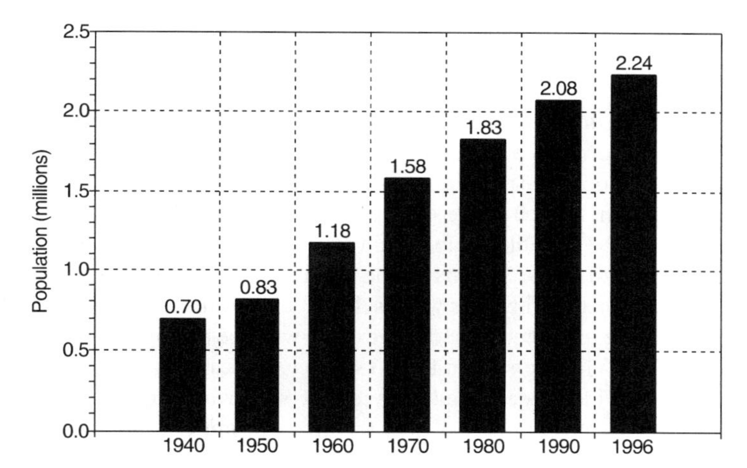

Figure 9-5 Long-term trends in population in the James estuary basin. *Sources:* Forstall, 1995; USDOC, 1998.

cluded in this case study. Figure 9-5 presents long-term population trends (1940–1996) for the counties listed in Table 9-1. From 1940 to 1996, the population in the James estuary case study area more than tripled (Forstall, 1995; USDOC, 1998).

HISTORICAL WATER QUALITY ISSUES

The estuarine system starts near Richmond, where the fall line is located, and extends approximately 100 miles from the mouth of the river. The historical water quality concerns in the estuarine system have been dissolved oxygen and increased nutrient loads. DO is affected by the carbon and nitrogen components of the wastewater effluents. It is also influenced indirectly by the phosphorus content of these sources insofar as the latter stimulates phytoplankton growth.

In 1947, the 14-mile stretch of the James River east of Richmond was described as "dead." In 1963, conditions had not improved despite growing public concern. The *Richmond News Leader* described the river as a sewer. After a powerboat tour of the river, the editor described the river as green with algae, septic, and laden with dead and dying fish. Even the hardy catfish, which normally tolerates severely polluted waters, was observed gasping for its last breath. The only birds in sight were circling turkey vultures, attracted by the floating offal. At that time, the sewage collection system for Richmond was only partially operational, and only 58 percent of the design flow of the city's sewage treatment plant was being used. Raw sewage was being discharged into the James through Gillies Creek, and it seemed doubtful that the river would ever meet the minimum standard of 4.0 mg/L of dissolved oxygen required to permit recreational river uses (Richmond News Leader, 1963).

LEGISLATIVE AND REGULATORY HISTORY

Concern over the severely degraded conditions in the James River prompted the General Assembly to establish the State Water Control Board (SWCB) in 1946. The Board used its authority to put pressure on the city of Richmond to expand its treatment facilities and on industries to cease their discharges into the river (Richmond News Leader, 1963). Although the city responded favorably, and hopes were raised that the river could be fishable again within 10 years, a brief inspection of the river in 1963 revealed that the expectations of the Game and Inland Fisheries Commission had been overoptimistic. The river was as dead as it had been in 1947.

The most significant impetus for change came with the passage of the federal Clean Water Act in 1972. This legislation forced states and localities to clean up municipal discharges and provided federal and state money with which to do it. Richmond upgraded its sewage treatment plant in 1974 to remove as much as 80 percent of the suspended solids (secondary treatment) (Epes, 1992). Later upgrades included a 500-million-gallon storm overflow basin in 1983, a $73 million filtering system in 1990, and an agreement in 1992 to spend $82 million for more improvements scheduled for completion in 1998 (Epes, 1992).

Water supply and wastewater treatment facilities have been developing at a rate commensurate with growth in the James River basin over the past few decades. As a result, the James River, including the Appomattox River, has received increased quantities of treated effluent from both municipal and industrial sources.

The Virginia SWCB realized the necessity of planning for waste treatment requirements many years ago. Between 1960 and 1962, several water quality studies were conducted to document the water quality conditions in the James River. These studies were among the earliest to quantitatively evaluate the natural assimilation capacity of the James River in the Hopewell and Richmond areas and to estimate the effect on stream quality of local industrial waste discharges.

Recognizing that proper planning must be implemented on a regional basis to protect the river system from impairment of its numerous desirable uses, SWCB entered into an agreement with the USEPA in 1971, under section 3(c) of the Federal Water Pollution Control Act of 1965, to study the James River. A principal outcome of this effort, completed in 1974, was the development of a James River ecosystem model by the Virginia Institute of Marine Science (VIMS). The SWCB used this model for wasteload allocations in the James River. Following the 3(c) study, the Richmond-Crater 208 study was funded, and a second detailed water quality management model, the James Estuary Model (JEM), was developed for the upper James River estuary. This model was found to be inconsistent with the VIMS model, and a review of both models was conducted by Hydroscience, Inc. The VIMS model was modified, and the revised James River model (JMSRV) was recalibrated for use in updating wasteload allocations (Hydroscience, 1980). The SWCB staff used the latter model to develop wasteload allocations, that is, the Upper James River Wasteload Allocation Plan, in 1982 (SWCB, 1982).

Nutrient reduction has also been considered, and control measures have been implemented as part of the effort to clean up the Chesapeake Bay. In 1987, the Virginia General Assembly took action to reduce nutrient enrichment by enacting a phosphate detergent ban. The next step was taken in March 1988, when the Virginia SWCB adopted the Policy for Nutrient-Enriched Waters and a water quality standard designating certain waters as nutrient-enriched. Under the policy, municipal and industrial wastewater treatment plants with flows higher than 1 mgd are required to remove phosphorus to meet a 2 mg/L limit. Facilities were given up to 3 years to complete plant modifications to meet this requirement.

IMPACTS OF WASTEWATER TREATMENT

Pollutant Loading and Water Quality Trends

Pollutant loads from POTWs have been reduced significantly over the past three decades. In 1971, a large number of the municipal wastewater treatment plants provided primary treatment. By 1984, there were more than 20 major point source (municipal and industrial) discharges in the James River estuary from Richmond to the mouth of the Chesapeake Bay. Table 9-2 lists the major municipal and industrial treatment facilities discharging to the James River during 1983. Figure 9-6 illustrates the locations of these point sources. Some of the municipal facilities were consolidated to form re-

TABLE 9-2 Major Point Source Loads to the James River Estuary in September 1983

Point Source Discharger	River Mile	Flow (mgd)	$CBOD_u$ (lb/day)
Richmond	97.8	56.5	4,512
DuPont	92.7	6.9	202
Falling Creek	92.2	7.2	714
Proctors Creek	86.9	3.1	2,602
Reynolds Metals	86.9	0.1	1
VEPCO	86.7	0.08	0
American Tobacco	81.5	0.07	84.8
ICI	80.6	0.06	8.9
Philip Morris	79.8	0.0	83.2
Allied–Chester	78.5	NA	3,859
Allied–Hopewell	77.2	0.0	4,809
Stone Container	76.8	−13.5	NA
Hopewell	76.1	35.8	16,200
Williamsburg	NA	9.2	229
James River	NA	12.5	436
Boat Harbor	NA	16.1	410
Nansemond	NA	6.8	770
Army Base	NA	12.1	413
Lambert's Points	NA	20.2	21,893
Petersburg	88.9	NA	NA

Source: Lung, 1991.

Notes: (1) River miles are the distance measured from confluence of James River with Chesapeake Bay. (2) Petersburg is 10.8 miles upstream from the James River.

gional treatment plants. In the early 1980s, all POTWs achieved secondary treatment levels except the Lambert's Point plant, which was considered at an advanced primary level (with phosphorus removal). Since the early 1980s, wasteload allocation studies have been prepared to recommend further reductions of the BOD_5 loads in the upper estuary. Some of them, such as those in the Hampton Roads Sanitation District, achieved BOD_5 concentrations in the effluent much lower than 30 mg/L.

A study by the Virginia SWCB showed that the phosphate detergent ban has resulted in reductions of total phosphorus concentrations of 34 percent for POTW influent and 50 percent for effluent (SWCB, 1990). The SWCB's analysis was based on the data collected from the POTWs operated in the Hampton Roads Sanitation District, which operates nine POTWs in the James River basin. The total phosphorus concentrations measured during different periods of the study are shown in Table 9-3.

It should be pointed out that the analysis shown in Table 9-3 was based on the POTWs that did not have phosphorus removal. The phosphate detergent ban would have no effect on the effluent phosphorus concentration from the POTWs that remove phosphorus. Eventually, when the POTWs remove phosphorus to meet the 2 mg/L requirement, the ban will reduce the costs of phosphorus removal by reducing the influent concentrations.

The upstream boundaries and tributaries of the watershed of the estuary account for approximately 94 percent of the drainage area measured below the confluence of the James and Chickahominy rivers. The area adjacent to the Appomattox and James

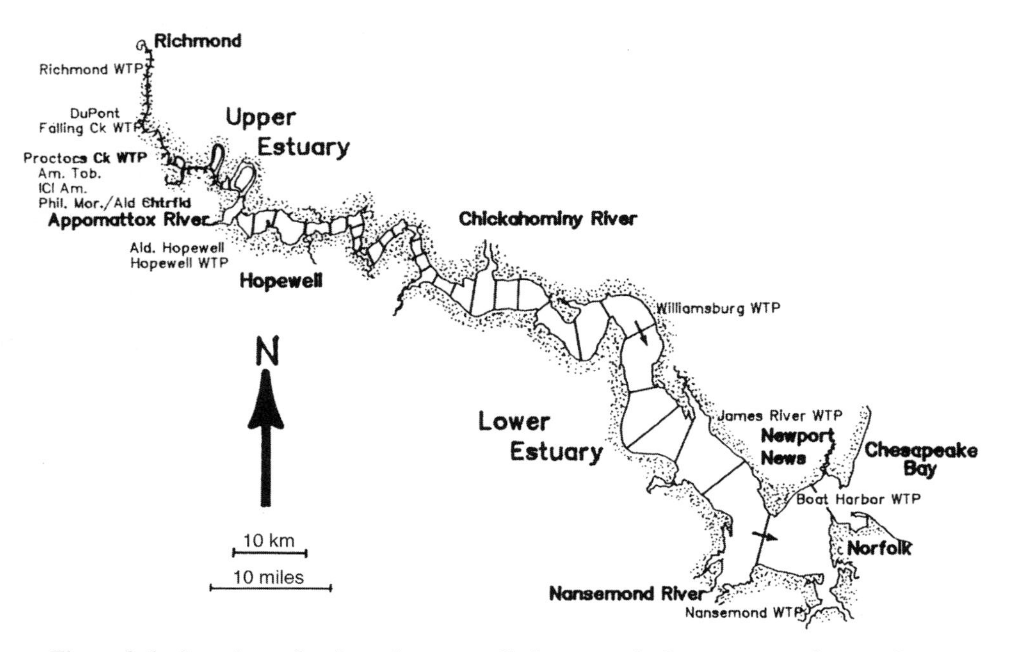

Figure 9-6 Locations of major point source discharges to the James estuary. *Source:* Lung and Testerman, 1989. Reprinted with permission of The American Society of Civil Engineers. Copyright 1989.

TABLE 9-3 Effect of Phosphate Detergent Ban: Hampton Roads Sanitation District

Time Period	Influent (mg/L)	Effluent (mg/L)
Pre-ban	7.4	5.3
Transition	5.6	3.7
Post-ban	4.9	2.5
Reduction	34%	53%

Source: Lung, 1991.

rivers below Richmond is thus a small fraction of the total area drained by this system. Runoff from the contiguous drainage area during the low-flow summer months represents a small fraction of the total river flow and has a negligible effect on the water quality in the watershed. The importance of the upstream pollutant loads was reported by HydroQual (1986). For example, in the James, the upstream ultimate BOD load is larger than any point source load, and the nitrogenous BOD (NBOD) is nearly equal in magnitude to several of the largest point source inputs. Similarly, the Appomattox River boundary load is significant relative to the Petersburg wastewater treatment plant discharge, the only significant point source input to this river. Further, the three point source inputs, the Richmond and Hopewell treatment plants and Allied–Hopewell, account for the major portion of the point source loads to the James. The nonpoint source runoff load was shown to be relatively small in comparison to the other inputs to the system (HydroQual, 1986).

It should be pointed out that CSO loads might be significant inputs to the river system during wet weather conditions and might also be a factor in the sediment interactions. In view of the purpose of this study, CSOs are not included in this analysis. The CSO impacts are indirectly incorporated into the modeling analysis to the degree that they are a component in the sediment oxygen demand rates determined by HydroQual (1986).

Figure 9-7 shows historical data of DO concentrations in the James estuary. The June 1971 survey shows that the river reach from Richmond to Hopewell was dominated by the waste discharges from and near Richmond. During that survey, the river was under a moderately high temperature and high flow. Consequently, the DO sag was carried downstream far enough (about 35 miles from Richmond) to merge with the Hopewell area discharges. Downstream from Hopewell, the DO concentrations started a slow recovery. In the lower estuary, from Mulberry Island (River Mile 27) to Old Point Comfort (River Mile 0), there were a number of large waste discharges. As a result of the strength of the tidal action combined with the massive amount of dilution water available, a rather steady DO level was measured. The DO levels seldom fell below 5.5 mg/L under the worst conditions, and the depression of DO due to waste stabilization by biological oxidation was usually less than 1 mg/L (Engineering Science, 1974).

The second survey in Figure 9-7 was conducted in September 1971, showing even lower DO concentrations below Richmond, compared with the data from the June 1971 survey. The DO sag was below 4 mg/L near River Mile 89, which was followed by a slow recovery. Also shown in Figure 9-7 is the DO profile measured in July

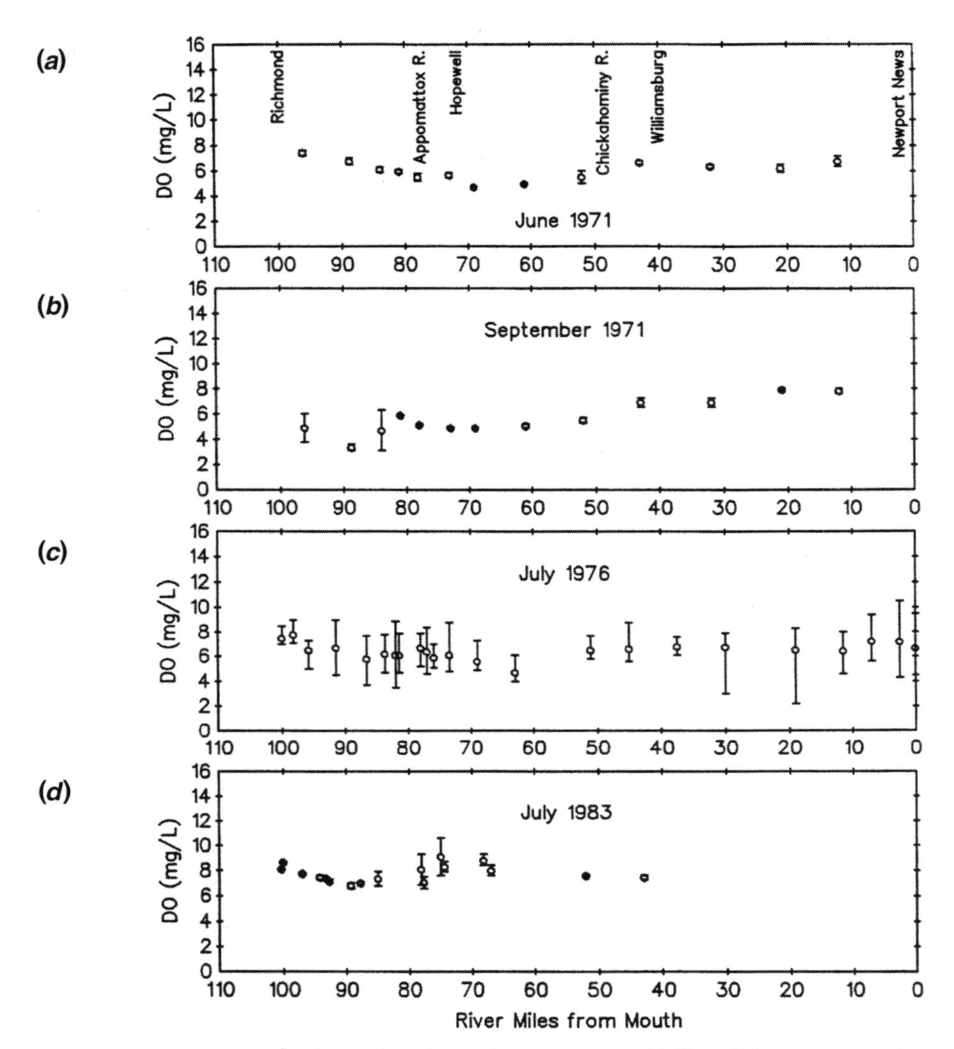

Figure 9-7 Spatial distribution of DO for the James estuary: (*a*) June 1971; (*b*) September 1971; (*c*) July 1976; and (*d*) July 1983. *Sources:* HydroQual, 1986; Lung, 1991.

1976. The DO sag level (below Richmond) improved slightly from the 1971 condition, although the sag was still below 5 mg/L. A mild recovery occurred until the wastes from the Hopewell area entered the river and depressed the DO concentration again, resulting in a second DO sag in the river. Such a two-sag DO profile has been consistently observed since the late 1970s. The low DO gradually increased downstream for a full recovery.

The DO condition observed in July 1983 is also presented in Figure 9-7. With continuing treatment upgrades beyond the secondary treatment for carbon removal, the DO condition in the James estuary continued to improve in the 1980s. The data indi-

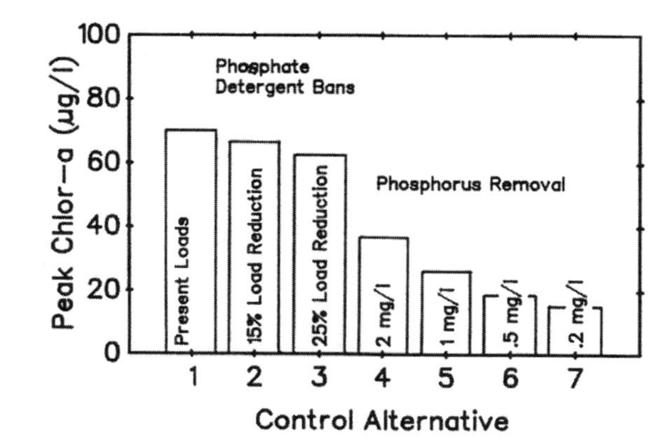

Figure 9-8 Projected impact of point source phosphorus controls. *Source:* Lung, 1986. Reprinted with permission of the American Society of Civil Engineers. Copyright 1986.

cate that the minimum DO level was above 6 mg/L in September 1983, a sign of continuing improvement of the water quality. The impact from the Richmond area discharges has been significantly reduced following the treatment plant upgrades.

Although the reduction of BOD_5 loads from the POTWs was measured in the last 30 years, no appreciable reduction of nutrient loads was detected until the phosphate detergent ban in 1988. Prior to the Virginia phosphate detergent ban, Lung (1986) conducted a modeling study assessing the water quality benefit of point source phosphorus control in the James River basin. The model results are summarized in Figure 9-8, showing the peak phytoplankton chlorophyll levels predicted in the upper James estuary for various control alternatives ranging from a phosphate detergent ban to phosphorus removal. The model suggests that the reduction of chlorophyll in the water column due to the phosphate detergent ban would be minimal, while phosphorus removal at POTWs would offer reasonable reductions in phytoplankton biomass in the upper estuary.

Evaluation of Water Quality Benefits Following Treatment Plant Upgrades

From a policy and planning perspective, the central question in water pollution control is simply: Would water quality standards be attained if primary treatment levels were considered acceptable? In addition to the qualitative assessment of historical data, water quality models can provide a quantitative approach to judge improvements in water quality achieved as a result of upgrades in wastewater treatment. The James River Model (JMSRV), originally developed by Hydroscience (1980) and subsequently enhanced by HydroQual (1986), Lung (1986), and Lung and Testerman (1989), and calibrated using data for September 1983 conditions (Figure 9-9), has been used to demonstrate the water quality benefits attained by the secondary treatment requirement of the 1972 CWA (Lung, 1991). Using the model, existing popu-

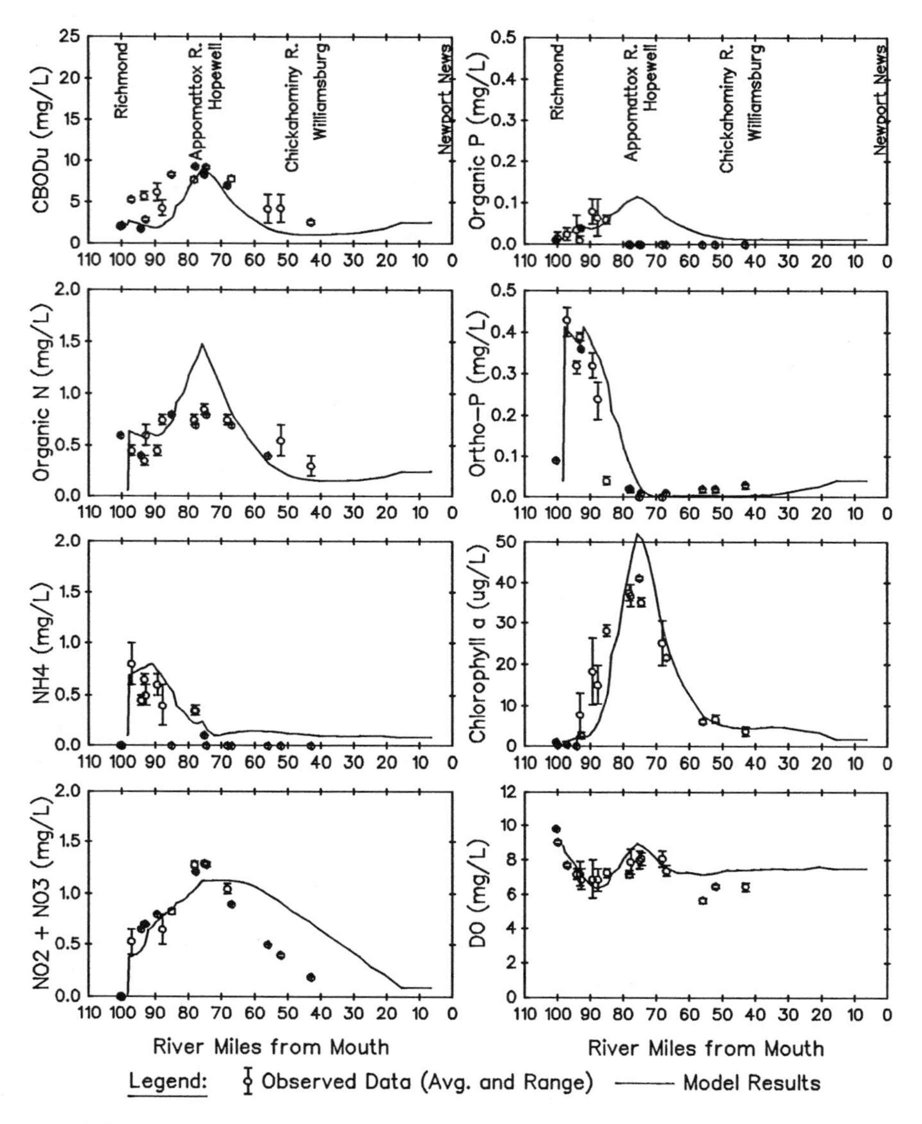

Figure 9-9 James River model calibrations for September 1983. *Source:* Lung and Testerman, 1989. Reprinted with permission of the American Society of Civil Engineers. Copyright 1989.

lation and wastewater flow data (ca. 1983) were used to compare water quality for summer low-flow and 7Q10 low-flow conditions simulated with three management scenarios: (1) primary effluent, (2) secondary effluent, and (3) existing wastewater loading. Water quality conditions for these alternatives were simulated using fresh-water and wastewater flow data for 1983, a year characterized by 66 percent of the summer average flow (see Figure 9-3) of the James River (Figure 9-10).

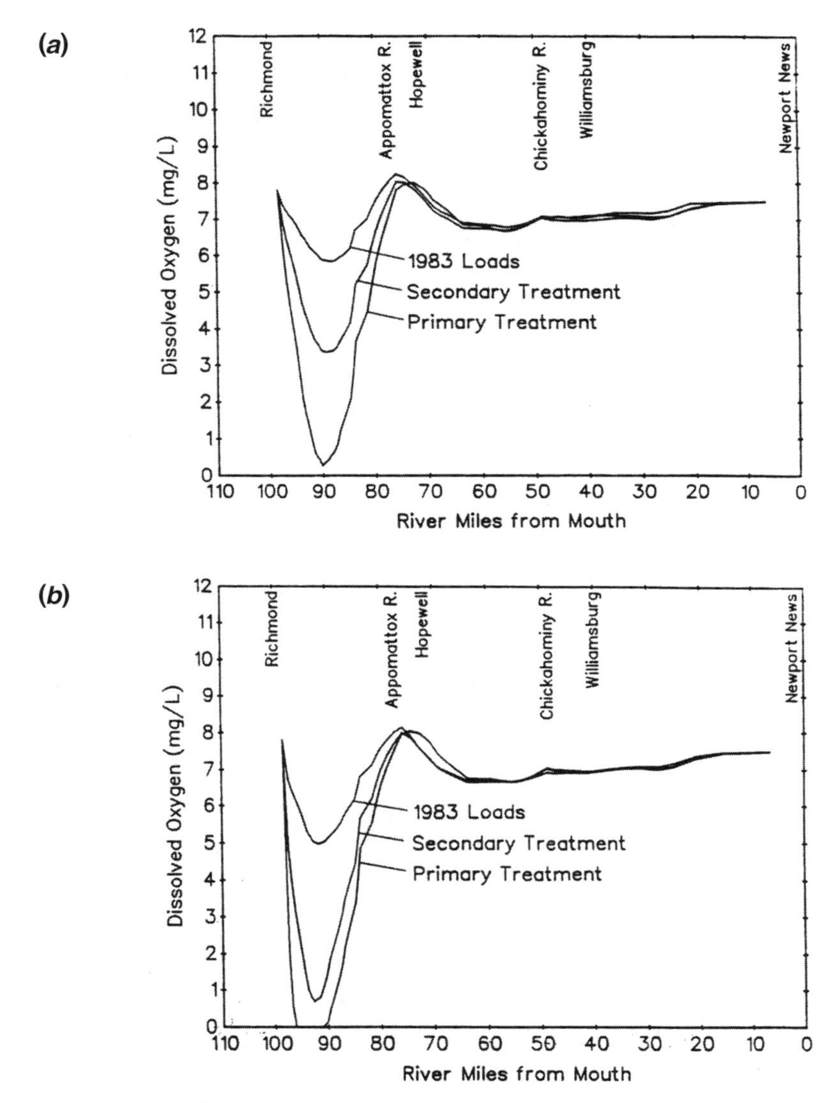

Figure 9-10 Comparison of simulated impact of primary, secondary, and existing 1983 effluent levels on DO: (*a*) summer 1983 conditions and (*b*) 7Q10 low-flow conditions. *Source:* Lung, 1991.

Using the primary effluent assumption, under summer low-flow conditions, water quality is noticeably deteriorated in comparison to the 1983 calibration results. DO concentrations downstream of Richmond (River Mile 90) are computed to be near zero under the primary scenario. Using the secondary assumption, the significant reduction in BOD_5 loading significantly improves DO between Richmond and Hopewell, Virginia. In comparison to the primary scenario, minimum monthly averaged oxygen levels increase to almost 3.5 mg/L from less than 0.5 mg/L under the secondary effluent scenario. As shown with both observed data (Figure 9-9) and model simulations (Figure 9-10), the implementation of secondary and better treatment has resulted in significant improvements in the DO status of the estuary.

As demonstrated with the model, better-than-secondary treatment is required to achieve compliance with the water quality standard of 5 mg/L under extreme 7Q10 low-flow conditions (Figure 9-10) for DO downstream of Richmond. In contrast to the 1950s and 1960s, the occurrence of low-oxygen conditions has been virtually eliminated within the upper James River estuary. Additional improvements in water quality, in terms of reduced algal biomass and still greater improvements in DO levels, have been achieved as a result of advanced secondary levels of wastewater treatment for the Upper James River.

Recreational and Living Resources Trends

Upgrades of wastewater treatment plants to secondary treatment in the 1970s and continued commitment to water quality–based pollution controls throughout the 1980s and 1990s have achieved a dramatic recovery for the James River. Instead of turkey vultures, residents of Richmond currently can gaze at blue herons, bald eagles, and ospreys circling overhead (Epes, 1992). Although passage of the Clean Water Act in 1972 was the most significant factor contributing to the comeback of the James, other factors contributing to improvements in wildlife habitat included the creation of a flood control reservoir in the early 1980s to stabilize flow, the ban of the insecticide DDT, and floods and hurricanes in the 1960s and 1970s.

The ban on DDT allowed certain birds affected by egg shell thinning, including eagles and ospreys, to recover. The floods and hurricanes contributed to habitat improvement by punching holes in several of the dams in the river, allowing migrating fish to pass through once more (Epes, 1992). Those holes, and subsequent man-made fish ladders, have allowed fish to swim farther upstream to spawn again.

Above the falls, the return of smallmouth bass has made the upper James River one of the best smallmouth bass fisheries in the country. Below Richmond, abundant largemouth bass attract the national Bassmasters fishing tournaments. Striped bass, an anadromous (saltwater-to-freshwater migrating) fish, has returned to the James due in part to a state harvesting moratorium in effect for several years in the Chesapeake Bay. In fact, a 25-pound striped bass was caught in 1992 near Williams Dam in Richmond (Epes, 1992).

Fish-eating birds have also returned to the James River. In the 1970s, there were no bald eagles or ospreys nesting on the James River. In 1992, three pairs of bald eagles and six pairs of ospreys had reclaimed their historical nesting sites on the James (Bradshaw, 1992). Great blue herons boast about 200 pairs (Bradshaw, 1992). Birds began to return in the mid-1980s. Cattle egrets and double-crested cormorants extended their ranges to colonize the James, possibly due to reduction in available habitat elsewhere. In 1992, there were about 250 pairs of each overwintering in the region from Richmond to the Benjamin Harris bridge (Bradshaw, 1992). Cattle egrets eat reptiles and eels, and double-crested cormorants eat fish. These birds are no doubt responding to the increase in the stream quality for fish and other aquatic life now that wastewater discharges of organic and nutrient loads to the James River have been controlled.

SUMMARY AND CONCLUSIONS

An analysis of the existing water quality data for the James River estuary has been conducted to document the historical changes in wasteloads and the water quality improvement in the estuary from 1971 to the mid-1990s. A water quality model for the upper James estuary was modified to include the lower portion of the estuary (Lung and Testerman, 1989). This modified model was calibrated and verified using three sets of water quality data. Finally, the verified model was used to evaluate the water quality improvement due to the treatment upgrades from primary to secondary at the POTWs. Altogether, six simulation scenarios, incorporating different ambient environmental conditions and wasteload levels, were developed for evaluation.

The analysis of POTW wasteloads indicated significant reduction of BOD_5 discharged into the James estuary starting in the early 1970s. By the mid-1980s, many POTWs had achieved high degrees of carbon removal with treatment levels beyond secondary. Nutrient reduction did not start until 1988, when the phosphate detergent ban became effective.

A review of the historical water quality data showed the improvement of DO conditions in the James estuary from a DO sag of much lower than 5 mg/L in 1971 to levels consistently above 5 mg/L in the 1980s. Nutrient concentrations in the water column of the James estuary have remained quite stable over the past 30 years. The model results showed a clear, progressive rise in DO levels in the estuary from primary treatment to secondary treatment, and to treatment beyond secondary at the POTWs. Based on the analyses of historical wasteload data, water quality data, and model results, it can be concluded that the treatment upgrades from primary to secondary and better levels of treatment at POTWs provided significant water quality improvement in the James River basin. With the cleanup of the James River, visitors to Richmond, Virginia, can enjoy a riverboat dinner cruise or a stroll along the refurbished 2-mile canal walk. More adventurous visitors can challenge themselves by rafting and kayaking on the only Class IV white water located in an urban river in the nation (McCulley, 1999). Birds and fish are also making a remarkable recovery in the James River basin in response to water quality improvements.

REFERENCES

Bradshaw, D. 1992. Virginia Department of Game and Inland Fisheries. Personal communication, November 4.

Engineering-Science. 1974. Lower James River basin comprehensive water quality management plan. Planning Bulletin 217-B. Final report prepared for Virginia State Water Control Board, Richmond, VA by Engineering-Science Inc., Fairfax, VA.

Epes, C. 1992. Honk if you spot a brown pelican. Richmond Times-Dispatch, Richmond, VA. June 15.

Forstall, R. L. 1995. Population by counties by decennial census: 1900 to 1990. U.S. Bureau of the Census, Population Division, Washington, DC. http://www.census.gov/population/www/censusdata/cencounts.html

HydroQual, Inc. 1986. Water quality analysis of the James and Appomattox Rivers. Report prepared for Richmond Regional Planning District Commission, Richmond, VA, by HydroQual, Inc., Mahwah, NJ. June.

Hydroscience, Inc. 1980. Water quality analysis of the upper James River estuary. Report prepared for Virginia Water Control Board by Hydroscience, Inc., Westwood, NJ. April.

Lung, W. S. 1986. Assessing phosphorus control in the James River basin. J. Environ. Eng., ASCE 112(1): 44–60.

Lung, W. S. 1991. Trends in BOD/DO modeling for wasteload allocations of the James River estuary. Technical memorandum prepared under sub-contract for Tetra Tech, Inc., Fairfax, VA, by Envirotech, Inc., Charlottesville, VA.

Lung, W. S. and N. Testerman. 1989. Modeling fate and transport of nutrients in the James estuary. J. Environ. Eng., ASCE 115(5): 978–991.

McCulley, C. 1999. Rapids transit. The Sun, Baltimore, MD. October 17, Section R.

OMB. 1999. OMB Bulletin No. 99-04. Revised statistical definitions of Metropolitan Areas (MAs) and guidance on uses of MA definitions. U.S. Census Bureau, Office of Management and Budget, Washington, DC. http://www.census.gov/population/www/estimates/metrodef.html

Richmond News Leader. 1963. Neglected asset: The James. Editorial, October 8.

SWCB. 1982. Upper James River estuary wasteload allocation plan. Final report prepared for Richmond Regional Planning District Commission, Richmond, VA, by Virginia State Water Control Board, Richmond, VA.

SWCB. 1990. Effects of phosphate detergent ban in Virginia. Final report prepared by the Chesapeake Bay Office, Richmond, VA 15 pp. Virginia State Water Control Board.

USDOC. 1998. Census of population and housing. U.S. Department of Commerce, Economics and Statistics Administration, Bureau of the Census—Population Division, Washington, DC.

USEPA (STORET). STOrage and RETrieval Water Quality Information System. U.S. Environmental Protection Agency, Office of Wetlands, Oceans, and Watersheds, Washington, DC.

USGS. 1999. Streamflow data downloaded from the U.S. Geological Survey's National Water Information System (NWIS)-W Data retrieval for historical streamflow daily values. http://waterdata.usgs.gov/nwis-w

CHAPTER 10

Upper Chattahoochee River Case Study

The Southeast Basin (Hydrologic Region 3), covering a drainage area of 278,523 square miles, includes the Chattahoochee-Flint-Apalachicola River, which has a length of 524 miles and a drainage area of 19,600 square miles (Iseri and Langbein, 1974). On the basis of a mean annual discharge (1941–1970) of 24,700 cfs, the Chattahoochee-Flint-Apalachicola River ranks twenty-third of the large rivers of the United States (Iseri and Langbein, 1974). Figure 10-1 highlights the location of the Upper Chattahoochee River case study watersheds (catalog units), and the city of Atlanta, Georgia, identified in this river basin as one of the urban-industrial waterways affected by severe water pollution problems during the 1950s and 1960s (see Table 4-2). In this chapter, information is presented to characterize long-term trends in population, municipal wastewater infrastructure and effluent loading of pollutants, ambient water quality, environmental resources, and uses of the Upper Chattahoochee River. Data sources include USEPA's national water quality database (STORET), published technical literature, and unpublished technical reports ("grey" literature) obtained from local agency sources.

The Chattahoochee River Basin constitutes almost 40 percent of the Chattahoochee-Flint-Apalachicola River Basin (Figure 10-2), which discharges into the Gulf of Mexico. The Chattahoochee River flows from northeast Georgia through metropolitan Atlanta to West Point Dam. From there, the river forms the Georgia-Alabama border and, for a short distance, the Georgia-Florida border. Near the southern border of Georgia, the Flint River joins the Chattahoochee River to form the Apalachicola River. Major urban centers in the Upper Chattahoochee River Basin include Atlanta, Gainesville, Marietta, Cornelia, and Alpharetta, Georgia. The Atlanta region represents only 3.6 percent of Georgia's total land area, but contains one-third of the state's population (ARC, 1984). The large volume of wastewater discharged in the Atlanta area has a far-reaching effect on water quality conditions in receiving waters. The Upper Chattahoochee River is by far the largest river in the Atlanta region. Other streams in the region include Sweetwater Creek, South River, Flint River, Yellow River, Peachtree Creek, and Line Creek.

The Chattahoochee River is Atlanta's major water supply source and receptacle for wastewater disposal. The Upper Chattahoochee River Basin provides numerous recreational areas and fish and wildlife habitats. Lake Sidney Lanier, for example, is

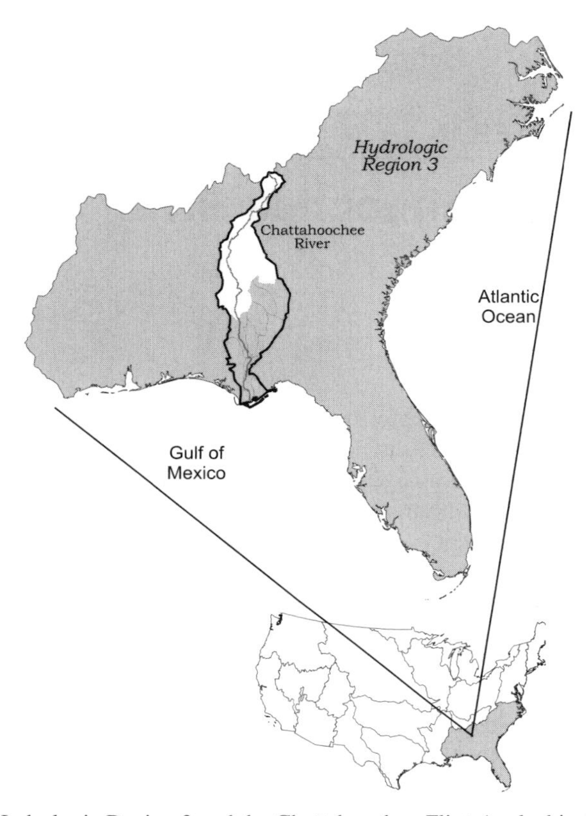

Figure 10-1 Hydrologic Region 3 and the Chattahoochee-Flint-Apalachicola River Basin.

a nationally popular water resort area. The area from Buford Dam to Peachtree Creek has been under intensive development pressures that threaten the water quality of the Chattahoochee River.

PHYSICAL SETTING AND HYDROLOGY

The Upper Chattahoochee River Basin covers 10,130 square miles from the southern slopes of the Blue Ridge mountains, in northeast Georgia, to the West Point Dam at the Georgia-Alabama state line. The flow length of this section is 250 river miles, generally to the southwest. The basin is narrow in relation to its length, the average width being less than 40 miles. Elevations in the Upper Chattahoochee Basin range from approximately 4,000 feet at the headwaters to approximately 635 feet at West Point Lake. Air temperature tends to be cooler in the mountains and warmer in the

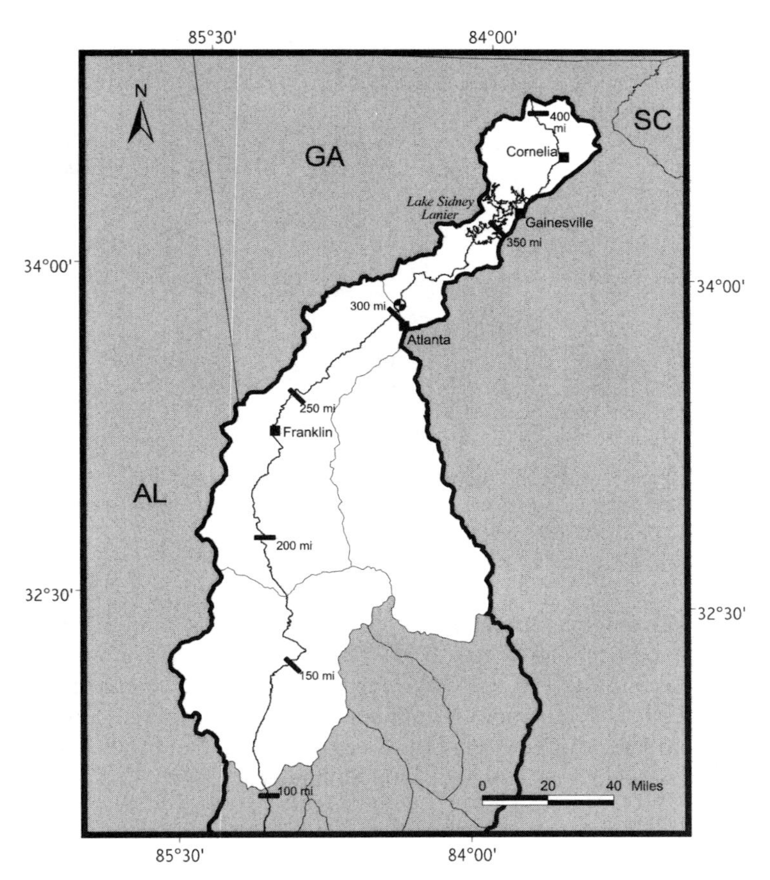

Figure 10-2 Location map of Upper Chattahoochee Basin. (River miles shown are distances from Gulf of Mexico.)

southern areas of the basin; the annual air temperature averages about 16°C. Average annual rainfall in the basin is about 54 inches over the basin area of 3,440 square miles. The rainfall tends to be greatest in upland areas and in the southern region of the basin (Cherry et al., 1980; Lium et al., 1979).

Flow in the river is dependent on rainfall and regulation by the hydroelectric generating facilities at Buford Dam and Morgan Falls Dam. High-flow conditions usually occur in the spring and low-flow conditions in late autumn (Figure 10-3). The most pronounced changes in regulated flow have occurred as a result of the construction and operation of the Buford Dam since 1957. In the mid-1960s, the city of Atlanta and the Georgia Power Company modified the Morgan Falls Dam and Reservoir, just upstream of Atlanta, to provide a minimum flow of 750 cfs from Morgan

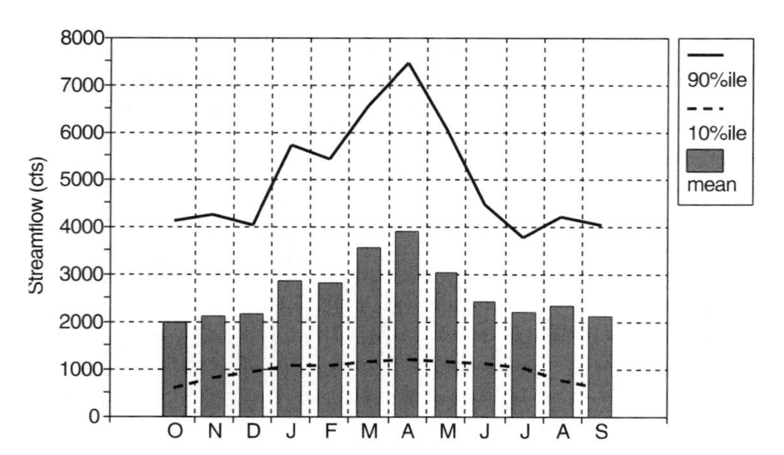

Figure 10-3 Monthly trends in streamflow for the Chattahoochee River. Monthly mean, tenth, and ninetieth percentile statistics computed for 1951–1980 (USGS Gage 02336000 at Atlanta, Georgia). *Source:* USGS, 1999.

Falls. Since 1965, minimum streamflows have been higher and more consistent as a result of those modifications (Figure 10-4). The average flow at Buford Dam, based on 35 years of record, is 2,168 cfs. The average flow near Atlanta, based on 43 years of record, is 2,603 cfs. Regulations for minimum streamflow volumes set in 1974 require a minimum release of 1,100 cfs from Morgan Falls, further increasing minimum streamflows near Atlanta (Cherry et al., 1980; Lium et al., 1979).

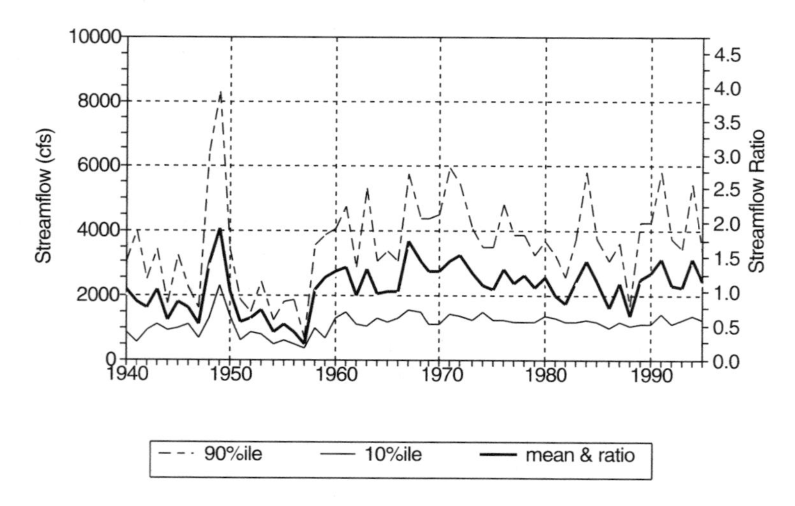

Figure 10-4 Long-term trends in mean, tenth, and ninetieth percentile statistics computed for summer (July–September) streamflow for the Chattahoochee River (USGS Gage 02336000 at Atlanta, Georgia). *Source:* USGS, 1999.

POPULATION, WATER, AND LAND USE TRENDS

The Upper Chattahoochee River case study area includes several counties that are defined by the Office of Management and Budget as Metropolitan Statistical Areas (MSAs) or Primary Metropolitan Statistical Areas (PMSAs). Table 10-1 lists MSA and counties included in this case study. Figure 10-5 presents long-term population trends (1940–1996) for the counties listed in Table 10-1.

From 1940 to 1996, the population in the Upper Chattahoochee River case study area increased dramatically (rising from 0.41 million in 1940 to 3.53 million in 1996). The U.S. Bureau of the Census reported the 1970 population of the Atlanta area to be 1.7 million. By 1990, this number had risen to 2.95 million (Forstall, 1995; USDOC, 1998). During the 1950s through the 1970s, population in the Atlanta region increased by 34 percent to 39 percent; the greatest growth rates were recorded in 1950–1960 (39 percent) and 1970–1980 (38 percent). During the 1980s and 1990s, the rate of growth slowed down considerably: the population increased by 22 percent from 1980 to 1990 and by only 19 percent from 1990 to 1996 (Forstall, 1995; USDOC, 1998). During the 1970s, population density in the area varied by about an order of magnitude from approximately 40 persons per square mile in the rural, headwater areas of the basin to 492 persons per square mile in the urban environs of Atlanta (Faye et al., 1980).

TABLE 10-1 Metropolitan Statistical Area (MSA) Counties in the Upper Chattahoochee Basin Case Study

County
Barrow County, GA
Bartow County, GA
Carroll County, GA
Cherokee County, GA
Clayton County, GA
Cobb County, GA
Coweta County, GA
DeKalb County, GA
Douglas County, GA
Fayette County, GA
Forsyth County, GA
Fulton County, GA
Gwinnett County, GA
Henry County, GA
Newton County, GA
Paulding County, GA
Pickens County, GA
Rockdale County, GA
Spalding County, GA
Walton County, GA

Source: OMB, 1999.

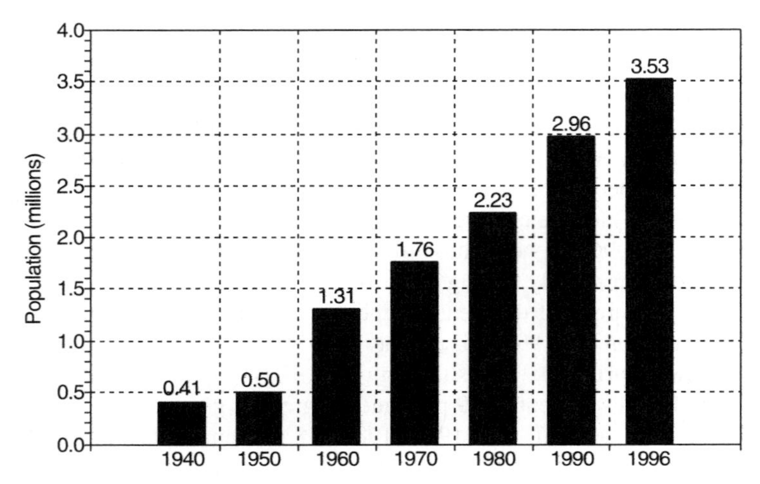

Figure 10-5 Long-term trends in population in the Upper Chattahoochee River basin. *Sources:* Forstall, 1995; USDOC, 1998.

Land in the Upper Chattahoochee Basin, upstream and downstream of Atlanta, is predominantly forest. The Atlanta area of the basin is predominantly residential. Agricultural activity is fairly evenly distributed through the basin. Table 10-2 shows the major land uses in the basin (Cherry et al., 1980; Lium et al., 1979; Stamer et al., 1979). Agricultural activities above the Buford Dam are concentrated in stream valleys and on the lower slopes. Crops and pastures occupy a significant portion of the agricultural areas, but poultry operations are the economically dominant agricultural activity. Urban areas are predominantly residential, but industrial activities are significant. Industrial activities include automobile assembly, food processing, and light manufacturing. Intense industrial land use dominates the area downstream of Interstate Highway 75 (Mauldin and McCollum, 1992).

Power generation, water supply, water-quality maintenance, and recreation are activities currently supported along the Chattahoochee River. Six power-generating facilities use the resources of the Chattahoochee River. The Buford Dam and Morgan

TABLE 10-2 Land Use in the Upper Chattahoochee River Basin

		Percentage Breakdown		
Location	Area (mi^2)	Urban	Agriculture	Forest
Above Buford Dam	1,040	4	16	81
Buford Dam–Atlanta	410	22	18	60
Atlanta–Fairburn	610	40	12	49
Fairburn–Whitesburg	370	6	17	77
Whitesburg–West Point Dam	1,010	4	17	79

Source: Lium et al., 1979.

Falls Dam are peak-power hydroelectric generating facilities. The other four are fossil-fuel thermoelectric power plants. The six plants have a combined generating capacity of approximately 3.8 million kilowatts. Two fossil-fuel plants near Atlanta discharge nearly 1,000 cfs of cooling water to the river.

As of 1998, 29 public water treatment plants process water withdrawn from rivers and lakes in the Atlanta region and 3 new treatment facilities were proposed for the Atlanta area. The largest water treatment plants in the region are operated by the city of Atlanta (Hemphill & Chattahoochee, design capacity 201 mgd), Dekalb County (Scott Candler, 128 mgd), Gwinnett County (Lake Lanier, 120 mgd), and Atlanta–Fulton County (Atlanta–Fulton County, 90 mgd). The Chattahoochee River and Lake Lanier are their main sources of raw water. The total capacity of the public water supply withdrawals from the 14 largest water treatment plants is 770.5 mgd (ARC, 1998). As of the late 1990s, approximately 443 mgd was withdrawn from water sources in the Upper Chattahoochee, primarily from surface water sources (ARC, 1998). During the mid-1970s, water use was estimated at 180 mgd with an increase in demand to 484 mgd fairly accurately projected for the year 2000 (Lium et al., 1979). The Chattahoochee River and Lake Lanier system and the Etowah River and Allatoona Lake system are the most important sources of public water, providing about 85 percent of the region's water supply. As of the late 1990s, residential and commercial water uses accounted for 54 percent and 23 percent of the total water demand, respectively. Government activities accounted for 6 percent and manufacturing uses for only 4 percent; approximately 14 percent could not be accounted for (Kundell and DeMeo, 1999). By the year 2020, regional water demand is expected to increase by approximately 46 percent of the withdrawals *ca.* 1998. The projected increase in water demand and the limited availability of surface water and groundwater supply sources in northern Georgia are a key factor in the need for regional cooperation to meet the challenges posed by water supply and water quality problems in the Atlanta region (Kundell and DeMeo, 1999).

Water-based recreational activities are abundant all along the Chattahoochee River. The headwaters are popular for trout fishing, camping, and hunting. Lake Sidney Lanier maintains numerous boat launches, campgrounds, marinas, yacht clubs, and cottages. The reach from Buford Dam to Atlanta supports fishing, canoeing, and rafting. The reach between Morgan Falls and Peachtree Creek, one of the most scenic on the river, is the site for an annual raft race that draws thousands of participants and onlookers to the area. West Point Lake, at the base of the Upper Chattahoochee River Basin, is an impoundment created by the construction of West Point Dam in 1974. This lake is widely used for fishing, boating, camping, and swimming.

HISTORICAL WATER QUALITY ISSUES

The poet Sidney Lanier, who praised the Chattahoochee in his "Song of the Chattahoochee," would not have been so inspired during the 1940s, 1950s, and 1960s. The Chattahoochee River was characterized by poor water quality for a reach of 70 miles below Atlanta. The first 40 miles were described as "grossly polluted," and responsibility was attributed to inadequately treated wastewater, particularly from Atlanta's R. M. Clayton sewage treatment plant, at the mouth of Peachtree Creek (EPD, 1981). Figure 10-6 shows the effect Atlanta's wastewater discharges historically have had on the water quality of the Chattahoochee River, with DO levels drastically depleted downstream of Atlanta near State Road (SR)-92 (River Mile 280). At Fairburn, an average of 13 percent of the river flow consisted of wastewater (Stamer et al., 1979). From July through October, heat and low flow placed the river in near septic conditions, with DO below 4 mg/L 64 percent of the time. During the period from 1968 to 1974, DO concentrations were 64 percent less in the summer months than in January and minimum DO levels were consistently below 1 mg/L (EPD, 1981). In 1973, DO concentrations dropped to zero during September. As of 1972, the R. M. Clayton plant was still releasing large quantities of wastewater receiving only primary treatment. Fecal coliform densities, ammonia-N, BOD_5, and suspended solids concentrations continued to be high above and below the discharge at Peachtree Creek. Fish kills caused by discharges of raw sanitary sewage and industrial chemicals were commonplace before 1976 (Mauldin and McCollum, 1992).

Rainfall in the area results in overflows from combined sewer systems (CSOs) and large amounts of urban runoff, contributing to large dissolved and suspended constituent loads to the river. Twelve CSOs have been identified in the watershed (Mauldin and McCollum, 1992). Low-flow periods result in less dilution of wastewater, resulting in low DO concentrations, high BOD_5, high fecal coliform densities, and other problems.

A severe drought in 1988 caused the DO level to dip below 4 mg/L in the study region from April to August (Mauldin and McCollum, 1992). A major fish kill occurred

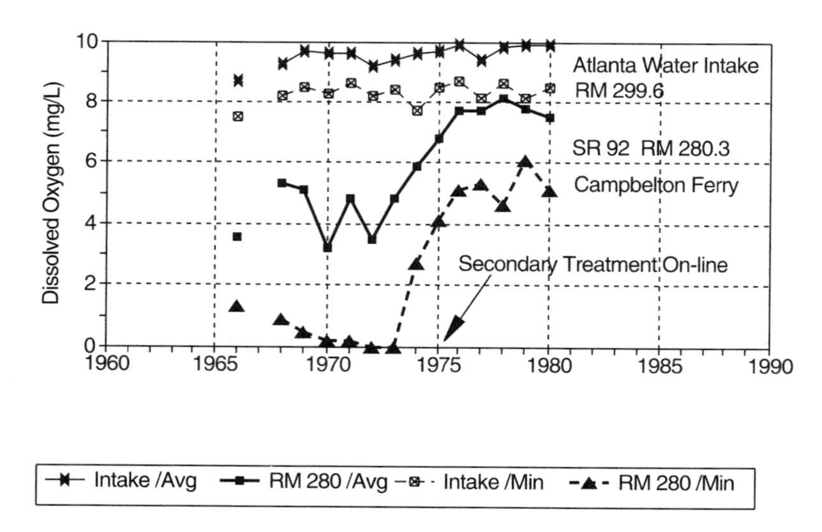

Figure 10-6 Long-term trends of DO concentration upstream and downstream of Atlanta wastewater discharges. *Source:* EPD, 1981.

during October of 1988, due to an unidentified agent (Mauldin and McCollum, 1992). The many impoundments along the river and releases of cooling water from fossil fuel plants, in excess of 1,000 cfs, contribute to water temperature increases, further reducing the waste assimilation capabilities of the river. Atlanta's population is served by 27 water pollution control plants, with designated flows greater than 0.01 mgd, located along the river and its tributaries. The 12 largest water pollution control plants in the Atlanta region have a total design capacity of 404 mgd. The largest facility, the R. M. Clayton plant, is operated by the city of Atlanta and has a capacity of 120 mgd. More than half of the total volume of wastewater enters the river near river mile 301 downstream of the city of Atlanta's water intake (Mauldin and McCollum, 1992).

LEGISLATIVE AND REGULATORY HISTORY

Concern for the coordination of water and sewer facility planning and operation has existed in Atlanta since the early 1930s. Construction of the metropolitan sewer system began in 1944 as a cooperative effort between Atlanta local governments. From 1950 to 1952, a major functional consolidation, The Plan of Improvement, was prepared to better define service functions between the city of Atlanta and Fulton County. Atlanta was the primary provider of sewage treatment at that time. During the 1960s, the near-septic conditions in the river concerned many people. Utility of the waters was greatly reduced, threatening water supplies, recreation, and aquatic habitats. Studies were conducted to identify problems and needs. Technology was available to remedy many of the problems identified, but funding was unavailable.

The Georgia Water Quality Control Act (enacted 1964, amended) was the first major state law to be applied to water quality management. The act gives the Georgia Environmental Protection Division (EPD) authority to control all types of pollution in the state's waters from both point and nonpoint sources. In the late 1960s the Atlanta Region Metropolitan Planning Commission (now the Atlanta Regional Commission or ARC) prepared several reports on the consolidation of water and sewer services. The Preliminary Water and Sewer Report, issued in 1968, provided elements of an Administrative Plan for water and sewers in the Atlanta region. The report called for a basinwide water and sewer authority, representing nine counties, to oversee water quality management on a basinwide scale. Unfortunately, local officials did not support the plan because of the large estimated cost (Hammer, Siler, George Associates, 1975).

The next state-level move toward regulation was the Metropolitan River Protection Act (MRPA) (enacted 1973, amended), which allows the ARC to advise local governments when proposed developments violate the Chattahoochee Corridor Plan. The plan establishes standards for development based on the carrying capacity of the land within 2,000 feet of impoundments or riverbank of the Chattahoochee or within the 100-year floodplain, whichever is greater (ARC, 1984). The Soil Erosion Act of 1975 also created controls over the effects of development in the area. This act requires local counties and municipalities to adopt and enforce local ordinances to control soil erosion from land-disturbing activities within their jurisdiction.

The 1972 CWA resulted in significant improvement of the water quality in the Upper Chattahoochee River Basin. Funding was provided under the CWA in the form of the Construction Grants Program. The state of Georgia received $117 million in 1976 under this program, but funding decreased steadily with only $41 million provided in 1983, despite the fact that Georgia reported needs of $300 million in 1983 (LMS, 1989). Beginning in 1988, funding for the Construction Grants Program was reallocated to the Clean Water State Revolving Fund (CWSRF) as a mechanism for providing financial assistance to municipalities. The CWA established secondary treatment as the minimum allowable level for municipal plants. The National Pollutant Discharge Elimination System (NPDES), a national permit program that regulates polluted discharges and requires permittees to monitor effluent quality, is also included in the CWA. States were called upon to develop water quality standards, water use classification, and effluent limits based on water quality criteria established by USEPA.

Attempts were made to improve water quality in the Chattahoochee River by regulating flow. The EPD set requirements for minimum flow of 750 cfs upstream of Atlanta (Cherry et al., 1980). A regulatory dam downstream from Buford Dam has been proposed and modeled (Zimmerman and Dortch, 1988). The dam would ensure Atlanta's water supply into the twenty-first century, aid in regulating river flow and eliminate the requirement for minimum releases from Buford Dam. It is not possible to greatly affect flow, since there is a limited amount of water available and water supply demands and wastewater flows continue to increase.

IMPACTS OF WASTEWATER TREATMENT

Pollutant Loading and Water Quality Trends

Major improvements in water quality occurred in the Chattahoochee Basin during the 1970s and early 1980s, resulting from implementation of secondary treatment. The effects of the increasing load of wastewater were diminished by better treatment technology. Figure 10-7 shows the increasing trends of effluent discharge rates for the area's larger wastewater treatment plants. By 1974, all Atlanta-area waste treatment facilities had been upgraded to provide secondary levels of treatment. Before implementation of secondary treatment, DO levels were severely reduced by wastewater discharges from Atlanta (Figure 10-8). Figures 10-6 and 10-8 show dramatic improvements beginning in 1974. The effects of secondary treatment on DO concentrations are particularly notable during the summer months (Figure 10-9). Water quality has improved despite a doubling of Atlanta's population over the period from 1970 to 1996 (Figure 10-5).

Many advances in improving water quality since 1974 can be attributed to continually improving operation and maintenance procedures. Figure 10-10 indicates improvements in suspended solids concentrations and BOD_5 in the effluent wastewater from the R. M. Clayton plant, the largest in the Atlanta region. These improvements resulted primarily from improved operator training and upgrading of the solids-handling facility. Similar changes took place at other area plants during this time. The R. M. Clayton plant operated at a primary level of treatment from the late 1930s to the mid-1960s. For much of this time, the capacity of the plant exceeded the design

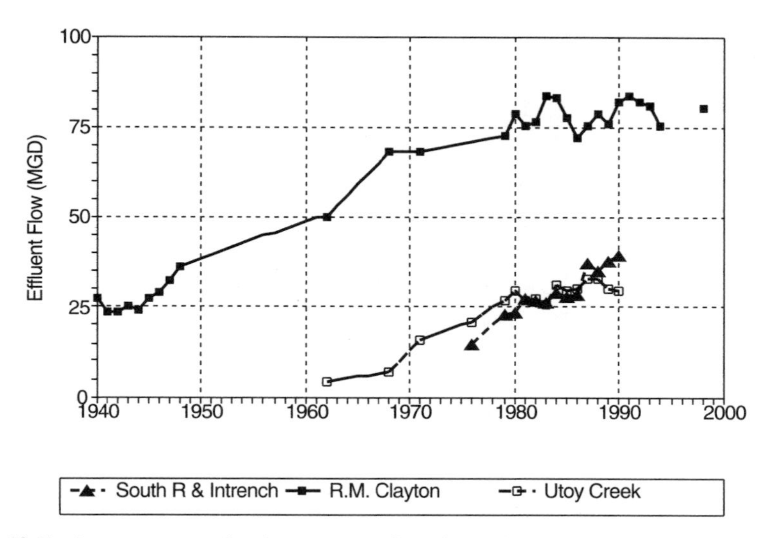

Figure 10-7 Long-term trends of wastewater flow for major wastewater treatment plants in the Atlanta area. *Sources:* ARC, 1984; USEPA, 1971; USPHS, 1963; Woodward, 1949; Richards, 1999.

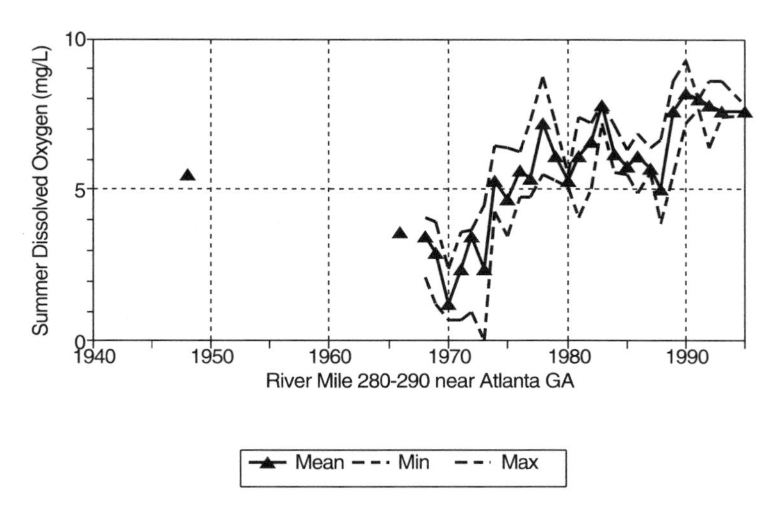

Figure 10-8 Long-term trends of mean, minimum, and maximum summer DO in the Chattahoochee River near Atlanta, Georgia (RF1–03130002066) (River Mile 280–290). *Source:* USEPA (STORET).

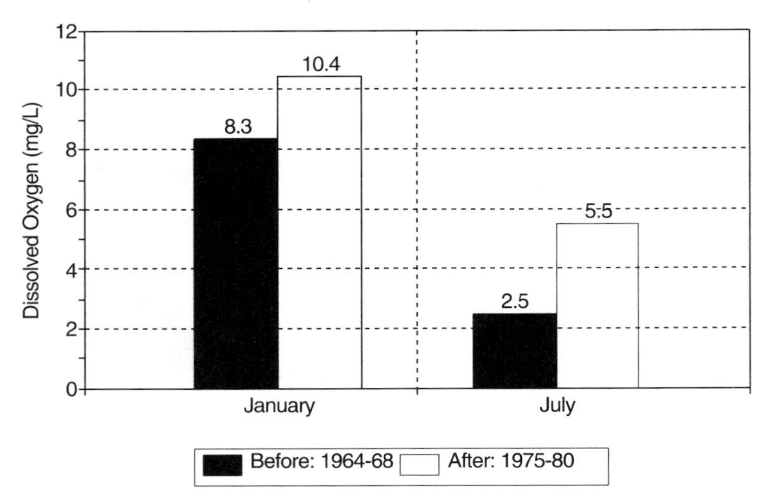

Figure 10-9 Comparison of January and July mean dissolved oxygen below Atlanta wastewater discharges before and after upgrade to secondary treatment. *Source:* EPD, 1981.

flow and treatment was below design level. When the plant was upgraded to provide secondary treatment, around 1968, the design flow was also increased to 120 mgd. A portion of the wastewater flow continued to receive only primary treatment into the early 1970s, when further improvements were made. In 1974, the R. M. Clayton Plant was providing secondary treatment to 100 percent of the plant's wastewater flow. In the early 1980s, operating and maintenance improvements further lowered BOD_5

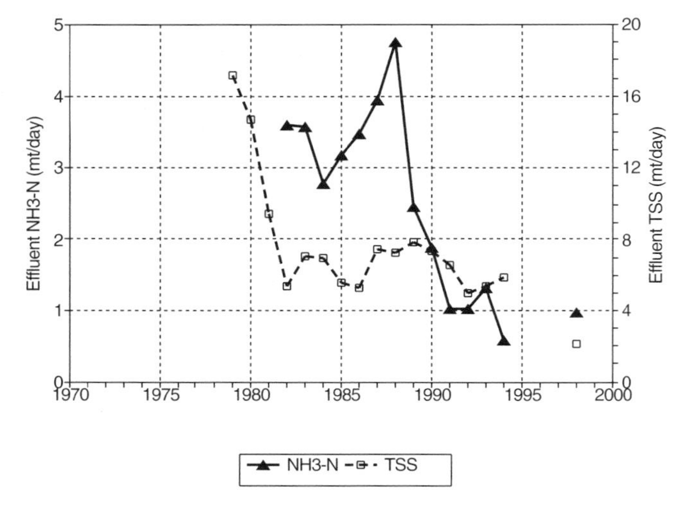

Figure 10-10 Performance of the R. M. Clayton wastewater treatment plant. *Sources:* ARC, 1984; USEPA, 1971; USPHS, 1963, Richards, 1999.

concentration in the effluent wastewater. The R. M. Clayton plant was upgraded to advanced secondary with ammonia removal in 1988. In June 2001 construction to upgrade and build new facilities at the R. M. Clayton plant, the Utoy plant, and the South River plant was completed. These facilities have state-of-the-art effluent filters, biological phosphorus removal, ultraviolet disinfection, and new headworks (Richards, 2001). Decreases in the BOD_5 loading of effluent at the R. M. Clayton Plant as a result of upgrading levels of treatment are shown in Figure 10-11.

All of the larger wastewater treatment plants in the Atlanta region must meet treatment requirements more stringent than secondary treatment. Phosphorus removal and restrictions on phosphates in detergents, for example, have resulted in a decline of ambient phosphorus concentrations downstream of Atlanta from approximately 1.0–1.2 mg/L in the early 1980s to approximately 0.1 mg/L a decade later (ARC, 1998). Land application of treated wastewater is also being used at several facilities in the region, with treated wastewater sprayed on forestland, golf courses, or other landscaped areas. At the 4,000-acre E. L. Huie Land Application site, the Clayton Water Authority operates the largest site, treating 18 mgd by reclaiming the treated effluent for its water supply since the water percolates through the soil and back to the raw water source (ARC, 1998).

A combined sewer system, originally constructed in Atlanta *ca.* 1900–1940, has historically contributed to water pollution problems in the Chattahoochee River. When constructed during the 1980s and 1990s, the seven combined sewer overflow control facilities were approved by the Georgia Department of Natural Resources Environmental Protection Division. As a consequence of a federal lawsuit, however,

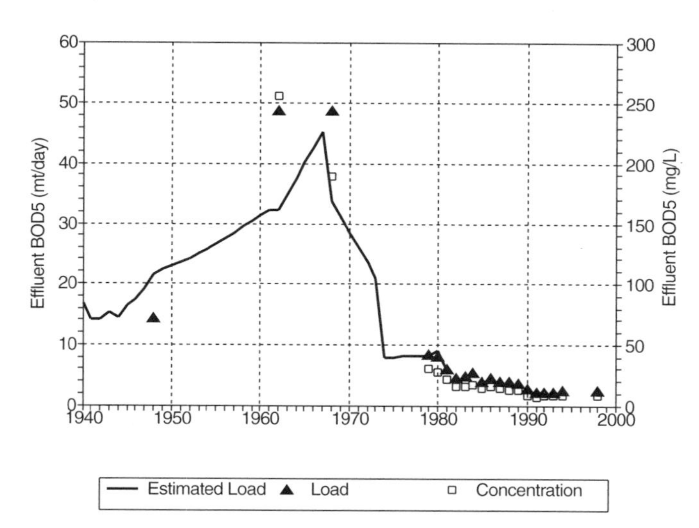

Figure 10-11 Long-term trends of effluent BOD_5 for the R. M. Clayton wastewater treatment plant. Data for estimated BOD_5 based on effluent flow data (Figure 10-7), 200 mg/L influent BOD_5, and removal rates of 35 percent (primary), 85 percent (secondary), and 95 percent (tertiary). *Sources:* ARC, 1984; USEPA, 1971; USPHS, 1963; Woodward, 1949, Richards, 1999.

the state agency and the presiding judge subsequently determined that the effluent from the CSO treatment facilities failed to meet current state and federal water quality standards. A consent decree signed in July 1998 required the City of Atlanta to develop a CSO remedial measures plan that would bring all the CSO control facilities into compliance by July 2007. In July 2001, the U.S. Environmental Protection Agency and the Georgia EPD authorized the City of Atlanta to proceed with the remedial measures plan developed for the following seven CSO treatment facilities: Clear Creek, Tanyard Creek, North Avenue, Greensferry, McDaniel, Custer and Intrenchment Creek. The plan includes: separation of additional areas of the existing combined sewer system, upgrade the Intrenchment Creek facility, dechlorination of disinfected discharges at existing CSO facilities, consolidate storage in deep tunnel sewer systems, treatment of stored combined wastewater at dedicated CSO facilities and develop site-specific permit limits. Implementation of the $950 million plan adopted by the city is expected to be completed by July 2007 and will result in the complete separation of all combined sewers within 25 years (see Mynhier et al., 2001).

Recreational and Living Resources Trends

Historical records of fish population in the Chattahoochee River below Atlanta are very limited. Conditions downstream of Atlanta's wastewater discharge were unsuitable for fish survival during the 1970s, and no fish surveys could be collected (Mauldin and McCollum, 1992). Shelton and Davies (1975) conducted a preimpoundment survey of the area to be flooded by the West Point Dam. The survey lasted from January 1972 to May 1974. The station closest to Atlanta on the Chattahoochee was at Franklin, Georgia. During the early 1970s study period, the Chattahoochee River was described as carrying a high organic load from municipal wastes, a high suspended solids load from agricultural and construction practices, and high chemical concentrations from industrial effluents. The relatively poor water quality in the Chattahoochee River affected the distribution and abundance of fish species sampled in the main stem versus the tributaries. Seventeen species of fish were collected in the Chattahoochee River at Franklin, which is less than half the number of species expected for Georgia rivers of similar size.

A fish survey of the Chattahoochee River conducted between July 1990 and June 1992 revealed the return of fish in great numbers to the portion of the river below the city of Atlanta. The number of species collected ranged from 14 or 15 at the sites in the direct vicinity of the wastewater treatment plant to 18 to 22 at the sampling sites located 63 and 23 km downstream, respectively. The diverse species collected represented a considerable improvement from conditions in the early 1970s, when only 17 species were sampled at Franklin, about 100 km downstream from the wastewater treatment plant, and no fish were present downstream of Atlanta's water supply intake (Mauldin and McCollum, 1992). The recent survey collected 12 gamefish species, compared to 8 collected by Shelton and Davies (1975); the most abundant of game species by weight were largemouth bass, bluegill, and channel catfish. Samples were analyzed using the Index of Biotic Integrity (IBI) (Karr, 1981; Karr et al., 1986). IBI scores for the four sampling sites (located 1 km upstream of the discharge, 1 km downstream of the discharge, 23 km downstream, and 63 km downstream) ranged from 22 to 32, which is 37 percent to 53 percent of the maximum score of 60. Scores in the 21 to 30 range indicated poor stream quality for fish and a population dominated by omnivorous, pollutant-tolerant forms. The Chattahoochee River below Atlanta's wastewater treatment plant discharge had a disproportionate segment of carp (75 percent), a higher proportion of bluegill to redbreast sunfish than is common in Georgia streams, and fewer gamefish than expected. A score of 32 measured 23 km downstream from the discharge indicated fair stream quality for fish. Overall, the fish sampled appeared to be healthy. Neoplasms were not observed in bluegill specimens, nor were gross external abnormalities observed in catfish.

The results of the 1990 to 1992 sampling show that water quality has improved immensely since 1972 when the river below Atlanta was described as "in near septic condition for a reach of 35 miles" (GADNR, 1991). The improvement is due to enhanced wastewater treatment (Mauldin and McCollum, 1992). Combined efforts of the state, communities, and industries, along with USEPA grants for municipal wastewater treatment systems, have put the Upper Chattahoochee River on the road to recovery. Fish kills have not been commonplace since 1976, except for one caused by an unidentified agent in 1988 (Mauldin and McCollum, 1992) (Table 10-3). Bloodworm-infested sludge beds no longer float in the shallows below Atlanta, sportfish populations are recovering, there is more DO in the water, macroinvertebrate fauna is more diverse, and fecal coliform bacteria levels dropped 82 percent in only 4 years (USEPA, 1980). The number of water quality violations has dropped dramatically since the 1970s even though standards have increased. Water-based and contact recreation are now fully supported along the Chattahoochee River reach from Buford Dam to Peachtree Creek. Fishing is generally supported along the entire river (GADNR,

TABLE 10-3 Fish Kills Due to Municipal Waste Discharges in the Greater Atlanta Region

Location	Date of Occurrence	Duration	Severity	Length of Stream Affected (miles)	Game Species (%)
Chattahoochee River, Atlanta	08/13/1964	1 day	Moderate	6	70
Proctor Creek, Atlanta	07/18/1976	1 day	Moderate	5	3
Chattahoochee River, Atlanta	07/29/1976	12 hours	Moderate	15	75
Nancy Creek, Chamblee	07/24/1981	12 hours	Severe	3	87
Marsh Creek, Sandy Springs	09/03/1981	1 day	Moderate	1	37
Little Nancy Creek, Atlanta	09/28/1984	Unknown	Moderate	1	63

Source: Mauldin and McCollum, 1992.

1991). As a result of the investments to upgrade water pollution control facilities in the Atlanta metropolitan region, the natural ecological balance of the river is beginning to be restored.

SUMMARY AND CONCLUSIONS

Results of legislation and regulations have been positive, due to active enforcement on all levels. Water quality monitoring by the EPD and under the NPDES program helps to evaluate progress and indicate violations. Water quality in the Upper Chattahoochee River, particularly in the vicinity of Atlanta, has improved dramatically with implementation of secondary waste treatment. Chemical, physical, and biological data all indicate a great improvement in water quality when compared to data from investigations done in the 1940s, 1950s, 1960s, and 1970s (LMS, 1989). Although total loading of pollutants to the Chattahoochee River, such as BOD5, suspended solids, and phosphorus, have been reduced significantly as a result of major capital improvements to the wastewater and water pollution control infrastructure of the Atlanta region during the 1970s and 1980s, the dramatic improvements in water quality of the river tended to level out during the 1990s.

Contemporary degradation of water quality is attributed to rapid urban development, the expanding area of the outer suburbs of Atlanta, and nonpoint source loading from stormwater runoff. The Georgia Department of National Resources (GADNR) listed more than 600 stream miles in the Atlanta area as impaired in the 1994–1995 305(b) report, with less than 20 percent of the degradation in stream miles attributed to point source pollution. As a result of increased sediment loading from watershed runoff to the Chattahoochee River and the reservoirs, water supply intakes are routinely shut down during and after rainstorms.

Contemporary water resource issues for Atlanta include the degradation of water quality in rivers and streams, the adverse impact of stormwater runoff on public water supplies and recreational lakes, and probable limits on future water supply allocations under the tristate river compacts that have sparked "water wars" between Georgia, Alabama, and Florida (Kundell and DeMeo, 1999). Despite the successes of past water pollution control efforts during the 1970s and 1980s, the Atlanta region is now confronted with serious water supply and water quality issues that will affect the future economic viability of the Atlanta metropolitan region. To achieve the solutions to contemporary water quality problems required by state and federal agencies, regional cooperation is essential for watershed management (Kundell and DeMeo, 1999).

REFERENCES

ARC. 1984. Status of water pollution control in the Atlanta region. Atlanta Regional Commission, Atlanta, GA.

ARC. 1998. Water resources of the Atlanta region. Atlanta Regional Commission, Atlanta, Georgia. January.
http://www.atlreg.com

Cherry, R. N., R. E. Faye, J. K. Stamer, and R. L. Kleckner. 1980. Summary of the river-quality assessment of the Upper Chattahoochee River basin, Georgia. U.S. Geological Survey, Circular 811, U.S. Department of Interior, Reston, VA.

EPD. 1981. Statement for the public hearing of the investigation and oversight subcommittee of the Public Works and Transportation Committee of the U.S. House of Representatives, May 18, 1981. Environmental Protection Division, Georgia Department of Natural Resources, Atlanta, GA.

Faye, R. E., W. P. Carey, J. K. Stamer, and R. L. Kleckner. 1980. Erosion, sediment discharge, and channel morphology in the Upper Chattahoochee River basin, Georgia. U.S. Geological Survey, Professional Paper 1107, U.S. Department of Interior, Reston, VA.

Forstall, R. L. 1995. Population by counties by decennial census: 1900 to 1990. U.S. Bureau of the Census, Population Division, Washington, DC. http://www.census.gov/population/www/censusdata/cencounts.html

GADNR. 1991. Rules and regulations for water quality control. Environmental Protection Division, Georgia Department of Natural Resources, Atlanta, GA.

Hammer, Siler, George Associates. 1975. Regional assessment study of the Chattahoochee-Flint-Apalachicola Basin. NTIS No. PB-252-318. National Commission on Water Quality, Washington, DC.

Iseri, K. T. and W. B. Langbein. 1974. Large rivers of the United States. U.S. Geological Survey Circular No. 686, U.S. Department of Interior, Reston, VA.

Karr, J. R. 1981. Assessment of biotic integrity using fish communities. Fisheries 6(6): 21–27.

Karr, J. R., K. D. Fausch, P. L. Angermeier, P. R. Yant, and I. J. Schlosser. 1986. Assessing biological integrity in running waters: A method and its rationale. Special Publication 5. Illinois Natural History Survey, Champaign, IL.

Kundell, J. E. and T. DeMeo. 1999. Cooperative regional water management alternatives for metropolitan Atlanta: A report of the Regional Water and Sewer Study Commission. Prepared for the Atlanta Regional Commission, Atlanta, GA, by the University of Georgia, The Carl Vinson Institute of Government, Athens, GA.

Lium, B. W., J. K. Stamer, T. A. Ehlke, R. E. Faye, and R. N. Cherry. 1979. Biological and microbiological assessment of the Upper Chattahoochee River basin, Georgia. U.S. Geological Survey, Circular 796, U.S. Department of Interior, Reston, VA.

LMS. 1989. Technical assistance for the development of a defensible water quality model of the Chattahoochee River. Report prepared for Georgia Department of Natural Resources, Environmental Protection Division, Atlanta, GA, by Lawler, Matusky, and Skelly (LMS) Engineers. Pearl River, NY.

Mauldin, A. C. and J. C. McCollum. 1992. Status of the Chattahoochee River fish population downstream of Atlanta, Georgia. Georgia Department of Natural Resources, Game and Fish Division, Atlanta, GA.

Mynhier, M. D., T. A. Richards, M. L. Griffin, and R. L. Wycoff. 2001. A separate peace. Atlanta considers three options for remediating its combined sewer systems. Water Environment & Technology, 13(10): 24–30, Water Environment Federation, Alexandria, Virginia, October.

OMB. 1999. OMB Bulletin No. 99-04. Revised statistical definitions of Metropolitan Areas (MAs) and guidance on uses of MA definitions. U.S. Census Bureau, Office of Management and Budget, Washington, DC. http://www.census.gov/population/www/estimates/metrodef.html

Richards, T. 1999. R. M. Clayton Wastewater Treatment Plant, Atlanta, GA. Personal communication. November 24.

Richards, T. 2001. Gwinnett County Department of Public Works, Lawrenceville, GA, Personal communication. October 22.

Shelton, W. L. and W. D. Davies. 1975. Preimpoundment survey of fishes in the West Point Reservoir Area (Chattahoochee River, Alabama and Georgia). Georgia Acad. Sci. 33: 221–230.

Stamer, J. K., R. N. Cherry, R. E. Faye, and R. L. Kleckner. 1979. Magnitudes, nature and effects of point and nonpoint discharges in the Chattahoochee River basin, Atlanta to West Point Dam, Georgia. U.S. Geological Survey, Water Supply Paper 2059, U.S. Department of Interior, Reston, VA, 65 p.

USDOC. 1998. Census of population and housing. U.S. Department of Commerce, Economics and Statistics Administration, Bureau of the Census—Population Division, Washington, DC.

USEPA (STORET). STOrage and RETrieval Water Quality Information System. U.S. Environmental Protection Agency, Office of Wetlands, Oceans, and Watersheds, Washington, DC.

USEPA. 1971. Inventory of municipal waste facilities, Region IV: Alabama, Florida, Georgia, Kentucky, Mississippi, North Carolina, South Carolina and Tennessee. U.S. Environmental Protection Agency, Washington, DC.

USEPA. 1980. National accomplishments in pollution control: 1970–1980, Some case histories. U.S. Environmental Protection Agency, Office of Planning and Management, Program Evaluation Division, Washington, DC.

USGS. 1999. Streamflow data downloaded from U.S. Geological Survey, United States National Water Information System (NWIS)-W. Data retrieval for historical streamflow daily values.
http://waterdata.usgs.gov/nwis-w/

USPHS. 1963. Inventory of municipal waste facilities, Region IV: Alabama, Florida, Georgia, Kentucky, Mississippi, South Carolina, and Tennessee. U.S. Public Health Service, Washington, DC.

Woodward, R. L. 1949. Flow requirements for pollution abatement below Atlanta, Georgia. U.S. Public Health Service, Environmental Health Center, Cincinnati, OH.

Zimmerman, M. J., and M. S. Dortch, 1988. Water Quality Modeling Study of Proposed Reregulation Dam Downstream from Buford Dam, Chattahoochee River, Georgia. Technical Report EL-88-14, US Army Corps of Engineers, Waterways Experiment Station, Vicksburg, MS.

CHAPTER 11

Ohio River Case Study

The Ohio River Basin, covering a drainage area of 204,000 square miles, extends 1,306 miles from the headwaters of the Alleghany River in Potter County, Pennsylvania, to the confluence of the Ohio River with the Mississippi River at Cairo, Illinois. With a length of 981 miles from the confluence of the Alleghany and Monangahela rivers with the Ohio River at Pittsburgh, Pennsylvania, to Cairo, Illinois, and a drainage area of 192,200 square miles, the Ohio River is the largest single tributary to the Mississippi River. In the United States, the Ohio River ranks tenth in length and third in mean annual discharge (258,000 cfs) (Iseri and Langbein, 1974).

Figure 11-1 highlights the location of the Ohio River case study watersheds (catalog units) identified along the Ohio River as major urban-industrial areas (e.g., Cincinnati, Ohio, and Louisville, Kentucky) affected by severe water pollution problems during the 1950s and 1960s (see Table 4-2). In this chapter, information is presented to characterize long-term trends in population, municipal wastewater infrastructure and effluent loading of pollutants, ambient water quality, environmental resources, and uses of the Ohio River. Data sources include USEPA's national water quality database (STORET), published technical literature, and unpublished technical reports ("grey" literature) obtained from the Ohio River Valley Sanitation Commission (ORSANCO) and other local agency sources.

The ORSANCO district encompasses three-quarters of the basin, accounting for 155,000 square miles of the Ohio River watershed. The district contains nearly one-tenth of the nation's population in one-twentieth of the nation's continental area. Ten percent of the people in the watershed receive their water supply from the Ohio River. Population densities in the ORSANCO district range from less than 50 people per square mile in the southwest to more than 600 in the eastern urban centers. Land use in the area is primarily agricultural, but concentrations of industry, coal mining, and oil and gas drilling are present throughout the region. In addition to agricultural and industrial uses, the Ohio River supports fish and wildlife habitats, water-based recreation, navigation, and power generation.

Utility of the Ohio River had significantly declined by the 1930s as the result of rising discharges of raw sewage and untreated industrial waste. Widespread public concern was spurred by drought-induced epidemics in 1930 and continually high levels of bacterial pollution. Citizens of the Ohio River Valley proposed a regional approach to water quality management in the form of an interstate compact. Eight states joined the Ohio River Valley Sanitation Commission in 1948, setting a precedent for

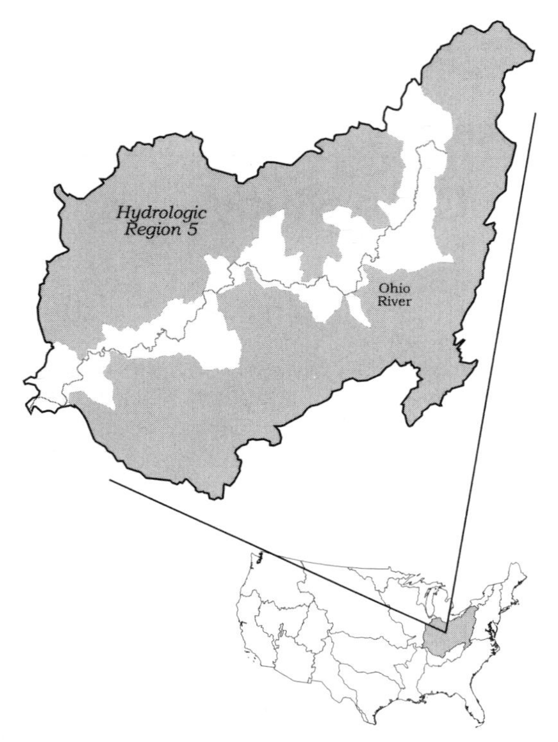

Figure 11-1 Hydrologic Region 5 and the Ohio River Basin.

cooperation among state, local, and private interests and the federal government for unifying waste management within individual watersheds. The benefits of pollution control standards implemented through this region-wide compact have been significant to the overall condition of waterways in the Ohio River Basin.

PHYSICAL SETTING AND HYDROLOGY

Nineteen major tributaries discharge to the Ohio River (Figure 11-2). The 155,000-square-mile ORSANCO district originates on the western slopes of the Appalachian Mountains, with the Allegheny River flowing into the Ohio River from the northwest and the Monangahela River from the south. The southwestern portion of the district is characterized by rolling hills and wide valleys, and the northwest is level or gently rolling. The elevation of the Ohio's riverbed drops 429 feet from the headwaters to the mouth at the confluence with the Mississippi River, with flow in the drainage basin generally toward the southwest. The ORSANCO district is approximately 700 miles long and has an average width of 220 miles. Rainfall in the basin averages 45 inches, and the average annual discharge of the Ohio River into the Mississippi River

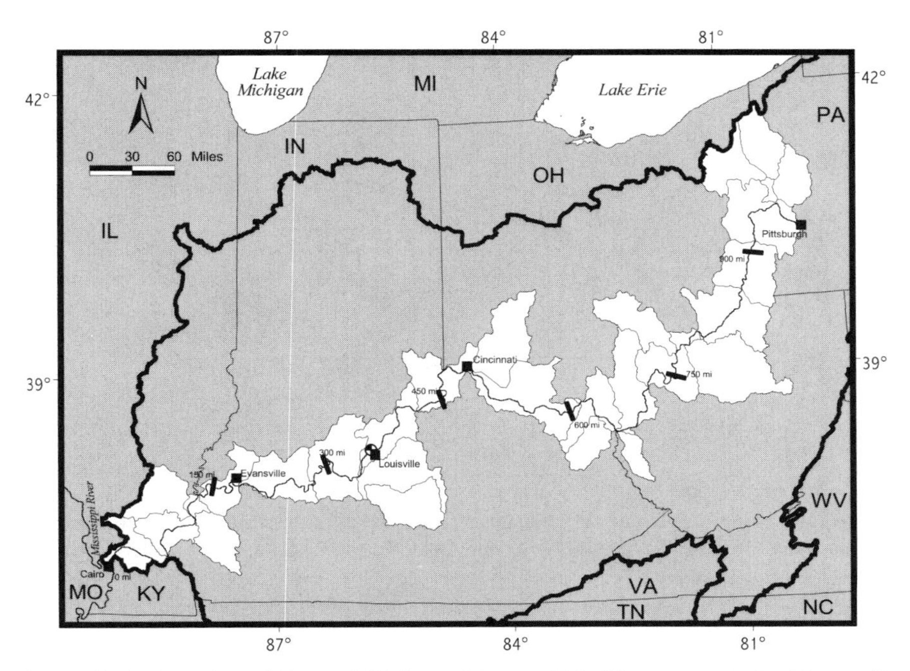

Figure 11-2 Location of Upper, Middle, and Lower Ohio River watersheds. (River miles shown are distances from confluence of Ohio River with Mississippi River at Cairo, Illinois.)

is 258,000 cfs. Variations in rainfall, temperature, vegetation coverage, and snow storage have historically caused wide ranges of runoff and streamflows. Low-flow conditions usually occur in July through November; the monthly average, taken at Louisville, Kentucky, ranges from 33,853 cfs in September to 239,613 cfs in March. Figures 11-3 and 11-4 show summer average flows (July–September) and monthly average flows over the 55-year period from 1940 to 1995.

Canalization of the entire Ohio River and some of its tributaries was achieved by 1929, converting the river into a series of backwater pools. The original system of submergible wicket dams has been almost completely replaced by high-lift permanent dams (Tennant, 1998).

POPULATION, WATER, AND LAND USE TRENDS

The Ohio River basin continues to be one of the most important agricultural and industrial centers of the nation. Population in the ORSANCO district has increased steadily over the past few decades, and use of the water resources has increased with the development of the basin. More than 3,700 municipalities, more than 1,800 industries, and three major cities—Louisville, Cincinnati, and Pittsburgh—depend on

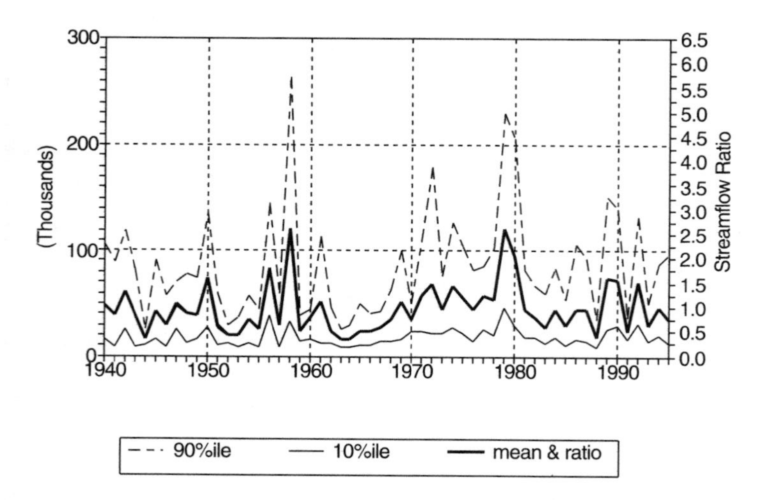

Figure 11-3 Long-term trends in mean, tenth, and ninetieth percentile statistics computed for summer (July–September) streamflow in the Ohio River. (USGS Gage 03294500 at Louisville, Kentucky.) *Source:* USGS, 1999.

the Ohio River Valley. The Ohio River case study area includes a number of counties identified by the Office of Management and Budget (OMB) as Metropolitan Statistical Areas (MSAs) or Primary Metropolitan Statistical Areas (PMSAs). Table 11-1 lists the MSAs and counties included in this case study. Figure 11-5 presents long-term population trends (1940–1996) for the counties listed in Table 11-1. From 1940

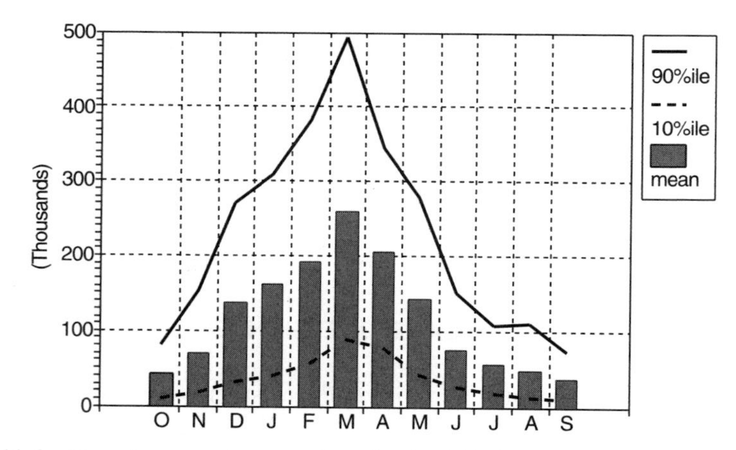

Figure 11-4 Monthly trends in streamflow for the Ohio River. Monthly mean, tenth, and ninetieth percentiles computed for 1951–1980. (USGS Gage 03294500 at Louisville, Kentucky.) *Source:* USGS, 1999.

TABLE 11-1 Metropolitan Statistical Areas (MSAs) and Counties in the Ohio River Basin Case Study

MSA	County
Wheeling, WV–OH MSA	Belmont County, OH
	Marshall County, WV
	Ohio County, WV
Steubenville–Weirton, OH–WV MSA	Jefferson County, OH
	Brooke County, WV
Huntington–Ashland, WV–KY–OH MSA	Boyd County, KY
	Carter County, KY
	Greenup County, KY
	Lawrence County, OH
	Cabell County, WV
	Hancock County, WV
	Wayne County, WV
Cincinnati–Hamilton, OH–KY–IN CMSA	Dearborn County, IN
	Ohio County, IN
	Boone County, KY
	Campbell County, KY
	Kenton County, KY
	Pendleton County, KY
	Brown County, OH
	Clermont County, OH
	Hamilton County, OH
	Warren County, OH
	Butler County, OH
Louisville, KY–IN MSA	Clark County, IN
	Floyd County, IN
	Harrison County, IN
	Scott County, IN
	Bullitt County, KY
	Jefferson County, KY
	Oldham County, KY
Evansville–Henderson, IN–KY MSA	Posey County, IN
	Vanderburgh County, IN
	Warrick County, IN
	Henderson County, KY

Source: OMB, 1999.

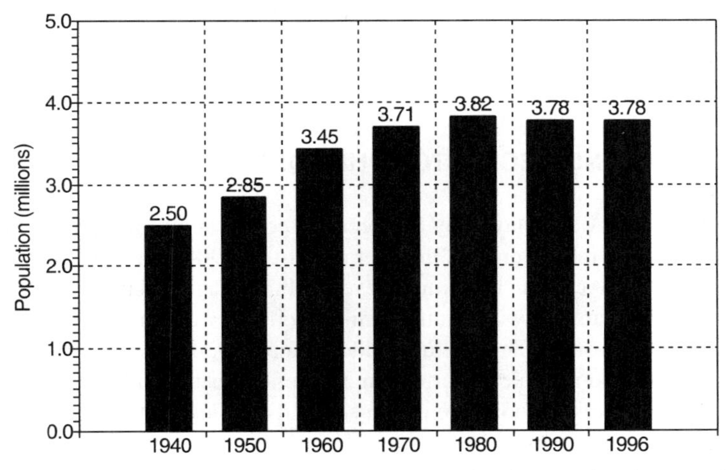

Figure 11-5 Long-term trends in population in the Ohio River Basin. *Sources:* Forstall, 1995; USDOC, 1998.

to 1996, the population in the Ohio River case study area increased by more than 50 percent (Forstall, 1995; USDOC, 1998). Agriculture continues to be the dominant land use in the area, although extensive mining is conducted in the watershed; 70 to 80 percent of the national total amount of bituminous coal and a significant amount of natural gas and oil are present in the basin.

The Ohio River supports navigation, power generation, industrial cooling and processing, warm-water aquatic habitats, public water supplies, and recreation. Because the river serves as a water source to industries, agricultural lands, and more than 3.5 million people, and as a waste receptacle for far larger numbers, the river's environment has been placed in a fragile balance.

HISTORICAL WATER QUALITY ISSUES

Growing concern for the deteriorating environmental conditions in the Ohio River peaked in the early 1930s, when serious drought turned many slackwater pools into virtual cesspools and a series of epidemics plagued cities along the Ohio River. Costs of water treatment increased dramatically from 1921 to 1934 as a result of an estimated 80-fold increase in the bacteria levels present in the river. In 1936, Congressman Brent Spence testified at a congressional hearing on the pollution of navigable waters that "the Ohio River is a cesspool." At the same hearing, the State Health Commissioner of Kentucky added that "the Ohio River, from Pittsburgh to Cairo, is an open sewer." In 1939, the city of Marietta, Ohio, was forced to change its water supply source from the Ohio River to wells and the Muskingum River as pollution levels in the river became untreatable. In 1951, only 39 percent of the sewered population was served by community treatment facilities. Sections of the Ohio River still suffered oxygen depletion so severe that aquatic life could not survive and pollution, bacteria levels, taste, and odor made large sections of the Ohio River unsuitable for most uses.

LEGISLATIVE AND REGULATORY HISTORY

Large-scale action was delayed by the need for cooperation throughout the basin to achieve significant improvements in water quality. In 1908, the Ohio state legislature adopted the Bense Act, which exempted every Ohio village and municipality from installing sewage treatment works until similar facilities were provided by all municipalities upstream from it. This attitude endured until 1924, when the Ohio River Valley Negotiating Committee reported an agreement between industries and state health commissioners to cooperate in carrying out a policy for the conservation of interstate streams. Congress authorized the states to negotiate the compact in 1936 and approved the resulting document in 1940. In June of 1948, the Federal Water Pollution Control Act (Public Law 80-845) was passed and the ORSANCO Compact was signed by Illinois, Indiana, Kentucky, New York, Ohio, Pennsylvania, Virginia, and

West Virginia, setting a precedent for cooperation among federal agencies, state governments, municipalities, and industries. Soon after, wastewater treatment standards were enacted for the Cincinnati pool. Bacterial quality objectives for the Ohio River were established in 1951, and an assessment of potential health hazards from trace constituents in wastewater was initiated. By 1954, municipal wastewater treatment standards for the Ohio River had been established. In relation to the industrial dischargers, a resolution adopted in 1959 placed responsibility on industries for reporting spills and accidental discharges to state agencies.

Following the 1965 Federal Water Quality Act, ORSANCO adopted stream water quality recommendations. In 1970, ORSANCO Pollution Control Standard 1-70 revised the pollution control standards established in 1954, making secondary treatment the minimum requirement for wastewater treatment plants and establishing equivalent treatment requirements for industry. From 1957 to 1965, \$82,786,500 in federal aid was allocated to 638 projects in the Ohio Valley. The communities matched every federal dollar with \$2.50 of local funds for a total of \$282,966,000 spent on improving conditions. The majority of treatment works, both in place and under construction during this time, were equipped for secondary treatment. For 3 years before federal aid was offered, Pennsylvania provided incentives for smaller communities to upgrade their treatment by offering funds to communities upon compliance with standards. Although the population served by municipal facilities has increased greatly under these programs (Figure 11-6), increasingly high water quality criteria and limited funds have caused a sharp increase in population served by facilities classified as inadequate between 1965 and 1990.

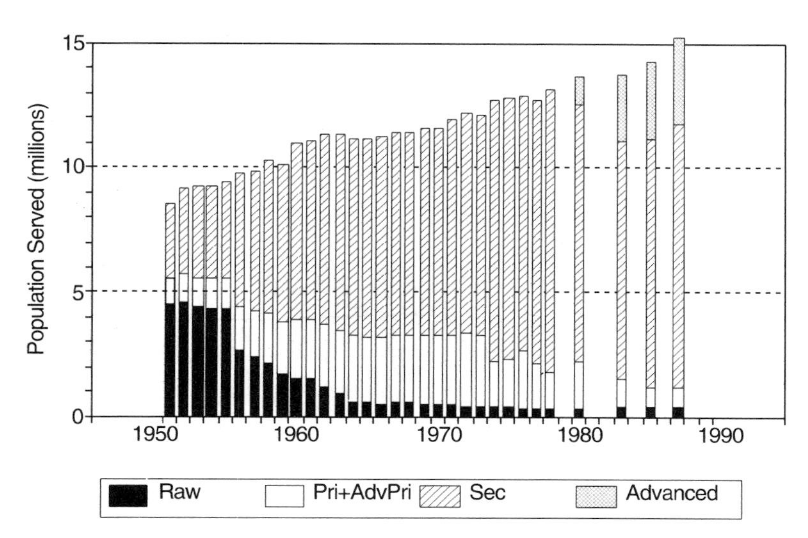

Figure 11-6 Long-term trends in population served by municipal wastewater treatment plants in the ORSANCO District. *Sources:* ORSANCO, 1978, 1988.

IMPACTS OF WASTEWATER TREATMENT

Pollutant Loading and Water Quality Trends

Following the 1948 advances in cooperative management, water quality conditions in the Ohio River began to improve. A dramatic decrease occurred in the discharge of raw sewage from 1950 to 1963 (Figure 11-6). As a result of the stringent permit requirements on dischargers and improvements in wastewater treatment facilities implemented in the late 1960s and 1970s, even more advances have been made to upgrade wastewater treatment plants. Levels of BOD_5 effluent loading have decreased significantly, even as the influent loading continues to increase as population increases (Figure 11-7). Corresponding to the decreasing levels of pollutant loading is the increased amount of DO available to support aquatic organisms. Figure 11-8 shows the typical oxygen sag curve observed during the mid-1960s downstream from Cincinnati, Ohio, while Figures 11-9, 11-10, and 11-11 indicate long-term trends of DO at various sampling locations. These data clearly illustrate an overall increase in oxygen following the 1972 CWA requirement for secondary treatment. A remarkable improvement in oxygen concentration occurs in the critical minimum occurring near North Bend/Fort Miami (River Mile 490) and at the pool formed by Markland Lock/Dam (River Mile 449–453). During the 1988 drought, for example, levels of DO continued to meet standards near Cincinnati and Louisville, in contrast to the mid-1960s when consistent low-flow conditions resulted in DO concentrations below water quality standards (see Figures 11-9 and 11-13). Using the data compiled for

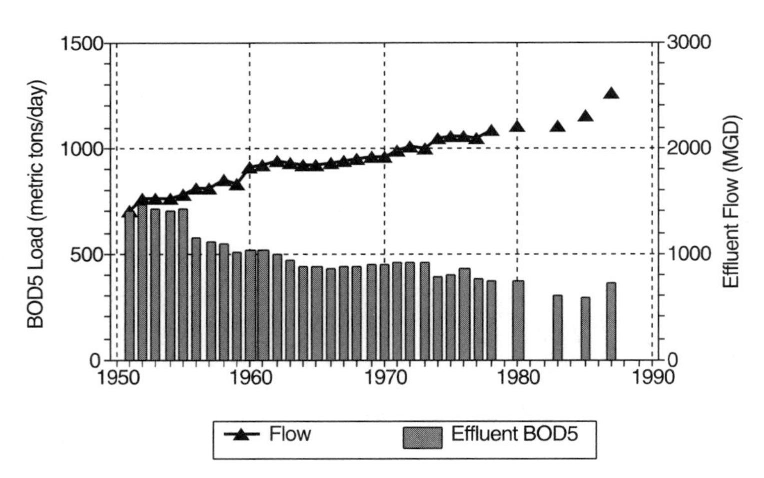

Figure 11-7 Long-term trends of wastewater flow, influent and effluent BOD_5 for the ORSANCO District. Data based on population served with 165 gallons per person per day, influent BOD_5 of 215 mg/L, and removal efficiencies of 36 percent (primary), 85 percent (secondary), and 95 percent (tertiary). *Sources:* ORSANCO, 1978, 1987.

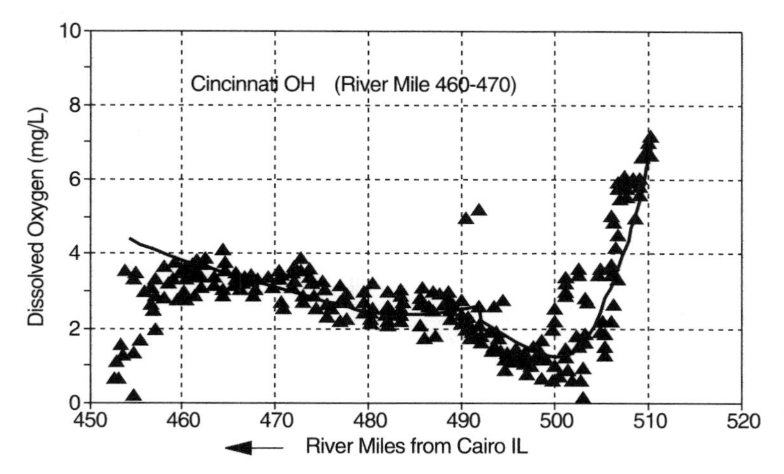

Figure 11-8 Spatial distribution of DO along the Ohio River downstream of Cincinnati: October–November 1963. *Source:* Hydroscience, 1969.

trends in DO near Cincinnati (Figure 11-9) and Louisville (Figure 11-12), the mean summer tenth percentile level of DO significantly improved after the CWA (1986–1995) in comparison to conditions before the CWA (1961–1970) (Figure 11-13).

Water quality data collected since the 1950s indicate increased compliance with federal and ORSANCO criteria for DO, BOD_5, turbidity, pH, and many other water

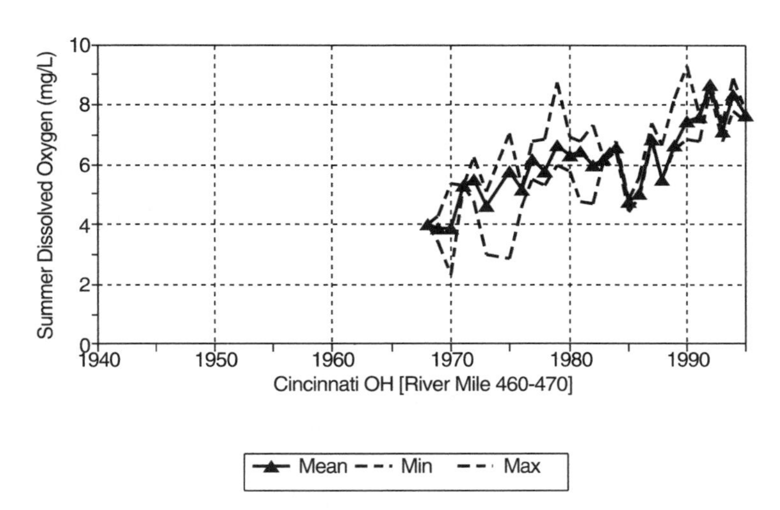

Figure 11-9 Long-term trends of DO near Cincinnati, Ohio (River Miles 460–470) (RF1–05090203002). *Source:* USEPA STORET.

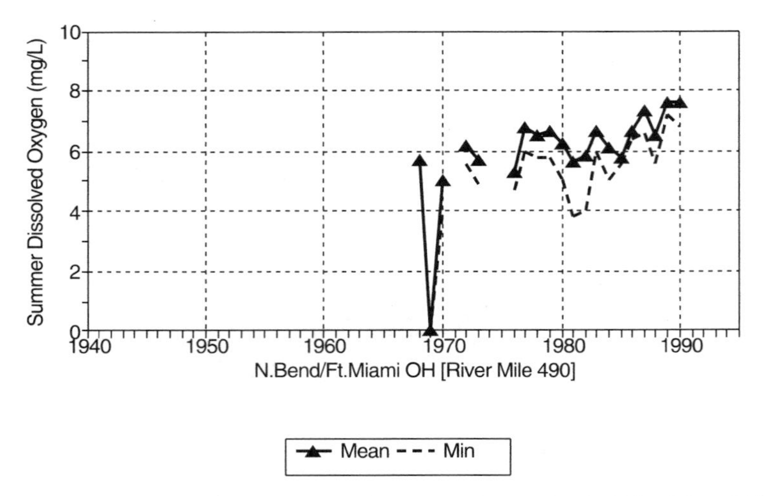

Figure 11-10 Long-term trends of DO at North Bend/Ft. Miami, Ohio (River Mile 490). (RF1–05090203012) *Source:* USEPA STORET.

quality factors (Cleary, 1978; Wolman, 1971). ORSANCO (1979) reports greater than 98.8 percent compliance for 15 out of 20 examined stream quality criteria. In 1990, ORSANCO published a statistical analysis of data resulting from water quality monitoring conducted over an 11-year period. Decreasing trends at individual sampling points were reported for a majority of the contaminants examined, and overall improving trends are indicated for total phosphorus, ammonia, nitrogen, copper, lead,

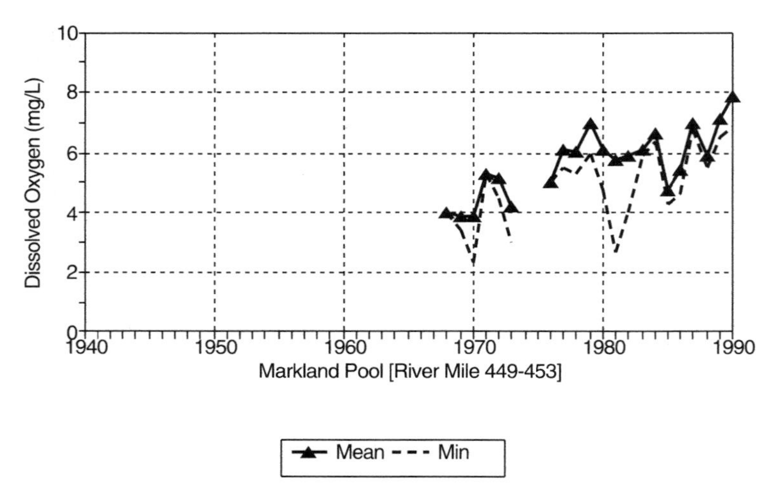

Figure 11-11 Long-term trends of DO at Markland Lock & Dam, Kentucky (River Miles 449–453) (RF1–05140101010). *Source:* USEPA STORET.

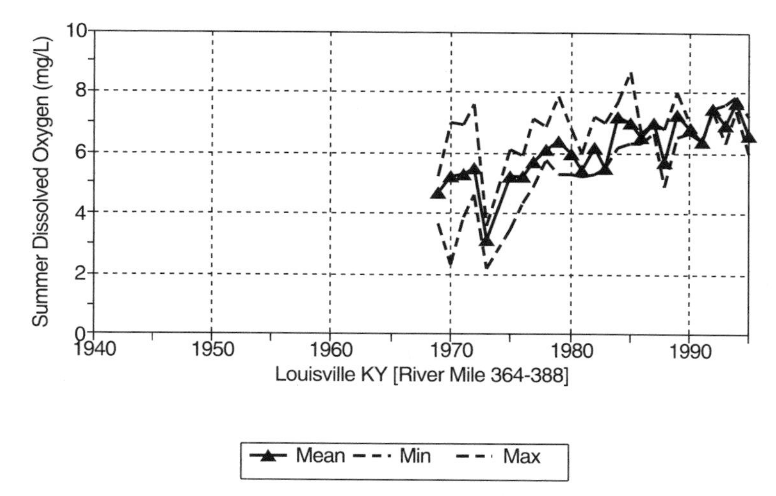

Figure 11-12 Long-term trends of DO at Louisville, Kentucky (River Miles 364–388) (RF1-05140101001). *Source:* USEPA STORET.

and zinc. An indication of the improving water quality in the Ohio River is the marked increase in diversity of fish species, with the greatest improvement seen in the upper reaches of the river (Figure 11-14). Increases are primarily noted in sport and commercially valuable species, which tend to be more pollution-sensitive than other fish species.

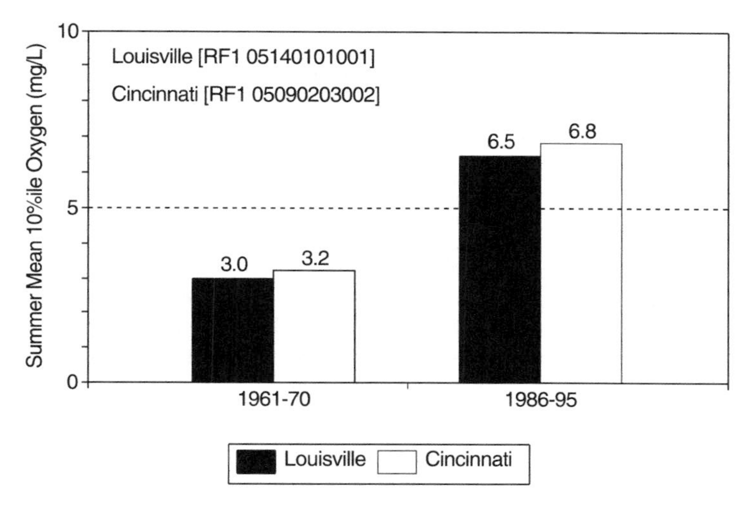

Figure 11-13 Before and after comparison of summer mean tenth percentile DO near Louisville, Kentucky (River Miles 364–368) and Cincinnati, Ohio (River Miles 460–470) during 1961–1970 and 1986–1995. *Source:* USEPA STORET.

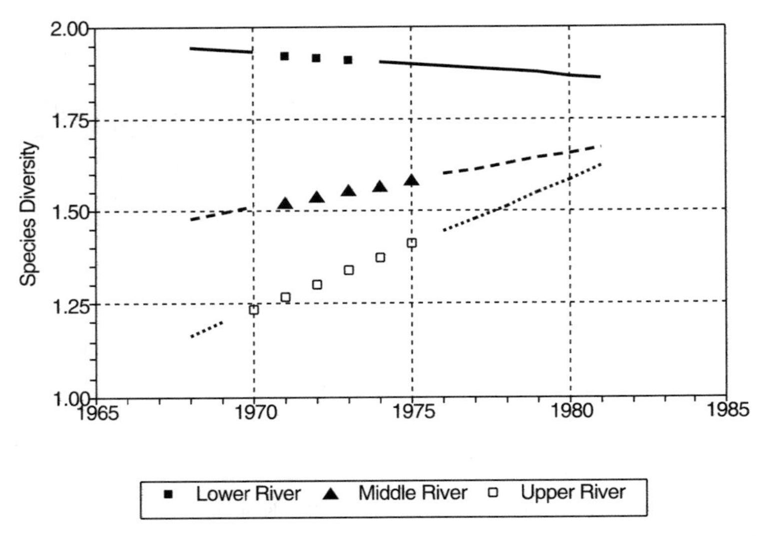

Figure 11-14 Long-term trends in fish diversity in the Ohio River. *Source:* ORSANCO, 1982.

Recreational and Living Resources Trends

There is little long-term information on biological trends in the Ohio River (Pearson, 1992). Information on plants, invertebrates, and plankton is scarce or nonexistent. The only historical population data are for mussels, which were diverse and abundant in the 1800s but are less so now, even with water quality improvements in the river.

Data on fish populations in the middle section of the Ohio River have been collected since the 1950s and indicate that the populations have responded more positively than mussels to improved water quality (Figure 11-15). The first comprehensive fish population study on the Ohio River was done by ORSANCO in 1957, and the study has continued almost yearly since then. The study reports fish data according to section of the river—upper, middle, and lower. Louisville and Cincinnati are located in the middle section of the river. Changes in fish diversity since the study began have been most dramatic in the upper river, where a 40 percent increase has been measured, but diversity has increased by 13 percent in the middle section as well (ORSANCO, 1982). Numbers of species and overall fish biomass are still increasing in the middle section of the river, though they have not returned to their original levels. ORSANCO attributes the improvements to increased DO concentrations and pH, and to decreased levels of toxic materials in the river (ORSANCO, 1982).

Other studies also indicate continuing improvements in the quality of the Ohio River habitat. Studies by Geo-Marine conducted in the early 1980s near North Bend, Ohio (about 30 miles downstream of Cincinnati), found increasing numbers of species of larval fish, a life stage generally sensitive to DO levels (Geo-Marine, 1986). A trend toward a more even distribution of the numbers of individuals among the spe-

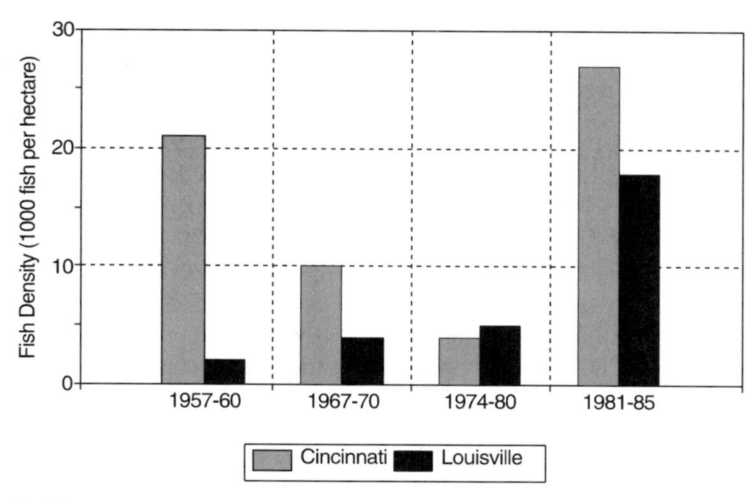

Figure 11-15 Long-term trends in Ohio River fish abundance at Cincinnati and Louisville. *Source:* W. D. Pearson, Water quality in North American river systems, eds. C. D. Becker and D. A. Neitzel, © 1992, Reprinted with permission: Battelle Press, Columbus, OH.

cies captured was found as well, indicating improved habitat quality. The Ohio EPA has also conducted fish studies along the river. Their studies have found an Index of Biotic Integrity (IBI) near North Bend between 46 and 48 (Ohio EPA, 1992). This is a fair to good rating, indicating habitats where tolerant and intolerant benthic species are both found. The IBI, developed by Karr (1981) for use in midwestern streams, has been adapted for use in other areas. This index is a composite of 12 ecological attributes of fish communities, including species richness, indicator taxa (both intolerant and tolerant), trophic guilds, fish abundance, and incidence of hybridization, disease, and abnormalities (Karr et al., 1986). IBI scores range from a low of 12 to a high of 60. A score of 12 is assigned to conditions where no fish are present even after repeated sampling; a score of 60 is assigned to conditions comparable to the best habitats without human disturbance (Karr et al., 1986). Benthos are particularly good indicators of long-term trends in water quality because the species are generally sedentary and have long life spans. For pollution studies, benthos are divided into three categories, and intolerant species are indicative of good water quality because of their inability to survive in, or intolerance of, low DO concentrations. Ohio EPA's sampling at North Bend in 1991 found a total of 23 species, with one intolerant species among them (Sanders, 1992a, 1992b; Plafkin et al., 1989).

Water quality improvements in the Ohio River have benefited both commercial and sport fisheries (Figure 11-15). Sportfishing, important recreationally and for tourism, began returning to the river in the mid-1980s. In 1982, the Bass Anglers Sportsman's Society held the Bass Champs Invitational at Cincinnati because of the reported bass catch in the river (ORSANCO, 1981). Such contests are now commonly held along the river.

SUMMARY AND CONCLUSIONS

Significant improvements have been accomplished throughout the Ohio River Basin through the combined efforts of federal, state, and local governments. The last half of the twentieth century has seen a reversal of the previous trend of river degradation. As of the mid-1990s, nearly 94 percent of the Ohio River Basin's sewered population was served by at least secondary treatment. This accomplishment, on such a large scale, has shown what regional cooperation can achieve. The Ohio River now supports many uses that had previously been seriously impaired. Support of use for public water supply and aquatic habitat is maintained along the entire river. Sportfishing has returned, and the dramatic improvement in water quality is reflected in the increasing number of fishing tournaments along the river, including the Bass Masters Classic at Cincinnati.

Much progress has been made, but there is a recognizable need for further action. Water-based recreation continues to be impaired by high bacteria levels in the river. As of 1988, contact recreation was not supported on 59 percent of the river and was fully supported on only 6.5 percent of the river. Fish consumption advisories were still in effect for Kentucky, Ohio, Pennsylvania, and West Virginia in 1989, due to high levels of PCBs and chlordane found in fish tissues (ORSANCO, 1989a). Certain metals, organic compounds, cyanide, phenol, copper, zinc, oxygen, and temperature also continue to pose a problem. ORSANCO is considering how to address these and other stream quality impairments by addressing nonpoint source pollution controls (Norman, 1991), combined sewer overflow controls (Tennant et al., 1990), control of toxic chemicals (Vicory and Tennant, 1994), and control of ecological effects of hydropower development. Continued improvements are seen in monitoring, detection, and regulation, as well as treatment and spill response (Vicory and Tennant, 1993). The combination of present efforts with past achievements has put the Ohio River on the road to recovery.

REFERENCES

Cleary, E. J. 1978. Perspective on river quality diagnosis. J. WPCF 50(5): 825–832.

Forstall, R. L. 1995. Population by counties by decennial census: 1900 to 1990. U.S. Bureau of the Census, Population Division, Washington, DC. http://www.census.gov/population/www/censusdata/cencounts.html

Geo-Marine. 1986. 1985 Ohio River ecological research program. Adult and juvenile fish and ichthyoplankton studies. Report prepared for Ohio Edison Company, Ohio Valley Electric Corporation by Geo-Marine, Plano, TX, December.

Hydroscience. 1969. Water quality analysis for the Markland Pool of the Ohio River. Technical report prepared for Malcolm Pirnie Engineers, White Plains, NY by Hydroscience, Westwood, NJ.

Iseri, K. T. and W. B. Langbein. 1974. Large rivers of the United States. Circular No. 686. U.S. Department of the Interior, U.S. Geological Survey, Reston, VA.

Karr, J. R. 1981. Assessment of biotic integrity using fish communities. Fisheries 6(6): 21–27.

Karr, J. R., K. D. Fausch, P. L. Angermeier, P. R. Yant, and I. J. Schlosser. 1986. Assessing biological integrity in running waters: A method and its rationale. Special Publication 5. Illinois Natural History Survey, Champaign, IL.

Norman, C. G. 1991. Urban runoff effects on Ohio River water quality. Water Environ. Technol. 3(6): 44–46.

Ohio EPA. 1992. Ohio Environmental Protection Agency water quality planning and assessment Fish Information System (FINS): Fish survey data. Ohio Environmental Protection Agency, Columbus, OH.

OMB. 1999. OMB Bulletin No. 99-04. Revised statistical definitions of Metropolitan Areas (MAs) and guidance on uses of MA definitions. U.S. Census Bureau, Office of Management and Budget, Washington, DC.
http://www.census.gov/population/www/estimates/metrodef.html

ORSANCO. 1978. ORSANCO in review. Ohio River Valley Sanitation Commission, Cincinnati, OH.

ORSANCO. 1979. ORSANCO in review. Ohio River Valley Sanitation Commission, Cincinnati, OH.

ORSANCO. 1981. Annual report. Ohio River Valley Sanitation Commission, Cincinnati, OH.

ORSANCO. 1982. Annual report. Ohio River Valley Sanitation Commission, Cincinnati, OH.

ORSANCO. 1985. Annual report. Ohio River Valley Sanitation Commission, Cincinnati, OH.

ORSANCO. 1986. Annual report. Ohio River Valley Sanitation Commission, Cincinnati, OH.

ORSANCO. 1987. 1987 status of wastewater facilities. Ohio River Valley Sanitation Commission, Cincinnati, OH.

ORSANCO. 1988. ORSANCO: Forty years of service. Ohio River Valley Sanitation Commission, Cincinnati, OH.

ORSANCO. 1989a. Water quality of the Ohio River, Biennial assessment: 1988–1989. Ohio River Valley Sanitation Commission, Cincinnati, OH.

ORSANCO. 1989b. Annual report. Ohio River Valley Sanitation Commission, Cincinnati, OH.

ORSANCO. 1990. Water quality trends, Ohio River and its tributaries. Statistical analyses of data resulting from water quality monitoring conducted by ORSANCO. Ohio River Valley Sanitation Commission, Cincinnati, OH.

Pearson, W. D. 1992. Historical changes in water quality and fishes of the Ohio River. In Water quality in North American river systems, ed. C. D. Becker and D. A. Neitzel, pp. 207–231. Battelle Press, Columbus, OH.

Plafkin, J. L., M. T. Barbour, K. D. Porter, S. K. Gross, and R. M. Hughes. 1989. Rapid bioassessment protocols for use in streams and rivers. Benthic macroinvertebrates and fish. Results of pilot studies in Ohio and Oregon. EPA 444/4–89/001. U.S. Environmental Protection Agency, Office of Water, Assessment and Watershed Protection Division, Washington, DC.

Sanders, R. E. 1992a. Day versus night electrofishing catches from near-shore waters of the Ohio and Muskingum Rivers. Ohio J. Sci. 92(3): 51–59.

Sanders, R. E. 1992b. Ohio's near-shore fishes of the Ohio River: 1991–2000. (Year one: 1991 results). Prepared for the Ohio Department of Natural Resources, Division of Wildlife, Ohio Nongame & Endangered Wildlife Program, Columbus, OH.

Tennant, P. 1998. Ohio River Valley Sanitation Commission, Cincinnati, OH. Personal communication, August 21.

Tennant, P., C. Norman, and P. McConocha. 1990. Toxic-substance control for the Ohio River. Water Environ. Technol. 2(10): 59–63.

USDOC. 1998. Census of population and housing. Prepared by U.S. Department of Commerce, Economics and Statistics Administration, Bureau of the Census—Population Division, Washington, DC.

USEPA (STORET). STOrage and RETrieval Water Quality Information System. U.S. Environmental Protection Agency, Office of Wetlands, Oceans, and Watersheds, Washington, DC.

USGS. 1999. Streamflow data downloaded from the U.S. Geological Survey's National Water Information System (NWIS)-W. Data retrieval for historical streamflow daily values. http://waterdata.usgs.gov/nwis-w

Vicory, A. H., Jr., and P. A. Tennant. 1993. The Ohio River Valley Water Sanitation Commission and its activities. Ohio J. Sci. 93(2): 11.

Vicory, A. H., Jr., and P. A. Tennant. 1994. A strategy for monitoring the impacts of combined sewer overflows on the Ohio River. Water Quality International '94. Part 1: Combined Sewer Overflows and Urban Storm Drainage. Biennial Conference of the International Association on Water Quality, July 24–30, 1994, Budapest, Hungary.

Wolman, M. G. 1971. The nation's rivers. Science (174): 905–915.

CHAPTER 12

Upper Mississippi River Case Study

The upper and lower watersheds of the Mississippi River extend 2,340 miles from the headwaters in Lake Itasca, Minnesota, to the Gulf of Mexico. With a drainage basin of 1.15 million square miles, the Mississippi River, known as the "Father of Waters," drains 40 percent of the continental United States and discharges an annual average flow of 640,000 cfs into the Gulf of Mexico. On the basis of drainage area and mean annual discharge, the Mississippi is the largest river in the United States (Iseri and Langbein, 1974) and is ranked by annual discharge as the sixth largest river in the world (Berner and Berner, 1996). Figure 12-1 highlights the location of the seven catalog units of Accounting Unit 070102 for the Upper Mississippi River case study in the vicinity of Minneapolis–St. Paul in Minnesota. The Twin Cities are one of the nation's many major urban areas characterized by water pollution problems during the 1950s and 1960s (see Table 4-2; FWPCA, 1966; USPHS, 1951; 1953). Federal enforcement conferences were convened in 1964 and 1967 to investigate water pollution problems in the Minnesota and Wisconsin sections of the Upper Mississippi River (Zwick and Benstock, 1971).

In this chapter, data and information are presented to characterize long-term trends in population, municipal wastewater infrastructure and effluent loading of pollutants, ambient water quality conditions, environmental resources, and uses of the Upper Mississippi River in the vicinity of the Twin Cities. Data sources included STORET, USEPA's national water quality database, USGS streamflow records (USGS, 1999a), published literature, unpublished data, newsletters, and technical reports obtained from the Metropolitan Council Environmental Services (MCES) in St. Paul and from other state, local, and federal agencies. Data have also been obtained from a validated water quality model of the Upper Mississippi River (Lung and Larson, 1995) to identify the progressive improvements in dissolved oxygen and other water quality parameters attributed to upgrades of the Metropolitan Wastewater Treatment Plant in St. Paul from primary to secondary and advanced secondary with nitrification (Lung, 1998).

The Twin Cities of Minneapolis and St. Paul are the major urban centers for more than 1,100 miles along the Mississippi River upstream of St. Louis, Missouri. About one-third of the population and a majority of the commercial and industrial activity of Minnesota are located within the Twin Cities metropolitan region. Outside the Twin Cities, the Upper Mississippi watershed is primarily rural and forested, with the population dispersed in small towns and farms. The glaciated topography of the wa-

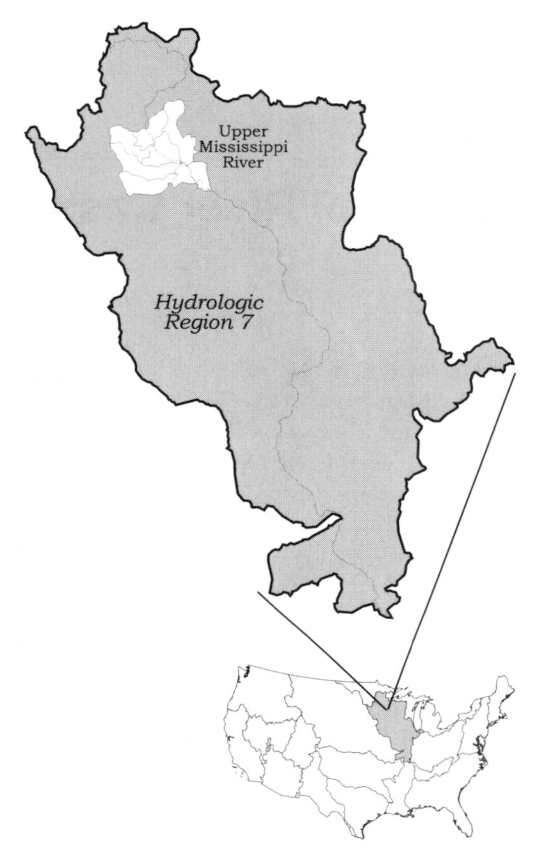

Figure 12-1 Hydrologic Region 7 and the Upper Mississippi River basin near Minneapolis–St. Paul, Minnesota.

tershed provides extensive habitat for fish and wildlife and also supports an economy historically based on agriculture and wood products. In addition to these economic sectors, industrial and manufacturing activities have become significant components of the overall economy.

PHYSICAL SETTING AND HYDROLOGY

The Upper Mississippi River basin (Hydrologic Region 7) covers a drainage area of 171,500 square miles over a reach of 1,170 miles from the headwaters in Lake Itasca to the confluence of the Missouri River with the Mississippi River at Alton, Illinois, just upstream of St. Louis, Missouri (Iseri and Langbein, 1974) (Figure 12-1). The water quality of the Upper Mississippi River has historically been dominated by wastewater loading from the Twin Cities, as well as sediments, nutrients, pesticides, oxidizable materials, and other pollutants from the Minnesota River basin. The watershed of the Upper Mississippi River basin (Catalog Unit 07010206) described in this case study includes a drainage area of 8,520 square miles, extending 83 miles from the confluence of the Crow River (UM milepoint 894) in Morrison County upstream of Anoka, Minnesota (UM milepoint 871) to the confluence of the St. Croix River downstream of Lock and Dam No. 2 at Prescott, Wisconsin (UM milepoint 811) (Figure 12-2).

Characterized by rolling hills and plains with numerous lakes, the basin topography reflects the effects of successive glacial advances over the region. Upstream of the Twin Cities, the major tributaries to the Upper Mississippi are the Minnesota River, the Rum River at Anoka, and the Crow River. Within the portion of the watershed influenced by wastewater loading from the Twin Cities, five locks and dams have been constructed for flood control, navigation, and hydropower purposes. Because of the flow-regulating nature of the series of locks and dams, the river essentially flows as a series of controlled backwater pools with relatively constant surface elevations. Over the 69-mile reach from Coon Rapids Dam upstream of Minneapolis

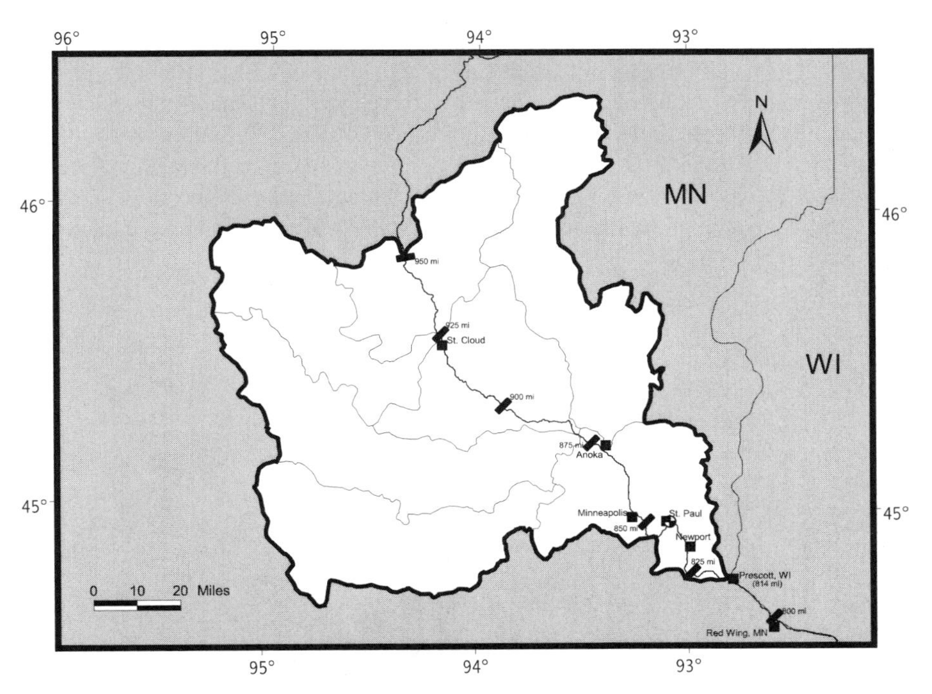

Figure 12-2 Location map of Upper Mississippi River (Accounting Unit 070102) near Minneapolis–St. Paul, Minnesota. (River miles shown are distances from confluence of Mississippi River with Ohio River at Cairo, Illinois.)

(UM river mile 866) to Lock and Dam No. 3 at Red Wing, Minnesota (UM river mile 797), the river drops from an elevation of 830 feet to 661 feet above mean sea level (Hydroscience, 1979).

The series of locks and dams, supplemented by dredging, maintain a 9-foot-deep navigation channel for commercial barge traffic. The navigation channel was authorized by the U.S. Congress in 1928, and the locks and dams were authorized in 1930. The U.S. Army Corps of Engineers is conducting a controversial environmental study assessing the impact of the lock and dam system on the ecological integrity of the Upper Mississippi River. In addition to the ecological effects of the flow control structures, the Great Flood of 1993 (Wahl et al., 1993) has generated investigations of the role that artificial drainage and flood-control structures might have played in actually increasing the extent of severe flooding in some areas of the watershed.

On a seasonal basis, streamflow of the Upper Mississippi River reflects peak precipitation during late spring snowmelt and early summer with severe subfreezing winter conditions (Figure 12-3). Although the minimum flow occurs during winter due to a reduction in watershed runoff as precipitation changes from rain to snow and ice, the critical period for water quality problems is during the low-flow, summer months. On the river itself, winter ice cover is intermittent, varying considerably both spatially and temporally. Ponded areas of the Upper Mississippi River, such as Lower Pool 2 and Lake Pepin (Pool 4), have permanent ice cover for about 3 months during the winter; the more riverine reaches freeze over only during extended periods of severe cold. DO levels are generally high during winter because of very low water temperature and open water conditions that allow oxygen exchange across the air-water interface. Reliable streamflow records from a USGS gage 300 feet upstream of the Roberts Street Bridge in St. Paul (UM milepoint 839.3) are available from 1892 to the present to characterize long-term monthly, annual, and extreme flow statistics over a drainage area of 36,800 square miles (USGS, 1999a). Based on the historical

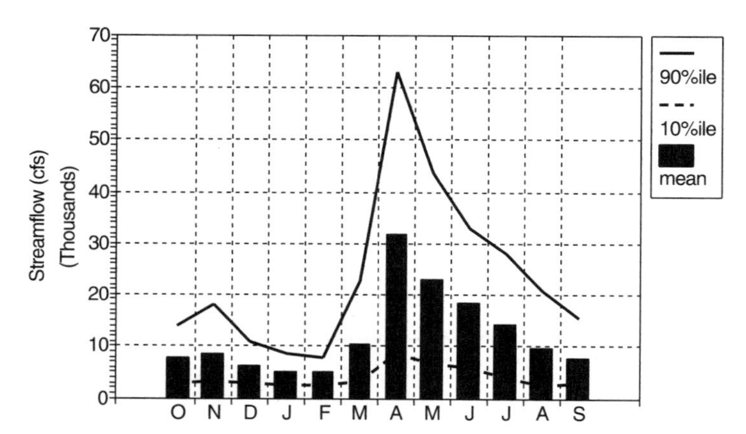

Figure 12-3 Monthly mean, tenth, and ninetieth percentile streamflow (1951–1980) at St. Paul, Minnesota (USGS Gage 05331000). *Source:* USGS, 1999a.

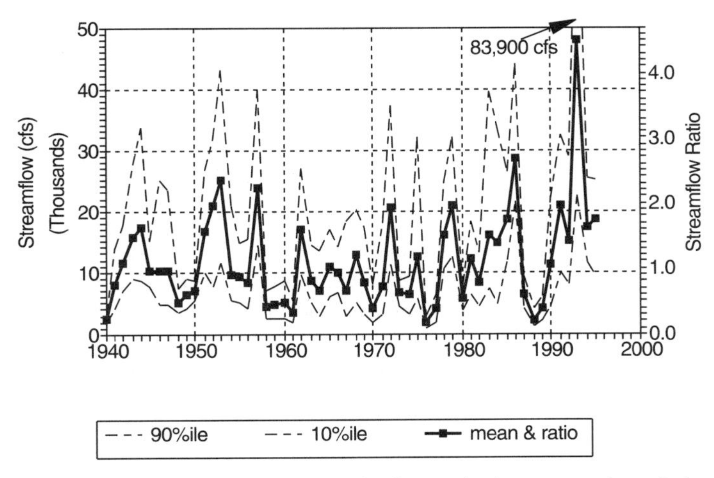

Figure 12-4 Long-term trends of summer (July–September) mean, tenth, and ninetieth percentile streamflow at St. Paul, Minnesota (USGS Gage 05331000). *Source:* USGS, 1999a.

data recorded for water years 1892–1998, monthly flow ranges from a maximum of 26,060 cfs in April to a winter minimum of 4,544 cfs during February and a summer low of 8,060 cfs during September. Over the period of record from 1892 to 1998, annual average discharge of the Upper Mississippi River at St. Paul has been 11,630 cfs, with the lowest daily mean flow of 632 cfs recorded on August 26, 1934, and the highest daily mean flow of 171,000 cfs observed on April 16, 1965 (USGS, 1999a). Using historical records from 1936–1979 to represent streamflow variability after the series of locks and dams were constructed on the Upper Mississippi River, the 7-day, 10-year flow (7Q10) for summer conditions (June–September) at St. Paul is reported as 1,633 cfs (MPCA, 1981).

The long-term (1940–1995) interannual variation of mean, tenth, and ninetieth percentile summer (July–September) streamflow is shown in Figure 12-4. The historical record exhibits pronounced year-to-year variability of summer streamflow. Based on data from 1951–1980, the long-term mean summer (July–September) flow of 10,659 cfs is used to compute a normalized streamflow ratio for each summer from 1940 to 1995 as dry (< 0.75), normal (0.75–1.50) or wet (> 1.50). For example, the summers of 1962, 1972, 1978–1979, 1983, 1985–1986, 1991, and 1993–1995 were all characterized by wet conditions, where the flow was greater than 150 percent of the long-term summer mean. The summers of 1960–1961, 1964, 1967, 1970–1971, 1973–1974, 1976–1977, 1980, and 1987–1989, in contrast, were characterized by dry conditions, where the flow was less than 75 percent of the long-term summer mean. The extreme droughts of 1976 (1,725 cfs, 16 percent of summer mean) and 1988 (2,334 cfs, 22 percent of summer mean) and the Great Flood of 1993 (47,789 cfs, 450 percent of summer mean) are particularly noticeable in the 55 years of historical records for the Upper Mississippi River at St. Paul.

POPULATION, WATER, AND LAND USE TRENDS

Beginning in 1838, when the Twin Cities area was first opened for settlement, the abundant land and water resources attracted homesteaders. The confluence of the Minnesota River and the Upper Mississippi River served as an important transportation link between military and trading posts and the growing towns and cities along the Mississippi River. St. Anthony's Falls provided a natural source of power for lumber and grist mills. The fertile soil supported an agricultural economy, and the vast forests provided resources for a growing wood products industry. Uses for the Upper Mississippi River have included municipal and industrial water supply, commercial navigation, log transport, commercial fishing, hydropower, and water-based recreational activities. Beginning with the construction of an urban sewer system in 1871, the Upper Mississippi River has also been used for wastewater disposal.

As a component of the Lake Pepin Phosphorus Study, conducted from 1994 to 1998, historical records of land uses, agricultural practices (e.g., manure applications and commercial fertilizer uses), and wastewater discharges, compiled beginning ca. 1860, were used to correlate long-term changes in land uses in the watersheds of the Upper Mississippi River, the Minnesota River, and the St. Croix River with long-term changes in sediment and phosphorus loading to Lake Pepin (Mulla et al., 1999). As of the mid-1990s, the USGS (1999b) had classified about 60 percent of the watershed (Upper Mississippi River, Minnesota River, and St. Croix River) as agriculture and 23 percent as forest. The remaining 17 percent of the drainage basin was classified as urban and suburban (5 percent), water (5 percent), and wetlands (7 percent) (USGS, 1999b).

The Upper Mississippi River case study area includes a number of counties identified by OMB (1999) as a Metropolitan Statistical Area (MSA) (Table 12-1). Long-term trends in the population of the 13-county Minneapolis–St. Paul MSA are shown in Figure 12-5. Resident population in this MSA increased by 150 percent, from 1.1 million

TABLE 12-1 Metropolitan Statistical Area (MSA) Counties in the Upper Mississippi River Case Study

County
Anoka County, MN
Carver County, MN
Chicago County, MN
Dakota County, MN
Hennepin County, MN
Isanti County, MN
Ramsey County, MN
Scott County, MN
Sherburne County, MN
Washington County, MN
Wright County, MN
Pierce County, WI
St. Croix County, WI

Source: OMB, 1999.

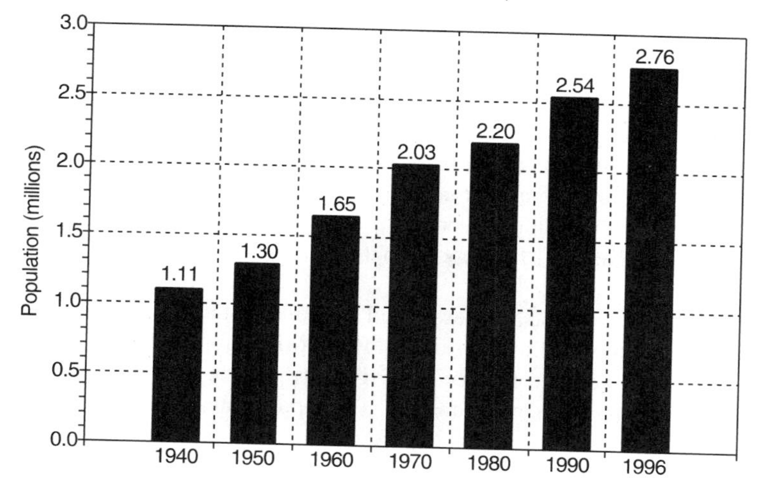

Figure 12-5 Long-term trends in population for the Minneapolis–St. Paul, Minnesota-Wisconsin MSA counties for the Upper Mississippi River case study area. *Sources:* Forstall, 1995; USDOC, 1998.

in 1940 to 2.76 million in 1996. After a small increase of population from 1940 to 1950, the greatest rate of growth occurred during the 1950s and 1960s, when population increased by 23 to 27 percent. The rate of population growth then declined during the 1970s to 8.5 percent, with an increase to 15 percent during the 1980s. During the period from 1990 to 1996, population increased by 9 percent (Forstall, 1995; USDOC, 1998). Reflecting population growth in the Twin Cities, the population served by the Metro wastewater plant increased from 1.04 million in 1962 to 1.68 million in 1997. By 2020, the plant is expected to provide service to 1.94 million people (Larson, 1999).

HISTORICAL WATER QUALITY ISSUES

As with many other urban areas of the United States, the Upper Mississippi River was grossly polluted early in the twentieth century because of growing urban populations and inadequately treated municipal and industrial wastewater discharges. Municipal officials simply relied on the natural flushing of rivers to dilute the human and industrial waste products of the growing metropolitan areas. City sewers, first constructed in 1871 in the Twin Cities, collected stormwater and sewage and discharged them directly into the river. By the early 1900s, the Upper Mississippi River was unable to biologically assimilate the untreated wastewater collected from the Twin Cities (MWCC, 1988).

Before construction of a lock and dam in Minneapolis in 1917, annual peak spring flows maintained a minimally acceptable degree of water quality by the physical removal of raw sewage and other waste materials accumulated during the previous year in the Twin Cities area. Construction of the lock and dam, however, drastically altered this natural cycle by slowing the current of the river and reducing the flushing effect of the peak spring flows. By 1920, 3 million cubic yards of sewage sludge had accumulated in the pool created by the lock and dam. Water quality was severely degraded by depletion of dissolved oxygen from decomposition of the sludge bed. Bacteria levels were extremely high, sewage sludge mats floated on the surface, and the river was noxious from hydrogen sulfide gas caused by septic conditions during the warm summer months. The Upper Mississippi River was grossly polluted for a distance of 30 miles from St. Anthony's Falls in Minneapolis to the St. Croix River at Prescott, Wisconsin (MWCC, 1988).

A 1928 joint report by the Minnesota and Wisconsin State Boards of Health stated that "a zone of heavy pollution extends from Minneapolis to the mouth of the St. Croix." The state report pronounced "the river in this zone . . . unfit for use as a water supply . . . fish life has been exterminated." The report stated that the river was "a potential danger from a health standpoint." Beginning with a river survey in 1926, the State Board of Health documented DO levels of less than 1 mg/L over a 25-mile reach from St. Paul to Hastings, Minnesota, that could not support a healthy aquatic ecosystem, including pollution-tolerant carp (Mockavak, 1990). From 1926 to 1937, minimum DO levels of 1 to 2 mg/L indicated less than 10 percent of oxygen saturation over a 20- to 25-mile reach downstream from St. Paul (Wolman, 1971). Bacteria levels were also extremely high, with total coliform concentrations of 10^5 to 10^6 MPN/100 mL measured downstream of St. Paul (MRI, 1976). The extent of the public health risk incurred from the discharge of raw sewage by the Twin Cities was made painfully clear in 1935, when a failure of the chlorination units at the public water supply plant resulted in a serious typhoid epidemic with 213 cases and 7 deaths (USPHS, 1953).

In adopting the 1928 Board of Health recommendations, the Twin Cities became the first major city on the Mississippi River to implement primary treatment and chlorination for its municipal water pollution control plant in 1938. Water quality quickly improved dramatically as the floating mats of sludge disappeared, and DO levels increased to better than 3 mg/L from 1942 through 1955 (Mockovak, 1990; Wolman, 1971). Within 2 years, fish returned and anglers reported catching walleye and other game fish in parts of the river that had been devoid of game fish prior to 1938. Maurice Robbins, a former deputy administrator of the Metropolitan Waste Control Commission (MWCC), recalled that "The impact [of waste treatment] on the river was tremendous . . . no more dead fish, no more sewage smell" (MWCC, 1988).

With increasing population (Figure 12-5), growth eventually overwhelmed the capacity of the river to assimilate the wastewater discharge from the primary Metro plant during the mid-1950s through the mid-1960s. Water quality once again deteriorated to conditions reminiscent of the 1920s and 1930s. During the summer of 1964, the Federal Water Pollution Control Administration (FWPCA) conducted a water pollution survey of the Upper Mississippi River that documented severe degradation of water quality (FWPCA, 1966). In contrast to an average of about 30,000 MPN/100 mL near St. Paul during the 1950s, total coliform densities ranged from 460,000 to 17,000,000 MPN/100 mL 9 miles downstream of St. Paul. Minimum DO levels of less than 1 mg/L were also recorded for 15 miles downstream of St. Paul. The biological health of the river abruptly changed, with a zone of degradation and decay extending 20 miles from St. Paul to Lock and Dam No. 2 at Hastings, Minnesota. The river bottom, thick with sewage sludge, was found to be devoid of the benthic organisms usually associated with clean waters (FWPCA, 1966; WRE, 1975).

In 1966, the Metro plant was upgraded to secondary treatment using the activated sludge process. Water quality once again improved, surpassing the 1928 guidelines. The rapidly growing suburban population, however, tended to generate more residual wasteload than could be removed by upgrading the plant to secondary treatment. Regardless of the Metro plant upgrades, annual high spring flows caused flooding of the plant, resulting in the discharge of raw sewage into the river. During the late 1960s, only 4 of the 33 suburban treatment plants provided adequate levels of treatment, thus contributing to the overall pollution loading of the river. Minneapolis and St. Paul further contributed to periodic pollution loading to the river through a network of combined stormwater and sewage collection sewers that discharged raw sewage during rainstorms.

In 1984, the Metro plant was upgraded once again to advanced secondary treatment with nitrification, designed to reduce effluent levels of ammonia. After implementation of secondary and advanced secondary waste treatment for the wastewater treatment plants of the Twin Cities area by the mid-1980s, water quality of the Upper Mississippi River routinely has been in compliance with water quality standards for dissolved oxygen and un-ionized ammonia. In contrast to the record of compliance for oxygen and un-ionized ammonia, turbidity levels have exceeded water quality objectives as a result of nonpoint source runoff of sediment from the Minnesota River basin (MWCC, 1994). Because the land uses of the Minnesota River basin are dominated by agricultural row crops and the fine-textured soils further contribute to sediment losses, the annual mean (1976–1996) sediment yield of 134 lb/acre-yr from the Minnesota River watershed is almost five times greater than the annual mean sediment yield of 28 lb/acre-yr estimated for the Upper Mississippi River basin upstream of Lock & Dam No. 1 (Meyer and Schellhaass, 1999). Fecal coliform levels also remained high, and often violated state water quality standards through the mid-1980s because of combined sewer overflows during rainstorms. Fecal coliform bacteria samples are in compliance with Minnesota water quality standards if the monthly geometric mean is less than 200 MPN/100 mL and any individual sample does not exceed 2,000 MPN/100 mL.

In 1984, it was estimated that 4.6 billion gallons per year of raw sewage and stormwater were discharged to the Upper Mississippi River. In response to this water quality problem and public pressure, the Twin Cities implemented an aggressive $320 million (1996 dollars) construction program from 1985 to 1995, intended to accelerate the completion of the ongoing project to separate the combined sewers (MCES, 1996). As a result of the separation of stormwater and raw sewage from the combined sewer system, fecal coliform bacteria levels have declined considerably, and compliance with state water quality standards has improved greatly at stations monitored at Lock and Dam No.1, St. Paul, Grey Cloud Island, and Pool 2 (Buttleman and Moore, 1999). Figure 12-6 shows the reduction in bacteria levels and the corresponding improvement in compliance with water quality standards. The monitoring station at St. Paul exhibits the greatest improvement, with compliance achieved at the

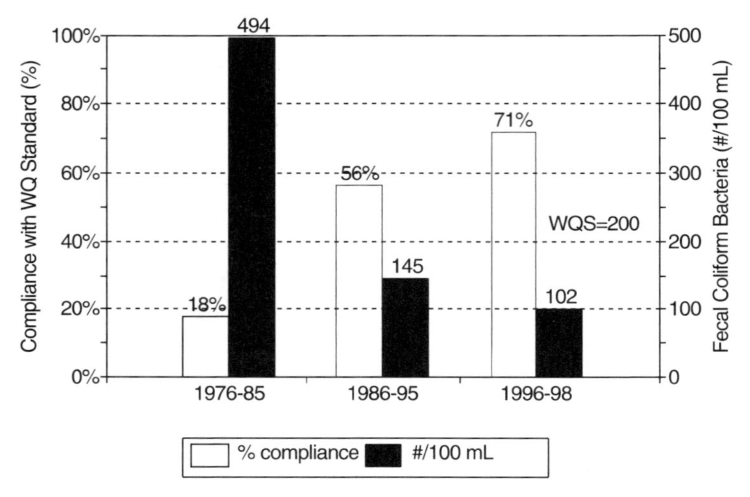

Figure 12-6 Pre- and post-CSO separation project trends in fecal coliform bacteria for the Upper Mississippi River at St. Paul. State standard for fecal coliform bacteria is 200 MPN/100 mL based on monthly geometric mean from May to October. *Source:* Buttleman and Moore, 1999.

71 percent level for samples collected from 1996 to 1998. High bacteria levels, however, do occasionally occur in the heavily urbanized area upstream of Lock and Dam No. 1; the high levels apparently are associated with urban stormwater runoff (Buttleman and Moore, 1999).

To remedy the periodic flooding of the Metro plant that resulted in the discharge of raw sewage to the river, flood protection projects were completed in 1975 and effluent pumps were installed in 1977. The pumps allowed the Metro plant to treat wastewater during the annual spring floods. The success of the flood control efforts at the Metro plant was dramatically demonstrated during the flood events of 1993 and 1997 when the plant recorded 100 percent compliance with NPDES permit limits during these two extreme events. Many other water pollution control plants in the region were forced to bypass waste treatment as a result of these extraordinary floods (Larson, 1999). In addition to their use for flood control, the effluent pumps are used during low-flow conditions when DO levels are depressed to aerate the effluent to increase ambient oxygen levels in the river.

Responding to federal industrial pretreatment requirements promulgated in 1979, the Twin Cities initiated a program to reduce discharges of heavy metals to the Upper Mississippi River. A comprehensive strategy was adopted in 1981 to reduce the discharge of heavy metals from municipal water pollution control plants contributed by sanitary sewer discharges from industrial sources. By 1992, a decade after beginning the program, the loading of heavy metals to the river had been reduced by an average of 82 percent, with declines in ambient levels of heavy metals. Using sediment cores collected in Lake Pepin, Balogh et al. (1999) have reconstructed historical loading rates of mercury from ca. 1800 to 1996 from the Upper Mississippi River watershed to Lake Pepin (Figure 12-7). Averaging the sediment core data by 10-year intervals, Balogh et al. estimated a loading rate of 3 kg/yr to characterize naturally occurring deposition of mercury under pristine conditions before European settlement began ca. 1830. Mercury deposition progressively increased during the nineteenth and twentieth centuries, with about one-half of the total mercury load deposited from 1940 to 1970 and the peak accumulation rate of 357 kg/yr identified during the 1960s. As a result of decreasing the discharges of mercury from municipal and industrial wastewater plants, the deposition rate in Lake Pepin has declined by almost 70 percent from the maximum loading during the 1960s to 110 kg/yr during 1990–1996. Although the investment in water pollution control has been very successful in reducing mercury in the Upper Mississippi River, ambient levels of mercury are still 30 times greater than the pristine conditions of the early 1800s (MCES, 2000). As of the late 1990s, the MCES is actively working to monitor and reduce even further the remaining sources of heavy metals, including mercury discharges to the river by wastewater treatment plants (MCES, 2000).

During the 1950s and 1960s, the depletion of dissolved oxygen in Pool 2 of the Upper Mississippi River near St. Paul adversely affected pollution-intolerant fish and other aquatic organisms. Studies during the early 1960s, for example, documented that burrowing mayflies (*Hexagenia*), an aquatic organism that is very sensitive to low DO conditions, were very scarce or absent from Pools 2 and 3 and Lake Pepin (Pool 4) of the river (Fremling, 1964). With the restoration of healthy levels of dissolved oxygen beginning in the mid-1980s, an abundance of mayflies once again col-

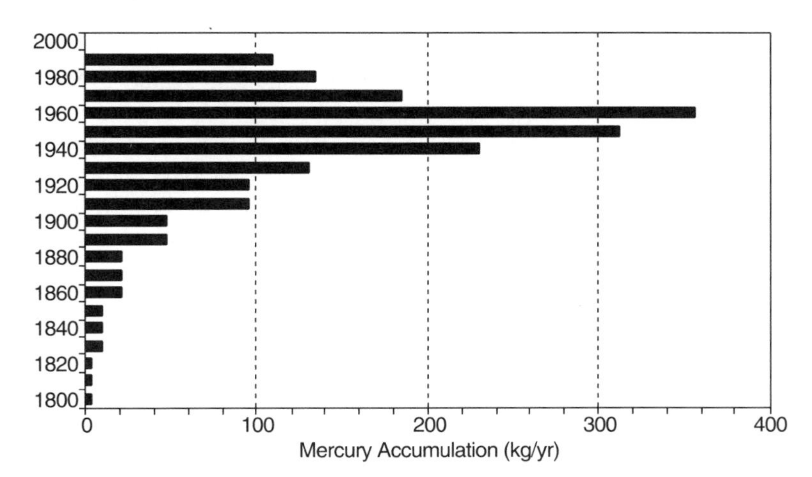

Figure 12-7 Historical mercury loading rates in the Upper Mississippi River reconstructed from sediments of Lake Pepin. Sediment core data averaged at 10-year intervals from 1800–1810 through 1980–1989 and 1990–1996. *Source:* Reprinted with permission from S. J. Balogh, D. E. Engstrom, J. E. Almendinger, M. L. Meyer, and D. K. Johnson, Environ. Sci. Technol. 33(19): 3297–3302, Copyright © 1999, American Chemical Society.

onized suitable habitats in the Upper Mississippi River from St. Paul to Lake Pepin after a 30-year absence from the river (Fremling and Johnson, 1990; MDNR, 1988). The resurgence of mayflies, significant improvements in ambient levels of DO and fecal coliform bacteria in Pool 2, and the reduction of mercury loading to the sediments of Lake Pepin demonstrate the successes of the water pollution control efforts implemented beginning in the 1980s. The Metro plant was upgraded to advanced secondary treatment with nitrification in 1984; the industrial pretreatment program was begun in 1982; and the accelerated CSO separation project, initiated in 1985 to jump-start an ongoing sewer separation project, was completed in 1995.

LEGISLATIVE AND REGULATORY HISTORY

The Minnesota State Legislature passed an act in 1885 to prevent the pollution of rivers and other water supply sources. For the next 60 years, the Minnesota Board of Health had responsibility for water pollution problems. By 1907, the State Board of Health realized that consumption of drinking water contaminated by raw sewage discharges posed a serious public health threat. Without any authority, the Board of Health attempted to pressure the Twin Cities communities to install wastewater treatment facilities. In 1917, the State Board of Health adopted regulations requiring towns to submit plans for sewers and wastewater treatment plants prior to construction. The Board also conducted water pollution surveys and made various recommendations for controlling pollution. Letters to the city councils of the Twin Cities urging action on controlling the discharge of raw sewage went unanswered in 1923 and 1925. At the request of the State Board of Health, the U.S. Public Health Service conducted the first water pollution survey of the Upper Mississippi River from the Twin Cities to Winona, Minnesota, in 1926.

During the 1920s, the Izaak Walton League, the Engineers Society of St. Paul, the Engineering Club of Minneapolis, and other private groups lobbied for immediate action on the problem of raw waste disposal into the river. In 1926 the Minneapolis Sanitary Commission was created to study "the condition of the river and the problems of sewage disposal" (MWCC, 1988). In 1927, when the Metropolitan Drainage Commission was formed, raw sewage was discharged through 84 outfalls over a network of 1,125 miles of sewers (MWCC, 1988). Maurice Robbins, a former deputy administrator of MWCC, remembering his experiences sampling the river during those years, stated that "It could get pretty awful down by the river. There were floating feces, dead fish and a terrible sewer smell" (MWCC, 1988).

In 1927, the State Board of Health was given the authority and the responsibility to administer and enforce all laws related to water pollution in Minnesota. The legislature directed the State Board of Health to form a Metropolitan Drainage Commission. The legislature, however, did not provide any substantial basis for managing waste disposal. In 1933, a decade after the Minnesota State Board of Health had begun to document the pollution problems of the Upper Mississippi River, the Minneapolis–St. Paul Sanitary District was finally created to oversee construction of the first primary wastewater treatment plant in the Twin Cities region. The primary treatment plant, located near Pig's Eye Lake in St. Paul, went online in 1938.

In 1945, the legislature passed the Water Pollution Control Act to establish the Water Pollution Control Commission for the regulation of the emerging problems of water pollution. The Minnesota Act, amended in 1951, 1959, and 1963, was regarded as one of the better state water pollution control acts in the United States (FWPCA, 1966). The main mission of the new Water Pollution Control Commission was to direct the construction of primary wastewater treatment plants for the smaller municipalities in the Twin Cities metropolitan area.

In 1967, the state legislature formed the Metropolitan Council as a regional coordination agency. In 1969, the Metropolitan Waste Control Commission (MWCC) was given the regional responsibility for wastewater collection and treatment systems for 33 plants within 200 political jurisdictions of the seven-county Twin Cities area. In 1967, the legislature also created the Minnesota Pollution Control Agency (MPCA) to replace the Water Pollution Control Commission. The new agency was soon given authority to regulate and enforce effluent limits for municipal and industrial treatment plants. The establishment of the U.S. Environmental Protection Agency in 1970 and the enactment of the 1972 Clean Water Act further strengthened the regulatory powers for requiring uniform effluent limits for wastewater dischargers. In July 1994, the MWCC and transit services were merged with the Metropolitan Council. The responsibility for operating municipal wastewater treatment plants was delegated to the Environmental Services Division of the Metropolitan Council (MCES).

Following the 1972 Clean Water Act, the MWCC, with federal (75 percent) and state (15 percent) funding assistance, spent more than $350 million to dramatically improve the technology of the Metro plant, upgrade other facilities, and build interceptor sewer systems (MWCC, 1988). During the 1970s and 1980s, MWCC phased out or upgraded old plants or constructed new plants for many of the suburban communities in the Twin Cities region. MCES now operates the Metro plant and eight other treatment plants in the Twin Cities area. The Metro plant and three other wastewater treatment plants discharge to the Upper Mississippi River; three plants discharge effluent to the Minnesota River; and the St. Croix River and the Vermilion River each receive effluent discharges from one municipal plant.

IMPACTS OF WASTEWATER TREATMENT

Pollutant Loading and Water Quality Trends

During the 1960s and 1970s, effluent loading from the Metro plant accounted for more than three-quarters of the total point source load of BOD_5 in the section of the Upper Mississippi River from the Twin Cities to the St. Croix River. Because this one wastewater treatment plant, the Metro plant, accounted for more than 75 percent of the total point source load, historical effluent data from the Metro plant can serve as an indicator to demonstrate the success of public investments to upgrade the plant in improving water quality in the Upper Mississippi River. Figures 12-8 through 12-11 present time-series trend data for population served, effluent flow, BOD_5, total suspended solids (TSS), total Kjeldahl nitrogen (TKN), and ammonia-N concentrations for the Metro plant from 1940 to 2000 (Larson, 1999, 2001; Johnson and Aasen, 1989).

During the early 1960s, the Metro plant served 1.05 million people and discharged 158 mgd to the Upper Mississippi River. By 1997, the population served by Metro had grown to 1.7 million with a corresponding increase in the effluent discharge rate to 225 mgd (Figure 12-8). Since enactment of the Clean Water Act in 1972, effluent BOD_5 loading from the Metro plant has been reduced greatly from the peak loading period of the mid-1960s. Before upgrading the Metro plant, effluent BOD_5 loading peaked at about 330,000 lb/day in 1968. After upgrading to secondary in 1966, effluent loading dropped to 114,000 lb/day by 1970 and 77,000 lb/day in 1973. Since the 1980s, effluent loading of BOD_5 has continued to decline as a result of additional upgrades (e.g., advanced secondary in 1984) and replacement, or abandonment, of 21 of the 33 suburban wastewater treatment plants that existed in 1969 when MWCC assumed responsibility for plant operations. BOD_5 loading from Metro declined again to 40,000 lb/day by 1980 and to 27,000 lb/day by 1990. Over a 30-year period, upgrades and improvements to the Metro plant have reduced effluent BOD_5 loading by 95 percent from the historical peak loading of 330,000 lb/day in 1968 to only 17,000 lb/day in 1998. Over the same period, the effluent concentration of BOD_5 has been reduced from 184 mg/L in 1968 to 9.7 mg/L in 1998 (Figure 12-9).

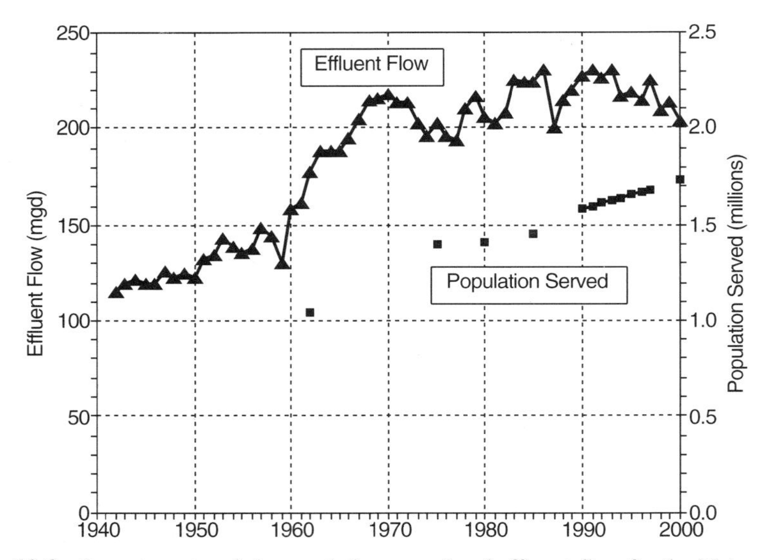

Figure 12-8 Long-term trends in population served and effluent flow for the Metro plant in St. Paul, Minnesota. *Source:* Larson, 1999, 2001; D. K. Johnson and P. W. Aasen, J. Minn. Acad. Sci. 55(1), 1989.

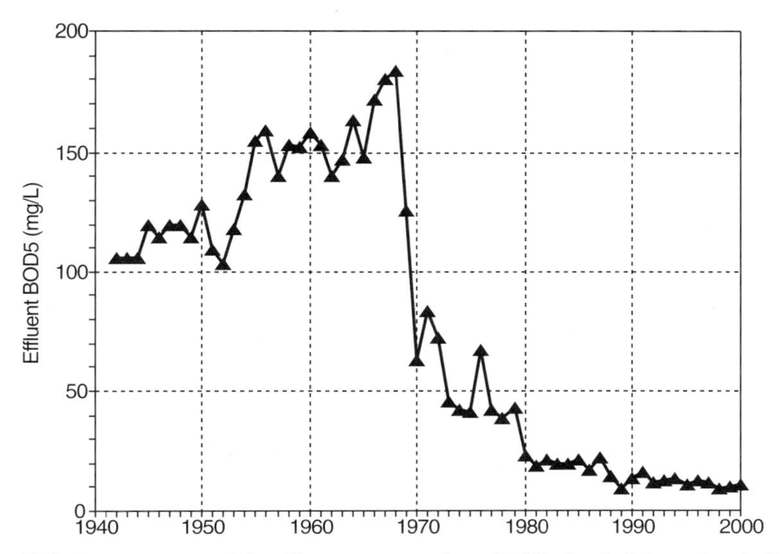

Figure 12-9 Long-term trends in effluent concentration of BOD_5 for the Metro plant in St. Paul, Minnesota. *Source:* Larson, 1999, 2001; D. K. Johnson and P. W. Aasen, J. Minn. Acad. Sci. 55(1), 1989.

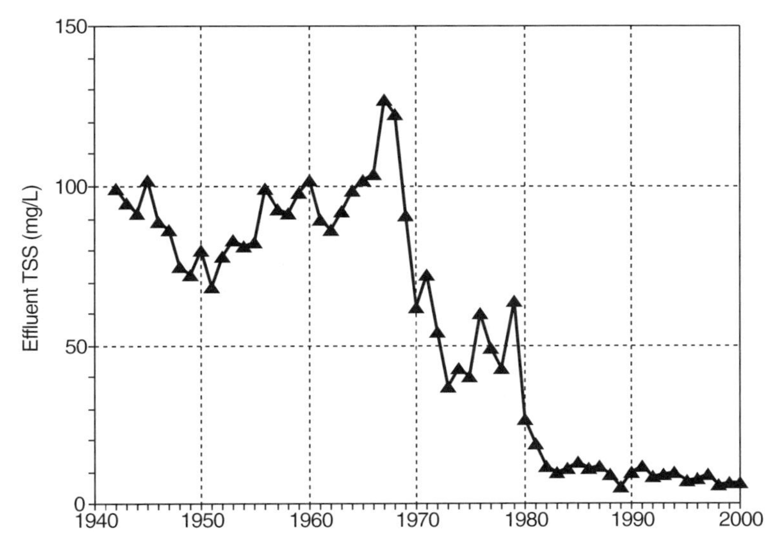

Figure 12-10 Long-term trends in effluent concentration of TSS for the Metro plant in St. Paul, Minnesota. *Source:* Larson, 1999, 2001; D. K. Johnson and P. W. Aasen, J. Minn. Acad. Sci. 55(1), 1989.

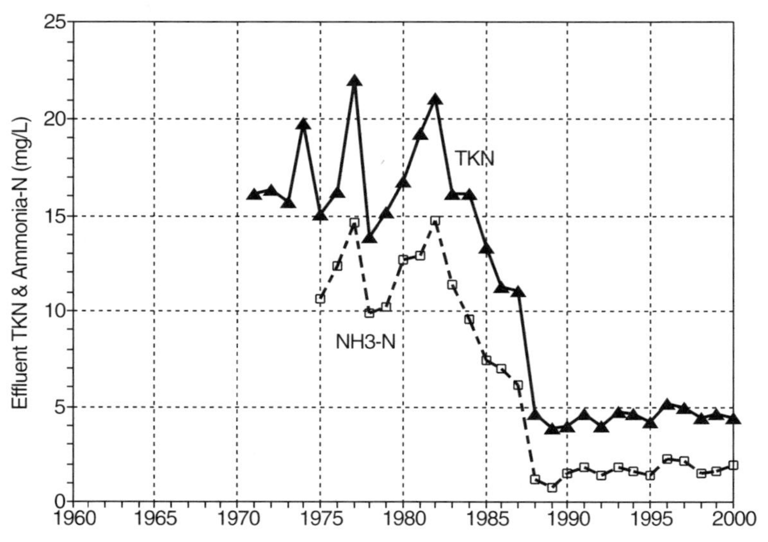

Figure 12-11 Long-term trends in effluent concentration of TKN and ammonia-N for the Metro plant in St. Paul, Minnesota. *Source:* Larson, 1999, 2001.

Upgrades and improvements to the Metro plant have also resulted in large reductions in effluent loading of suspended solids and nitrogen. TSS loading has dropped by 95 percent from the peak loading rate of 219,000 lb/day in 1968 to 10,000 lb/day by 1998; effluent concentration declined from 122 mg/L in 1968 to 5.7 mg/L by 1998 (Figure 12-10). Based on effluent data from monitoring that began in 1971, TKN loading has dropped by 78 percent from the peak loading rate of 36,500 lb/day in 1982 to 7,800 lb/day by 1998; effluent concentration has been reduced from 21 mg/L in 1982 to 4.5 mg/L by 1998 (Figure 12-11). Prior to the upgrade to advanced secondary with nitrification, toxicity-based water quality standards for the un-ionized portion of ammonia were frequently violated in the Upper Mississippi River. After upgrading the plant to nitrification with ammonia removal in 1984, effluent discharges of ammonia-N declined considerably. Using effluent data collected since 1975, ammonia-N loading has dropped by 90 percent from the peak loading rate of 25,500 lb/day in 1982 to 2,600 lb/day by 1998. The effluent concentration of ammonia-N has been reduced from 14.7 mg/L in 1982 to 1.5 mg/L by 1998 (Figure 12-11).

Beginning in the 1920s through the 1970s, the major water quality issues for the Upper Mississippi River have been bacterial contamination and depletion of DO from sewage discharges and combined sewer overflows. Historical DO data sets collected since 1926 illustrate the dramatic change in long-term trends in the spatial distribution of DO recorded 5 miles downstream of the confluence with the Minnesota River near St. Paul (UM milepoint 840) to Lock and Dam No. 3 at Red Wing, Minnesota (UM milepoint 797) (Figure 12-12). These historical data sets clearly illustrate the adverse impacts of wastewater loading and the effectiveness of upgrades in waste-

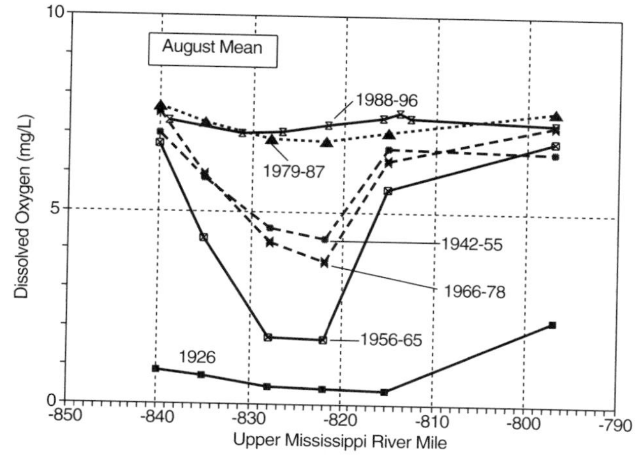

Figure 12-12 Spatial trends of August DO in the Upper Mississippi River from 1926 to 1988–1996 from St. Paul, Minnesota (UM milepoint 840), to Lock & Dam No. 3 at Redwing, Minnesota (UM milepoint 797). *Sources:* Larson, 1999; Mockovak, 1990; MWCC, 1989; D. K. Johnson and P. W. Aasen, J. Minn. Acad. Sci. 55(1), 1989.

water treatment implemented in 1938, 1966, and the early 1970s at the Metro plant. Because of the hydraulic characteristics of the Upper Mississippi River, minimum DO levels have been consistently observed in a zone 5 to 15 miles downstream of the Metro plant discharge, within the oxygen sag region from Newport, Minnesota (UM milepoint 820) to Grey Cloud, Minnesota (UM milepoint 830).

Using historical data available from USEPA's STORET water quality database, the long-term trend of summer DO and BOD_5 (1940–1995) has been compiled from monitoring station records extracted for RF1 reach 07010206001 from the Minnesota River (UM milepoint 844.7) to the St. Croix River (UM milepoint 811). Although DO is characterized by a high degree of interannual variability because of temporal variability in streamflow and the spatial gradient over this 34-mile-long reach, there has been a definite improvement in this long reach between the 1960s, when summer mean oxygen levels ranged from approximately 4 to 7 mg/L, to the period from the mid-1970s through the mid-1990s, when summer mean oxygen levels consistently ranged from approximately 7 to 8 mg/L even during the drought conditions of 1987–1988 (Figure 12-13). The trend of improvement in DO during the 1980s and 1990s is consistent with the long-term trend of improvement in ambient BOD_5 extracted for the same reach (Figure 12-14). During the 1960s and 1970s, summer mean BOD_5 ranged from approximately 4 to 8 mg/L. During the 1980s, mean BOD_5 ranged from approximately 2.5 to 4.5 mg/L. In the period 1990–1995, mean ambient BOD_5 declined even further to levels ranging from approximately 2 to 3.5 mg/L as a result of upgrading the Metro plant to advanced secondary with nitrification in the late 1980s.

In interpreting the year-to-year variability of the long-term DO data from 1940 through 1995, it is important to understand the influence of streamflow on summer

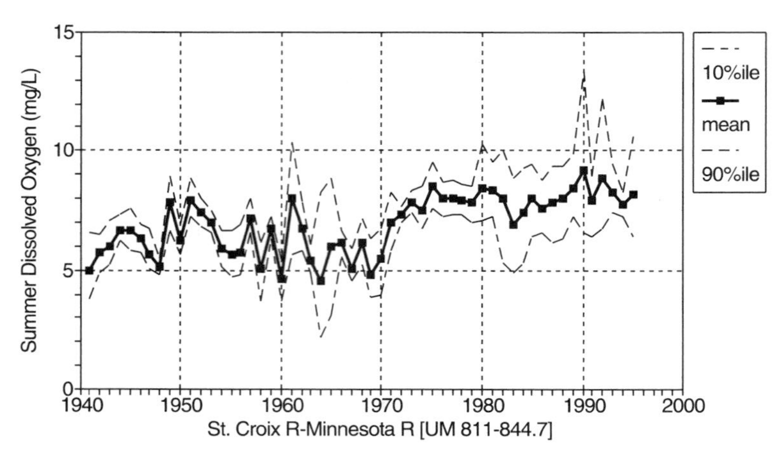

Figure 12-13 Long-term trends of mean, tenth percentile, and ninetieth percentile summer DO in the Upper Mississippi River for RF1 reach 07010206001 from the Minnesota River (UM milepoint 844.7) to the St. Croix River (UM milepoint 811). *Source:* USEPA STORET.

oxygen levels under the peak effluent loading conditions of the 1960s and early 1970s compared to the greatly reduced effluent loading conditions that have characterized the Twin Cities area since the mid-1970s. Under conditions of similar effluent loading rates, DO decreases during low-flow conditions in contrast to a relative increase during higher summer flow conditions. Over years of comparable effluent BOD_5 loading, the interannual cycles that appear to show fluctuating trends of either "im-

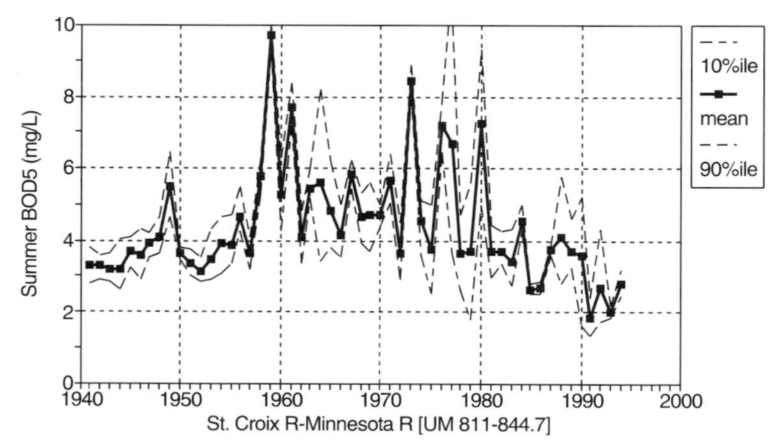

Figure 12-14 Long-term trends of mean, tenth percentile, and ninetieth percentile summer BOD_5 in the Upper Mississippi River for RF1 reach 07010206001 from the Minnesota River (UM milepoint 844.7) to the St. Croix River (UM milepoint 811). *Source:* USEPA STORET.

provement" or "degradation" in DO (Figure 12-13) are caused primarily by year-to-year variability of summer streamflow (see Figure 12-4). An accurate evaluation of the long-term trend in improvement of DO is possible only by filtering the time series of oxygen records to extract only those summers that are characterized by dry streamflow conditions.

Figure 12-15 shows long-term trends in DO conditions for "dry" summers for a 10-mile subreach of the RF1 reach (07010206001) for the critical oxygen sag location from Newport, Minnesota (UM milepoint 820), to Grey Cloud, Minnesota (UM milepoint 830). The time series record of DO data in Figure 12-15 is extracted to highlight the trend in improvement for summers of comparable "dry" streamflow conditions when the flow at the St. Paul USGS gage was less than 75 percent of the long-term (1951–1980) summer mean. During the 1960s, low-flow summer mean DO levels violated water quality standards with concentrations as low as less than 1 mg/L in 1961 to approximately 4 mg/L in 1964. After the upgrade of the Metro plant to advanced secondary with nitrification in the late 1980s, mean summer DO levels in the critical subreach had improved to levels as high as approximately 6 to 7 mg/L even during the extreme drought conditions of 1987–1988. Using before and after data in a post-audit model applied to the low-flow summers of 1976 and 1988, Lung (1996a) has clearly demonstrated that the improvements in DO can be directly related to upgrades of the Metro plant.

As shown by the historical records for fecal coliform bacteria (Figure 12-6), DO (Figure 12-15), and levels of sediment mercury in Lake Pepin (Figure 12-7), investments in water pollution control programs of the 1970s and 1980s have succeeded in

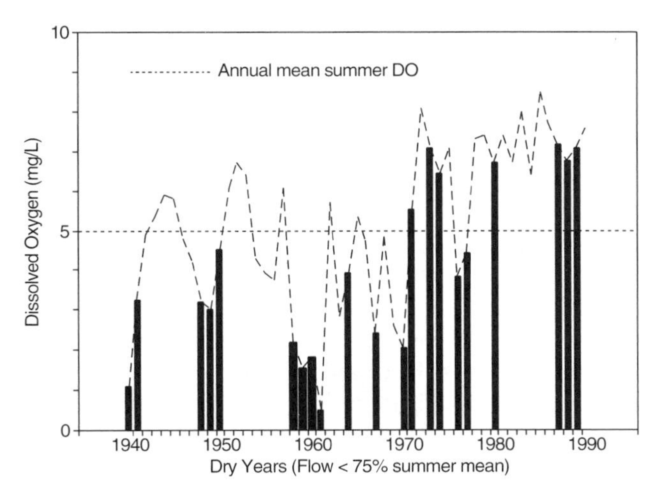

Figure 12-15 Long-term trends of mean summer DO in the Upper Mississippi River for 1940–1990 and years characterized by "dry" streamflow less than 75 percent of long-term (1951–1980) summer mean streamflow. Data extracted for subreach from Newport, Minnesota to Grey Cloud, Minnesota (UM milepoint 820–830). *Source:* USEPA STORET.

improving water quality conditions for these historical problems of the 1950s, 1960s, and 1970s in the Upper Mississippi River. During the 1980s and 1990s, water quality and comprehensive ecological investigations in the Upper Mississippi River have identified a number of contemporary chemical and nonchemical problems in the basin. Nonchemical issues identified as threats to the ecological processes of the river and floodplain ecosystem include, for example, loss of habitat and wetlands and man-made alterations from flood control and navigation projects.

Contemporary chemical problems include inputs of nutrients, sediments, heavy metals, pesticides, and other toxic chemicals. For example, the loading of phosphorus and suspended solids influences water quality in Lake Pepin, a natural impoundment located about 50 miles downstream of St. Paul. Lake Pepin is eutrophic, with high annual mean concentrations of total phosphorus (0.16 mg/L) and soluble reactive phosphorus (SRP) (0.07 mg/L) (James et al., 1996) recorded at the inlet to the lake (UM milepoint 797) during the average flow years of 1994–1996. Eutrophic conditions in the lake are caused by excessive loading of nutrients from point and nonpoint sources in the watershed. When physical and hydrological conditions are favorable, such as during low-flow summers, nuisance algal blooms (i.e., viable chlorophyll-*a* greater than 30 μg/L) occur. Concerns related to the need for controls on phosphorus loading arose after severe algal blooms and fish kills in Lake Pepin occurred under the drought conditions of 1987–1988 (Johnson, 1999).

The Lake Pepin Phosphorus Study, conducted from 1994 to 1998, compiled historical and contemporary data sets to evaluate the human impact on (1) long-term records of sediment and phosphorus loading to Lake Pepin and (2) the corresponding water quality responses to changes in loading to the lake. Since European settlement *ca.* 1830s, the contemporary (1990–1996) annual input of approximately 850,000 metric tons/year (mt/year) of sediment is about ten times greater than the loading rates estimated for the presettlement era. Analysis of data from the three basins included in the study, the Upper Mississippi River, the Minnesota River, and the St. Croix River, indicates that 90 percent of the increased sediment load to Lake Pepin is contributed by erosion of fine-textured soils from the Minnesota River basin. The record of sediment deposition in Lake Pepin also indicates that the most rapid rates of sediment input to the lake occurred during the 1940s and 1950s. If current sedimentation rates continue from erosion in the Minnesota River basin, Lake Pepin could be completely filled in about 340 years (Engstrom and Almendinger, 1998).

Over the past two centuries, phosphorus concentrations in the sediments of Lake Pepin have increased twofold, while water column concentrations (inferred from diatom assemblages in the sediments) appear to have increased by a factor of 4 since European settlement *ca.* 1830s. Increased phosphorus levels in the sediments and water column are the result of an increase in phosphorus loads to Lake Pepin by a factor of 5 to 7 since the 1830s to the contemporary estimated loading rate of approximately 4,000 to 5,000 mt/year for 1990–1996. Wastewater discharges and agricultural applications of manure and commercial fertilizer are most likely the key factors controlling historical phosphorus loads to Lake Pepin, and the statewide ban on phosphates in detergents contributed to a reduction in phosphorus loading from municipal wastewater plants by approximately 40 percent over the period from 1970 to 1980.

Since the 1830s era, the progressive increase in phosphorus loading has resulted in a shift in assemblages of diatoms from clear water benthic algae and mesotrophic water column species in the presettlement era to planktonic species exclusively characteristic of highly eutrophic conditions in the 1990s (Engstrom and Almendinger, 1998).

In evaluating strategies to reduce phosphorus loads to Lake Pepin, the significant differences in the relative contributions of point and nonpoint sources of flow, solids, and nutrient loads under a range of flow conditions need to be considered over a time scale of decades. Point and nonpoint source loading data for suspended solids and total phosphorus have been compiled for low-flow (1988), average-flow (1994–1996), and high-flow (1993) conditions for the Upper Mississippi River (upstream of Lock and Dam No. 1), the Minnesota River, and the St. Croix River (Meyer and Schellhaass, 1999). Based on 21 years of data (1976–1996), the mean yield of total phosphorus from the agriculturally dominated Minnesota River (0.33 lb/acre-year) is twice as great as the mean yield from the Upper Mississippi River basin upstream of Lock and Dam No. 1 (0.16 lb/acre-year) and the St. Croix River basin (0.14 lb/acre-year). Figure 12-16 presents a comparison of the magnitude of point source loads and nonpoint source loads of total phosphorus from the Upper Mississippi River, Minnesota River, and St. Croix River basins for 1988 (drought), 1993 (flood), and 1994–1996 (average conditions) (Meyer and Schellhaass, 1999).

Under the extreme flood conditions of 1993, nonpoint source loadings of total phosphorus from the Minnesota River and the Upper Mississippi River watersheds have been shown to account for 58 percent and 15 percent, respectively, of the total phosphorus load of 6,030 mt/year estimated for 1993, while point sources from the Metro plant accounted for 15 percent of the total phosphorus load. During the severe

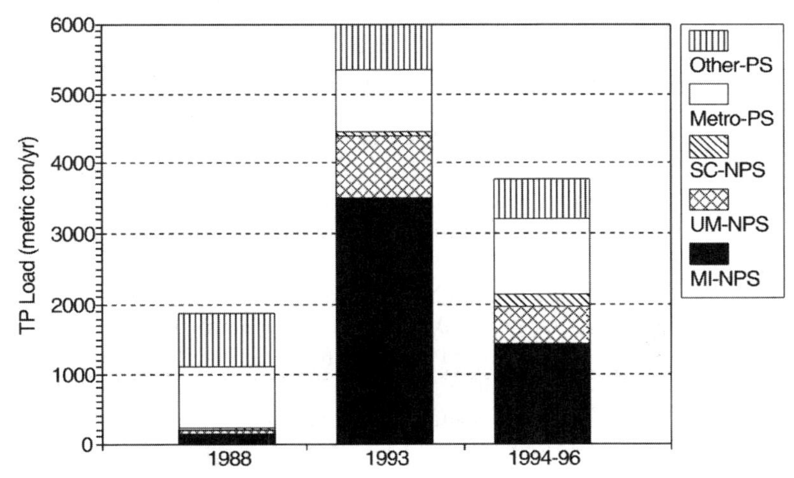

Figure 12-16 Comparison of total phosphorus loadings from nonpoint sources (NPS) in the Upper Mississippi River (UM, to Lock & Dam No. 1), Minnesota River (MI), and St. Croix River (SC) basins and point source (PS) loadings from the Metro plant and other facilities in the three river basins. *Source:* Meyer and Schellhaass, 1999.

drought conditions of 1988, the total phosphorus load of 1,900 mt/year was only about one-third of the 1993 load. Under the drought conditions, the Metro plant accounted for 47 percent of the total phosphorus load, and nonpoint source loading from the Minnesota River and the Upper Mississippi River contributed only 6 percent and 3 percent, respectively, of the total phosphorus load of 1,900 mt/year. During the average flow conditions of 1994–1996, the total phosphorus load of 3,800 mt/year was two times greater than the 1988 drought load. Under average flow conditions, the Metro plant accounted for 28 percent of the total phosphorus load, and nonpoint source loading from the Minnesota River and the Upper Mississippi River contributed 38 percent and 14 percent, respectively, of the total phosphorus load of 3,800 mt/year.

Meyer and Schellhaass (1999) have used this data set to develop summary budgets of the relative contributions of point source and nonpoint source loadings of total phosphorus to the three river basins during 1988, 1993, and 1994–1996. Under the drought conditions of 1988, the contribution from point sources (88.5 percent) dominated the total inputs of phosphorus, compared to the 11.5 percent accounted for by nonpoint sources. During the extreme flood conditions of 1993, nonpoint source loads accounted for about three-quarters (74.5 percent) of the total input of phosphorus, with point sources accounting for about one-quarter (25.5 percent). During the average flow conditions of 1994–1996, the relative contribution of point sources (56.2 percent) and nonpoint sources (43.8 percent) was almost comparable.

These point source loading and nonpoint source loading data sets for suspended sediments and phosphorus and a number of other field studies (e.g., James et al., 1999) have been used to support the development of an advanced model of hydrodynamics, sediment transport, and eutrophication for the Upper Mississippi River and Lake Pepin (HydroQual, 1999a, 1999b; Garland et al., 1999). As of 2000, the MCES is using the model to evaluate the effectiveness of alternative strategies to control point and nonpoint phosphorus loading to the Upper Mississippi River to achieve the water quality objectives established for Lake Pepin. Evaluations of sediment loading contributed primarily from agricultural runoff in the Minnesota River basin have also been a key issue in the Minnesota River Assessment Project (MPCA, 1994).

On the much larger scale of the entire Mississippi River basin, nitrogen loading from the Mississippi River has been identified as a major cause of the algal blooms and hypoxia that occur over a 16,000-square-kilometer area of the inner Gulf of Mexico known as the "Dead Zone" (Christen, 1999; Malakoff, 1998; Moffat, 1998; Rabelais et al., 1996; Vitousek et al., 1997). Based on technical assessments of the "Dead Zone" problem, a USEPA and NOAA Action Plan, released in January 2001, recommended that efforts be undertaken to reduce inputs of nitrogen from wastewater treatment plants and agricultural land uses (e.g., fertilizer applications and confined animal feedlots) over the entire Mississippi River basin, which drains 40 percent of the land area of the continental United States (Christen, 1999; Mississippi River/Gulf of Mexico Watershed Nutrient Task Force, 2001).

The series of locks and dams and maintained navigation channel have been an integral physical feature of the Upper Mississippi River since the early 1930s when the U.S. Congress authorized the U.S. Army Corps of Engineers to maintain the river for navigation purposes. Concerns have been raised about the disposal of dredged sediments, often contaminated with heavy metals and toxic chemicals, to maintain the navigation channel and the loss of ecologically critical backwater habitats to sediment deposits. The devastation caused by the Great Flood of 1993 (Wahl et al., 1993) in the upper Midwest has also triggered debates about the failure of flood control measures intended to protect river communities from floods. As the key federal agency responsible for inland waterways, the U.S. Army Corps of Engineers has initiated controversial studies to evaluate the ecological impact of maintenance dredging, flood control structures, and widening the series of locks and dams (Phillips, 1999).

Evaluation of Water Quality Benefits Following Treatment Plant Upgrades

From a policy and planning perspective, the central question related to the effectiveness of the secondary treatment requirement of the 1972 CWA is simply: Would water quality standards for DO be attained if primary treatment levels were considered acceptable? In addition to the qualitative assessment of historical data, water quality models can provide a quantitative approach to evaluate improvements in dissolved oxygen and other water quality parameters achieved as a result of upgrades to secondary and greater levels of wastewater treatment. Since the 1970s, increasingly complex models have been developed to determine wasteload allocation requirements for municipal and industrial dischargers to meet the needs of decision-makers for the Upper Mississippi River.

During the mid-1970s, the National Commission on Water Quality (see Appendix A-6) funded Water Resources Engineers (WRE, 1975) to develop a steady-state, one-dimensional water quality model (QUAL-II) of DO, BOD_5, nutrients, and fecal coliform bacteria, using data collected in the Upper Mississippi River in 1964–1965 (FWPCA, 1966). The model was applied to evaluate the effectiveness of the technology-based requirements of the 1972 Clean Water Act for municipal and industrial dischargers. With funding available from the CWA Section 208 program, Hydroscience (1979) developed a water quality model (AESOP) of DO, BOD_5, nutrients, algae, and bacteria using data collected in 1973, 1976, and 1977. The model, further validated by the Minnesota Pollution Control Agency using data obtained in 1980, was used for a wasteload allocation study of the Metro plant's impact on DO and un-ionized ammonia in Pool 2 (MPCA, 1981).

As a result of the severe algal blooms and fish kills that occurred in Lake Pepin during the extreme drought of 1988, a time-variable water quality model (WASP5-EUTRO5) of DO, BOD_5, nutrients, and algae was developed using data collected during 1988 (MWCC, 1989), 1990, and 1991 (EnviroTech, 1992, 1993; Lung and Larson, 1995). The validated model was used to evaluate alternatives for phosphorus controls at the Metro plant and to perform a postaudit of the Hydroscience (1979) AESOP model using low-flow data collected during 1988 (Lung, 1996a). The model was also applied to track the fate and transport of phosphorus and the relative impact of the point and nonpoint sources on eutrophication in Lake Pepin (Lung, 1996b).

Following completion of the model by EnviroTech (1992, 1993), a number of uncertainty issues were identified related to: (1) fate and transport of phosphorus from point and nonpoint sources; (2) interaction of suspended solids with phosphorus transport; and (3) interaction of nonpoint source phosphorus inputs generated under low-flow and high-flow hydrologic conditions with interannual variation in the benthic release of phosphorus. To address these issues, a three-dimensional hydrodynamic, sediment transport, and advanced eutrophication model was developed and calibrated using data collected over 12 years from 1985 through 1996 (Garland et al., 1999; HydroQual, 1999a, 1999b). The calibrated model was used to simulate the long-term (24-year) water quality response in the Upper Mississippi River and Lake Pepin to a number of alternative control scenarios over a range of hydrologic (e.g., dry and wet years) and loading conditions for point source and nonpoint source discharges of phosphorus.

To evaluate the incremental improvements in water quality conditions that have been achieved by upgrading municipal wastewater plants from primary to secondary and from secondary to advanced secondary levels of wastewater treatment, Lung (1998) used the WASP5-EUTRO5 model developed by EnviroTech (1992, 1993) to demonstrate the water quality benefits attained by the secondary treatment requirements of the 1972 CWA. Using the model, municipal and industrial wastewater flow and effluent loading data were used with boundary flow and loading data describing the Upper Mississippi River and Minnesota River to compare water quality conditions for three summers (1964, 1976, and 1988) characterized by comparable low-flow conditions and primary (1964), secondary (1976), and advanced secondary (1988) levels of wastewater treatment at the Metro plant. The model was applied to evaluate the water quality impact of three different treatment levels for Metro and the other municipal plants: (1) primary, (2) secondary, and (3) advanced secondary with nitrification. CBOD oxidation rates were calibrated for each of these three different data sets to reflect differences in the proportion of labile and refractory oxidizable material discharged from the Metro plant.

A comparison of the results of the model runs and observed data sets is presented in Figure 12-17. Spatial distributions of CBOD-ultimate, ammonia-N, nitrate + nitrite-N, algal chlorophyll-*a*, and DO are presented from St. Paul (UM milepoint 840) to Lock & Dam No. 2 (UM milepoint 815) for 1964 (primary), 1976 (secondary), and 1988 (advanced secondary with nitrification) loading and flow conditions. The upgrade of the Metro plant from primary to secondary and the corresponding reduction of effluent BOD_5 loading (see Figure 12-9) is reflected in the decrease in ambient CBOD from a peak of approximately 20 mg/L in 1964 to approximately 7 to 8 mg/L in 1976 at UM milepoint 835 near the Metro plant. As shown in the simulation results, the distributions of ammonia-N and nitrate + nitrite-N are similar under the primary and secondary treatment scenarios because upgrading from primary to secondary treatment does not change the effluent concentration of ammonia. The progressive reduction in ambient ammonia-N and corresponding increase in ambient nitrate + nitrite-N for the 1988 simulation, however, reflect the impact of the upgrade from secondary to advanced secondary with nitrification and the drop in effluent loading of ammonia-N at the Metro plant (see Figure 12-11).

During the 1960s, when Metro discharged primary effluent, a large section of the river was hypoxic or anoxic, with the worst conditions (< 2 mg/L) observed over ap-

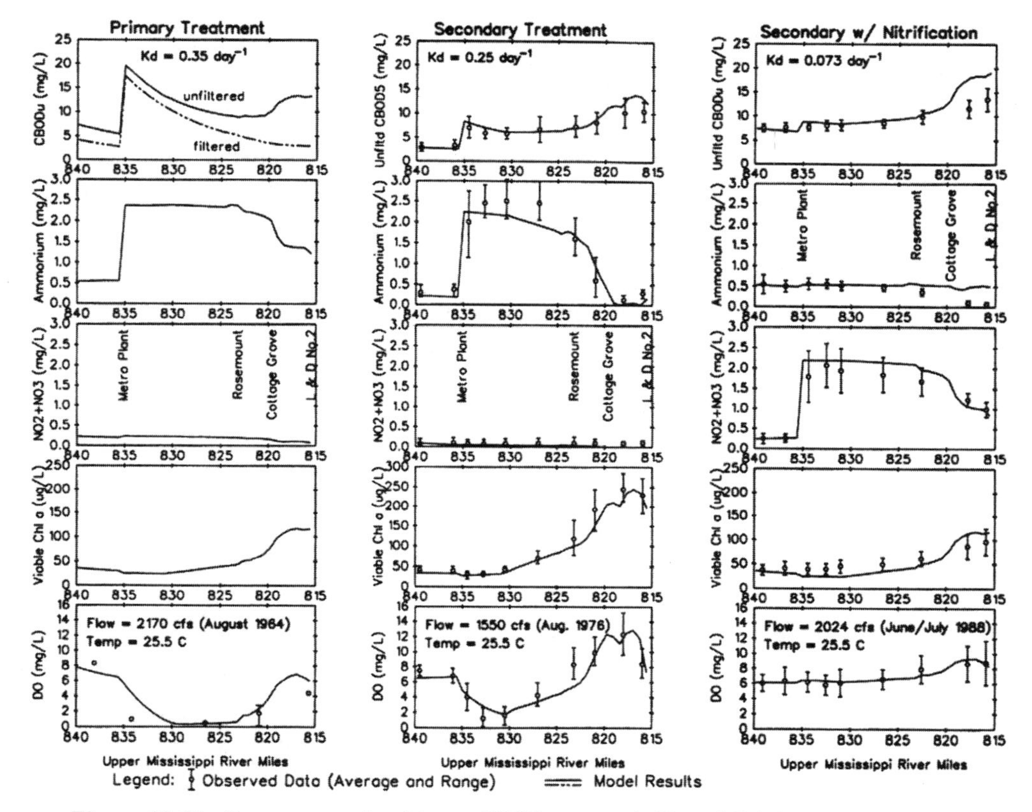

Figure 12-17 Improvement in ultimate CBOD, ammonia-N, and DO levels in the Upper Mississippi River related to Metro treatment plant upgrades from primary to secondary and advanced secondary with nitrification. *Source:* Lung, 1998. Reprinted with permission of the American Society of Civil Engineers. Copyright 1998.

proximately 15 miles from UM milepoint 820 to UM milepoint 835. The observed data and the model results indicate the elimination of anoxic conditions and a nominal improvement in DO conditions under the extreme low-flow conditions of August 1976. The minimum DO level is increased from approximately 0.5 mg/L in 1964 to approximately 2 mg/L in 1976 as a result of the upgrade from primary to secondary treatment. Even with secondary treatment at Metro, however, compliance with the water quality standard for dissolved oxygen of 5 mg/L was not achieved, and a distinct oxygen sag is observed in the 1976 data set. Compliance with the DO standard was finally achieved, even under the extreme drought conditions of 1988, after Metro was upgraded from secondary to advanced secondary treatment with nitrification.

The model results demonstrate very clearly the progressive increase in DO levels in the river following the upgrades at Metro to secondary and advanced secondary treatment. The model results also demonstrate the ability of a well-calibrated model to match observed water quality distributions that are directly related to changes in effluent loading from Metro under the three different treatment levels. The data used to define the effluent flow and loading characteristics for the primary, secondary, and advanced secondary treatment levels for the 1964, 1976, and 1988 simulations are given in Lung (1998). The data used to define effluent flow and loads from the other municipal and industrial point sources and the boundary inputs from the Upper Mississippi River and the Minnesota River are summarized by WRE (1975) for 1964 and by EnviroTech (1992, 1993) for the 1976 and 1988 simulations.

In generating the simulation results for the three different treatment scenarios, all model coefficients, except the CBOD oxidation rate, are based on the same numerical values for each of the three model runs. The in-stream oxidation rate for CBOD (K_d) is assigned different values for primary (0.35/day), secondary (0.25/day), and advanced secondary with nitrification (0.07/day) treatment levels, since this kinetic reaction rate is dependent upon stabilization of the effluent and the quantity of labile and refractory components of oxidizable organic matter in the effluent (Chapra, 1997; Thomann and Mueller, 1987). Using effluent loading rates that are representative of the three different treatment levels for Metro, the model results confirm that the improvement in water quality observed in the Upper Mississippi River can be attributed to investments made to upgrade the Metro plant.

Recreational and Living Resources Trends

Long-term trends in recreational uses, private investments along the riverfront, and biological resources dependent on the integrity of aquatic ecological conditions are meaningful nonchemical indicators of water quality conditions in the Upper Mississippi River. One very simple indicator is the use of the river for recreational boating. If water quality conditions are very poor, as was the case during the 1950s and 1960s, the noxious conditions are not desirable for boating as a recreational activity. If water quality is not degraded, the river might be considered desirable for boating. As shown in the long-term trend of recreational boat traffic through Locks 1 through 4 of the river (Figure 12-18), annual recreational vessel usage of the river ranged from approximately 25,000 vessels to approximately 30,000 vessels from the mid-1970s through the mid-1980s. Beginning in the mid-1980s, the improvements in water quality in the Upper Mississippi River suggest a strong correlation with the dramatic increase in annual recreational vessel traffic on the river to approximately 45,000 to approximately 53,000 boats (Erickson, 2000), with the recreational vessel traffic in Locks 1 through 4 increasing by about two-thirds between 1986 and 1998 (Figure 12-18). Note that traffic in 1993 dropped by about one-half because of the extreme flood conditions of that year.

Recreational boats require marina space, and in 1990 about 2,700 new marina slips were in various planning stages—enough to double marina capacity. The number of permit applications received by the St. Paul District U.S. Army Corps of Engineers for docks, marinas, boathouses, boat ramps, and beach and wildlife improvements soared from only 3 in 1981 to 22 by 1989 (Figure 12-19). In the late 1970s, nobody would have considered investing in a marina in Pool 2 because of poor water quality conditions in the vicinity of St. Paul. Apparently related to improvements in

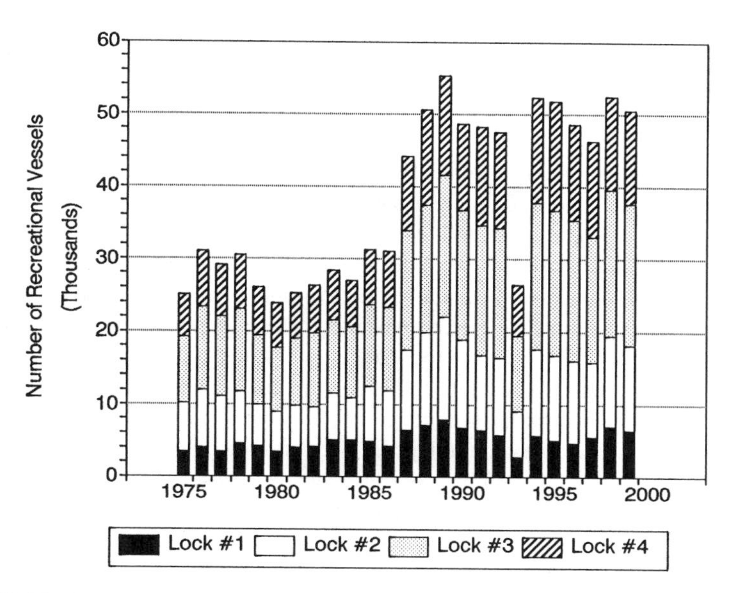

Figure 12-18 Recreational vessel traffic in Locks 1 through 4 of the Upper Mississippi River. *Source:* Erickson, 2000.

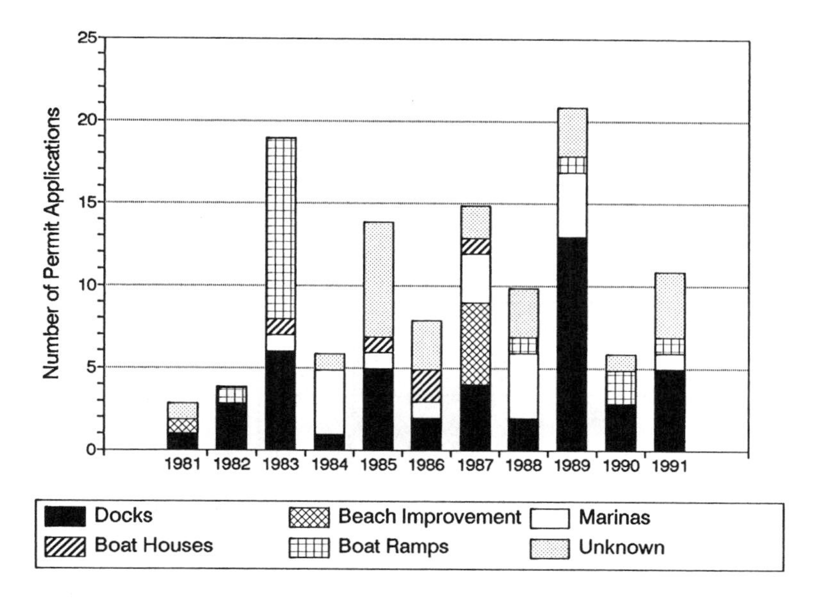

Figure 12-19 Recreational permit applications for Wabasha, Dakota, Washington, Goodhue, Pierce, and Pepin counties along the Upper Mississippi River. *Source:* Erickson, 2000.

water quality conditions, several marinas were proposed and constructed for this area of the river beginning in the mid-1980s. Lake City, located on Lake Pepin, for example, obtained a permit for a new marina in 1984; within a year, several hundred spaces were added for sailboats.

Increases in recreational uses of the river prompted eight agencies to form a partnership agreement in 1990 to study recreational trends and resolve conflicts over river and parkland use. The agencies included two park services, three state departments of natural resources, the U.S. Fish and Wildlife Service, the U.S. Army Corps of Engineers, and the Minnesota-Wisconsin Boundary Area Commission. They conducted a study to sort out the issues, uses, and resource management conflicts related to the rediscovery of the delights of a cleaned-up river by boaters, fishermen, and hikers (MPCA, 1993).

Partly because of the ban on DDT, the establishment of wildlife reserves, and reduced loadings of industrial pollutants from the pretreatment program, populations of water birds have increased in the Upper Mississippi River. Peregrine falcons, bald eagles, mallard ducks, and great blue herons have been observed in the Minneapolis–Saint Paul metropolitan area and in the floodplain wetlands located on the Upper Mississippi River near the Metro plant. Black crowned night herons have been observed feeding below the Ford Dam (Galli, 1992). Animals sensitive to the bioaccumulation of PCBs in their aquatic food, such as fish-eating mink, are also making a comeback (Smith, 1992). The number of great egrets and great blue herons nesting in Pig's Eye Lake has increased since the late 1970s and early 1980s, and cormorants has been observed nesting in the lake since 1983 (Galli, 1992) (Figure 12-20).

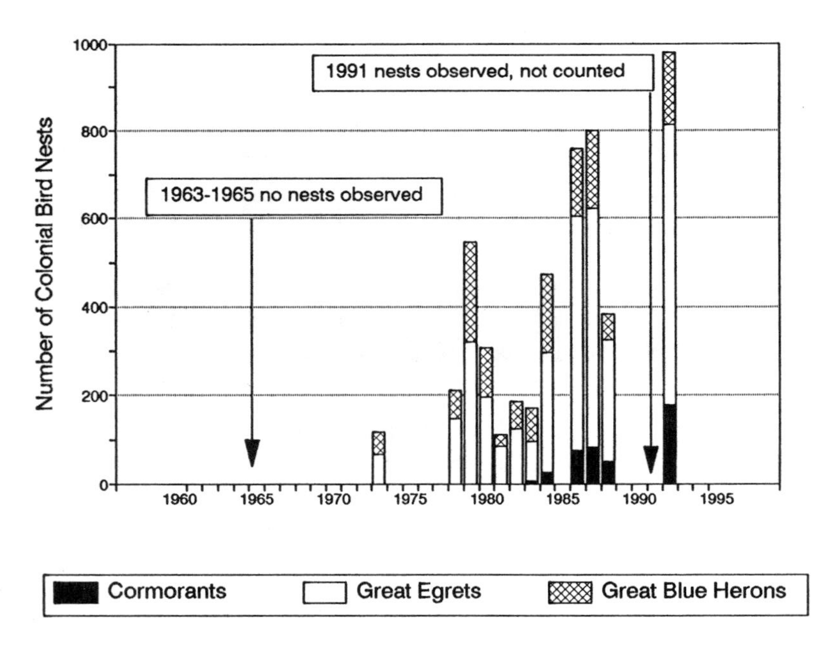

Figure 12-20 Colonial bird nest counts for Pigs Eye Lake. *Source:* Galli, 1992.

Electrofishing samples from Spring Lake, a backwater area affected by the Metro plant, were collected in 1981, 1986, and 1991. These samples showed an increase in species diversity and abundance of certain species (Gilbertson, 1992). The ecological quality of Spring Lake, as expressed by the Index of Biotic Integrity (IBI) (Karr, 1981; Karr et al., 1986), has improved since the mid-1980s (Figure 12-21). Species that have returned to Pool 2 include blue sucker and paddle fish; this is particularly noteworthy because paddle fish had not been observed in Pool 2 since the 1950s.

As in many other urban waterways of the United States, detectable levels of PCBs, a toxic organic chemical that adsorbs to sediment particles, have been identified in fish tissue and sediments as a result of contamination from industrial sources, transport of contaminated sediments, atmospheric deposition, stormwater runoff, and wastewater discharges. In 1975, PCB residues found in common carp and other fish species taken from the Upper Mississippi River exceeded the FDA action level of 5 mg/kg. The Minnesota Department of Health issued fish consumption advisories for a number of species including common carp, catfish, walleye, and smallmouth buffalo (MDH, 1998). Since the ban on production of PCBs in 1979, the level of PCBs in fish tissue in the Upper Mississippi River, as well as in many other rivers and lakes in the United States, has been declining. In the Upper Mississippi River, median PCB levels in common carp and walleye dropped by over 80 percent during the period between 1975–1979 and 1988–1995. Dramatic decreases have been recorded in fish tissue levels of PCBs from common carp collected in Pool 2 and Pool 4 (Lake Pepin) of the river. Lipid-normalized median PCB concentrations have declined in Pool 2 from 121 μg/g in 1975–1976 to 18 μg/g in 1987–1988 and in Lake Pepin from 62 μg/g in 1973–1974 to 16 μg/g in 1987–1988 (Biedron and Helwig, 1991)

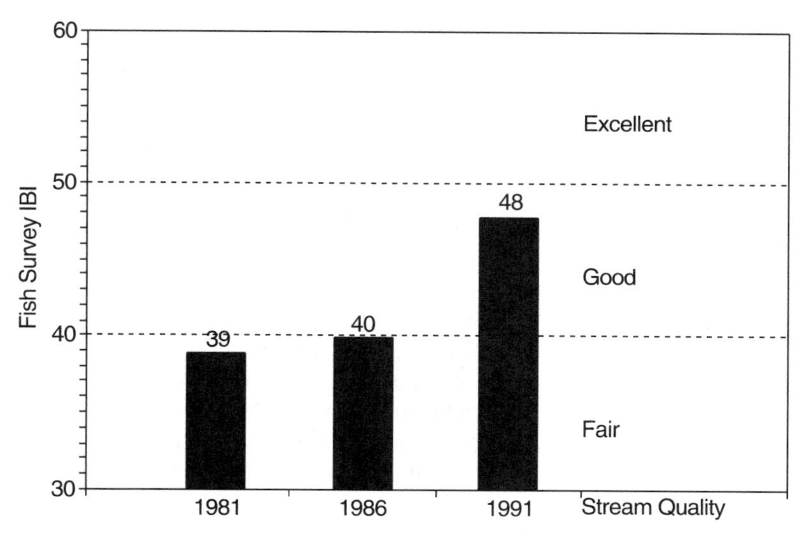

Figure 12-21 Fish survey IBI results for Spring Lake, backwater to Pool 2, which receives discharges from the Metro wastewater treatment plant. *Source:* Gilbertson, 1992.

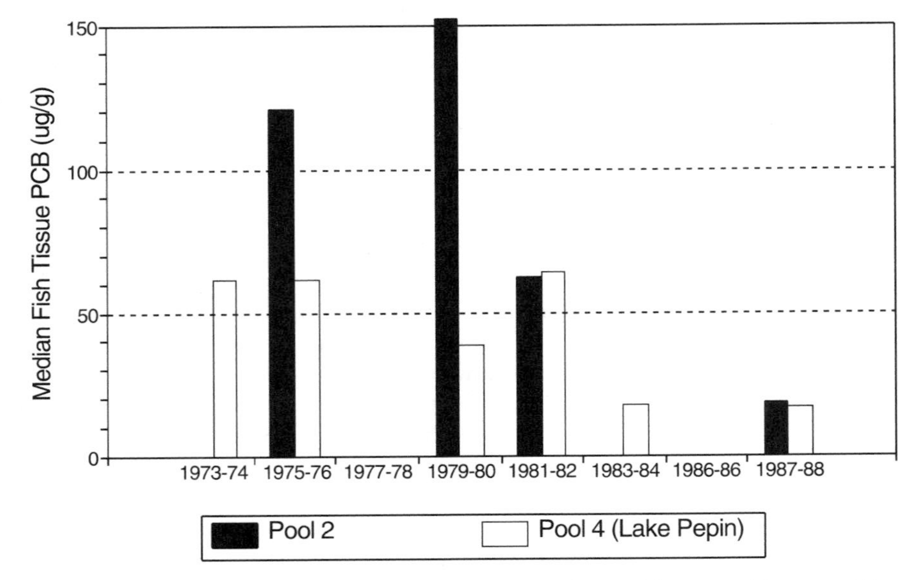

Figure 12-22 Lipid-normalized median PCB concentrations in fillet tissue (skin on) of 20 to 24.9-inch carp from Pool 2 and Pool 4 (Lake Pepin) of the Upper Mississippi River, 1973–1988. *Source:* Biedron and Helwig, 1991.

(Figure 12-22). Low levels of PCBs still persist, however, in fish tissue and other chemical pathways in the aquatic environment of the Upper Mississippi River, despite the PCB ban (Lee and Anderson, 1998).

SUMMARY AND CONCLUSIONS

As a result of strong state, local, and federal legislative actions with overwhelming public support, the cleanup of the Upper Mississippi River in the Twin Cities area is a national environmental success story. Comprehensive water pollution surveys dating back to 1926 documented the magnitude of the problems and provided the technical basis for the implementation of effective engineering proposals for abatement of water pollution in the river. Since enactment of the Clean Water Act in 1972, Minnesota has increased the water quality standard for DO to 5 mg/L and invested in upgrades to obtain better-than-secondary levels of wastewater treatment for the Metro plant and the other wastewater treatment plants operated by the MCES.

In contrast to the excessive effluent loading from the Metro plant during the 1960s, the investment in upgrades to the Metro plant during the 1980s, including nitrification, have succeeded in reducing effluent discharges of BOD_5 from 1968 to 1998 by 95 percent (Figure 12-9), suspended solids from 1968 to 1998 by 95 percent (Figure 12-10) and ammonia-N from 1982 to 1998 by 90 percent (Figure 12-11). As a direct result of these upgrades, compliance with water quality standards for DO has been achieved even under the low-flow conditions of the drought of 1988 (Lung, 1998). The accelerated program to separate stormwater and sanitary flow succeeded in achieving compliance with state standards (200 MPN/100 mL monthly geometric mean and 2,000 MPN/100 mL for individual samples) for fecal coliform bacteria at the 71 percent level for samples collected during 1996–1998. As a result of the industrial waste pretreatment program initiated in 1982, the discharge of heavy metals to the Metro plant (and the Upper Mississippi River) has been reduced by about 90 percent (MCES, 1999) and mercury loading from the Upper Mississippi River to Lake Pepin in 1990–1996 declined by almost 70 percent from the 1960s level (Balogh et al., 1999). Despite these significant improvements, MCES has targeted toxic chemicals (e.g., PCBs) and heavy metals (e.g., mercury) as contaminants of concern for monitoring, identification of sources, and reduction of the load discharged to the river.

In contrast to the degraded environmental conditions during the 1950s through 1970s, the Upper Mississippi River is no longer a place to avoid. Parks, trails, and marinas have been developed along the river in areas where no one would have considered making such investments in the 1970s. A thriving riverfront corridor increases the value of both commercial and residential properties along the waterfront. The city of St. Paul, for example, through the St. Paul Riverfront Corporation, has invested nearly $500 million (as of the mid-1990s) for land acquisitions and infrastructure development along the riverfront (Donlan et al., 1995). In the late 1980s, private developers began to respond to riverfront infrastructure investments by obtaining more than $7 million in tax increment financing for development along the riverfront (Donlan et al., 1995). In addition to private development, in December 1999, the Science Museum of Minnesota completed a new museum along the riverfront in St. Paul that features exhibits on the Upper Mississippi River.

The record clearly shows that the Clean Water Act of 1972 profoundly affected every community in Minnesota, including the Twin Cities. The CWA accelerated the cleanup of the Upper Mississippi River by providing federal funds for the construction of new wastewater collection and treatment systems and the upgrading of existing sewage treatment plants. Since the mid-1980s, the resurgence of mayflies and the record of greatly improved compliance with water quality standards for dissolved oxygen and fecal coliform bacteria are key indicators of the effectiveness of the water pollution control efforts accomplished by state, federal, and local governments in the Twin Cities.

By the end of 2005, the Metro plant will have implemented biological removal of phosphorus to meet an annual effluent level of 1 mg/L for phosphorus (MCES, 1998a). As of 1999, "BioP" had been successfully implemented in a portion of the Metro plant and at suburban plants discharging to the Minnesota River. Under the Metro Environment Partnership, the control of urban and rural runoff will be addressed by a $7.5 million commitment from the MCES to reduce pollution from the seven-county Twin Cities metropolitan region (MCES, 1998a). State-of-the-art technology for solids processing, approved in July 1998, will reduce mercury emissions by 70 percent along with other pollutants and odors (MCES, 1998a). Further reductions in mercury discharges to the river from the Metro plant will be accomplished as a result of a partnership between the MCES and the Minnesota Dental Association to test and evaluate new technologies to filter dental amalgam from wastewater (MCES, 1998b, 2000).

Although significant accomplishments have been made to improve water quality and ecological conditions in the Upper Mississippi River, continued investments are needed to address contemporary issues for continued restoration and maintenance of the ecological integrity of the river. The designation of the Upper Mississippi River as an American Heritage River in July 1998 recognizes both the significant environmental improvements that have been accomplished and the continuing need to address the key ecological issues identified in the 1990s. The key water quality and resource management issues identified for the Upper Mississippi River (USGS, 1999b) for the twenty-first century include the following:

- Point and nonpoint source loading of nutrients, sediments, heavy metals, and toxic chemicals in the Minnesota River and Upper Mississippi River from agricultural and urban land uses.
- Point and nonpoint source loading of nutrients and pesticides to aquifer systems from agricultural land uses.
- Contamination of groundwater with toxic chemicals from industrial activities and leachate from landfills.
- Contamination of surface waters and groundwater in areas characterized by rapid urbanization.
- Degradation of biological communities by riparian and bottom habitat losses, river channel modifications, construction of locks and dams, increasing backwater sedimentation rates and loss of wetlands, effects of reservoir operations on fisheries, and eutrophication.
- Contamination of bottom sediments in the river with toxic chemicals and subsequent benthic release and bioaccumulation of toxic substances within the aquatic food chain.

Water quality in the Upper Mississippi River, as measured by indicators presented in this chapter such as DO, ammonia, fecal coliform bacteria, and sediment levels of mercury, has improved greatly since the 1960s and 1970s as a result of upgrades to wastewater treatment plants required by the 1972 CWA. Despite these improvements, contaminant loading from municipal (nutrients) and industrial (heavy metals, toxic chemicals) dischargers and runoff from urban (heavy metals) and agricultural (nutrients, pesticides, sediments) watersheds continue to adversely affect the ecological integrity of the Upper Mississippi River. In addition to chemical inputs to the river, the Upper Mississippi River Conservation Committee has warned that the ecosystem of the Upper Mississippi River is threatened by structural alterations of the river such as continued stream channelization, flood control levees that separate the river from the floodplain, and the proposed expansion of the commercial navigation infrastructure (UMRCC, 1994). If the current ecological benefits are to be maintained and degraded ecological conditions restored, an ongoing effort will be needed to maintain environmental monitoring and research programs to document the status and trends of the Upper Mississippi River to provide the scientific data needed for effective resource management decisions (USGS, 1998).

REFERENCES

Balogh, S. T., D. R. Engstrom, J. E. Almendinger, M. L. Meyer, and D. K. Johnson. 1999. History of mercury loading in the Upper Mississippi River reconstructed from the sediments of Lake Pepin. Environ. Sci. Technol. 33(19): 3297–3302.

Berner, E. K. and R. A. Berner. 1996. Global environment water, air and geochemical cycles. Prentice Hall, Upper Saddle River, NJ.

Biedron, C. J. and D. D. Helwig. 1991. PCBs in common carp of the Upper Mississippi River: Investigation of trends from 1973–1988 and the design of a long term fish monitoring program. Technical report. Minnesota Pollution Control Agency, Water Quality Division, St. Paul, MN.

Buttleman, K. and W. Moore. 1999. Update on elevated levels of fecal coliform bacteria in the Mississippi River above Lock and Dam No. 1. Technical memorandum to chair and members of the Environment Committee from Keith Buttleman, Director, Environmental Planning and Evaluation Department, MCES, and Bill Moore, Director, Wastewater Services Department, MCES. Metropolitan Council Environmental Services, St. Paul, MN. January 18.

Chapra, S. C. 1997. Surface water quality modeling. McGraw-Hill, New York.

Christen, K. 1999. Gulf dead zone grows, perplexes scientists. Environ. Sci. Technol. News & Research Notes 4(10): 396a–397a.

Donlan, M. C., M. D. Ewen, T. B. Peterson, and B. G. Morrison. 1995. The costs and benefits of municipal wastewater treatment: Upper Mississippi and Potomac River case studies. Final technical report prepared for Tetra Tech, Inc., Fairfax, VA, and U.S. Environmental Protection Agency, Office of Water, Policy and Resource Management Office, Washington, DC, by Industrial Economics, Inc., Cambridge, MA. September.

Engstrom, D. R. and J. E. Almendinger. 1998. Historical changes in sediment and phosphorus loading to the Upper Mississippi River: Mass-balance reconstruction from the sediments of Lake Pepin. Final research report prepared for the Metropolitan Council Environmental Services, St. Paul, MN by St. Croix Watershed Research Station, Science Museum of Minnesota, Marine on St. Croix, MN, June.

EnviroTech. 1992. Post-audit of the Upper Mississippi River Model using 1988 summer low flow water quality data. Final technical report submitted to Metropolitan Waste Control Commission, St. Paul, MN by EnviroTech Associates, Inc., Charlottesville, VA. November.

EnviroTech. 1993. Water quality modeling of the Upper Mississippi River and Lake Pepin. Mississippi River phosphorus study, Section 6: Part A. Technical report submitted to Metropolitan Waste Control Commission, St. Paul, MN by EnviroTech Associates, Inc., Charlottesville, VA. March.

Erickson, D. 2000. U.S. Army Corps of Engineers, St. Paul District. Data extracted from general database that includes data from annual reports of "Lock Performance Monitoring System (LPMS), Summary of Lock Statistics." Navigation Data Center, U.S. Army Corps of Engineers, Water Resources Support Center, Alexandria, VA. Personal communication. January 24.

Forstall, R. L. 1995. Population by counties by decennial census: 1900 to 1990. Population Division, U.S. Bureau of the Census, Washington, DC. http://www.census.gov/population/www/censusdata/cencounts.html

Fremling, C. R. 1964. Mayfly distribution indicates water quality on the Upper Mississippi River. Science 146: 1164–1166.

Fremling, C. R. and D. K. Johnson. 1990. Recurrence of *Hexagenia* mayflies demonstrates improved water quality in pool 2 and Lake Pepin, Upper Mississippi River. In: Mayflies and stoneflies: Life Histories and Biology. ed. I. C. Campbell, pp. 243–248. Kluwer Academic Publishers. Dordrecht, The Netherlands.

FWPCA. 1966. Pollution of the Upper Mississippi River and major tributaries. NTIS No. PB-217-267. U.S. Department of the Interior, Federal Water Pollution Control Administration, Great Lakes Region, Twin Cities Upper Mississippi River Project, Chicago, IL.

Galli, J. 1992. Minnesota Department of Natural Resources. St. Paul, MN. Personal communication, November 24.

Garland, E., J. J. Szydlik, D. D. Di Toro, and C. E. Larson. 1999. Framework for point and non-point source nutrient control evaluations. Presented at WEFTEC'99, New Orleans, LA, October 9–13. Water Environment Federation, Alexandria, VA.

Gilbertson, B. 1992. Minnesota Department of Natural Resources, Metro Region Headquarters, Division of Fish and Wildlife, St. Paul, MN. Personal communication, September 11.

HydroQual. 1999a. Advanced eutrophication model of the Upper Mississippi River, Summary report. Draft report prepared for Metropolitan Council Environmental Services, St. Paul, MN, by HydroQual, Inc., Mahwah, NJ. April.

HydroQual. 1999b. Advanced eutrophication model of Pools 2 to 4 of the Upper Mississippi River. Draft report prepared for Metropolitan Council Environmental Services, St. Paul, MN, by HydroQual, Inc., Mahwah, NJ. April.

Hydroscience. 1979. Upper Mississippi River 208 Grant water quality modeling study. Technical report prepared for Metropolitan Waste Control Commission, St. Paul, MN, by Hydroscience, Westwood, NJ. August.

Iseri, K. T. and W. B. Langbein. 1974. Large rivers of the United States. Circular No. 686. U.S. Department of the Interior, U.S. Geological Survey, Reston, VA.

James, W. F., J. W. Barko, and H. L. Eakin. 1996. Analysis of nutrient/seston fluxes and phytoplankton dynamics in Lake Pepin (Upper Mississippi River), 1996: Third annual report. Technical report submitted to Metropolitan Council Environmental Services, St. Paul, MN.

James, W. F., J. W. Barko, and H. L. Eakin. 1999. Diffusive and kinetic fluxes of phosphorus from sediments in relation to phosphorus dynamics in Lake Pepin, Upper Mississippi River. Misc. paper No. W-99–1. U.S. Army Corps of Engineers, Engineering Research and Development Center, Vicksburg, MS.

Johnson, D. K. 1999. Environmental studies of phosphorus in the Upper Mississippi River, 1994–1998. Presented at Mississippi River Research Consortium, Annual Meeting, La Crosse, WI, April 22–23.

Johnson, D. K. and P. W. Aasen. 1989. The metropolitan wastewater treatment plant and the Mississippi River: 50 years of improving water quality. J. Minn. Acad. Sci. 55(1): 134–138.

Karr, J. R. 1981. Assessment of biotic integrity using fish communities. Fisheries 6(6): 21–27.

Karr, J. R., K. D. Fausch, P. L. Angermeier, P. R. Yant, and I. J. Schlosser. 1986. Assessing biological integrity in running waters: A method and its rationale. Special Publication 5. Illinois Natural History Survey, Champaign, IL.

Larson, C. E. 1999. Metropolitan Council Environmental Services, St. Paul, Minnesota. Personal communication, March 12.

Larson, C. E. 2001. Metropolitan Council Environmental services, St. Paul. Minnesota. Personal communication. August 30.

Lee, K. E., and J. P. Anderson. 1998. Water quality assessment of the Upper Mississippi River Basin, Minnesota and Wisconsin—Polychlorinated biphenyls in common carp and walleye fillets, 1975–95. U.S. Geological Survey. Originally published as Water Resources Investigations Report 98-4126. Available on USGS Water Resources in Minnesota web site. http://wwwmn.cr.usgs.gov/pcb.htm

Lung, W. 1996a. Post-audit of Upper Mississippi River BOD/DO model. J. Environ. Eng., ASCE 122(5): 350–358.

Lung, W. 1996b. Fate and transport modeling using numerical tracers. Water Resources Res. 32(1): 171–178.

Lung, W. 1998. Trends in BOD/DO modeling for waste load allocations. J. Environ. Eng., ASCE 124(10): 1004–1007.

Lung, W. and C. Larson. 1995. Water quality modeling of Upper Mississippi River and Lake Pepin. J. Environ. Eng., ASCE 121(10): 691–699.

Malakoff, D. 1998. Death by suffocation in the Gulf of Mexico. Science 281 (July 10): 190–192.

MCES. 1996. Separating combined sewers to improve and protect Mississippi River water quality: A ten-year commitment. Annual progress report to the public, Year ten. Legislative and regulatory history. Cities of Minneapolis, St. Paul, and South St. Paul, and Metropolitan Council Environmental Services, St. Paul, MN.

MCES. 1998a. Update (newsletter). Metropolitan Council Environmental Services, St. Paul, MN. August 3 and November 25 issues.

MCES. 1998b. Protecting our future. Biennial report 1996–98. Metropolitan Council Environmental Services, St. Paul, MN. September.

MCES. 1999. 100 years of improvements in the Twin Cities. Metropolitan Council Environmental Services, St. Paul, MN. March.

MCES. 2000. Council directions (newsletter). Metropolitan Council Environmental Services, St. Paul, MN. January/February.

MDH. 1998. Minnesota fish consumption advisory. Minnesota Department of Health, St. Paul, MN. May.

MDNR. 1988. Fifty years to a cleaner Mississippi River. Minnesota Volunteer. Minnesota Department of Natural Resources, St. Paul, MN. July–August.

Meyer, M. L. and S. M. Schellhaass. 1999. Phosphorus sources and Upper Mississippi River water quality. Presented at Mississippi River Research Consortium, Annual Meeting, La Crosse, WI, April 22–23.

Mississippi River/Gulf of Mexico Watershed Nutrient Task Force. 2001. Action plan for reducing, mitigating, and controlling hypoxia in the northern Gulf of Mexico. U.S. Environmental Protection Agency, Office of Water, Washington, DC. January 18.

http://www.epa.gov/msbasin/actionplan.html

Mockovak, C. 1990. Mississippi River water quality mirrors metro area history. Minnesota Environment. Minnesota Pollution Control Agency, St. Paul, MN. March/April.

Moffat, A. 1998. Global nitrogen overload problem grows critical. Science 279: 988–989. Research News, (February 13).

MPCA. 1981. Mississippi River waste load allocation study. Technical report prepared by Division of Water Quality, Minnesota Pollution Control Agency, St. Paul, MN.

MPCA. 1993. Mississippi River phosphorus study, Section 8, Benefit inventory report. Technical report prepared by Water Quality Division, Minnesota Pollution Control Agency, St. Paul, MN. March.

MPCA. 1994. Minnesota River assessment project report. Executive summary report to the Legislative Commission on Minnesota Resources. Prepared by Minnesota Pollution Control Agency, St. Paul, MN. January.

MRI. 1976. Water Pollution Control Act of 1972 environmental impact assessment Upper Mississippi River Basin. NTIS No. PB-251-222. Technical report NCWQ 75/74 prepared for National Commission on Water Quality, Washington, DC, by Midwest Research Institute, Minneapolis, MN.

Mulla, D. J., A. Sekely, D. Wheeler, and J. C. Bell. 1999. Historical trends affecting accumulation of sediment and phosphorus in Lake Pepin. Draft technical report prepared for the Metropolitan Council Environmental Services by Department of Soil, Water and Climate, University of Minnesota, St. Paul, MN.

MWCC. 1988. 50 years treating the Mississippi right. Metropolitan Waste Control Commission, St. Paul, MN.

MWCC. 1989. 1988 Mississippi River low flow survey report. Report No. QC-88-162. Water Quality Monitoring and Analysis Division, Quality Control Department, Metropolitan Waste Control Commission, St. Paul, MN. August.

MWCC. 1994. 1992 river quality report. Report No. QC-92-284. Water Quality Division, Quality Control Department, Metropolitan Waste Control Commission, St. Paul, MN. June.

OMB. 1999. OMB Bulletin No. 99-04. Revised statistical definitions of Metropolitan Areas (MAs) and guidance on uses of MA definitions. U.S. Census Bureau, Office of Management and Budget, Washington, DC.

http://www.census.gov/population/www/estimates/metrodef.html

Phillips, D. 1999. Life on the Mississippi: A voyage of commerce and conflict. The Washington Post, Business Section, pp. H1–H8. December 19.

Rabelais, N., R. E. Turner, D. Justic, Q. Dortch, W. J. Wiseman, Jr., and B. K. Sen Gupta. 1996. Nutrient changes in the Mississippi River and system responses on the adjacent continental shelf. Estuaries 19: 386–407.

Smith, S. 1992. U.S. Fish and Wildlife Service, Fort Snelling, MN. Personal communication. September 14.

Thomann, R. V. and J. A. Mueller. 1987. Principles of surface water quality modeling and control. Harper & Row Publishers, New York.

UMRCC. 1994. Facing the threat: An ecosystem management strategy for the Upper Mississippi River system. Upper Mississippi River Conservation Committee, Rock Island, IL. January.

USDOC. 1998. Census of population and housing. U.S. Department of Commerce, Economics and Statistics Administration, Bureau of the Census—Population Division, Washington, DC.

USEPA (STORET). STOrage and RETrieval Water Quality Information System. U.S. Environmental Protection Agency, Office of Wetlands, Oceans, and Watersheds, Washington, DC.

USGS. 1998. Ecological status and trends of the Upper Mississippi River System 1998: A report of the Long Term Resource Monitoring Program. Press release "Old Man River Gets Health Checkup." U.S. Geological Survey, Biological Resources Division, Upper Midwest Environmental Sciences Center, La Crosse, WI.

http://www.umesc.usgs.gov/reports_publications/status_and_trends.html

USGS. 1999a. Mississippi River at St. Paul, MN (05331000). U.S. Geological Survey, National Water Information Service (NWIS-W). Daily flow data, Minnesota NWIS-S. http://waterdata.usgs.gov/nwis-s/MN/data.components/hist.cgi?statnum=05331000 Summary statistics of streamflow from USGS Water Resources in Minnesota. http://wwwmn.cr.usgs.gov/umis/descript1.html

USGS. 1999b. Upper Mississippi River National Water Quality Assessment Study, Study unit description, General information. U.S. Geological Survey. USGS Water Resources in Minnesota.
http://wwwmn.cr.usgs.gov/umis/descript1.html

USPHS. 1951. Upper Mississippi drainage basin: A cooperative state-federal report on water pollution. NTIS No. PB-215–584. Federal Security Agency, U.S. Public Health Service, Chicago, IL.

USPHS. 1953. Upper portion Upper Mississippi River drainage basin: A cooperative state-federal report on water pollution. NTIS No. PB-215–864. Water Pollution Series No. 57, U.S. Public Health Service, Chicago, IL.

Vitousek, P., J. D. Aber, R. W. Howarth, G. E. Likens, P. A. Matson, D. W. Schindler, W. H. Schlesinger, and D. Tilman. 1997. Human alterations of the global nitrogen cycle: Sources and consequences. Ecol. Applic. 7(3): 737–750.

Wahl, K. L., K. C. Vining, and G. J. Wiche. 1993. Precipitation in the Upper Mississippi River Basin, January 1 through July 31, 1993: Floods in the Upper Mississippi River Basin, 1993. U.S. Geological Survey Circular 1120-B. U.S. Department of the Interior, U.S. Geological Survey, Denver, CO.

Wolman, M. G. 1971. The nation's rivers. Science 174: 905–917.

WRE. 1975. Water quality analysis of the Upper Mississippi River Basin. NTIS No. PB-250-982, Technical report NCWQ 75/64 prepared for National Commission on Water Quality, Washington, DC, by Water Resources Engineers, Walnut Creek, CA.

Zwick, D. and M. Benstock. 1971. Water wasteland: Ralph Nader's study group report on water pollution. The Center for Study of Responsive Law, Grossman Publishers, New York.

CHAPTER 13

Willamette River Case Study

The Pacific Northwest basin, covering a drainage area of 277,612 square miles, includes the "mighty" Columbia River. Based on its mean annual discharge (262,000 cfs, 1941–1970), the Columbia is the second largest river in the continental United States (Iseri and Langbein, 1974). With a length of 270 miles, a drainage area of 11,200 square miles, and a mean annual discharge of 35,660 cfs (1941–1970), the Willamette River is the fifteenth largest waterway in the United States ranked on the basis of annual discharge (Iseri and Langbein, 1974). Figure 13-1 highlights the location of the Willamette River case study watersheds (catalog units), identified in the Pacific Northwest basin as major urban-industrial areas affected by severe water pollution problems during the 1950s and 1960s (see Table 4-2). In this chapter, information is presented to characterize long-term trends in population, municipal wastewater infrastructure and effluent loading of pollutants, ambient water quality, environmental resources, and uses of the Willamette River. Data sources include USEPA's national water quality database (STORET), published technical literature, and unpublished technical reports ("grey" literature) obtained from local agency sources.

The Willamette River extends for 270 miles from its headwaters in the southern Cascade Mountains in Douglas County, Oregon, to the city of Portland, Oregon, where it meets the tidal Columbia River (Figure 13-2) (Iseri and Langbein, 1974). More than two-thirds of Oregon's population lives within the major urban centers that have developed in the valley. The basin provides extensive natural habitat for fish and wildlife and supports a prosperous economy based on agriculture, timber and wood products, and recreation.

The Willamette River was once one of the nation's most grossly polluted waterways because of raw sewage discharges and inadequate levels of municipal and industrial waste treatment. Since the late 1920s, when a survey found that nearly half of the citizens of Portland were in favor of antipollution laws, public opinion in Oregon has strongly favored regulatory controls on wastewater discharges to clean up the Willamette River. As a result of strong legislative actions with overwhelming public support, the cleanup has become a major national environmental success. In particular, Oregon's legislative actions mandating a minimum level of secondary waste treatment have played an important role in restoring the ecological balance of the Willamette.

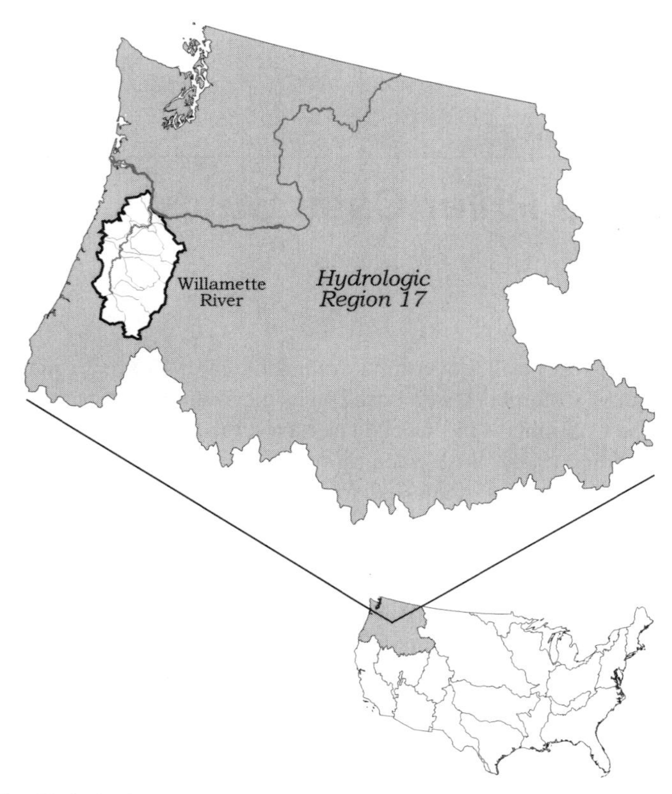

Figure 13-1 Hydrologic Region 17 and Willamette watersheds.

PHYSICAL SETTING AND HYDROLOGY

With a watershed of 11,200 square miles, the Willamette River basin in northwestern Oregon is bounded by the Coast (west) and Cascade (east) mountain ranges, which have a north-south length of 150 miles and an east-west width of 75 miles (Figure 13-2). Elevations range from less than 10 feet at the mouth near the Columbia River to 450 feet in the valley near Eugene to greater than 10,000 feet in the headwaters of the Cascade mountain range. Physical transport in the river can be described in terms of three distinctive physiographic reaches and characterized by the key physical parameters that strongly influence water quality—length, summer low-flow velocity, and travel time (Table 13-1). The longer travel time in the tidal portion of the Willamette River (10 days) can lead to decreased water quality.

Seasonal variation in the river flow is the result of the region's heavy winter rains and spring snowmelt from November through March. Low-flow conditions occur during the summer months of July through September, with the seasonal minimum occurring during August. Based on data from 1940–1990, monthly average flows

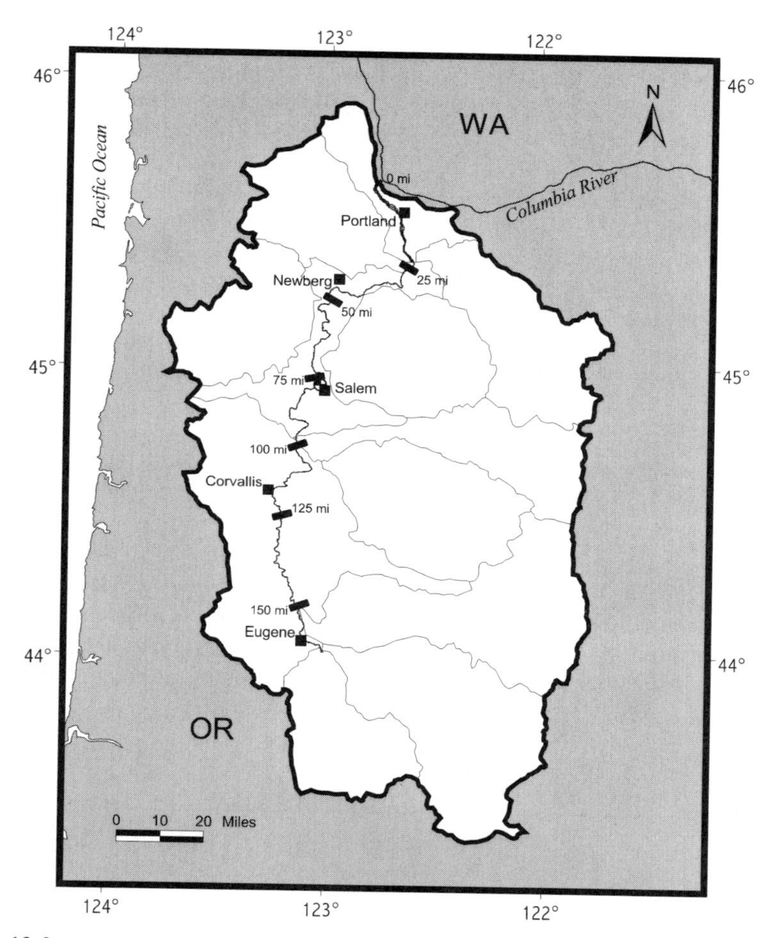

Figure 13-2 Location map of Willamette River basin. (River miles shown are distances from the confluence of the Willamette River with the Columbia River at Portland, OR.)

TABLE 13-1 Physical Characteristics of Willamette River at 6,000 cfs

Reach	Length (miles)	Average Velocity (cm/sec)	Travel Time (days)
Upstream	135.0	60	2.8
Newberg Pool	25.5	8	3.9
Tidal	26.5	3	10.0

Source: Rickert et al., 1976.

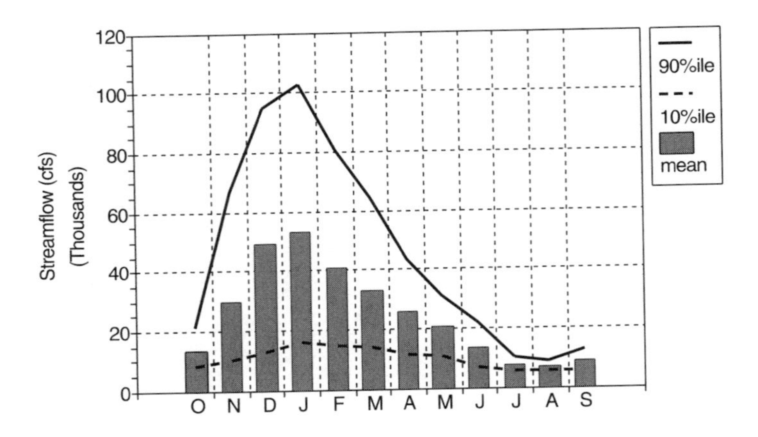

Figure 13-3 Monthly mean, tenth, and ninetieth percentile streamflow of the Willamette River at Salem, Oregon (USGS Gage 14191000), 1951–1980. *Source:* USGS, 1999.

range from 6,246 cfs in August to 48,060 cfs in January (Figure 13-3). Before 1953, the natural summer low flow ranged from 2,500 cfs to 5,000 cfs at Salem. Since 1953, flow augmentation by 14 U.S. Army Corps of Engineers (USACE) reservoirs has been used to maintain a summer low flow of about 6,000 cfs at Salem (Hines et al., 1976) (Figure 13-4).

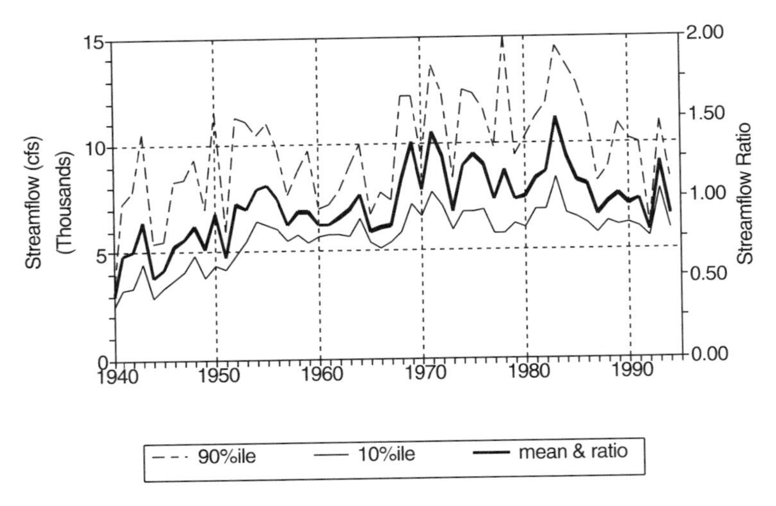

Figure 13-4 Long-term trends of summer mean, tenth, and ninetieth percentile streamflow of the Willamette River at Salem, Oregon (USGS Gage 14191000), July–September. *Source:* USGS, 1999.

POPULATION, WATER, AND LAND USE TRENDS

Because of abundant natural resources, the river has played a key historical role in the agricultural and industrial development of the valley. The Willamette River, a major source for the basin's municipal (20 cities) and industrial (600 facilities) water supply, also provides irrigation water for the rich fruit and vegetable farms of the valley. Other major uses include commercial navigation, hydroelectric power production, commercial and recreational fisheries, and water-based recreational activities, including aesthetic enjoyment of the Greenway Trail along the length of the river. As the region has grown, the river has also been used, and misused, for municipal and industrial waste disposal, including the disposal of wastewater generated by the pulp and paper industry since the 1920s.

Oregon's three largest cities—Salem, Portland, and Eugene—with a total population of 1.8 million (nearly 70 percent of the state's population) are within the Willamette River basin. The population of the basin has steadily increased since World War II. With a significant wood products and agricultural economy, the Willamette basin accounts for about 70 percent of the total industrial production of Oregon. Industrial production, like the population of the basin, has steadily increased over the past several decades.

The Willamette River case study area includes a number of counties identified by the Office of Management and Budget (OMB) as Metropolitan Statistical Areas (MSAs) or Primary Metropolitan Statistical Areas (PMSAs). Table 13-2 lists the MSAs and counties included in this case study. Figure 13-5 presents long-term population trends (1940–1996) for the counties listed in Table 13-2. From 1940 to 1996 the population in the area more than tripled (Forstall, 1995; USDOC, 1998).

TABLE 13-2 Metropolitan Statistical Areas (MSAs) and Counties in the Willamette River Case Study

Portland–Salem, OR–WA CMSA
Clackmas County, OR
Columbia County, OR
Marion County, OR
Multnomah County, OR
Polk County, OR
Washington County, OR
Yamhill County, OR
Clark County, WA
Corvallis, OR MSA
Benton County, OR

Source: OMB, 1999.

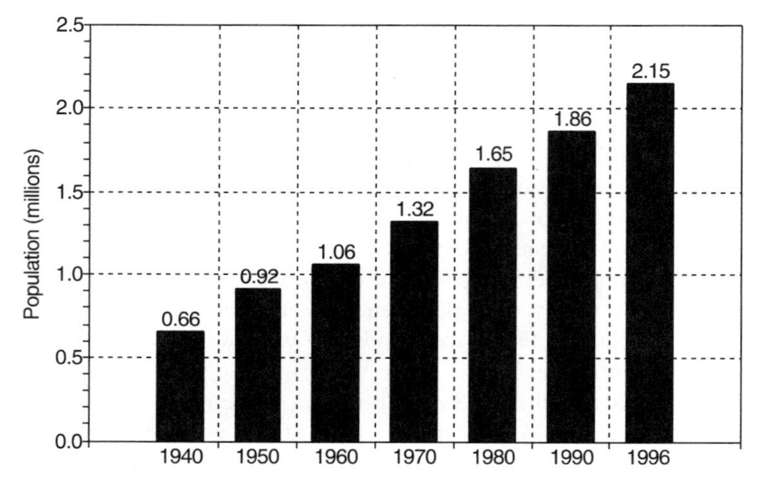

Figure 13-5 Long-term trends in population in the Willamette River basin. *Sources:* Forstall, 1995; USDOC, 1998.

HISTORICAL WATER QUALITY ISSUES

In the early 1920s, the Oregon Board of Health determined that the Lower Willamette River near Portland was grossly polluted as a result of raw waste discharges from municipal and industrial sources. In 1927, the Portland City Club declared the Willamette "ugly and filthy" with "intolerable" conditions. The first comprehensive water quality survey in 1929 found severely declining oxygen levels downstream of Newberg with an estimated concentration of 0.5 mg/L at the confluence with the Columbia River. Not surprisingly, bacteria levels were also found to be significantly increased downstream of each major city along the river. Industrial disposal from pulp and paper mills had resulted in extensive bottom sludge deposits that frequently surfaced during summer low-flow conditions as noxious, unsightly floating mats of sludge. By 1930, the municipal waste from the 300,000 inhabitants of Portland flowed untreated into Portland Harbor, resulting in severe oxygen depletion during the summer (Oregon State Sanitary Authority, 1964; Gleeson, 1972).

During the 1950s, Kessler Cannon, a state official, described the Willamette River from Eugene to the Columbia River as the "filthiest waterway in the Northwest and one of the most polluted in the nation." Gross water pollution conditions resulted in high bacteria counts, oxygen depletion, and fish kills (e.g., Gleeson and Merryfield, 1936; Merryfield et al., 1947; Merryfield and Wilmot, 1945). Cannon recounted the noxious conditions in the Willamette: "As the bacteria count rose, oxygen levels dropped—to near zero in some places. Fish died. The threat of disease put a stop to safe swimming. Rafts of sunken sludge, surfacing in the heat of summer, discouraged water-skiing and took the pleasure out of boating" (Starbird and Georgia, 1972). In

1967, the Izaak Walton League described the Lower Willamette River as a "stinking slimy mess, a menace to public health, aesthetically offensive, and a biological cesspool" (USEPA, 1980).

LEGISLATIVE AND REGULATORY HISTORY

After more than a decade of public concern about the polluted conditions of the Willamette River, the citizens of Oregon passed a referendum in 1938 setting water quality standards and establishing the Oregon State Sanitary Authority. With the establishment of the Sanitary Authority, it became Oregon's public policy to restore and maintain the natural purity of all public waters. As a result of regulatory actions by the Sanitary Authority, all municipalities discharging into the Willamette implemented primary treatment during the period from 1949 to 1957, with all costs borne by the municipalities. Beginning in 1952, industrial waste discharges from the pulp and paper mills were controlled by required lagoon diversions during summer months. In 1953, the new U.S. Army Corps of Engineers dams began to operate, resulting in augmentation of the natural summer low flow. Although not originally planned for water quality management, summer reservoir releases have become a significant factor in maintaining water quality and enabling salmon migration during the fall.

Although tremendous accomplishments had been made in controlling water pollution in the Willamette basin, large increases in industrial production and in the population served by municipal wastewater plants exceeded the assimilative capacity of the river. By 1960, the Sanitary Authority required that all municipalities discharging to the Willamette River achieve a minimum of secondary treatment (85 percent removal of BOD_5). In 1964, the pulp and paper mills were directed to implement primary treatment, with secondary treatment during the summer months. In 1967, industrial secondary treatment was required on a year-round basis. The Sanitary Authority had thus established a minimum policy of secondary treatment for all municipal and industrial waste dischargers with the option of requiring tertiary treatment if needed to maintain water quality. The state initiated the issuance of discharge permits for wastewater plants in 1968, 4 years before the 1972 CWA established the National Pollutant Discharge Elimination System (NPDES). The policy adopted in 1967 remains the current water pollution control policy of the state of Oregon for the Willamette River (ODEQ, 1970).

In response to the 1965 Federal Water Quality Act, Oregon established intrastate and interstate water quality standards in 1967 that were among the first new state water quality standards to be approved by the federal government. The 1972 CWA provided even further authority for Oregon to issue discharge permits limiting the pollutant loading from municipal and industrial facilities.

From 1956 to 1972, Federal Construction Grants to Oregon totaled $33.4 million for municipal wastewater facilities (CEQ, 1973). Since 1974, the cities of Salem, Corvallis, and Portland have received Construction Grants under the 1972 CWA to build and upgrade secondary wastewater treatment facilities.

IMPACTS OF WASTEWATER TREATMENT

Pollutant Loading and Water Quality Trends

As a result of the stringent regulatory requirements for municipal and industrial wastewater treatment, total pollutant loading has decreased substantially over the past 30–40 years (Figure 13-6), while total wastewater flow has increased over the same period. By 1972, when the CWA was passed, the total oxygen demand of wastewater discharges to the Willamette had been decreased to 25 percent of the demand of the pollutant load discharged in 1957 (CEQ, 1973). Following the implementation of basinwide secondary treatment for municipal and industrial wastewater sources, water quality model budgets have shown that about 46 percent of the oxygen demand in the Willamette River during the critical summer months results from upstream nonpoint source loads from rural tributary basins. The remaining half of the total oxygen demand is accounted for by municipal (22 percent) and industrial (32 percent) point source loads (Rickert and Hines, 1978).

Severe summer oxygen depletion has been the key historical water quality problem in the Willamette River. Since the 1970s, however, summer oxygen levels have increased significantly as a result of: (1) the implementation of basinwide secondary treatment for municipal and industrial point sources, and (2) low-flow augmentation from reservoir releases. Based on data obtained from the earliest water quality survey in 1929 to the most recently available monitoring programs, the dramatic improvements in summer oxygen levels in the river are clearly shown in the spatial distribution of oxygen from Salem to Portland Harbor (Figure 13-7) and the long-term historical trend for oxygen in the Lower Willamette River near Portland Harbor (Figure 13-8). These historical data sets document the grossly polluted water quality condi-

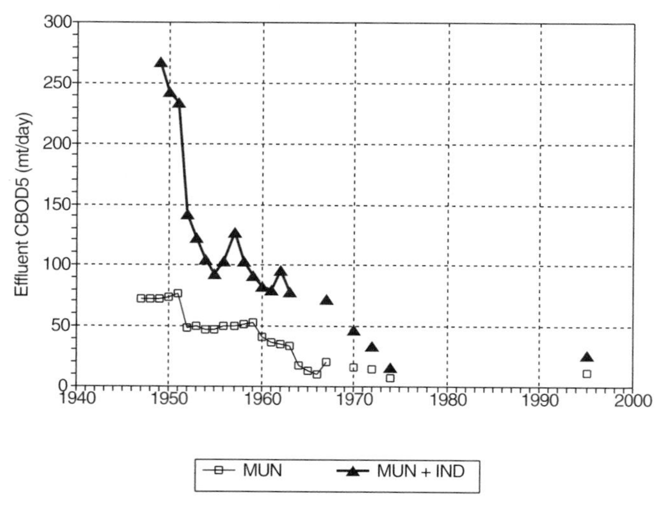

Figure 13-6 Long-term trends in municipal and industrial effluent BOD_5 loading to the Willamette River. *Sources:* Gleeson, 1972; ODEQ, 1970; Bondelid et al., 2000.

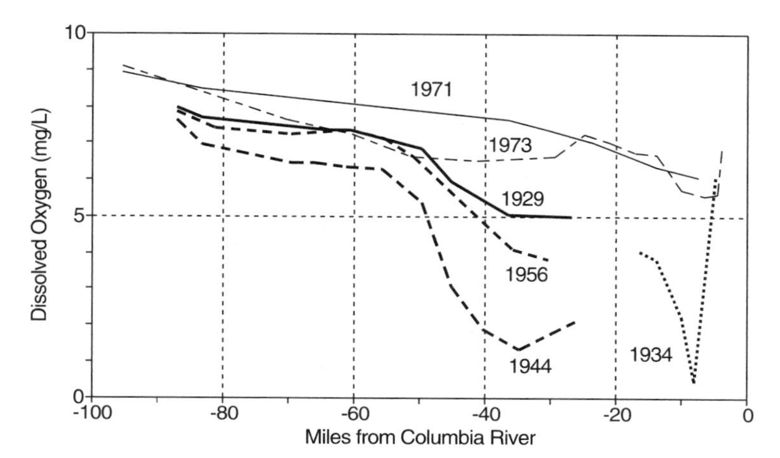

Figure 13-7 Long-term trends in the spatial distribution of DO in the Willamette River. *Source:* Reprinted with permission from D. A. Rickert and W. G. Hines, River quality assessment: Implications of a prototype project, Science 200 (June 9, 1978): 1113–1118, Copyright © 1978 American Association for the Advancement of Science.

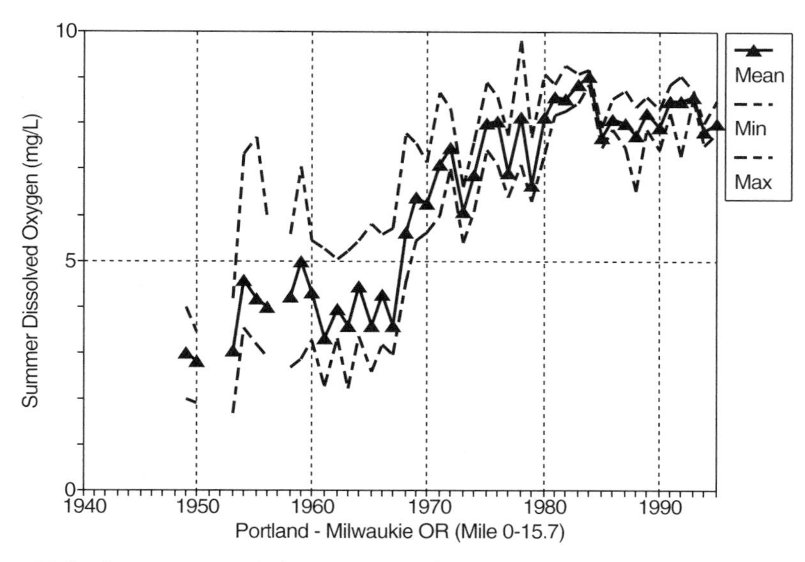

Figure 13-8 Long-term trends in summer DO in the Lower Willamette River at Portland, OR, for RF1 reach 17090012017 (River Mile 0–15.7). *Source:* USEPA STORET.

tions that existed prior to implementation of a minimum level of secondary treatment for municipal and industrial discharges to the river.

Although the current status of the river is visibly much improved and water contact sports and salmon migration are once again possible in most of the river, there are still concerns about the levels of toxic contamination. Oregon's 1990 water quality status assessment report (ODEQ, 1990a) classified the river as "water quality limited" as a result of seven contaminants exceeding USEPA draft sediment guidelines (arsenic, chromium, lead, zinc, and DDT), state water quality standards (arsenic), or both (2,3,7,8-TCDD). Surveys have found levels of toxic chemicals in water, sediments, and fish tissue at various locations in the river basin (ODEQ, 1994). Surveys conducted by ODEQ in 1994 indicated that levels of metals (arsenic, barium, cadmium, chromium, copper, lead, mercury, nickel, silver, and zinc), pesticides (chlordane and DDT), other organic chemicals (carbon tetrachloride, creosote, dichloroethylene, dioxin, PAHs, PCBs, phenol, pentachlorophenol, phenanthrene, phthalates, trichloroethane, trichloroethylene, and trichlorophenol), and bacteria exceed regulatory or guidance criteria for the protection of aquatic life and human health in at least one location of the river.

As a result of these findings, in 1990, the Oregon legislature directed ODEQ to develop a comprehensive study that would generate a technical and regulatory understanding and an information base on the river system that could be used to protect and enhance its water quality. To meet this directive, ODEQ developed and implemented a comprehensive, multiphase investigation known as the Willamette River Basin Water Quality Study (WRBWQS) (ODEQ, 1990b; Tetra Tech, 1995).

Recreational and Living Resources Trends

The first comprehensive study of the Willamette River biota was conducted by Dimick and Merryfield (1945) in the summer of 1944. Their study was specifically intended to assess the impact of water pollution on fish and benthic invertebrates in the river. Benthos are particularly good indicators of long-term trends in water quality because most benthic species are sedentary and have long life spans. Their state of health is therefore a gauge of both past and present water quality. Reactions to even occasional toxic discharges are measurable as variances in the species assemblages of benthic invertebrates. For pollution studies, benthos are divided into three categories: (1) intolerant species (e.g., stoneflies, mayflies, caddisflies) are indicative of good water quality because of their inability to survive in or tolerate low DO concentrations; (2) facultative species are indicative of a transition between good and poor water quality because they can survive under a wide range of DO conditions; and (3) tolerant species (e.g., sludgeworms), which are adapted to low DO levels, become dominant where poor water quality is prevalent.

Dimick and Merryfield (1945) found very different biological conditions in different stretches of the river. Upstream of Salem, where pollutant sources to the river were few, they found an abundance of healthy fish and populations of intolerant caddisfly, mayfly, and stonefly nymphs (Figure 13-9). Downstream of Salem to Portland, where pollutant loadings to the river were greatest, they found few to no fish, dead

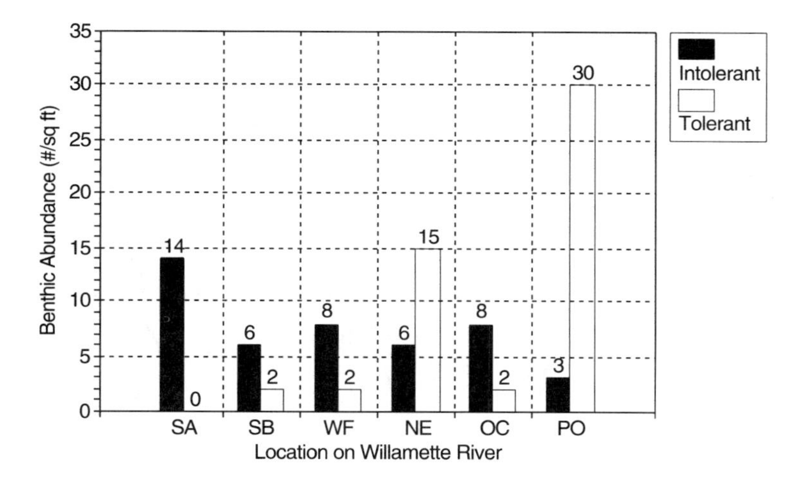

SA=Salem (1 mile above city); SB=Salem (2 miles below city); WF=Wheatland Ferry; NE=Newberg (0.8 miles above city); OC=Oregon City (above and below Willamette Falls); PO=Portland (0.3 miles upstream of Sellwood Bridge).

Figure 13-9 Spatial distribution of tolerant and intolerant benthic organisms in the Willamette River upstream and downstream of municipal waste discharges in 1945. *Source:* Dimick and Merryfield, 1945.

fish in or on the banks of the river, and a total absence of stoneflies and mayflies. They further noted that the biomass of insect larvae downstream of Salem was less than that upstream, and that largemouth bass collected below Salem were generally smaller than normal and in poor physical condition. Both of these conditions are indicative of poor water quality.

Dimick and Merryfield attributed the poor biological condition below Salem to the effects of pollution, but it is uncertain whether fish were directly affected or whether their populations were diminished because of the lack of their invertebrate foodstuffs (Dimick and Merryfield, 1945). Regardless, the study demonstrated that pollution was a major factor in the decline of the river's commercial and sport fisheries.

In 1983, the study was repeated to assess the changes that had occurred in the river since its cleanup began. Hughes and Gammon (1987) sampled the same sites that Dimick and Merryfield had sampled in 1944. Although the 1983 study showed some signs of a pollution-stressed river downstream from Salem, the differences between the findings of the studies demonstrated a marked improvement in water quality. Where Dimick and Merryfield had found only tolerant species associated with sluggish, warm water and muddy or sandy substrates, Hughes and Gammon found many intolerant species suited to fast-moving, cold water and rubble and gravel bottoms.

The improvements in the fish communities of the Willamette River between 1944 and 1983 (Figure 13-10) were not solely due to water quality improvements. Historically, the river provided important spawning and nursery grounds for salmon and steelhead, but dams built along the river prevented these fish from reaching their

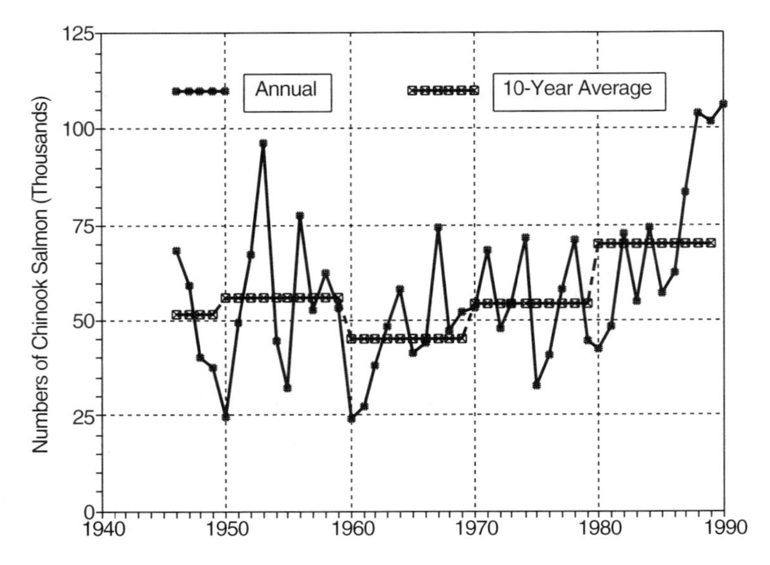

Figure 13-10 Long-term trends of spring Chinook salmon runs at Willamette Falls. *Source:* Bennett, 1991.

spawning grounds. Corrections to this situation have accompanied water quality improvements. Fish ladders have been built at dams, and four large fish hatcheries have been put into operation, producing 3.8 million salmon per year (Bennett, 1991). The dams also provide flow augmentation during autumn low-flow periods, thereby providing faster moving, oxygenated water to running fall chinook salmon (Starbird and Georgia, 1972).

Water quality has nevertheless played an important role in the survival and return of both natural-born and hatchery-reared salmon in the Willamette River. In 1965, only 79 chinook salmon were counted in the fall run. That number increased to 5,000 in 1971 (Starbird and Georgia, 1972). A record high of 106,300 spring chinook salmon were counted in the 1990 run, up 30 percent from the 1985–1989 average of 81,900. The 1990 catch of chinook salmon of 27,700 was 39 percent greater than the 1980–1989 average of 20,000 (Bennett, 1991). With the recent and continuing population growth in the Portland area (where most of the salmon are caught) and water quality improvements, interest in angling in the river has increased dramatically. The Willamette River is once again able to support important commercial and recreational fisheries.

SUMMARY AND CONCLUSIONS

The cleanup of the Willamette River has been accomplished because of overwhelming public support; strong commitment by federal, state, and local governments; comprehensive water quality studies that documented the extent of the problems; and the implementation of sound engineering proposals for controlling water pollution. Public pressure and responsive political leadership have resulted in the basinwide implementation of secondary treatment requirements with a minimum of legal actions needed to ensure compliance with the regulations. Water quality studies of the Willamette (e.g., Rickert, 1984; Rickert et al., 1976) have demonstrated the importance of the minimum requirement of secondary waste treatment for municipal and industrial dischargers, as well as the significance of background water quality and summer low-flow augmentation from USACE reservoirs, in achieving Oregon's water quality goals.

Vast improvements in the water quality of the Willamette River, facilitated by stringent regulatory controls, have led to remarkable improvements in the integrity of the river's biological communities. Of major importance, both recreationally and economically, is the continuing recovery of the fisheries. Salmon and steelhead on their migratory spawning runs are no longer precluded from reaching their spawning grounds in the Willamette River basin because of severely depressed or nonexistent concentrations of DO. Recreational anglers are once again able to enjoy pursuing these valuable gamefish as the fish make their way up the river to their spawning grounds. Another significant improvement is the return of viable populations of resident species of gamefish, including bass, catfish, perch, sturgeon, and crappies.

Although the severe water quality problems that have plagued the Willamette River in the past are clearly gone, there are still reasons for concern about the river's overall health (Tetra Tech, 1995). Until the continued presence of toxic contaminants in the water and sediments, the loads of suspended sediment and nutrients, and the alteration of the habitat can be abated, the overall ecological conditions of the Willamette River will continue to suffer.

For four decades, beginning in the 1920s, the Lower Willamette River near Portland, Oregon, was considered one of the most polluted urban-industrial rivers in the United States. Three decades after enactment of strict water pollution control regulations by the state of Oregon in the late 1960s and the federal Clean Water Act in 1972, the remarkable improvements in water quality and the ecological health of the river now provide important recreational and commercial benefits to the citizens of the Willamette valley. Salmon and steelhead fisheries, once blocked by dams without fish ladders and constrained by low dissolved oxygen conditions, are now sustained by migratory populations that can safely reach upriver spawning grounds. The local economies of major cities on the Willamette River are thriving, and upscale developments are attracted to riverfront locations by the aesthetics of a clean river that was once considered noxious with an unsightly riverfront. Although significant accomplishments have been made to improve water quality and ecological conditions in the Willamette River, continued investments are needed to address contemporary issues for continued restoration and maintenance of the river's ecological integrity. The designation of the Willamette River as an American Heritage River in 1998 recognizes both the significant environmental improvements that have been accomplished and the continuing need to address the key ecological issues identified in the 1990s. Nutrient enrichment, sediment loading, and the lingering presence of toxic chemicals in the river, sediment bed, and biota are ecological problems that remain. With cooperative efforts by state, local and federal agencies, they will be addressed in the early decades of the twenty-first century.

REFERENCES

Bennett, D. E. 1991. 1990 Willamette River spring Chinook salmon run. Oregon Department of Fish and Wildlife, Columbia River Management. Portland, OR. November.

Bondelid, T., S. Unger, and A. Stoddard. 2000. National water pollution control assessment model (NWPCAM) Version 1.1. Final report prepared by Research Triangle Institute, Research Triangle Park, NC for U.S. Environmental Protection Agency, Office of Policy, Economics and Innovation, Washington, DC, November, RTI Project Number 92U-7640-031.

CEQ. 1973. Chapter 2 in: Cleaning up the Willamette. Fourth annual report on environmental quality. Council on Environmental Quality, Washington, DC.

Dimick, R. E. and F. Merryfield. 1945. The fishes of the Willamette River system in relation to pollution. Bulletin Series No. 20. Oregon State College Engineering Experiment Station, Corvallis, OR. June.

Forstall, R. L. 1995. Population by counties by decennial census: 1900 to 1990. Population Division, U.S. Bureau of the Census, Washington, DC.
http://www.census.gov/population/www/censusdata/cencounts.html

Gleeson, G. W. 1972. The return of a river, The Willamette River, Oregon. Oregon State University, Corvallis, OR.

Gleeson, G. W. and F. Merryfield. 1936. Industrial and domestic wastes of the Willamette Valley. Engineering Experiment Station Bulletin No. 7. Oregon State Agricultural College, Corvallis, OR.

Hines, W. G., D. A. Rickert, and S. W. McKenzie. 1976. Hydrologic analysis and river quality data programs. U.S. Geological Survey Circular 715-D. U.S. Department of the Interior, Arlington, VA.

Hughes, R. M. and J. R. Gammon. 1987. Longitudinal changes in fish assemblages and water quality in the Willamette River, Oregon. Trans. Amer. Fish. Soc. 116: 196–209.

Iseri, K. T. and W. B. Langbein. 1974. Large rivers of the United States. Circular No. 686. U.S. Department of the Interior, U.S. Geological Survey, Reston, VA.

Merryfield, F. and W. G. Wilmot. 1945. 1945 progress report on pollution in Oregon streams. Engineering Experiment Station Bulletin No. 19. Oregon State University, Corvallis, OR.

Merryfield, F., W. B. Bollen, and F. C. Kachelhoffer. 1947. Industrial and city wastes. Engineering Experiment Station Bulletin No. 22. Oregon State University, Corvallis, OR.

ODEQ. 1970. Water quality control in Oregon. Oregon Department of Environmental Quality, Portland, OR. December.

ODEQ. 1990a. Oregon 1990 water quality status assessment report: 305(b) report. Prepared for U.S. Environmental Protection Agency, Washington, DC, by Oregon Department of Environmental Quality, Portland, OR.

ODEQ. 1990b. Willamette River basin water quality study: Rationale for a draft work plan. Oregon Department of Environmental Quality, Portland, OR.

ODEQ. 1994. Oregon's 1994 water quality status assessment report: 305(b) report. Prepared for U.S. Environmental Protection Agency, Washington, DC, by Oregon Department of Environmental Quality, Portland, OR.

OMB. 1999. OMB Bulletin No. 99-04. Revised statistical definitions of Metropolitan Areas (MAs) and guidance on uses of MA definitions. U.S. Census Bureau, Office of Management and Budget, Washington, DC.
http://www.census.gov/population/www/estimates/metrodef.html

Oregon State Sanitary Authority. 1964. Report on water quality and waste treatment needs for the Willamette River. Division of the State Board of Health, Portland, OR. May.

Rickert, D. A. 1984. Use of dissolved oxygen modeling results in the management of river quality. J. WPCF 56(1): 94–101.

Rickert, D. A., W. G. Hines, and S. W. McKenzie. 1976. Methodology for river quality assessment with application to the Willamette River Basin, Oregon. U.S. Geological Survey Circular 715-M. U.S. Department of the Interior, Arlington, VA.

Rickert, D. A. and W. G. Hines. 1978. River quality assessment: Implications of a prototype project. Science 200 (June 9): 1113–1118.

Starbird, E. A. and L. J. Georgia. 1972. A river restored: Oregon's Willamette. National Geographic 141(6): 816–835.

Tetra Tech. 1995. Willamette River basin water quality study: Summary and synthesis of study findings. Prepared for Oregon Department of Environmental Quality, Water Quality Division, Portland, OR, by Tetra Tech, Inc., Redmond, WA.

USDOC. 1998. Census of population and housing. U.S. Department of Commerce, Economics and Statistics Administration, Bureau of the Census—Population Division, Washington, DC.

USEPA (STORET). STOrage and RETrieval Water Quality Information System. U.S. Environmental Protection Agency, Office of Wetlands, Oceans, and Watersheds, Washington, DC.

USEPA. 1980. National accomplishments in pollution control: 1970–1980. U.S. Environmental Protection Agency, Office of Planning and Evaluation, Washington, DC.

USGS. 1999. Streamflow data downloaded from the U.S. Geological Survey's National Water Information System (NWIS)-W. Data retrieval for historical streamflow daily values. http://waterdata.usgs.gov/nwis-w

APPENDIX A

United States Waterways Identified with Water Pollution Problems Before the 1972 Clean Water Act

A-1 1963 List of Potential Enforcement Actions
A-2 Joint State-Federal Water Pollution Enforcement Conferences, January 1957–February 1971
A-3 National Archive Records of Joint State-Federal Enforcement Conference Transcripts, 1957–1972
A-4 National Archive Records of Interstate Water and Conference Files
A-5 National Archive Records of 180-Day Notice Transcripts
A-6 National Commission on Water Quality (NCWQ) Reports

INTRODUCTION

The collection of State-Federal Enforcement Conference proceedings and reports, and technical and administrative files related to enforcement notices are excellent primary sources of data and information related to water pollution problems in the United States during the 1950s and 1960s. Federal agency documents related to water pollution and enforcement activities are located in the National Archives, 8601 Adelphi Road, Adelphi, Maryland. A shuttle bus is available from the National Archives building in Washington, DC. Records are available for review only from 9:00 A.M. through 5 P.M. at the National Archives. The contact telephone number for the National Archives in Adelphi, Maryland (as of October 3, 1996) is (301) 713-7250, ext. 331.

Since the number of boxes included in the set of enforcement documents is quite large, anyone interested in examining the records should review the detailed listing of the contents of document boxes listed in Appendix A-3, Appendix A-4, and Appendix A-5. Each appendix provides the National Archives tracking and accession numbers required to retrieve document box numbers for specific waterways or enforcement activities of interest.

Appendix A-3 National Archive Records of Joint State-Federal Enforcement Conference Transcripts, 1957–1972

Accession Number: 412-74-0027
Archives Tracking Number: TN28/C3/7A
Number of Document Boxes: 47 (#1–#47)

Appendix A-4 National Archive Records of Interstate Water and Conference Enforcement Action Files

Accession Number: 412-74-0022
Archives Tracking Number: TN28/C3/7A
Number of Document Boxes: 117 (#1–62, #62a, #62b, #62c, #63–#114)

Appendix A-5 National Archive Records of 180-Day Notice Transcripts

Accession Number: 412-74-0026
Archives Tracking Number: TN28/C3/9A
Number of Document Boxes: 8 (#1–#8)

REFERENCE

Felton, B. 1996. Retrieval of documents from National Archives. Memorandum to Robert K. Bastian, USEPA, Office of Water, Municipal Technology Branch from Barbara Felton, Contract Support, National Records Management Program, Enterprise Information Management Division, Washington, DC. October 3.

APPENDIX A-1 1963 LIST OF POTENTIAL ENFORCEMENT ACTIONS[1]

From testimony of Murray Stein before the U.S. Senate Special Subcommittee on Air and Water Pollution of the Committee on Public Works, June 17, 1963.

1. Alleghany River (New York, Pennsylvania)
2. Applegate River–Eliot River (California, Oregon)
3. Arkansas River, area I (Colorado, Kansas)
4. Arkansas River, area II (Kansas, Oklahoma)
5. Arkansas River, area III (Oklahoma, Arkansas)
6. Batten Kill (Vermont, New York)
7. Green River (Kentucky, Tennessee)

[1] *Source:* Zwick and Benstock, 1971.

8. Big Sandy River (Kentucky, West Virginia, Virginia)
9. Big Horn River (Wyoming, Montana)
10. Big Sioux River (South Dakota, Iowa)
*11. Blackstone River (Massachusetts, Rhode Island)
12. Bodcau River (Arkansas, Louisiana)
13. Buntings Branch (Delaware, Maryland)
14. Byram River (Connecticut, New York)
15. Connecticut River, upper (New Hampshire, Vermont)
*16. Connecticut River, lower (Massachusetts, Connecticut) [25]
17. Catawba and Wateree Rivers (North Carolina, South Carolina)
*18. Chattahoochee River, upper (Alabama, Georgia) [38]
19. Chattahoochee River, lower (Alabama, Georgia, Florida)
*20. Coosa River (Alabama, Georgia) [21]
21. Delaware River (New York, Pennsylvania, New Jersey, Delaware)
22. Des Moines River (Minnesota, Iowa, Missouri)
23. Delores River (Colorado, Utah)
24. French River (Massachusetts, Connecticut)
25. French Broad River (Tennessee, North Carolina)
*26. Grand Calumet/Little Calumet Rivers (Indiana, Illinois) [34]
27. Grand (Neosho) River (Kansas, Oklahoma)
28. Green River (Wyoming, Utah)
29. Hoosic River (Vermont, Massachusetts, New York)
30. Kanab Creek (Utah, Arizona)
31. Kanawha River, tributary to Ohio River (West Virginia, Ohio)
32. Klamath River (Oregon, California)
33. Leviathan Creek (California, Nevada)
34. Little Blue River (Nebraska, Kansas)
35. Lost River (Oregon, California)
*36. Mahoning River (Ohio, Pennsylvania) [33]
37. Malad Rivers (Idaho, Utah)
38. Marais Des Cygnes River (Kansas, Missouri)
39. McElmo Creek (Colorado, Utah)
*40. Menominee River (Wisconsin, Michigan) [24]
*41. Merrimack River (New Hampshire, Massachusetts) [29]
*42. Mississippi River, area I (Minnesota, Wisconsin) [28]
*43. Mississippi River, area IX (Memphis, Tennessee–Vicksburg, Missssippi; Tennessee, Arkansas, Mississippi, Louisiana) [30]
*44. Mississippi River, area X (Vicksburg, Mississippi–mouth; Mississippi, Louisiana) [30]

45. Missouri River, including lower Yellowstone River (Montana, North Dakota, South Dakota)
*46. Monongahela River (West Virginia, Maryland, Pennsylvania) [26]
47. Montreal River (Michigan, Wisconsin)
*48. Nashua River (Massachusetts, New Hampshire) [29]
49. Nolichucky River (North Carolina, Tennessee)
50. Ochlockonee River (Georgia, Florida)
51. Ohio River, area I (Pittsburgh, Pennsylvania–Pennsylvania State Line; Ohio, Pennsylvania)
52. Ohio River, area II (Pittsburgh, Pennsylvania–Pennsylvania State Line–Huntington, West Virginia; West Virginia, Ohio)
53. Ohio River, area III (Huntington, West Virginia–above Cincinnati, Ohio; Kentucky, West Virginia, Ohio)
54. Ohio River, area IV (Indiana, Kentucky, Ohio)
55. Ohio River, area V (Illinois, Kentucky, Ohio)
56. Ouachita River (Arkansas, Louisiana)
57. Pigeon River (North Carolina, Tennessee)
58. Pawcatuck River (Rhode Island, Connecticutt)
59. Pea and Choctawhatchee Rivers (Alabama, Florida)
*60. Pearl River (Louisiana, Mississippi) [22]
61. Piscataqua River (New Hampshire, Maine)
62. Potomac River, Luke–Cumberland, Maryland area (Maryland, West Virginia)
63. Quinebaug River (Massachusetts, Connecticut)
64. Red River, upper (Arkansas, Oklahoma, Texas)
65. Red River, lower (Arkansas, Louisiana)
*66. Red River of the North (Minnesota, North Dakota, South Dakota) [36]
67. Rio Grande River (Texas, New Mexico)
68. Roanoke River (Virginia, North Carolina)
69. Rock River (Illinois, Wisconsin)
70. Saco River (New Hampshire, Maine)
71. St. Croix River (Minnesota, Wisconsin)
*72. Hudson and East Rivers (tributary to Raritan Bay enforcement area; New Jersey, New York) [37]
73. Snake River, area II (Idaho, Oregon)
*74. Snake River, area III (Washington, Idaho) [27]
75. St. Louis River (Wisconsin, Minnesota)
76. St. Mary's River (Georgia, Florida)
77. Chattoga, Tugaloo, Seneca, and upper Savannah Rivers (South Carolina, Georgia)

*78. Savannah River, lower (South Carolina, Georgia) [32]
*79. Savannah River, mouth (South Carolina, Georgia) [32]
*80. South Platte (Colorado, Nebraska) [23]
81. Suwannee River (Georgia, Florida)
82. Susquehanna River, north branch (Pennsylvania, New York)
*83. Ten Mile River (Rhode Island, Massachusetts) [31]
84. Tennessee River (Georgia, Tennessee, Alabama)
85. Verdigris River (Kansas, Oklahoma)
86. Virgin River (Utah, Arizona, Nevada)
87. Wabash River (Illinois, Indiana)
88. Warm Springs Run (West Virginia, Maryland)
89. Yellowstone River, lower (Montana, North Dakota)
*90. Youghiogheny River (West Virginia, Maryland, Pennsylvania) [26]

Notes. An asterisk (*) indicates that enforcement action has taken place (as of February 1971). Numbers in brackets (e.g., [99]) refer to identifying numbers assigned to Enforcement Conferences by Zwick and Benstock (1971) (e.g., *90. Youghiogheny River . . . [26] was included in the Enforcement Conference for the Monongahela River from December 17–18, 1963). A list of the Conferences follows in Appendix A-2.

REFERENCE

Zwick, D. and M. Benstock. 1971. Water wasteland. Ralph Nader's study group report on water pollution. Center for Study of Responsive Law, Washington, DC. Grossman Publishers, New York.

APPENDIX A-2 JOINT STATE-FEDERAL WATER POLLUTION ENFORCEMENT CONFERENCES, JANUARY 1957–FEBRUARY 1971[1]

1. Corney Creek Drainage System (AR, LA)
Hearing: January 16–17, 1957. Initiated by Surgeon General, U.S. Public Health Service.

2. Big Blue River (NE, KS)
Conference: May 3, 1957. Initiated by Surgeon General, U.S. Public Health Service.

3. Missouri River—St. Joseph, Missouri Area (MO, KS)
Conference: June 11, 1957; Hearing: July 27–30,1959; Suit Filed: September 29,

[1]*Source:* Zwick and Benstock, 1971.

1960; Court Order: October 31, 1961. Initiated by Surgeon General, U.S. Public Health Service.

4. Missouri River—Omaha, Nebraska Area (NE, KS, MO, IA)
Conference: (Session 1) June 14, 1957; (Session 2) July 21, 1964. Initiated by Surgeon General, U.S. Public Health Service.

5. Potomac River—Washington Metropolitan Area (DC, MD, VA)
Conference: (Session 1) August 22, 1957; (Session 2) February 13, 1958; (Session 3) April 2–4, May 8, 1969; (Session 3 reconvened) May 21–22, 1970, October 13, 1970; Progress meeting: December 8–9, 1970. Initiated by Surgeon General, U.S. Public Health Service.

6. Missouri River—Kansas Cities Metropolitan Area (KS, MO)
Conference: December 3, 1967; Hearing: June 13–17, 1960. Initiated by Surgeon General, U.S. Public Health Service.

7. Mississippi River—St. Louis Metropolitan Area (MO, IL)
Conference: March 4, 1958. Initiated by Missouri Health Division; Illinois Sanitary Water Board; Bi-State Development Agency.

8. Animas River (CO, NM)
Conference: (Session 1) April 29, 1958; (Session 2) June 24, 1959. Initiated by New Mexico Department of Public Health.

9. Missouri River—Sioux City Area (SD, IA, NE, KS, MO)
Conference: July 24, 1958; Hearing: March 23–27, 1959. Initiated by Iowa Commissioner of Public Health.

10. Lower Columbia River (WA, OR)
Conference: (Session 1) September 10–11, 1958; (Session 2) September 3–4, 1959; (Session 3) September 8–9, 1965. Initiated by Surgeon General, U.S. Public Health Service.

11. Bear River (ID, WY, UT)
Conference: (Session 1) October 8, 1958; (Session 2) July 19, 1960. Initiated by Utah Water Pollution Control Board.

12. Colorado River and All Tributaries (CO, UT, AZ, NV, CA, NM, WY)
Conference: (Session 1) January 13, 1960; (Session 2) May 11, 1961; (Session 3) May 9–10, 1962; (Session 4) May 27–28,1963; (Session 5) May 26, 1964; (Session 6) July 26, 1967. Initiated by New Mexico Department of Public Health; Arizona State Department of Health; Nevada State Board of Health; Colorado Department of Public Health; Utah Water Pollution Control Board; California State Water Pollution Control Board.

13. North Fork of the Holston River (TN, VA)
Conference: (Session 1) September 28, 1960;(Session 2) June 19, 1962; (Session 3) Called and postponed. Initiated by Tennessee Stream Pollution Control Board.

14. Raritan Bay (NJ, NY)
Conference: (Session 1) August 22, 1961; (Session 2) May 9, 1963; (Session 3) June 13–14, 1967. Initiated by Surgeon General, U.S. Public Health Service.

15. North Platte River (NE, WY)
Conference: (Session 1) September 21, 1961; (Session 2) March 21, 1962; (Session 3) November 20, 1963. Initiated by Nebraska Department of Health.

16. Puget Sound (WA)
Conference: (Session 1) January 16–17,1962; (Session 2) September 6–7, October 6, 1963. Initiated by Governor of Washington.

17. Mississippi River—Clinton, Iowa, Area (IL, IA)
Conference: March 8, 1962. Initiated by Secretary of Health, Education and Welfare.

18. Detroit River (MI)
Conference: (Session 1) March 27–28, 1962; (Session 2) June 15–18, 1965. Initiated by Governor of Michigan.

19. Androscoggin River (NH, ME)
Conference: (Session 1) September 24, 1962; February 6, 1963; (Session 2) October 21, 1969. Initiated by Secretary of Health, Education and Welfare.

20. Escambia River (AL, FL)
Conference: October 24, 1962. Initiated by Florida State Board of Health.

21. Coosa River (GA, AL)
Conference: (Session 1) August 27, 1963; (Session 2) April 11, 1968. Initiated by Secretary of Health, Education and Welfare.

22. Pearl River (MS, LA)
Conference: (Session 1) October 22, 1963; (Session 2) November 7, 1968. Initiated by Secretary of Health, Education and Welfare.

23. South Platte River (CO)
Conference: (Session 1) October 29, 1963; (Session 2) Apri1 27–28, 1966; (Session 2 reconvened) November 10, 1966. Initiated by Governor of Colorado.

24. Menominee River (MI, WI)
Conference: November 6–8, 1963. Initiated by Secretary of Health, Education and Welfare.

25. Lower Connecticut River (MA, CT)
Conference: (Session 1) December 2, 1963; (Session 2) September 27, 1967. Initiated by Secretary of Health, Education and Welfare.

26. Monongahela River (WV, VA, PA, MD)
Conference: December 17–18, 1963. Initiated by Secretary of Health, Education and Welfare.

27. Snake River—Lewiston, Idaho—Clarkston, Washington Area (ID, WA)
Conference: January 15, 1964. Initiated by Secretary of Health, Education and Welfare.

28. Upper Mississippi River (MI, WI)
Conference: (Session 1) February 7–8, 1964; (Session 2) February 28, March 1 and 20, 1967. Initiated by Secretary of Health, Education and Welfare; Governors of Minnesota and Wisconsin.

29. Merrimack and Nashua Rivers (NH, MA)
Conference: (Session 1) February 11, 1964; (Session 2) December 18, 1968; Work-

shops: October 20–21, 1970. Initiated by Secretary of Health, Education and Welfare; Governor of Massachusetts.

30. Lower Mississippi River (AR, TN, MS, LA)
Conference: May 5–6, 1964. Initiated by Secretary of Health, Education and Welfare.

31. Blackstone and Ten Mile Rivers (MA, RI)
Conference: (Session 1) January 26, 1965; (Session 2) May 28, 1968. Initiated by Secretary of Health, Education and Welfare.

32. Lower Savannah River (SC, GA)
Conference: (Session 1) February 2, 1965; (Session 2) October 29, 1969. Initiated by Secretary of Health, Education and Welfare.

33. Mahoning River (OH, PA)
Conference: February 16–17, 1965. Initiated by Secretary of Health, Education and Welfare.

34. Grand Calumet River, Little Calumet River, Calumet River, Wolf Lake, Lake Michigan, and Their Tributaries (IL, IN)
Conference: (Session 1) March 2–9, 1965; (Technical Session) January 4, 5, 31, February 1, 1966; (Reconvened) August 26, 1969. Initiated by Secretary of Health, Education and Welfare.

35. Lake Erie (MI, IN, OH, PA, NY)
Conference: (Session 1) August 3–5, 1965; (Session 2) August 10–12,1965; (Session 3) March 22, 1967; (Session 4) October 4, 1968; (Session 5) June 3–4,1970. Initiated by Secretary of Health, Education and Welfare; Governor of Ohio.

36. Red River of the North (MN, ND)
Conference: September 14–15, 1965; January 18, March 4, 1966. Initiated by Secretary of Health, Education and Welfare.

37. Hudson River (NY, NJ)
Conference: (Session 1) September 28–30, 1965; (Session 2) September 20–21, 1967; (Session 3) June 18–19, 1969; (Session 3 reconvened) November 25, 1969. Initiated by Secretary of Health, Education and Welfare; Governors of New York and New Jersey.

38. Chattahoochee River and its Tributaries (GA, AL)
Conference: (Session 1) July 14–15, 1966; (Session 2) February 17, 1970. Initiated by Secretary of the Interior.

39. Lake Tahoe (CA, NV)
Conference: July 18–20, 1966. Initiated by Secretary of the Interior.

40. Moriches Bay and Eastern Section of Great South Bay and Their Tributaries (NY)
Conference: (Session 1) September 20–21, 1966; (Session 2) June 21, 1967. Initiated by Secretary of the Interior.

41. Penobscot River and Upper Penobscot Bay and Their Tributaries (ME)
Conference: April 20, 1967. Initiated by Secretary of the Interior.

42. Eastern New Jersey Shore—from Shark River to Cape May (NJ)
Conference: November 1,1967. Initiated by Secretary of the Interior.

43. Lake Michigan (MI, IN, IL, WI)
Conference: (Session 1) January 31, February 1–2, 5–7; March 7–8, 12, 1968; (Session 2) February 25, 1969; (Session 3) March 31, April 1, May 7, 1970; Workshops held September 28–October 2, 1970; (Session 3) reconvened March 23–24, 1971; (Session 4) Unknown dates. Initiated by Governor of Illinois; Secretary of the Interior.

44. Boston Harbor (MA)
Conference: (Session 1) May 20, 1968; (Session 2) April 30, 1969. Initiated by Secretary of the Interior.

45. Lake Champlain (NY, VT)
Conference: (Session 1) November 13, December 19–29, 1968; (Session 2) June 25, 1970. Initiated by Secretary of Interior; Vermont Department of Water Resources.

46. Lake Superior and its Tributary Basin (WI, MN, MI)
Conference: (Session 1) May 13–15, September 30, October 1, 1969; (Session 2) April 29–30, August 12–13, 1970; (Session 2 reconvened) January 14–15, 1971. Initiated by Secretary of the Interior.

47. Escambia River Basin (AL, FL)
Conference: January 20–21, 1970. Initiated by Governor of Florida.

48. Perdido Bay (FL, AL)
Conference: January 22, 1970. Initiated by Governor of Alabama.

49. Mobile Bay (AL)
Conference: January 27–28, 1970. Initiated by Secretary of the Interior.

50. Biscayne Bay (FL)
Conference: February 24–26, 1970. Initiated by Governor of Florida.

51. Dade County (FL)
Conference: October 20–21, 1970. Initiated by Governor of Florida.

REFERENCE

Zwick, D. and M. Benstock. 1971. Water wasteland. Ralph Nader's study group report on water pollution. Center for Study of Responsive Law, Washington, DC. Grossman Publishers, New York.

APPENDIX A-3 NATIONAL ARCHIVE RECORDS OF JOINT STATE-FEDERAL ENFORCEMENT CONFERENCE TRANSCRIPTS, 1957–1972[1]

Archives Tracking Number: TN28/C3/7A
Accession Number: 412-74-22
Record Group Number: 412

[1]*Source:* Standard Form 135, Records Transmittal and Receipt.

FROM: U.S. Environmental Protection Agency
Enforcement Division, OEGC
Room 3600, Waterside Mall (WSM), Washington, DC 20460

TO: Federal Records Center, GSA
Washington National Records Center
Accession Section
Washington, DC 20409, Stop Code 386

Agency Custodian of Records: Frank Hall
Agency Official: Harold Masters, Records Management Officer
Number of Boxes of Documents: 47
Date: April 16, 1974

U.S. Environmental Protection Agency

Enforcement Compliance Division

Documents pertaining to conferences of an interstate body of water, and a state and federal effort to make or keep it clean. Complete verbatim record (certified conference reporter typed copy) of conference including charts, pictures, reports, and other exhibits.

ENFORCEMENT CONFERENCE TRANSCRIPTS — WATER (1957–1972)

1 Androscoggin River, 1st Session through Biscayne Bay, lst Session
2 Biscayne Bay through Boston Harbor, lst Session
3 Boston Harbor, 2nd Session through Chattahoochee River, lst Session
4 Chattahoochee River, 2nd Session through Colorado River, 7th Session
5 Colorado River, 7th Session reconvened through Connecticut River, 2nd Session
6 Coosa River through Corney Creek, Exhibits
7 Dade County through Dade County, 3rd Session with exhibits
8 Dade County, Progress Meeting through Detroit River, 2nd Session
9 Detroit River, 2nd Session through Lake Erie, 1st Session
10 Lake Erie, 2nd Session through Lake Erie, Progress Meeting
11 Lake Erie, Technical Session through Lake Erie, 5th Session
12 Lake Erie, 5th Session Workshop through Lake Erie, 5th Session Workshop (All)
13 Escambia River through Conecuh-Escambia River
14 Galveston Bay, 1st Session through Galveston Bay, Executive Session
15 Galveston Bay, Follow-up through Hudson River
16 Hudson River, 2nd Session through Long Island Sound, 1st Session

17 Mahoning River, lst Session through Merrimack River–Nashua River
18 Merrimack River–Nashua River, 2nd Session through Lake Michigan, lst Session
19 Lake Michigan, Technical Session through Lake Michigan, 2nd Progress Meeting
20 Lake Michigan (Calumet Area), lst Session through Lake Michigan, 2nd Session
21 Lake Michigan (4-States), 1st Session All
22 Lake Michigan (4-States), 1st Session Exhibits through Lake Michigan, 3rd Session
23 Lake Michigan (4-States), 3rd Session Workshop A11
24 Lake Michigan (4-States), 3rd Session Workshop through Lake Michigan, 4th Session
25 Lake Michigan, 4th Session through 4th Session Exhibits
26 Lake Michigan, 4th Session Exhibits through Lake Michigan, Public Session
27 Mississippi River (Upper), 1st Session through Mississippi River (Upper), Progress Meeting
28 Mississippi River (Lower) through Mississippi River (St. Louis)
29 Missouri River (Omaha), Progress Meeting through Missouri River (Kansas City)
30 Missouri River (Kansas City), Exhibits #1 through Hearings, Missouri River (Sioux City)
31 Missouri River (Sioux City), Exhibits through Mobil Bay, Legal Exhibits
32 Monongahela River, lst Session through Moriches Bay, 2nd Session
33 Moriches Bay, Exhibits through New Jersey, Exhibits
34 Pearl Harbor, Technical Session through Ohio River (Pittsburgh Area), 1st Session
35 Public Meeting Pearl River through Peridido Bay, 2nd Session
36 North Platte River, 1st, 2nd, 3rd Sessions through Potomac River
37 Potomac River, Exhibits WC-10-11 through 3rd Session
38 Potomac River, 3rd Session through Potomac River, Reconvened 3rd Session
39 Potomac River, Exhibits Legal through Potomac River, Progress Meeting
40 Potomac River, Progress Meeting through Potomac River, Progress Meeting June 20, 1972
41 Puget Sound, 1st Session through Puget Sound, Litigation
42 Puget Sound, Exhibits through Raritan Bay, 2nd Session
43 Raritan Bay, 3rd Session through Savannah River, Georgia
44 Snake River through Lake Superior, Executive Session October, 1969
45 Lake Superior, 2nd Session through Lake Superior, 2nd Session Reconvened
46 Lake Superior, 2nd Session Reconvened through Lake Superior, 2nd Session Reconvened for second time
47 Lake Superior, 2nd Session Reconvened for third time through Lake Tahoe, 1st Session

List of Conference Transcripts (1957–1972) in Box Numbers 1–47

Androscoggin River, lst Session and reconvened
Androscoggin River, lst Session
Androscoggin River, 2nd Session
Biscayne Bay, 1st Session Vol. I
Biscayne Bay, 1st Session Vol. II
Biscayne Bay, lst Session Exhibits
Biscayne Bay, lst Session Exhibits
Blackstone River, 1st and 2nd Sessions
Boston Harbor
Boston Harbor, 2nd Session Vol. I
Boston Harbor, 3rd Session
Lake Champlain, lst Session
Lake Champlain, 2nd Session
Chattahoochee River, 1st Session
Chattahoochee River, 2nd Session
Colorado River through 6th Session
Colorado River Exhibits
Colorado River, 7th Session Vol. I
Colorado River, 7th Session Vol. II
Colorado River, 7th Session Reconvened
Lower Columbia River, 1st through 3rd Sessions plus slides
Lower Columbia River, Atomic Energy Facilities
Connecticut River, 2nd Session
Coosa River, lst and 2nd Session
Coosa River, 2nd Session, Georgia River
Corny Creek, Maps & charts
Corney Creek, Maps & photos
Corney Creek, Exhibits
Dade County, 1st Session
Dade County, 2nd Session
Dade County, 3rd Session
Dade County, 3rd Session with exhibits
Dade County, Progress Meeting
Detroit River, Legal #2
Detroit River, 1st Session Vol. I & II
Detroit River, 2nd Session Vol. I, II, & III
Detroit River, 2nd Session
Lake Erie, 1st Session

Lake Erie, 1st Session
Lake Erie, 2nd Session
Lake Erie, 3rd Session
Lake Erie, 3rd Session
Lake Erie, Progress Meeting
Lake Erie, Technical Session
Lake Erie, 4th Session
Progress Evaluation Meeting
Lake Erie, 5th Session
Lake Erie, 5th Session Workshop
Escambia River Basin
Escambia River Basin, 1st Session
Conecuh-Escambia River
Galveston Bay, 1st Session
Galveston Bay, lst Session Exhibits
Galveston Bay, lst Session
Galveston Bay, 1st Session Exhibits
Galveston Bay, Executive Session
Galveston Bay, Follow-up
Galveston Bay, Follow-up
Holston River, 2nd Session
Hudson River, Vol. I & II
Hudson River, 2nd Session
Hudson River, 3rd Session
Hudson River, 3rd Session Reconvened
Long Island Sound, 1st Session
Mahoning River, 1st Session
Menominee River, 1st Session
Merrimack River, lst Session
Merrimack River & Nashua River, 2nd Session
Merrimack River & Nashua River, 2nd Session
Lake Michigan, 1st Session
Lake Michigan, 1st Session
Lake Michigan, Technical Session
Lake Michigan, 1st Progress Session
Lake Michigan, 2nd Progress Session
Lake Michigan (Calumet Area), lst Session
Lake Michigan (Calumet Area), 2nd Session
Lake Michigan (4-States), 1st Session

Lake Michigan (4-States), 1st Session Exhibits
Lake Michigan (4-States), 2nd Session
Lake Michigan (4-States), 3rd Session
Lake Michigan (4-States), 3rd Session Workshop
Lake Michigan (4-States), 3rd Session Workshop
Lake Michigan (4-States), 3rd Session Reconvened
Lake Michigan (4-States), 4th Session
Lake Michigan, 4th Session Vol. II
Lake Michigan, 4th Session Vol. III
Lake Michigan, Exhibits
Lake Michigan, Exhibits
Lake Michigan, 4th Session Exhibits
Lake Michigan, 4th Session Exhibits
Lake Michigan, 4th Session Exhibits
Lake Michigan, 4th Session Exhibits
Lake Michigan, 4th Session, Public Session
Mississippi River (Upper), 1st Session
Mississippi River (Upper), 1st Session
Mississippi River (Upper), 2nd Session
Mississippi River (Upper), Legal
Mississippi River (Upper), Progress Meeting
Mississippi River (Lower), 1st Session
Upper Mississippi River (Wisconsin & Minnesota)
Mississippi River (Clinton)
Mississippi River (St. Louis)
Missouri River (Omaha), Progress Meeting
Missouri River (Omaha)
Missouri River (Omaha), 1st Session & pictures
Missouri River (Kansas City), Progress Meeting
Missouri River (Kansas City), 1st Session
Missouri River (Kansas City), Exhibits 411
Missouri River (Kansas City), Exhibits 412
Missouri River (Sioux City), Slides
Missouri River (Kansas City), Exhibits #3
Missouri River Hearings (Sioux City)
Missouri River Hearings (Sioux City)
Missouri River (Sioux City), Exhibits
Mobile Bay, 1st Session
Mobile Bay, Legal Exhibits

Monongahela River, 1st Session
Monongahela River, 2nd Session
Moriches Bay, 2nd Session
Moriches Bay, Exhibits
Moriches Bay, Great South Bay
Mount Hope Bay, 1st Session
Mount Hope Bay, 1st Session Reconvened
New Jersey Shore
Pearl River, Technical Session
Pearl River, lst Session
Ohio River, Wheeling Area, lst Session
Ohio River, Pittsburgh Area, lst Session
Public Meeting Pearl River
Perdido Bay, Florida, Alabama
Perdido Bay, 2nd Session
Penobscot River
North Platte River
South Platte River, 1st Session
South Platte River, 2nd Session
Potomac River, Exhibits WC 9
Potomac River, Exhibits WC 10–11
Potomac River, Exhibits
Potomac River, Exhibits
Potomac River, Exhibits
Potomac River, Exhibits
Potomac River, 2nd Session
Potomac River, 3rd Session
Potomac River, 3rd Session
Potomac River, 3rd Session
Potomac River, Progress Meeting
Potomac River, Work Session
Potomac River Reconvened, 3rd Session
Potomac River, Exhibits, Legal
Potomac River, 3rd Session Reconvened and Executive Session
Potomac River, Progress Meeting
Potomac River, Progress Meeting
Potomac River, Progress Meeting
Potomac River, Progress Meeting
Potomac River, Progress Meeting

Potomac River, Progress Meeting Reconvened
Potomac River, Progress Meeting Back-up Material
Potomac River, Progress Meeting June 20–21, 1972
Puget Sound, lst Session
Puget Sound, Exhibits
Puget Sound, Legal
Puget Sound, Litigation
Puget Sound, Exhibits
Red River of the North, lst & 2nd Sessions
Raritan Bay, 2nd Session
Raritan Bay, 3rd Session
Raritan Bay, 3rd Session & Appendix
Savannah River, lst Session
Savannah River, Georgia & South Carolina
Snake River
Western South Dakota River, 1st Session
Lake Superior, May 1969
Lake Superior
Lake Superior
Lake Superior, Executive Session October 1969
Lake Superior, 2nd Session
Lake Superior, Exhibits #1
Lake Superior, 2nd Session & Exhibits #2–5
Lake Superior, 2nd Session Reconvened
Lake Superior, 2nd Session Reconvened Exhibits
Lake Superior, 2nd Session for Second Time
Lake Superior, 2nd Session, 3rd Meeting
Lake Superior, 2nd Session, 3rd Meeting Reconvened
Lake Tahoe, 1st Session

APPENDIX A-4 NATIONAL ARCHIVE RECORDS OF INTERSTATE WATER AND CONFERENCE ENFORCEMENT ACTION FILES[1]

Archives Tracking Number: TN28/C3/7A
Accession Number: 412-74-22
Record Group Number: 412

[1] *Source:* Standard Form 135, Records Transmittal and Receipt.

FROM: U.S. Environmental Protection Agency
Permit Assistance and Evaluation Division
Crystal Mall #2, Rm. 706, Arlington, VA

TO: Federal Records Center, GSA
Washington National Records Center
Accession Section
Washington, DC 20409, Stop Code 386

Agency Custodian of Records: Robert Hardeker
Agency Official: Harold Masters, Records Management Officer
Number of Document Boxes: 117 (#1–62, #62a, #62b, #62c, #63–114)
Date: undated

U.S. Environmental Protection Agency

Permit Assistance and Evaluation Division

Documents pertaining to conferences of an interstate body of water, and a state and federal effort to make or keep it clean. Documents and dates contained in the file concern the administrative aspects of enforcement, the social and economic forces acting both for and against clean-up, the development and application of rules for establishing/maintaining clean water conditions, and the monitoring and testing of the water.

Each water body section is divided into 8 parts which contain the following types: Administrative, Compliance, Correspondence, Legal, Legislation, Litigation, News clippings and Reports.

INTERSTATE WATER AND CONFERENCE FILES

1 Alabama (State) through Androscoggin River (Correspondence)
2 Androscoggin River (Legislative) through Applegate River (Reports)
3 Arizona (State) through Arkansas River, Area III (Clips #3)
4 Arkansas River, Area III (Reports #1) through Bear River (Clips #1)
5 Bear River (Reports) through Big Sioux River (Reports #1)
6 Black Fork River (Administrative) through Boston Harbor (Reports #1)
7 Brood River (Administrative) through California (State) (Clips)
8 California (State) (Reports #1) through California (State) (Reports #3)
9 California (State) (Reports #4) through Canadian River (Reports #1)
10 Carson River (Administrative) through Upper Chattahoochee River (Clips #1)
11 Upper Chattahoochee River (Clips #2) through Chicago Sanitary and Ship Canal (Legislative #1)
12 Chicago Sanitary and Ship Canal (Legislative #2) through Chicago Sanitary and Ship Canal (Reports #1)

13 Chicago Sanitary and Ship Canal (Reports #2) through Colorado River (Administrative #3)
14 Colorado River (Administrative #4) through Colorado River (Administrative #13A)
15 Colorado River (Compliance) through Colorado River (Reports #5)
16 Colorado River (Reports #6) through Colorado River (Reports #8)
17 Colorado River (Reports #9) through Colorado River (Reports #12)
18 Colorado River, Area I (Administrative #1) through Co1orado River (General Clips #1)
19 Co1orado River (General Reports #1) through Columbia River (General Reports #5)
20 Columbia River, General (Reports #6) through Columbia River, Area III (Clips #1)
21 Columbia River, Area III (Clips #2) through Connecticut River, General (Clips #1)
22 Connecticut River, General (Reports #1) through Coosa River (Compliance #1)
23 Coosa River (Correspondence) #1 through Corney Creek (Reports #1)
24 Corney Creek (Reports #2) through Delaware River, General (Correspondence #1)
25 Delaware River, General (Clips #1) through Delaware River, General (Reports #3)
26 Delaware River, General (Clips #2) through Delaware River, General (Reports #3)
27 Detroit River (Administrative #2) through Detroit River (Litigation #1)
28 Detroit River (Litigation #1A) through Detroit River (Reports #2)
29 Detroit River (Reports #3) through Lake Erie (Administrative #2)
30 Lake Erie (Administrative #3) through Lake Erie (Correspondence #1)
31 Lake Erie (Correspondence #2) through Lake Erie (Clips #8)
32 Lake Erie (Clips #9) through Escambia River (Clips #1)
33 Escambia River (Reports #1) through Florida State (General #4)
34 Florida (State) (Reports #1) through Georgia (State) (General #2)
35 Lake Gibson (General #1) through Great Lakes (Correspondence #1)
36 Great Lakes (Legal #1) through Gypsum Creek (Reports #1)
37 Hackensack River, Upper (Administrative #1) through Holston River, North Fork (Reports #1)
38 Holston River, North Fork (Reports #2) through Hudson River (Correspondence #1)
39 Hudson River (Correspondence #2) through Idaho (State) (Reports #1)
40 Illinois (State) (General #1) through Iowa (State)(Reports #1)
41 James River (Administrative #1) through Kewaunee Harbor (Reports #1)
42 Keya Paha River (General) through Long Island Sound (Correspondence #1)
43 Long Island Sound (Correspondence #1) through Mahoning River (Clips #1)
44 Mahoning River (Reports #1) through Massachusetts (State) (Reports #1)
45 Maumee River (Administrative) through Merrimack River (Compliance #1)

46 Merrimack River (Compliance #2) through Gulf of Mexico (General #2)
47 Gulf of Mexico (Reports #1) through Lake Michigan (Calumet) (Administrative #2)
48 Lake Michigan (Calumet) (Compliance #1) through Lake Michigan (Calumet) (Clips #3)
49 Lake Michigan (Calumet) (Clips #4) through Lake Michigan (Calumet) (Reports #2)
50 Lake Michigan (Calumet) (Reports #3) through Lake Michigan (4-States)(Compliance #1)
51 Lake Michigan (4-States) (Compliance 1A) through Lake Michigan (4-States) (Correspondence #5)
52 Lake Michigan (4-States) (Correspondence #6) through Minnesota River (Clips #1)
53 Minnesota River (Reports #1) through Mississippi River (Upper) (Correspondence #1)
54 Mississippi River (Upper) (Correspondence #2) through Mississippi River (Upper)(Reports #2)
55 Mississippi River (Upper) (Reports #3) through Mississippi River (St. Louis)(Administrative #3)
56 Mississippi River (St. Louis) (Compliance #1) through Mississippi River (St. Louis) (Clips #4)
57 Mississippi River (St. Louis) (Clips #5) through Mississippi River (Lower) (Administrative #l)
58 Mississippi River (Lower) (Administrative #2) through Mississippi River (Lower) (Reports #l)
59 Mississippi River (Lower) (Reports #2) through Missouri (State) (Reports #1)
60 Missouri River (General #1) through Missouri River (St. Joseph)(Clips #4)
61 Missouri River (Kansas City) (Administrative #1) through Missouri River (Kansas City) (Clips #3)
62 Missouri River (Kansas City) (Clips #4) through Mohawk River (Clips #1)
62a Missouri River (Reports #1) through Missouri River (Sioux City) (Administrative #2)
62b Missouri River (Sioux City) (Compliance #1) through Missouri River (Omaha) (Correspondence #1)
62c Missouri River (Omaha) (Legislative #1) through Missouri River (St. Joseph) (Compliance #3)
63 Mohawk River (Reports #1) through Monongahela River (Reports #2)
64 Monongahela River (Reports #3) through Monongahela River (Reports #6)
65 Montana (State) (General #1) through Nashua River (Correspondence #1)
66 Nashua River (Reports #1) through New Jersey (State), Eastern (Compliance #l)
67 New Jersey (State), Eastern (Compliance #1) through New York (State) (Reports #3)
68 New York (State) (Reports #4) through North Carolina (State) (Reports #l)

69 North Carolina (State)(Reports #2) through Ohio (State) (General #1)
70 Ohio (State) (General #2) through Ohio River (Reports #1)
71 Ohio River (Reports #2) through Ohio River, Area V (Administrative #1)
72 Ohio River, Area V (Correspondence #1) through Oregon (State) (Reports #1)
73 Ouachita River (Administrative #1) through Pascagoula River (Reports #1)
74 Pawcatuck River (General #1) through Pecos River (Clips #1)
75 Pecos River (Reports #1) through Pigeon River (Clips #1)
76 Pigeon River (Reports #1) through Puget Sound (Administrative #1)
77 Puget Sound (Compliance #1) through Puget Sound (Reports #2)
78 Puget Sound (Reports #3) through Raritan Bay (Administrative #3)
79 Raritan Bay (Compliance #l) through Raritan Bay (Reports #1)
80 Raritan Bay (Reports #2) through Red River (Upper) (Clips #l)
81 Red River (Upper) (Reports #1) through Republican River (Lower) (Administrative #1)
82 Republican River, Upper (Compliance #1) through Sabine River (Administrative #1)
83 Sabine River (Correspondence #2) through St. Lawrence River (Clips #1)
84 St. Lawrence River (Reports #1) through San Francisco Bay (Reports #1)
85 San Francisco Bay (Reports #2) through Savannah River, Upper (Clips #1)
86 Savannah River, Upper (Reports #1) through Savannah River, Lower (Clips #2)
87 Savannah River, Lower (Reports #l) through Snake River, Area III (Reports #1)
88 Souris River (Reports #1) through South Platte River (Clips #1)
89 South Platte River (Reports #1) through South Platte River (Reports #5)
90 South Platte River (Reports #6) through Spokane River (Administrative #1)
91 Spokane River (Correspondence #1) through Susquehanna River (Reports #2)
92 Suwannee and Withlacoochee Rivers (Administrative #l) through 10 Mile River (Reports #1)
93 Tennessee (State) (Reports #1) through Texas (State) (Reports #2)
94 Texas (State) (Reports #3) through Trinity River (Clips #1)
95 Trinity River (Reports #1) through Virgin Islands (Reports #1)
96 Virgin River (General #1) through Washita River (Reports #1)
97 Walker River (Administrative #1) through Wichita River (General #1)
98 West Virginia (State) (General #1) through Wisconsin (State) (General #4)
99 Wisconsin (State) (General #5) through Youghiogheny River (General)
100 Region, Great Lakes #l through Region, Northeast #3
101 Region, Northeast (Reports #1) through Region, Ohio Basin (Reports #2)
102 Region, Ohio Basin (Reports #3) through Region, Southeast (General #1)
103 Region, Southeast (General #2) through Region, All (Reports #1)
104 Region, All (Reports #2) through Laboratory, Grosse Ile
105 Laboratory, National Marine Quality Laboratory through General, Intra-Headquarters #5

106 General, Newsclips #1 through General, Reports #2
107 Program, Acid Mine Drainage through Program, Coastal Waters Monitoring
108 Program, Congressional Records through Program, Impact Statements
109 Program, Industrial Wastes through Program, Oil Pollution Control #3
110 Program, Oil Pollution Control #4 through Program, Sewage and Waste Treatment Reports
111 Program, Shellfish Sanitation through Program, U.S. Water Resources Council
112 Program, Water Law (General #7) through Program, Water Resources #2
113 Program, Water Resources #4 through Program, Water Resources, New England #1
114 Program, Water Resources, New York through Program, Water Resources, West Virginia

LIST OF ENFORCEMENT ACTION FILES CONTAINED IN DOCUMENT BOXES #1–#114

1	Alabama (State)	[AL]
2	Alabama River	[AL]
3	Alaska (State)	[AK]
4	Alleghany River	[NY PA]
5	Altahama River	[GA]
6	Amite River	[MS LA]
7	Androscoggin River	[ME NH]
8	Animas River	[NM CO]
9	Applegate River	[OR CA]
10	Arizona (State)	[AZ]
11	Arkansas (State)	[AR]
12	Arkansas River (General)	[ ]
13	Arkansas River I	[KS CO]
14	Arkansas River II	[KS OK]
15	Arkansas River III	[AR OK]
16	Aucilla River	[GA FL]
17	Barren River	[KY TN]
18	Batten Kill	[NY VT]
19	Bear Lake	[UT ID]
20	Bear River	[UT WY ID]
21	Beaver Brook	[MA NH]
22	Belle Fourche River	[WY SD]
23	Big Blue River	[KS NE]
24	Big Creek	[MO LA]
25	Bighorn River	[MT WY]
26	Big Muddy River	[IL]
27	Big Sandy River	[KY WV VA]
28	Big Sioux River	[IA SD]
29	Biscayne Bay	[FL]

30	Black Fork River	[UT WY]
31	Blackstone River	[RI MA]
32	Blue Earth River	[IA MN]
33	Badcau Bayou	[LA AR]
34	Boston Harbor	[MA]
35	Brandywine Creek	[PA DE]
36	Broad River (Congaree River)	[NC SC]
37	Bruneau River	[NV ID]
38	Buntings Branch	[MD DE]
39	Butler Creek (General)	[ ]
40	Byram River	[CT NY]
41	California (State)	[CA]
42	Canada (General)	[ ]
43	Canadian River	[OK TX NM]
44	Carson River	[NV CA]
45	Catawba-Wateree River	[NC SC]
46	Cedar River	[MN IA]
47	Champlain, Lake	[VT NY]
48	Chariton River	[IA MO]
49	Charles River	[MA]
50	Charleston Harbor	[SC]
51	Chattahoochee River (Upper)	[AL GA]
52	Chattahoochee River (Lower)	[AL GA FL]
53	Chautauqua, Lake	[NY]
54	Chesapeake Bay	[MD DE VA]
55	Cheyenne River	[WY SD]
56	Chicago Sanitary & Ship Canal	[IL]
57	Chowan River Basin	[VA NC]
58	Cimarron River	[NM OK CO KS]
59	Clarks Fork River	[MT WY]
60	Colorado (State)	[CO]
61	Colorado River (General)	[CO UT NV AZ CA]
62	Colorado River I	[UT CO]
63	Colorado River II	[UT AZ]
64	Colorado River III	[AZ NV]
65	Colorado River IV	[AZ NV]
66	Colorado River of Texas	[TX]
67	Columbia River (General)	[WA OR]
68	Columbia River I	[WA OR]
69	Columbia River II	[WA OR]
70	Columbia River III	[WA OR]
71	Connecticut (State)	[CT]
72	Connecticut (General)	[CT]
73	Connecticut River (Upper)	[VT NY]
74	Connecticut River (Lower)	[MA CT]

75	Cooper River	[SC]
76	Coosa River	[AL GA]
77	Corney Creek	[AR LA]
78	Crow Creek	[WY CO]
79	Cumberland River	[TN KY]
80	Cuyahoga River	[OH]
81	Deep Creek	[ID UT]
82	Deerfield River	[MA VT]
83	Delaware (State)	[DE]
84	Delaware (General)	[DE]
85	Delaware River I	[NY PA NJ]
86	Delaware River II	[NY PA NJ]
87	Delaware River III	[PA DE]
88	Delaware River IV	[PA DE]
89	Des Moines River	[MO IA]
90	Detroit River	[MI]
91	District of Columbia	[DC]
92	Delores River	[UT CO]
93	Erie, Lake	[NY PA MI IN OH]
94	Escambia River	[AL FL]
95	Escambia River Basin	[AL FL]
96	Fabius River	[IA MO]
97	Flint River	[FL GA]
98	Florida (State)	[FL]
99	Fox River	[IA MO]
100	Fox River	[WI IL]
101	French River	[CT MA]
102	French Broad River	[TN NC]
103	Gallatin River	[WY ID]
104	Galveston Bay	[TX]
105	Georgia (State)	[GA]
106	Gibson, Lake	[OK]
107	Gila & Salt Rivers	[AZ]
108	Goose Creek	[NV ID UT]
109	Goose, Lake	[OR CA]
110	Grand River	[IA MO]
111	Grand River	[SD ND]
112	Grand & Little Calumet Rivers	[IL IN]
113	Grand (Neosho) River	[OK KS]
114	Grants-Bluewaters	[NM]
115	Great Basin Drainage Basin	[CA NV]
116	Great Lakes	[MN WI MI IL OH IN]
117	Green Bay	[MI WI]
118	Green River	[UT WY]
119	Guadalupe River	[TX]

120	Guam Territory	[ ]
121	Gypsum Creek	[AZ UT]
122	Hackensack River (Upper)	[NY NJ]
123	Harris Fork Creek	[TN KY]
124	Hawaii (State)	[HI]
125	Henry's Fork	[WY UT]
126	Henry's Fork	[WY ID]
127	Hocking River	[OH]
128	Holston River (General)	[TN VA]
129	Holston River (North Fork)	[TN VA]
130	Holston River (South Fork)	[TN VA]
131	Hoosic River	[NY VT]
132	Housatonic River (Upper)	[MA CT]
133	Houston Ship Canal	[TX]
134	Hudson River (Lower)	[NY NJ]
135	Hudson River (Upper)	[NY]
136	Huron, Lake	[MI]
137	Huron River	[MI]
138	Idaho (State)	[ID]
139	Illinois (State)	[IL]
140	Illinois River	[IL]
141	Imperial Beach	[CA]
142	Indiana (State)	[IN]
143	Indian Creek	[KS MO]
144	Iowa (State)	[IA]
145	James River	[SD ND]
146	Jarbridge River	[NV ID]
147	Juniata River	[PA]
148	Kalamazoo River	[MI]
149	Kanab Creek	[AZ UT]
150	Kanawha River	[WV OH]
151	Kansas (State)	[KS]
152	Kansas River	[KS]
153	Kentucky (State)	[KY]
154	Kewaunee Harbor	[WI]
155	Key Paha River	[SD NE]
156	Klamath River	[CA OR]
157	Kootenai River	[MT ID]
158	La Plata River	[NM CA]
159	Laramie River	[CO WY]
160	Lehigh River	[PA DE]
161	Leviathan Creek	[NV CA]
162	Little Beaver Creek	[OH PA]
163	Little Blue River	[KS NE]
164	Little Missouri River	[WY MT SD ND]

165	Little River	[OK AR]
166	Little Sioux River	[MN IA]
167	Little Snake River	[WY CO]
168	Little Tennessee River	[TN NC GA]
169	Long Island Sound	[NY]
170	Los Pinos River	[NM CO]
171	Lost River	[OR CA]
172	Louisiana (State)	[LA]
173	Madison River	[WY MT]
174	Mahoning River	[PA OH]
175	Maine (State)	[ME]
176	Malad River	[UT ID]
177	Maple River	[ND SD]
178	Marais Des Cynes River	[KS MO]
179	Maryland (State)	[MD]
180	Massachusetts (State)	[MA]
181	Maumee River	[OH IN]
182	McElmo Creek	[UT CO]
183	Mead, Lake	[NV AZ]
184	Menominee River	[MI WI]
185	Meramec River	[MO]
186	Merrimack River	[MA NH]
187	Mexico, Gulf of	[LA TX AL FL]
188	Michigan (State)	[MI]
189	Michigan, Lake (Calumet) (III-IV)	[IN]
190	Michigan, Lake (Tributaries)	[MI WI IL IN]
191	Mill River	[MA RI]
192	Minnesota (State)	[MN]
193	Minnesota River	[MN]
194	Mississippi (State)	[MS]
195	Mississippi River (General)	[ ]
196	Mississippi River (Upper) I	[MN WI]
197	Mississippi River (Prairie du Chien) II	[IA WI]
198	Mississippi River (Clinton) III	[IL IA]
199	Mississippi River (Davenport) IV	[IL IA]
200	Mississippi River (Burlington) V	[IL IA]
201	Mississippi River (Hannibal) VI	[MO IL]
202	Mississippi River (St. Louis) VII	[MO IL]
203	Mississippi River (Cairo-Memphis) VIII	[KY MO AR TN]
204	Mississippi River (Lower) (Memphis-Vicksburg) IX	[TN AR LA MS]
205	Mississippi River (Lower) (Vicksburg-Mouth) X	[LA MS]
206	Mississippi–Missouri Rivers (General)	[ ]
207	Missouri (State)	[MO]

208	Missouri River (General)	[ ]
209	Missouri River (Headwaters–Ft. Randall Dam Site) I	[MT ND SD]
210	Missouri River (Sioux City) II	[SD NE IA]
211	Missouri River (Omaha) III	[NE IA]
212	Missouri River (St. Joseph) IV	[KS MO]
213	Missouri River (Kansas City) V	[KS MO]
214	Mobile Bay	[AL]
215	Mohawk River	[NY]
216	Monangahela River	[PA MD WV]
217	Montana (State)	[MT]
218	Montezeuma Creek	[CO UT]
219	Montreal River	[MI WI]
220	Mt. Hope Bay Drainage Area	[MA RI]
221	Moriches Bay	[NY]
222	Mullica River	[NJ]
223	Muskingum River	[OH WV]
224	Nanticoke River	[MD DE]
225	Narragansett Bay	[RI]
226	Nashua River	[NH MA]
227	Nebraska (State)	[NE]
228	Nevada (State)	[NV]
229	New England (States)	[ME VT NH MA CT]
230	New Hampshire (State)	[NH]
231	New Jersey (State)	[NJ]
232	New Jersey (Eastern Shore)	[NJ]
233	New Mexico (State)	[NM]
234	New River	[WV VA OH]
235	New York (State)	[NY]
236	New York Bight	[NY NJ]
237	NY-NJ-CT-Tri-State Area	[NY NJ CT]
238	Niagra River	[NY]
239	Nishnabota River	[IA MO]
240	Nodaway River	[IA MO]
241	Nolichucky River	[TN NC]
242	North Canadian River	[NM OK KS TX]
243	North Carolina (State)	[NC]
244	North Dakota (State)	[ND]
245	North Platte River (Upper)	[CO WY]
246	North Platte River (Lower)	[NE WY]
247	Norwalk River	[CT]
248	Oconto River	[WI]
249	Ohio (State)	[OH]
250	Ohio River (General)	[OH]
251	Ohio River I	[OH PA WV]

252	Ohio River II	[OH WV]
253	Ohio River III	[OH WV KY]
254	Ohio River IV	[OH IN KY]
255	Ohio River V	[IL IN KY]
256	Oklahoma (State)	[OK]
257	Ontario, Lake	[NY]
258	Oregon (State)	[OR]
259	Ouachita River	[LA AR]
260	Pacific Northwest Drainage Basin	[WA OR CA]
261	Palouse River	[WA ID]
262	Pascagoula River	[MS AL]
263	Pea & Choctawhatchee Rivers	[FL AL]
264	Peach River	[FL]
265	Pearl Harbor	[HA]
266	Pearl River	[LA MS]
267	Pecos River	[TX NM]
268	Pend Oreille River	[ID WA]
269	Pennsylvania (State)	[PA]
270	Penobscot River	[ME]
271	Permian Basin	[ ]
272	Perdido Bay	[FL AL]
273	Peshtigo River	[WI]
274	Pigeon River	[NC TN]
275	Piscataqua River	[NH ME]
276	Platte River	[IA MO]
277	Potomac River (Luke–Cumberland)	[MD WV]
278	Potomac River (DC Metro)	[MD VA DC]
279	Potomac River (General)	[MD VA DC WV]
280	Powder River	[WY MT]
281	Presumpscot River	[ME]
282	Price River	[UT]
283	Puerto Rico	[PR]
284	Puget Sound	[WA]
285	Pymatuning Creek	[OH PA]
286	Quinebaug River	[MA CT]
287	Raft River	[UT ID]
288	Rainy River	[MN]
289	Raritan Bay	[NY NJ]
290	Red River (General)	[AR OK TX]
291	Red River (Upper)	[AR OK TX]
292	Red River of the North	[ND SD]
293	Republican River (Upper)	[KS NE]
294	Republican River (Lower)	[CO KS NE]
295	Rhode Island (State)	[RI]
296	Rio Grande	[TX NM]

297	Roanoke River	[NC VA]
298	Rock River	[IL WI]
299	Rock River	[IA MN]
300	Sabine Lake	[LA TX]
301	Sabine River	[LA TX]
302	Saco River	[ME NH]
303	St. Croix River	[WI MN]
304	St. Francis River	[AR MO]
305	St. Johns River	[FL]
306	St. Joseph River	[IN MI]
307	St. Lawrence River	[NY]
308	St. Louis River	[WI MN]
309	St. Marys River	[FL GA]
310	Salmon Falls Creek	[NV ID]
311	Salmon Falls River	[ME NH]
312	Salton Sea	[CA]
313	San Francisco Bay	[CA]
314	San Juan River (General)	[CO NM UT]
315	San Juan River I	[CO NM]
316	San Juan River II	[CO NM UT]
317	Santa Ana River	[CA]
318	Santee River	[NC SC]
319	Stilla River	[GA]
320	Savannah River (General)	[GA SC]
321	Savannah River (Upper)	[GA SC]
322	Savannah River (Lower)	[GA SC]
323	Savannah River (Mouth)	[GA SC]
324	Shell Rock River	[IA MN]
325	Shenandoah River	[VA]
326	Smith River	[NC VA]
327	Smoky Hill River	[CO KS]
328	Snake River (General)	[WY ID OR WA]
329	Snake River I	[WY IED]
330	Snake River II	[OR ID]
331	Snake River III	[ID WA]
332	Souris River	[ND]
333	South Carolina (State)	[SC]
334	South Dakota (State)	[SD]
335	South Platte River	[CO NE]
336	Southeast Drainage Basin	[ ]
337	Split Rock Creek	[SD MN]
338	Spokane River	[WA ID]
339	Spring Creek	[PA]
340	Spring River	[MO KS]

341	Superior, Lake	[MN WI]
342	Susquehanna River (Upper)	[PA NY]
343	Suwanee River	[FL GA]
344	Tahoe, Lake	[CA NV]
345	Tangipahoa River	[LA MS]
346	Ten Mile River	[RI MA]
347	Tennessee (State)	[TN]
348	Tennessee River	[AL GA TN]
349	Teton River	[WY ID]
350	Texas (State)	[TX]
351	Tombigee River	[AL MS]
352	Tongue River	[WY MT]
353	Trinity River	[TX]
354	Truckee Carson Irrigation District	[ ]
355	Truckee River	[NV CA]
356	Umpqua River	[OR]
357	Utah (State)	[UT]
358	Verdigris River	[OK KS]
359	Vermilion River	[IL]
360	Vermont (State)	[VT]
361	Virgin Islands	[VI]
362	Virgin River	[NV AZ UT]
363	Virginia (State)	[VA]
364	Wabash River	[IL IN]
365	Walker River	[NV CA]
366	Walloomsac River	[NY VT]
367	Warm Springs Run	[MD WV]
368	Washington (State)	[WA]
369	Washita River	[OK TX]
370	West Virginia (State)	[WV]
371	White River	[AR MO]
372	White River	[NE SD]
373	White River	[UT CO]
374	Whitewood Creek	[SD]
375	Wichita River	[TX]
376	Whitewater River	[OH IN]
377	Willamette River	[OR]
378	Winooski River	[VT]
379	Wisconsin (State)	[WI]
380	Wisconsin River	[WI MI]
381	Withlacoochee River	[GA FL]
382	Wyoming (State)	[WY]
383	Yadkin-Pee Dee River	[NC SC]
384	Yakima River	[WA]

385	Yampa River	[CO]
386	Yankee Creek	[OH PA]
387	Yellowstone River	[MT ND]
388	Yellowstone River	[WY MT]
389	Youghiogheny River	[WV MD PA]

APPENDIX A-5 NATIONAL ARCHIVE RECORDS OF 180-DAY NOTICE TRANSCRIPTS[1]

National Archives Tracking Number: TN28/C3/9A
Accession Number: 412-74-26
Record Group Number: 412

FROM: U.S. Environmental Protection Agency
Enforcement Division, OEGC
Rm. 3600, Waterside Mall, Washington, DC

TO: Federal Records Center, GSA
Washington National Records Center
Accession Section
Washington, DC 20409, Stop Code 386

Agency Custodian of Records: Frank Hall
Agency Official: Harold Masters, Records Management Officer
Number of Document Boxes: 8
Date: April 16, 1974

U.S. Environmental Protection Agency

Enforcement Division

Documents pertaining to conferences on an interstate body of water, and a state and federal effort to make or keep it clean. Complete verbatim transcripts (certified conference reporter typed copy) of informal hearings conducted by EPA under Sec. 10(c)(5)of the Federal Water Pollution Control Act. (33 U.S.C. 466 *et seqn.*)

180-DAY NOTICE TRANSCRIPTS

1 Alton Box Board Co., Lafayette, Indiana Mill through Cleveland, Ohio
2 Cleveland, Ohio, Exhibits through City of Fargo, North Dakota
3 City of Franklin, New Hampshire, through Town of Darien, Connecticut

[1] *Source:* Standard Form 135, Records Transmittal and Receipt.

4 Jones & Laughlin Steel Corporation through GAF
5 Lake Mead & the Colorado River through Piel Bros., Inc.
6 Republic Steel Corporation through City of Toledo, Ohio
7 Stamford, Connecticut, through Vincennes, Indiana
8 WSSC (Washington Suburban Sanitary Commission) through Willoughby-Eastlake, Ohio

List of 180-Day Notice Transcripts in Document Boxes #1 through #8

Alton Box Board Co., Lafayette, Indiana Mill
Riverside Paper Corp. and Consolidated Papers, Inc., Appleton, Wisconsin
City of Chicopee, Massachusetts
City of Atlanta, Georgia
Banies Co., Inc., East Pepperell, Massachusetts
City of Bogalusa, Louisiana
Cleveland, Ohio
Cleveland, Ohio, Exhibits
2nd Meeting of City of Cleveland, Ohio
City of Cleveland and its 32 Suburbs
Cuyahoga County, Cleveland, Ohio
Covington, Indiana
Fairfax Drainage District, Wyandotte County, Kansas
City of Fargo, North Dakota
City of Franklin, New Hampshire
Euclid, Ohio
Willoughby-Eastlake, Ohio
Eagle-Pitcher Industries, Baxter Springs, Kansas
City of Detroit, Michigan
Town of Darien, Connecticut
Jones & Laughlin Steel Corp., Cleveland, Ohio
Interlake Steel Corp., Toledo, Ohio
City of Hammond, Indiana, and the Hammond Sanitary District
Hurley, Wisconsin
Holly Sugar Co., Torrington, Wyoming
Green Bay Metro Sewerage District, Charmin Paper Products Co., and American Can Company, Green Bay, Wisconsin
Granite, Illinois
Gary Sanitary District, Gary, Indiana
GAF Corp., Linden, New Jersey
Lake Mead & Colorado River, Las Vegas, Nevada

City of Logansport, Indiana
Town of Montezeuma, Indiana
Neenah-Menasha Sewerage Comm., et al.
Penn Central Transportation Corp., Harmon Yards, New York
Pepperell Paper Co., East Pepperell, Massachusetts
Piel Bros., Inc., Hampden-Harvard Div., Chicopee, Massacusetts
Republic Steel Corp., Cleveland, Ohio
Reserve Mining Co., Silver Bay, Minnesota
City of Riverview, Michigan
Santa Fe Land Development Co.
Village of Sauget, Illinois
City of Sheboygan, Wisconsin
East-Side Levee and Sanitary District
Tahoe-Douglas District & Kingsbury General Improvement District
City of Toledo, Ohio
City of Stamford, Connecticut
Superior, Wisconsin
Superior Fiber Products, Superior, Wisconsin
Toledo, Ohio
U.S. Steel Corp., Cleveland, Ohio
Vincennes, Indiana
WSSC (Washington Suburban Sanitary Commission), Washington DC
County of Wayne, Michigan
Whiting, Indiana
Willoughby-Eastlake, Ohio

APPENDIX A-6 NATIONAL COMMISSION ON WATER QUALITY (NCWQ) REPORTS[1]

In the course of its work evaluating progress of water pollution control programs under PL 92–500 [Clean Water Act, CWA], the National Commission on Water Quality contracted for the preparation of a large number of technical reports. A complete listing of all the technical reports (with NTIS accession numbers and cost as of 1976) prepared for the NCWQ has been compiled by Mitchell (1976). Appendix A-6 documents the technical reports prepared for the NCWQ that were related to water quality and environmental assessments of site-specific waterways. The reports are available for purchase from the National Technical Information Service (NTIS), U.S. Department of Commerce, Springfield, Virginia 22161.

[1]*Source:* Mitchell, 1976.

NCWQ STAFF TECHNICAL VOLUMES

1. Technological Assessment . . . . PB 252 298
2. Economic and Social Impacts . . . . PB 253 037
3. Water Quality Analysis, Environmental Impact Assessment . . . . PB 252 167
4. Regional Assessment . . . . PB 252 099

CONTRACTOR REPORTS: REGIONAL IMPACTS

Delaware River Basin; Betz Environmental Engineers. . . . . PB 249 910
San Francisco Bay/Central Valley; Arthur D. Little Inc.. . . . . PB 249 730
Ohio River Basin; Dames & Moore, Inc. . . . . PB 249 680
Colorado River Basin; Utah State University . . . . PB 249 660
Merrimack-Nashua River Basin; Abt Associates . . . . PB 250 060
Kanawha River; Dames & Moore, Inc. . . . . PB 250 105
Lake Erie; Dalton, Dalton, Little & Newport . . . . PB 251 009
Yellowstone River Basin; Stevens, Thompson & Runyan, Inc. . . . . PB 251 072
Puget Sound/Lake Washington; Stevens, Thompson & Runyan, Inc.. . . PB 251 319
Houston Ship Channel/Galveston Bay; Bernard Johnson, Inc.. . . . . PB 252 460
Chattahoochee-Flint-Apalachicola; Hammer, Siler, George Assoc. . . . . PB 252 318

CONTRACTOR REPORTS: ENVIRONMENTAL IMPACT ASSESSMENT, WATER QUALITY ANALYSIS

Puerto Rico; Tetra Tech, Inc.. . . . . PB 251 323
Gulf of Alaska; Tetra Tech, Inc.. . . . . PB 251 322
Hawaii; Tetra Tech, Inc.. . . . . PB 251 320
St. John River (Florida); Atlantis Scientific . . . . PB 251 225
S. California Bight; Tetra Tech, Inc. . . . . PB 251 435
South Platte; Tetra Tech, Inc.. . . . . PB 251 446
Escambia River & Bay; Atlantis Scientific . . . . PB 251 447
Hudson River; Lawler, Matusky & Skelly . . . . PB 251 099
J. Percy Priest Reservoir; Vanderbilt University . . . . PB 251 098
St. John River (Maine); Meta Systems. . . . . PB 250 942
Yadkin Peedee River Basin; TRW, Inc. . . . . PB 250 933
Connecticut River; The Center for Environment and Man . . . . PB 250 924
Susquehanna; Lawler, Matusky & Skelly . . . . PB 250 925
Housatonic River; Lawler, Matusky & Skelly . . . . PB 250 926
Missouri River Basin; Midwest Research Inst.. PB 250 930
Biscayne Bay; Water Resource Engineers. . . . . PB 252 095
Trinity River; Water Resource Engineers . . . . PB 252 097
Charles River/Boston Harbor; Process Research . . . . PB 252 098
Guadalupe/San Antonio River Basin; Water Resource Engineers. . . . . . PB 252 430
Chesapeake Bay; Virginia Inst. of Marine Sciences . . . . PB 252 096
Utah Lake/Jordan River, Utah; Environmental Dynamics, Inc. . . . . PB 250 932

CONTRACTOR REPORTS: WATER QUALITY ANALYSIS

Upper Mississippi River Basin; Water Resource Engineers .... PB 250 982
Snake River; Tetra Tech, Inc. .... PB 250 929
Columbia River; Tetra Tech, Inc. .... PB 250 927
Santee River Basin; Water Resource Engineers .... PB 250 928
Upper Rio Grande; Water Resource Engineers .... PB 250 981
Iowa-Cedar River Basin; Water Resource Engineers .... PB 250 931
The Potomac River; GKY Associates .... PB 252 038

CONTRACTOR REPORTS: ENVIRONMENTAL IMPACT ASSESSMENT

The Potomac River; Academy of Natural Sciences .... PB 250 934
Iowa-Cedar River Basin; Midwest Research Inst. .... PB 251 227
Snake River; Parametrix .... PB 251 226
Upper Mississippi River Basin; Midwest Research Inst. .... PB 251 222
Santee River Basin; Academy of Natural Sciences .... PB 251 354
Upper Rio Grande; Academy of Natural Sciences .... PB 251 228
Columbia River; Parametrix .... PB 251 324
Ocean Discharges, "Environmental Effects of the Disposal of Municipal Effluents, Following Various Levels of Wastewater Treatment, into Nearshore Marine Waters"; Southern California Coastal Water Research Project and Engineering-Science, Inc.. .... PB 254 396
Disposal of Wastewater Residuals; Environmental Quality Systems. .... PB 251 371
National Residuals Discharge Inventory; National Research Council, National Academy of Sciences, Committee on Water Quality Policy .... PB 252 288

REFERENCE

Mitchell, D. 1976. NCWQ reports available. J. WPCF 48(10): 2427–2433.

APPENDIX B

National Municipal Wastewater Inventory and Infrastructure, 1940–2016

TABLES

B-1 POTW facility inventory
B-2 Resident population served by POTWs
B-3 Wastewater flow
B-4 Influent wastewater flow normalized to population served
B-5 Influent $CBOD_5$ load
B-6 Effluent $CBOD_5$ load
B-7 $CBOD_5$ removal efficiency
B-8 Influent $CBOD_5$ load normalized to population served
B-9 Influent $CBOD_5$ concentration
B-10 Effluent $CBOD_5$ concentration
B-11 Influent $CBOD_u$ load
B-12 Effluent $CBOD_u$ load
B-13 $CBOD_u$ removal efficiency
B-14 Effluent $CBOD_u$ concentration
B-15 Influent NBOD load
B-16 Effluent NBOD load
B-17 NBOD removal efficiency
B-18 Effluent NBOD concentration
B-19 Influent BOD_u load
B-20 Effluent BOD_u load
B-21 Removal efficiency BOD_u
B-22 Effluent BOD_u concentration
B-23 Influent TSS load
B-24 Effluent TSS load
B-25 TSS removal efficiency
B-26 Effluent TSS concentration
B-27 Influent POC load
B-28 Effluent POC load

B-29 POC removal efficiency
B-30 Effluent POC concentration
B-31 Influent and effluent characteristics for 5-day carbonaceous BOD: $CBOD_5$
B-32 Influent and effluent characteristics for ultimate carbonaceous BOD: $CBOD_u$
B-33 Influent and effluent characteristics for total Kjedhal nitrogen: TKN
B-34 Influent and effluent characteristics for nitrogenous BOD: NBOD
B-35 Influent and effluent characteristics for ultimate BOD: BOD_u
B-36 Influent and effluent characteristics for total suspended solids (TSS)
B-37 Influent and effluent characteristics for particulate organic carbon (POC)
B-38 Middle population projections of POTW loads of $CBOD_5$, $CBOD_u$, NBOD and BOD_u, 1996–2050
B-39 Inventory of POTW facilities in 1972 in the United States
B-40 Population served (millions) by POTW facilities in 1972 in the United States
B-41 Municipal wastewater treatment categories presented in the tables of Appendix B

Note: Values of −9.0 in the tables indicate that data were not available for the particular year and category of waste treatment.

DATA SOURCES FOR TRENDS IN MUNICIPAL WASTEWATER TREATMENT, 1940–2016

1940 Population served data from FWPCA (1970). Inventory of municipal wastewater facilities from NCWQ (1976). Data not available for no-discharge category; assumed zero for calculation of totals.

1950 Population served data and inventory of municipal wastewater facilities from USPHS (1951). Data not available for no-discharge category; assumed zero for calculation of totals.

1962 Population served data and inventory of municipal wastewater facilities from USEPA (1974). Data not available for no-discharge category; assumed zero for calculation of totals.

1968 Population served data and inventory of municipal wastewater facilities from USEPA (1974) and FWQA (1970). Data not available for no-discharge category; assumed zero for calculation of totals.

1972 Population served data from ASIWPCA (1984). Inventory of municipal wastewater facilities from USEPA (1972). Significant differences in population served data between USEPA (1972) (102.3 million) and ASIWPCA (1984) (76.3 million) for secondary treatment. USEPA (1972) includes population served by oxidation ponds (7.5 million, 4,467 facilities) and land application (0.5 million, 142 facilities) in the total population served of 102.3 million for "secondary" treatment. ASIWPCA (1984) reports 76.2 million served by secondary with no breakdown of categories of facilities given. To be consistent with trends in population served by secondary plants reported for 1968 (85.6 million) and 1978 (56.3 million), the population served data from ASIWPCA (1984) is used to estimate wastewater flow, influent and effluent loading of BOD_5 for 1972.

1973 Population served data and inventory of municipal wastewater facilities from USEPA (1974). Data not available for no-discharge category; assumed zero for calculation of totals.

1974 Inventory of municipal wastewater facilities from NCWQ (1976). Population served data not available for any category.

1976 Population served data and inventory of municipal wastewater facilities from USEPA (1976). Data not available for raw population served and effluent flow. Data not available for no-discharge category; assumed zero for calculation of totals.

1978 Population served data, inventory of municipal wastewater facilities, wastewater effluent flow (as 1,000 m^3/day), influent and effluent BOD_5 loading data from USEPA (1978). Effluent flow from USEPA (1978) converted from 1,000 m^3/day to mgd using conversion factor of 0.2642. Population served by raw discharge reported as 364,000 persons in Table 10 of the USEPA (1978) Needs Survey. To be consistent with trends in population served by raw discharge for data reported for 1968 (10.1 million), 1972 (4.9 million), 1980 (2.3 million), and 1982 (1.9 million), the value for 1978 given in Table 10 appears to be in error by factor of 10. The population served by raw discharge for 1978 is increased by a factor of 10 to 3.64 million for this study. Data reported in USEPA (1978) for effluent flow, influent and effluent BOD_5 loading for raw discharge, computed from erroneous population served data, is not used. The number of facilities reported in USEPA (1978) as raw discharge systems ($n = 91$) is not consistent with trend of 2,265 raw facilities reported for 1972 and 237 raw facilities reported for 1982. There is a possible factor of 10 error in USEPA Needs Survey data table since $n = 910$ raw facilities would be consistent with the decreasing trend in the number of raw facilities from 1972 to 1982.

1980 Population served data, inventory of municipal wastewater facilities, wastewater effluent flow (as 1,000 m^3/day), influent and effluent BOD_5 loading data, and percent BOD_5 removal from USEPA (1980). Effluent flow from USEPA (1980) converted from 1,000 m^3/day to mgd by conversion factor of 0.2642. Data not available for raw effluent flow.

1982 Population served data, inventory of municipal wastewater facilities, wastewater effluent flow (as 1,000 m^3/day), influent and effluent BOD_5 loading data, and percent BOD_5 removal from USEPA (1982). Effluent flow from USEPA (1982) converted from 1,000 m^3/day to mgd by conversion factor of 0.2642. Data not available for raw effluent flow.

1984 Population served data, inventory of municipal wastewater facilities, wastewater effluent flow (as mgd), influent and effluent BOD_5 loading data, and percent BOD_5 removal from USEPA (1984). Data not available for raw effluent flow.

1986 Population served data, inventory of municipal wastewater facilities, wastewater effluent flow (as mgd), influent and effluent BOD_5 loading data, and percent BOD_5 removal from USEPA (1986). Data not available for raw effluent flow.

1988 Population served data and inventory of municipal wastewater facilities from USEPA (1989).

1992 Population served data and inventory of municipal wastewater facilities from USEPA (1993).

1996 Population served data and inventory of municipal wastewater facilities from USEPA (1997).

2016 Projection of population served and inventory of municipal wastewater facilities for the year 2016 from USEPA (1997).

Data Sources The primary data sources for the analysis of POTW wastewater trends included the municipal wastewater inventories published by the USPHS from 1940 through 1968 (USPHS, 1951; NCWQ, 1976; USEPA, 1974) and USEPA's Clean Water Needs Surveys (CWNS) conducted from 1973 through 1996 (USEPA, 1976, 1978, 1980, 1982, 1984, 1986, 1989, 1993, 1997). Each data source reported the number of municipal wastewater treatment plants and the population served by raw, primary, advanced primary, secondary, advanced secondary and advanced treatment levels of wastewater treatment.

Where data were available to differentiate primary from advanced primary and advanced secondary from advanced treatment, the data were added to define "less than secondary" and "greater than secondary" categories. For some of the USEPA Needs Surveys, facility inventories and population served data were compiled as "less than secondary" and "better than secondary"; data were not available to differentiate primary and advanced primary or advanced secondary and advanced treatment levels of wastewater treatment. The USEPA Clean Water Needs Surveys also compiled effluent flow, influent and effluent loading and percent removal of BOD_5 in the reports for 1978, 1980, 1982, 1984 and 1986. For the years in which these data were not available, wastewater effluent flow, influent and effluent BOD_5, $CBOD_u$, TKN, NBOD, BOD_u, TSS, and POC loading data were estimated based on: (1) population served; (2)constant normalized flow rate of 165 gallons per capita per day (gpcd); (3) influent BOD_5 of 215 mg/L and effluent BOD_5 removal efficiencies; (4) effluent ratios of $CBOD_u$/ BOD_5; (5) influent TKN of 30.3 mg/L and effluent TKN removal efficiencies; (6) influent TSS of 215 mg/L and effluent TSS removal efficiencies; and (7) effluent ratios of the particulate organic fraction of TSS and a constant ratio of carbon to dry weight (C/DW).

Constant Per Capita Flow Rate The constant per capita flow rate of 165 gpcd is based on the mean ($n = 5$) of the total population served and total wastewater flow data compiled in the USEPA Clean Water Needs Surveys for 1978 through 1986 (see Table B-4). The rate of per capita flow, ranging from 160 to 173 gpcd, includes residential (55 percent), commercial and industrial (20 percent), stormwater (4 percent), and infiltration and inflow (20 percent) components of wastewater flow (see AMSA, 1997). The constant per capita flow rate of 165 gpcd used in this study to estimate trends of municipal wastewater flow and loading is identical to the typical United

States average (165 gpcd) within the wide range (65 to 290 gpcd) of municipal water use that accounts for residential, commercial and industrial, and public water uses in the United States (see Metcalf and Eddy, 1991). Public wastewater flow, obviously related to public water withdrawals, can range from 70 to 130 percent of the rate of water withdrawal, with a reasonable assumption being that wastewater flow is approximately equal to withdrawals by water supplies (Steel, 1960).

Influent BOD_5 The influent BOD_5 concentration of 215 mg/L, consistent with many other estimates of raw wastewater strength (e.g., AMSA, 1997; Tetra Tech, 1999; Metcalf and Eddy, 1991), is based on the mean ($n = 5$) nationally aggregated ratio of the total influent BOD_5 loading rate normalized to total wastewater flow reported in the USEPA Clean Water Needs Surveys for 1978 through 1986 (see Table B-9). The influent concentration, ranging from 209 to 229 mg/L for these years, includes the residential, commercial and industrial, and infiltration and inflow contributions to the total influent BOD_5 load to municipal wastewater treatment plants. Using the influent BOD_5 concentration of 215 mg/L and the normalized flow rate of 165 gpcd, the normalized influent BOD_5 loading rate (0.296 lb BOD_5 per person per day) accounts for the nationally aggregated mixture of domestic, commercial and industrial, and infiltration and inflow components of wastewater. The loading rate used in this analysis is almost a factor of two greater than the typical textbook value for the "population equivalent" (PE) of 1 PE = 0.17 lb BOD_5 per person per day. Typical textbook values account only for the average per capita residential load contributed by combined stormwater and domestic wastewater; the industrial and commercial component is not included (Fair et al., 1971).

Effluent BOD_5 The effluent BOD_5 loading rates, estimated using removal efficiencies typically assigned for NPDES permit limits and wastewater treatment plant design assumptions, are based on an influent concentration of 215 mg/L and removal efficiencies (as percentage) assigned to each level of treatment. The BOD_5 removal efficiencies assumed for primary (35 percent), advanced primary (50 percent), and less than secondary (42.5 percent), although somewhat lower than removal efficiencies reported for primary (41 percent), advanced primary (64 percent), and less than secondary (54 to 57 percent) based on PCS data compiled for the USEPA Clean Water Needs Surveys for 1976, 1978, and 1982, are consistent with typical textbook values reported for BOD_5 removal efficiency for primary treatment plants (see Metcalf and Eddy, 1991).

The BOD_5 removal efficiencies assumed for secondary (85 percent), advanced secondary (90 percent), advanced waste treatment (95 percent), and better than secondary (92.5 percent) are comparable to removal efficiencies for secondary (82 to 86 percent), advanced secondary (89 to 92 percent), advanced waste treatment (87 to 94 percent), and better than secondary (94 percent) based on PCS data compiled in the Clean Water Needs Surveys for 1976, 1978, and 1982. The removal efficiencies for secondary, advanced secondary, and advanced waste treatment used in this study, consistent with textbook design values, are based on an influent concentration of 215 mg/L and the ranges of removal efficiencies and effluent concentrations of BOD_5

used to define these treatment categories in the USEPA Clean Water Needs Survey for 1978 (USEPA, 1978).

Secondary treatment was defined in the 1978 Needs Survey by a BOD_5 effluent concentration of 30 mg/L with the removal efficiency ranging from 84 to 89 percent (USEPA, 1978). Advanced secondary was defined by an effluent BOD_5 range of 10 to 30 mg/L, and advanced waste treatment was defined by an effluent BOD_5 concentration less than, or equal to, 10 mg/L (USEPA, 1978). Assuming a mean influent concentration of 215 mg/L, a midrange effluent concentration of 20 mg/L for advanced secondary, and 10 mg/L for advanced waste treatment, BOD_5 removal efficiencies were assigned as 90 percent for advanced secondary and 95 percent for advanced waste treatment.

Ultimate BOD Influent and effluent loading rates of the ultimate carbonaceous BOD ($CBOD_u$) were estimated from the BOD_5 loads and conversion ratios of ultimate to 5–day BOD ($CBOD_u/CBOD_5$) defined for each level of municipal treatment (Leo et al., 1984; Thomann and Mueller, 1987). Since it is impossible to determine if historical BOD_5 effluent loads included the suppression of nitrification (see Hall and Foxen, 1984), it is assumed that BOD_5 is approximately equal to $CBOD_5$ (see Lung, 1998).

Loading rates for the nitrogenous component of oxygen demand (NBOD) are estimated from the influent concentration and removal efficiencies of oxidizable nitrogen (TKN) for each level of treatment and the oxygen/nitrogen ratio of 4.57 g O_2 per g N. Removal efficiencies for TKN are based on data compiled by: (1) Gunnerson et al. (1982) for primary (22 percent); (2) advanced primary (22 percent) assumed the same as primary, and (3) secondary (36 percent). TKN removal efficiencies assigned for (4) advanced secondary (78 percent); and (5) advanced treatment (83 percent) are within the twenty-fifth and seventy-fifth percentile ranges of data compiled for advanced secondary (72 to 92 percent) and advanced treatment (79 to 95 percent) from AMSA (1997). Less than secondary removal efficiencies are assigned as the mean of primary and advanced primary removal percentages. Better than secondary removal efficiencies are assigned as the mean of advanced secondary and advanced treatment removal percentages.

The total ultimate BOD (BOD_u) load is calculated as the sum of the ultimate carbonaceous ($CBOD_u$) and nitrogenous (NBOD) components of the effluent load of oxygen demanding substances . Table B-31 summarizes the influent and effluent concentrations and removal efficiencies assumed for calculation of the loads of BOD_5, $CBOD_u$, TKN, NBOD, and BOD_u.

Projection of Effluent Load Trends from 1996–2050 Projections of effluent loading trends for ultimate BOD are based on: (1) U.S. Census Bureau "middle" population projections from 1996–2050 (U.S. Census, 1996); (2) constant wastewater inflow rate of 165 gpcd; (3) constant BOD_u influent concentration of 396.5 mg/L; and (4) linear extrapolations of the percentage of projected population served by POTWs and removal efficiencies for BOD_5, $CBOD_u$, NBOD, and BOD_u estimated for 1996 and 2016 using data obtained from the 1996 Clean Water Needs Survey (USEPA, 1997). It is assumed that the design BOD_u removal efficiency (65 percent in 1996 and 71 percent in 2016) and the proportion of the United States population served by POTWs (72 percent in 1996 and 88 percent in 2016) increases linearly from 1996 to 2016. After 2016, it is assumed that these percentages remain constant over time from 2016 to 2050 as 71 percent removal efficiency for BOD_u and 88 percent of the projected population served by POTWs.

Influent TSS and POC The influent TSS concentration of 215 mg/L, consistent with other estimates of raw wastewater strength (e.g., Tetra Tech, 1999; Metcalf and Eddy, 1991), is based on the mean influent concentration from 60 wastewater facilities reported in AMSA (1997). The influent concentration of POC (70.95 mg/L) was estimated from the particulate organic matter (POM) (volatile suspended solids) fraction of influent TSS (POM/TSS = 0.75) for "medium" strength raw wastewater and a C DW ratio of 0.44 mg C/mg DW (Metcalf and Eddy, 1991).

Effluent TSS and POC Removal efficiencies for TSS are based on data compiled by: (1) Gunnerson et al. (1982) for primary (50 percent); (2) NRC (1993) for advanced primary (70 percent); and mean removal efficiencies computed from data reported by AMSA (1997) for (3) secondary (92 percent); (4) advanced secondary (97 percent); and (5) advanced treatment (98 percent). Less than secondary removal efficiencies are assigned as the mean of primary and advanced primary removal percentages. Better than secondary removal efficiencies are assigned as the mean of advanced secondary and advanced treatment removal percentages. Effluent concentrations of POC are estimated from particulate organic matter (volatile suspended solids) fractions of TSS for primary and advanced primary (0.83), secondary and advanced secondary (0.67) obtained from Clark et al.(1977). Data were not available to define the organic matter fraction of TSS for effluent from advanced treatment; a value of 0.5 was assumed as a reasonable characterization of the organic matter fraction of effluent TSS. The organic component of effluent TSS was converted to POC using a constant C/DW ratio of 0.44 mg C/mg DW (Metcalf and Eddy, 1991). Table B-32 summarizes the influent and effluent concentrations and removal efficiencies assumed for calculation of the loads of TSS and POC.

REFERENCES

AMSA. 1997. The AMSA financial survey. American Metropolitan Sewerage Association, Washington, DC.

ASIWPCA. 1984. America's clean water: The states evaluation of progress 1972–1982. Association of State and Interstate Water Pollution Control Administrators (ASIWPCA), Washington, D. C. Executive Summary and technical appendix. Technical appendix provides state inventories of population served by municipal wastewater treatment category and influent and effluent BOD_5 loadings for 1972 and 1982.

Clark, J. W., W. Viessman, and M. L. Hammer. 1977. Elements of water supply and wastewater disposal, 3rd ed. Harper & Row Publishers, New York.

Fair, G. M., J. C. Geyer, and D. A. Okun. 1971. Elements of water supply and wastewater disposal, 2nd ed. John Wiley & Sons, New York. 752 pp.

FWPCA. 1970. The economics of clean water. Vol. I, Detailed analysis. U.S. Department of the Interior, Federal Water Pollution Control Administration (FWPCA), Washington, DC. March. Population served data for 1940 raw, primary and intermediate, and secondary treatment digitized from Figure 4, "Growth of public waste handling services," p. 75.

FWQA. 1970. Municipal waste facilities in the U.S.: Statistical summary, 1968 inventory. Publication No. CWT-6. U.S. Department of the Interior, Federal Water Quality Administration (FWQA), Washington, DC.

Gunnerson, C. G. et al. 1982. Management of domestic waste. In: Ecological stress and the New York Bight: Science and management, ed. G. F. Mayer, pp. 91–112. Estuarine Research Foundation, Columbia, SC.

Hall, J. C. and R. J. Foxen. 1984. Nitrification in BOD_5 test increases POTW noncompliance. J. WPCF 55(12): 1461–1469.

Leo, W.M, R. V. Thomann, and T. W. Gallagher. 1984. Before and after case studies: Comparisons of water quality following municipal treatment plant improvements. EPA 430/9–007. U.S. Environmental Protection Agency, Office of Water, Program Operations, Washington, DC.

Lung, W. 1998. Trends in BOD/DO modeling for waste load allocations. J. Environ. Eng., ASCE 124(10):1004–1007.

Metcalf and Eddy. 1991. Wastewater engineering: Treatment, disposal and reuse. 3rd ed. McGraw-Hill Series in Water Resources and Environmental Engineering, New York. 1334 pp.

MWCOG. 1989. Potomac River water quality trends and issues in the Washington metropolitan area, 1982–1986. Metropolitan Washington Council of Governments, Washington, DC. April.

NCWQ. 1976. Staff report to the National Commission on Water Quality. U.S. Government Printing Office, Washington, DC. p. II-5. Municipal facility inventory data for 1940 and 1974 taken from numerical data presented in FWQA (1970) figure "Trends in municipal wastewater treatment" for 1940 (2,938 primary, 2,630 secondary for a total of 5,568) and 1974 (3,032 primary, 16,987 secondary, 992 tertiary for a total of 21,011 facilities).

NRC. 1993. Managing wastewater in coastal urban areas. Committee on Wastewater Management for Coastal Urban Areas, Water Science and Technology Board, Commission on Engineering and Technical Systems, National Research Council. National Academy Press, Washington, DC.

Steel, E. W. 1960. Water supply and sewerage, 4th ed. McGraw-Hill, New York. 655 pp.

Tetra Tech. 1999. Improving point source loadings data for reporting national water quality indicators. Final Technical Report prepared for U.S. Environmental Protection Agency, Office of Wastewater Management, Washington, DC, by Tetra Tech, Inc. Fairfax, VA.

Thomann, R. V. and J. A. Mueller. 1987. Principles of surface water quality modeling and control. Harper & Row, Inc., New York.

U.S. Census. 1996. Population projections of the United States by age, sex, race and Hispanic origin: 1995–2050. Current Population Reports Series, pp.25–1130. Population Division, U.S. Bureau of Census, Washington, DC.

USEPA. 1972. Working file notes from EPA Office of Municipal Pollution Control, Washington, DC. Handwritten tables compiled from November 1972 EPA STORET data extraction and tabulation of population served and facilities inventory for raw, primary, intermediate (advanced primary), secondary, and tertiary treatment. Detailed types of waste treatment facilities are compiled. Secondary treatment category (total of 14,035 facilities serving 102.3 million persons) included subcategories of: other; activated sludge; extended aeration; high rate trickling filters; standard rate trickling filter; intermediate sand filtration; land application; oxidation pond; unknown filter; and unknown. Primary data sources reproduced in Tables B-39 and B-40.

USEPA. 1974. National water quality inventory, 1974. Report to Congress. EPA-440/9-74-001, Vols. I and II. Office of Water Planning and Standards, U.S. Environmental Protection Agency, Washington, D. C. Population served and number of facilities for raw, primary, secondary, and tertiary for 1962, 1968, and 1973 compiled on p. 278. Data from 1962 and 1968 are from USPHS Municipal Waste Inventories. Data for 1973 are from STORET database.

USEPA. 1976. 1976 NEEDS survey, Conveyance and treatment of municipal wastewater. Summaries of technical data. Water Program Operations, U.S. Environmental Protection Agency, Washington, DC. Data not available for population served by no discharge, raw discharges, or tertiary treatment. Data not available for facilities inventory.

USEPA. 1978. 1978 NEEDS survey, Conveyance and treatment of municipal wastewater. Summaries of technical data. Water Program Operations, U.S. Environmental Protection Agency, Washington, DC. Data not available for raw effluent flow.

USEPA. 1980. 1980 NEEDS survey, Conveyance and treatment of municipal wastewater. Summaries of technical data. Office of Water Program Operations, U.S. Environmental Protection Agency, Washington, DC. Data not available for raw effluent flow.

USEPA. 1982. 1982 NEEDS survey, Conveyance, treatment, and control of municipal wastewater, combined sewer overflows and stormwater runoff. Summaries of technical data. Office of Water Program Operations, U.S. Environmental Protection Agency, Washington, DC. Data not available for raw effluent flow.

USEPA. 1984. 1984 NEEDS survey, Conveyance, treatment, and control of municipal wastewater, combined sewer overflows and stormwater runoff. Summaries of technical data. Office of Water Program Operations, U.S. Environmental Protection Agency, Washington, DC. Data not available for raw effluent flow.

USEPA. 1986. 1986 NEEDS survey, Conveyance, treatment, and control of municipal wastewater, combined sewer overflows and stormwater runoff. Summaries of technical data. Office of Water Program Operations, U.S. Environmental Protection Agency, Washington, DC. Data not available for raw effluent flow.

USEPA. 1989. 1988 NEEDS survey, Conveyance, treatment, and control of municipal wastewater, combined sewer overflows and stormwater runoff. Summaries of technical data. Office of Water Program Operations, U.S. Environmental Protection Agency, Washington, DC.

USEPA. 1993. 1992 Clean Water Needs Survey (CWNS), Conveyance, treatment, and control of municipal wastewater, combined sewer overflows and stormwater runoff. Summaries of technical data. EPA-832-R-93-002. Office of Water Program Operations, U.S. Environmental Protection Agency, Washington, DC.

USEPA. 1997. 1996 Clean Water Needs Survey (CWNS), Conveyance, treatment, and control of municipal wastewater, combined sewer overflows and stormwater runoff. Summaries of technical data. EPA-832-R-97-003. Office of Water Program Operations, U.S. Environmental Protection Agency, Washington, DC.

USPHS. 1951. Water pollution in the United States. A report on the polluted conditions of our waters and what is needed to restore their quality. NTIS No. PB-218-308/BA. US Federal Security Agency, Public Health Service, Washington, DC. Population served and facilities inventory data for raw, primary, and secondary treatment compiled for 1950 (pp.31–33) by major river basins.

TABLE B-1 POTW Facility Inventory (Count)

Year	Total	No-Discharge	Raw	Primary	Advanced Primary	Secondary	Advanced Secondary	Advanced Treatment	Less than Secondary	Greater than Secondary
1940	−9.0	−9.0	−9.0	2,938.0	0.0	2,630.0	0.0	0.0	2,938.0	0.0
1950	11,784.0	0.0	5,156.0	3,099.0	0.0	3,529.0	0.0	0.0	3,099.0	0.0
1962	11,698.0	0.0	2,262.0	2,717.0	0.0	6,719.0	0.0	0.0	2,717.0	0.0
1968	14,051.0	0.0	1,564.0	2,435.0	0.0	10,042.0	0.0	10.0	2,435.0	10.0
1972	19,355.0	142.0	2,265.0	2,530.0	64.0	13,893.0	0.0	461.0	2,594.0	461.0
1973	−9.0	−9.0	1,532.0	2,723.0	−9.0	16,015.0	−9.0	795.0	−9.0	−9.0
1974	−9.0	−9.0	−9.0	3,032.0	−9.0	16,987.0	−9.0	992.0	−9.0	−9.0
1976	−9.0	−9.0	−9.0	−9.0	−9.0	2,838.0	−9.0	2,719.0	2,451.0	−9.0
1978	14,850.0	985.0	91.0	1,306.0	2,972.0	6,608.0	2,187.0	701.0	4,278.0	2,888.0
1980	15,522.0	1,361.0	272.0	1,043.0	2,300.0	7,852.0	2,443.0	251.0	3,343.0	2,694.0
1982	15,662.0	1,600.0	237.0	1,036.0	2,083.0	7,946.0	2,529.0	231.0	3,119.0	2,760.0
1984	15,580.0	1,726.0	202.0	−9.0	−9.0	8,070.0	−9.0	−9.0	2,617.0	2,965.0
1986	15,541.0	1,762.0	149.0	−9.0	−9.0	8,403.0	−9.0	−9.0	2,112.0	3,115.0
1988	15,708.0	1,854.0	117.0	−9.0	−9.0	8,536.0	−9.0	−9.0	1,789.0	3,412.0
1992	15,613.0	1,981.0	0.0	−9.0	−9.0	9,086.0	−9.0	−9.0	868.0	3,678.0
1996	16,024.0	2,032.0	0.0	−9.0	−9.0	9,388.0	−9.0	−9.0	176.0	4,428.0
2016	18,303.0	2,369.0	0.0	−9.0	−9.0	9,738.0	−9.0	−9.0	61.0	6,135.0

TABLE B-2 Resident Population Served by POTWs

Year	Total	No-Discharge	Raw	Primary	Advanced Primary	Secondary	Advanced Secondary	Advanced Treatment	Less than Secondary	Greater than Secondary
1940	70,800,000.0	0.0	32,200,000.0	18,500,000.0	0.0	20,100,000.0	0.0	0.0	18,500,000.0	0.0
1950	91,762,001.0	0.0	35,268,437.0	24,599,743.0	0.0	31,893,821.0	0.0	0.0	24,599,743.0	0.0
1962	118,300,000.0	0.0	14,600,000.0	42,200,000.0	0.0	61,500,000.0	0.0	0.0	42,200,000.0	0.0
1968	140,100,000.0	0.0	10,100,000.0	44,100,000.0	0.0	85,600,000.0	0.0	300,000.0	44,100,000.0	300,000.0
1972	141,722,242.0	825,000.0	4,939,928.0	50,502,813.0	1,376,059.0	76,270,812.0	2,257,101.0	5,550,529.0	51,878,872.0	7,807,630.0
1973	−9.0	−9.0	3,200,000.0	54,600,000.0	−9.0	105,000,000.0	−9.0	2,700,000.0	−9.0	−9.0
1974	−9.0	−9.0	−9.0	−9.0	−9.0	−9.0	−9.0	−9.0	−9.0	−9.0
1976	−9.0	−9.0	−9.0	−9.0	−9.0	32,523,000.0	−9.0	45,733,000.0	40,271,000.0	−9.0
1978	155,227,000.0	2,197,000.0	3,640,000.0	18,747,000.0	25,333,000.0	56,256,000.0	30,937,000.0	18,117,000.0	44,080,000.0	49,054,000.0
1980	158,337,000.0	3,599,000.0	2,307,000.0	19,101,714.0	18,214,286.0	62,680,000.0	47,518,000.0	4,917,000.0	37,316,000.0	52,435,000.0
1982	163,525,000.0	4,172,000.0	1,876,000.0	−9.0	−9.0	67,609,000.0	50,853,000.0	5,411,000.0	33,604,000.0	56,264,000.0
1984	170,643,000.0	5,514,000.0	1,273,000.0	−9.0	−9.0	70,656,000.0	−9.0	−9.0	33,675,000.0	59,525,000.0
1986	163,319,000.0	5,679,000.0	1,605,000.0	−9.0	−9.0	72,285,000.0	−9.0	−9.0	28,815,000.0	54,935,000.0
1988	177,536,335.0	6,079,611.0	1,367,172.0	−9.0	−9.0	77,954,544.0	−9.0	−9.0	26,484,096.0	65,650,912.0
1992	180,614,290.0	7,764,363.0	0.0	−9.0	−9.0	82,907,949.0	−9.0	−9.0	21,712,715.0	68,229,263.0
1996	189,710,899.0	7,660,876.0	0.0	−9.0	−9.0	81,944,349.0	−9.0	−9.0	17,177,492.0	82,928,182.0
2016	274,722,315.0	14,163,722.0	0.0	−9.0	−9.0	102,321,429.0	−9.0	−9.0	5,513,147.0	152,724,017.0

TABLE B-3 Wastewater Flow (million gallons per day, mgd)

Year	Total	No-Discharge	Raw	Primary	Advanced Primary	Secondary	Advanced Secondary	Advanced Treatment	Less than Secondary	Greater than Secondary
1940	11,682.0	0.0	5,313.0	3,052.5	0.0	3,316.5	0.0	0.0	3,052.5	0.0
1950	15,140.7	0.0	5,819.3	4,059.0	0.0	5,262.5	0.0	0.0	4,059.0	0.0
1962	19,519.5	0.0	2,409.0	6,963.0	0.0	10,147.5	0.0	0.0	6,963.0	0.0
1968	23,116.5	0.0	1,666.5	7,276.5	0.0	14,124.0	0.0	49.5	7,276.5	49.5
1972	23,384.2	136.1	815.1	8,333.0	227.0	12,584.7	372.4	915.8	8,560.0	1,288.3
1973	−9.0	−9.0	528.0	9,009.0	−9.0	17,325.0	−9.0	445.5	−9.0	−9.0
1974	−9.0	−9.0	−9.0	−9.0	−9.0	−9.0	−9.0	−9.0	−9.0	−9.0
1976	−9.0	−9.0	−9.0	−9.0	−9.0	5,437.5	−9.0	7,545.9	12,653.6	7,007.1
1978	26,799.5	362.5	600.6	3,415.0	3,737.4	10,138.7	4,732.1	3,813.2	7,152.4	8,545.3
1980	25,510.0	438.3	380.7	3,035.4	2,895.1	9,882.1	8,654.9	751.6	6,157.1	8,651.8
1982	27,202.7	490.9	309.5	2,474.5	2,825.4	11,009.5	9,377.5	714.7	5,300.6	10,092.2
1984	27,305.0	600.0	210.0	−9.0	−9.0	11,047.0	−9.0	−9.0	5,335.0	10,113.0
1986	27,956.8	608.0	264.8	−9.0	−9.0	12,140.0	−9.0	−9.0	4,580.0	10,364.0
1988	29,293.5	1,003.1	225.6	−9.0	−9.0	12,862.5	−9.0	−9.0	4,369.9	10,832.4
1992	29,801.4	1,281.1	0.0	−9.0	−9.0	13,679.8	−9.0	−9.0	3,582.6	11,257.8
1996	31,302.3	1,264.0	0.0	−9.0	−9.0	13,520.8	−9.0	−9.0	2,834.3	13,683.1
2016	45,329.2	2,337.0	0.0	−9.0	−9.0	16,883.0	−9.0	−9.0	909.7	25,199.5

TABLE B-4 **Influent Wastewater Flow Normalized to Population Served (gallons per person per day, gpcd) (Default = 165)**

Year	Total	No-Discharge	Raw	Primary	Advanced Primary	Secondary	Advanced Secondary	Advanced Treatment	Less than Secondary	Greater than Secondary
1940	165.0	−9.0	165.0	165.0	−9.0	165.0	−9.0	−9.0	165.0	−9.0
1950	165.0	−9.0	165.0	165.0	−9.0	165.0	−9.0	−9.0	165.0	−9.0
1962	165.0	−9.0	165.0	165.0	−9.0	165.0	−9.0	−9.0	165.0	−9.0
1968	165.0	−9.0	165.0	165.0	−9.0	165.0	−9.0	165.0	165.0	165.0
1972	165.0	165.0	165.0	165.0	165.0	165.0	165.0	165.0	165.0	165.0
1973	−9.0	−9.0	165.0	165.0	−9.0	165.0	−9.0	165.0	−9.0	−9.0
1974	−9.0	−9.0	−9.0	−9.0	−9.0	−9.0	−9.0	−9.0	−9.0	−9.0
1976	−9.0	−9.0	−9.0	−9.0	−9.0	167.2	−9.0	165.0	314.2	−9.0
1978	172.6	165.0	165.0	182.2	147.5	180.2	153.0	210.5	162.3	174.2
1980	161.1	121.8	165.0	158.9	158.9	157.7	182.1	152.9	165.0	165.0
1982	166.4	117.7	165.0	−9.0	−9.0	162.8	184.4	132.1	157.7	179.4
1984	160.0	108.8	165.0	−9.0	−9.0	156.3	−9.0	−9.0	158.4	169.9
1986	171.2	107.1	165.0	−9.0	−9.0	167.9	−9.0	−9.0	158.9	188.7
1988	165.0	165.0	165.0	−9.0	−9.0	165.0	−9.0	−9.0	165.0	165.0
1992	165.0	165.0	−9.0	−9.0	−9.0	165.0	−9.0	−9.0	165.0	165.0
1996	165.0	165.0	−9.0	−9.0	−9.0	165.0	−9.0	−9.0	165.0	165.0
2016	165.0	165.0	−9.0	−9.0	−9.0	165.0	−9.0	−9.0	165.0	165.0

TABLE B-5 Influent $CBOD_5$ Load (metric tons per day, mt/day)

Year	Total	No-Discharge	Raw	Primary	Advanced Primary	Secondary	Advanced Secondary	Advanced Treatment	Less than Secondary	Greater than Secondary
1940	9,507.5	0.0	4,324.0	2,484.3	0.0	2,699.2	0.0	0.0	2,484.3	0.0
1950	12,322.5	0.0	4,736.1	3,303.4	0.0	4,282.9	0.0	0.0	3,303.4	0.0
1962	15,886.2	0.0	1,960.6	5,666.9	0.0	8,258.7	0.0	0.0	5,666.9	0.0
1968	18,813.6	0.0	1,356.3	5,922.1	0.0	11,495.0	0.0	40.3	5,922.1	40.3
1972	19,031.5	110.8	663.4	6,781.9	184.8	10,242.2	303.1	745.4	6,966.7	1,048.5
1973	−9.0	−9.0	429.7	7,332.1	−9.0	14,100.1	−9.0	362.6	−9.0	−9.0
1974	−9.0	−9.0	−9.0	−9.0	−9.0	−9.0	−9.0	−9.0	−9.0	−9.0
1976	−9.0	−9.0	−9.0	−9.0	−9.0	3,764.0	−9.0	6,141.4	10,177.0	5,392.0
1978	21,252.8	295.0	488.8	2,511.0	3,210.0	8,222.0	3,419.0	3,107.0	5,721.0	6,526.0
1980	20,528.9	356.7	309.8	4,700.0	2,356.2	7,810.0	6,498.0	598.0	5,011.1	7,041.3
1982	21,170.4	399.5	251.9	−9.0	−9.0	8,623.0	7,030.0	586.0	4,280.0	7,616.0
1984	23,395.3	488.3	170.9	−9.0	−9.0	9,448.0	−9.0	−9.0	4,917.0	8,371.0
1986	23,927.4	494.8	215.5	−9.0	−9.0	10,378.0	−9.0	−9.0	4,325.0	8,514.0
1988	23,840.8	816.4	183.6	−9.0	−9.0	10,468.3	−9.0	−9.0	3,556.5	8,816.1
1992	24,254.2	1,042.7	0.0	−9.0	−9.0	11,133.5	−9.0	−9.0	2,915.7	9,162.3
1996	25,475.7	1,028.8	0.0	−9.0	−9.0	11,004.1	−9.0	−9.0	2,306.7	11,136.2
2016	36,891.7	1,902.0	0.0	−9.0	−9.0	13,740.4	−9.0	−9.0	740.3	20,508.9

TABLE B-6 Effluent $CBOD_5$ Load (metric tons per day, mt/day)

Year	Total	No-Discharge	Raw	Primary	Advanced Primary	Secondary	Advanced Secondary	Advanced Treatment	Less than Secondary	Greater than Secondary
1940	6,343.7	0.0	4,324.0	1,614.8	0.0	404.9	0.0	0.0	1,614.8	0.0
1950	7,525.8	0.0	4,736.1	2,147.2	0.0	642.4	0.0	0.0	2,147.2	0.0
1962	6,882.9	0.0	1,960.6	3,683.5	0.0	1,238.8	0.0	0.0	3,683.5	0.0
1968	6,931.9	0.0	1,356.3	3,849.3	0.0	1,724.2	0.0	2.0	3,849.3	2.0
1972	6,767.9	0.0	663.4	4,408.2	92.4	1,536.3	30.3	37.3	4,500.6	67.6
1973	−9.0	−9.0	429.7	4,765.8	−9.0	2,115.0	−9.0	18.1	−9.0	−9.0
1974	−9.0	−9.0	−9.0	−9.0	−9.0	−9.0	−9.0	−9.0	−9.0	−9.0
1976	−9.0	−9.0	−9.0	−9.0	−9.0	518.0	−9.0	307.1	4,360.0	324.0
1978	5,509.8	0.0	488.8	1,487.0	1,167.0	1,596.0	362.0	409.0	2,654.0	771.0
1980	5,188.3	0.0	309.8	2,316.0	1,178.1	1,469.0	752.0	75.0	2,881.4	528.1
1982	4,379.9	0.0	251.9	−9.0	−9.0	1,539.0	582.0	32.0	1,975.0	614.0
1984	3,943.9	0.0	170.9	−9.0	−9.0	1,135.0	−9.0	−9.0	2,030.0	608.0
1986	3,923.5	0.0	215.5	−9.0	−9.0	1,279.0	−9.0	−9.0	1,834.0	595.0
1988	4,460.0	0.0	183.6	−9.0	−9.0	1,570.2	−9.0	−9.0	2,045.0	661.2
1992	4,033.7	0.0	0.0	−9.0	−9.0	1,670.0	−9.0	−9.0	1,676.5	687.2
1996	3,812.2	0.0	0.0	−9.0	−9.0	1,650.6	−9.0	−9.0	1,326.4	835.2
2016	4,024.9	0.0	0.0	−9.0	−9.0	2,061.1	−9.0	−9.0	425.7	1,538.2

TABLE B-7 $CBOD_5$ Removal Efficiency (Percent)

Year	Total	No-Discharge	Raw	Primary	Advanced Primary	Secondary	Advanced Secondary	Advanced Treatment	Less than Secondary	Greater than Secondary
1940	33.3	−9.0	−9.0	35.0	−9.0	85.0	−9.0	−9.0	35.0	−9.0
1950	38.9	−9.0	−9.0	35.0	−9.0	85.0	−9.0	−9.0	35.0	−9.0
1962	56.7	−9.0	−9.0	35.0	−9.0	85.0	−9.0	−9.0	35.0	−9.0
1968	63.2	−9.0	−9.0	35.0	−9.0	85.0	−9.0	95.0	35.0	95.0
1972	64.4	−9.0	−9.0	35.0	50.0	85.0	90.0	95.0	35.4	93.6
1973	−9.0	−9.0	−9.0	35.0	−9.0	85.0	−9.0	95.0	−9.0	−9.0
1974	−9.0	−9.0	−9.0	−9.0	−9.0	−9.0	−9.0	−9.0	−9.0	−9.0
1976	−9.0	−9.0	−9.0	−9.0	−9.0	86.2	−9.0	95.0	57.2	94.0
1978	74.1	100.0	0.0	40.8	63.6	80.6	89.4	86.8	53.6	88.2
1980	74.7	−9.0	−9.0	50.7	50.0	81.2	88.4	87.5	42.5	92.5
1982	79.3	100.0	0.0	−9.0	−9.0	82.2	91.7	94.5	53.9	91.9
1984	83.1	−9.0	−9.0	−9.0	−9.0	88.0	−9.0	−9.0	58.7	92.7
1986	83.6	−9.0	−9.0	−9.0	−9.0	87.7	−9.0	−9.0	57.6	93.0
1988	81.3	−9.0	−9.0	−9.0	−9.0	85.0	−9.0	−9.0	42.5	92.5
1992	83.4	−9.0	−9.0	−9.0	−9.0	85.0	−9.0	−9.0	42.5	92.5
1996	85.0	−9.0	−9.0	−9.0	−9.0	85.0	−9.0	−9.0	42.5	92.5
2016	89.1	−9.0	−9.0	−9.0	−9.0	85.0	−9.0	−9.0	42.5	92.5

TABLE B-8 Influent $CBOD_5$ Load Normalized to Population Served (lb $CBOD_5$ per person per day) (Default = 0.296)

Year	Total	No-Discharge	Raw	Primary	Advanced Primary	Secondary	Advanced Secondary	Advanced Treatment	Less than Secondary	Greater than Secondary
1940	0.296	−9.000	0.296	0.296	−9.000	0.296	−9.000	−9.000	0.296	−9.000
1950	0.296	−9.000	0.296	0.296	−9.000	0.296	−9.000	−9.000	0.296	−9.000
1962	0.296	−9.000	0.296	0.296	−9.000	0.296	−9.000	−9.000	0.296	−9.000
1968	0.296	−9.000	0.296	0.296	−9.000	0.296	−9.000	0.296	0.296	0.296
1972	0.296	0.296	0.296	0.296	0.296	0.296	0.296	0.296	0.296	0.296
1973	−9.000	−9.000	0.296	0.296	−9.000	0.296	−9.000	0.296	−9.000	−9.000
1974	−9.000	−9.000	−9.000	−9.000	−9.000	−9.000	−9.000	−9.000	−9.000	−9.000
1976	−9.000	−9.000	−9.000	−9.000	−9.000	0.255	−9.000	0.296	0.557	−9.000
1978	0.302	0.296	0.296	0.295	0.279	0.322	0.244	0.378	0.286	0.293
1980	0.286	0.219	0.296	0.542	0.285	0.275	0.301	0.268	0.296	0.296
1982	0.285	0.211	0.296	−9.000	−9.000	0.281	0.305	0.239	0.281	0.298
1984	0.302	0.195	0.296	−9.000	−9.000	0.295	−9.000	−9.000	0.322	0.310
1986	0.323	0.192	0.296	−9.000	−9.000	0.317	−9.000	−9.000	0.331	0.342
1988	0.296	0.296	0.296	−9.000	−9.000	0.296	−9.000	−9.000	0.296	0.296
1992	0.296	0.296	−9.000	−9.000	−9.000	0.296	−9.000	−9.000	0.296	0.296
1996	0.296	0.296	−9.000	−9.000	−9.000	0.296	−9.000	−9.000	0.296	0.296
2016	0.296	0.296	−9.000	−9.000	−9.000	0.296	−9.000	−9.000	0.296	0.296

TABLE B-9 Influent $CBOD_5$ Concentration (mg/L) (Default = 215.0 mg/L)

Year	Total	No-Discharge	Raw	Primary	Advanced Primary	Secondary	Advanced Secondary	Advanced Treatment	Less than Secondary	Greater than Secondary
1940	215.0	−9.0	215.0	215.0	−9.0	215.0	−9.0	−9.0	215.0	−9.0
1950	215.0	−9.0	215.0	215.0	−9.0	215.0	−9.0	−9.0	215.0	−9.0
1962	215.0	−9.0	215.0	215.0	−9.0	215.0	−9.0	−9.0	215.0	−9.0
1968	215.0	−9.0	215.0	215.0	−9.0	215.0	−9.0	215.0	215.0	215.0
1972	215.0	215.0	215.0	215.0	215.0	215.0	215.0	215.0	215.0	215.0
1973	−9.0	−9.0	215.0	215.0	−9.0	215.0	−9.0	215.0	−9.0	−9.0
1974	−9.0	−9.0	−9.0	−9.0	−9.0	−9.0	−9.0	−9.0	−9.0	−9.0
1976	−9.0	−9.0	−9.0	−9.0	−9.0	182.9	−9.0	215.0	212.5	203.3
1978	209.5	215.0	215.0	194.2	226.9	214.2	190.9	215.2	211.3	201.7
1980	212.6	215.0	215.0	409.0	215.0	208.8	198.3	210.2	215.0	215.0
1982	205.6	215.0	215.0	−9.0	−9.0	206.9	198.0	216.6	213.3	199.4
1984	226.3	215.0	215.0	−9.0	−9.0	225.9	−9.0	−9.0	243.5	218.7
1986	226.1	215.0	215.0	−9.0	−9.0	225.8	−9.0	−9.0	249.5	217.0
1988	215.0	215.0	215.0	−9.0	−9.0	215.0	−9.0	−9.0	215.0	215.0
1992	215.0	215.0	−9.0	−9.0	−9.0	215.0	−9.0	−9.0	215.0	215.0
1996	215.0	215.0	−9.0	−9.0	−9.0	215.0	−9.0	−9.0	215.0	215.0
2016	215.0	215.0	−9.0	−9.0	−9.0	215.0	−9.0	−9.0	215.0	215.0

TABLE B-10 Effluent $CBOD_5$ Concentration (mg/L)

Default:		0.0	215.0	139.75	107.5	32.25	21.5	10.75	123.63	16.12
Year	Total	No-Discharge	Raw	Primary	Advanced Primary	Secondary	Advanced Secondary	Advanced Treatment	Less than Secondary	Greater than Secondary
1940	143.5	−9.0	215.0	139.8	−9.0	32.3	−9.0	−9.0	139.8	−9.0
1950	131.3	−9.0	215.0	139.8	−9.0	32.3	−9.0	−9.0	139.8	−9.0
1962	93.2	−9.0	215.0	139.8	−9.0	32.3	−9.0	−9.0	139.8	−9.0
1968	79.2	−9.0	215.0	139.8	−9.0	32.3	−9.0	10.8	139.8	10.8
1972	76.5	0.0	215.0	139.8	107.5	32.3	21.5	10.8	139.8	10.8
1973	−9.0	−9.0	215.0	139.8	−9.0	32.3	−9.0	10.8	138.9	13.9
1974	−9.0	−9.0	−9.0	−9.0	−9.0	−9.0	−9.0	10.8	−9.0	−9.0
1976	−9.0	−9.0	−9.0	−9.0	−9.0	25.2	−9.0	−9.0	−9.0	−9.0
1978	54.3	0.0	215.0	115.0	82.5	41.6	20.2	10.8	91.0	12.2
1980	53.7	0.0	215.0	201.6	107.5	39.3	23.0	28.3	98.0	23.8
1982	42.5	0.0	215.0	−9.0	−9.0	36.9	16.4	26.4	123.6	16.1
1984	38.2	0.0	215.0	−9.0	−9.0	27.1	−9.0	11.8	98.4	16.1
1986	37.1	0.0	215.0	−9.0	−9.0	27.8	−9.0	−9.0	100.5	15.9
1988	40.2	0.0	215.0	−9.0	−9.0	32.3	−9.0	−9.0	105.8	15.2
1992	35.8	0.0	−9.0	−9.0	−9.0	32.3	−9.0	−9.0	123.6	16.1
1996	32.2	0.0	−9.0	−9.0	−9.0	32.3	−9.0	−9.0	123.6	16.1
2016	23.5	0.0	−9.0	−9.0	−9.0	32.3	−9.0	−9.0	123.6	16.1

TABLE B-11 Influent $CBOD_u$ Load (metric tons per day, mt/day)

Year	Total	No-Discharge	Raw	Primary	Advanced Primary	Secondary	Advanced Secondary	Advanced Treatment	Less than Secondary	Greater than Secondary
1940	11,409.0	0.0	5,188.9	2,981.2	0.0	3,239.0	0.0	0.0	2,981.2	0.0
1950	14,786.9	0.0	5,683.3	3,964.1	0.0	5,139.5	0.0	0.0	3,964.1	0.0
1962	19,063.4	0.0	2,352.7	6,800.3	0.0	9,910.4	0.0	0.0	6,800.3	0.0
1968	22,576.3	0.0	1,627.6	7,106.5	0.0	13,794.0	0.0	48.3	7,106.5	48.3
1972	22,837.8	132.9	796.0	8,138.2	221.7	12,290.6	363.7	894.4	8,360.0	1,258.2
1973	−9.0	−9.0	515.7	8,798.5	−9.0	16,920.2	−9.0	435.1	−9.0	−9.0
1974	−9.0	−9.0	−9.0	−9.0	−9.0	−9.0	−9.0	−9.0	−9.0	−9.0
1976	−9.0	−9.0	−9.0	−9.0	−9.0	4,516.8	−9.0	7,369.6	12,212.4	6,470.4
1978	25,503.4	354.0	586.6	3,013.2	3,852.0	9,866.4	4,102.8	3,728.4	6,865.2	7,831.2
1980	24,634.7	428.1	371.8	5,640.0	2,827.5	9,372.0	7,797.6	717.6	6,013.3	8,449.6
1982	25,404.5	479.4	302.3	−9.0	−9.0	10,347.6	8,436.0	703.2	5,136.0	9,139.2
1984	28,074.3	586.0	205.1	−9.0	−9.0	11,337.6	−9.0	−9.0	5,900.4	10,045.2
1986	28,712.8	593.8	258.6	−9.0	−9.0	12,453.6	−9.0	−9.0	5,190.0	10,216.8
1988	28,609.0	979.7	220.3	−9.0	−9.0	12,561.9	−9.0	−9.0	4,267.8	10,579.3
1992	29,105.0	1,251.2	0.0	−9.0	−9.0	13,360.2	−9.0	−9.0	3,498.9	10,994.8
1996	30,570.9	1,234.5	0.0	−9.0	−9.0	13,204.9	−9.0	−9.0	2,768.1	13,363.4
2016	44,270.0	2,282.4	0.0	−9.0	−9.0	16,488.5	−9.0	−9.0	888.4	24,610.6

TABLE B-12 Effluent $CBOD_u$ Load (metric tons per day, mt/day)

Year	Total	No-Discharge	Raw	Primary	Advanced Primary	Secondary	Advanced Secondary	Advanced Treatment	Less than Secondary	Greater than Secondary
1940	8,922.4	0.0	5,188.9	2,583.7	0.0	1,149.8	0.0	0.0	2,583.7	0.0
1950	10,943.4	0.0	5,683.3	3,435.6	0.0	1,824.5	0.0	0.0	3,435.6	0.0
1962	11,764.5	0.0	2,352.7	5,893.6	0.0	3,518.2	0.0	0.0	5,893.6	0.0
1968	12,689.4	0.0	1,627.6	6,158.9	0.0	4,896.9	0.0	6.0	6,158.9	6.0
1972	12,558.1	0.0	796.0	7,053.1	147.8	4,363.2	86.1	111.8	7,201.0	197.9
1973	−9.0	−9.0	515.7	7,625.4	−9.0	6,006.7	−9.0	54.4	−9.0	−9.0
1974	−9.0	−9.0	−9.0	−9.0	−9.0	−9.0	−9.0	−9.0	−9.0	−9.0
1976	−9.0	−9.0	−9.0	−9.0	−9.0	1,471.1	−9.0	921.2	6,976.0	939.6
1978	11,620.7	0.0	586.6	2,379.2	1,867.2	4,532.6	1,028.1	1,227.0	4,246.4	2,255.1
1980	10,685.4	0.0	371.8	3,705.6	1,885.0	4,172.0	2,135.7	225.0	4,610.2	1,531.5
1982	9,581.9	0.0	302.3	−9.0	−9.0	4,370.8	1,652.9	96.0	3,160.0	1,748.9
1984	8,439.7	0.0	205.1	−9.0	−9.0	3,223.4	−9.0	−9.0	3,248.0	1,763.2
1986	8,550.9	0.0	258.6	−9.0	−9.0	3,632.4	−9.0	−9.0	2,934.4	1,725.5
1988	9,869.3	0.0	220.3	−9.0	−9.0	4,459.5	−9.0	−9.0	3,272.0	1,917.5
1992	9,418.1	0.0	0.0	−9.0	−9.0	4,742.9	−9.0	−9.0	2,682.5	1,992.8
1996	9,232.0	0.0	0.0	−9.0	−9.0	4,687.7	−9.0	−9.0	2,122.2	2,422.1
2016	10,995.2	0.0	0.0	−9.0	−9.0	5,853.4	−9.0	−9.0	681.1	4,460.7

TABLE B-13 $CBOD_u$ Removal Efficiency (Percent)

Default:		100%	0%	13.3%	33.3%	64.5%	76.3%	87.5%	23.3%	81.9%
Year	Total	No-Discharge	Raw	Primary	Advanced Primary	Secondary	Advanced Secondary	Advanced Treatment	Less than Secondary	Greater than Secondary
1940	21.8	−9.0	−9.0	13.3	−9.0	64.5	−9.0	−9.0	13.3	−9.0
1950	26.0	−9.0	−9.0	13.3	−9.0	64.5	−9.0	−9.0	13.3	−9.0
1962	38.3	−9.0	−9.0	13.3	−9.0	64.5	−9.0	−9.0	13.3	−9.0
1968	43.8	−9.0	−9.0	13.3	−9.0	64.5	−9.0	87.5	13.3	87.5
1972	45.0	−9.0	−9.0	13.3	33.3	64.5	76.3	87.5	13.9	84.3
1973	−9.0	−9.0	−9.0	13.3	−9.0	64.5	−9.0	87.5	−9.0	−9.0
1974	−9.0	−9.0	−9.0	−9.0	−9.0	−9.0	−9.0	−9.0	−9.0	−9.0
1976	−9.0	−9.0	−9.0	−9.0	−9.0	67.4	−9.0	87.5	42.9	85.5
1978	54.4	100.0	0.0	21.0	51.5	54.1	74.9	67.1	38.1	71.2
1980	56.6	−9.0	−9.0	34.3	33.3	55.5	72.6	68.6	23.3	81.9
1982	62.3	100.0	0.0	−9.0	−9.0	57.8	80.4	86.3	38.5	80.9
1984	69.9	−9.0	−9.0	−9.0	−9.0	71.6	−9.0	−9.0	45.0	82.4
1986	70.2	−9.0	−9.0	−9.0	−9.0	70.8	−9.0	−9.0	43.5	83.1
1988	65.5	−9.0	−9.0	−9.0	−9.0	64.5	−9.0	−9.0	23.3	81.9
1992	67.6	−9.0	−9.0	−9.0	−9.0	64.5	−9.0	−9.0	23.3	81.9
1996	69.8	−9.0	−9.0	−9.0	−9.0	64.5	−9.0	−9.0	23.3	81.9
2016	75.2	−9.0	−9.0	−9.0	−9.0	64.5	−9.0	−9.0	23.3	81.9

TABLE B-14 Effluent $CBOD_u$ Concentration (mg/L) [$CBOD_u = BOD_5 \times (CBOD_u/BOD_5)$]

Default $CBOD_u/CBOD_5$		0.0	1.2	1.6	1.6	2.84	2.84	3.0	1.6	2.9
Default: $CBOD_u$		0.0	258.0	223.60	172.0	91.59	61.06	32.25	197.80	46.76
Year	Total	No-Discharge	Raw	Primary	Advanced Primary	Secondary	Advanced Secondary	Advanced Treatment	Less than Secondary	Greater than Secondary
1940	201.8	−9.0	258.0	223.6	−9.0	91.6	−9.0	−9.0	223.6	−9.0
1950	190.9	−9.0	258.0	223.6	−9.0	91.6	−9.0	−9.0	223.6	−9.0
1962	159.2	−9.0	258.0	223.6	−9.0	91.6	−9.0	−9.0	223.6	−9.0
1968	145.0	−9.0	258.0	223.6	−9.0	91.6	−9.0	32.3	223.6	32.3
1972	141.9	0.0	258.0	223.6	172.0	91.6	61.1	32.3	222.2	40.6
1973	−9.0	−9.0	258.0	223.6	−9.0	91.6	−9.0	32.3	−9.0	−9.0
1974	−9.0	−9.0	−9.0	−9.0	−9.0	−9.0	−9.0	−9.0	−9.0	−9.0
1976	−9.0	−9.0	−9.0	−9.0	−9.0	71.5	−9.0	32.3	145.6	35.4
1978	114.5	0.0	258.0	184.0	132.0	118.1	57.4	85.0	156.8	69.7
1980	110.7	0.0	258.0	322.5	172.0	111.5	65.2	79.1	197.8	46.8
1982	93.1	0.0	258.0	−9.0	−9.0	104.9	46.6	35.5	157.5	45.8
1984	81.7	0.0	258.0	−9.0	−9.0	77.1	−9.0	−9.0	160.8	46.1
1986	80.8	0.0	258.0	−9.0	−9.0	79.0	−9.0	−9.0	169.3	44.0
1988	89.0	0.0	258.0	−9.0	−9.0	91.6	−9.0	−9.0	197.8	46.8
1992	83.5	0.0	−9.0	−9.0	−9.0	91.6	−9.0	−9.0	197.8	46.8
1996	77.9	0.0	−9.0	−9.0	−9.0	91.6	−9.0	−9.0	197.8	46.8
2016	64.1	0.0	−9.0	−9.0	−9.0	91.6	−9.0	−9.0	197.8	46.8

TABLE B-15 Influent NBOD Load (metric tons per day, mt/day) (NBOD = 4.57 × TKN)

Year	Total	No-Discharge	Raw	Primary	Advanced Primary	Secondary	Advanced Secondary	Advanced Treatment	Less than Secondary	Greater than Secondary
1940	6,123.3	0.0	2,784.9	1,600.0	0.0	1,738.4	0.0	0.0	1,600.0	0.0
1950	7,936.3	0.0	3,050.3	2,127.6	0.0	2,758.4	0.0	0.0	2,127.6	0.0
1962	10,231.5	0.0	1,262.7	3,649.8	0.0	5,319.0	0.0	0.0	3,649.8	0.0
1968	12,116.9	0.0	873.5	3,814.1	0.0	7,403.4	0.0	25.9	3,814.1	25.9
1972	12,257.2	71.4	427.2	4,367.9	119.0	6,596.5	195.2	480.1	4,486.9	675.3
1973	−9.0	−9.0	276.8	4,722.2	−9.0	9,081.2	−9.0	233.5	−9.0	−9.0
1974	−9.0	−9.0	−9.0	−9.0	−9.0	−9.0	−9.0	−9.0	−9.0	−9.0
1976	−9.0	−9.0	−9.0	−9.0	−9.0	2,850.2	−9.0	3,955.3	6,632.6	−9.0
1978	14,047.4	190.0	314.8	1,790.1	1,959.0	5,314.4	2,480.4	1,998.8	3,749.1	4,479.2
1980	13,371.5	229.7	199.5	1,591.1	1,517.5	5,179.9	4,536.6	394.0	3,227.4	4,535.0
1982	14,258.8	257.3	162.3	−9.0	−9.0	5,770.8	4,915.4	374.6	2,778.4	5,290.0
1984	14,312.4	314.5	110.1	−9.0	−9.0	5,790.5	−9.0	−9.0	2,796.4	5,300.9
1986	14,654.1	318.7	138.8	−9.0	−9.0	6,363.4	−9.0	−9.0	2,400.7	5,432.5
1988	15,354.7	525.8	118.2	−9.0	−9.0	6,742.1	−9.0	−9.0	2,290.5	5,678.0
1992	15,620.9	671.5	0.0	−9.0	−9.0	7,170.5	−9.0	−9.0	1,877.9	5,901.0
1996	16,407.7	662.6	0.0	−9.0	−9.0	7,087.2	−9.0	−9.0	1,485.6	7,172.3
2016	23,760.1	1,225.0	0.0	−9.0	−9.0	8,849.5	−9.0	−9.0	476.8	13,208.8

TABLE B-16 Effluent NBOD Load (metric tons per day, mt/day) [NBOD = 4.57 × TKN)

Year	Total	No-Discharge	Raw	Primary	Advanced Primary	Secondary	Advanced Secondary	Advanced Treatment	Less than Secondary	Greater than Secondary
1940	5,145.5	0.0	2,784.9	1,248.0	0.0	1,112.6	0.0	0.0	1,248.0	0.0
1950	6,475.2	0.0	3,050.3	1,659.5	0.0	1,765.4	0.0	0.0	1,659.5	0.0
1962	7,513.7	0.0	1,262.7	2,846.8	0.0	3,404.2	0.0	0.0	2,846.8	0.0
1968	8,591.1	0.0	873.5	2,975.0	0.0	4,738.1	0.0	4.4	2,975.0	4.4
1972	8,273.3	0.0	427.2	3,406.9	92.8	4,221.8	42.9	81.6	3,499.8	124.6
1973	−9.0	−9.0	276.8	3,683.3	−9.0	5,812.0	−9.0	39.7	−9.0	−9.0
1974	−9.0	−9.0	−9.0	−9.0	−9.0	−9.0	−9.0	−9.0	−9.0	−9.0
1976	−9.0	−9.0	−9.0	−9.0	−9.0	1,824.1	−9.0	672.4	5,173.4	−9.0
1978	7,525.8	0.0	314.8	1,396.2	1,528.0	3,401.2	545.7	339.8	2,924.3	885.5
1980	6,916.3	0.0	199.5	1,241.0	1,183.7	3,315.1	998.1	67.0	2,517.4	884.3
1982	7,167.8	0.0	162.3	−9.0	−9.0	3,693.3	1,081.4	63.7	2,167.2	1,145.1
1984	7,030.9	0.0	110.1	−9.0	−9.0	3,705.9	−9.0	−9.0	2,181.2	1,033.7
1986	7,143.3	0.0	138.8	−9.0	−9.0	4,072.6	−9.0	−9.0	1,872.5	1,059.3
1988	7,327.0	0.0	118.2	−9.0	−9.0	4,315.0	−9.0	−9.0	1,786.6	1,107.2
1992	7,204.6	0.0	0.0	−9.0	−9.0	4,589.1	−9.0	−9.0	1,464.7	1,150.7
1996	7,093.2	0.0	0.0	−9.0	−9.0	4,535.8	−9.0	−9.0	1,158.8	1,398.6
2016	8,611.3	0.0	0.0	−9.0	−9.0	5,663.7	−9.0	−9.0	371.9	2,575.7

TABLE B-17 NBOD Removal Efficiency (Percent)

Default:		100%	0%	22%	22%	36%	78%	83%	22%	80.5%
Year	Total	No-Discharge	Raw	Primary	Advanced Primary	Secondary	Advanced Secondary	Advanced Treatment	Less than Secondary	Greater than Secondary
1940	16.0	−9.0	−9.0	22.0	−9.0	36.0	−9.0	−9.0	22.0	−9.0
1950	18.4	−9.0	−9.0	22.0	−9.0	36.0	−9.0	−9.0	22.0	−9.0
1962	26.6	−9.0	−9.0	22.0	−9.0	36.0	−9.0	−9.0	22.0	−9.0
1968	29.1	−9.0	−9.0	22.0	−9.0	36.0	−9.0	83.0	22.0	83.0
1972	32.5	−9.0	−9.0	22.0	22.0	36.0	78.0	83.0	22.0	81.6
1973	−9.0	−9.0	−9.0	22.0	−9.0	36.0	−9.0	83.0	−9.0	−9.0
1974	−9.0	−9.0	−9.0	−9.0	−9.0	−9.0	−9.0	−9.0	−9.0	−9.0
1976	−9.0	−9.0	−9.0	−9.0	−9.0	36.0	−9.0	83.0	22.0	−9.0
1978	46.4	−9.0	−9.0	22.0	22.0	36.0	78.0	83.0	22.0	80.2
1980	48.3	−9.0	−9.0	22.0	22.0	36.0	78.0	83.0	22.0	80.5
1982	49.7	100.0	0.0	−9.0	−9.0	36.0	78.0	83.0	22.0	78.4
1984	50.9	−9.0	−9.0	−9.0	−9.0	36.0	−9.0	−9.0	22.0	80.5
1986	51.3	−9.0	−9.0	−9.0	−9.0	36.0	−9.0	−9.0	22.0	80.5
1988	52.3	−9.0	−9.0	−9.0	−9.0	36.0	−9.0	−9.0	22.0	80.5
1992	53.9	−9.0	−9.0	−9.0	−9.0	36.0	−9.0	−9.0	22.0	80.5
1996	56.8	−9.0	−9.0	−9.0	−9.0	36.0	−9.0	−9.0	22.0	80.5
2016	63.8	−9.0	−9.0	−9.0	−9.0	36.0	−9.0	−9.0	22.0	80.5

TABLE B-18 Effluent NBOD Concentration (mg/L) (NBOD = 4.57 × TKN)

Default:		0.0	138.47	108.0	108.0	88.62	30.46	23.54	108.0	27.0
Year	Total	No-Discharge	Raw	Primary	Advanced Primary	Secondary	Advanced Secondary	Advanced Treatment	Less than Secondary	Greater than Secondary
1940	116.4	−9.0	138.5	108.0	−9.0	88.6	−9.0	−9.0	108.0	−9.0
1950	113.0	−9.0	138.5	108.0	−9.0	88.6	−9.0	−9.0	108.0	−9.0
1962	101.7	−9.0	138.5	108.0	−9.0	88.6	−9.0	−9.0	108.0	−9.0
1968	98.2	−9.0	138.5	108.0	−9.0	88.6	−9.0	23.5	108.0	23.5
1972	93.5	0.0	138.5	108.0	108.0	88.6	30.5	23.5	108.0	25.5
1973	−9.0	−9.0	138.5	108.0	−9.0	88.6	−9.0	23.5	−9.0	−9.0
1974	−9.0	−9.0	−9.0	−9.0	−9.0	−9.0	−9.0	−9.0	−9.0	−9.0
1976	−9.0	−9.0	−9.0	−9.0	−9.0	88.6	−9.0	23.5	108.0	−0.3
1978	74.2	0.0	138.5	108.0	108.0	88.6	30.5	23.5	108.0	27.4
1980	71.6	0.0	138.5	108.0	108.0	88.6	30.5	23.5	108.0	27.0
1982	69.6	0.0	138.5	−9.0	−9.0	88.6	30.5	23.5	108.0	30.0
1984	68.0	0.0	138.5	−9.0	−9.0	88.6	−9.0	−9.0	108.0	27.0
1986	67.5	0.0	138.5	−9.0	−9.0	88.6	−9.0	−9.0	108.0	27.0
1988	66.1	0.0	138.5	−9.0	−9.0	88.6	−9.0	−9.0	108.0	27.0
1992	63.9	0.0	−9.0	−9.0	−9.0	88.6	−9.0	−9.0	108.0	27.0
1996	59.9	0.0	−9.0	−9.0	−9.0	88.6	−9.0	−9.0	108.0	27.0
2016	50.2	0.0	−9.0	−9.0	−9.0	88.6	−9.0	−9.0	108.0	27.0

TABLE B-19 Influent BOD_u Load (metric tons per day, mt/day) ($BOD_u = CBOD_u + NBOD$)

Year	Total	No-Discharge	Raw	Primary	Advanced Primary	Secondary	Advanced Secondary	Advanced Treatment	Less than Secondary	Greater than Secondary
1940	17,532.4	0.0	7,973.8	4,581.2	0.0	4,977.4	0.0	0.0	4,581.2	0.0
1950	22,723.2	0.0	8,733.6	6,091.7	0.0	7,897.9	0.0	0.0	6,091.7	0.0
1962	29,294.9	0.0	3,615.4	10,450.1	0.0	15,229.4	0.0	0.0	10,450.1	0.0
1968	34,693.3	0.0	2,501.1	10,920.6	0.0	21,197.3	0.0	74.3	10,920.6	74.3
1972	35,095.0	204.3	1,223.3	12,506.1	340.8	18,887.1	558.9	1,374.5	12,846.9	1,933.4
1973	−9.0	−9.0	792.4	13,520.7	−9.0	26,001.4	−9.0	668.6	−9.0	−9.0
1974	−9.0	−9.0	−9.0	−9.0	−9.0	−9.0	−9.0	−9.0	−9.0	−9.0
1976	−9.0	−9.0	−9.0	−9.0	−9.0	7,367.0	−9.0	11,325.0	18,845.0	−9.0
1978	39,550.8	544.0	901.4	4,803.3	5,811.0	15,180.8	6,583.2	5,727.2	10,614.3	12,310.4
1980	38,006.2	657.8	571.3	7,231.1	4,345.0	14,551.9	12,334.2	1,111.6	9,240.6	12,984.6
1982	39,663.3	736.7	464.6	−9.0	−9.0	16,118.4	13,351.4	1,077.8	7,914.4	14,429.2
1984	42,386.8	900.5	315.2	−9.0	−9.0	17,128.1	−9.0	−9.0	8,696.8	15,346.1
1986	43,366.9	912.5	397.4	−9.0	−9.0	18,817.0	−9.0	−9.0	7,590.7	15,649.3
1988	43,963.7	1,505.5	338.6	−9.0	−9.0	19,304.1	−9.0	−9.0	6,558.3	16,257.3
1992	44,725.9	1,922.7	0.0	−9.0	−9.0	20,530.7	−9.0	−9.0	5,376.8	16,895.8
1996	46,978.5	1,897.1	0.0	−9.0	−9.0	20,292.1	−9.0	−9.0	4,253.7	20,535.7
2016	68,030.1	3,507.4	0.0	−9.0	−9.0	25,338.1	−9.0	−9.0	1,365.2	37,819.4

TABLE B-20 Effluent BOD_u Load (metric tons per day, mt/day) ($BOD_u = CBOD_u + NBOD$)

Year	Total	No-Discharge	Raw	Primary	Advanced Primary	Secondary	Advanced Secondary	Advanced Treatment	Less than Secondary	Greater than Secondary
1940	14,067.9	0.0	7,973.8	3,831.7	0.0	2,262.4	0.0	0.0	3,831.7	0.0
1950	17,418.6	0.0	8,733.6	5,095.1	0.0	3,589.9	0.0	0.0	5,095.1	0.0
1962	19,278.2	0.0	3,615.4	8,740.4	0.0	6,922.3	0.0	0.0	8,740.4	0.0
1968	21,280.5	0.0	2,501.1	9,133.9	0.0	9,635.0	0.0	10.5	9,133.9	10.5
1972	20,831.4	0.0	1,223.3	10,460.1	240.7	8,584.9	129.0	193.4	10,700.8	322.4
1973	−9.0	−9.0	792.4	11,308.7	−9.0	11,818.6	−9.0	94.1	−9.0	−9.0
1974	−9.0	−9.0	−9.0	−9.0	−9.0	−9.0	−9.0	−9.0	−9.0	−9.0
1976	−9.0	−9.0	−9.0	−9.0	−9.0	3,295.2	−9.0	1,593.6	12,149.4	−9.0
1978	19,146.5	0.0	901.4	3,775.4	3,395.2	7,933.8	1,573.8	1,566.8	7,170.7	3,140.6
1980	17,601.7	0.0	571.3	4,946.6	3,068.6	7,487.1	3,133.7	292.0	7,127.5	2,415.8
1982	16,749.8	0.0	464.6	−9.0	−9.0	8,064.1	2,734.3	159.7	5,327.2	2,893.9
1984	15,470.6	0.0	315.2	−9.0	−9.0	6,929.3	−9.0	−9.0	5,429.2	2,796.9
1986	15,694.2	0.0	397.4	−9.0	−9.0	7,704.9	−9.0	−9.0	4,806.9	2,784.8
1988	17,196.3	0.0	338.6	−9.0	−9.0	8,774.4	−9.0	−9.0	5,058.6	3,024.7
1992	16,622.7	0.0	0.0	−9.0	−9.0	9,332.0	−9.0	−9.0	4,147.2	3,143.5
1996	16,325.2	0.0	0.0	−9.0	−9.0	9,223.5	−9.0	−9.0	3,281.0	3,820.7
2016	19,606.6	0.0	0.0	−9.0	−9.0	11,517.1	−9.0	−9.0	1,053.0	7,036.4

TABLE B-21 Removal Efficiency BOD_u (Percent)

Default:		100%	0%	16.4%	29.4%	54.5%	76.9%	85.9%	22.9%	81.4%
Year	Total	No-Discharge	Raw	Primary	Advanced Primary	Secondary	Advanced Secondary	Advanced Treatment	Less than Secondary	Greater than Secondary
1940	19.8	−9.0	−9.0	16.4	−9.0	54.5	−9.0	−9.0	16.4	−9.0
1950	23.3	−9.0	−9.0	16.4	−9.0	54.5	−9.0	−9.0	16.4	−9.0
1962	34.2	−9.0	−9.0	16.4	−9.0	54.5	−9.0	−9.0	16.4	−9.0
1968	38.7	−9.0	−9.0	16.4	−9.0	54.5	−9.0	85.9	16.4	85.9
1972	40.6	−9.0	−9.0	16.4	29.4	54.5	76.9	85.9	16.7	83.3
1973	−9.0	−9.0	−9.0	16.4	−9.0	54.5	−9.0	85.9	−9.0	−9.0
1974	−9.0	−9.0	−9.0	−9.0	−9.0	−9.0	−9.0	−9.0	−9.0	−9.0
1976	−9.0	−9.0	−9.0	−9.0	−9.0	55.3	−9.0	85.9	35.5	−9.0
1978	51.6	−9.0	−9.0	21.4	41.6	47.7	76.1	72.6	32.4	74.5
1980	53.7	−9.0	−9.0	31.6	29.4	48.5	74.6	73.7	22.9	81.4
1982	57.8	−9.0	−9.0	−9.0	−9.0	50.0	79.5	85.2	32.7	79.9
1984	63.5	−9.0	−9.0	−9.0	−9.0	59.5	−9.0	−9.0	37.6	81.8
1986	63.8	−9.0	−9.0	−9.0	−9.0	59.1	−9.0	−9.0	36.7	82.2
1988	60.9	−9.0	−9.0	−9.0	−9.0	54.5	−9.0	−9.0	22.9	81.4
1992	62.8	−9.0	−9.0	−9.0	−9.0	54.5	−9.0	−9.0	22.9	81.4
1996	65.2	−9.0	−9.0	−9.0	−9.0	54.5	−9.0	−9.0	22.9	81.4
2016	71.2	−9.0	−9.0	−9.0	−9.0	54.5	−9.0	−9.0	22.9	81.4

TABLE B-22 Effluent BOD_u Concentration (mg/L) (BOD_u = $CBOD_u$ + NBOD)

Default:		0.0	396.5	331.6	280.0	180.2	91.5	55.8	305.8	73.8
Year	Total	No-Discharge	Raw	Primary	Advanced Primary	Secondary	Advanced Secondary	Advanced Treatment	Less than Secondary	Greater than Secondary
1940	318.1	−9.0	396.5	331.6	−9.0	180.2	−9.0	−9.0	331.6	−9.0
1950	303.9	−9.0	396.5	331.6	−9.0	180.2	−9.0	−9.0	331.6	−9.0
1962	260.9	−9.0	396.5	331.6	−9.0	180.2	−9.0	−9.0	331.6	−9.0
1968	243.2	−9.0	396.5	331.6	−9.0	180.2	−9.0	55.8	331.6	55.8
1972	235.3	0.0	396.5	331.6	280.0	180.2	91.5	55.8	330.2	66.1
1973	−9.0	−9.0	396.5	331.6	−9.0	180.2	−9.0	55.8	−9.0	−9.0
1974	−9.0	−9.0	−9.0	−9.0	−9.0	−9.0	−9.0	−9.0	−9.0	−9.0
1976	−9.0	−9.0	−9.0	−9.0	−9.0	160.1	−9.0	55.8	253.6	−0.3
1978	188.7	0.0	396.5	292.1	240.0	206.7	87.9	108.5	264.8	97.1
1980	182.3	0.0	396.5	430.5	280.0	200.1	95.7	102.6	305.8	73.8
1982	162.7	0.0	396.5	−9.0	−9.0	193.5	77.0	59.0	265.5	75.8
1984	149.7	0.0	396.5	−9.0	−9.0	165.7	−9.0	−9.0	268.8	73.1
1986	148.3	0.0	396.5	−9.0	−9.0	167.7	−9.0	−9.0	277.3	71.0
1988	155.1	0.0	396.5	−9.0	−9.0	180.2	−9.0	−9.0	305.8	73.8
1992	147.4	0.0	−9.0	−9.0	−9.0	180.2	−9.0	−9.0	305.8	73.8
1996	137.8	0.0	−9.0	−9.0	−9.0	180.2	−9.0	−9.0	305.8	73.8
2016	114.3	0.0	−9.0	−9.0	−9.0	180.2	−9.0	−9.0	305.8	73.8

TABLE B-23 Influent TSS Load (metric tons per day, mt/day)

Year	Total	No-Discharge	Raw	Primary	Advanced Primary	Secondary	Advanced Secondary	Advanced Treatment	Less than Secondary	Greater than Secondary
1940	9,507.5	0.0	4,324.0	2,484.3	0.0	2,699.2	0.0	0.0	2,484.3	0.0
1950	12,322.5	0.0	4,736.1	3,303.4	0.0	4,282.9	0.0	0.0	3,303.4	0.0
1962	15,886.2	0.0	1,960.6	5,666.9	0.0	8,258.7	0.0	0.0	5,666.9	0.0
1968	18,813.6	0.0	1,356.3	5,922.1	0.0	11,495.0	0.0	40.3	5,922.1	40.3
1972	19,031.5	110.8	663.4	6,781.9	184.8	10,242.2	303.1	745.4	6,966.7	1,048.5
1973	−9.0	−9.0	429.7	7,332.1	−9.0	14,100.1	−9.0	362.6	−9.0	−9.0
1974	−9.0	−9.0	−9.0	−9.0	−9.0	−9.0	−9.0	−9.0	−9.0	−9.0
1976	−9.0	−9.0	−9.0	−9.0	−9.0	4,425.4	−9.0	6,141.4	10,298.3	−9.0
1978	21,811.1	295.0	488.8	2,779.4	3,041.7	8,251.5	3,851.3	3,103.4	5,821.1	6,954.7
1980	20,761.6	356.7	309.8	2,470.4	2,356.2	8,042.7	7,043.9	611.7	5,011.1	7,041.3
1982	22,139.2	399.5	251.9	−9.0	−9.0	8,960.2	7,632.0	581.6	4,314.0	8,213.6
1984	22,222.5	488.3	170.9	−9.0	−9.0	8,990.7	−9.0	−9.0	4,341.9	8,230.6
1986	22,753.0	494.8	215.5	−9.0	−9.0	9,880.3	−9.0	−9.0	3,727.5	8,434.9
1988	23,840.8	816.4	183.6	−9.0	−9.0	10,468.3	−9.0	−9.0	3,556.5	8,816.1
1992	24,254.2	1,042.7	0.0	−9.0	−9.0	11,133.5	−9.0	−9.0	2,915.7	9,162.3
1996	25,475.7	1,028.8	0.0	−9.0	−9.0	11,004.1	−9.0	−9.0	2,306.7	11,136.2
2016	36,891.7	1,902.0	0.0	−9.0	−9.0	13,740.4	−9.0	−9.0	740.3	20,508.9

TABLE B-24 Effluent TSS Load (metric tons per day, mt/day)

Year	Total	No-Discharge	Raw	Primary	Advanced Primary	Secondary	Advanced Secondary	Advanced Treatment	Less than Secondary	Greater than Secondary
1940	5,782.1	0.0	4,324.0	1,242.2	0.0	215.9	0.0	0.0	1,242.2	0.0
1950	6,730.4	0.0	4,736.1	1,651.7	0.0	342.6	0.0	0.0	1,651.7	0.0
1962	5,454.7	0.0	1,960.6	2,833.5	0.0	660.7	0.0	0.0	2,833.5	0.0
1968	5,237.7	0.0	1,356.3	2,961.0	0.0	919.6	0.0	0.8	2,961.0	0.8
1972	4,953.1	0.0	663.4	3,390.9	55.4	819.4	9.1	14.9	3,446.4	24.0
1973	−9.0	−9.0	429.7	3,666.0	−9.0	1,128.0	−9.0	7.3	−9.0	−9.0
1974	−9.0	−9.0	−9.0	−9.0	−9.0	−9.0	−9.0	−9.0	−9.0	−9.0
1976	−9.0	−9.0	−9.0	−9.0	−9.0	354.0	−9.0	122.8	4,119.3	−9.0
1978	3,628.7	0.0	488.8	1,389.7	912.5	660.1	115.5	62.1	2,302.2	177.6
1980	3,133.7	0.0	309.8	1,235.2	706.9	643.4	211.3	12.2	2,004.4	176.0
1982	2,934.9	0.0	251.9	−9.0	−9.0	716.8	229.0	11.6	1,725.6	240.6
1984	2,832.7	0.0	170.9	−9.0	−9.0	719.3	−9.0	−9.0	1,736.8	205.8
1986	2,707.8	0.0	215.5	−9.0	−9.0	790.4	−9.0	−9.0	1,491.0	210.9
1988	2,664.0	0.0	183.6	−9.0	−9.0	837.5	−9.0	−9.0	1,422.6	220.4
1992	2,286.0	0.0	0.0	−9.0	−9.0	890.7	−9.0	−9.0	1,166.3	229.1
1996	2,081.4	0.0	0.0	−9.0	−9.0	880.3	−9.0	−9.0	922.7	278.4
2016	1,908.1	0.0	0.0	−9.0	−9.0	1,099.2	−9.0	−9.0	296.1	512.7

TABLE B-25 TSS Removal Efficiency (Percent)

Default:		100%	0%	50%	70%	92%	97%	98%	60%	97.5%
Year	Total	No-Discharge	Raw	Primary	Advanced Primary	Secondary	Advanced Secondary	Advanced Treatment	Less than Secondary	Greater than Secondary
1940	39.2	−9.0	−9.0	50.0	−9.0	92.0	−9.0	−9.0	50.0	−9.0
1950	45.4	−9.0	−9.0	50.0	−9.0	92.0	−9.0	−9.0	50.0	−9.0
1962	65.7	−9.0	−9.0	50.0	−9.0	92.0	−9.0	−9.0	50.0	−9.0
1968	72.2	−9.0	−9.0	50.0	−9.0	92.0	−9.0	98.0	50.0	98.0
1972	74.0	−9.0	−9.0	50.0	70.0	92.0	97.0	98.0	50.5	97.7
1973	−9.0	−9.0	−9.0	50.0	−9.0	92.0	−9.0	98.0	−9.0	−9.0
1974	−9.0	−9.0	−9.0	−9.0	−9.0	−9.0	−9.0	−9.0	−9.0	−9.0
1976	−9.0	−9.0	−9.0	−9.0	−9.0	92.0	−9.0	98.0	60.0	−9.0
1978	83.4	−9.0	−9.0	50.0	70.0	92.0	97.0	98.0	60.5	97.4
1980	84.9	−9.0	−9.0	50.0	70.0	92.0	97.0	98.0	60.0	97.5
1982	86.7	100.0	0.0	−9.0	−9.0	92.0	97.0	98.0	60.0	97.1
1984	87.3	−9.0	−9.0	−9.0	−9.0	92.0	−9.0	−9.0	60.0	97.5
1986	88.1	−9.0	−9.0	−9.0	−9.0	92.0	−9.0	−9.0	60.0	97.5
1988	88.8	−9.0	−9.0	−9.0	−9.0	92.0	−9.0	−9.0	60.0	97.5
1992	90.6	−9.0	−9.0	−9.0	−9.0	92.0	−9.0	−9.0	60.0	97.5
1996	91.8	−9.0	−9.0	−9.0	−9.0	92.0	−9.0	−9.0	60.0	97.5
2016	94.8	−9.0	−9.0	−9.0	−9.0	92.0	−9.0	−9.0	60.0	97.5

TABLE B-26 Effluent TSS Concentration (mg/L) (Default Influent TSS = 215.0 mg/L)

Default:		0.0	215.0	107.5	64.5	17.2	6.45	4.3	86.0	5.38
Year	Total	No-Discharge	Raw	Primary	Advanced Primary	Secondary	Advanced Secondary	Advanced Treatment	Less than Secondary	Greater than Secondary
1940	130.8	−9.0	215.0	107.5	−9.0	17.2	−9.0	−9.0	107.5	−9.0
1950	117.4	−9.0	215.0	107.5	−9.0	17.2	−9.0	−9.0	107.5	−9.0
1962	73.8	−9.0	215.0	107.5	−9.0	17.2	−9.0	−9.0	107.5	−9.0
1968	59.9	−9.0	215.0	107.5	−9.0	17.2	−9.0	−9.0	107.5	−9.0
1972	56.0	0.0	215.0	107.5	−9.0	17.2	−9.0	4.3	107.5	4.3
1973	−9.0	−9.0	215.0	107.5	64.5	17.2	6.5	4.3	106.4	4.9
1974	−9.0	−9.0	−9.0	−9.0	−9.0	17.2	−9.0	4.3	−9.0	−9.0
1976	−9.0	−9.0	−9.0	−9.0	−9.0	−9.0	−9.0	−9.0	−9.0	−9.0
1978	35.8	0.0	215.0	107.5	−9.0	17.2	−9.0	4.3	86.0	−0.3
1980	32.5	0.0	215.0	107.5	64.5	17.2	6.5	4.3	85.0	5.5
1982	28.5	0.0	215.0	−9.0	64.5	17.2	6.5	4.3	86.0	5.4
1984	27.4	0.0	215.0	−9.0	−9.0	17.2	6.5	4.3	86.0	6.3
1986	25.6	0.0	215.0	−9.0	−9.0	17.2	−9.0	−9.0	86.0	5.4
1988	24.0	0.0	215.0	−9.0	−9.0	17.2	−9.0	−9.0	86.0	5.4
1992	20.3	0.0	−9.0	−9.0	−9.0	17.2	−9.0	−9.0	86.0	5.4
1996	17.6	0.0	−9.0	−9.0	−9.0	17.2	−9.0	−9.0	86.0	5.4
2016	11.1	0.0	−9.0	−9.0	−9.0	17.2	−9.0	−9.0	86.0	5.4

TABLE B-27 Influent POC Load (metric tons per day, mt/day) [POC = TSS × (POM/TSS) × (C/DW)]

Year	Total	No-Discharge	Raw	Primary	Advanced Primary	Secondary	Advanced Secondary	Advanced Treatment	Less than Secondary	Greater than Secondary
1940	3,137.5	0.0	1,426.9	819.8	0.0	890.7	0.0	0.0	819.8	0.0
1950	4,066.4	0.0	1,562.9	1,090.1	0.0	1,413.4	0.0	0.0	1,090.1	0.0
1962	5,242.4	0.0	647.0	1,870.1	0.0	2,725.4	0.0	0.0	1,870.1	0.0
1968	6,208.5	0.0	447.6	1,954.3	0.0	3,793.3	0.0	13.3	1,954.3	13.3
1972	6,280.4	36.6	218.9	2,238.0	61.0	3,379.9	100.0	246.0	2,299.0	346.0
1973	−9.0	−9.0	141.8	2,419.6	−9.0	4,653.0	−9.0	119.6	−9.0	−9.0
1974	−9.0	−9.0	−9.0	−9.0	−9.0	−9.0	−9.0	−9.0	−9.0	−9.0
1976	−9.0	−9.0	−9.0	−9.0	−9.0	1,460.4	−9.0	2,026.6	3,398.4	−9.0
1978	7,197.6	97.4	161.3	917.2	1,003.8	2,723.0	1,270.9	1,024.1	1,921.0	2,295.0
1980	6,851.3	117.7	102.2	815.2	777.5	2,654.1	2,324.5	201.9	1,653.6	2,323.6
1982	7,305.9	131.8	83.1	−9.0	−9.0	2,956.9	2,518.6	191.9	1,423.6	2,710.5
1984	7,333.4	161.1	56.4	−9.0	−9.0	2,966.9	−9.0	−9.0	1,432.8	2,716.1
1986	7,508.5	163.3	71.1	−9.0	−9.0	3,260.5	−9.0	−9.0	1,230.1	2,783.5
1988	7,867.5	269.4	60.6	−9.0	−9.0	3,454.5	−9.0	−9.0	1,173.6	2,909.3
1992	8,003.9	344.1	0.0	−9.0	−9.0	3,674.0	−9.0	−9.0	962.2	3,023.6
1996	8,407.0	339.5	0.0	−9.0	−9.0	3,631.3	−9.0	−9.0	761.2	3,674.9
2016	12,174.2	627.7	0.0	−9.0	−9.0	4,534.3	−9.0	−9.0	244.3	6,767.9

TABLE B-28 Effluent POC Load (metric tons per day, mt/day) [POC = TSS × (POM/TSS) × (C/DW)]

Year	Total	No-Discharge	Raw	Primary	Advanced Primary	Secondary	Advanced Secondary	Advanced Treatment	Less than Secondary	Greater than Secondary
1940	1,944.0	0.0	1,426.9	453.4	0.0	63.7	0.0	0.0	453.4	0.0
1950	2,266.9	0.0	1,562.9	602.9	0.0	101.1	0.0	0.0	602.9	0.0
1962	1,876.1	0.0	647.0	1,034.2	0.0	194.9	0.0	0.0	1,034.2	0.0
1968	1,799.8	0.0	447.6	1,080.8	0.0	271.3	0.0	0.2	1,080.8	0.2
1972	1,724.5	0.0	218.9	1,237.7	20.2	241.7	2.7	3.3	1,257.9	6.0
1973	−9.0	−9.0	141.8	1,338.1	−9.0	332.8	−9.0	1.6	−9.0	−9.0
1974	−9.0	−9.0	−9.0	−9.0	−9.0	−9.0	−9.0	−9.0	−9.0	−9.0
1976	−9.0	−9.0	−9.0	−9.0	−9.0	104.4	−9.0	27.0	1,503.5	−9.0
1978	1,244.1	0.0	161.3	507.2	333.1	194.7	34.1	13.7	840.3	47.7
1980	1,069.0	0.0	102.2	450.8	258.0	189.8	62.3	2.7	731.6	45.3
1982	994.5	0.0	83.1	−9.0	−9.0	211.5	67.5	2.6	629.8	70.1
1984	955.5	0.0	56.4	−9.0	−9.0	212.2	−9.0	−9.0	633.9	53.0
1986	902.8	0.0	71.1	−9.0	−9.0	233.2	−9.0	−9.0	544.2	54.3
1988	883.6	0.0	60.6	−9.0	−9.0	247.1	−9.0	−9.0	519.2	56.7
1992	747.4	0.0	0.0	−9.0	−9.0	262.7	−9.0	−9.0	425.7	59.0
1996	668.1	0.0	0.0	−9.0	−9.0	259.7	−9.0	−9.0	336.8	71.7
2016	564.3	0.0	0.0	−9.0	−9.0	324.3	−9.0	−9.0	108.1	132.0

TABLE B-29 POC Removal Efficiency (Percent)

Default:		100%	0%	44.7%	66.8%	92.8%	97.3%	98.7%	55.8%	98.0%
Year	Total	No-Discharge	Raw	Primary	Advanced Primary	Secondary	Advanced Secondary	Advanced Treatment	Less than Secondary	Greater than Secondary
1940	38.0	−9.0	−9.0	44.7	−9.0	92.8	−9.0	−9.0	44.7	−9.0
1950	44.3	−9.0	−9.0	44.7	−9.0	92.8	−9.0	−9.0	44.7	−9.0
1962	64.2	−9.0	−9.0	44.7	−9.0	92.8	−9.0	−9.0	44.7	−9.0
1968	71.0	−9.0	−9.0	44.7	−9.0	92.8	−9.0	98.7	44.7	98.7
1972	72.5	−9.0	−9.0	44.7	66.8	92.8	97.3	98.7	45.3	98.3
1973	−9.0	−9.0	−9.0	44.7	−9.0	92.8	−9.0	98.7	−9.0	−9.0
1974	−9.0	−9.0	−9.0	−9.0	−9.0	−9.0	−9.0	−9.0	−9.0	−9.0
1976	−9.0	−9.0	−9.0	−9.0	−9.0	92.8	−9.0	98.7	55.8	−9.0
1978	82.7	−9.0	−9.0	44.7	66.8	92.8	97.3	98.7	56.3	97.9
1980	84.4	−9.0	−9.0	44.7	66.8	92.8	97.3	98.7	55.8	98.0
1982	86.4	100.0	0.0	−9.0	−9.0	92.8	97.3	98.7	55.8	97.4
1984	87.0	−9.0	−9.0	−9.0	−9.0	92.8	−9.0	−9.0	55.8	98.0
1986	88.0	−9.0	−9.0	−9.0	−9.0	92.8	−9.0	−9.0	55.8	98.0
1988	88.8	−9.0	−9.0	−9.0	−9.0	92.8	−9.0	−9.0	55.8	98.0
1992	90.7	−9.0	−9.0	−9.0	−9.0	92.8	−9.0	−9.0	55.8	98.0
1996	92.1	−9.0	−9.0	−9.0	−9.0	92.8	−9.0	−9.0	55.8	98.0
2016	95.4	−9.0	−9.0	−9.0	−9.0	92.8	−9.0	−9.0	55.8	98.0

TABLE B-30 Effluent POC Concentration (mg/L) [POC = TSS × (POM/TSS) × (C/DW)] (Default Influent POC = 70.95 mg/L; C/DW = 0.44 mg C/mg DW)

Default POM/TSS (mg/mg)		0.00	0.75	0.83	0.83	0.67	0.67	0.50	0.83	0.585
Default Effluent POC (mg/L)		0.0	70.95	39.24	23.54	5.07	1.90	0.95	31.39	1.38
Year	Total	No-Discharge	Raw	Primary	Advanced Primary	Secondary	Advanced Secondary	Advanced Treatment	Less than Secondary	Greater than Secondary
1940	44.0	−9.0	71.0	39.2	−9.0	5.1	−9.0	−9.0	39.2	−9.0
1950	39.6	−9.0	71.0	39.2	−9.0	5.1	−9.0	−9.0	39.2	−9.0
1962	25.4	−9.0	71.0	39.2	−9.0	5.1	−9.0	−9.0	39.2	−9.0
1968	20.6	−9.0	71.0	39.2	−9.0	5.1	−9.0	0.9	39.2	0.9
1972	19.5	0.0	71.0	39.2	23.5	5.1	1.9	0.9	38.8	1.2
1973	−9.0	−9.0	71.0	39.2	−9.0	5.1	−9.0	0.9	−9.0	−9.0
1974	−9.0	−9.0	−9.0	−9.0	−9.0	−9.0	−9.0	−9.0	−9.0	−9.0
1976	−9.0	−9.0	−9.0	−9.0	−9.0	5.1	−9.0	0.9	31.4	−0.3
1978	12.3	0.0	71.0	39.2	23.5	5.1	1.9	0.9	31.0	1.5
1980	11.1	0.0	71.0	39.2	23.5	5.1	1.9	0.9	31.4	1.4
1982	9.7	0.0	71.0	−9.0	−9.0	5.1	1.9	0.9	31.4	1.8
1984	9.2	0.0	71.0	−9.0	−9.0	5.1	−9.0	−9.0	31.4	1.4
1986	8.5	0.0	71.0	−9.0	−9.0	5.1	−9.0	−9.0	31.4	1.4
1988	8.0	0.0	71.0	−9.0	−9.0	5.1	−9.0	−9.0	31.4	1.4
1992	6.6	0.0	−9.0	−9.0	−9.0	5.1	−9.0	−9.0	31.4	1.4
1996	5.6	0.0	−9.0	−9.0	−9.0	5.1	−9.0	−9.0	31.4	1.4
2016	3.3	0.0	−9.0	−9.0	−9.0	5.1	−9.0	−9.0	31.4	1.4

TABLE B-31 Influent and Effluent Characteristics for 5–day Carbonaceous BOD: $CBOD_5$

Category	Parameter	Influent (mg/L)	Effluent (mg/L)	Removal (Percent)	Conversion Factor
1 No-discharge	$CBOD_5$	215.00	0.00	100.0	1.000
2 Raw	$CBOD_5$	215.00	215.00	0.0	1.000
3 Primary	$CBOD_5$	215.00	139.75	35.0	1.000
4 Advanced primary	$CBOD_5$	215.00	107.50	50.0	1.000
5 Secondary	$CBOD_5$	215.00	32.25	85.0	1.000
6 Advanced secondary	$CBOD_5$	215.00	21.50	90.0	1.000
7 Advanced treatment	$CBOD_5$	215.00	10.75	95.0	1.000
8 Less than secondary (primary + advanced primary)	$CBOD_5$	215.00	123.63	42.5	1.000
9 Greater than secondary (advanced secondary + advanced treatment)	$CBOD_5$	215.00	16.12	92.5	1.000

TABLE B-32 Influent and Effluent Characteristics for Ultimate Carbonaceous BOD: $CBOD_u$ (Conversion Factor = $CBOD_u/BOD_5$ Ratios)

Category	Parameter	Influent (mg/L)	Effluent (mg/L)	Removal (Percent)	Conversion Factor
1 No discharge	$CBOD_u$	258.00	0.00	100.0	1.000
2 Raw	$CBOD_u$	258.00	258.00	0.0	1.200
3 Primary	$CBOD_u$	258.00	223.60	13.3	1.600
4 Advanced primary	$CBOD_u$	258.00	172.00	33.3	1.600
5 Secondary	$CBOD_u$	258.00	91.59	64.5	2.840
6 Advanced secondary	$CBOD_u$	258.00	61.06	76.3	2.840
7 Advanced treatment	$CBOD_u$	258.00	32.25	87.5	3.000
8 Less than secondary (primary + advanced primary)	$CBOD_u$	258.00	197.80	23.3	1.600
9 Greater than secondary (advanced secondary and advanced treatment)	$CBOD_u$	258.00	46.76	81.9	2.900

TABLE B-33 Influent and Effluent Characteristics for Total Kjedhal Nitrogen: TKN = Organic $-N + NH_3-N$

Category	Parameter	Influent (mg/L)	Effluent (mg/L)	Removal (Percent)	Conversion Factor
1 No-discharge	TKN	30.30	0.00	100.0	1.000
2 Raw	TKN	30.30	30.30	0.0	1.000
3 Primary	TKN	30.30	23.63	22.0	1.000
4 Advanced primary	TKN	30.30	23.63	22.0	1.000
5 Secondary	TKN	30.30	19.39	36.0	1.000
6 Advanced secondary	TKN	30.30	6.67	78.0	1.000
7 Advanced treatment	TKN	30.30	5.15	83.0	1.000
8 Less than secondary (primary + advanced primary)	TKN	30.30	23.63	22.0	1.000
9 Greater than secondary (advanced secondary and advanced treatment)	TKN	30.30	5.91	80.5	1.000

TABLE B-34 Influent and Effluent Characteristics for Nitrogenous BOD: NBOD = 4.57 × TKN (Conversion factor = 4.57 mg O_2 per mg N)

Category	Parameter	Influent (mg/L)	Effluent (mg/L	Removal (Percent)	Conversion Factor
1 No-discharge	NBOD	138.47	0.00	100.0	4.570
2 Raw	NBOD	138.47	138.47	0.0	4.570
3 Primary	NBOD	138.47	108.01	22.0	4.570
4 Advanced primary	NBOD	138.47	108.01	22.0	4.570
5 Secondary	NBOD	138.47	88.62	36.0	4.570
6 Advanced secondary	NBOD	138.47	30.46	78.0	4.570
7 Advanced treatment	NBOD	138.47	23.54	83.0	4.570
8 Less than secondary (primary + advanced primary)	NBOD	138.47	108.01	22.0	4.570
9 Greater than secondary (advanced secondary + advanced treatment)	NBOD	138.47	27.00	80.5	4.570

TABLE B-35 Influent and Effluent Characteristics for Ultimate BOD: $BOD_u = CBOD_u + NBOD$

Category	Parameter	Influent (mg/L)	Effluent (mg/L)	Removal (Percent)	Conversion Factor
1 No-discharge	$CBOD_u$ + NBOD	396.47	0.00	100.0	1.000
2 Raw	$CBOD_u$ + NBOD	396.47	396.47	0.0	1.000
3 Primary	$CBOD_u$ + NBOD	396.47	331.61	16.4	1.000
4 Advanced primary	$CBOD_u$ + NBOD	396.47	280.01	29.4	1.000
5 Secondary	$CBOD_u$ + NBOD	396.47	180.21	54.5	1.000
6 Advanced secondary	$CBOD_u$ + NBOD	396.47	91.52	76.9	1.000
7 Advanced treatment	$CBOD_u$ + NBOD	396.47	55.79	85.9	1.000
8 Less than secondary (primary + advanced primary)	$CBOD_u$ + NBOD	396.47	305.81	22.9	1.000
9 Greater than secondary (advanced secondary + advanced treatment)	$CBOD_u$ + NBOD	396.47	73.76	81.4	1.000

TABLE B-36 Influent and Effluent Characteristics for Total Suspended Solids (TSS): TSS = Volatile (Organic) SS + Nonvolatile (Inorganic) SS

Category	Parameter	Influent (mg/L)	Effluent (mg/L)	Removal (Percent)	Conversion Factor
1 No-discharge	TSS	215.00	0.00	100.0	1.000
2 Raw	TSS	215.00	215.00	0.0	1.000
3 Primary	TSS	215.00	107.50	50.0	1.000
4 Advanced primary	TSS	215.00	64.50	70.0	1.000
5 Secondary	TSS	215.00	17.20	92.0	1.000
6 Advanced secondary	TSS	215.00	6.45	97.0	1.000
7 Advanced treatment	TSS	215.00	4.30	98.0	1.000
8 Less than secondary (primary + advanced primary)	TSS	215.00	86.00	60.0	1.000
9 Greater than secondary (advanced secondary + advanced treatment)	TSS	215.00	5.38	97.5	1.000

TABLE B-37 Influent and Effluent Characteristics for Particulate Organic Carbon (POC): POC = TSS × [(POM/TSS) × (C/DW)]

Category		POM/TSS	Influent (mg/L)	Effluent (mg/L)	Removal (Percent)	Conversion Factor
1 No-discharge	POC	0.00	70.95	0.00	100.0	0.330
2 Raw	POC	0.75	70.95	70.95	0.0	0.330
3 Primary	POC	0.83	70.95	39.24	44.7	0.365
4 Advanced primary	POC	0.83	70.95	23.54	66.8	0.365
5 Secondary	POC	0.67	70.95	5.07	92.8	0.295
6 Advanced secondary	POC	0.67	70.95	1.90	97.3	0.295
7 Advanced treatment	POC	0.50	70.95	0.95	98.7	0.220
8 Less than secondary (primary + advanced primary)	POC	0.83	70.95	31.39	55.8	0.365
9 Greater than secondary (advanced secondary + advanced treatment)	POC	0.585	70.95	1.38	98.0	0.257

POM/TSS = Fraction of particulate organic matter (volatile SS/TSS).
C/DW = Carbon-dry weight = 0.44 mg C/mg DW
Conversion factor = (POM/TSS) × (C/DW)

TABLE B-38 Middle Population Projections of POTW Loads of $CBOD_5$, $CBOD_u$, NBOD, and BOD_u: 1996–2050

Year	Population	Population Served (Percent)	Effluent $CBOD_5$ (mt/day)	$CBOD_5$ (% removal)	Effluent $CBOD_u$ (mt/day)	$CBOD_u$ (% removal)	Effluent NBOD (mt/day)	NBOD (% removal)	Effluent BOD_u (mt/day)	BOD_u (% removal)
1940	132,164,569	53.5	6,344	33.3	8,922	21.8	5,146	16.0	14,068	19.8
1950	151,325,798	60.6	7,526	38.9	10,943	26.0	6,475	18.4	17,419	23.3
1962	NA	NA	6,883	56.7	11,765	38.3	7,514	26.6	19,278	34.2
1968	NA	NA	6,932	63.2	12,689	43.8	8,591	29.1	21,281	38.7
1972	NA	NA	6,768	64.4	12,558	45.0	8,273	32.5	20,831	40.6
1978	219,555,000	70.7	5,510	74.1	11,621	54.4	7,526	46.4	19,147	51.6
1982	230,075,000	71.7	4,380	79.3	9,582	62.3	7,168	49.7	16,750	57.8
1988	227,258,220	78.1	4,460	81.3	9,869	65.5	7,327	52.3	17,196	60.9
1992	246,928,467	73.1	4,034	83.4	9,418	67.6	7,205	53.9	16,623	62.8
1996	NA	NA	3,812	85.0	9,232	69.8	7,093	56.8	16,325	65.2
1997	267,645,000	72.6	3,853	85.2	9,375	70.1	7,208	57.1	16,583	65.5
1998	270,002,000	73.4	3,877	85.4	9,478	70.3	7,294	57.5	16,772	65.8
1999	272,330,000	74.3	3,899	85.6	9,579	70.6	7,378	57.8	16,957	66.1
2000	274,634,000	75.1	3,919	85.8	9,678	70.9	7,460	58.2	17,138	66.4
2001	276,918,000	75.9	3,938	86.0	9,774	71.1	7,541	58.5	17,315	66.7
2002	279,189,000	76.7	3,954	86.3	9,868	71.4	7,620	58.9	17,489	67.0
2003	281,452,000	77.5	3,970	86.5	9,960	71.7	7,698	59.2	17,658	67.3
2004	283,713,000	78.4	3,983	86.7	10,050	71.9	7,775	59.6	17,825	67.6
2005	285,981,000	79.2	3,996	86.9	10,139	72.2	7,851	59.9	17,990	67.9
2006	288,269,000	80.0	4,006	87.1	10,226	72.5	7,926	60.3	18,152	68.2

TABLE B-38 *Continued*

Year	Population	Population Served (Percent)	Effluent $CBOD_5$ (mt/day)	$CBOD_5$ (% removal)	Effluent $CBOD_u$ (mt/day)	$CBOD_u$ (% removal)	Effluent NBOD (mt/day)	NBOD (% removal)	Effluent BOD_u (mt/day)	BOD_u (% removal)
2007	290,583,000	80.8	4,016	87.3	10,313	72.8	8,000	60.6	18,313	68.5
2008	292,928,000	81.6	4,025	87.5	10,398	73.0	8,074	61.0	18,472	68.8
2009	295,306,000	82.5	4,032	87.7	10,482	73.3	8,148	61.3	18,631	69.1
2010	297,716,000	83.3	4,037	87.9	10,566	73.6	8,221	61.7	18,787	69.4
2011	300,157,000	84.1	4,042	88.1	10,648	73.8	8,294	62.0	18,942	69.7
2012	302,624,000	84.9	4,045	88.3	10,730	74.1	8,366	62.4	19,096	70.0
2013	305,112,000	85.7	4,046	88.5	10,809	74.4	8,437	62.7	19,247	70.3
2014	307,617,000	86.6	4,046	88.7	10,887	74.6	8,507	63.1	19,395	70.6
2015	310,134,000	87.4	4,044	88.9	10,963	74.9	8,576	63.4	19,540	70.9
2016	312,658,000	88.2	4,040	89.1	11,037	75.2	8,644	63.8	19,681	71.2
2017	315,185,000	88.2	4,073	89.1	11,126	75.2	8,714	63.8	19,840	71.2
2018	317,711,000	88.2	4,106	89.1	11,215	75.2	8,784	63.8	19,999	71.2
2019	320,231,000	88.2	4,138	89.1	11,304	75.2	8,853	63.8	20,158	71.2
2020	322,742,000	88.2	4,171	89.1	11,393	75.2	8,923	63.8	20,316	71.2
2021	325,239,000	88.2	4,203	89.1	11,481	75.2	8,992	63.8	20,473	71.2
2022	327,720,000	88.2	4,235	89.1	11,569	75.2	9,060	63.8	20,629	71.2
2023	330,183,000	88.2	4,267	89.1	11,656	75.2	9,129	63.8	20,784	71.2
2024	332,626,000	88.2	4,298	89.1	11,742	75.2	9,196	63.8	20,938	71.2
2025	335,050,000	88.2	4,330	89.1	11,827	75.2	9,263	63.8	21,090	71.2
2026	337,454,000	88.2	4,361	89.1	11,912	75.2	9,330	63.8	21,242	71.2
2027	339,839,000	88.2	4,391	89.1	11,996	75.2	9,396	63.8	21,392	71.2
2028	342,208,000	88.2	4,422	89.1	12,080	75.2	9,461	63.8	21,541	71.2
2029	344,560,000	88.2	4,452	89.1	12,163	75.2	9,526	63.8	21,689	71.2

TABLE B-38 *Continued*

Year	Population	Population Served (Percent)	Effluent $CBOD_5$ (mt/day)	$CBOD_5$ (% removal)	Effluent $CBOD_u$ (mt/day)	$CBOD_u$ (% removal)	Effluent NBOD (mt/day)	NBOD (% removal)	Effluent BOD_u (mt/day)	BOD_u (% removal)
2030	346,899,000	88.2	4,483	89.1	12,246	75.2	9,591	63.8	21,836	71.2
2031	349,227,000	88.2	4,513	89.1	12,328	75.2	9,655	63.8	21,983	71.2
2032	351,544,000	88.2	4,543	89.1	12,410	75.2	9,719	63.8	22,129	71.2
2033	353,853,000	88.2	4,573	89.1	12,491	75.2	9,783	63.8	22,274	71.2
2034	356,157,000	88.2	4,602	89.1	12,573	75.2	9,847	63.8	22,419	71.2
2035	358,457,000	88.2	4,632	89.1	12,654	75.2	9,910	63.8	22,564	71.2
2036	360,756,000	88.2	4,662	89.1	12,735	75.2	9,974	63.8	22,709	71.2
2037	363,056,000	88.2	4,691	89.1	12,816	75.2	10,037	63.8	22,853	71.2
2038	365,358,000	88.2	4,721	89.1	12,897	75.2	10,101	63.8	22,998	71.2
2039	367,666,000	88.2	4,751	89.1	12,979	75.2	10,165	63.8	23,144	71.2
2040	369,980,000	88.2	4,781	89.1	13,060	75.2	10,229	63.8	23,289	71.2
2041	372,303,000	88.2	4,811	89.1	13,142	75.2	10,293	63.8	23,435	71.2
2042	374,636,000	88.2	4,841	89.1	13,225	75.2	10,358	63.8	23,582	71.2
2043	376,981,000	88.2	4,871	89.1	13,308	75.2	10,422	63.8	23,730	71.2
2044	379,339,000	88.2	4,902	89.1	13,391	75.2	10,488	63.8	23,878	71.2
2045	381,713,000	88.2	4,933	89.1	13,475	75.2	10,553	63.8	24,028	71.2
2046	384,106,000	88.2	4,964	89.1	13,559	75.2	10,619	63.8	24,178	71.2
2047	386,522,000	88.2	4,995	89.1	13,644	75.2	10,686	63.8	24,331	71.2
2048	388,962,000	88.2	5,026	89.1	13,731	75.2	10,754	63.8	24,484	71.2
2049	391,431,000	88.2	5,058	89.1	13,818	75.2	10,822	63.8	24,640	71.2
2050	393,931,000	88.2	5,090	89.1	13,906	75.2	10,891	63.8	24,797	71.2

TABLE B-39 Inventory of POTW Facilities by Flow Range (as mgd) in 1972 in the United States

	Flow Range (mgd)						
Facility Type	0–1	1–5	5–10	10–25	>25	Unknown	Total
None (Raw)	141	9	0	0	4	2,111	2,265
Minor	17	4	3	0	0	20	44
Primary	1,523	312	60	1	76	558	2,530
Settling tank (no details)	92	16	0	0	6	17	131
Septic tank	252	3	0	0	1	394	650
Imhoff tanks	624	19	1	0	2	36	682
Mechanical cleaned tanks	474	264	59	1	60	40	898
Plain, hop-bottom tanks	48	3	0	0	3	2	56
Unknown	33	7	0	0	4	69	113
Intermediate	18	21	4	0	14	7	64
Secondary	10,839	1,220	172	2	156	1,646	14,035
Other	547	19	4	0	3	285	858
Activated sludge	1,151	306	80	1	82	123	1,743
Extended aeration	1,568	44	4	1	4	102	1,723
High rate trickling filter	1,231	417	37	0	36	77	1,798
Standard rate trickling filter	1,414	201	22	0	8	93	1,738
Intermediate sand filter	483	4	1	0	0	77	565
Land application	70	6	1	0	0	65	142
Oxidation pond	3,638	115	6	0	3	705	4,467
Unknown filter	249	37	7	0	2	39	334
Unknown	488	71	10	0	18	80	667
Tertiary	373	44	6	1	5	32	461
Unknown category	18	21	4	0	14	7	64
Total	12,911	1,610	245	4	255	4,374	19,399

Source: USEPA, 1972.

TABLE B-40 Population Served (Millions) by Flow Range of POTW Facilities (as mgd) in 1972 in the United States

	Flow Range (mgd)						
Facility Type	0–1	1–5	5–10	10–25	>25	Unknown	Total
None (Raw)	0.2	0.2	0.0	0.0	1.8	3.1	5.3
Minor	0.1	0.1	0.2	0.0	0.0	0.7	1.1
Primary	3.5	5.4	2.6	0.2	21.8	12.4	45.9
Settling tank (no details)	0.2	0.3	0.0	0.0	1.0	0.6	2.1
Septic tank	0.2	0.1	0.0	0.0	0.4	0.4	1.1
Imhoff tanks	1.1	0.2	0.0	0.0	0.5	0.1	1.9
Mechanical cleaned tanks	1.7	4.5	2.6	0.2	16.1	4.7	29.8
Plain, hop-bottom tanks	0.2	0.2	0.0	0.0	1.8	0.0	2.2
Unknown	0.1	0.1	0.0	0.0	2.0	6.6	8.8
Intermediate	0.1	0.4	0.2	0.0	5.1	0.1	5.9
Secondary	22.0	19.2	8.6	1.4	46.1	5.0	102.3
Other	0.3	0.2	0.1	0.0	0.2	1.4	2.2
Activated sludge	4.1	5.4	4.0	0.3	29.1	0.4	43.3
Extended aeration	1.4	0.5	0.6	1.1	0.4	0.1	4.1
High rate trickling filter	4.9	7.0	1.7	0.0	4.8	1.2	19.6
Standard rate trickling filter	4.4	2.9	1.0	0.0	0.7	0.3	9.3
Intermediate sand filter	0.3	0.1	0.1	0.0	0.0	0.1	0.6
Land application	0.1	0.1	0.1	0.0	0.0	0.2	0.5
Oxidation pond	4.7	1.3	0.2	0.0	0.3	0.8	7.3
Unknown filter	0.6	0.7	0.3	0.0	2.8	0.1	4.5
Unknown	1.2	1.0	0.5	0.0	7.8	0.4	10.9
Tertiary	0.5	0.7	0.2	0.2	1.1	0.1	2.8
Unknown category	0.1	0.4	0.2	0.0	5.1	0.1	5.9
Total	26.4	26.0	11.8	1.8	75.9	21.4	163.3

Source: USEPA, 1972.

TABLE B-41 Municipal Wastewater Treatment Categories Presented in the Tables of Appendix B

Category	Description
Total	Sum of all treatment categories
No-Discharge	Municipal wastewater treatment facilities that do not discharge effluent to surface waters; most no-discharge facilities are oxidation or stabilization ponds designed for evaporation and or infiltration; no-discharge also includes facilities designed for recycling and reuse or spray irrigation systems
Raw	Collection system only; no treatment provided; effluent = influent
Primary	Primary treatment
Advanced Primary	Advanced primary treatment
Less than Secondary	Less than secondary is the sum of primary + advanced primary
Secondary	Secondary treatment
Advanced Secondary	Advanced secondary treatment
Advanced Treatment	Advanced treatment (tertiary, AWT)
Greater than Secondary	Greater than secondary is the sum of advanced secondary + advanced treatment

Note: See Table 2–2 for descriptions of treatment categories.

APPENDIX C

National Public and Private Sector Investment in Water Pollution Control

TABLES

C-1 U.S. Environmental Protection Agency Construction Grants Program and Clean Water State Revolving Fund expenditures for municipal water pollution control

C-2 Water pollution control abatement, current year dollars

C-3 Water pollution control abatement (constant 1995 dollars)

C-4 Gross domestic product index and plant construction index used for inflation adjustment of O&M and capital expenditures

REFERENCES

CE. 1995. Economic indicators: Chemical engineering plant cost index. Chemical Engineering, The McGraw-Hill Companies, Vol. 102, No. 7, p. 192.

Council of Economic Advisors. 1997. Table B-3, Chain Type price indexes for gross domestic product, 1959–96, Appendix B, Statistical Tables Relating to Income, Employment and Production. In: Economic Report to the President, Transmitted to the Congress, February, 1997 together with the Annual Report of the Council of Economic Advisors, U. S. Government Printing Office, Washington, DC.

Vogan, C. R. 1996. Pollution abatement and control expenditures, 1972–94. Survey of current business. Vol. 76, No. 9, pp. 48–67. U. S. Department of Commerce, Bureau of Economic Analysis, Washington, DC.

TABLE C-1 U. S. Environmental Protection Agency Construction Grants Program and Clean Water State Revolving Fund Expenditures for Municipal Water Pollution Control

	Construction Grants				Clean Water State Revolving Fund			
Year	Annual, Current $ ($1,000)	Cumulative, Current $ ($1,000)	Annual, 1995 $ ($1,000)	Cumulative, 1995 $ ($1,000)	Annual, Current $ ($1,000)	Cumulative, Current $ ($1,000)	Annual, 1995 $ ($1,000)	Cumulative, 1995 $ ($1,000)
1970	139	139	512	512	0	0	0	0
1971	31	170	100	612	0	0	0	0
1972	132,080	132,250	364,950	365,562	0	0	0	0
1973	3,043,502	3,175,752	8,006,876	8,372,438	0	0	0	0
1974	2,519,179	5,694,931	5,774,008	14,146,446	0	0	0	0
1975	4,343,443	10,038,374	9,027,401	23,173,847	0	0	0	0
1976	4,598,985	14,637,359	9,075,865	32,249,712	0	0	0	0
1977	7,272,400	21,909,759	13,507,911	45,757,623	0	0	0	0
1978	2,832,399	24,742,158	4,907,498	50,665,121	0	0	0	0
1979	5,112,276	29,854,434	8,119,248	58,784,369	0	0	0	0
1980	3,807,997	33,662,431	5,526,847	64,311,216	0	0	0	0
1981	3,605,439	37,267,870	4,602,088	68,913,304	0	0	0	0
1982	2,250,355	39,518,225	2,716,928	71,630,232	0	0	0	0
1983	3,988,124	43,506,349	4,770,899	76,401,131	0	0	0	0
1984	4,565,966	48,072,315	5,363,493	81,764,624	0	0	0	0
1985	2,129,228	50,201,543	2,481,372	84,245,996	0	0	0	0
1986	2,319,335	52,520,878	2,761,491	87,007,487	0	0	0	0
1987	2,442,281	54,963,159	2,859,376	89,866,863	0	0	0	0
1988	3,062,053	58,025,212	3,389,277	93,256,140	252,227	252,227	279,181	279,181
1989	1,295,664	59,320,876	1,382,068	94,638,208	1,153,764	1,405,991	1,230,703	1,509,883

TABLE C-1 *Continued*

	Construction Grants				Clean Water State Revolving Fund			
Year	Annual, Current $ ($1,000)	Cumulative, Current $ ($1,000)	Annual, 1995 $ ($1,000)	Cumulative, 1995 $ ($1,000)	Annual, Current $ ($1,000)	Cumulative, Current $ ($1,000)	Annual, 1995 $ ($1,000)	Cumulative, 1995 $ ($1,000)
1990	945,677	60,266,553	1,002,530	95,640,738	1,368,882	2,774,873	1,451,183	2,961,067
1991	279,960	60,546,513	293,754	95,934,492	1,971,827	4,746,700	2,068,972	5,030,039
1992	284,006	60,830,519	300,577	96,235,069	1,891,382	6,638,082	2,001,739	7,031,778
1993	118,912	60,949,431	125,501	96,360,570	1,890,597	8,528,678	1,995,337	9,027,115
1994	105,055	61,054,486	108,193	96,468,763	1,270,191	9,798,870	1,308,148	10,335,264
1995	11,065	61,065,551	11,065	96,479,828	1,318,463	11,117,332	1,318,463	11,653,726
1996	0	61,065,551	0	96,479,828	1,714,318	12,831,650	NA	NA
1997	0	61,065,551	0	96,479,828	792,498	13,624,148	NA	NA
1998	0	61,065,551	0	96,479,828	1,240,294	14,864,442	NA	NA
1999	0	61,065,551	0	96,479,828	1,299,931	16,164,373	NA	NA
2000	0	61,065,551	0	96,479,828	0	16,164,373	NA	NA
		1,321,677		1,962,605	Grants data matched to River Basins >18			
		587,221		842,962	Grants data not matched to any CU			
		59,156,653		93,674,326	Grants data matched to River Basins 1–18			
		0		−65	Rounding error			
		61,065,551		96,479,828	Grants data for all grants, USA total			

Sources: Construction Grants Program expenditures extracted August, 1995 from USEPA "GICS" database; Clean Water State Revolving Fund (CWSRF) expenditures obtained from USEPA Office of Wastewater Management, CWSRF Program, April 2000.

TABLE C-2 Water Pollution Control Abatement, Current Year Dollars (Million $)

Year	Private Capital, p&e	Private O&M, p&e[a]	Private, Total	Public Capital, Sewer	Public Capital, Elec. Ut.	Public Capital, Total K[b]	Public O&M, Sewer	Public O&M, Elec. Ut.	Public O&M, Other	Public O&M, Total O&M	Public Capital + O&M, Total Public	Public + Private, Capital O&M, Total
1972	1,501	789	2,290	2,260	29	2,289	1,125	3	0	1,128	3,416	5,706
1973	1,770	972	2,742	2,534	22	2,556	1,308	4	1	1,313	3,869	6,611
1974	1,765	1,188	2,953	3,105	29	3,134	1,567	5	1	1,573	4,706	7,659
1975	2,145	1,409	3,554	3,762	32	3,794	1,838	7	0	1,845	5,639	9,193
1976	2,607	1,726	4,333	4,082	36	4,118	2,156	9	1	2,166	6,283	10,616
1977	2,827	2,064	4,891	4,287	52	4,339	2,553	10	0	2,563	6,902	11,793
1978	2,683	2,357	5,040	4,992	63	5,055	2,977	10	1	2,988	8,043	13,083
1979	2,873	2,788	5,661	5,945	81	6,026	3,399	12	1	3,412	9,438	15,099
1980	2,795	2,985	5,780	6,592	61	6,653	3,915	13	0	3,928	10,581	16,361
1981	2,848	3,210	6,058	6,404	57	6,461	4,556	18	1	4,575	11,035	17,093
1982	2,937	3,466	6,403	5,851	86	5,937	5,168	17	2	5,187	11,123	17,526
1983	2,422	3,753	6,175	5,735	73	5,808	5,643	19	2	5,664	11,472	17,647
1984	2,730	4,052	6,782	5,794	54	5,848	6,057	20	2	6,079	11,927	18,709
1985	2,670	4,350	7,020	6,193	63	6,256	6,554	10	3	6,567	12,823	19,843
1986	2,534	4,741	7,275	6,884	40	6,924	7,201	10	3	7,214	14,138	21,413
1987	2,614	5,088	7,702	7,803	37	7,840	7,792	13	3	7,808	15,648	23,350
1988	2,581	5,427	8,008	8,322	28	8,350	8,363	12	3	8,378	16,727	24,735
1989	3,196	5,767	8,963	8,350	49	8,399	9,325	11	3	9,339	17,737	26,700
1990	4,430	6,492	10,922	8,730	77	8,807	10,262	8	3	10,273	19,080	30,002
1991	4,666	6,223	10,889	9,015	64	9,079	10,995	14	3	11,012	20,091	30,980
1992	4,532	6,522	11,054	9,589	14	9,603	11,929	10	3	11,942	21,545	32,599
1993	4,335	6,513	10,848	5,126	11	5,137	6,220	11	3	6,234	11,371	22,219
1994	4,720	7,057	11,777	0	10	10	0	10	3	13	23	11,800
Total	68,181	88,939	157,120	131,350	1,068	132,418	120,897	256	42	121,195	253,612	410,732

Source: Vogan, 1996.

[a]p&e = industrial plant and equipment

[b]Total K = total capital = public + public electric utility

TABLE C-3 Water Pollution Control Abatement, Constant 1995 Dollars (1995 $, GDP Index for O&M; PCI for Capital) (Million $)

Year	Private Capital, p&e	Private O&M, p&e[a]	Private, Total	Public Capital, Sewer	Public Capital, Elec. Ut.	Public Capital, Total K[b]	Public O&M, Sewer	Public O&M, Elec. Ut.	Public O&M, Other	Public O&M, Total O&M	Public Capital + O&M, Total Public	Public + Private, Capital O&M, Total
1972	4,147	2,642	6,789	6,243	80	6,323	3,765	10	0	3,775	10,098	16,887
1973	4,657	3,057	7,714	6,666	58	6,724	4,114	12	3	4,130	10,854	18,568
1974	4,045	3,437	7,482	7,116	66	7,182	4,532	15	3	4,550	11,732	19,214
1975	4,458	3,720	8,178	7,819	67	7,885	4,851	20	0	4,871	12,757	20,935
1976	5,145	4,287	9,432	8,055	71	8,126	5,354	23	3	5,380	13,505	22,937
1977	5,251	4,796	10,047	7,962	97	8,058	5,933	24	0	5,956	14,015	24,062
1978	4,649	5,078	9,726	8,649	109	8,758	6,412	22	2	6,437	15,195	24,921
1979	4,563	5,521	10,084	9,442	129	9,570	6,730	25	2	6,757	16,327	26,411
1980	4,057	5,408	9,465	9,567	89	9,655	7,093	26	0	7,118	16,774	26,238
1981	3,635	5,285	8,920	8,174	73	8,246	7,500	33	2	7,535	15,781	24,701
1982	3,546	5,373	8,919	7,063	104	7,167	8,010	28	3	8,042	15,209	24,128
1983	2,897	5,591	8,488	6,860	87	6,947	8,406	29	3	8,438	15,385	23,874
1984	3,207	5,784	8,991	6,807	63	6,870	8,645	29	3	8,677	15,547	24,538
1985	3,112	5,986	9,097	7,217	73	7,291	9,018	14	4	9,036	16,326	25,424
1986	3,017	6,356	9,373	8,196	48	8,243	9,653	13	4	9,670	17,914	27,287
1987	3,060	6,609	9,670	9,136	43	9,179	10,122	17	4	10,142	19,321	28,991
1988	2,857	6,785	9,642	9,211	31	9,242	10,455	15	4	10,474	19,715	29,357
1989	3,409	6,904	10,314	8,906	52	8,959	11,164	13	4	11,180	20,139	30,453
1990	4,696	7,443	12,140	9,255	82	9,337	11,766	9	3	11,778	21,115	33,254
1991	4,896	6,874	11,770	9,459	67	9,526	12,144	16	3	12,163	21,690	33,459
1992	4,796	7,008	11,804	10,148	15	10,163	12,817	11	3	12,831	22,994	34,798
1993	4,575	6,851	11,426	5,410	12	5,422	6,542	11	3	6,557	11,978	23,404
1994	4,861	7,270	12,131	0	10	10	0	10	3	13	23	12,154
Total	93,537	128,063	221,600	177,361	1,525	178,886	175,027	424	59	175,510	354,396	575,996

Source: Vogan, 1996.

[a]p&e = industrial plant and equipment

[b]Total K = total capital = public + public electric utility

TABLE C-4 Gross Domestic Product (GDP) Index and Plant Construction Index (PCI) Used for Inflation Adjustment of O&M and Capital Expenditures

Fiscal Year	O&M GDP	Capital PCI	Fiscal Year	O&M GDP	Capital PCI
1955	—	—	1976	52.3	192.1
1956	—	—	1977	55.9	204.1
1957	—	—	1978	60.3	218.8
1958	—	—	1979	65.6	238.7
1959	25.6	—	1980	71.7	261.2
1960	26.0	—	1981	78.9	297.0
1961	26.3	—	1982	83.8	314.0
1962	26.9	—	1983	87.2	316.9
1963	27.2	—	1984	91.0	322.7
1964	27.7	—	1985	94.4	325.3
1965	28.4	—	1986	96.9	318.4
1966	29.4	—	1987	100.0	323.8
1967	30.3	—	1988	103.9	342.5
1968	31.8	—	1989	108.5	355.4
1969	33.4	—	1990	113.3	357.6
1970	35.2	—	1991	117.6	361.3
1971	37.1	—	1992	120.9	358.2
1972	38.8	137.2	1993	123.5	359.2
1973	41.3	144.1	1994	126.1	368.1
1974	44.9	165.4	1995	129.9	379.1
1975	49.2	182.4			

Source: Council of Economic Advisors (1997) and CE (1995).

APPENDIX D

Before and After CWA Changes in Tenth Percentile Dissolved Oxygen and Ninetieth Percentile BOD_5 at the Catalog Unit Level

TABLES

D-1 Before and after CWA changes in tenth percentile dissolved oxygen at the catalog unit level (mg/L)

D-2 Before and after CWA changes in ninetieth percentile BOD_5 at the catalog unit level (mg/L)

TABLE D-1 Before and After CWA Changes in Tenth Percentile Dissolved Oxygen at the Catalog Unit Level (mg/L)

Rank	Catalog Unit	Catalog Unit Name	Mean Before CWA	Mean After CWA	After-Before CWA	Number of Stations Before CWA	Number of Stations After CWA	Number of Dry Years Before CWA	Number of Dry Years After CWA
1	04030204	Lower Fox	0.160	7.205	7.045	1	6	2	3
2	04120102	Cattaraugus	1.323	7.600	6.277	3	2	4	2
3	04110002	Cuyahoga	0.295	6.501	6.206	2	25	5	1
4	17010307	Lower Spokane	3.500	9.700	6.200	1	1	1	2
5	07070002	Lake Dubay	0.880	6.683	5.803	1	3	2	3
6	18060005	Salinas	3.180	8.750	5.570	4	1	4	4
7	02050306	Lower Susquehanna	0.880	6.196	5.316	1	10	4	1
8	04030104	Oconto	0.500	5.800	5.300	1	2	1	3
9	05080002	Lower Great Miami	1.185	6.468	5.282	6	4	4	1
10	08030204	Coldwater	0.000	5.208	5.208	1	11	2	3
11	10170203	Lower Big Sioux	0.000	5.143	5.143	1	9	3	2
12	04040002	Pike-Root	0.940	5.940	5.000	1	1	2	1
13	08030203	Yocona	0.000	4.854	4.854	1	16	2	4
14	04040003	Milwaukee	2.180	6.957	4.777	2	3	3	1
15	06010104	Holston	0.157	4.869	4.712	1	5	2	2
16	08030205	Yalobusha	0.000	4.629	4.629	1	11	3	4
17	06010205	Upper Clinch	1.614	6.082	4.468	1	7	3	2
18	02040204	Delaware Bay	0.530	4.910	4.380	1	8	5	3
19	04100002	Raisin	4.059	8.340	4.281	17	2	4	1
20	11070207	Spring	1.600	5.625	4.025	1	4	2	2
21	04040001	Little Calumet-Galien	0.570	4.555	3.985	2	11	2	1
22	18090208	Mojave	4.020	7.977	3.957	2	3	4	4
23	07120007	Lower Fox	3.780	7.576	3.796	2	11	3	1
24	07130011	Lower Illinois	1.940	5.722	3.783	1	20	3	3
25	04100009	Lower Maumee	2.068	5.847	3.780	8	7	5	1

TABLE D-1 *Continued*

Rank	Catalog Unit	Catalog Unit Name	Mean Before CWA	Mean After CWA	After-Before CWA	Number of Stations Before CWA	Number of Stations After CWA	Number of Dry Years Before CWA	Number of Dry Years After CWA
26	04130003	Lower Genesee	1.043	4.710	3.668	4	2	4	1
27	06010102	South Fork Holston	2.623	6.183	3.560	3	10	2	3
28	05050008	Lower Kanawha	1.463	5.013	3.550	3	7	3	2
29	02040203	Schuylkill	3.830	7.367	3.537	1	9	4	1
30	04110001	Black-Rocky	1.688	4.910	3.222	6	6	5	1
31	07090001	Upper Rock	2.760	5.980	3.220	1	8	4	1
32	03010106	Roanoke Rapids	1.423	4.608	3.186	4	6	3	3
33	07130006	Upper Sangamon	0.000	3.130	3.130	1	21	2	2
34	07120004	Des Plaines	1.477	4.605	3.128	18	49	3	1
35	06010105	Upper French Broad	4.100	7.033	2.933	1	7	1	2
36	05030103	Mahoning	2.627	5.540	2.913	6	2	5	1
37	05080001	Upper Great Miami	3.533	6.443	2.909	3	4	4	1
38	02040202	Lower Delaware	1.298	4.172	2.874	16	24	4	1
39	03100204	Alafia	2.598	5.387	2.789	4	3	2	4
40	14010005	Colorado Headwaters-Plateau	4.880	7.560	2.680	1	2	2	3
41	05090101	Raccoon-Symmes	3.200	5.835	2.635	1	3	4	3
42	05050003	Greenbrier	3.233	5.720	2.487	3	1	4	1
43	07010204	Crow	4.200	6.680	2.480	1	2	3	3
44	04030101	Manitowoc-Sheboygan	5.010	7.490	2.480	3	5	1	1
45	07140201	Upper Kaskaskia	3.260	5.700	2.440	6	11	3	3
46	02040201	Crosswicks-Neshaminy	3.446	5.815	2.369	7	4	4	1
47	03170006	Pascagoula	2.266	4.541	2.275	5	13	2	2
48	09020301	Sandhill-Wilson	4.100	6.347	2.247	2	3	2	3
49	03100101	Peace	4.030	6.250	2.220	4	2	3	4
50	02070003	Cacapon-Town	4.600	6.800	2.200	1	1	5	2

TABLE D-1 *Continued*

Rank	Catalog Unit	Catalog Unit Name	Mean Before CWA	Mean After CWA	After-Before CWA	Number of Stations Before CWA	Number of Stations After CWA	Number of Dry Years Before CWA	Number of Dry Years After CWA
51	05030101	Upper Ohio	4.530	6.702	2.172	1	39	5	1
52	05030202	Upper Ohio-Shade	3.765	5.830	2.065	2	3	4	3
53	07130001	Lower Illinois-Senaca	4.174	6.186	2.012	9	14	3	1
54	05090202	Little Miami	4.005	5.999	1.994	2	15	3	1
55	07120006	Upper Fox	4.595	6.550	1.955	2	39	3	1
56	07090004	Sugar	5.900	7.800	1.900	1	2	2	1
57	01080205	Lower Connecticut	4.300	6.163	1.863	1	7	4	1
58	02070008	Middle Potomac-Catoctin	5.100	6.937	1.837	1	14	5	2
59	17050115	Middle Snake-Payette	6.470	8.300	1.830	1	1	1	3
60	06010207	Lower Clinch	5.260	7.081	1.821	2	4	1	3
61	18070203	Santa Ana	3.833	5.633	1.800	6	9	5	3
62	04090001	St. Clair	5.213	7.000	1.787	6	1	4	1
63	05090201	Ohio Brush-Whiteoak	4.400	6.170	1.770	1	3	3	3
64	05120201	Upper White	3.802	5.552	1.750	37	14	3	1
65	08020401	Lower Arkansas	5.650	7.390	1.740	2	2	2	4
66	07060005	Apple-Plum	3.946	5.582	1.636	5	6	2	2
67	03050109	Saluda	2.645	4.270	1.625	6	68	1	4
68	17060107	Lower Snake-Tucannon	7.600	9.200	1.600	1	1	1	2
69	02040205	Brandywine-Christina	4.600	6.126	1.526	1	23	4	1
70	07010206	Twin Cities	3.851	5.373	1.522	23	64	3	3
71	07070003	Castle Rock	5.000	6.515	1.515	1	4	2	3
72	03050103	Lower Catawba	2.273	3.771	1.498	13	22	2	2
73	10190006	Big Thompson	5.443	6.901	1.458	8	11	2	1
74	10200203	Salt	3.750	5.195	1.445	2	20	3	1
75	05120114	Little Wabash	2.340	3.754	1.414	1	11	3	2

TABLE D-1 *Continued*

Rank	Catalog Unit	Catalog Unit Name	Mean Before CWA	Mean After CWA	After-Before CWA	Number of Stations Before CWA	Number of Stations After CWA	Number of Dry Years Before CWA	Number of Dry Years After CWA
76	03180004	Lower Pearl	3.777	5.185	1.408	7	4	1	1
77	04140203	Oswego	4.300	5.700	1.400	1	1	4	2
78	05020002	West Fork	5.325	6.691	1.366	4	2	4	2
79	17080005	Lower Cowlitz	7.564	8.930	1.366	5	4	2	2
80	05030203	Little Kanawha	4.403	5.743	1.340	6	6	4	2
81	07120005	Upper Illinois	5.077	6.412	1.334	4	10	3	1
82	07030005	Lower St. Croix	5.731	7.057	1.326	9	7	3	3
83	06010108	Nolichucky	5.580	6.900	1.320	1	2	1	2
84	07090006	Kishwaukee	5.000	6.280	1.280	1	12	4	1
85	07010207	Rum	5.800	7.080	1.280	1	3	3	3
86	17020006	Okanogan	6.460	7.725	1.265	2	2	2	3
87	05040005	Wills	4.220	5.480	1.260	1	3	4	3
88	18080003	Honey-Eagle Lakes	6.820	8.040	1.220	1	1	3	4
89	06030005	Pickwick Lake	4.800	5.968	1.168	1	5	1	4
90	18020103	Sacramento-Lower Thomes	8.385	9.550	1.165	4	2	3	2
91	03170008	Escatawpa	2.078	3.232	1.153	6	12	2	2
92	14070006	Lower Lake Powell	6.380	7.500	1.120	1	1	1	1
93	11070103	Middle Verdigris	5.010	6.117	1.107	1	4	2	3
94	06030001	Guntersville Lake	4.850	5.935	1.085	2	13	1	3
95	11140208	Saline Bayou	4.160	5.240	1.080	1	1	4	4
96	07140106	Big Muddy	2.290	3.366	1.076	1	36	3	2
97	05010009	Lower Allegheny	6.060	7.132	1.072	1	19	4	1
98	08050003	Tensas	3.100	4.168	1.068	1	5	3	4
99	03170001	Chunky-Okatibbee	4.433	5.500	1.067	3	1	4	3
100	05050002	Middle New	5.000	6.066	1.066	2	7	3	2

TABLE D-1 *Continued*

Rank	Catalog Unit	Catalog Unit Name	Mean Before CWA	Mean After CWA	After-Before CWA	Number of Stations Before CWA	Number of Stations After CWA	Number of Dry Years Before CWA	Number of Dry Years After CWA
101	02070007	Shenandoah	5.260	6.307	1.047	1	3	5	2
102	05120202	Lower White	4.725	5.750	1.025	2	3	3	1
103	08090203	Eastern Louisiana Coastal	4.550	5.548	0.998	1	4	2	2
104	07040001	Rush-Vermillion	4.857	5.853	0.996	10	6	3	2
105	03140305	Escambia	5.120	6.100	0.980	2	1	2	2
106	02050103	Owego-Wappasening	6.080	7.047	0.967	1	3	4	1
107	17110013	Duwamish	6.001	6.957	0.956	9	7	2	3
108	08030100	L. Mississippi-Greenville	5.100	6.020	0.920	1	1	3	4
109	05130108	Caney	4.200	5.110	0.910	1	1	1	3
110	05120103	Mississinewa	5.840	6.715	0.875	1	6	3	1
111	17060110	Lower Snake	7.170	8.030	0.860	1	1	3	3
112	17080001	Lower Columbia-Sandy	7.800	8.625	0.825	1	2	2	1
113	05120111	Middle Wabash-Busser	5.100	5.892	0.792	1	23	3	1
114	11110207	Lower Arkansas-Maumelle	6.360	7.140	0.780	2	7	2	4
115	04090005	Huron	5.000	5.750	0.750	2	2	4	1
116	11110104	Robert S. Kerr Reservoir	5.620	6.367	0.747	1	3	4	3
117	06020001	Mid. Tennessee-Chickamauga	4.340	5.071	0.730	15	34	1	3
118	08040303	Dugdemona	3.000	3.725	0.725	1	2	4	4
119	17110007	Lower Skagit	9.420	10.117	0.697	2	3	3	3
120	17010305	Upper Spokane	7.200	7.895	0.695	2	11	1	2
121	08040304	Little	3.250	3.938	0.688	4	2	4	3
122	03050208	Broad-St. Helena	2.287	2.970	0.683	4	11	1	2
123	17110019	Puget Sound	8.043	8.720	0.677	7	1	1	2
124	10190007	Cache La Poudre	6.391	7.038	0.647	12	6	2	1
125	17110011	Snohomish	8.360	8.987	0.627	2	3	3	4

TABLE D-1 *Continued*

Rank	Catalog Unit	Catalog Unit Name	Mean Before CWA	Mean After CWA	After-Before CWA	Number of Stations Before CWA	Number of Stations After CWA	Number of Dry Years Before CWA	Number of Dry Years After CWA
126	07090005	Lower Rock	5.253	5.866	0.613	3	23	3	1
127	05050005	Gauley	7.090	7.685	0.595	1	5	4	2
128	07130002	Vermilion	3.775	4.370	0.595	2	4	2	2
129	10190018	Lower South Platte	6.220	6.800	0.580	3	1	1	1
130	04110003	Ashtabula-Chagrin	5.225	5.793	0.568	5	3	5	1
131	10200101	Middle Platte-Buffalo	6.800	7.305	0.505	1	2	3	1
132	17020015	Lower Crab	6.680	7.180	0.500	3	1	1	2
133	07070005	Lower Wisconsin	6.365	6.830	0.465	2	2	2	2
134	04110004	Grand	5.665	6.120	0.455	3	3	5	1
135	04080204	Flint	5.400	5.850	0.450	3	2	4	1
136	08040302	Castor	3.440	3.870	0.430	1	2	4	4
137	08030206	Upper Yazoo	5.520	5.948	0.428	1	5	2	4
138	07130003	Low.Illinois-Lk. Chautauqua	4.078	4.489	0.412	5	14	3	2
139	07120003	Chicago	3.416	3.824	0.407	13	19	2	1
140	05120106	Tippecanoe	6.135	6.525	0.390	2	2	3	1
141	03050202	South Carolina Coast	3.553	3.932	0.379	3	9	2	3
142	14080105	Middle San Juan	4.790	5.160	0.370	1	1	2	1
143	04030108	Menominee	6.440	6.750	0.310	1	2	1	3
144	07080209	Lower Iowa	6.540	6.845	0.305	1	2	2	2
145	10270205	Lower Big Blue	5.880	6.170	0.290	1	2	2	1
146	10200202	Lower Platte	6.900	7.173	0.273	1	2	2	1
147	18030012	Tulare-Buena Vista Lake	7.240	7.500	0.260	1	2	3	4
148	05020006	Youghiogheny	7.197	7.446	0.249	11	5	4	1
149	18040005	Lower Cosumnes-Lower	8.460	8.700	0.240	2	1	3	3
150	06010103	Watauga	6.095	6.333	0.238	2	3	1	3

TABLE D-1 *Continued*

Rank	Catalog Unit	Catalog Unit Name	Mean Before CWA	Mean After CWA	After-Before CWA	Number of Stations Before CWA	Number of Stations After CWA	Number of Dry Years Before CWA	Number of Dry Years After CWA
151	10180009	Mid. N. Platte-Scotts Bluff	7.000	7.233	0.233	1	8	2	2
152	05080003	Whitewater	6.880	7.100	0.220	1	1	3	1
153	04080206	Saginaw	4.899	5.117	0.218	19	3	4	1
154	08040207	Lower Ouachita	4.000	4.200	0.200	2	1	4	4
155	05050007	Elk	6.400	6.577	0.177	1	7	4	2
156	08070205	Tangipahoa	6.200	6.353	0.153	2	3	3	2
157	03040201	Lower Pee Dee	4.120	4.270	0.150	19	39	2	3
158	02070001	South Branch Potomac	6.300	6.450	0.150	1	1	4	1
159	17020011	Wenatchee	9.300	9.450	0.150	1	2	1	4
160	05070101	Upper Guyandotte	6.400	6.545	0.145	2	4	3	1
161	07020012	Lower Minnesota	5.439	5.580	0.141	17	8	2	3
162	05140101	Silver-Little Kentucky	3.400	3.530	0.130	1	5	2	1
163	03180005	Bogue Chitto	6.000	6.100	0.100	2	2	2	2
164	03050205	Edisto	4.006	4.053	0.047	5	7	1	3
165	17090003	Upper Willamette	7.657	7.699	0.042	6	7	1	1
166	05120112	Embarras	5.080	5.102	0.022	1	15	3	2
167	02070004	Conococheague-Opequon	6.040	6.060	0.020	1	14	5	2
168	03160203	Lower Tambigbee	5.988	5.967	−0.020	8	15	2	2
169	05050006	Upper Kanawha	7.050	6.990	−0.060	2	2	3	2
170	09030004	Upper Rainy	7.200	7.135	−0.065	1	4	1	2
171	05020003	Upper Monongahela	6.325	6.250	−0.075	6	19	4	2
172	05020001	Tygart Valley	7.028	6.926	−0.102	6	4	4	2
173	07090003	Pecatonica	5.500	5.388	−0.112	1	6	2	2
174	06030002	Wheeler Lake	5.486	5.368	−0.119	2	13	1	3
175	05070204	Big Sandy	5.720	5.550	−0.170	1	2	3	4

TABLE D-1 *Continued*

Rank	Catalog Unit	Catalog Unit Name	Mean Before CWA	Mean After CWA	After-Before CWA	Number of Stations Before CWA	Number of Stations After CWA	Number of Dry Years Before CWA	Number of Dry Years After CWA
176	02070010	Middle Potomac-Anacostia	4.340	4.160	−0.180	1	81	5	2
177	03040204	Little Pee Dee	3.686	3.480	−0.205	9	11	2	3
178	05050001	Upper New	5.973	5.734	−0.240	3	17	1	2
179	08040301	Lower Red	5.450	5.200	−0.250	1	2	4	1
180	07140204	Lower Kaskaskia	4.130	3.864	−0.266	1	19	3	4
181	07040007	Black	7.450	7.180	−0.270	1	3	2	3
182	10240011	Independence-Sugar	5.420	5.125	−0.295	1	2	3	2
183	17110008	Stillaguamish	8.675	8.360	−0.315	2	1	3	3
184	03050101	Upper Catawba	4.667	4.321	−0.345	3	36	1	2
185	02080206	Lower James	6.140	5.774	−0.366	1	20	3	3
186	04080201	Tittabawassee	6.035	5.660	−0.375	2	3	3	1
187	04030202	Wolf	7.205	6.825	−0.380	2	2	1	3
188	07080208	Middle Iowa	3.827	3.445	−0.382	6	6	2	2
189	08050002	Bayou Macon	3.700	3.287	−0.413	2	3	3	4
190	08040202	Lower Ouachita-Bayou	2.635	2.220	−0.415	4	7	4	4
191	05120105	Middle Wabash-Deer	6.860	6.435	−0.425	1	1	2	2
192	08080203	Upper Calcasieu	4.590	4.162	−0.428	6	5	4	3
193	05020004	Cheat	6.886	6.432	−0.454	8	4	4	2
194	06040006	Lower Tennessee	5.400	4.930	−0.470	1	4	3	2
195	05070102	Lower Guyandotte	6.387	5.850	−0.537	3	2	3	3
196	07040006	La Crosse-Pine	6.740	6.200	−0.540	1	2	3	2
197	10300101	Lower Missouri-Crooked	5.150	4.586	−0.564	1	5	3	1
198	07040002	Cannon	7.260	6.662	−0.598	1	6	2	3
199	07020007	Middle Minnesota	6.803	6.197	−0.607	6	3	3	3
200	07010203	Clearwater-Elk	6.200	5.541	−0.660	1	18	3	3

TABLE D-1 *Continued*

Rank	Catalog Unit	Catalog Unit Name	Mean Before CWA	Mean After CWA	After-Before CWA	Number of Stations Before CWA	Number of Stations After CWA	Number of Dry Years Before CWA	Number of Dry Years After CWA
201	05120104	Eel	7.000	6.290	−0.710	1	2	3	1
202	17100103	Upper Chehalis	7.900	7.18	−0.715	2	2	2	1
203	03040202	Lynches	4.917	4.18	−0.728	17	18	2	3
204	03160106	Middle Tombigbee-Lubbub	5.520	4.776	−0.744	1	7	2	4
205	07140101	Cahokia-Joachim	4.490	3.658	−0.832	1	16	2	4
206	03050105	Upper Broad	6.700	5.867	−0.833	1	29	1	3
207	07050005	Lower Chippewa	5.980	5.100	−0.880	3	1	4	3
208	04030105	Peshtigo	6.240	5.300	−0.940	1	1	1	3
209	10170101	Lewis And Clark Lake	7.500	6.548	−0.952	1	3	1	3
210	05120108	Mid.Wabash-Little Vermilion	6.800	5.802	−0.998	3	7	3	2
211	08040206	Bayou D'arbonne	2.880	1.867	−1.013	1	3	3	3
212	05010006	Middle Allegheny-Redbank	8.000	6.977	−1.023	2	26	4	1
213	03050207	Salkehatchie	3.805	2.770	−1.035	2	7	1	3
214	11140202	Middle Red-Coushatta	7.120	6.050	−1.070	2	2	4	3
215	05070201	Tug	6.173	5.100	−1.073	3	1	3	2
216	05030204	Hocking	4.860	3.717	−1.143	1	4	4	2
217	10270207	Lower Little Blue	7.330	6.070	−1.260	1	2	1	1
218	05020005	Lower Monongahela	7.937	6.661	−1.275	3	31	4	1
219	05120101	Upper Wabash	6.450	5.173	−1.277	3	8	3	1
220	10240008	Big Nemaha	7.340	6.060	−1.280	1	1	2	1
221	10240006	Little Nemaha	6.820	5.520	−1.300	1	1	2	1
222	05090203	Middle Ohio-Laughery	5.400	4.043	−1.357	1	6	2	2
223	03040205	Black	3.950	2.581	−1.369	2	17	2	2
224	16050101	Lake Tahoe	7.190	5.818	−1.372	1	4	2	4
225	03050201	Cooper	6.045	4.566	−1.479	1	7	2	3

TABLE D-1 *Continued*

Rank	Catalog Unit	Catalog Unit Name	Mean Before CWA	Mean After CWA	After-Before CWA	Number of Stations Before CWA	Number of Stations After CWA	Number of Dry Years Before CWA	Number of Dry Years After CWA
226	16050102	Truckee	7.780	6.264	−1.516	3	13	3	4
227	08050001	Boeuf	4.347	2.629	−1.717	3	14	4	4
228	05140202	Highland-Pigeon	5.140	3.388	−1.753	1	4	5	1
229	12010004	Toledo Bend Reservoir	4.960	3.200	−1.760	1	1	4	4
230	08090201	Liberty Bayou-Tchefuncta	6.400	4.620	−1.780	2	2	2	2
231	05030201	Little Muskingum-Mid.Island	7.100	5.301	−1.799	1	10	4	3
232	08030201	Little Tallahatchie	4.810	3.006	−1.804	2	5	3	4
233	08080202	Mermentau	2.260	0.430	−1.830	2	2	1	1
234	04120103	Buffalo-Eighteen Mile	1.880	0.000	−1.880	2	1	3	3
235	08010100	Lower Mississippi-Memphis	5.800	3.830	−1.970	1	1	3	3
236	04010302	Bad-Montreal	7.460	5.463	−1.997	2	4	3	3
237	07080202	Shell Rock	8.100	6.080	−2.020	1	1	2	3
238	05140206	Lower Ohio	5.800	3.642	−2.158	1	5	3	4
239	04030201	Upper Fox	6.760	4.400	−2.360	1	1	2	2
240	05120113	Lower Wabash	4.990	2.550	−2.440	1	3	4	1
241	03060101	Seneca	7.920	5.416	−2.504	1	21	1	4
242	12010005	Lower Sabine	6.580	3.836	−2.744	2	6	3	1
243	03060106	Middle Savannah	8.210	4.957	−3.253	1	13	1	3
244	05110003	Middle Green	5.850	1.942	−3.908	6	4	5	4
245	11140203	Loggy Bayou	5.300	1.006	−4.294	2	5	3	3
246	11140304	Cross Bayou	6.640	1.253	−5.387	2	3	4	2

TABLE D-2 Before and After CWA Changes in Ninetieth Percentile BOD_5 at the Catalog Unit Level (mg/L)

Rank	Catalog Unit	Catalog Unit Name	Mean Before CWA	Mean After CWA	After-Before CWA	Number of Stations Before CWA	Number of Stations After CWA	Number of Dry Years Before CWA	Number of Dry Years After CWA
1	03050109	Saluda	64.105	4.848	−59.257	4	58	1	4
2	03050103	Lower Catawba	48.153	8.831	−39.322	8	15	2	2
3	10270205	Lower Big Blue	43.400	6.510	−36.890	1	2	2	1
4	10190006	Big Thompson	36.303	5.700	−30.603	8	2	2	1
5	05120201	Upper White	34.823	6.869	−27.955	19	14	3	1
6	03170008	Escatawpa	21.990	1.880	−20.110	6	1	2	2
7	04100002	Raisin	21.440	6.240	−15.200	1	1	4	1
8	05050008	Lower Kanawha	13.663	2.550	−11.113	3	1	3	2
9	04110001	Black-Rocky	16.406	8.000	−8.406	5	2	5	1
10	04040002	Pike-Root	12.400	6.100	−6.300	1	1	2	1
11	05120104	Eel	8.300	2.100	−6.200	1	1	3	1
12	07040006	La Crosse-Pine	9.400	3.745	−5.655	1	2	3	2
13	02040204	Delaware Bay	5.200	0.038	−5.162	1	8	5	3
14	07010207	Rum	9.000	4.340	−4.660	1	1	3	3
15	07040001	Rush-Vermillion	6.779	2.160	−4.619	6	1	3	2
16	07010206	Twin Cities	10.143	5.580	−4.563	21	5	3	3
17	04030104	Oconto	6.860	2.450	−4.410	1	2	1	3
18	04080206	Saginaw	10.876	6.560	−4.316	14	2	4	1
19	04090005	Huron	9.780	5.500	−4.280	2	2	4	1
20	02070007	Shenandoah	6.490	2.500	−3.990	1	2	5	2
21	05070204	Big Sandy	5.100	1.180	−3.920	1	1	3	4
22	05120103	Mississinewa	8.240	4.337	−3.903	1	3	3	1
23	05090202	Little Miami	7.500	3.605	−3.895	1	2	3	1
24	05030103	Mahoning	9.600	5.740	−3.860	6	1	5	1
25	03050208	Broad-St. Helena	7.900	4.122	−3.778	1	11	1	2

TABLE D-2 *Continued*

Rank	Catalog Unit	Catalog Unit Name	Mean Before CWA	Mean After CWA	After-Before CWA	Number of Stations Before CWA	Number of Stations After CWA	Number of Dry Years Before CWA	Number of Dry Years After CWA
26	08040301	Lower Red	6.620	3.210	−3.410	1	1	4	1
27	08020401	Lower Arkansas	6.610	3.230	−3.380	1	2	2	4
28	16050102	Truckee	5.500	2.317	−3.183	1	7	3	4
29	04030204	Lower Fox	9.660	6.543	−3.117	1	3	2	3
30	05070201	Tug	4.693	1.612	−3.081	3	1	3	2
31	05010009	Lower Allegheny	4.470	1.780	−2.690	1	1	4	1
32	10190007	Cache La Poudre	8.192	5.700	−2.492	10	2	2	1
33	05050006	Upper Kanawha	2.450	0.000	−2.450	2	1	3	2
34	07070002	Lake Dubay	8.840	6.430	−2.410	1	3	2	3
35	05140202	Highland-Pigeon	4.970	2.660	−2.310	1	2	5	1
36	02040203	Schuylkill	5.840	3.587	−2.253	1	3	4	1
37	02040202	Lower Delaware	5.792	3.580	−2.211	16	19	4	1
38	05120101	Upper Wabash	7.763	5.562	−2.201	3	5	3	1
39	05050002	Middle New	7.550	5.350	−2.200	2	2	3	2
40	05090203	Middle Ohio-Laughery	6.100	3.945	−2.155	1	2	2	2
41	11110207	Lower Arkansas-Maumelle	5.300	3.366	−1.934	1	7	2	4
42	07030005	Lower St. Croix	4.238	2.340	−1.898	5	1	3	3
43	0316020	Lower Tambigbee	2.969	1.251	−1.718	8	9	2	2
44	03040202	Lynches	5.781	4.326	−1.456	13	17	2	3
45	07080202	Shell Rock	16.890	15.500	−1.390	1	1	2	3
46	07010203	Clearwater-Elk	5.000	3.650	−1.350	1	2	3	3
47	03050205	Edisto	3.916	2.720	−1.196	5	6	1	3
48	03140305	Escambia	2.935	2.050	−0.885	2	1	2	2
49	07020007	Middle Minnesota	7.733	6.850	−0.883	5	3	3	3
50	03100204	Alafia	2.693	1.810	−0.883	3	2	2	4

TABLE D-2 *Continued*

Rank	Catalog Unit	Catalog Unit Name	Mean Before CWA	Mean After CWA	After-Before CWA	Number of Stations Before CWA	Number of Stations After CWA	Number of Dry Years Before CWA	Number of Dry Years After CWA
51	08040202	Lower Ouachita-Bayou	4.500	3.773	−0.727	1	3	4	4
52	03080103	Lower St. Johns	3.340	2.642	−0.698	1	26	1	4
53	07020012	Lower Minnesota	8.272	7.810	−0.462	16	2	2	3
54	04030101	Manitowoc-Sheboygan	5.550	5.290	−0.260	3	2	1	1
55	16020204	Jordan	16.843	16.594	−0.249	7	25	2	1
56	11110104	Robert S. Kerr Reservoir	6.300	6.060	−0.240	1	2	4	3
57	03050105	Upper Broad	4.200	3.980	−0.220	1	26	1	3
58	07060005	Apple-Plum	5.896	5.700	−0.196	5	4	2	2
59	17080001	Lower Columbia-Sandy	1.480	1.300	−0.180	1	1	2	1
60	09030004	Upper Rainy	1.560	1.390	−0.170	1	4	1	2
61	09020301	Sandhill-Wilson	4.955	4.830	−0.125	2	2	2	3
62	04110002	Cuyahoga	8.400	8.300	−0.100	2	6	5	1
63	17090003	Upper Willamette	2.345	2.283	−0.062	6	6	1	1
64	03040205	Black	3.130	3.208	0.078	2	16	2	2
65	04080201	Tittabawassee	4.000	4.100	0.100	1	2	3	1
66	05090101	Raccoon-Symmes	1.600	1.700	0.100	1	2	4	3
67	05140101	Silver-Little Kentucky	2.880	3.000	0.120	1	2	2	1
68	03160106	Middle Tombigbee-Lubbub	2.520	2.700	0.180	1	1	2	4
69	05140206	Lower Ohio	1.990	2.230	0.240	1	1	3	4
70	04090001	St. Clair	3.707	4.000	0.293	6	1	4	1
71	0503010	Upper Ohio	3.200	3.700	0.500	1	2	5	1
72	04030108	Menominee	2.760	3.420	0.660	1	1	1	3
73	03050201	Cooper	2.540	3.239	0.699	1	6	2	3
74	06020001	Mid. Tennessee-Chickamauga	1.692	2.455	0.763	11	6	1	3
75	03060101	Seneca	3.720	4.739	1.019	1	14	1	4

TABLE D-2 *Continued*

Rank	Catalog Unit	Catalog Unit Name	Mean Before CWA	Mean After CWA	After-Before CWA	Number of Stations Before CWA	Number of Stations After CWA	Number of Dry Years Before CWA	Number of Dry Years After CWA
76	03040204	Little Pee Dee	3.487	5.054	1.566	7	11	2	3
77	05120113	Lower Wabash	5.540	7.170	1.630	1	1	4	1
78	02040201	Crosswicks-Neshaminy	4.527	6.215	1.688	7	4	4	1
79	05030202	Upper Ohio-Shade	1.900	3.640	1.740	1	2	4	3
80	02070010	Middle Potomac-Anacostia	3.880	5.762	1.882	1	38	5	2
81	05110003	Middle Green	2.236	4.154	1.918	5	2	5	4
82	02070004	Conococheague-Opequon	2.200	4.200	2.000	1	2	5	2
83	02070008	Middle Potomac-Catoctin	3.580	5.607	2.027	1	14	5	2
84	05050001	Upper New	2.073	4.251	2.178	3	11	1	2
85	07080209	Lower Iowa	6.550	9.000	2.450	2	1	2	2
86	03040201	Lower Pee Dee	3.979	6.731	2.753	18	36	2	3
87	07040002	Cannon	2.410	5.255	2.845	1	2	2	3
88	05120111	Middle Wabash-Busser	5.970	9.640	3.670	1	1	3	1
89	02050306	Lower Susquehanna	2.980	6.800	3.820	1	2	4	1
90	03050207	Salkehatchie	4.475	9.056	4.581	2	5	1	3
91	05090201	Ohio Brush-Whiteoak	1.900	6.810	4.910	1	2	3	3
92	07070003	Castle Rock	3.000	8.470	5.470	1	3	2	3
93	10240011	Independence-Sugar	3.000	8.750	5.750	1	1	3	2
94	10300101	Lower Missouri-Crooked	6.500	12.278	5.778	1	5	3	1
95	05120202	Lower White	7.000	13.293	6.293	1	3	3	1
96	10200203	Salt	9.900	20.000	10.100	2	1	3	1
97	07120003	Chicago	7.922	19.000	11.078	13	6	2	1

APPENDIX E

Before and After CWA Changes in Tenth Percentile Dissolved Oxygen at the RF1 Reach Level

TABLE

E-1 Before and after CWA changes in tenth percentile dissolved oxygen at the RF1 reach level (mg/L)

TABLE E-1 Before and After CWA Changes in Tenth Percentile Dissolved Oxygen at the RF1 Reach Level (mg/L)

Rank	RF1 Reach ID	RF1 Reach Name	Oxygen Before CWA	Oxygen After CWA	Oxygen After–Before	Number of Stations Before CWA	Number of Stations After CWA
1	10170203037	Big Sioux River	0.0000	7.2200	7.2200	1	1
2	04100002001	River Raisin	1.6000	8.3400	6.7400	2	2
3	04110002001	Cuyahoga River	0.2950	6.4967	6.2017	2	24
4	05030103007	Mahoning River	1.0900	7.1600	6.0700	1	1
5	07070002034	Wisconsin River	0.8800	6.8400	5.9600	1	1
6	05120201004	White River	0.6900	6.4240	5.7340	5	1
7	05080002008	Great Miami River	0.2000	5.8600	5.6600	1	1
8	07120004018	Du Page River, E. Branch	0.5750	5.9200	5.3450	4	3
9	07090001004	Rock River	2.7600	8.0500	5.2900	1	1
10	05020006031	Casselman River	2.9600	8.0000	5.0400	1	1
11	04040002005	Root River	0.9400	5.9400	5.0000	1	1
12	02040201011	Neshaminy River	2.6000	7.5600	4.9600	1	1
13	04030101012	Manitowoc River	5.9500	10.9000	4.9500	1	1
14	03170006007	Pascagoula River	0.0000	4.9200	4.9200	1	7
15	06010102004	Holston River, S. Fork	1.6000	6.4800	4.8800	1	2
16	08030203006	Enid Lake	0.0000	4.8673	4.8673	1	3
17	04040003001	Milwaukee River	2.1800	6.9567	4.7767	2	3
18	04030104002	Oconto River	0.5000	5.2000	4.7000	1	1
19	08030205018	Grenada Lake	0.0000	4.6160	4.6160	1	4
20	05050008006	Kanawha River	0.0000	4.5667	4.5667	2	3
21	04120102002	Cattaraugus Creek	3.3000	7.6000	4.3000	1	2
22	03050109053	Reedy River	1.9500	6.2270	4.2770	4	10
23	07120004002	Des Plains River	1.7620	6.0000	4.2380	2	1
24	05120201013	White River	2.2267	6.3750	4.1483	3	2
25	03050103037	Catawba River	1.6780	5.8000	4.1220	5	1

TABLE E-1 *Continued*

Rank	RF1 Reach ID	RF1 Reach Name	Oxygen Before CWA	Oxygen After CWA	Oxygen After–Before	Number of Stations Before CWA	Number of Stations After CWA
26	03170006025	Escatawpa River	0.0000	4.0983	4.0983	1	6
27	18090208001	Mojave River	4.0200	7.9767	3.9567	2	3
28	06010104007	Cherokee Lake	0.1570	4.1007	3.9437	1	3
29	04130003001	Genesee River	1.0425	4.7100	3.6675	4	2
30	18020103014	Sacramento River	5.9200	9.5500	3.6300	1	2
31	02040202045	Delaware River	1.2000	4.7000	3.5000	2	1
32	07010206001	Mississippi River	2.5246	5.9924	3.4678	13	26
33	07140201014	Kaskaskia River	2.8933	6.2000	3.3067	3	2
34	07120007006	Fox River	4.4000	7.6800	3.2800	1	4
35	02040205007	Brandywine Creek, E. Branch	4.6000	7.7800	3.1800	1	1
36	03050109068	Salada River	0.5200	3.6925	3.1725	1	4
37	02040202035	Delaware River	0.7000	3.8600	3.1600	1	2
38	03180004009	Pearl River	2.0000	5.0000	3.0000	2	1
39	05080001019	Great Miami River	3.5000	6.4800	2.9800	1	1
40	05090101004	Ohio River	3.2000	6.1600	2.9600	1	1
41	07120003006	Little Calumet River	0.0000	2.9214	2.9214	1	7
42	03100204003	Alafia River	2.7400	5.6000	2.8600	2	1
43	02040202043	Delaware River	1.6000	4.4500	2.8500	1	1
44	05120201011	White River	3.3800	6.2200	2.8400	3	1
45	04100009001	Maumee River	0.9835	3.8000	2.8165	4	2
46	03100204001	Alafia River	2.4550	5.2600	2.8050	2	1
47	02040202027	Delaware River	0.2000	3.0000	2.8000	1	1
48	08030204015	Coldwater River	0.0000	2.7990	2.7990	1	1
49	07120004010	Des Plains River	0.3000	3.0600	2.7600	3	4
50	02040201004	Delaware River	2.8800	5.6300	2.7500	3	1
51	18070203005	Santa Ana River	4.1400	6.8500	2.7100	2	2

TABLE E-1 *Continued*

Rank	RF1 Reach ID	RF1 Reach Name	Oxygen Before CWA	Oxygen After CWA	Oxygen After–Before	Number of Stations Before CWA	Number of Stations After CWA
52	05090202001	Little Miami River	4.0050	6.6800	2.6750	2	1
53	07130001005	Illinois River	3.1900	5.8500	2.6600	2	2
54	02040202085	Delaware River	0.4000	3.0300	2.6300	1	1
55	01080205033	Connecticut River	4.3000	6.8980	2.5980	1	1
56	04100009005	Maumee River	4.0690	6.6660	2.5970	3	5
57	07140201013	Kaskaskia River	2.7400	5.3367	2.5967	2	3
58	06010205001	Clinch River	1.6140	4.1965	2.5825	1	2
59	03040201049	Jeffries Creek	2.4400	5.0100	2.5700	2	2
60	07060005028	Mississippi River	3.3450	5.8900	2.5450	4	1
61	02040202030	Delaware River	0.5000	3.0000	2.5000	2	3
62	06010105021	French Broad River	4.1000	6.6000	2.5000	1	1
63	07120006001	Fox River	3.6600	6.1591	2.4991	1	11
64	06020001020	Tennessee River	3.2200	5.7067	2.4867	2	3
65	04110001004	Black River	1.3775	3.8400	2.4625	4	2
66	02040203002	Wassahickon Creek	3.8300	6.2750	2.4450	1	2
67	05090201001	Ohio River	4.4000	6.8400	2.4400	1	1
68	04040001010	*B	0.5700	3.0056	2.4356	2	5
69	03180004027	Bogue Lusa Creek	4.1333	6.5600	2.4267	3	1
70	07130006003	Sangamon River	0.0000	2.3875	2.3875	1	4
71	03040201045	Big Black Creek	2.1000	4.4500	2.3500	1	1
72	14010005007	Colorado River	4.8800	7.2200	2.3400	1	1
73	03010106018	Roanoke River	0.3633	2.5700	2.2067	3	1
74	03170001001	Okatibbee Creek	3.3000	5.5000	2.2000	2	1
75	06010207003	Clinch River	5.2600	7.4390	2.1790	2	2
76	09020301004	Red River	4.1000	6.2600	2.1600	2	1
77	07140106002	Big Muddy River	2.2900	4.3988	2.1088	1	8

TABLE E-1 *Continued*

Rank	RF1 Reach ID	RF1 Reach Name	Oxygen Before CWA	Oxygen After CWA	Oxygen After–Before	Number of Stations Before CWA	Number of Stations After CWA
78	07120005013	Illinois River	3.9320	6.0400	2.1080	1	2
79	05030103001	Mahoning River	1.9225	3.9200	1.9975	4	1
80	07040001001	Mississippi River	3.9390	5.9000	1.9610	2	2
81	11070207018	*A	1.6000	3.5000	1.9000	1	1
82	07120004017	Du Page River	4.0000	5.8620	1.8620	2	5
83	03050101086	Crowders Creek	1.9000	3.7600	1.8600	1	2
84	07040001012	Vermilion River	4.2205	6.0800	1.8595	2	1
85	05120202031	White River	3.4200	5.2700	1.8500	1	1
86	05140101002	Ohio River	3.4000	5.2500	1.8500	1	1
87	05030101014	Ohio River	4.5300	6.3760	1.8460	1	5
88	07090004004	Sugar River	5.9000	7.7000	1.8000	1	1
89	03050208037	Sanders Brook	1.9167	3.7033	1.7866	3	3
90	07120004019	Du Page River, W. Branch	2.9333	4.7033	1.7700	3	9
91	07120004007	Chicago Ship Canal	0.0250	1.7800	1.7550	1	2
92	08020401001	Arkansas River	5.6500	7.3900	1.7400	2	2
93	05080001001	Mad River	2.8200	4.5400	1.7200	1	1
94	02040202048	Delaware River	3.3100	5.0000	1.6900	2	2
95	05030203050	Little Kanawha River	5.1400	6.8250	1.6850	1	4
96	07130001025	Illinois River	4.2700	5.9000	1.6300	1	2
97	17110013003	Elliot Bay	4.3500	5.9600	1.6100	2	1
98	02040201002	Delaware River	2.8400	4.4200	1.5800	2	1
99	03050103013	Cane Creek	0.0000	1.5800	1.5800	1	1
100	06020001030	Tennessee River	3.9000	5.4717	1.5717	1	11
101	17080005007	Cowlitz River	8.1800	9.7300	1.5500	1	2
102	17110013004	Duwamish Waterway	5.6940	7.2400	1.5460	5	3
103	07120004012	Des Plains River	2.1500	3.6835	1.5335	1	2

TABLE E-1 *Continued*

Rank	RF1 Reach ID	RF1 Reach Name	Oxygen Before CWA	Oxygen After CWA	Oxygen After–Before	Number of Stations Before CWA	Number of Stations After CWA
104	07140201004	Lake Shelbyville	5.4000	6.9100	1.5100	1	2
105	17090003063	Willamette River	7.5200	8.9800	1.4600	3	1
106	04080204005	Flint River	4.2350	5.6800	1.4450	2	1
107	07090006003	Kishwaukee River	5.0000	6.4225	1.4225	1	2
108	06010108010	Nolichucky River	5.5800	7.0000	1.4200	1	1
109	05140202016	Ohio River	5.1400	6.5400	1.4000	1	1
110	07130001026	Illinois River	1.2000	2.6000	1.4000	1	1
111	02070007003	Shenandoah River	5.2600	6.6600	1.4000	1	2
112	04140203001	Oswego River	4.3000	5.7000	1.4000	1	1
113	06010102018	S. Holston Lake	3.7380	5.1150	1.3770	1	2
114	17080005002	Coweman River	5.9900	7.3400	1.3500	2	1
115	04030101020	Sheboygan River	4.5000	5.8200	1.3200	1	1
116	10190006002	Thompson River	5.6500	6.9620	1.3120	3	5
117	07030005018	St. Croix River	5.7580	7.0500	1.2920	2	1
118	11110207005	Arkansas River	6.2000	7.4900	1.2900	1	2
119	07120005001	Illinois River	5.2170	6.4800	1.2630	2	3
120	04030101008	E. Twin River	4.5800	5.8300	1.2500	1	1
121	03040202022	Little Lynches River	4.6417	5.8850	1.2433	3	2
122	05020003026	Monongahela River	5.8100	7.0480	1.2380	1	3
123	03040204015	Little Pee Dee River	4.6400	5.8600	1.2200	2	1
124	18040005002	Mokelumne River	7.4800	8.7000	1.2200	1	1
125	18080003022	Susan River	6.8200	8.0400	1.2200	1	1
126	02050306013	Susquehanna River	0.8800	2.0700	1.1900	1	1
127	03040201038	Black Creek	4.5350	5.7100	1.1750	2	1
128	03040201005	Pee Dee River	4.6133	5.7800	1.1667	3	1
129	03010106001	Roanoke River	4.6000	5.7600	1.1600	1	1

TABLE E-1 *Continued*

Rank	RF1 Reach ID	RF1 Reach Name	Oxygen Before CWA	Oxygen After CWA	Oxygen After–Before	Number of Stations Before CWA	Number of Stations After CWA
130	05120114001	Little Wabash River	2.3400	3.5000	1.1600	1	2
131	05010009001	Allegheny River	6.0600	7.1850	1.1250	1	14
132	05020003016	Monongahela River	6.3000	7.4120	1.1120	1	1
133	11140208006	Saline Bayou	4.1600	5.2400	1.0800	1	1
134	03180005003	Bogue Chito River	6.0000	7.0000	1.0000	2	1
135	05020002007	West Fork River	4.8000	5.8000	1.0000	1	1
136	05050002030	New River	5.7000	6.7000	1.0000	1	2
137	07070003013	Castle Rock Flowage	5.0000	6.0000	1.0000	1	1
138	08080203011	Calcasieu River	4.3200	5.3200	1.0000	2	1
139	10190007003	Cache La Poudre River	6.2578	7.2500	0.9922	9	1
140	07020012013	Minnesota River	5.6690	6.6600	0.9910	1	1
141	03140305004	Escambia River	5.1200	6.1000	0.9800	2	1
142	07130003018	Illinois River	4.3920	5.3650	0.9730	2	2
143	03100101010	Peace River	4.1300	5.0800	0.9500	1	1
144	17110011002	Snohomish River	8.3600	9.3000	0.9400	2	1
145	05120103010	Mississinewa River	5.8400	6.7800	0.9400	1	1
146	05130108022	Cumberland River, Caney	4.2000	5.1100	0.9100	1	1
147	03170008001	Escatawpa River	1.5800	2.4900	0.9100	3	8
148	10200203040	Salt Creek	4.4000	5.3007	0.9007	1	9
149	03040202014	Lynches River	5.5825	6.4600	0.8775	2	1
150	05020002001	West Fork River	6.7200	7.5820	0.8620	1	1
151	17060110001	Snake River	7.1700	8.0300	0.8600	1	1
152	17010305001	Spokane River	7.2000	8.0587	0.8587	2	3
153	03040202012	Lynches River	5.1833	6.0100	0.8267	3	2
154	07120004009	Chicago Ship Canal	0.4460	1.2650	0.8190	2	2
155	03040201008	Pee Dee River	5.7250	6.5400	0.8150	4	1

TABLE E-1 *Continued*

Rank	RF1 Reach ID	RF1 Reach Name	Oxygen Before CWA	Oxygen After CWA	Oxygen After–Before	Number of Stations Before CWA	Number of Stations After CWA
156	18070203008	Santa Ana River	3.6267	4.4333	0.8066	3	3
157	07120006007	Fox River	5.5300	6.3343	0.8043	1	3
158	05120113006	Wabash River	4.9900	5.7800	0.7900	1	1
159	10190006001	Thompson River	6.2400	7.0000	0.7600	1	1
160	07130002001	Vermilion River	3.7750	4.4750	0.7000	2	2
161	07020012001	Minnesota River	4.7316	5.4257	0.6941	9	7
162	17110019081	*W	8.0429	8.7200	0.6771	7	1
163	04110003008	Ashtabula River	5.1517	5.7700	0.6183	3	2
164	03050202006	Ashley River	3.7200	4.3300	0.6100	1	1
165	08040304014	Little River	3.2400	3.8500	0.6100	1	1
166	08030206007	Black Creek	5.5200	6.1100	0.5900	1	4
167	03040204018	Little Pee Dee River	4.7300	5.3000	0.5700	1	1
168	17090003058	Willamette River	7.2700	7.8400	0.5700	1	1
169	05030202005	Ohio River	5.7000	6.2600	0.5600	1	1
170	17020011001	Wenatchee River	9.3000	9.8600	0.5600	1	1
171	04080201001	Tittabawassee River	6.0350	6.5800	0.5450	2	1
172	05120201032	Eagle Creek	4.7600	5.2967	0.5367	1	3
173	08090203007	Intracoastal Waterway	4.5500	5.0750	0.5250	1	2
174	17110007006	Skagit River	9.4200	9.9350	0.5150	2	2
175	04110003005	Chagrin River	5.3350	5.8400	0.5050	2	1
176	08040304013	Little River	3.5200	4.0250	0.5050	1	1
177	17020015001	Crab Creek	6.6800	7.1800	0.5000	3	1
178	03040204016	Little Pee Dee River	3.4320	3.9187	0.4867	5	4
179	05120101004	Wabash River	6.3400	6.8050	0.4650	1	2
180	04110004001	Grand River	5.6650	6.1200	0.4550	3	3
181	03160203006	Tombigbee River	5.6040	6.0480	0.4440	5	5

TABLE E-1 *Continued*

Rank	RF1 Reach ID	RF1 Reach Name	Oxygen Before CWA	Oxygen After CWA	Oxygen After–Before	Number of Stations Before CWA	Number of Stations After CWA
182	07010207005	Rum River	5.8000	6.2400	0.4400	1	1
183	08040302001	Castor Creek	3.4400	3.8700	0.4300	1	2
184	03050103018	Catawba River	4.6725	5.1000	0.4275	2	1
185	07010206002	Mississippi River	4.4700	4.8606	0.3906	5	36
186	05020003003	Monongahela River	5.5400	5.9000	0.3600	2	1
187	03050205015	Polk Swamp	3.3000	3.6500	0.3500	1	1
188	04010302018	Montreal River	7.8000	8.1500	0.3500	1	1
189	03040201039	Black Creek	3.5450	3.8888	0.3438	3	8
190	08070205005	Tangipahoa River	6.2000	6.5400	0.3400	2	1
191	11110207011	Arkansas River	6.5200	6.8600	0.3400	1	2
192	04030108001	Menominee River	6.4400	6.7500	0.3100	1	2
193	05050005005	Gauley River	7.0900	7.4000	0.3100	1	1
194	07040007002	Black River	7.4500	7.7400	0.2900	1	1
195	06030002052	Tennessee River	5.4790	5.7500	0.2710	1	1
196	18030012014	Kings River	7.2400	7.5000	0.2600	1	2
197	06030005051	Wilson Lake	4.8000	5.0500	0.2500	1	2
198	03160203007	Tombigbee River	6.2400	6.4850	0.2450	1	2
199	06010103019	Watauga River	6.0950	6.3333	0.2383	2	3
200	03050103021	Twelvemile Creek	3.7600	3.9900	0.2300	1	2
201	03050205021	Edisto River	5.4000	5.6200	0.2200	1	1
202	04080206001	Saginaw River	4.8989	5.1167	0.2178	19	3
203	03050202011	Ashley River	3.6200	3.8350	0.2150	1	2
204	10190007004	Cache La Poudre River	6.7900	6.9960	0.2060	3	5
205	06020001032	Tennessee River	4.4000	4.6000	0.2000	1	1
206	06020001001	Tennessee River	4.6140	4.8067	0.1927	3	3
207	05120106002	Tippecanoe River	6.1350	6.3000	0.1650	2	1

TABLE E-1 *Continued*

Rank	RF1 Reach ID	RF1 Reach Name	Oxygen Before CWA	Oxygen After CWA	Oxygen After–Before	Number of Stations Before CWA	Number of Stations After CWA
208	07040001008	Mississippi River	5.5850	5.7467	0.1617	4	3
209	02070001001	Potomac River, S. Branch	6.3000	6.4500	0.1500	1	1
210	17080005011	Toutle River	8.8000	8.9200	0.1200	1	1
211	05020006024	Casselman River	8.6100	8.7200	0.1100	2	1
212	03040202015	Little Fork Creek	6.0933	6.1567	0.0634	3	3
213	17090003009	Willamette River	8.3600	8.4000	0.0400	1	1
214	09030004013	Rainy River	7.2000	7.2333	0.0333	1	3
215	05020004001	Cheat River	6.5033	6.5300	0.0267	3	2
216	05050001044	New River	6.6000	6.5700	−0.0300	1	1
217	05120108018	Wabash River	6.5000	6.4700	−0.0300	1	1
218	10180009016	N. Platte River	7.0000	6.9500	−0.0500	1	1
219	08030201005	Sardis Lake	3.4200	3.3680	−0.0520	1	1
220	02070010033	Potomac River	4.3400	4.2520	−0.0880	1	5
221	05020005030	Dunkard Creek	7.7100	7.6000	−0.1100	1	1
222	03160203015	Tombigbee River	6.8200	6.6950	−0.1250	2	2
223	06020001033	Tennessee River	4.4400	4.3000	−0.1400	1	1
224	03170008002	Escatawpa River	5.3700	5.2250	−0.1450	1	2
225	05070204034	Big Sandy River	5.7200	5.5500	−0.1700	1	2
226	05010006005	Allegheny River	7.2000	7.0290	−0.1710	1	10
227	07030005003	St. Croix River	5.4125	5.2200	−0.1925	4	1
228	08050001011	Boeuf River	4.1600	3.9560	−0.2040	1	2
229	03050207015	Lemon Creek	3.8050	3.5950	−0.2100	2	2
230	04010302002	White River	7.1200	6.8800	−0.2400	1	1
231	07050005019	Chippewa River	5.3400	5.1000	−0.2400	1	1
232	08040301020	Red River	5.4500	5.2000	−0.2500	1	2
233	05120201007	White River	5.5271	5.2750	−0.2521	7	2

TABLE E-1 *Continued*

Rank	RF1 Reach ID	RF1 Reach Name	Oxygen Before CWA	Oxygen After CWA	Oxygen After–Before	Number of Stations Before CWA	Number of Stations After CWA
234	10190018001	S. Platte River	7.0600	6.8000	−0.2600	1	1
235	05050006007	Kanawha River	7.0500	6.7800	−0.2700	2	1
236	03050105019	Broad River	6.7000	6.4200	−0.2800	1	1
237	06040006017	Tennessee River	5.4000	5.1133	−0.2867	1	3
238	17110008007	Stillaguamish River	8.6750	8.3600	−0.3150	2	1
239	06020001019	Tennessee River	5.0460	4.7300	−0.3160	1	2
240	02080206006	James River	6.1400	5.7930	−0.3470	1	1
241	05050007001	Elk River	6.4000	6.0500	−0.3500	1	1
242	08080203008	Calcasieu River	6.0000	5.6000	−0.4000	2	1
243	03050202010	Ashley River	3.3200	2.9000	−0.4200	1	1
244	05120105009	Wabash River	6.8600	6.4350	−0.4250	1	1
245	07020007002	Minnesota River	6.9480	6.5200	−0.4280	1	1
246	07080208002	Lake McBride	3.2840	2.8525	−0.4315	5	4
247	05120101006	Wabash River	7.0000	6.5600	−0.4400	1	1
248	08090201013	Tchefuncta River	6.4000	5.9600	−0.4400	2	1
249	05030204009	Hocking River	4.8600	4.3850	−0.4750	1	2
250	04090005001	Huron River	5.0000	4.5000	−0.5000	2	1
251	05070102002	Guyandotte River	6.6200	6.1000	−0.5200	1	1
252	05120201009	White River	4.6700	4.1400	−0.5300	1	1
253	05120201006	Fall Creek	5.9600	5.4250	−0.5350	1	2
254	03050205016	Edisto River	4.9000	4.3500	−0.5500	1	1
255	03050201018	Cooper River, W. Branch	6.0450	5.4650	−0.5800	1	2
256	03060101024	Golden Creek	7.9200	7.3400	−0.5800	1	2
257	03050109075	Georges Creek	7.5500	6.9250	−0.6250	1	2
258	06010102009	Boone Lake	2.5300	1.9000	−0.6300	1	1
259	08040202006	Bayou De Loutre	1.5000	0.8600	−0.6400	2	1

TABLE E-1 *Continued*

Rank	RF1 Reach ID	RF1 Reach Name	Oxygen Before CWA	Oxygen After CWA	Oxygen After–Before	Number of Stations Before CWA	Number of Stations After CWA
260	03040202024	Hanging Rock Creek	4.9800	4.3067	−0.6733	2	3
261	05020001027	Tygart Valley River	7.2500	6.5500	−0.7000	1	1
262	05120104001	Eel River	7.0000	6.1300	−0.8700	1	1
263	08050002003	Bayou Macon	3.7000	2.8300	−0.8700	2	2
264	06030002001	Tennessee River	5.4940	4.6000	−0.8940	1	1
265	03040201003	Catfish Creek	2.4900	1.5575	−0.9325	2	6
266	04030105002	Peshtigo River	6.2400	5.3000	−0.9400	1	1
267	03050103054	Rocky Creek	2.9950	2.0087	−0.9863	2	4
268	08050001024	Bayou Lafourche	3.6800	2.6670	−1.0130	1	2
269	12010005005	Sabine River	6.5800	5.5400	−1.0400	1	1
270	05020001001	Tygart Valley River	8.1000	7.0513	−1.0487	1	3
271	03050101007	Catawba River	4.8000	3.7500	−1.0500	1	2
272	11140202005	Red River	7.1200	6.0500	−1.0700	2	2
273	05020006001	Youghiogheny River	8.1200	7.0400	−1.0800	1	1
274	05070201010	Big Sandy River, Tug Fork	6.2500	5.1000	−1.1500	1	1
275	05020005001	Monongahela River	7.6000	6.4060	−1.1940	1	9
276	03160106031	Tombigbee River	5.5200	4.3000	−1.2200	1	1
277	05120101003	Wabash River	6.0100	4.7650	−1.2450	1	1
278	03050205009	Edisto River	4.6000	3.3350	−1.2650	1	2
279	08080203001	Calcasieu River	3.4500	2.1700	−1.2800	2	2
280	08050001026	Bayou Boeuf	5.2000	3.9000	−1.3000	1	1
281	10240006001	Little Nemaha River	6.8200	5.5200	−1.3000	1	1
282	16050102006	Truckee River	7.7050	6.3852	−1.3198	2	5
283	05020004003	Cheat River	7.0800	5.7500	−1.3300	2	1
284	05030201007	Ohio River	7.1000	5.7500	−1.3500	1	1
285	07090005004	Rock River	7.0600	5.7000	−1.3600	1	2

TABLE E-1 *Continued*

Rank	RF1 Reach ID	RF1 Reach Name	Oxygen Before CWA	Oxygen After CWA	Oxygen After–Before	Number of Stations Before CWA	Number of Stations After CWA
286	16050101004	Lake Tahoe	7.1900	5.8175	−1.3725	1	4
287	05020003001	Monongahela River	7.6000	6.2260	−1.3740	1	5
288	03050208004	Ashepoo River	3.4000	1.9600	−1.4400	1	1
289	08050003005	Tensas River	3.1000	1.6400	−1.4600	1	1
290	05120108005	Wabash River	7.0000	5.4640	−1.5360	1	1
291	05020005006	Monongahela River	8.5000	6.8383	−1.6617	1	6
292	12010004010	Toledo Bend Reservoir	4.9600	3.2000	−1.7600	1	1
293	05010006001	Allegheny River	8.8000	7.0150	−1.7850	1	4
294	07120003002	Chicago Sanitary Ship Canal	6.4122	4.6180	−1.7942	5	10
295	05030203014	Little Kanawha River	5.3900	3.5800	−1.8100	1	2
296	08080202035	Mermentau River	2.2600	0.3200	−1.9400	2	1
297	08040202002	Ouachita River	3.7700	1.7900	−1.9800	2	2
298	03040202004	Lynches River	6.0400	4.0367	−2.0033	1	3
299	03040205003	Black River	4.7000	2.6890	−2.0110	1	5
300	10270207005	Little Blue River	7.3300	5.3000	−2.0300	1	1
301	03050101005	Lake Wylie	7.3000	5.1025	−2.1975	1	4
302	07090005006	Rock River	5.5000	3.1250	−2.3750	1	4
303	07140204001	Kaskaskia River	4.1300	1.7150	−2.4150	1	2
304	05120112010	Embarras River, N. Fork	5.0800	2.6600	−2.4200	1	2
305	08030201003	Little Tallahatchie River	6.2000	3.6400	−2.5600	1	1
306	03060106055	Savannah River	8.2100	5.6150	−2.5950	1	3
307	05120111001	Wabash River	5.1000	2.5000	−2.6000	1	1
308	17110013005	Green River	8.4200	5.6300	−2.7900	1	2
309	11140203012	Bayou Dorcheat	5.3000	2.4300	−2.8700	2	1
310	03040201050	Jeffries Creek	4.6200	0.4300	−4.1900	1	2
311	11140304013	*B	6.6400	0.6300	−6.0100	2	2

APPENDIX F

Hydrologic Conditions of the 48 Contiguous States, Summer (July–September) from 1961 through 1995

Annual streamflow ratio is estimated for each catalog unit as a percentage of long-term mean summer streamflow computed from data extracted from ~ 5,000 USGS gages for the months of July through September for the years from 1951 to 1980. Hydrologic conditions for each summer from 1961 through 1995 are characterized with respect to the long-term summer mean from 1951–1980 as follows:

Dry summer	0–75 percent of long-term summer mean
Normal summer	75–150 percent of long-term summer mean
Wet summer	>150 percent of long-term summer mean

Normalized summer runoff is estimated as the ratio of summer streamflow to the watershed drainage area (as cfs/square mile) based on the long-term (1951–1980) average annual summer streamflow and drainage area for gage stations in a catalog unit.

FIGURES

F-1 Ratio of streamflow to watershed area for July–September, 1951–1980 (as cfs/square mile).

F-2 Persistence of dry hydrologic conditions during the summer for July–September for the period 1961–1965.

F-3 Persistence of dry hydrologic conditions during the summer for July – September for the period 1966–1970.

F-4 Persistence of dry hydrologic conditions during the summer for July–September for the period 1971–1975.

F-5 Persistence of dry hydrologic conditions during the summer for July – September for the period 1976–1980.

F-6 Persistence of dry hydrologic conditions during the summer for July–September for the period 1981–1985.

F-7 Persistence of dry hydrologic conditions during the summer for July–September for the period 1986–1990.

F-8 Persistence of dry hydrologic conditions during the summer for July–September for the period 1991–1995.
F-9 Hydrologic conditions during the summer (July–September) 1961.
F-10 Hydrologic conditions during the summer (July–September) 1962.
F-11 Hydrologic conditions during the summer (July–September) 1963.
F-12 Hydrologic conditions during the summer (July–September) 1964.
F-13 Hydrologic conditions during the summer (July–September) 1965.
F-14 Hydrologic conditions during the summer (July–September) 1966.
F-15 Hydrologic conditions during the summer (July–September) 1967.
F-16 Hydrologic conditions during the summer (July–September) 1968.
F-17 Hydrologic conditions during the summer (July–September) 1969.
F-18 Hydrologic conditions during the summer (July–September) 1970.
F-19 Hydrologic conditions during the summer (July–September) 1971.
F-20 Hydrologic conditions during the summer (July–September) 1972.
F-21 Hydrologic conditions during the summer (July–September) 1973.
F-22 Hydrologic conditions during the summer (July–September) 1974.
F-23 Hydrologic conditions during the summer (July–September) 1975.
F-24 Hydrologic conditions during the summer (July–September) 1976.
F-25 Hydrologic conditions during the summer (July–September) 1977.
F-26 Hydrologic conditions during the summer (July–September) 1978.
F-27 Hydrologic conditions during the summer (July–September) 1979.
F-28 Hydrologic conditions during the summer (July–September) 1980.
F-29 Hydrologic conditions during the summer (July–September) 1981.
F-30 Hydrologic conditions during the summer (July–September) 1982.
F-31 Hydrologic conditions during the summer (July–September) 1983.
F-32 Hydrologic conditions during the summer (July–September) 1984.
F-33 Hydrologic conditions during the summer (July–September) 1985.
F-34 Hydrologic conditions during the summer (July–September) 1986.
F-35 Hydrologic conditions during the summer (July–September) 1987.
F-36 Hydrologic conditions during the summer (July–September) 1988.
F-37 Hydrologic conditions during the summer (July–September) 1989.
F-38 Hydrologic conditions during the summer (July–September) 1990.
F-39 Hydrologic conditions during the summer (July–September) 1991.
F-40 Hydrologic conditions during the summer (July–September) 1992.
F-41 Hydrologic conditions during the summer (July–September) 1993.
F-42 Hydrologic conditions during the summer (July–September) 1994.
F-43 Hydrologic conditions during the summer (July–September) 1995.

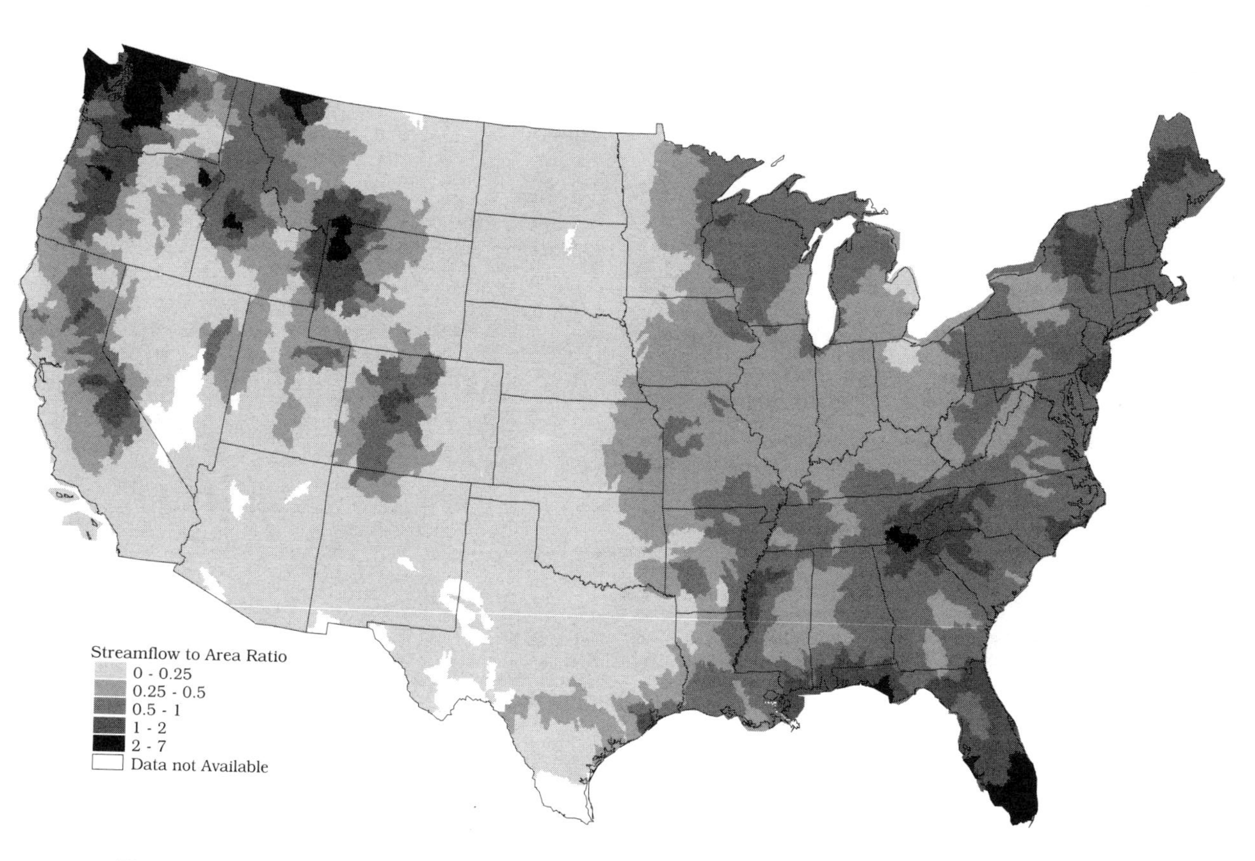

Figure F-1 Ratio of streamflow to watershed area for July–September, 1951–1980 (as cfs/square mile).

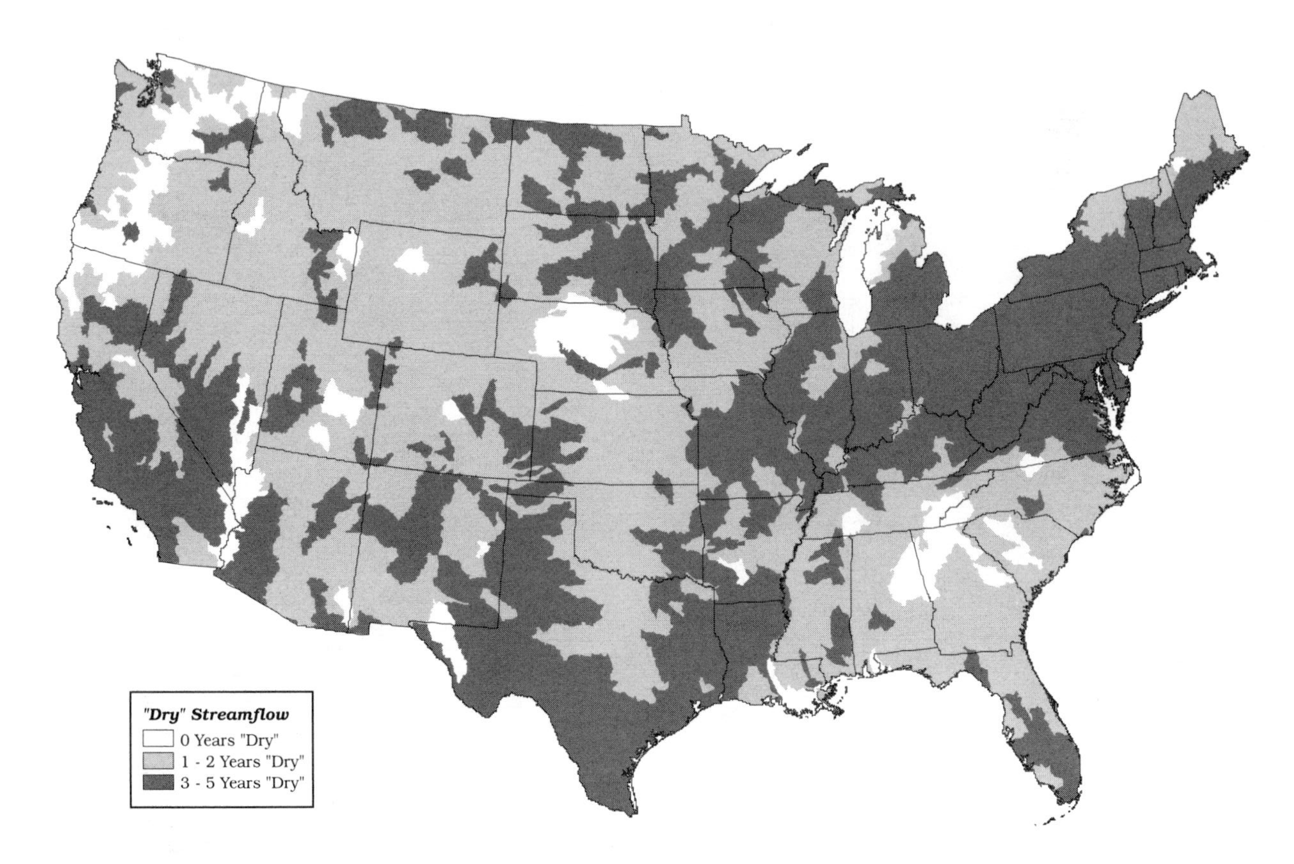

Figure F-2 Persistence of dry hydrologic conditions during the summer for July–September for the period 1961–1965.

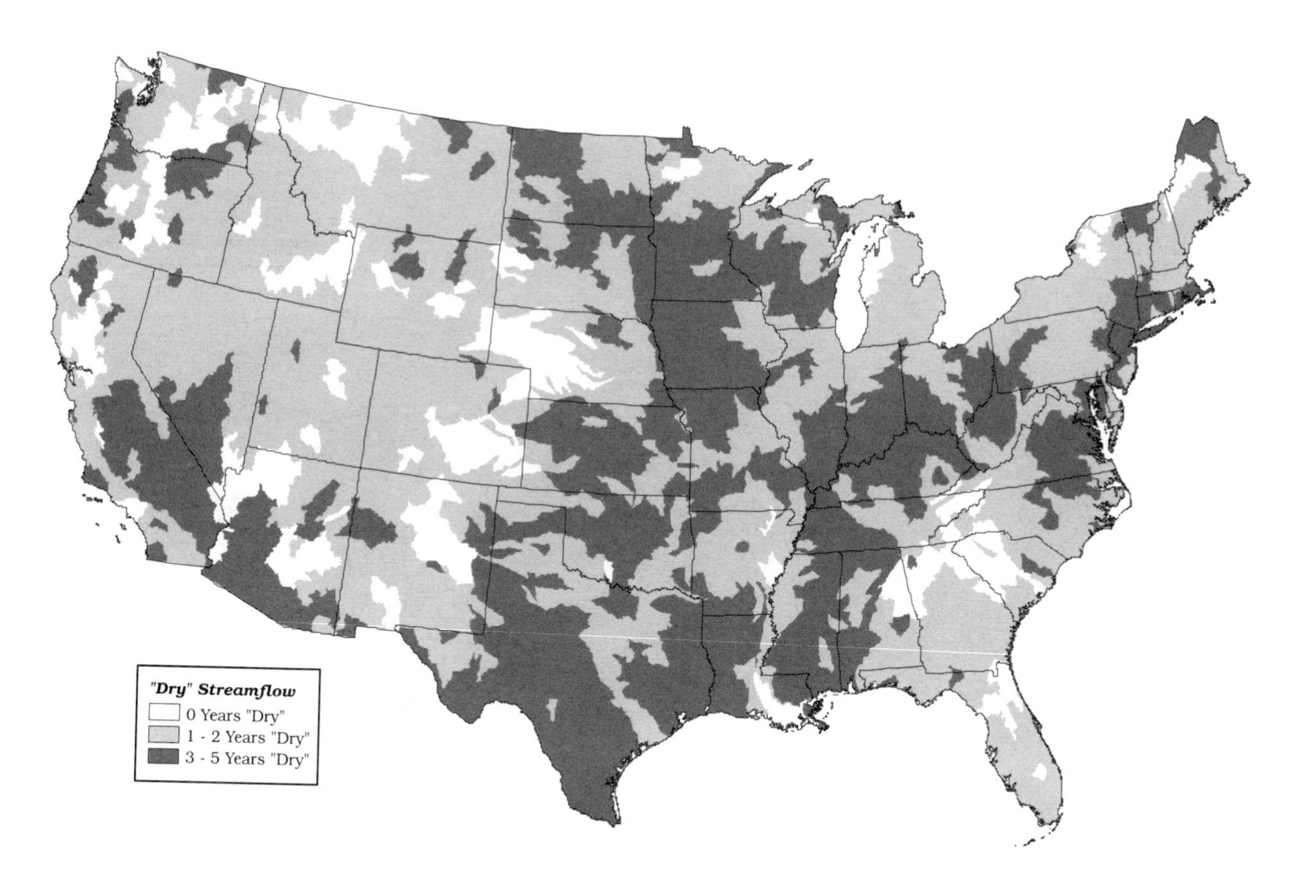

Figure F-3 Persistence of dry hydrologic conditions during the summer for July–September for the period 1966–1970.

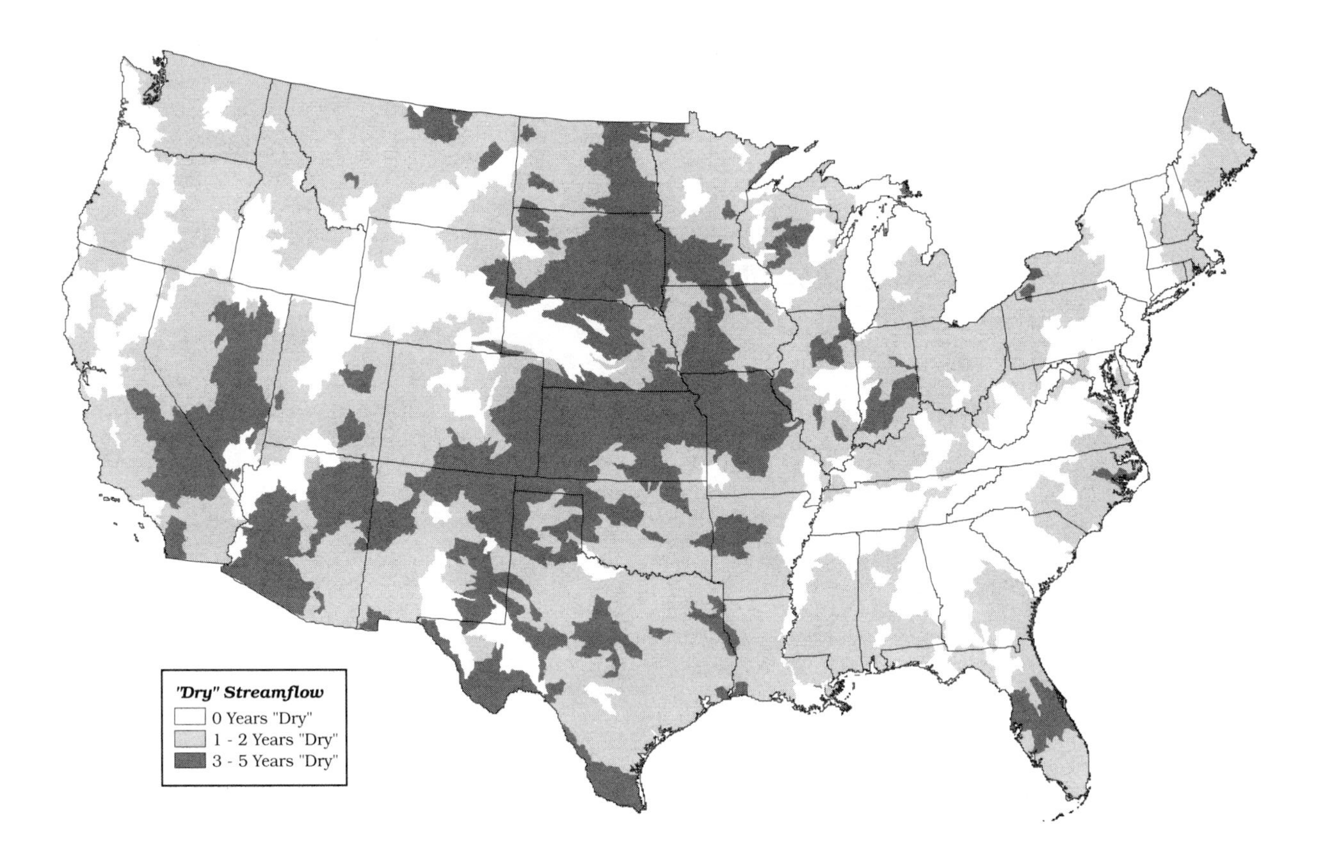

Figure F-4 Persistence of dry hydrologic conditions during the summer for July–September for the period 1971–1975.

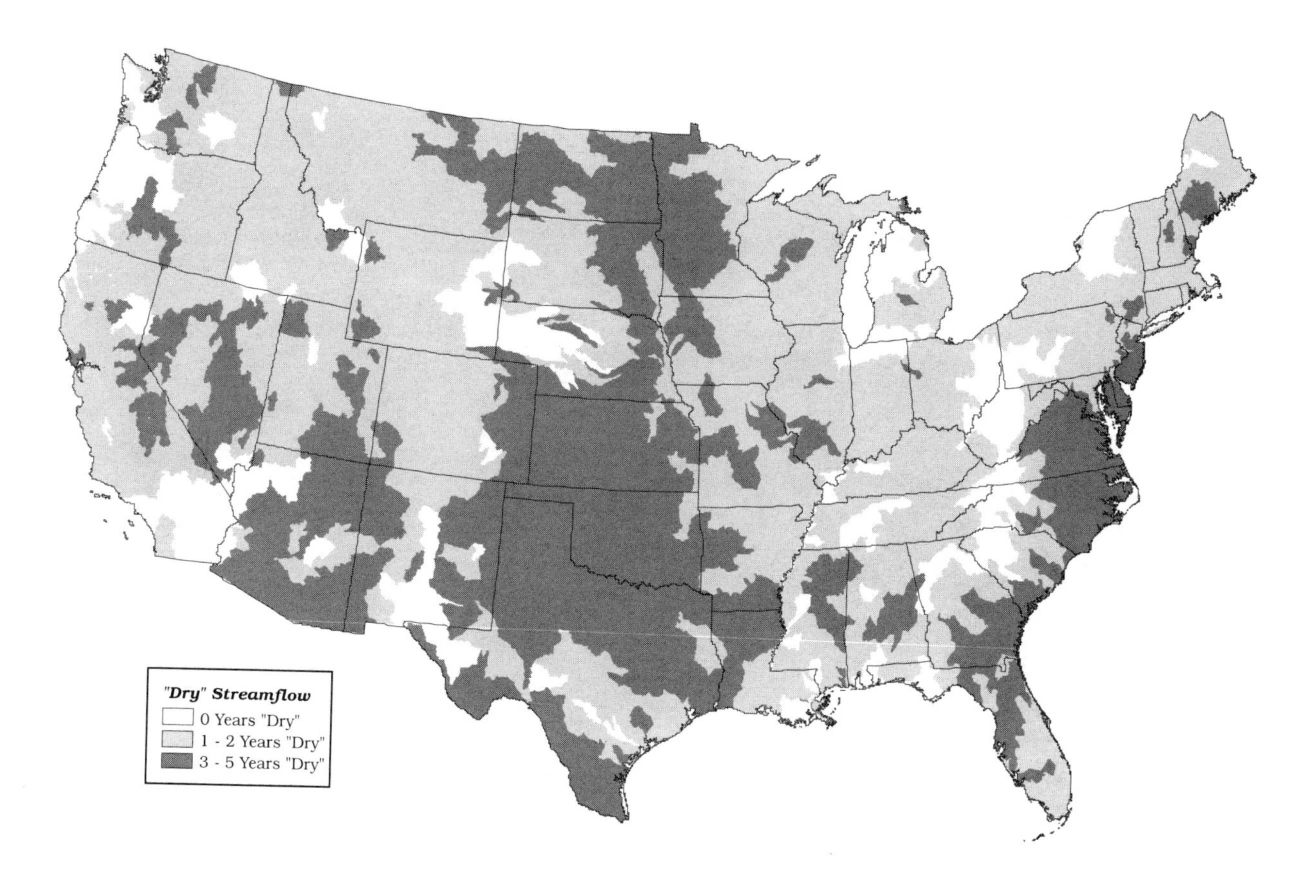

Figure F-5 Persistence of dry hydrologic conditions during the summer for July–September for the period 1976–1980.

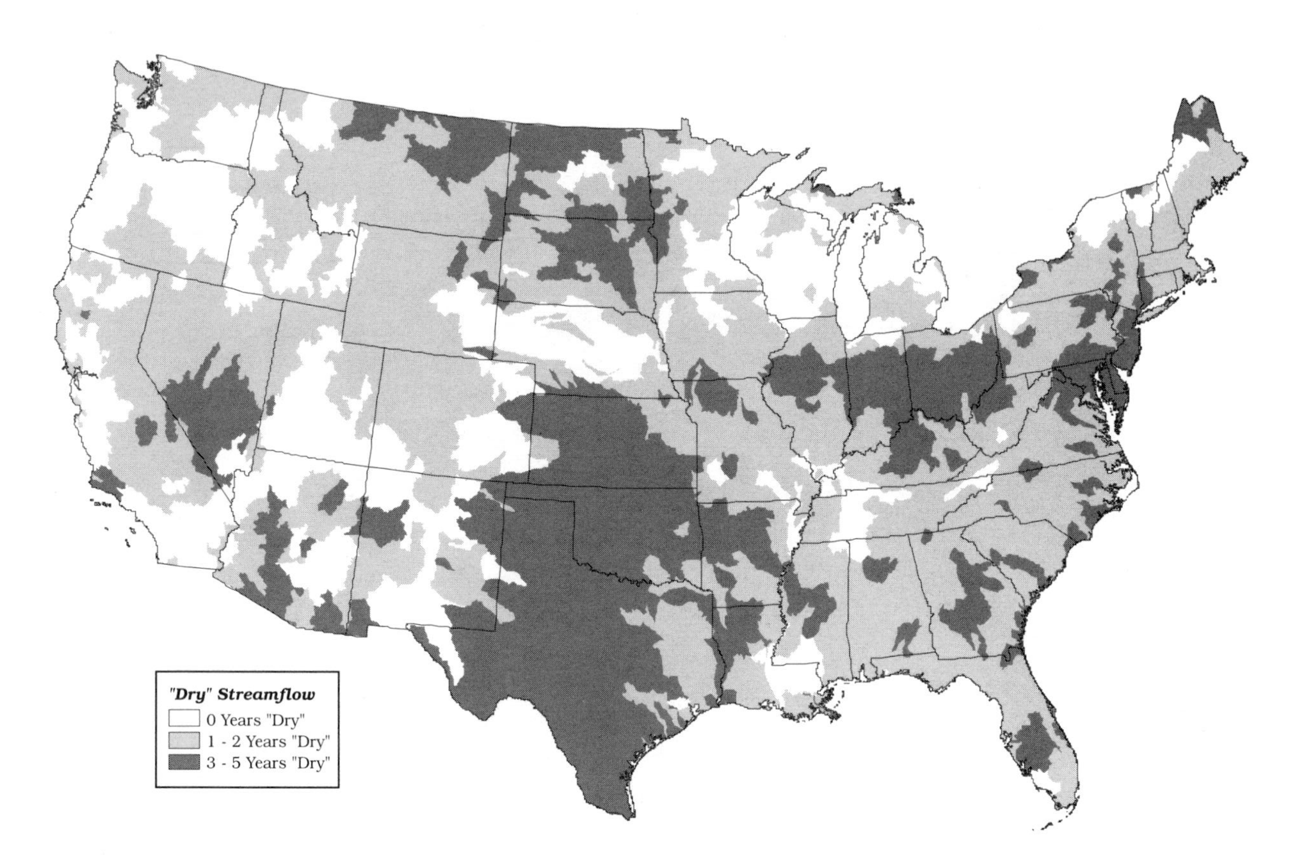

Figure F-6 Persistence of dry hydrologic conditions during the summer for July–September for the period 1981–1985.

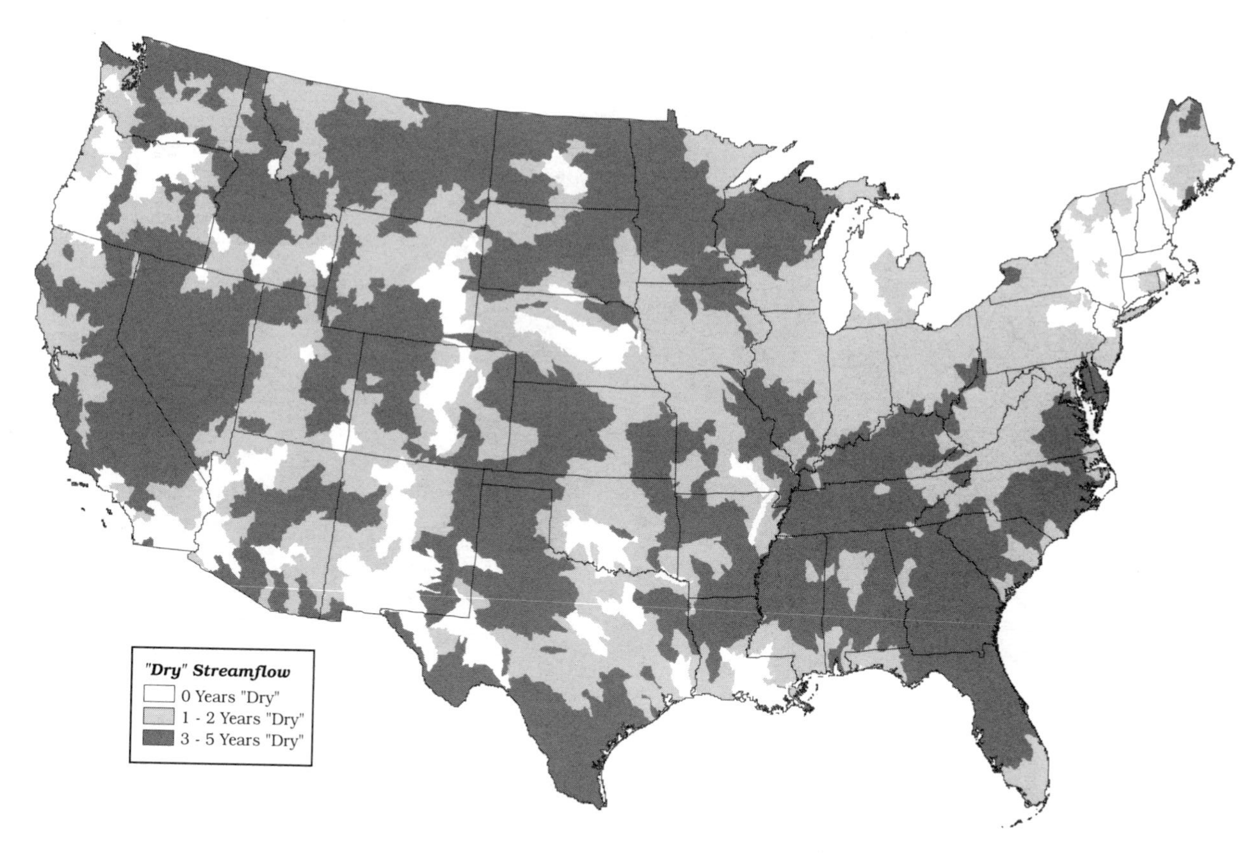

Figure F-7 Persistence of dry hydrologic conditions during the summer for July–September for the period 1986–1990.

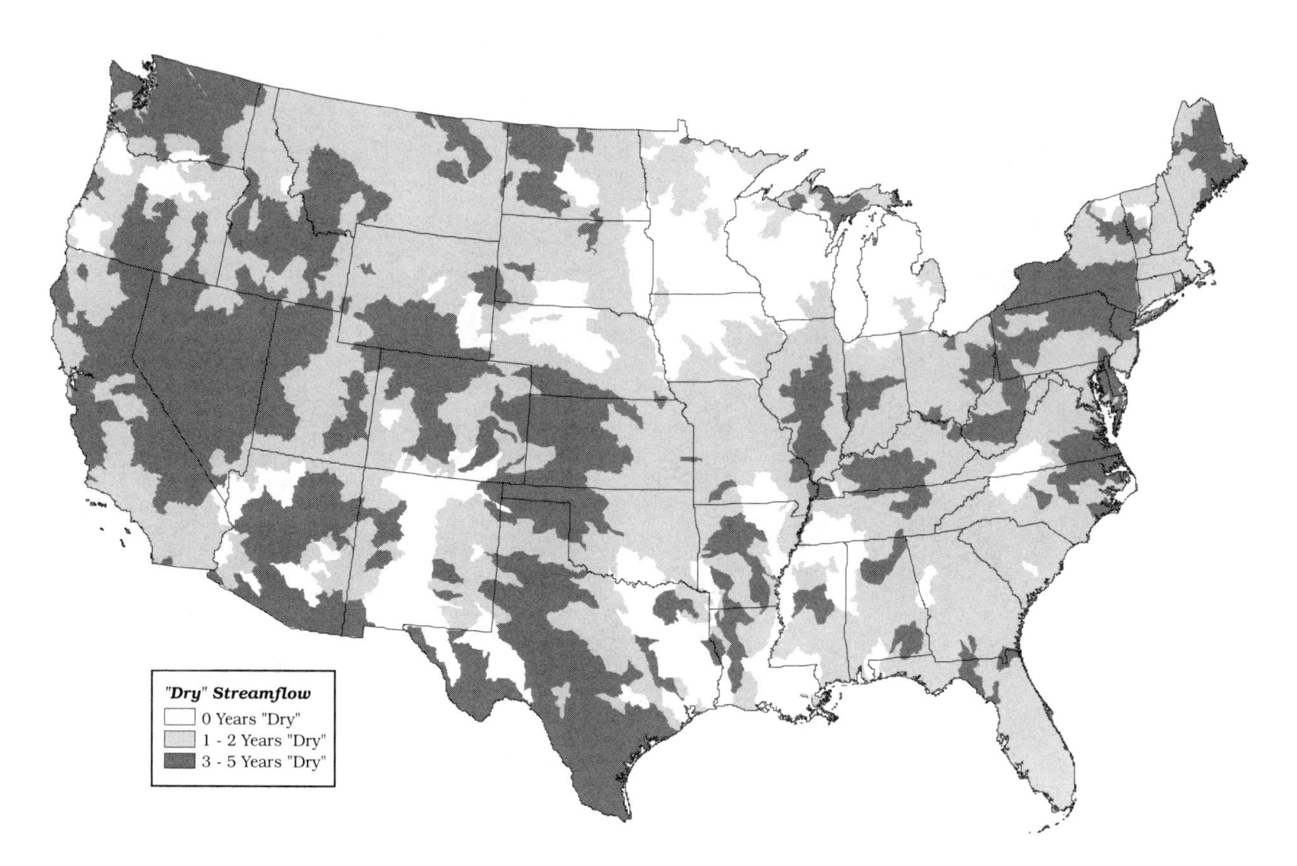

Figure F-8 Persistence of dry hydrologic conditions during the summer for July–September for the period 1991–1995.

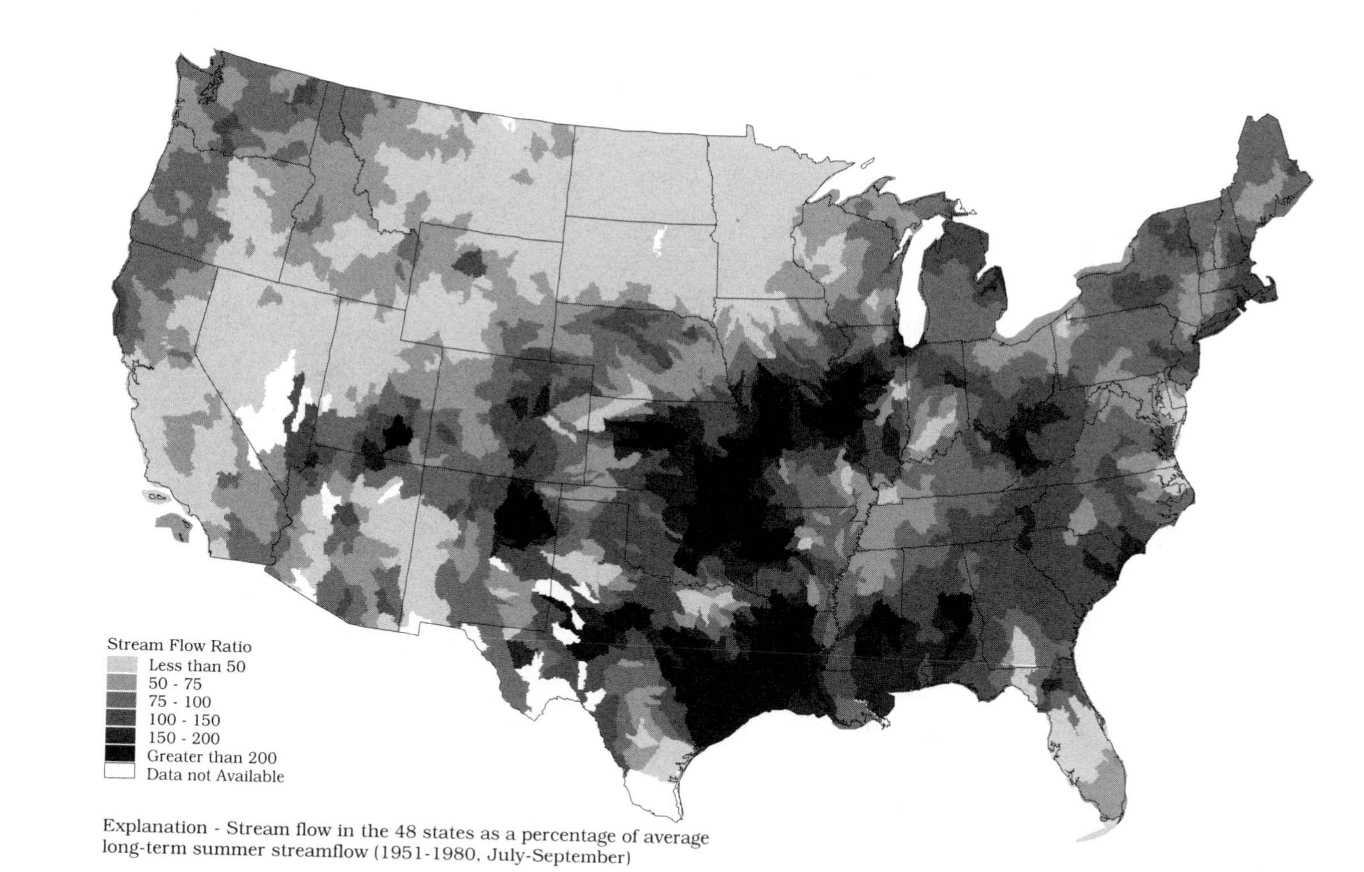

Figure F-9 Hydrologic conditions during the summer (July–September) 1961.

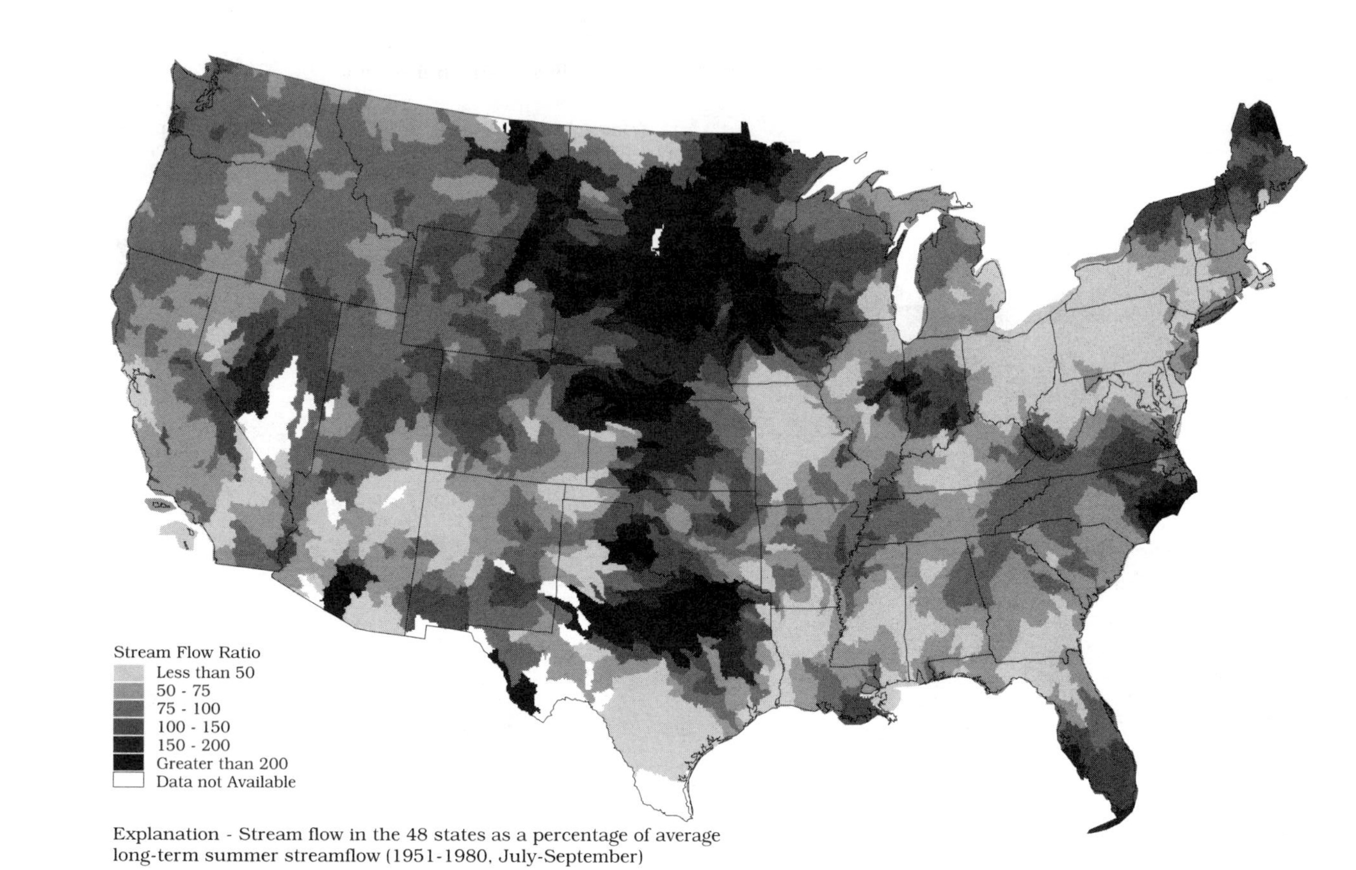

Figure F-10 Hydrologic conditions during the summer (July–September) 1962.

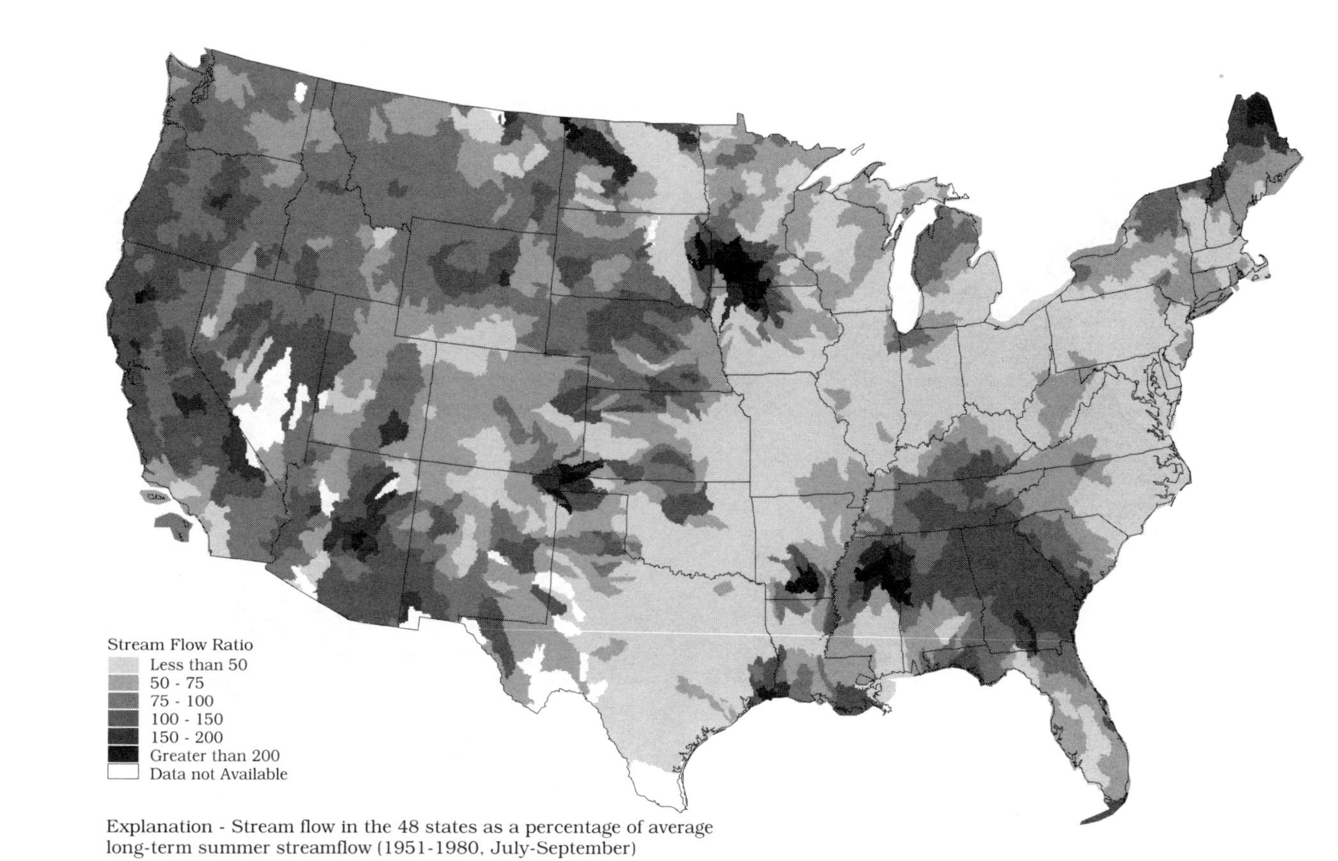

Figure F-11 Hydrologic conditions during the summer (July–September) 1963.

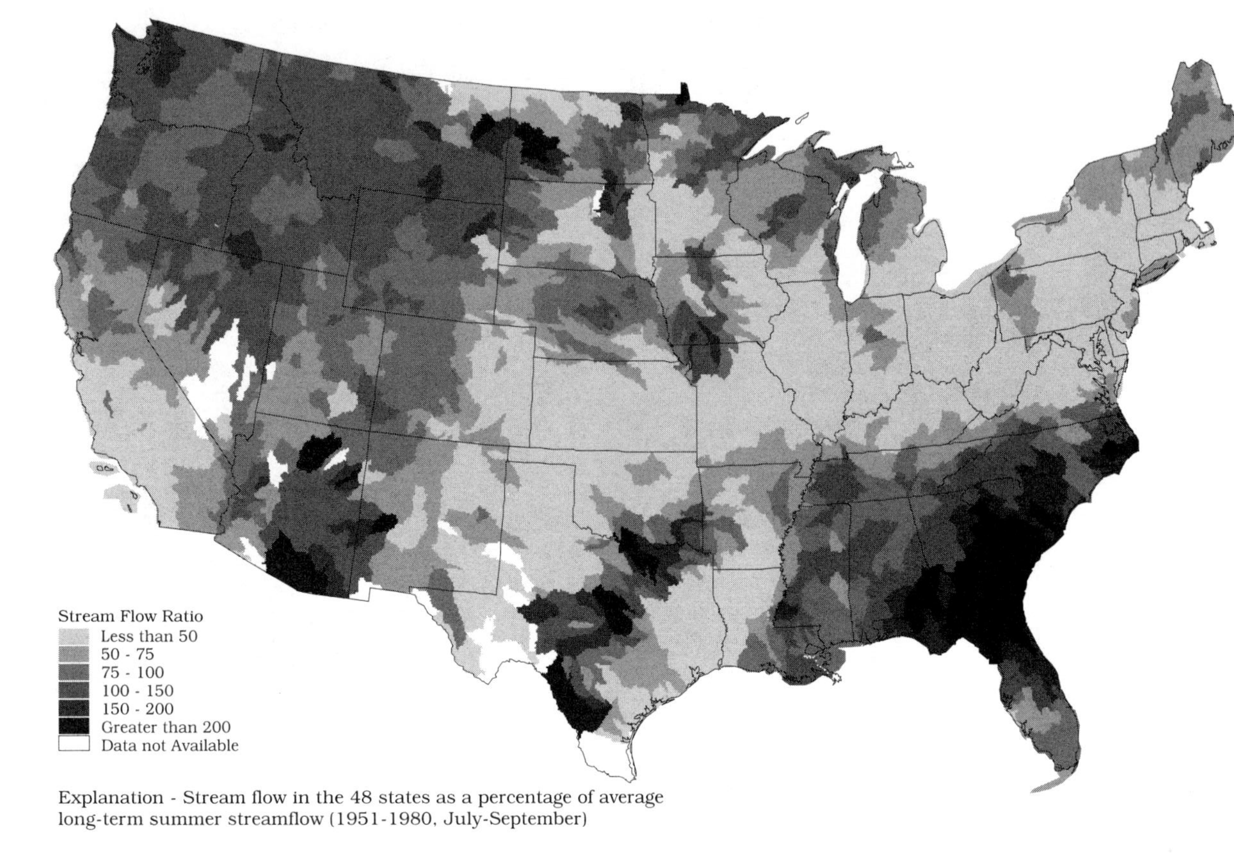

Figure F-12 Hydrologic conditions during the summer (July–September) 1964.

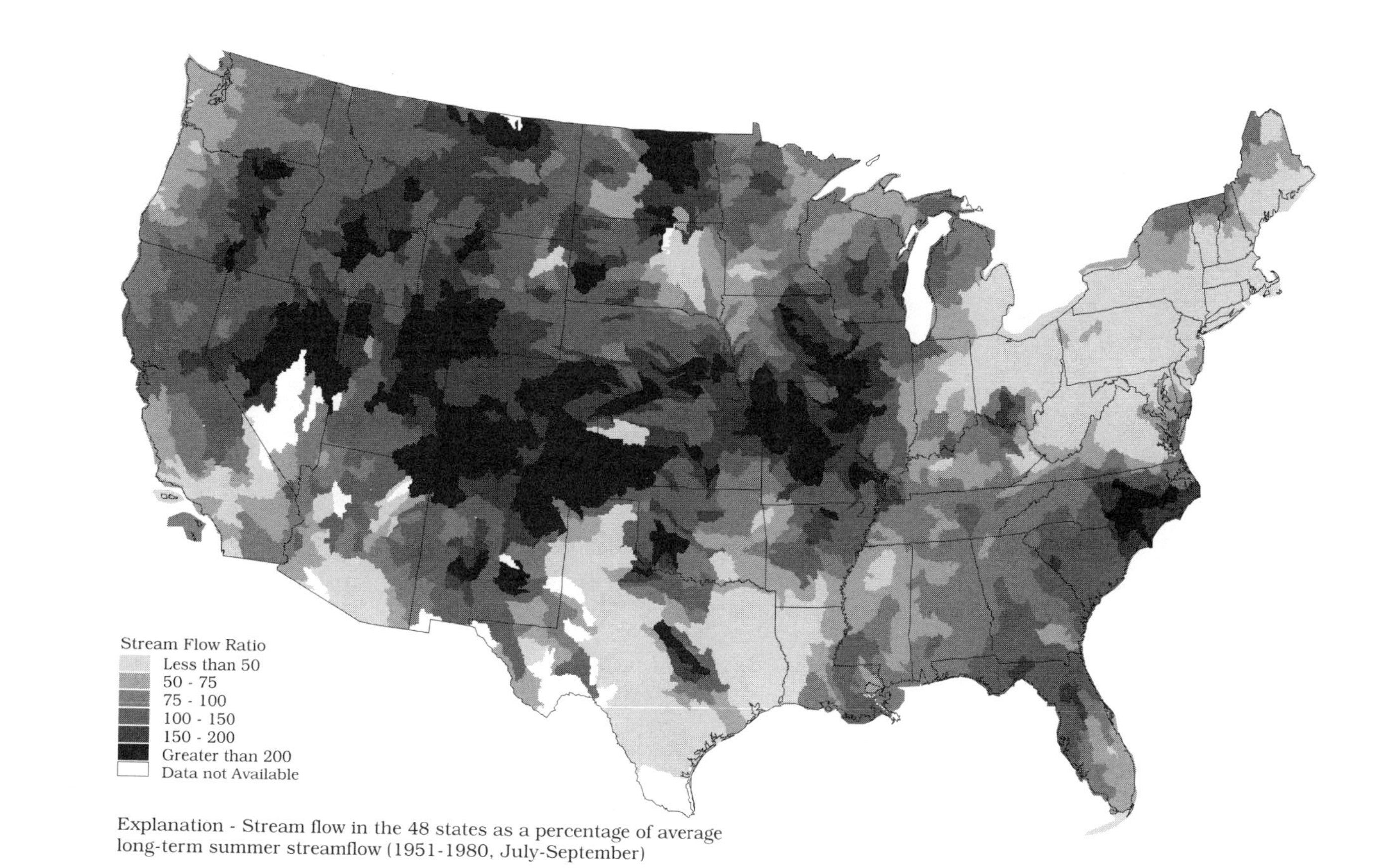

Figure F-13 Hydrologic conditions during the summer (July–September) 1965.

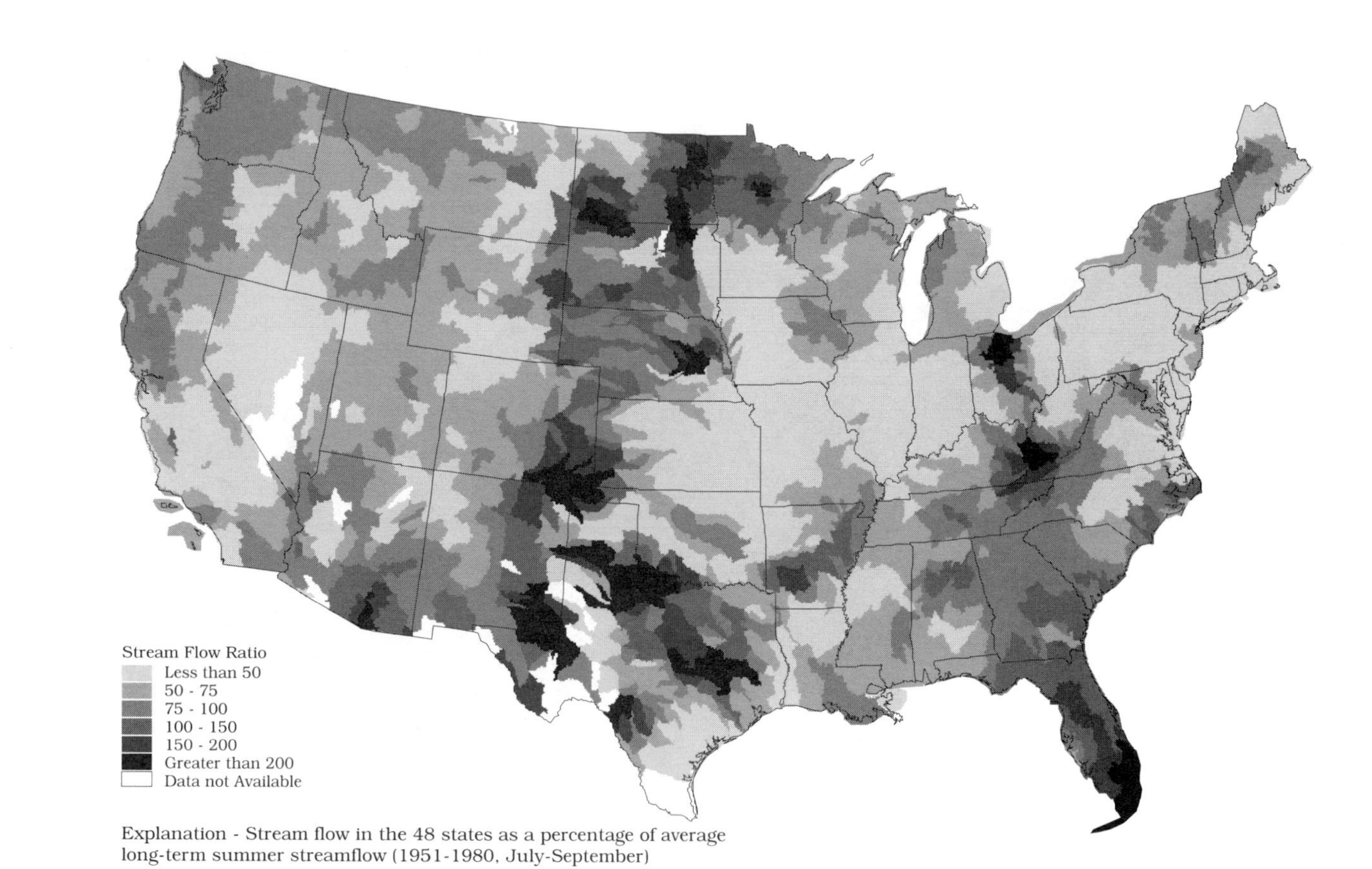

Figure F-14 Hydrologic conditions during the summer (July–September) 1966.

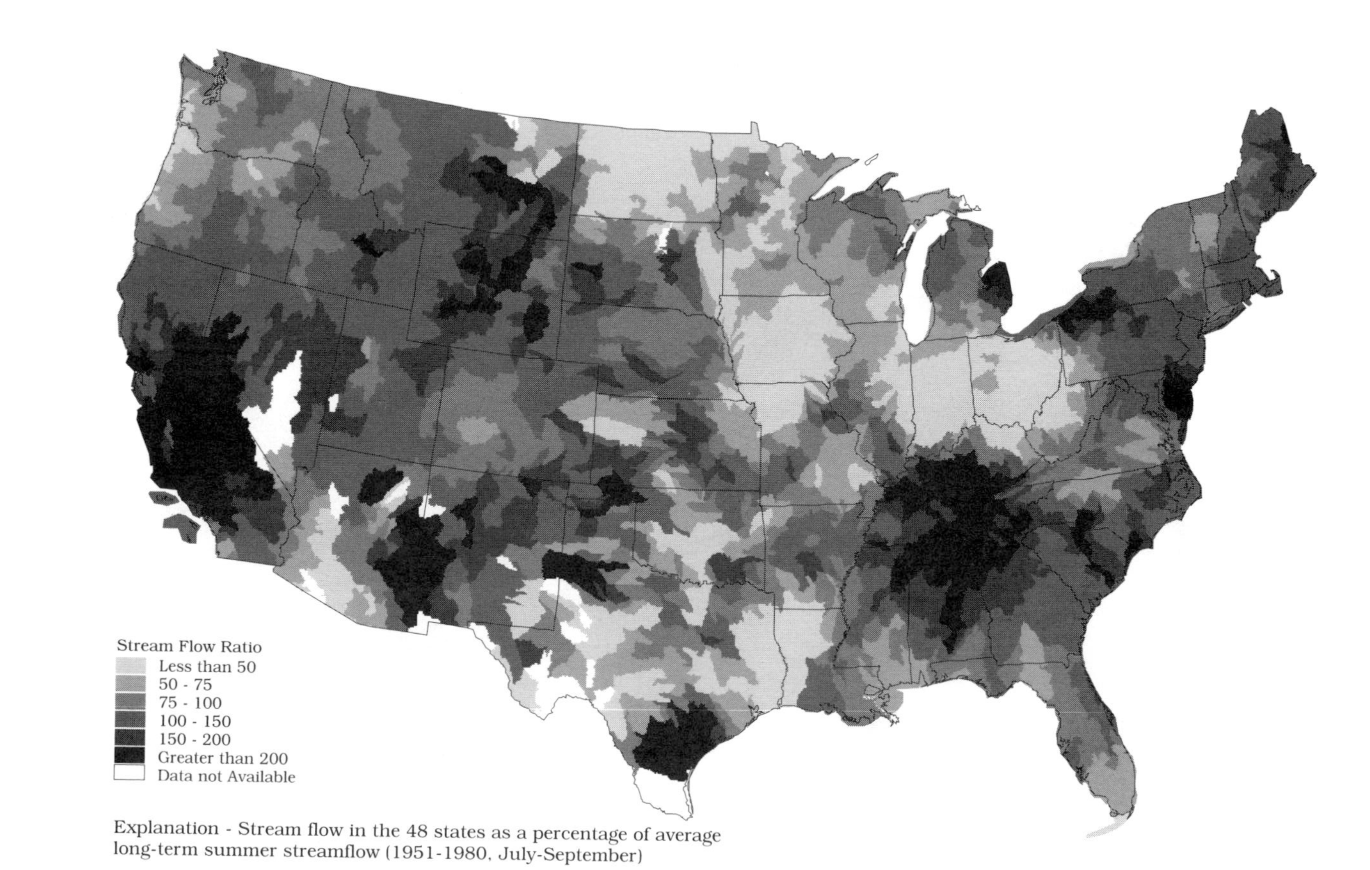

Figure F-15 Hydrologic conditions during the summer (July–September) 1967.

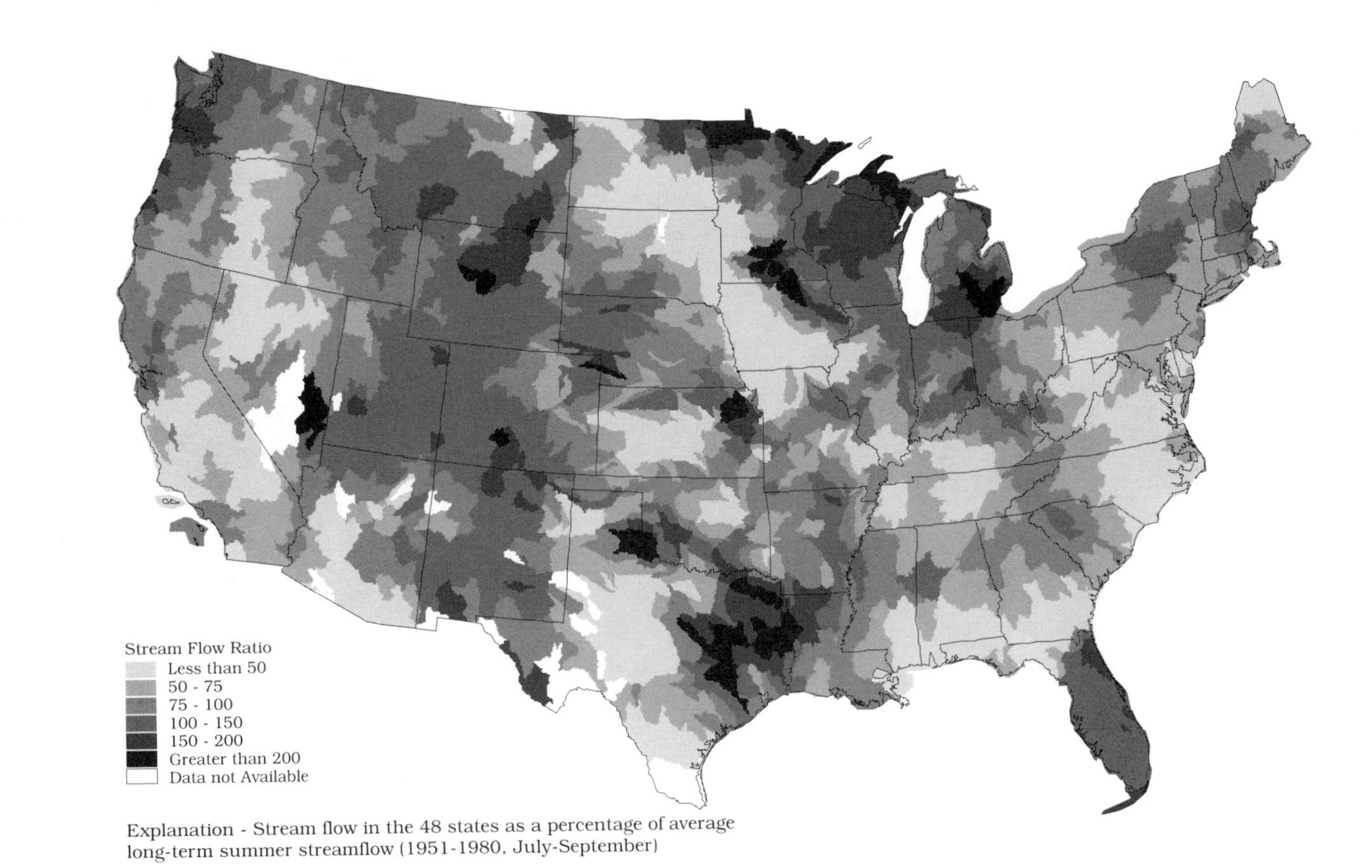

Figure F-16 Hydrologic conditions during the summer (July–September) 1968.

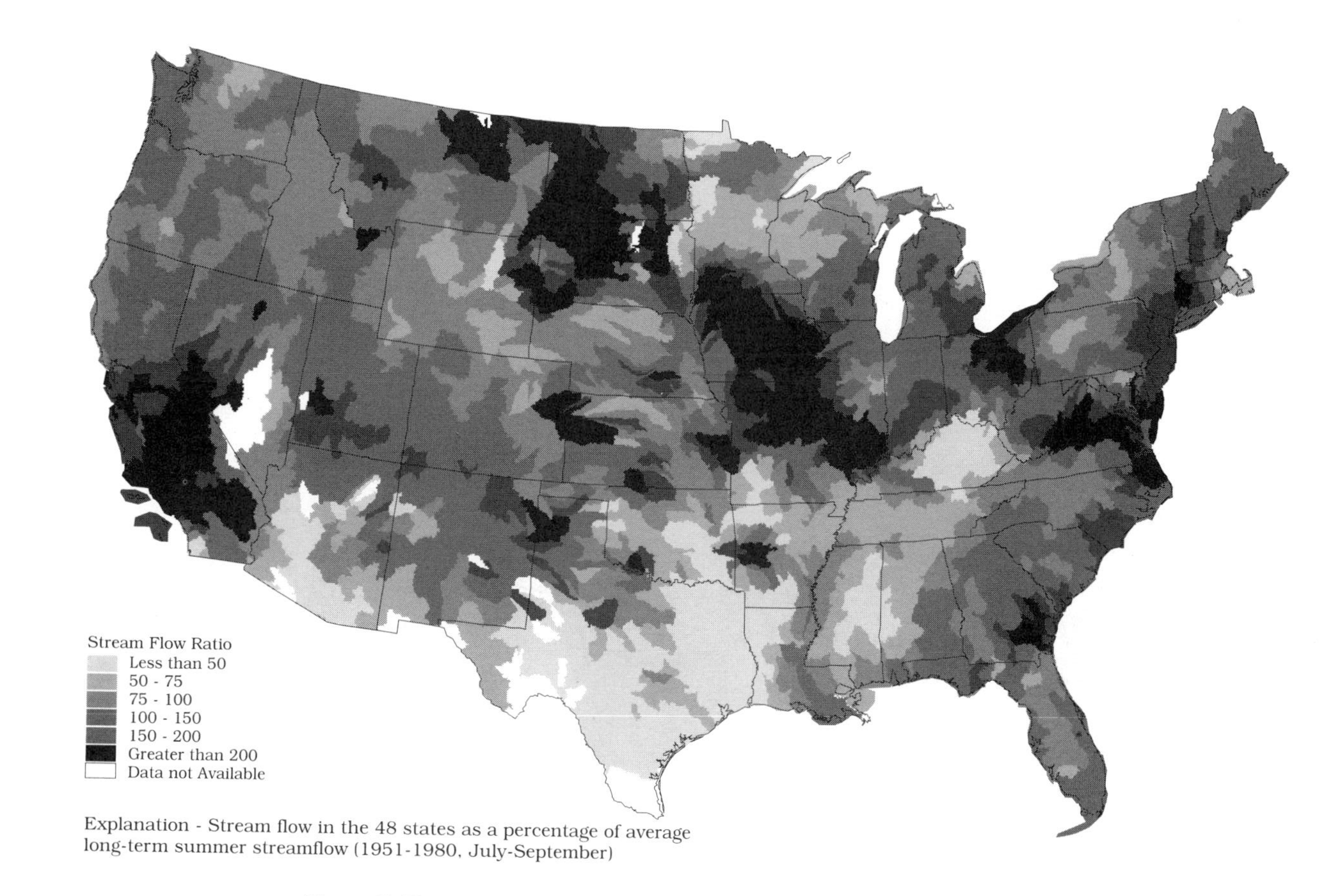

Figure F-17 Hydrologic conditions during the summer (July–September) 1969.

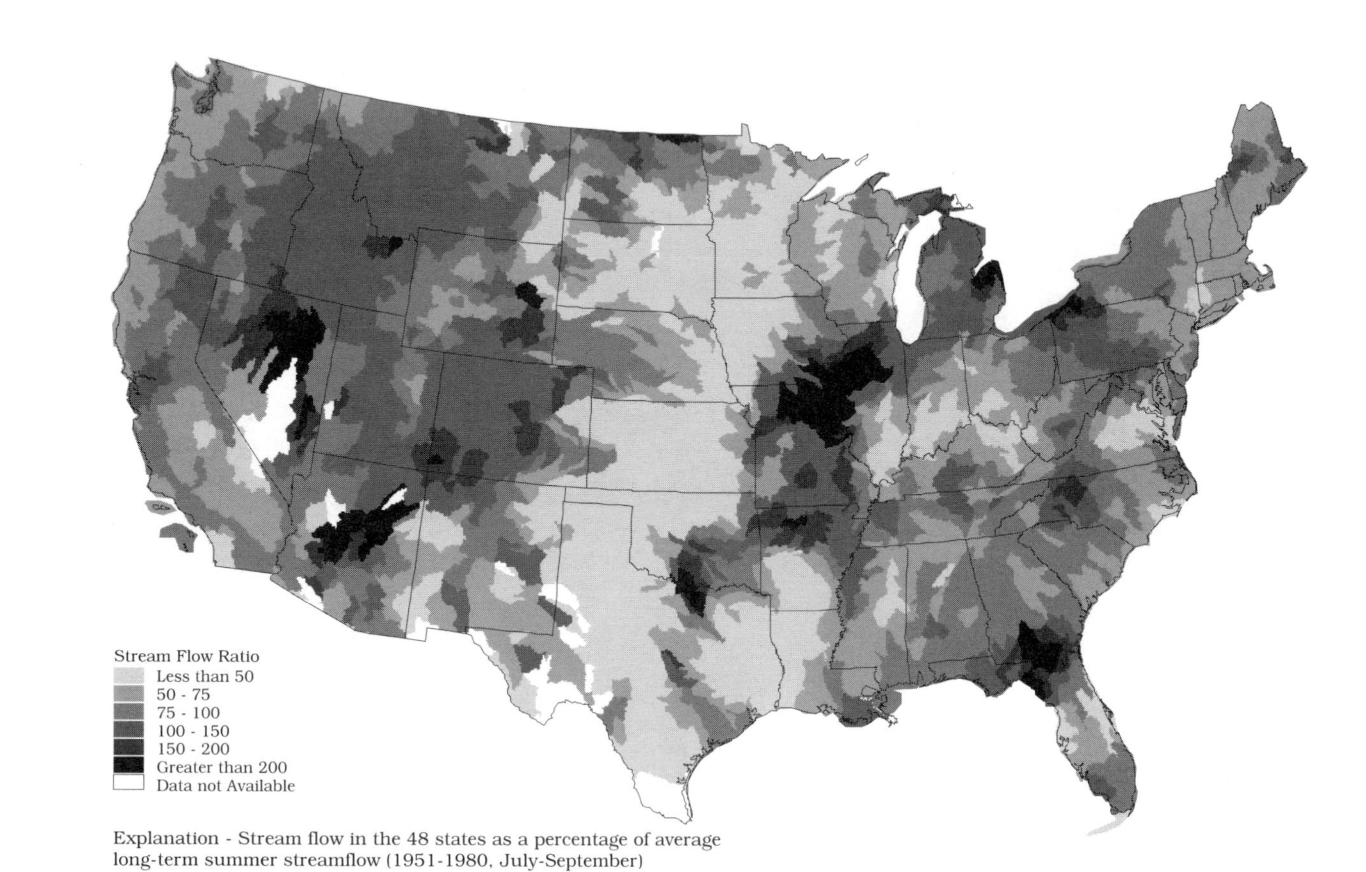

Figure F-18 Hydrologic conditions during the summer (July–September) 1970.

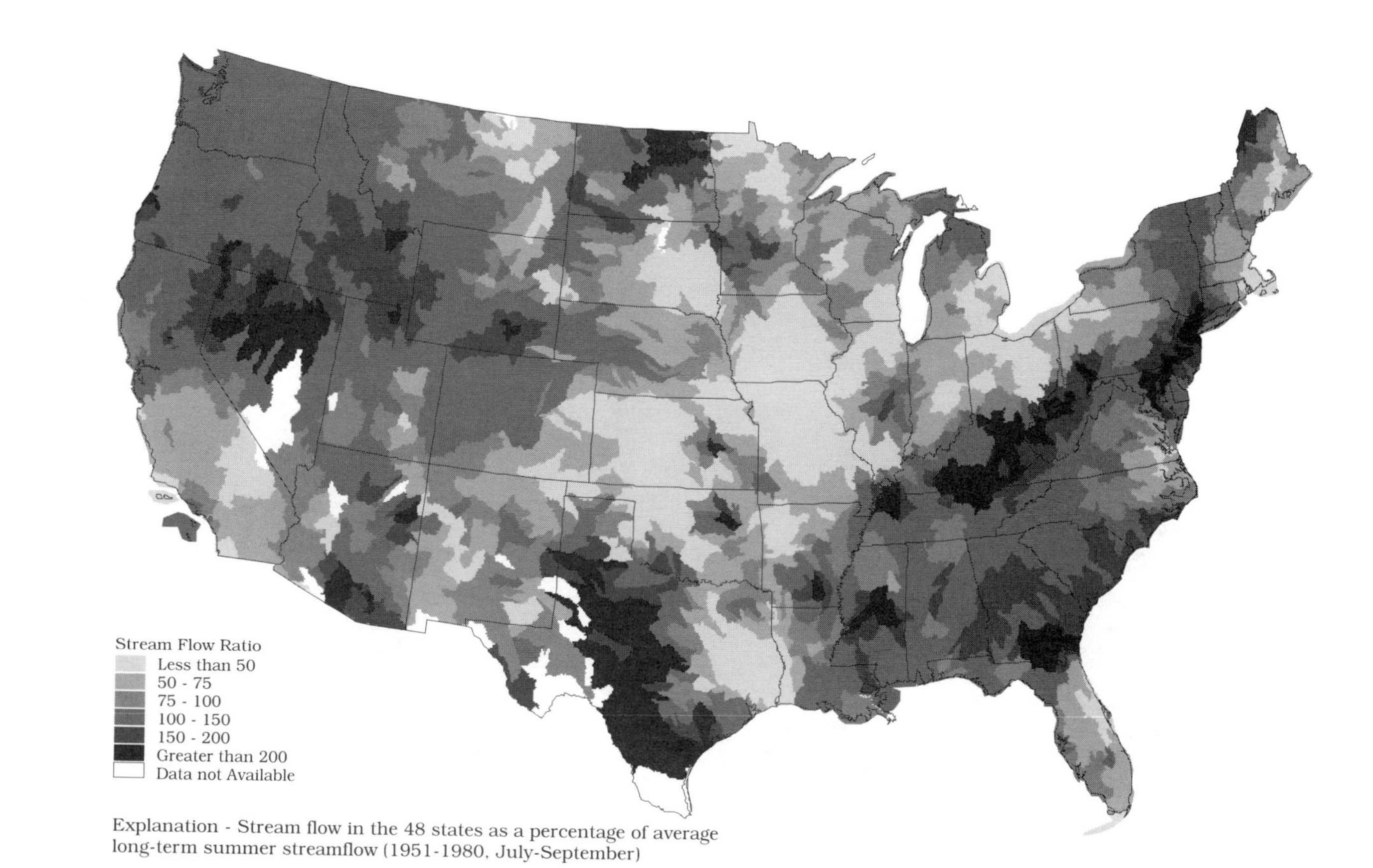

Figure F-19 Hydrologic conditions during the summer (July–September) 1971.

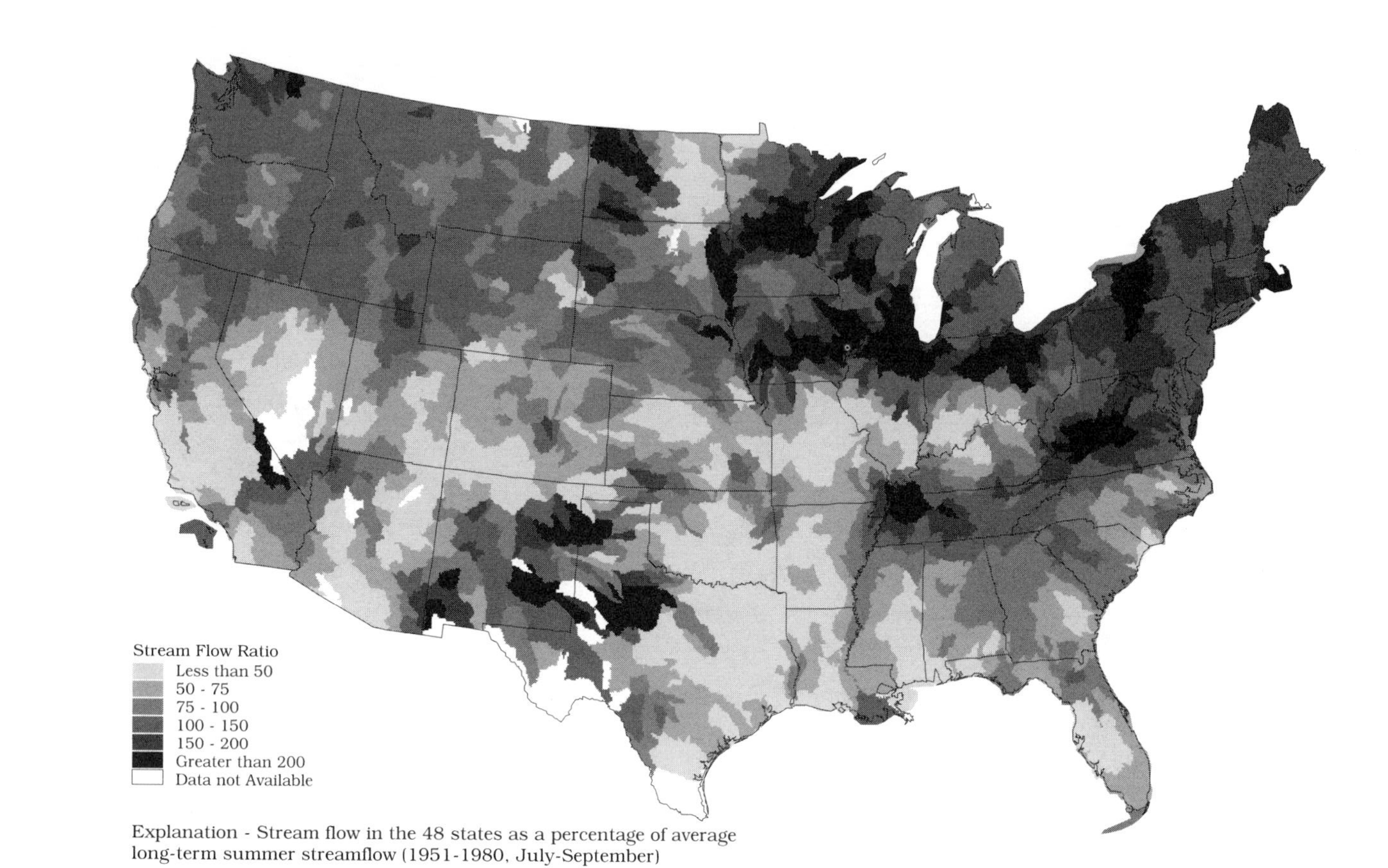

Figure F-20 Hydrologic conditions during the summer (July–September) 1972.

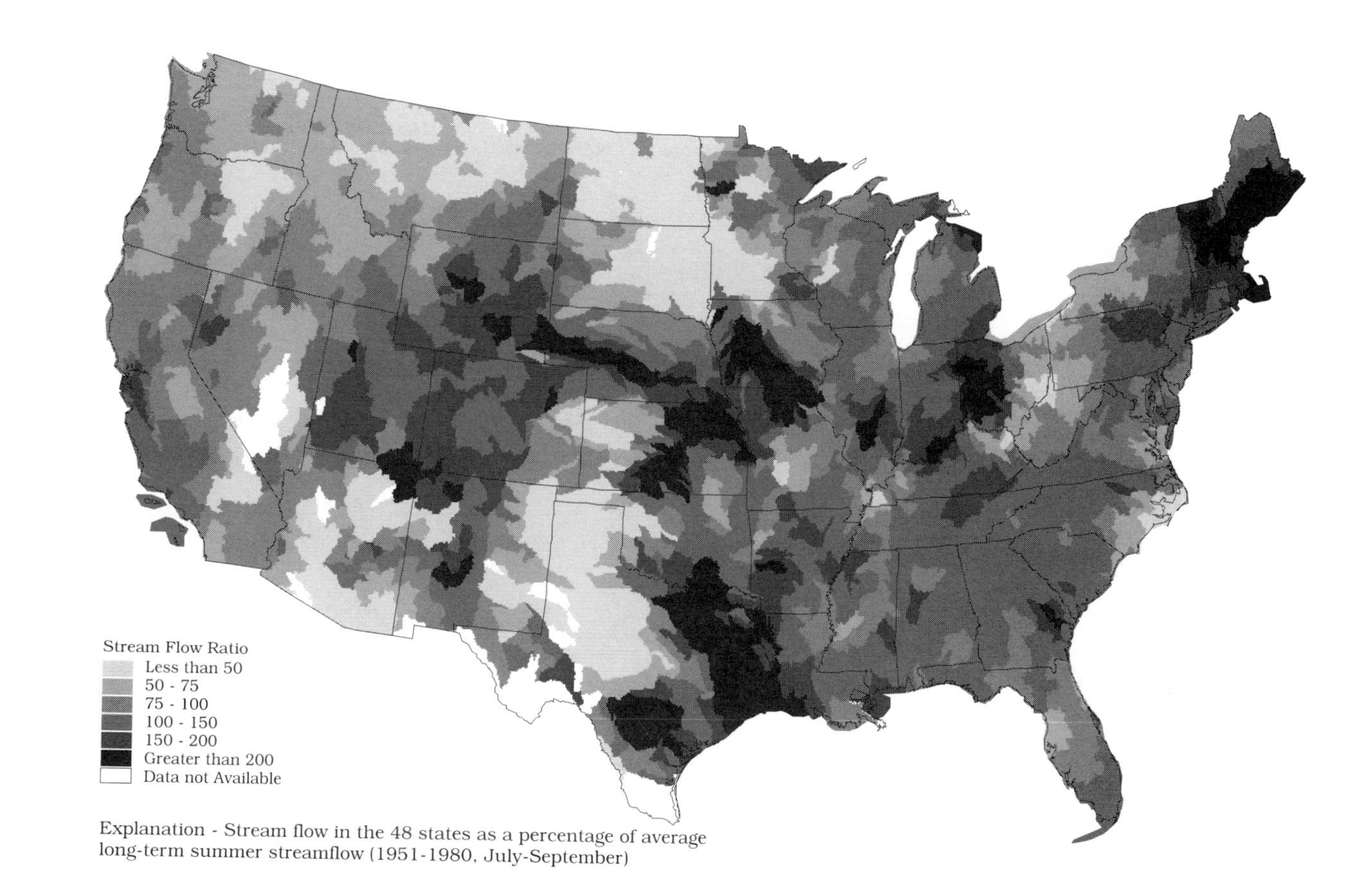

Figure F-21 Hydrologic conditions during the summer (July–September) 1973.

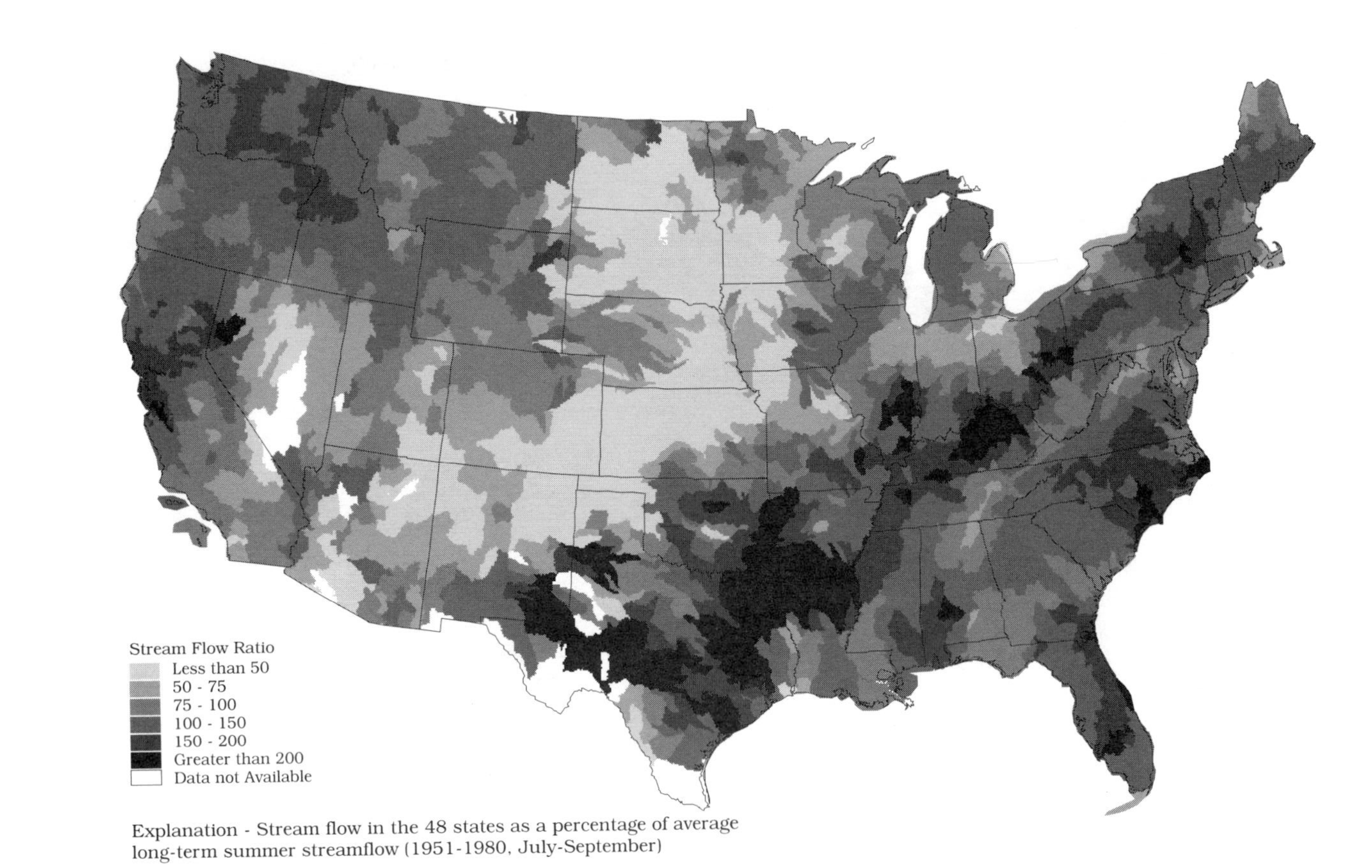

Figure F-22 Hydrologic conditions during the summer (July–September) 1974.

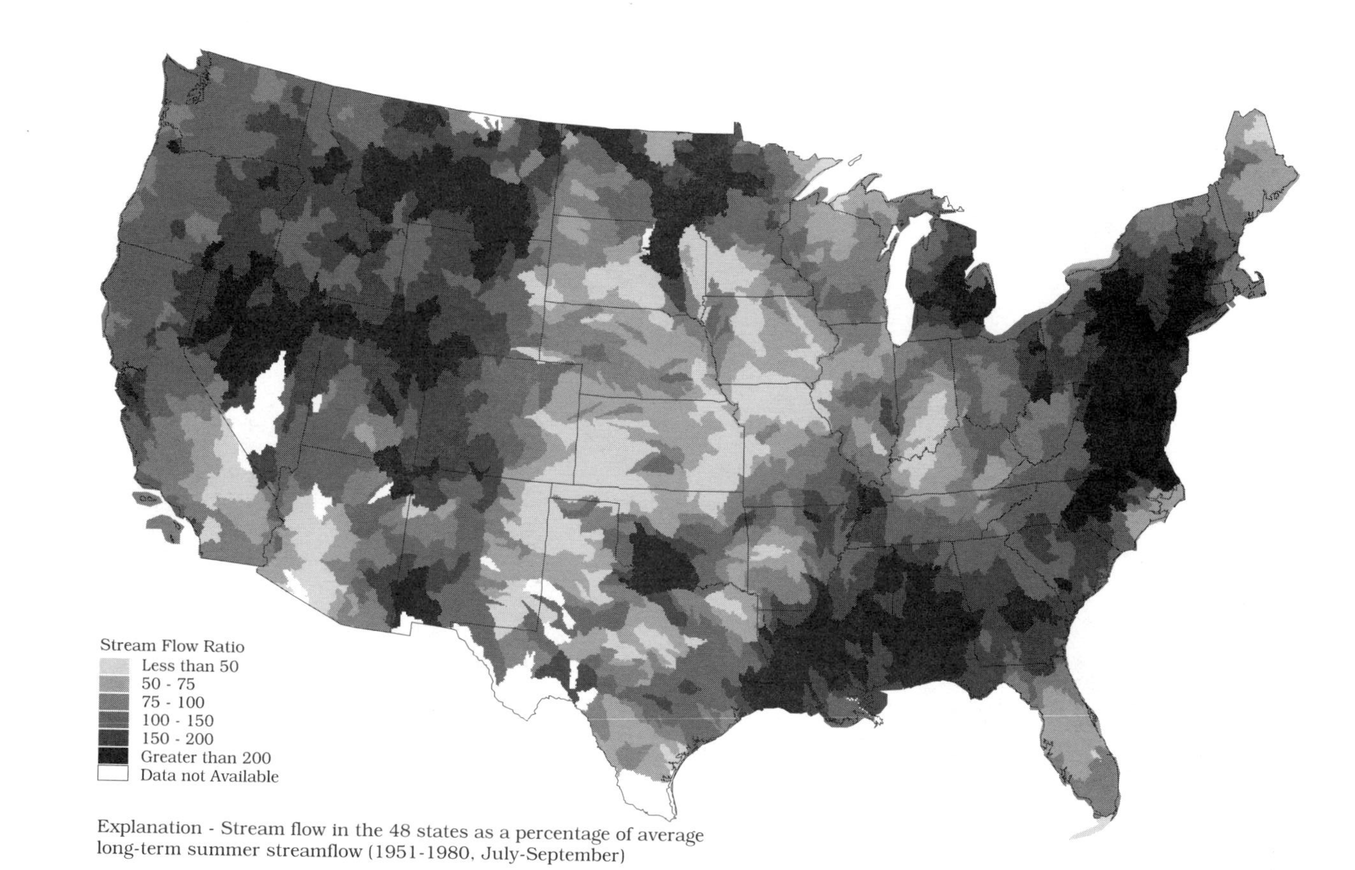

Figure F-23 Hydrologic conditions during the summer (July–September) 1975.

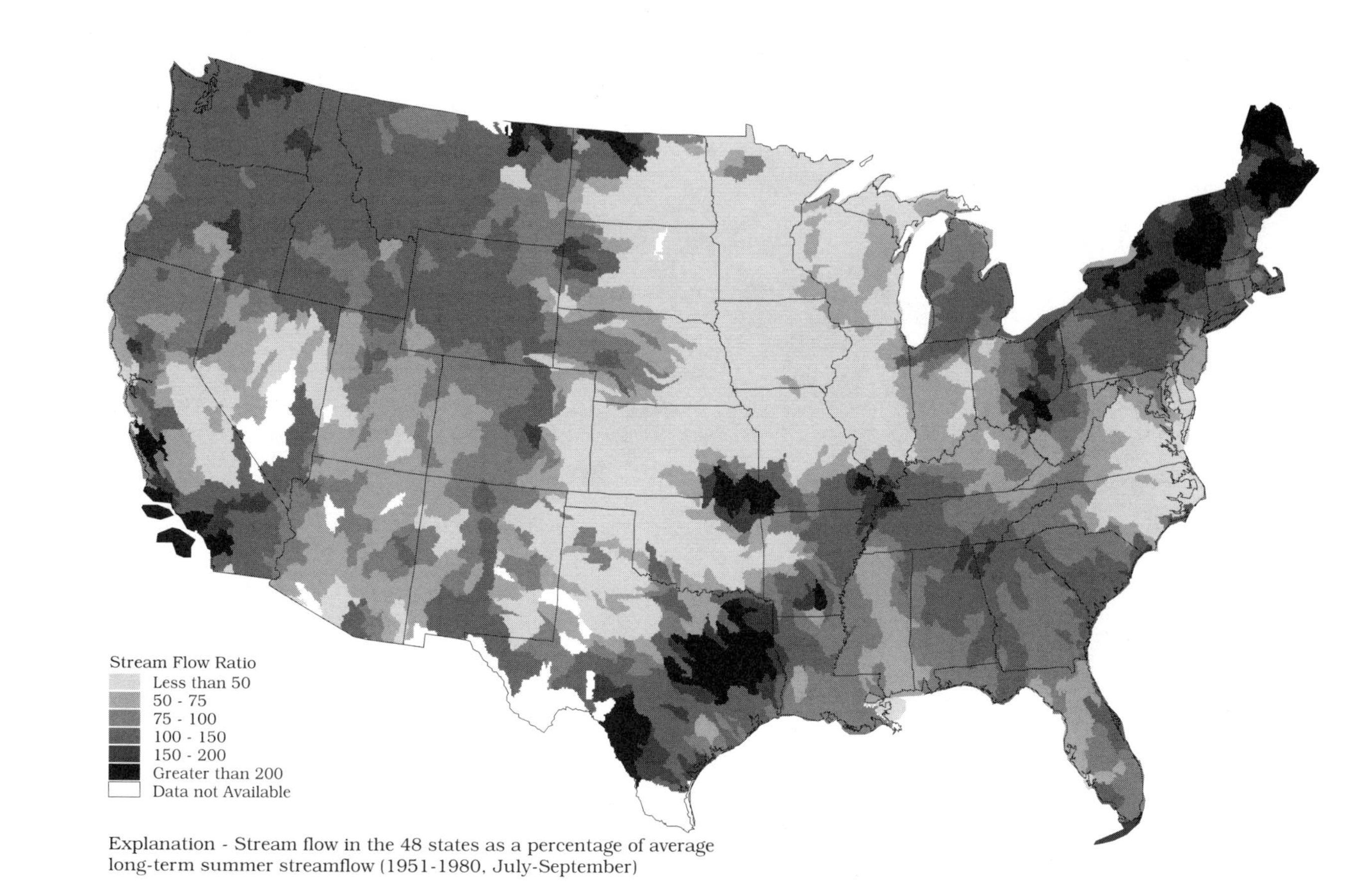

Figure F-24 Hydrologic conditions during the summer (July–September) 1976.

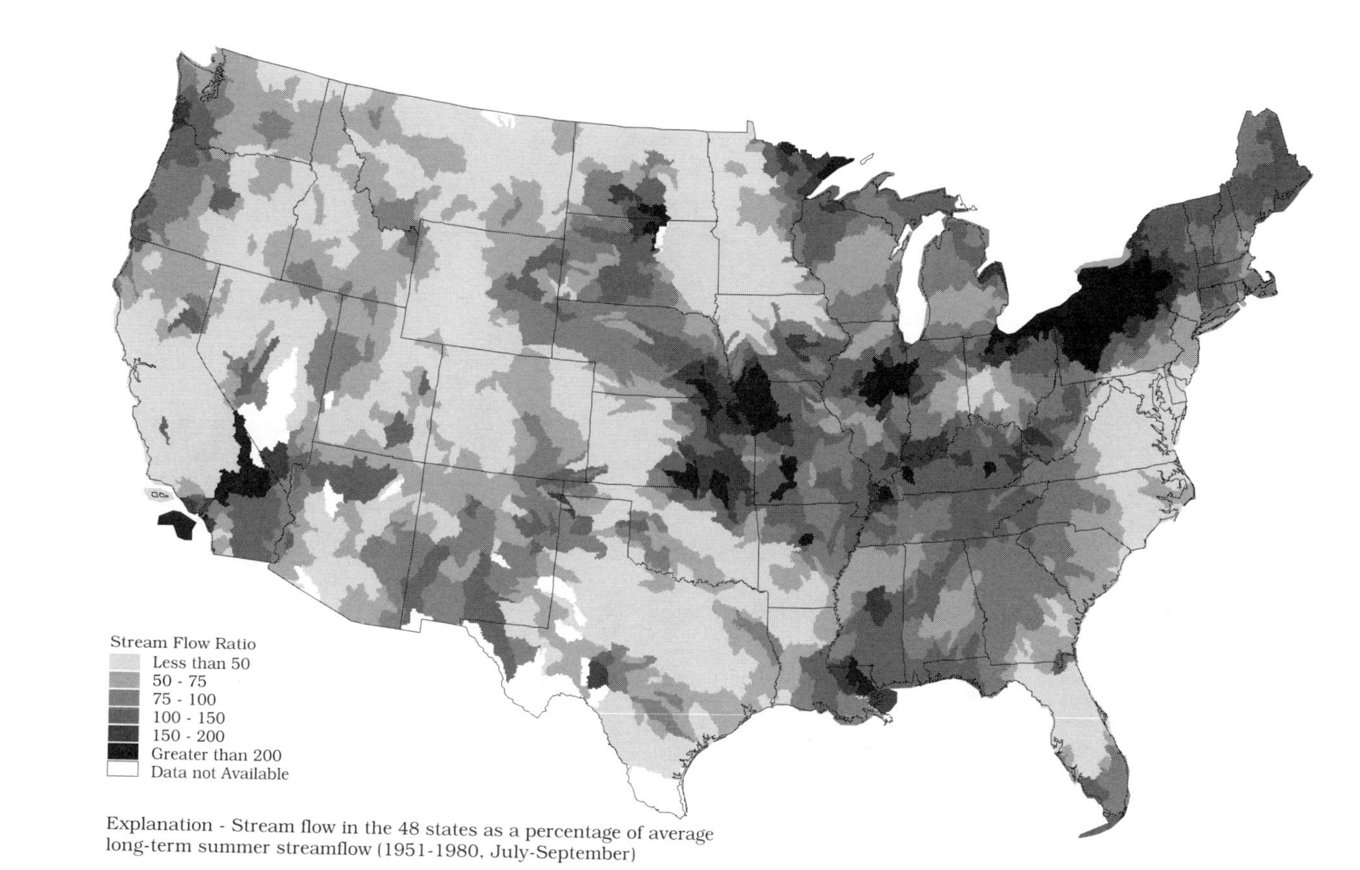

Figure F-25 Hydrologic conditions during the summer (July–September) 1977.

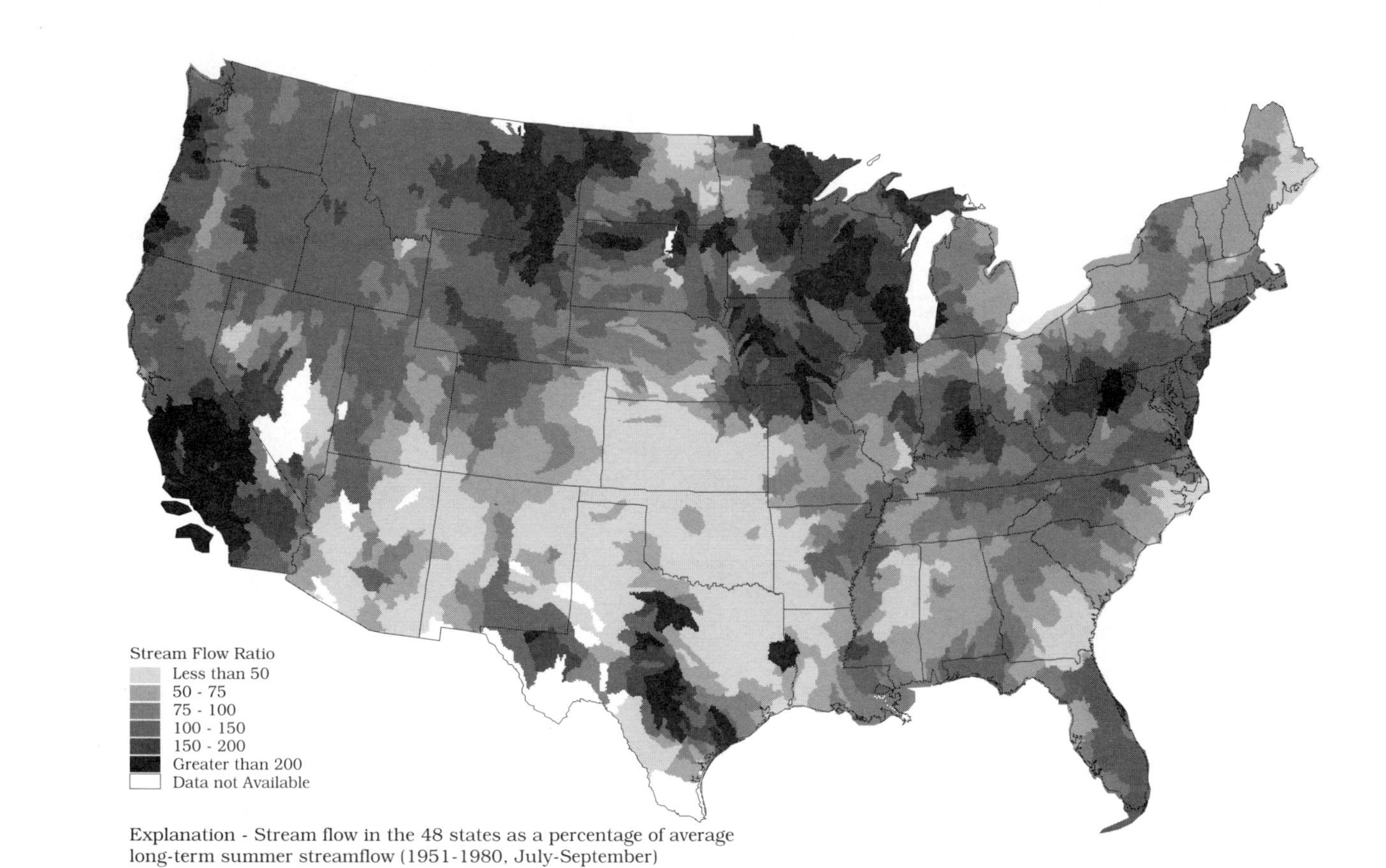

Figure F-26 Hydrologic conditions during the summer (July–September) 1978.

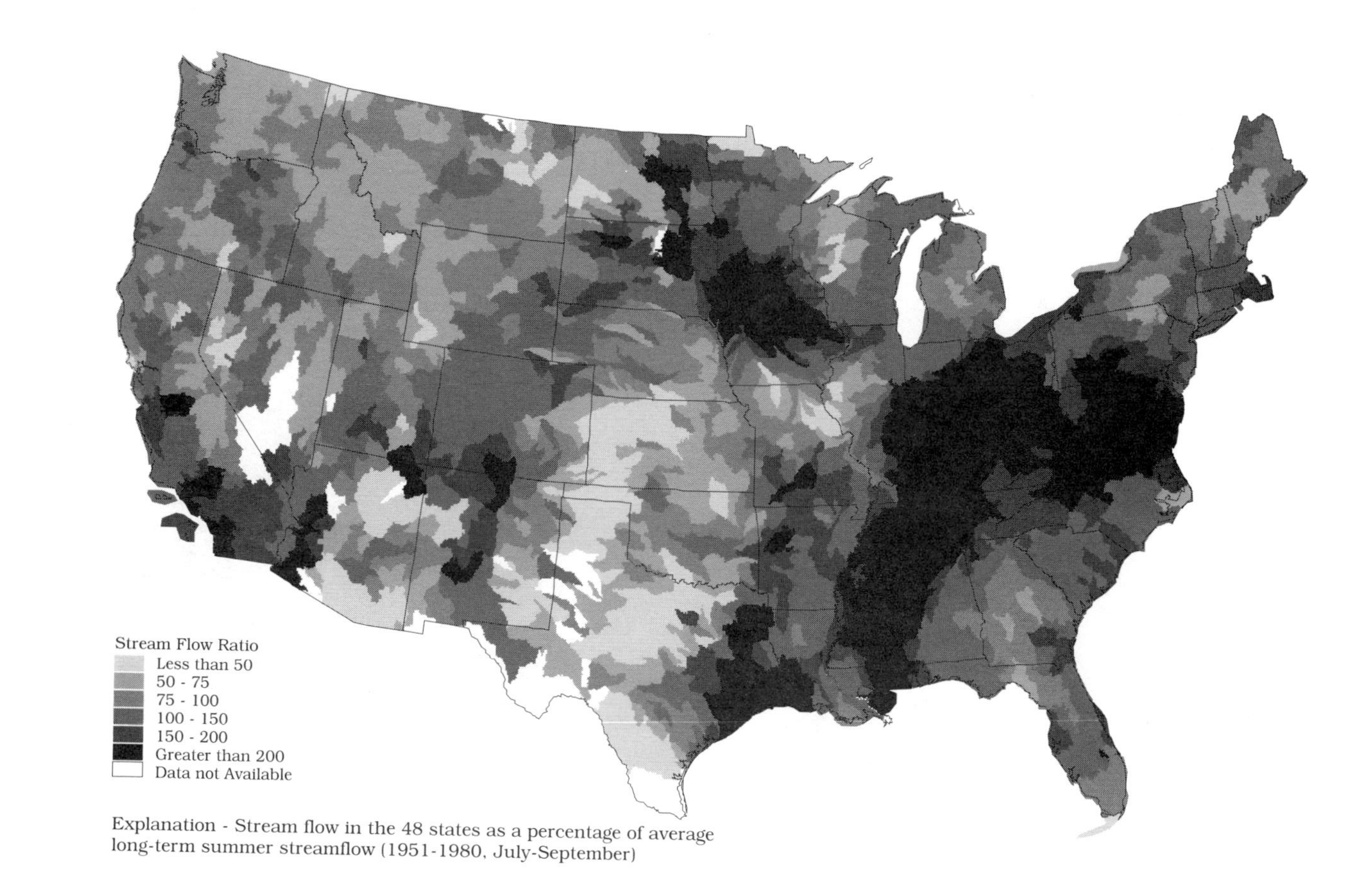

Figure F-27 Hydrologic conditions during the summer (July–September) 1979.

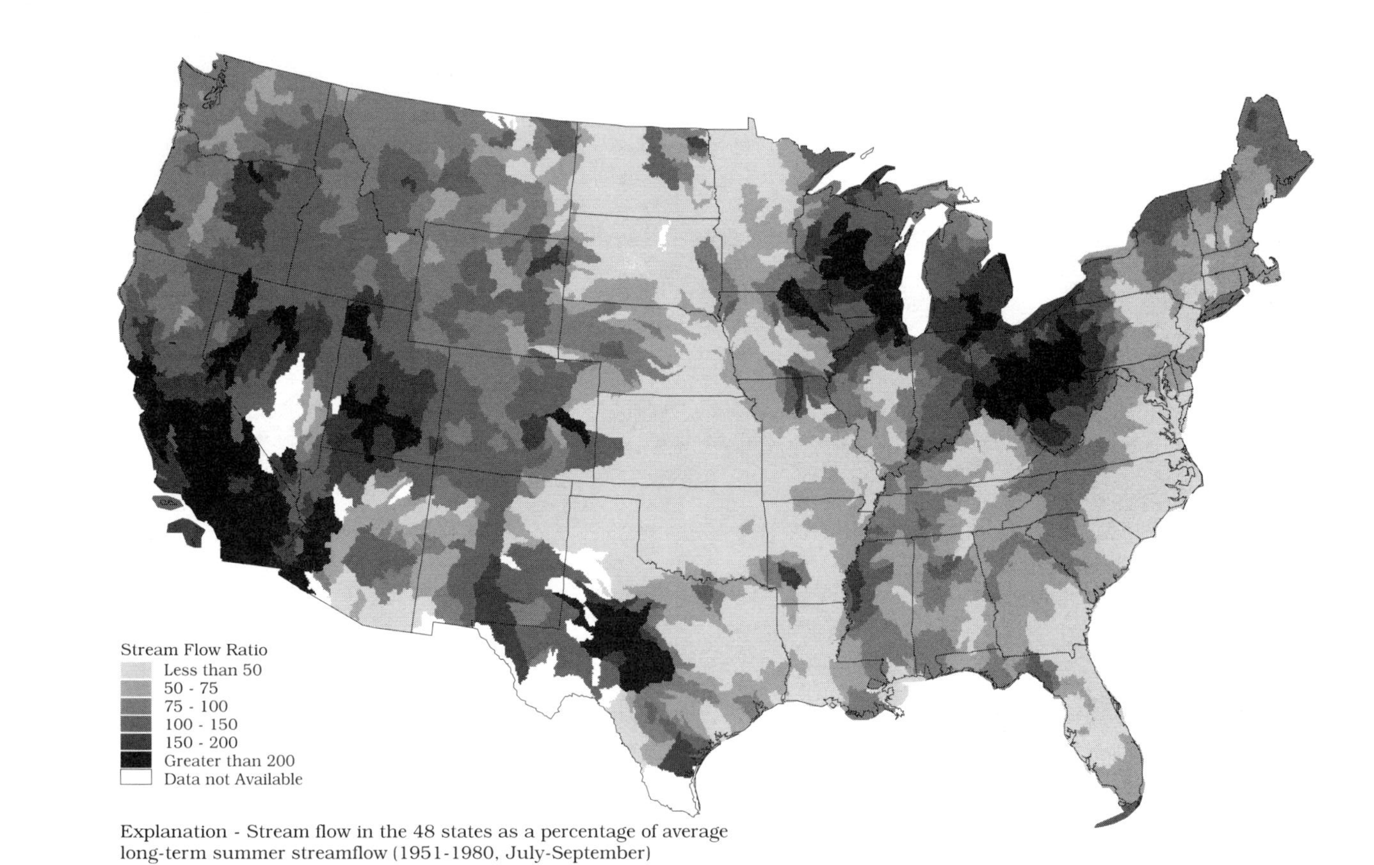

Figure F-28 Hydrologic conditions during the summer (July–September) 1980.

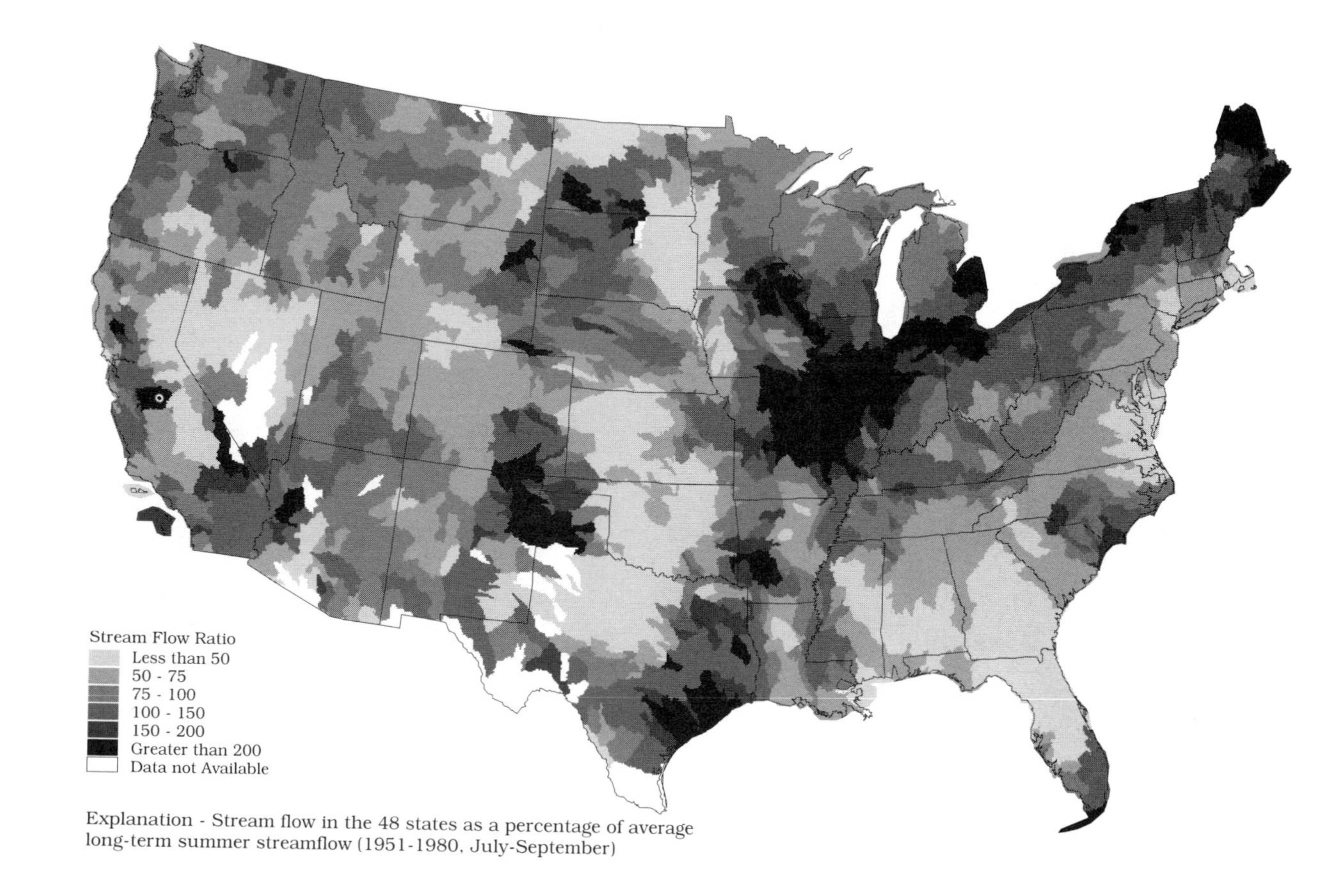

Explanation - Stream flow in the 48 states as a percentage of average long-term summer streamflow (1951-1980, July-September)

Figure F-29 Hydrologic conditions during the summer (July–September) 1981.

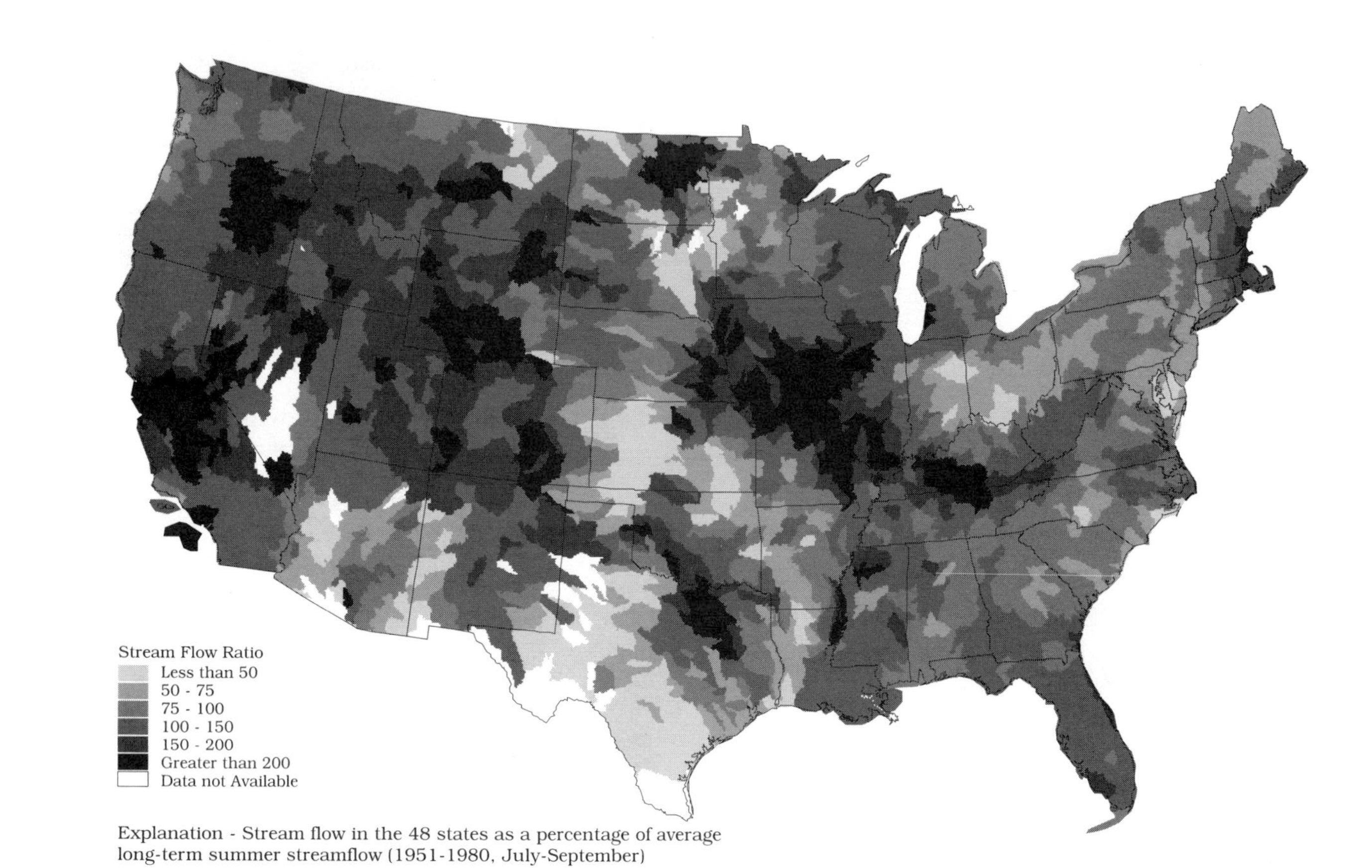

Figure F-30 Hydrologic conditions during the summer (July–September) 1982.

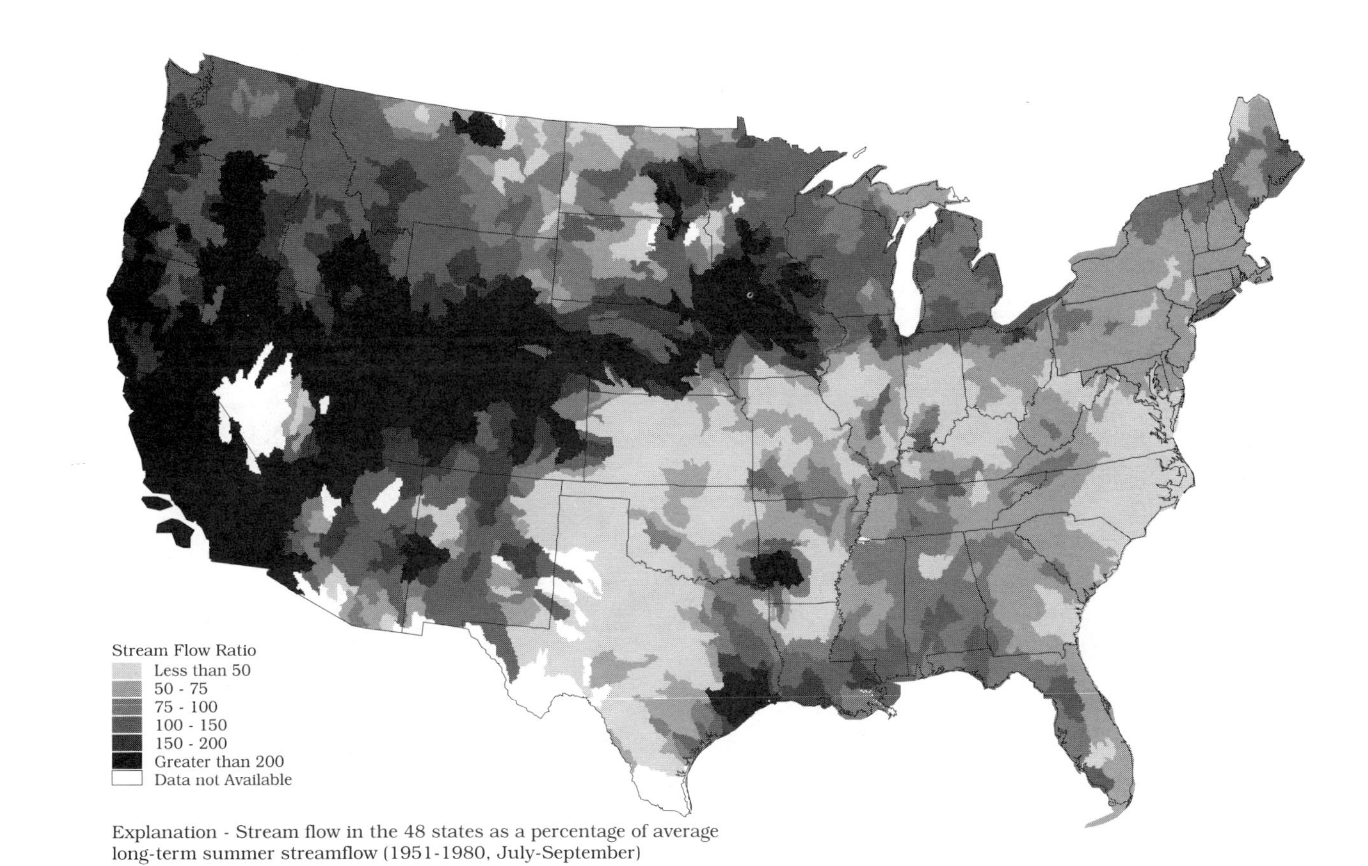

Figure F-31 Hydrologic conditions during the summer (July–September) 1983.

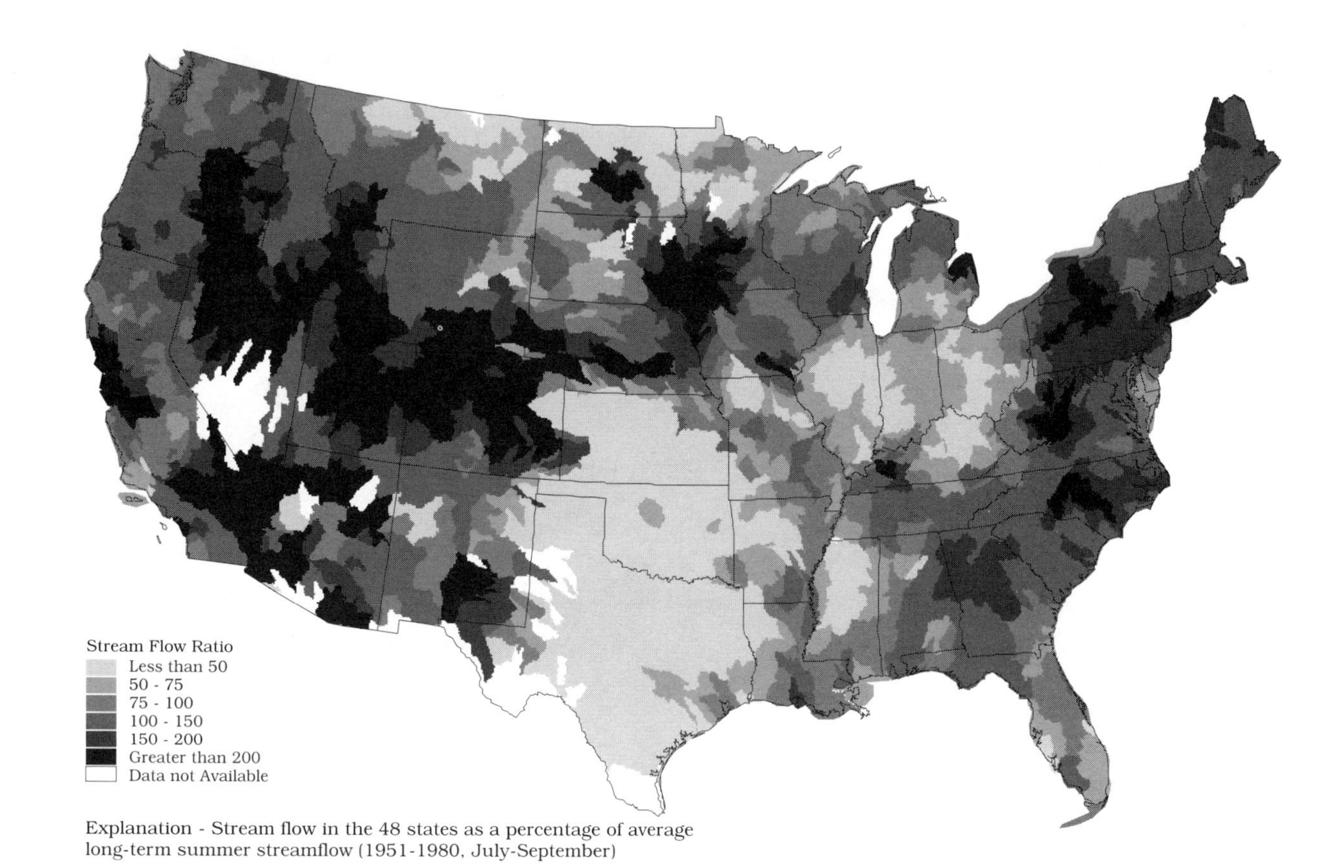

Figure F-32 Hydrologic conditions during the summer (July–September) 1984.

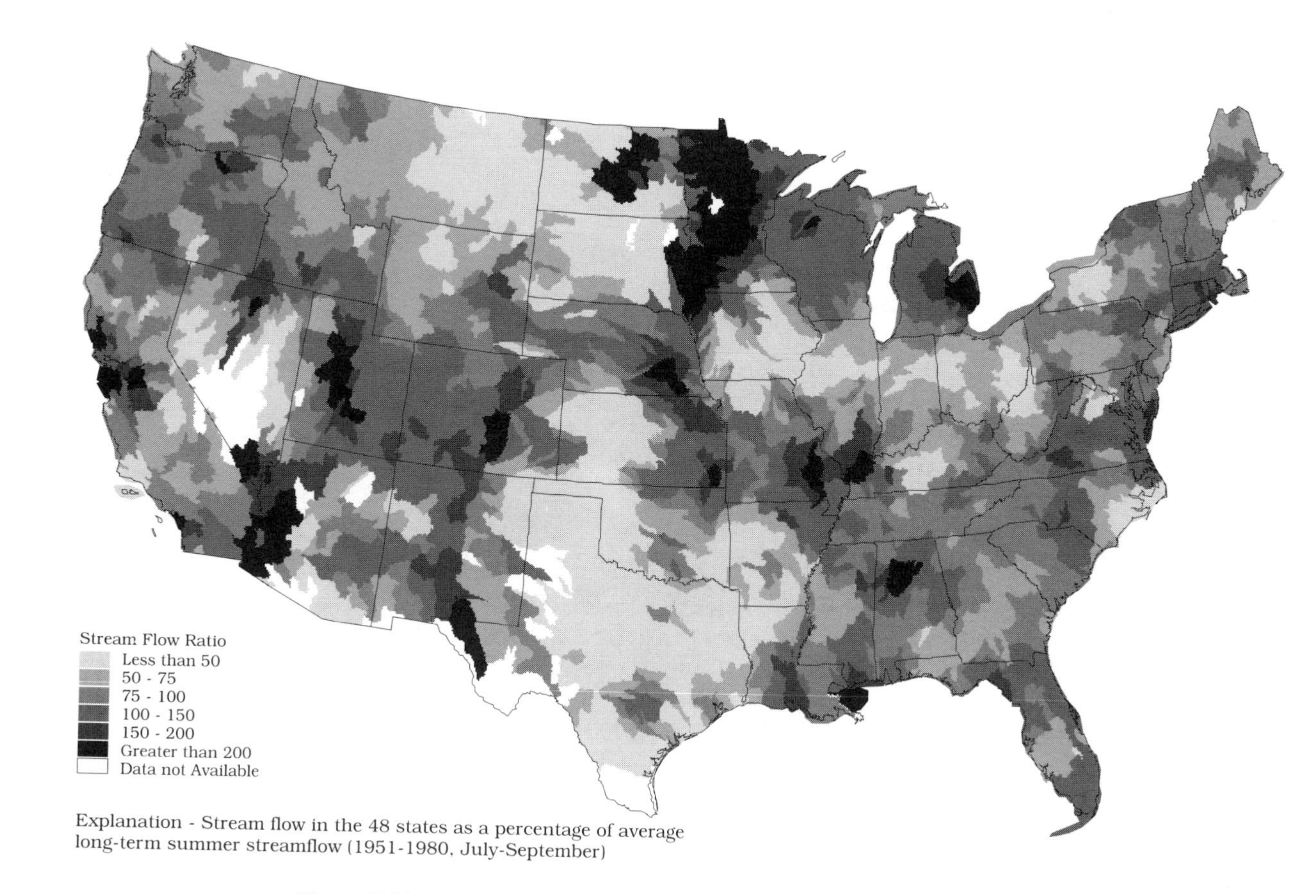

Figure F-33 Hydrologic conditions during the summer (July–September) 1985.

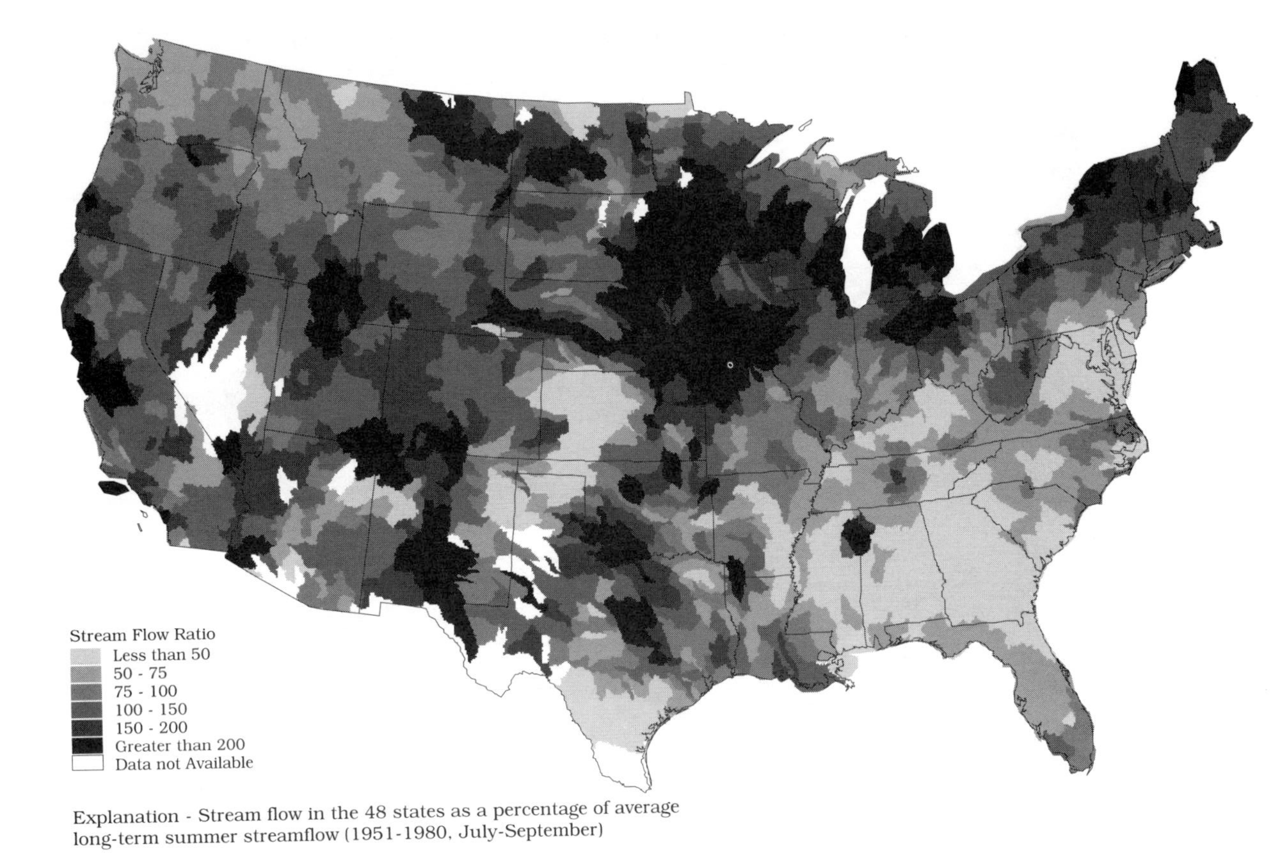

Figure F-34 Hydrologic conditions during the summer (July–September) 1986.

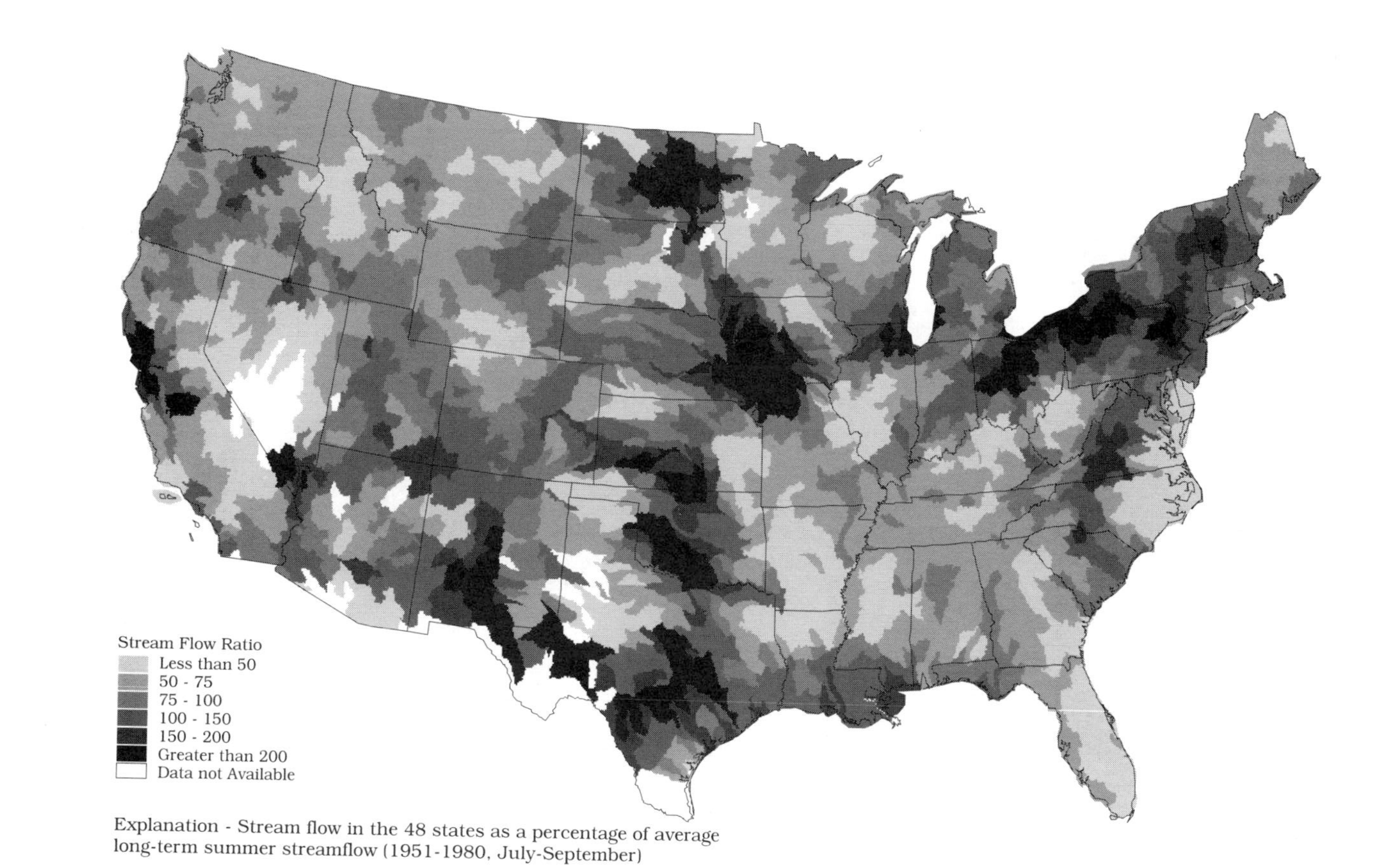

Figure F-35 Hydrologic conditions during the summer (July–September) 1987.

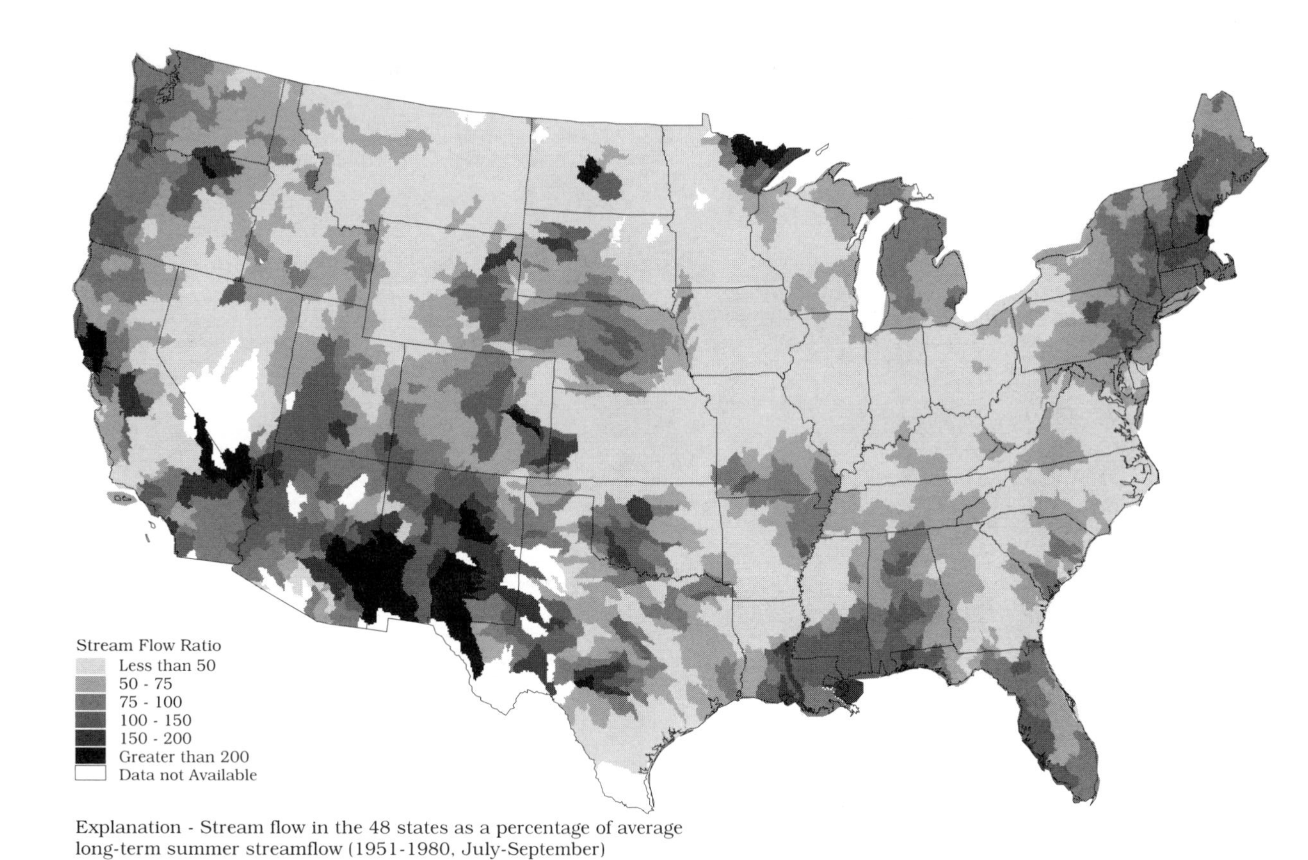

Figure F-36 Hydrologic conditions during the summer (July–September) 1988.

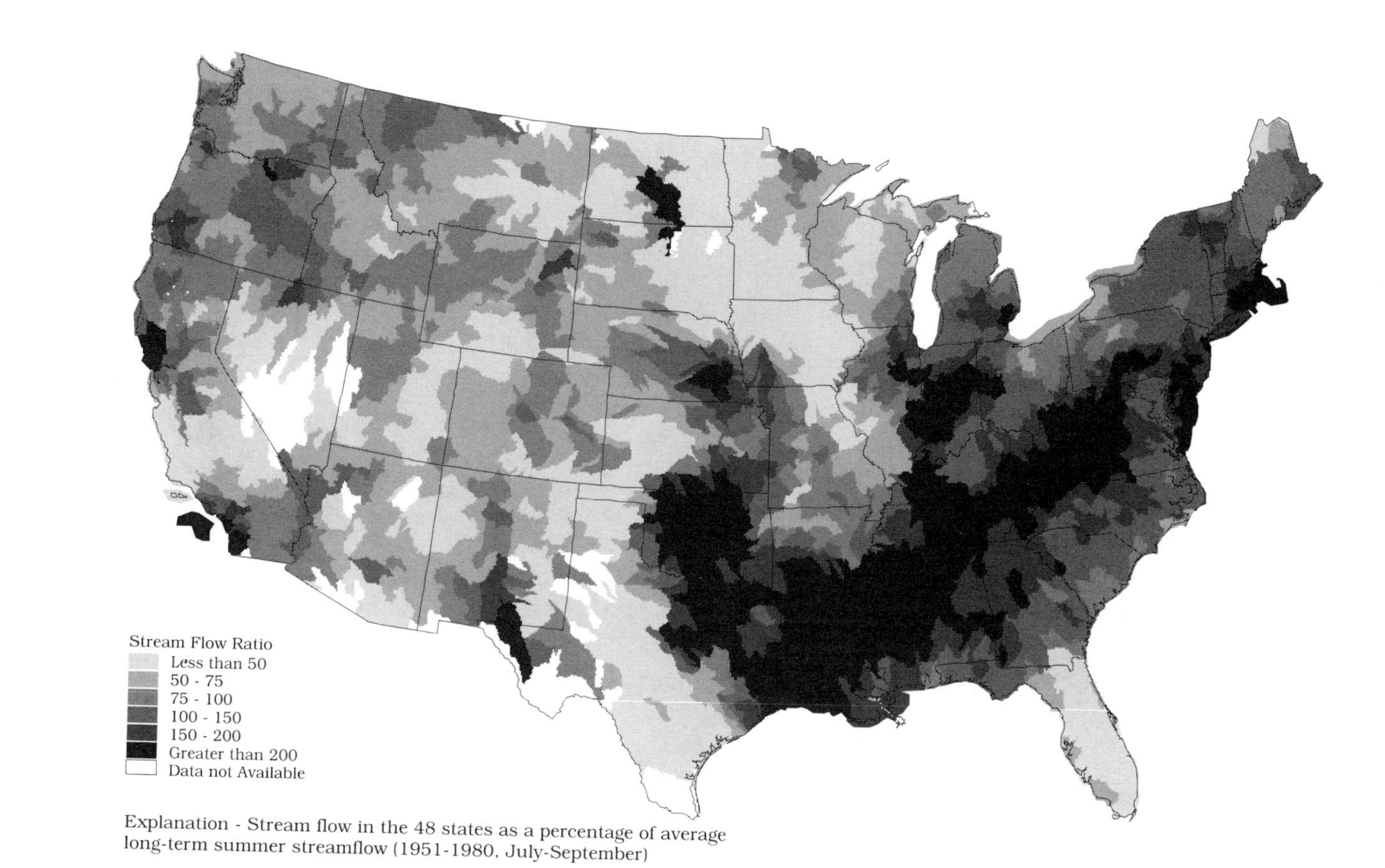

Figure F-37 Hydrologic conditions during the summer (July–September) 1989.

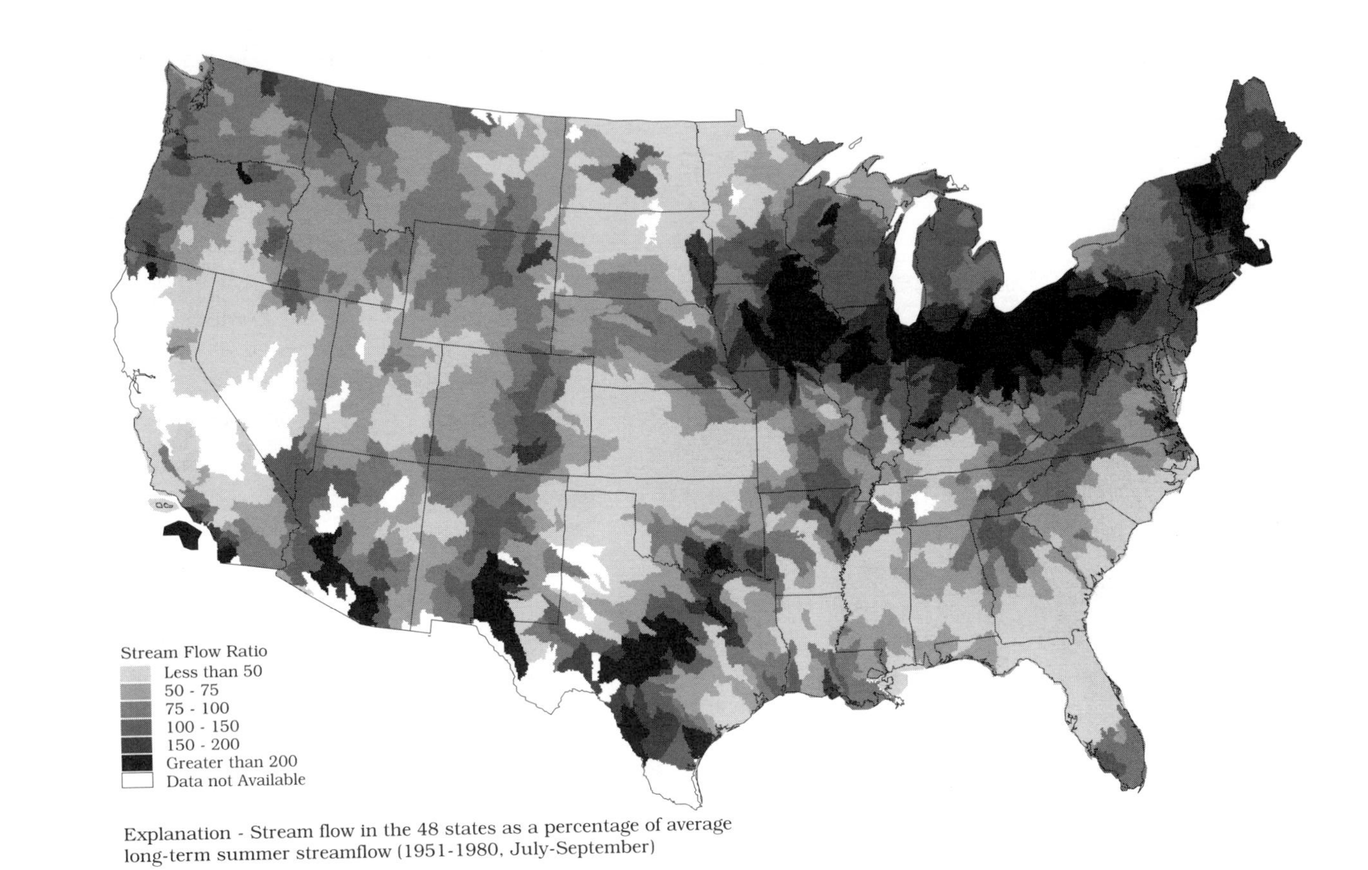

Explanation - Stream flow in the 48 states as a percentage of average long-term summer streamflow (1951-1980, July-September)

Figure F-38 Hydrologic conditions during the summer (July–September) 1990.

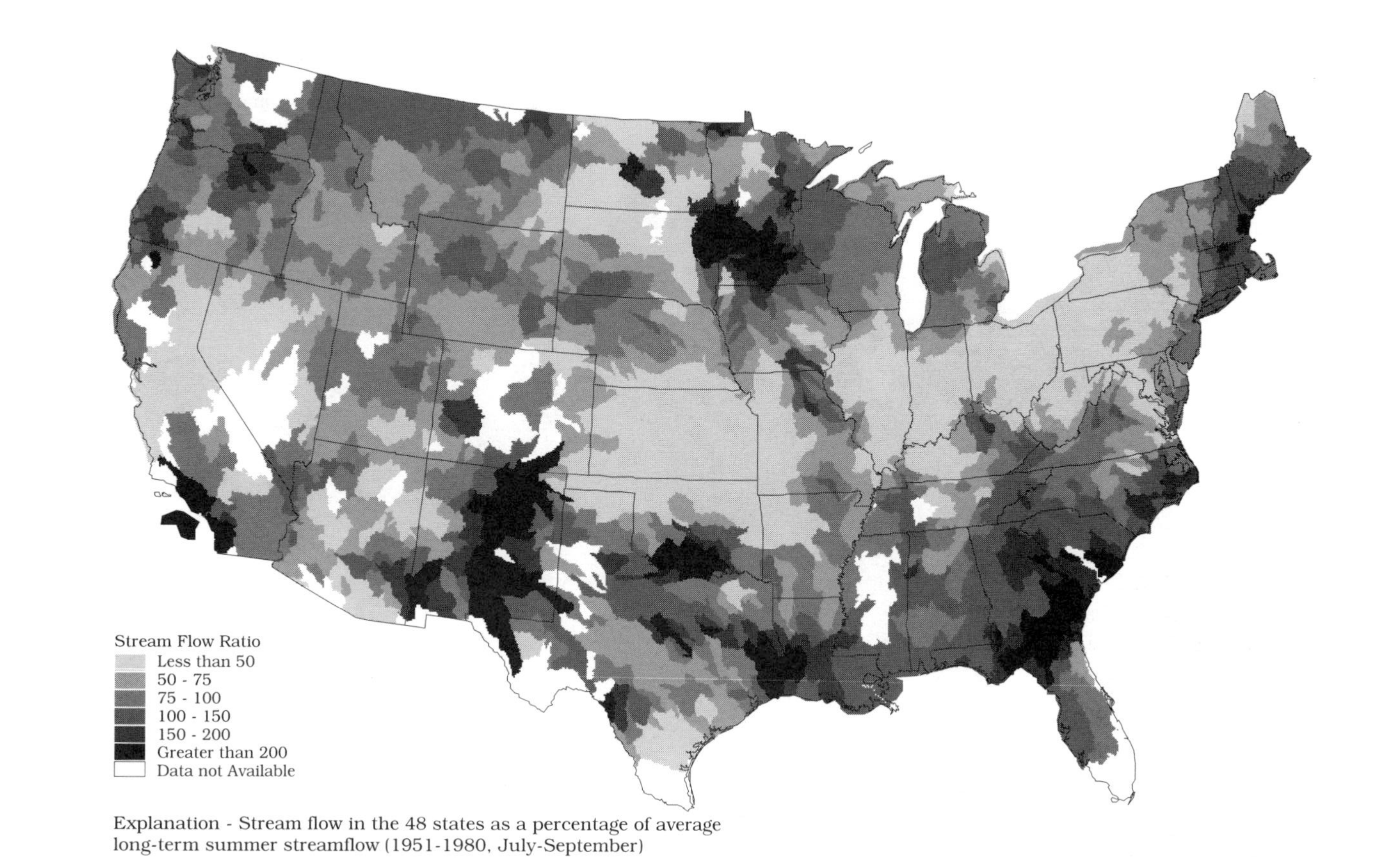

Explanation - Stream flow in the 48 states as a percentage of average long-term summer streamflow (1951-1980, July-September)

Figure F-39 Hydrologic conditions during the summer (July–September) 1991.

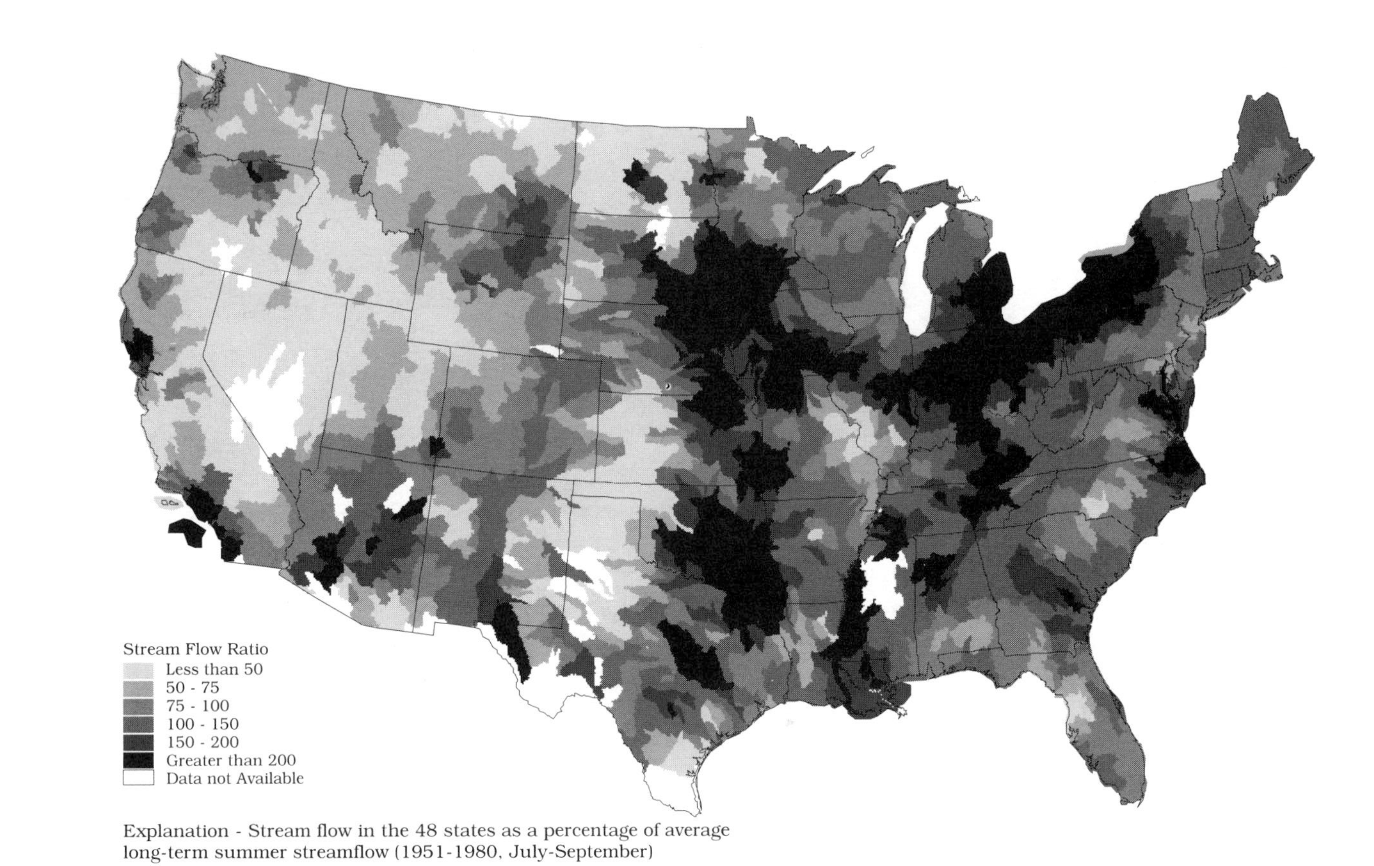

Figure F-40 Hydrologic conditions during the summer (July–September) 1992.

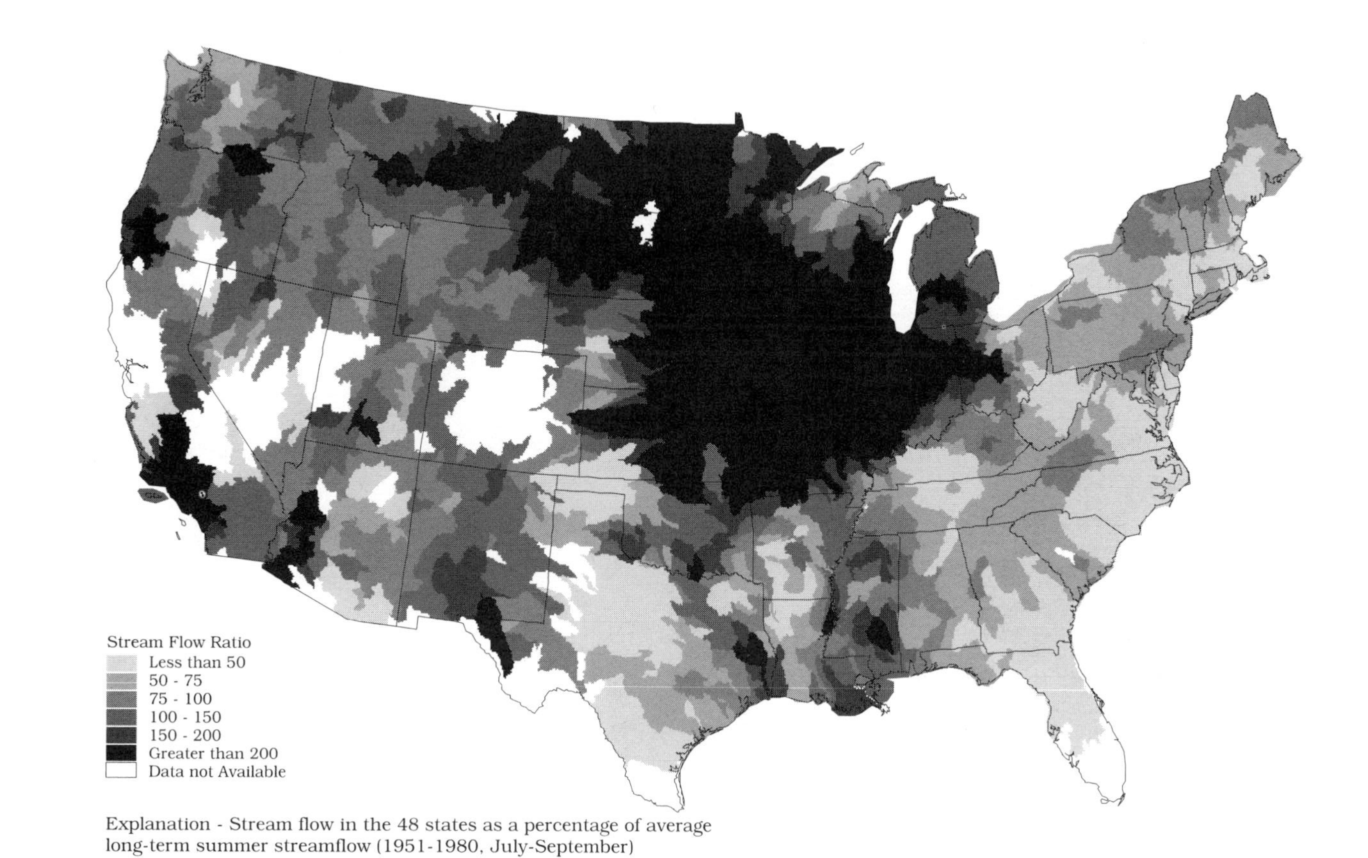

Figure F-41 Hydrologic conditions during the summer (July–September) 1993.

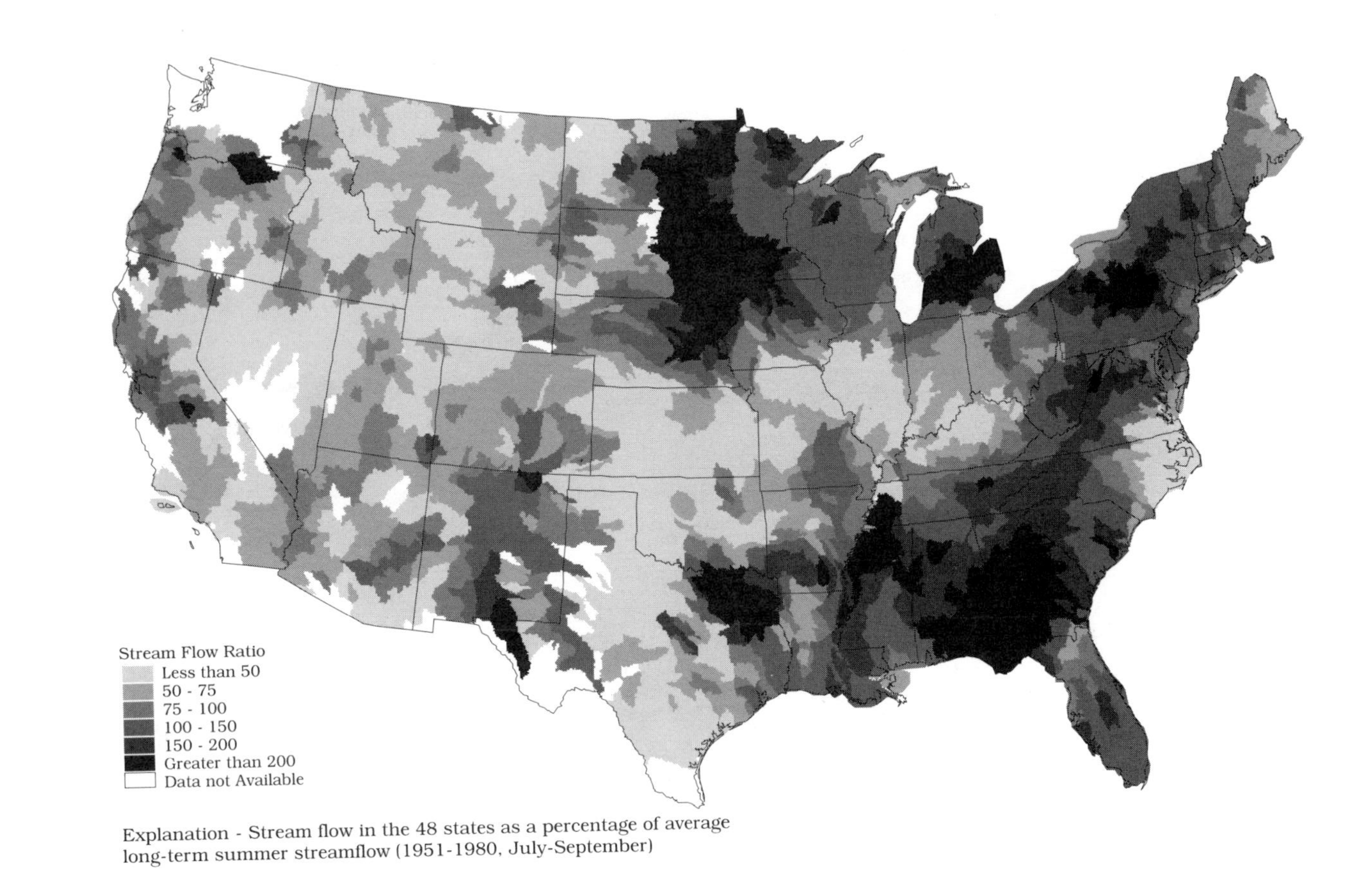

Explanation - Stream flow in the 48 states as a percentage of average long-term summer streamflow (1951-1980, July-September)

Figure F-42 Hydrologic conditions during the summer (July–September) 1994.

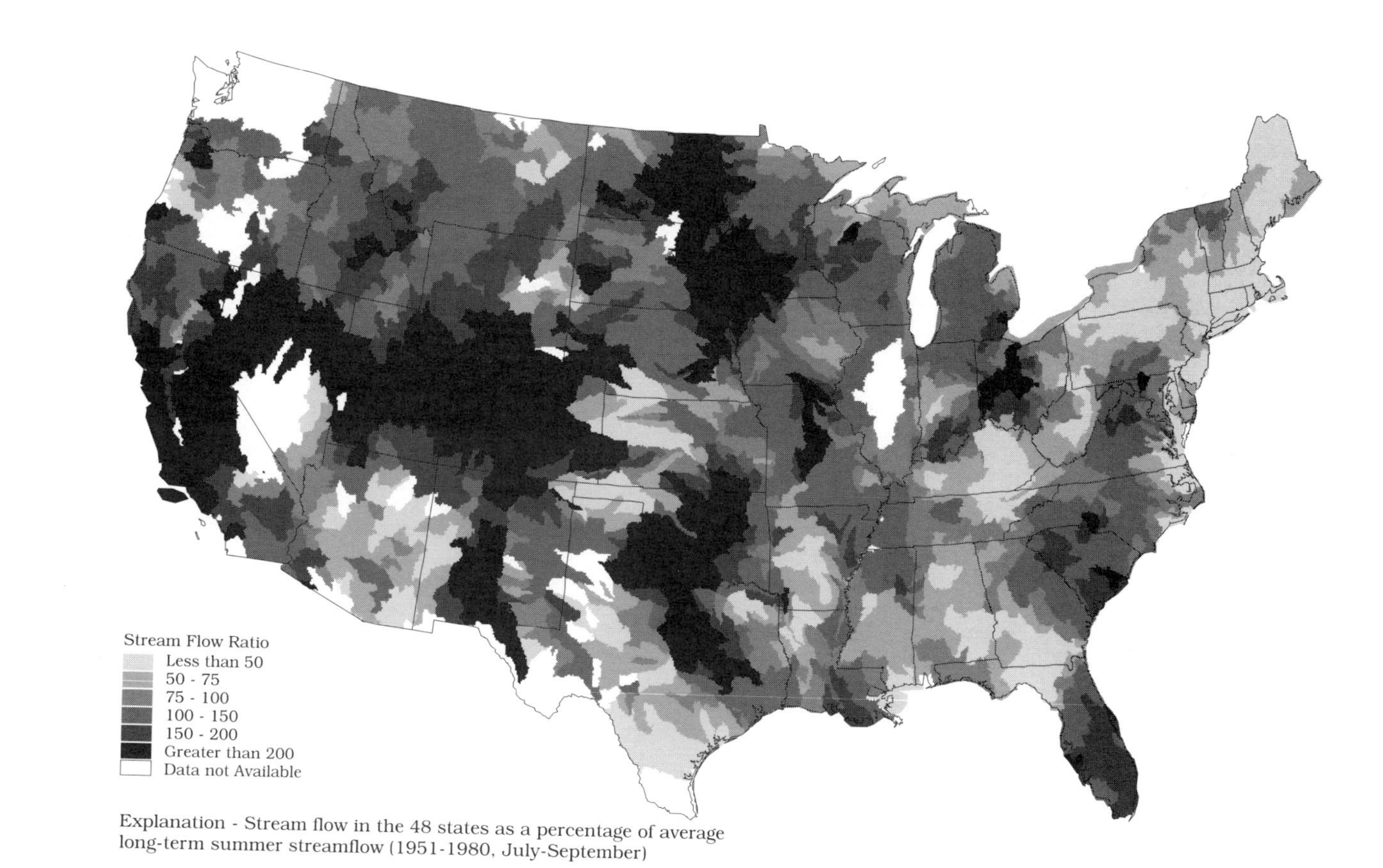

Figure F-43 Hydrologic conditions during the summer (July–September) 1995.

APPENDIX G

Municipal and Industrial Wastewater Loads by Major River Basin Before and After the Clean Water Act: 1950, 1973, and *ca.* 1995

TABLES

G-1 River basin municipal and industrial effluent BOD_5 loads: 1950
G-2 River basin municipal and industrial effluent BOD_5 loads: 1973
G-3 River basin municipal and industrial effluent TSS loads: 1973
G-4 River basin municipal and industrial effluent BOD_5 loads: *ca.* 1995
G-5 River basin municipal and industrial effluent TSS loads: *ca.* 1995

DATA SOURCES FOR MUNICIPAL AND INDUSTRIAL LOADS

1950

Population served and facilities inventory data compiled for 1950 for raw, primary, and secondary plants by major river basins (see USPHS, 1951; pp. 31–33). BOD_5 loads reported by USPHS (1951) as population equivalents (PE) for municipal and industrial sources. PE conversion factor (0.347 lb BOD_5 per person per day) assigned to match estimated municipal BOD_5 load (7,526 metric tons per day, mt/day) computed from unit wastewater flow (165 gpcd) and population served (91.76 million) by raw (35.26 million), primary (24.6 million), and secondary (31.89 million) wastewater treatment plants in 1950.

Data not available for Souris–Red Rainy Basin; municipal and industrial loads included in total for Missouri River–Hudson Bay Basin (see MacKichan, 1957).

Data not available for Upper and Lower Colorado; municipal and industrial loads included as total for Colorado Basin; Colorado Basin loads assigned to Upper Colorado Basin (see MacKichan, 1957).

Data not available for Rio Grande basin; Rio Grande Basin included in Western Gulf [Texas-Gulf]; data not available for Arkansas-Red-White and Texas-Gulf (see MacKichan, 1957).

1973

Industrial and municipal loads of BOD_5 and TSS estimated as million pounds per year for 1973 for the 48 States (see Table B-1 in Luken et al., 1976) for the National Commission on Water Quality.

Industrial Loads Influent and effluent loads for numerous industrial categories were estimated based on residual generation coefficients that defined industrial wastewater flow and residual loads per unit of product output or per unit of raw material processed. Industrial categories that accounted for the majority of process water use were investigated in greater detail (in-depth) than other categories (general). Industries evaluated in-depth included pulp and paper; petroleum refining; textiles; iron and steel; plastics and synthetics; organic chemicals; and inorganic chemicals. Effluent loads for the 1973 industrial inventory were based on removal efficiencies of BOD and TSS assigned for unit processes associated with primary, secondary, and tertiary industrial waste treatment. Total residual generation loads were reduced to total effluent discharge loads of 4,350 million lb BOD per year and 722,610 million lb TSS per year for all industrial categories.

Municipal Loads Population served in 1973 (ca. 156 million) by municipal wastewater treatment plants (24,209) is based on data obtained from the USEPA 1974 Needs Survey (USEPA, 1975). National municipal wastewater flow (~20,300 mgd; ~130 gpcd) was estimated as the sum of industrial flow reported to municipal facilities and population served multiplied by a constant 97.85 gpcd that accounted for residential, commercial, and institutional discharges. National influent loads were based on 1974 Needs Survey data where population served data were available for municipal facilities; otherwise, influent levels were assigned as follows: $BOD_5 = 200$ mg/L; TSS = 230 mg/L; TN = 40 mg/L, and TP = 10 mg/L.

Estimates of national municipal effluent loads of 5,740 million lb BOD_5 per year and 6,060 million lb TSS per year were based on removal efficiencies given in Luken et al. (1976) as:

	Removal Effiency (%)			
Type of Treatment	BOD_5	TSS	TN	TP
Conventional Treatment				
Primary	33	52	10	9
Activated sludge	78.6	73	20	20
Trickling filter	78.6	68.2	20	10
Lagoon	78.6	68.2	20	10
Disinfection	0	0	0	0
Advanced Treatment				
P removal	67	71	0	94
Nitrification	70	57	0	0
Nitrification and denitrification	0	0	90	0
Organics removal	60	60	0	20
Polishing lagoon	78.6	68	20	10
Carbon adsorption	50	72	0	50

1995

Industrial and municipal loads of BOD_5 and TSS estimated as million pounds per year *ca.* 1995 for the 48 States by Bondelid et al. (2000) for the National Water Pollution Control Assessment Model Version 1.1 (NWPCAM). The primary USEPA data sources used to identify, locate, and estimate annual effluent flow and loading rates of BOD_5 and TSS for municipal and industrial point sources were the Permit Compliance System (PCS) and the Clean Water Needs Survey (CWNS) national databases.

The PCS database provides information to characterize pollutant loads from both municipal and industrial wastewater dischargers. PCS defines a wastewater facility as either "major" or "minor," based on the population served (major $>$10,000), effluent flow rate (major $>$ 1 mgd), and/or the potential threat to human health or the aquatic environment. PCS further defines the type of wastewater facility using the Standard Industrial Classification (SIC) code for municipal wastewater facilities (SIC = 4952) and industrial facilities (e.g., inorganic chemicals; pulp and paper; iron and steel).

The CWNS database is used by the states and USEPA to track the technical status and needs of municipal wastewater facilities in the United States. The CWNS database provides information about the population served and the level of wastewater treatment for each facility (e.g., primary, secondary, advanced treatment). Data taken from the 1996 Clean Water Needs Survey (USEPA, 1997) were used to characterize municipal wastewater loads *ca.* 1995.

Industrial Loads The point source inventory compiled by Bondelid et al. (2000) included ~27,000 major and minor industrial facilities. Effluent loads were assigned using composite categories of SIC industrial classifications as compiled by NOAA (1994). Where effluent loading data were not available from the PCS database, typical pollutant concentrations assigned for industrial categories as default values were used to estimate pollutant loads.

Municipal Loads Population served in 1996 (189.7 million) by municipal wastewater treatment plants is based on data obtained from the USEPA 1996 Needs Survey (USEPA, 1997). The municipal point source inventory compiled by Bondelid et al. (2000), using data extracted from the PCS and CWNS databases, included ~15,500 municipal facilities. Effluent loads were estimated using either effluent flow and effluent concentration data extracted from PCS for a discharger or the level of wastewater treatment assigned for the discharger in the CWNS database. Where effluent loading data were not available in the PCS database, influent concentrations of 215 mg/L BOD_5 and and 215 mg/L TSS were assumed with the following typical removal efficiencies used to estimate the pollutant load of BOD_5 and TSS for a discharger:

	Removal Efficiency (%)	
Type of Treatment	BOD_5	TSS
Raw (untreated)	0	0
Primary	35	50
Advanced primary	50	70
Secondary	85	92
Advanced secondary	90	97
Advanced treatment	95	98

CONVERSION FACTORS

1 PE	=	Population equivalent
1 PE	=	0.347 lb BOD_5 per person per day for 1950 waste loads
1 metric ton (mt)	=	2204.6 lb
1 mt/day	=	1.2427 million lb per year

REFERENCES

Bondelid, T., S. Unger, and A. Stoddard. 2000. National water pollution control assessment model (NWPCAM) Version 1.1. RTI Project Number 92U-7640-031. Final report prepared for U. S. Environmental Protection Agency, Office of Policy, Economics and Innovation, Washington, DC, by Research Triangle Institute, Research Triangle Park, NC. November.

Luken, R. A., D. J. Basta, and E. H. Pechan. 1976. The national residuals discharge inventory. An analysis of the generation, discharge, cost of control and regional distribution of liquid wastes to be expected in achieving the requirements of Public Law 92–500.NCWQ 75/104. NTIS Accession No. PB-252–288. National Commission on Water Quality, Washington, DC.

MacKichan, K. A. 1957. Estimated use of water in the United States, 1955. U.S. Geological Survey, Circular 398, Reston, VA. 18 pp.

NOAA. 1994. Gulf of Maine point source inventory; A summary by watershed for 1991. National Coastal Pollutant Discharge Inventory. National Oceanic Atmospheric Administration, Strategic Environmental Assessments Division, Pollution Sources Characterization Branch, Silver Spring, MD.

USEPA. 1975. Procedural Guidance: 1974 joint State-EPA survey of needs for municipal wastewater treatment facilities. U.S. Environmental Protection Agency, Washington, DC.

USEPA. 1997. 1996 Clean Water Needs Survey (CWNS), conveyance, treatment, and control of municipal wastewater, combined sewer overflows and stormwater runoff: Summaries of technical data. EPA-832–R-97–003. U.S. Environmental Protection Agency, Office of Water Program Operations, Washington, DC.

USPHS. 1951. Water pollution in the United States. A report on the polluted conditions of our waters and what is needed to restore their quality. NTIS No. PB-218–308/BA. U.S. Federal Security Agency, U.S. Public Health Service, Washington, DC.

TABLE G-1 River Basin Municipal and Industrial Effluent BOD_5 Loads: 1950

Major River Basin	Municipal (PE)	Industrial (PE)	Municipal + Industrial (PE)	Municipal (mt/day)	Industrial (mt/day)	Municipal + Industrial (mt/day)	Municipal (%)	Industrial (%)
1 New England Basin	4,295,668	1,495,300	5,790,968	676	235	912	74	26
2 Mid-Atlantic Basin	6,240,831	3,597,598	9,838,429	982	566	1,549	63	37
3 South Atlantic-Gulf	3,257,344	3,535,814	6,793,158	513	557	1,069	48	52
4 Great Lakes Basin	2,029,507	6,169,338	8,198,845	319	971	1,291	25	75
5 Ohio River Basin	8,767,395	4,464,602	13,231,997	1,380	703	2,083	66	34
6 Tennessee River Basin	1,069,445	1,412,730	2,482,175	168	222	391	43	57
7 Upper Mississippi Basin	5,296,520	2,257,040	7,553,560	834	355	1,189	70	30
8 Lower Mississippi Basin	1,232,545	1,357,244	2,589,789	194	214	408	48	52
9 Souris-Red-Rainy Basin	0	0	0	0	0	0	NA	NA
10 Missouri River Basin	3,271,173	8,486,503	11,757,676	515	1,336	1,851	28	72
11 Arkansas-White-Red Basin	0	0	0	0	0	0	NA	NA
12 Texas Gulf Basin	0	0	0	0	0	0	NA	NA
13 Rio Grande River Basin	0	0	0	0	0	0	NA	NA
14 Upper Colorado Basin	4,375	279,458	283,833	1	44	45	2	98
15 Lower Colorado Basin	0	0	0	0	0	0	NA	NA
16 Great Basin	623,620	1,201,150	1,824,770	98	189	287	34	66
17 Pacific Northwest Basin	3,002,187	12,825,773	15,827,960	473	2,019	2,491	19	81
18 California Basin	8,723,140	2,797,000	11,520,140	1,373	440	1,813	76	24
USA (48 States)	47,813,750	49,879,550	97,693,300	7,526	7,851	15,377	49	51

Data Source: USPHS (1951)

TABLE G-2 River Basin Municipal and Industrial Effluent BOD_5 Loads: 1973

Major River Basin	Municipal (10^6 lb/year)	Industrial (10^6 lb/year)	Municipal + Industrial (10^6 lb/year)	Municipal (mt/day)	Industrial (mt/day)	Municipal + Industrial (mt/day)	Municipal (%)	Industrial (%)
1 New England Basin	340	150	490	423	186	609	69	31
2 Mid-Atlantic Basin	1,260	570	1,830	1,566	708	2,274	69	31
3 South Atlantic-Gulf	610	640	1,250	758	795	1,553	49	51
4 Great Lakes Basin	610	420	1,030	758	522	1,280	59	41
5 Ohio River Basin	450	310	760	559	385	944	59	41
6 Tennessee River Basin	90	160	250	112	199	311	36	64
7 Upper Mississippi Basin	300	270	570	373	336	708	53	47
8 Lower Mississippi Basin	80	310	390	99	385	485	21	79
9 Souris-Red-Rainy Basin	10	10	0	12	12	25	50	50
10 Missouri River Basin	250	40	290	311	50	360	86	14
11 Arkansas-White-Red Basin	120	130	250	149	162	311	48	52
12 Texas Gulf Basin	160	620	780	199	770	969	21	79
13 Rio Grande River Basin	10	20	30	12	25	37	33	67
14 Upper Colorado Basin	10	0	10	12	0	12	100	0
15 Lower Colorado Basin	60	10	70	75	12	87	86	14
16 Great Basin	20	10	30	25	12	37	67	33
17 Pacific Northwest Basin	100	470	570	124	584	708	18	82
18 California Basin	1,260	210	1,470	1,566	261	1,827	86	14
USA (48 States)	5,740	4,350	10,070	7,133	5,406	12,539	57	43

Data Source: Lukens et al. (1976)

TABLE G-3 River Basin Municipal and Industrial Effluent TSS Loads: 1973

Major River Basin	Municipal (10^6 lb/year)	Industrial (10^6 lb/year)	Municipal + Industrial (10^6 lb/year)	Municipal (mt/day)	Industrial (mt/day)	Municipal + Industrial (mt/day)	Municipal (%)	Industrial (%)
1 New England Basin	330	1,900	2,230	410	2,361	2,771	15	85
2 Mid-Atlantic Basin	1,280	9,970	11,250	1,591	12,390	13,981	11	89
3 South Atlantic-Gulf	670	3,170	3,840	833	3,939	4,772	17	83
4 Great Lakes Basin	690	27,600	28,290	857	34,299	35,157	2	98
5 Ohio River Basin	490	4,740	5,230	609	5,891	6,499	9	91
6 Tennessee River Basin	100	9,530	9,630	124	11,843	11,968	1	99
7 Upper Mississippi Basin	300	7,100	7,400	373	8,823	9,196	4	96
8 Lower Mississippi Basin	110	1,500	1,610	137	1,864	2,001	7	93
9 Souris-Red-Rainy Basin	10	120	0	12	149	162	8	92
10 Missouri River Basin	250	392,000	392,250	311	487,151	487,461	0	100
11 Arkansas-White-Red Basin	120	5,610	5,730	149	6,972	7,121	2	98
12 Texas Gulf Basin	190	2,570	2,760	236	3,194	3,430	7	93
13 Rio Grande River Basin	10	16,490	16,500	12	20,493	20,505	0	100
14 Upper Colorado Basin	10	3,790	3,800	12	4,710	4,722	0	100
15 Lower Colorado Basin	80	94,620	94,700	99	117,587	117,687	0	100
16 Great Basin	30	86,800	86,830	37	107,869	107,906	0	100
17 Pacific Northwest Basin	100	39,600	39,700	124	49,212	49,336	0	100
18 California Basin	1,290	15,500	16,790	1,603	19,262	20,865	8	92
USA (48 States)	6,060	722,610	728,540	7,531	898,010	905,541	1	99

Data Source: Lukens et al. (1976)

TABLE G-4 River Basin Municipal and Industrial Effluent BOD_5 Loads: *ca.* 1995

Major River Basin	Municipal (10^6 lb/year)	Industrial (10^6 lb/year)	Municipal + Industrial (10^6 lb/year)	Municipal (mt/day)	Industrial (mt/day)	Municipal + Industrial (mt/day)	Municipal (%)	Industrial (%)
1 New England Basin	205	105	310	255	130	385	66	34
2 Mid-Atlantic Basin	260	456	717	324	567	891	36	64
3 South Atlantic-Gulf	114	333	447	142	414	555	25	75
4 Great Lakes Basin	194	355	549	241	441	682	35	65
5 Ohio River Basin	134	97	232	167	121	288	58	42
6 Tennessee River Basin	36	54	90	45	67	111	40	60
7 Upper Mississippi Basin	121	63	184	150	79	229	66	34
8 Lower Mississippi Basin	56	228	284	70	283	353	20	80
9 Souris-Red-Rainy Basin	2	6	9	3	8	11	28	72
10 Missouri River Basin	72	90	162	90	112	202	45	55
11 Arkansas-White-Red Basin	33	80	112	41	99	139	29	71
12 Texas Gulf Basin	63	527	589	78	654	732	11	89
13 Rio Grande River Basin	7	1	8	9	1	10	93	7
14 Upper Colorado Basin	2	3	5	3	3	6	48	52
15 Lower Colorado Basin	16	4	20	20	5	24	81	19
16 Great Basin	9	9	18	12	11	22	52	48
17 Pacific Northwest Basin	58	124	182	72	155	226	32	68
18 California Basin	228	76	304	284	95	378	75	25
USA (48 States)	1,612	2,609	4,221	2,004	3,243	5,246	38	62

Data Source: Bondelid et al. (2000)

Notes: Industrial load is sum of minor and major industrial dischargers.

Population Served in 1996 = 189.7 million

1 metric ton per day = 1.2427 × (million lb per year)

1 metric ton (mt) = 2204.6 lb

TABLE G-5 River Basin Municipal and Industrial Effluent TSS Loads: *ca.* 1995

Major River Basin	Municipal (10^6 lb/year)	Industrial (10^6 lb/year)	Municipal + Industrial (10^6 lb/year)	Municipal (mt/day)	Industrial (mt/day)	Municipal + Industrial (mt/day)	Municipal (%)	Industrial (%)
1 New England Basin	167	616	783	207	766	973	21	79
2 Mid-Atlantic Basin	283	2,330	2,613	352	2,896	3,247	11	89
3 South Atlantic-Gulf	132	2,335	2,467	164	2,902	3,066	5	95
4 Great Lakes Basin	221	2,432	2,653	275	3,022	3,297	8	92
5 Ohio River Basin	139	1,129	1,267	172	1,403	1,575	11	89
6 Tennessee River Basin	42	162	204	53	202	254	21	79
7 Upper Mississippi Basin	115	1,264	1,378	142	1,570	1,713	8	92
8 Lower Mississippi Basin	68	558	626	84	694	778	11	89
9 Souris-Red-Rainy Basin	2	29	0	2	36	38	6	94
10 Missouri River Basin	71	381	452	88	473	561	16	84
11 Arkansas-White-Red Basin	36	152	188	45	189	234	19	81
12 Texas Gulf Basin	73	612	685	91	760	851	11	89
13 Rio Grande River Basin	11	1	12	13	1	14	92	8
14 Upper Colorado Basin	3	4	7	3	5	9	37	63
15 Lower Colorado Basin	18	9	27	22	11	33	66	34
16 Great Basin	10	15	25	12	19	31	39	61
17 Pacific Northwest Basin	64	203	267	80	253	332	24	76
18 California Basin	231	619	850	287	769	1,057	27	73
USA (48 States)	1,684	12,851	14,504	2,092	15,971	18,063	12	88

Data Source: Bondelid et al. (2000)

Notes: Industrial load is sum of minor and major industrial dischargers.

Population Served in 1996 = 189.7 million

1 metric ton per day = 1.2427 × (million lb per year)

1 metric ton (mt) = 2204.6 lb

APPENDIX H

Municipal and Industrial Water Withdrawals by Major River Basin: 1940–1995

TABLES

H-1 Public water supply freshwater withdrawals (million gallons per day, mgd)
H-2 Self-supplied industrial water use (million gallons per day, mgd)

FIGURES

H-1 Trends in public water supply withdrawals and per capita use, 1940–1995. *Sources:* USDOC, 1975; Solley et al. 1998.
H-2 Trends in Self-Supplied Industrial Water Withdrawals and Industrial Production Index, 1940–1995. *Sources:* USDOC (1996); Solley et al. (1998); Board of Governors of Federal Reserve System.

DATA SOURCES FOR MUNICIPAL AND INDUSTRIAL WATER USE

1940 Municipal water use river basin subtotals not available; USA total of 10,000 mgd from USDOC (1975).
1940 Industrial water use; USA total of 29,000 mgd from USDOC (1996), p. 233.
1945 Municipal water use river basin subtotals not available; USA total of 12,000 mgd from USDOC (1975); industrial water use data not available.
1950 MacKichan (1951); municipal water use river basin subtotals not available; USA total of 13,640 mgd from MacKichan (1951); industrial water use USA total of 77,216 mgd reported in MacKichan (1951)as 5,525 mgd (groundwater) and 71,691 mgd (surface water) included cooling water withdrawals; data adjusted by 40,000 mgd as given in 1980 USGS Water Use report (Solley et al., 1983).
1955 MacKichan (1957); river basins (regions) defined as:
- 01 New England = North Atlantic
- 02 Mid-Atlantic = Upper Hudson + Lower Hudson + Delaware + Chesapeake

03 South Atlantic–Gulf = Southeast
04 Great Lakes = Eastern Great Lakes/St.Lawrence + Western Great Lakes
05 Ohio
06 Tennessee = Tennessee-Cumberland
07 Upper Mississippi
08 Lower Mississippi
09 Souris Red Rainy = N/A, included in total for Missouri–Hudson Bay
10 Missouri = Missouri–Hudson Bay + Souris Red Rainy
11 Arkansas-White-Red
12 Texas-Gulf = Western Gulf
13 Rio Grande = N/A, included in Western Gulf
14 Upper Colorado = N/A, included as total for Colorado Basin
15 Lower Colorado = N/A, included as total for Colorado Basin
16 Great Basin
17 Pacific Northwest
18 California = South Pacific

1960 MacKichan and Kammerer (1957); river basins (regions) defined as:
01 New England
02 Mid-Atlantic = Delaware-Hudson + Chesapeake
03 South Atlantic–Gulf = South Atlantic + Eastern Gulf
04 Great Lakes = Eastern Great Lakes/St.Lawrence + Western Great Lakes
05 Ohio
06 Tennessee = Tennessee–Cumberland
07 Upper Mississippi
08 Lower Mississippi
09 Souris Red Rainy = Hudson Bay
10 Missouri = Upper Missouri + Lower Missouri
11 Arkansas-White-Red = Upper Arkansas–Red + Lower Arkansas–Red
12 Texas–Gulf = Western Gulf
13 Rio Grande = N/A, included in Western Gulf
14 Upper Colorado = N/A, included as total for Colorado Basin
15 Lower Colorado = N/A, included as total for Colorado Basin
16 Great Basin
17 Pacific Northwest
18 California = South Pacific

1965 Murray (1968); river basins (regions) defined as:
01 New England
02 Mid-Atlantic = Delaware-Hudson + Chesapeake
03 South Atlantic–Gulf = South Atlantic + Eastern Gulf
04 Great Lakes = Eastern Great Lakes/St.Lawrence + Western Great Lakes
05 Ohio
06 Tennessee = Tennessee-Cumberland

07 Upper Mississippi
08 Lower Mississippi
09 Souris Red Rainy = Hudson Bay
10 Missouri = Upper Missouri + Lower Missouri
11 Arkansas-White-Red = Upper Arkansas–Red + Lower Arkansas–Red
12 Texas-Gulf = Western Gulf
13 Rio Grande = N/A, included in Western Gulf
14 Upper Colorado = N/A, included as total for Colorado Basin
15 Lower Colorado = N/A, included as total for Colorado Basin
16 Great Basin
17 Pacific Northwest
18 California = South Pacific

1970 Murray and Reeves (1972)
1975 Murray and Reeves (1977)
1980 Solley et al.(1983)
1985 Solley et al. (1988); industrial water use data taken from Table 11, total = fresh and saline withdrawals + fresh and saline consumptive uses
1990 Solley et al.(1993); industrial water use data taken from Table 3, total withdrawals (groundwater + surface) = commercial + industrial + mining (fresh and saline)
1995 Solley et al.(1998); industrial water use data taken from Table 3, total withdrawals (groundwater + surface) = commercial + industrial + mining (fresh and saline)

Public Supply Public (water) supply refers to water withdrawn (from freshwater sources) by public and private water suppliers and delivered to a variety of users for domestic or household use, public uses, industrial and commercial uses. Domestic uses include drinking, food preparation, bathing, washing clothes and dishes, toilet flushing, watering lawns and gardens. Public uses include firefighting, street washing, municipal parks and swimming pools. Industrial and commercial establishments use public water supplies as either their primary source or as a supplemental source. Commercial users include hotels, restaurants, laundry services, public and private office facilities, and civilian and military institutions (Solley et al., 1983).

Self-Supplied Industrial Self-supplied industrial water use is categorized as "other" self-supplied water uses; thermoelectric power (electric utility) water withdrawals are not included in Table H-2. "Other" self-supplied water using industries include, but are not limited to, steel, chemical and allied products, paper and allied products, mining and petroleum refining (Solley et al., 1983). Beginning with the 1985 report, mining and commercial water uses were excluded from the statistics for "other" self-supplied industrial uses (Solley et al., 1988).

Industrial Production Index Monthly Industrial Production Index (IPI) data from 1936–2000 averaged over 5-year periods (1936–1940, 1941–1945, . . . , 1986–1990, 1991–1995). IPI data published by Board of Governors of U. S. Federal

Reserve System, Washington, DC. IPI time series data downloaded from http://www.economagic.com, U. S. Government, Federal Reserve, Board of Governors, Industrial Production by Market Group, Total Index (SA, NSA) (Economagic, 2001).

REFERENCES

Economagic. 2001. Total Industrial Production Index (SA, NSA)from 1936–2000. Board of Governors of Federal Reserve System, Washington, DC. IPI time series data downloaded from http://www.economagic.com, U. S. Government, Federal Reserve, Board of Governors, Industrial Production by Market Group, Total Index (SA, NSA).

MacKichan, K. A. 1951. Estimated use of water in the United States, 1950. USGS Circular 115, 13 pp. U.S. Geological Survey, Washington, DC.

MacKichan, K. A. 1957. Estimated use of water in the United States, 1955. USGS Circular 398, 18 pp. U.S. Geological Survey, Washington, DC.

MacKichan, K. A. and J. C. Kammerer. 1957. Estimated use of water in the United States, 1960. USGS Circular 456, 26 pp. U.S. Geological Survey, Washington, DC.

Murray, C. R. 1968. Estimated use of water in the United States, 1965. USGS Circular 556, 53 pp. U.S. Geological Survey, Washington, DC.

Murray, C. R. and E. B. Reeves. 1972. Estimated use of water in the United States, 1970. USGS Circular 676, 37 pp. U.S. Geological Survey, Washington, DC.

Murray, C. R. and E. B. Reeves. 1977. Estimated use of water in the United States, 1975. USGS Circular 765, 37 pp. U.S. Geological Survey, Washington, DC.

Solley, W. B., E. B. Chase, and W. B. Mann. 1983. Estimated use of water in the United States, 1980. USGS Circular 1001, 56 pp. U.S. Geological Survey, Alexandria, VA.

Solley, W. B., C. F. Merk, and R. R. Pierce. 1988. Estimated use of water in the United States, 1985. USGS Circular 1004, 82 pp. U.S. Geological Survey, Denver, CO.

Solley, W. B., R. R. Pierce, and H. A. Perlman. 1993. Estimated use of water in the United States, 1990. USGS Circular 1081, 76 pp. U.S. Geological Survey, Denver, CO.

Solley, W. B., R. R. Pierce, and H. A. Perlman. 1998. Estimated use of water in the United States, 1995. USGS Circular 1200, 71 pp. U.S. Geological Survey, Denver, CO.

USDOC. 1975. Historical Statistics of the United States, Colonial Times to 1970, Series J 92-103. U. S. Department of Commerce, Bureau of Census, Washington, DC.

USDOC. 1996. Statistical Abstract of the United States, No. 375, U. S. Department of Commerce, Bureau of Census, Washington, DC.

TABLE H-1 Public Water Supply Freshwater Withdrawals (million gallons per day, mgd)

Major River Basin	1940	1945	1950	1955	1960	1965	1970	1975	1980	1985	1990	1995
1 New England Basin	NA	NA	NA	950	1,000	5,500	1,400	1,400	1,500	1,453	1,400	1,440
2 Mid-Atlantic Basin	NA	NA	NA	3,499	3,880	0	5,200	5,300	5,400	6,042	5,980	6,000
3 South Atlantic-Gulf	NA	NA	NA	1,340	1,790	2,000	2,700	3,100	3,800	4,200	4,850	5,470
4 Great Lakes Basin	NA	NA	NA	3,290	3,400	3,800	4,400	3,100	3,900	4,082	4,340	4,420
5 Ohio River Basin	NA	NA	NA	1,620	1,500	1,800	2,100	2,200	2,200	2,414	2,530	2,680
6 Tennessee River Basin	NA	NA	NA	241	320	250	300	330	410	469	511	574
7 Upper Mississippi Basin	NA	NA	NA	580	1,000	1,200	1,600	2,900	1,900	1,876	1,890	1,880
8 Lower Mississippi Basin	NA	NA	NA	500	380	500	610	750	920	673	1,040	1,070
9 Souris-Red-Rainy Basin	NA	NA	NA	0	33	35	48	48	57	64	72	66
10 Missouri River Basin	NA	NA	NA	800	810	970	1,000	1,200	1,400	1,585	1,620	1,570
11 Arkansas-White-Red Basin	NA	NA	NA	610	630	730	730	930	1,600	1,318	1,400	1,550
12 Texas-Gulf Basin	NA	NA	NA	1,050	1,200	970	1,100	1,400	3,000	2,453	2,520	2,840
13 Rio Grande Basin	NA	NA	NA	0	0	250	310	350	320	455	533	487
14 Upper Colorado Basin	NA	NA	NA	190	270	53	58	77	120	127	118	141
15 Lower Colorado Basin	NA	NA	NA	0	0	290	390	510	720	829	1,070	1,170
16 Great Basin	NA	NA	NA	220	280	270	320	380	810	529	610	605
17 Pacific Northwest Basin	NA	NA	NA	820	1,200	1,200	1,300	1,200	1,300	1,607	1,580	1,910
18 California Basin	NA	NA	NA	1,270	2,700	4,000	3,400	3,700	4,100	5,301	5,750	5,610
USA (48 States)	10,000	12,000	13,640	16,980	20,393	23,818	26,966	28,875	33,457	35,477	37,814	39,483
19 Alaska	NA	NA	NA	NA	23	32	60	81	53	76	92	81
20 Hawaii	NA	NA	NA	NA	85	110	140	180	200	204	238	214
21 Puerto Rico	NA	NA	NA	NA	69	140	200	290	350	395	411	437
USA (Total)	10,000	12,000	13,640	16,980	20,570	24,100	27,366	29,426	34,060	36,152	38,555	40,215
Population of USA (millions)	—	—	150.7	164.0	179.3	193.8	205.9	216.4	229.6	242.4	252.3	267.1
Population served (millions)	—	—	—	115.0	136.0	152.6	165.0	175.0	186.0	197.4	210.0	225.0
Per capita use (gpcd)	—	—	—	147.7	151.3	157.9	165.9	168.1	183.1	183.1	183.6	178.7

TABLE H-2 Self-Supplied Industrial Water Use (million gallons per day, mgd)

Major River Basin	1940	1945	1950	1955	1960	1965	1970	1975	1980	1985	1990	1995
1 New England Basin	NA	NA	NA	2,097	1,410	1,560	1,520	1,670	1,578	1,004	700	267
2 Mid-Atlantic Basin	NA	NA	NA	7,286	7,080	7,490	8,000	6,200	5,500	4,525	3,750	2,569
3 South Atlantic-Gulf	NA	NA	NA	2,448	3,490	3,330	3,810	4,710	6,230	4,064	3,484	3,308
4 Great Lakes Basin	NA	NA	NA	5,776	7,725	9,025	9,000	7,600	6,120	4,790	4,558	4,723
5 Ohio River Basin	NA	NA	NA	6,841	7,234	8,627	5,858	6,020	5,024	3,740	3,481	4,210
6 Tennessee River Basin	NA	NA	NA	1,920	1,500	1,100	1,400	1,600	2,000	1,989	1,338	1,103
7 Upper Mississippi Basin	NA	NA	NA	1,875	1,720	1,618	1,719	1,815	3,315	1,199	1,385	1,334
8 Lower Mississippi Basin	NA	NA	NA	1,770	1,425	2,640	4,780	4,500	4,530	2,501	2,819	2,931
9 Souris-Red Rainy Basin	NA	NA	NA	0	91	104	83	33	9	50	58	24
10 Missouri River Basin	NA	NA	NA	1,399	478	466	720	556	706	280	527	530
11 Arkansas-Red-White Basin	NA	NA	NA	777	1,064	915	685	1,070	937	638	898	893
12 Texas-Gulf Basin	NA	NA	NA	4,060	1,317	4,010	4,900	3,070	1,620	3,560	2,787	2,619
13 Rio Grande Basin	NA	NA	NA	0	0	213	360	101	17	16	137	144
14 Upper Colorado Basin	NA	NA	NA	236	133	40	68	96	594	11	91	50
15 Lower Colorado Basin	NA	NA	NA	0	0	140	210	280	250	151	374	243
16 Great Basin	NA	NA	NA	160	322	211	335	316	563	43	317	352
17 Pacific Northwest Basin	NA	NA	NA	1,163	2,148	1,830	1,837	3,441	3,742	1,235	1,799	2,223
18 California Basin	NA	NA	NA	856	1,000	1,250	1,080	1,190	1,300	1,007	758	1,202
USA (48 States)	NA	NA	NA	38,664	38,137	44,569	46,365	44,268	44,035	30,803	29,261	28,725
19 Alaska	NA	NA	NA	NA	82	100	110	90	130	125	511	208
20 Hawaii	NA	NA	NA	NA	145	136	286	205	52	26	85	66
21 Puerto Rico	NA	NA	NA	NA	320	310	580	380	1,050	27	65	39
USA (Total)	29,000	NA	37,216	38,664	38,684	45,115	47,341	44,943	45,267	30,981	29,922	29,038
Industrial Production Index	11.8	22.0	21.9	29.1	34.3	42.6	57.3	65.7	77.0	82.4	93.9	104.8

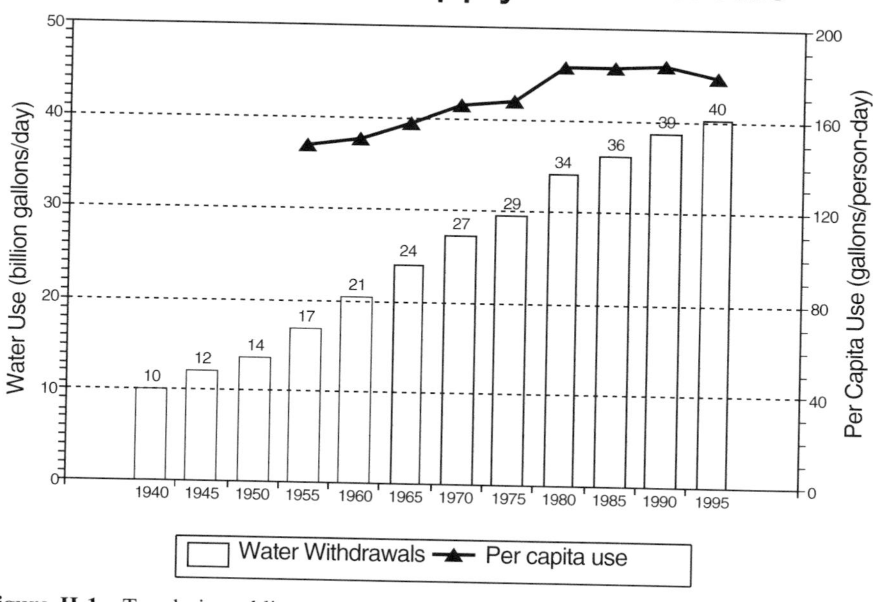

Figure H-1 Trends in public water supply withdrawals and per capita use, 1940–1995. *Sources:* USDOC, 1975; Solley et al., 1998.

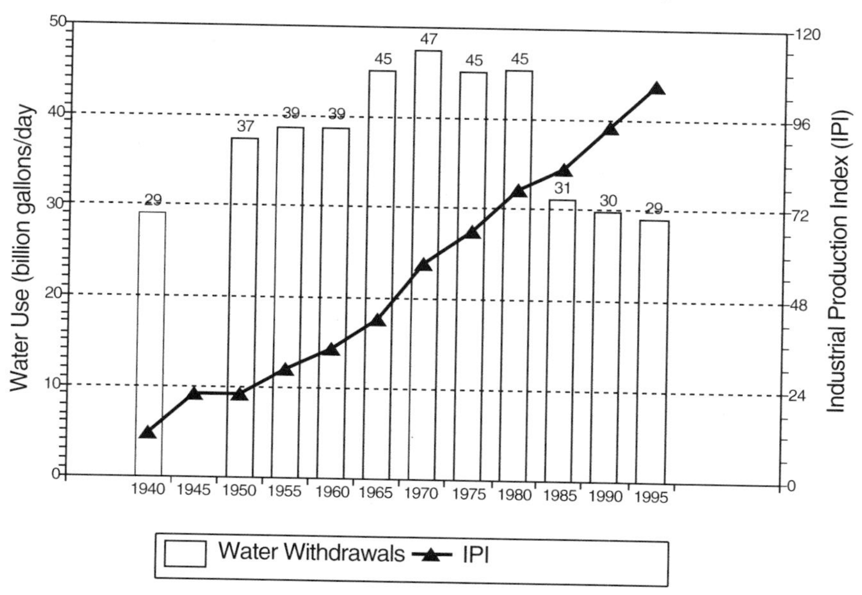

Figure H-2 Trends in self-supplied industrial water withdrawals and industrial production index, 1940–1995. *Sources:* USDOC, 1996; Solley et al., 1998; Economagic, 2001.

GLOSSARY

Activated sludge A secondary wastewater treatment process that removes organic matter by mixing air and recycled sludge bacteria with sewage to promote decomposition.

Advanced primary treatment Waste treatment process that incorporates primary sedimentation of suspended solids with chemical addition and flocculation to increase the overall removal of organic solids. Advanced primary treatment typically achieves about 50% removal of suspended solids and BOD.

Advanced secondary treatment Biological or chemical treatment processes added to a secondary treatment plant including a conventional activated sludge to increase the removal of solids and BOD. Typical removal rates for advanced secondary plants are on the order of 90% removal of solids and BOD.

Advanced wastewater treatment (AWT) Wastewater treatment process that includes combinations of physical and chemical operation units designed to remove nutrients, toxic substances, or other pollutants. Advanced, or tertiary, treatment processes treat effluent from secondary treatment facilities using processes such as nutrient removal (nitrification, denitrification), filtration, or carbon adsorption. Tertiary treatment plants typically achieve about 95% removal of solids and BOD in addition to removal of nutrients or other materials.

Aerobic Environmental conditions characterized by the presence of dissolved oxygen; used to describe biological or chemical processes that occur in the presence of oxygen.

Algae Any organisms of a group of chiefly aquatic microscopic nonvascular plants; most algae have chlorophyll as the primary pigment for carbon fixation. As primary producers, algae serve as the base of the aquatic food web, providing food for zooplankton and fish resources. An overabundance of algae in natural waters is known as eutrophication. See **chlorophyll**, **plankton**, and **phytoplankton**.

Algal bloom Rapidly occurring growth and accumulation of algae within a body of water. It usually results from excessive nutrient loading and/or sluggish circulation regime with a long residence time. Persistent and frequent blooms can result in low oxygen conditions.

Ambient water quality Natural concentration of water quality constituents prior to mixing of either point or nonpoint source load of contaminants. Reference ambient concentration is used to indicate the concentration of a chemical that will not cause adverse impact to human health.

Ammonia Inorganic form of nitrogen; product of hydrolysis of organic nitrogen and denitrification. Ammonia is preferentially used by phytoplankton over nitrate for uptake of inorganic nitrogen.

Ammonia toxicity Under specific conditions of temperature and pH, the un-ionized component of ammonia can be toxic to aquatic life. The un-ionized component of ammonia increases with pH and temperature.

Anadromous Characteristic of fish that live in the ocean but spawn in freshwater. Examples: salmon and steelhead.

Anaerobic Environmental condition characterized by zero oxygen levels. Describes biological and chemical processes that occur in the absence of oxygen.

Anoxic Aquatic environmental conditions containing zero or little dissolved oxygen. See also **anaerobic**.

Anthropogenic Pertains to the (environmental) influence of human activities.

Assimilative capacity The amount of contaminant load (expressed as mass per unit time) that can be discharged to a specific river, lake, or estuary without exceeding water quality standards or criteria. Assimilative capacity is used to define the capacity of a waterbody to naturally absorb and use waste matter and organic materials without impairing water quality or harming aquatic life.

Bacterial decomposition Breakdown by oxidation, or decay, of organic matter by heterotrophic bacteria. Bacteria use the organic carbon in organic matter as the energy source for cell synthesis.

Baseflow Sustained, low-flow discharge rate in a stream derived from groundwater discharge into the stream channel. During extended periods of low precipitation, baseflow may account for most, or all, of the streamflow.

Benthic Refers to material, especially sediment, at the bottom of an aquatic ecosystem. It can be used to describe the organisms that live on, or in, the bottom of a waterbody.

Biochemical oxygen demand (BOD) The amount of oxygen per unit volume of water required to bacterially or chemically oxidize (stabilize) the oxidizable matter in water. Biochemical oxygen demand measurements are usually conducted over specific time intervals (5, 10, 20, 30 days). The term BOD_5 generally refers to standard 5-day BOD test.

Carbonaceous Pertaining to or containing organic carbon derived from plant and animal residues.

Carbonaceous BOD Biochemical oxygen demand accounted for by decomposition of organic carbon derived from plant and animal residues.

Chlorophyll A group of green photosynthetic pigments that occur primarily in the chloroplast of plant cells. The amount of chlorophyll-*a*, a specific pigment, is frequently used as a measure of algal biomass in natural waters.

Clean Water State Revolving Fund (CWSRF) Passed under the Amendments to the Clean Water Act (CWA) in 1987, the Clean Water State Revolving Fund (CWSRF) program replaced the long-running federal Construction Grants program. Under the CWSRF program, each state and Puerto Rico created revolving loan funds to provide independent and permanent sources of low-cost financing for a range of environmental water quality projects. As payments are made on loans, funds are recycled to fund additional water protection projects. While traditionally used to build or improve wastewater treatment plants, loans are used increasingly for agricultural, rural, and urban runoff control; wet weather flow control, including stormwater and sewer overflows; alternative treatment technologies; small decentralized systems; brownfields remediation; and estuary improvements projects. Also called State Revolving Fund (SRF).

Funds to establish SRF programs are provided through federal government grants (83% of total capitalization) and state matching funds (17% of total capitalization). To augment the federal and state capitalization, states may use the assets of the fund to support the issuance of bonds. At their option, some states choose to transfer some Construction Grant funds into their CWSRF.

From the beginning of capitalization in 1988 through 1999, federal contributions to the CWSRF program grew to $16.1 billion. Additional state match, state leveraged bonds, loan repayments, and fund earnings have increased CWSRF assets to over $30 billion since 1988.

Coliform bacteria A group of bacteria that normally live within the intestines of mammals, including humans. Coliform bacteria are used as an indicator of the presence of sewage in natural waters.

Combined sewer overflows (CSOs) A combined sewer carries both wastewater and stormwater runoff. CSOs discharged to receiving waters can result in contamination problems that may prevent the attainment of water quality standards.

Commercial water use Water used for motels, hotels, restaurants, office buildings, and other commercial operations.

Concentration Mass amount of a substance or material in a given unit volume of solution. Usually measured in milligrams per liter (mg/L) or parts per million (ppm).

Confluence The physical location where a lower order stream or river flows into a higher order stream or river as a tributary. See **mouth**.

Constituent A chemical or biological substance in water, sediments, or biota that can be measured by an analytical method (e.g., nitrate-N, organic carbon, or chlorophyll).

Construction Grants Program Federal funding authorized by amendments (1956 through 1987) to 1948 Federal Water Pollution Control Act to provide technical assistance and construction money to aid municipalities in building and upgrading sewerage collection systems and municipal wastewater treatment plants. After 1972 amendments, $61.1 billion (current year dollars) in federal funding provided under Clean Water Act to upgrade municipal wastewater facilities to a minimum of secondary treatment.

Consumptive use That part of water withdrawn that is evaporated, transpired, or incorporated into a manufactured product, or consumed by humans or animals, or otherwise removed from the immediate waterbody environment.

Contamination Act of polluting or making impure; any indication of chemical, sediment, or biological impurities.

Conventional pollutants As specified under the Clean Water Act, conventional pollutants include suspended solids, coliform bacteria, biochemical oxygen demand, pH, and oil and grease.

Decay Gradual decrease in the amount of a given substance in a given system due to various loss/sink processes including chemical and biological transformation, dissipation to other environmental media, or deposition into storage areas.

Decomposition Metabolic breakdown of organic materials; the by-products formation releases energy and simple organics and inorganic compounds See also **respiration**.

Denitrification Describes the decomposition of ammonia compounds, nitrites, and nitrates (by bacteria) that results in the eventual release of nitrogen gas into the atmosphere.

Dilution Addition of a volume of less concentrated liquid (water) that results in a decrease in the original concentration.

Discharge The volume of water that passes a given point within a given period of time. It is an all-inclusive outflow term, describing a variety of flows such as from a pipe to a stream, or from a stream to a river, lake, or ocean.

Discharge permits (NPDES) A permit issued by the USEPA or a state regulatory agency that sets specific limits on the type and amount of pollutants that a municipality or industry can discharge to a receiving water; it also includes a compliance schedule for achieving those limits. It is called the NPDES because the permit process was established under the National Pollutant Discharge Elimination System, under provisions of the Federal Clean Water Act.

Dispersion The turbulent mixing and spreading of chemical or biological constituents, including pollutants, in various directions from a point source, at varying velocities depending on the differential in-stream flow characteristics.

Dissolved oxygen (DO) The amount of oxygen gas that is dissolved in water. It also refers to a measure of the amount of oxygen available for biochemical activity in a waterbody, and as indicator of the quality of that water.

Diurnal Actions or processes that have a period or a cycle of approximately one tidal-day or that are completed within a 24-hour period and that recur every 24 hours.

Domestic wastewater Also called sanitary wastewater, consists of wastewater discharged from residences and from commercial, institutional, and similar facilities.

Domestic water use Water used for household purposes such as drinking, food preparation, bathing, washing clothes and dishes, watering lawns and gardens, flushing toilets, and so on. Also called residential water use.

Drainage basin A part of the land area enclosed by a topographic divide from which direct surface runoff from precipitation normally drains by gravity into a receiving water. Also referred to as watershed, river basin, or hydrologic unit.

Dynamic model A mathematical formulation describing the physical behavior of a system or a process and its temporal variability.

Dynamic simulation Modeling of the behavior of physical, chemical, and/or biological phenomena and their variation over time.

Ecosystem An interactive system that includes the organisms of a natural community association together with their abiotic physical, chemical, and geochemical environment.

Effluent Municipal sewage or industrial liquid waste (untreated, partially treated, or completely treated) that flows out of a wastewater treatment plant, septic system, pipe, and so on.

Enforcement conference Joint state-federal water pollution conference convened by U.S. Public Health Service under authority of 1956 amendments to the 1948 Federal Water Pollution Control Act. Federal regulatory authority was restricted to enforcement of water pollution problems only in interstate waters because of the Commerce clause of the U.S. Constitution. Fifty-one enforcement conferences were convened from 1957 to 1971.

Estuary Brackish-water areas influenced by the ocean tides where the mouth of the river meets the sea.

Eutrophication Enrichment of an aquatic ecosystem with nutrients (nitrogen, phosphorus) that accelerate biological productivity (growth of algae, periphyton, and macrophytes/weeds) and an undesirable accumulation of plant biomass.

Factor of safety Coefficient used to account for uncertainties in representing, simulating, or designing a system.

Fecal coliform bacteria Coliform bacteria that are present in the intestines or feces of warm-blooded animals including humans. They are often used as indicators of the sanitary quality of water. See **coliform bacteria**.

Flocculation The process by which suspended colloidal or very fine particles are assembled into larger masses or floccules that eventually settle out of suspension.

Flushing characteristics Measure of the displacement of water from a riverine or estuarine system as controlled by the combined actions of freshwater inflow and tidal mixing and exchange.

Frequency analysis A numerical determination of the distribution of values for a parameter within a data set. See **median**, **ninetieth percentile**, and **tenth percentile**.

Freshwater Water that contains less than 1,000 mg/L of dissolved solids. Water that contains more than 500 mg/L of dissolved solids is undesirable for drinking water and for many industrial uses.

Gaging station A specific location on a stream, river, canal, lake, or reservoir where systematic measurements of hydrologic data, such as stage height and streamflow, are collected. The USGS maintains and operates a network of stream gaging stations to collect hydrologic data for many streams and rivers. Historical streamflow and stage height data are available from the USGS streamflow database (www.waterdata.usgs.gov/nwis-w). Earliest historical records are available from the late nineteenth century for some rivers.

"Grey" literature Unpublished technical reports and memoranda, data reports, or other documents prepared by academic researchers, federal, state, or local agencies, or other institutions and organizations. Typically, limited distribution makes "grey" literature difficult to obtain except from agency or institutional sources.

Hydrologic Accounting Unit Geographical subdivision of watersheds within each Hydrologic Sub-Region. There are a total of 352 Accounting Units in the United States with 334 Accounting Units located in the 48 contiguous states. Accounting Units are identified by a six-digit code where the first two digits identify the Hydrologic Sub-Region as the larger hydrologic units. Example: 070102 is the Accounting Unit for the Platte-Spunk basin of the Upper Mississippi River basin.

Hydrologic Catalog Unit Geographical subdivision of watersheds within each Hydrologic Accounting Unit. There are a total of 2,150 Catalog Units in the United States with 2,111 Catalog Units located in the 48 contiguous states. Catalog Units are identified by an eight-digit code where the first two digits identify the Hydrologic Accounting Unit. Example: 07010206 is the Catalog Unit for the Twin Cities area of the Upper Mississippi River.

Hydrologic cycle The representation of the cycle of water on earth based on all hydrologic processes and the interactions of water between the atmosphere, surface waters, polar ice, glaciers, and groundwater.

Hydrologic Region Largest geographical subdivision of the United States into a hierarchical succession of hydrologic units based on drainage area. There are a total of 21 Hydrologic Regions in the United States, with 18 Hydrologic Regions located within the 48 contiguous states. Hydrologic regions are identified by a two-digit numerical code from 01 to 21. Example: 07 is the Hydrologic Region for the Upper Mississippi River basin.

Hydrologic Sub-Region Geographical subdivision of watersheds within each Hydrologic Region. There are a total of 222 Sub-Regions in the United States, with 204 Sub-Regions located in the 48 contiguous states. Sub-Regions are identified by a four-digit code with the first two digits used to identify the larger Hydrologic Region. Example: 0701 is the Hydrologic Sub-Region for the Mississippi Headwaters of the Upper Mississippi River basin.

In situ Latin for "in place"; *in situ* measurements consist of measurement of component or processes in a full-scale system or in the field rather than in a laboratory.

Industrial water use Water used for industrial purposes such as fabricating, processing, washing, and cooling. Industries that use water include steel, chemical and allied products, paper and allied products, mining, and petroleum refining.

Influent Water volume flow rate or mass loading of a pollutant or other constituent into a waterbody or wastewater treatment plant.

Inorganic Pertaining to matter that is neither living nor immediately derived from living matter.

Loading, Load, Loading rate The total amount of material (pollutants) entering the system from one or multiple sources; measured as a rate in weight (mass) per unit time.

Low-flow (7Q10) Low-flow (7Q10) is the seven-day average low-flow occurring once in 10 years; this probability-based statistic is used in determining stream design flow conditions and for evaluating the water quality impact of effluent discharge limits.

Major river basin Alternative terminology for **Hydrologic Region**.

Margin of Safety (MOS) A required component of the total maximum daily load (TMDL) that accounts for the uncertainty about the relationship between the pollutant load and the quality of the receiving waterbody.

Mass balance An equation that accounts for the flux of mass going into a defined area and the flux of mass leaving the defined area. The flux in must equal the flux out to achieve a mass balance.

Mathematical model A system of mathematical expressions that describe the spatial and temporal distribution of water quality constituents resulting from fluid transport and the one, or more, individual processes and interactions within some prototype aquatic ecosystem. A mathematical water quality model is used as the basis for waste load allocation and TMDL evaluations.

Mean The numerical average of a set of observations; computed as the sum of the observations divided by the number of observations in the data set.

Median (fiftieth percentile) A middle statistic based on ranking a data set from the minimum to the maximum value. The median value divides the data set into two parts so that one-half of the values are lower than the median and one-half of the values are greater than the median. The median is also defined as the fiftieth percentile value.

Milligrams per liter (mg/L) A unit of measurement expressing the concentration of a constituent in solution as the weight (mass) of solute (1 milligram) per unit volume (1 liter) of water; equivalent to 1 part per million (ppm) for a water density ~1 gram per cubic centimeter. 1 mg/L = 1,000 ug/L; 1 g/L = 1,000 mg/L.

Million gallons per day (mgd) Rate of water volume discharge representing a volume of 1 million gallons of water passing across a given location in a time interval of one day. A flow rate of 1 mgd = 1.54723 cubic feet per second (cfs) = 0.04381 cubic meters per second (cms).

Mineralization The biochemical transformation of organic matter into a mineral or an inorganic compound.

Mixing characteristics Refers to the tendency for natural waters to blend; that is, for dissolved and particulate substances to disperse into adjacent waters.

Most probable number (MPN) Measure of concentration, or abundance, of total and fecal coliform bacteria based on incubation results and a statistical interpretation of the results.

Municipal wastewater inventory U.S. Public Health Service compilations of inventory of municipal wastewater plants, population served, and influent flow by different categories of municipal wastewater treatment facilities. Inventories were compiled in 1950, 1962, and 1968.

N/P ratio The ratio of nitrogen to phosphorus in an aquatic system. The ratio is used as an indicator of the nutrient-limiting conditions for algal growth; also used as indicator for the analysis of trophic levels of receiving waters.

Natural waters Flowing waterbody within a physical system that has developed without human intervention, in which natural processes continue to take place; streams, rivers, lakes, bays, estuaries, and coastal and open ocean are examples of natural waters.

Needs Survey USEPA Clean Water Needs Surveys (CWNS) compiled from 1976 through 1996 at 2- to 4-year intervals. Needs Surveys document inventory of wastewater plants, population served, influent flow and effluent load of 5-day biochemical oxygen demand and total suspended solids by six different categories of municipal wastewater treatment facilities. Information is compiled for existing conditions (e.g., 1996) and 20-year projections (e.g., 2016).

Ninetieth percentile A high-end statistic based on ranking a data set from the minimum to the maximum value. The ninetieth percentile defines the value for which 90% of the ranked data set is less than the statistic and 10% of the data set is greater than the statistic. Used in this study to define "worst-case" conditions for evaluation of "before and after" trends of BOD_5.

Nitrate (NO_3) and Nitrite (NO_2) Oxidized nitrogen species. Nitrate is the form of nitrogen used by aquatic plants for photosynthesis.

Nitrification Biologically mediated process of the oxidation of ammonium salts to nitrites (via ***nitrosomonas*** bacteria) and the further oxidation of nitrite to nitrate via ***nitrobacter*** bacteria.

Nitrifier organisms Bacterial organisms that mediate the biochemical oxidative processes of nitrification.

Nitrobacter Type of bacteria responsible for the conversion of nitrite to nitrate.

Nitrogenous BOD (NBOD) Refers to the biochemical oxygen demand associated with the oxidation of ammonia to nitrite and nitrate.

Nitrosomonas Type of bacteria responsible for the oxidation of ammonia to the intermediate product nitrite.

Nonpoint source Pollution that is not released through pipes but rather originates as surface runoff over a relatively large drainage area. Nonpoint sources can be divided into source activities related to either land or water use including failing septic tanks, improper animal-keeping practices, forest practices, and urban and rural runoff from a drainage basin.

Numerical model Mathematical models that approximate a solution of governing partial differential equations that describe a natural process. The approximation uses a numerical discretization of the space and time components of the system or process. See **mathematical model**.

Nutrient A primary element necessary for the growth of living organisms. Carbon dioxide, nitrogen, and phosphorus, for example, are required nutrients for phytoplankton (algae) and other aquatic plants' growth.

Organic matter The organic fraction that includes plant and animal residue at various stages of decomposition, cells and tissues of soil organisms, and substances synthesized by the soil population. Commonly determined as the amount of organic material contained in a soil or water sample.

Organic nitrogen Organic form of nitrogen bound to organic matter.

Outfall Location point where wastewater or stormwater flows from a conduit, stream, or drainage ditch into natural waters.

Oxidation The chemical union of oxygen with metals or organic compounds accompanied by a removal of hydrogen or another atom. It is an important factor for soil formation and permits the release of energy from cellular fuels.

Oxygen demand Measure of the amount of dissolved oxygen used by microorganisms and or chemical compounds in the oxidation of organic matter. See also **biochemical oxygen demand**.

Oxygen depletion Deficit of dissolved oxygen in a natural waters system due to oxidation of natural and anthropogenic organic matter.

Oxygen sag Description of characteristic spatial trend of the concentration of dissolved oxygen in a stream or river downstream of high loading rate of oxygen-demanding materials from tributaries, municipal or industrial wastewater dischargers, or urban stormwater and combined sewer overflow systems.

Oxygen saturation The maximum amount of oxygen gas that can be dissolved in natural waters by transfer of oxygen from the atmosphere across the air-water interface. The concentration of oxygen saturation is influenced by water temperature, salinity or chlorides concentration, elevation above mean sea level, and other water characteristics.

Parameters Constituents and pollutants measured in water quality monitoring programs. Examples: dissolved oxygen, 5-day biochemical oxygen demand, total suspended solids, water temperature.

Parts per million (ppm) Measure of concentration of 1 part solute to 1 million parts water (by weight). See **milligrams per liter**.

Parts per thousand (ppt) Measure of concentration of 1 part solute to 1,000 parts water (by weight). See **milligrams per liter**.

Pathogens Microorganisms capable of producing disease. Pathogens are of great concern to protect human health relative to drinking water, swimming beaches, and shellfish beds.

Per capita use The quantity of water used per person averaged over a time interval of 1 day; expressed as gallons per capita per day (gpcd).

pH A measure of acidity indicated by the logarithm of the reciprocal of the hydrogen ion concentration (activity) of a solution. pH values less than 7 are acidic; values greater than 7 are basic; pH of 7 is neutral. pH of natural waters typically ranges from $\sim$6 to $\sim$8.

Phosphorus A nutrient essential for plant growth that can play a key role in stimulating the growth of aquatic plants in streams, rivers, and lakes.

Photosynthesis The biochemical synthesis of carbohydrate-based organic compounds from water and carbon dioxide using light energy in the presence of chlorophyll. Photosynthesis occurs in all plants, including aquatic organisms such as algae and macrophytes. Photosynthesis also occurs in primitive bacteria such as blue-green algae (cyanobacteria).

Phytoplankton A group of generally unicellular microscopic plants characterized by passive drifting within the water column. See **algae**.

Plankton Group of generally microscopic plants and animals passively floating, drifting, or swimming weakly. Plankton include the phytoplankton (plants) and zooplankton (animals).

Point source Pollutant loads discharged at a specific location from pipes, outfalls, and conveyance channels from either municipal wastewater treatment plants or industrial waste treatment facilities. Point sources can also include pollutant loads contributed by urban stormwater systems or tributaries to the main receiving water stream or river.

Pollutant A contaminant in a concentration or amount that adversely alters the physical, chemical, or biological properties of a natural environment. The term include pathogens, toxic metals, carcinogens, oxygen demanding substances, or other harmful substances.

Pretreatment The treatment of wastewater to remove or reduce contaminants prior to discharge into another municipal treatment system or a receiving water.

Primary productivity A measure of the rate at which new organic matter is formed and accumulated through photosynthesis and chemosynthesis activity of producer organisms (chiefly, green plants). The rate of primary production is estimated by measuring the amount of oxygen released (oxygen method) or the amount of carbon assimilated by the plant (carbon method).

Primary treatment plant Wastewater treatment process where solids are removed from raw sewage primarily by physical settling. The process typically removes about 25 to 35% of solids and related organic matter (BOD_5).

Publicly owned treatment works (POTW) Municipal wastewater treatment plant owned and operated by a public governmental entity such as a town or city.

Public-supply withdrawals Water withdrawn from surface water or groundwater by public or private water suppliers for use within a community. Water is used for domestic, commercial, industrial, and public water uses such as fire fighting.

Range Statistical measure expressing the difference between the minimum and maximum values recorded for a given constituent in time and space.

Raw sewage Untreated municipal sewage that is discharged to a water body.

Reach (of a river) A longitudinal section of a stream or river defined by the upstream and downstream locations of lower order stream tributaries flowing into a higher order stream.

Reach File Version 1 (RF1) USEPA hydrologic database designed to define the downstream hydraulic routing of a connecting network of streams, rivers, lakes and reservoirs, bays, and tidal waters. Version 1 of the Reach File (RF1) includes a network of 64,902 RF1 reaches that includes 632,552 miles of waterbodies in the 48 contiguous states. RF1 reaches are indexed using an 11-digit code.

Reach File Version 3 USEPA hydrologic database that defines the downstream hydraulic routing of a more complex connecting network of streams, rivers, lakes and reservoirs, bays, and tidal waters than the RF1 database. Version 3 of the Reach File (RF3) includes a network of 1,821,245 RF3 reaches that includes about 3.5 million miles of waterbodies in the 48 contiguous states. RF3 reaches are indexed using a 17-digit code.

Reaction rate coefficient Coefficient describing the rate of transformation of a substance in an environmental medium characterized by a set of physical, chemical, and biological conditions such as temperature and dissolved oxygen level.

Reaeration The net flux of oxygen transfer occurring from the atmosphere to a body of water with a free surface.

Receiving waters Creeks, streams, rivers, lakes, estuaries, groundwater formations, or other bodies of water into which surface water and/or treated or untreated wastewater are discharged, either naturally or in man-made systems.

Removal efficiency A measure of how much of a pollutant is removed from raw sewage prior to discharge into a receiving water after completion of wastewater treatment processes. Expressed as percentage, calculated as Removal % = [(influent − effluent) /(influent)] × 100.

Residential water use See **domestic water use**.

Respiration Biochemical process by means of which cellular fuels are oxidized with the aid of oxygen to permit the release of the energy required to sustain life; during respiration, oxygen is consumed and carbon dioxide is released.

River basin See **drainage basin**.

Rotating biological contactors (RBCs) A wastewater treatment process consisting of a series of closely spaced rotating circular disks of polystyrene or polyvinyl chloride. Attached biological growth is promoted on the surface of the disks. The rotation of the disks allows contact with the wastewater and the atmosphere to enhance oxygenation.

Salinity The total amount of dissolved salts, measured by weight as parts per thousand (ppt) or g/L. Salinity concentrations range from ~0.5 to ~1 ppt for tidal freshwaters; from ~20 to ~25 ppt for estuarine waters; ~30 ppt for coastal waters to ~35 ppt for the open ocean.

Submersed or submerged aquatic vegetation (SAV) Rooted aquatic plants that grow in shallow clear water.

Secchi depth A measure of the light penetration into the water column. Light penetration is influenced by turbidity of the water body.

Secondary treatment plant Waste treatment process where oxygen-demanding organic materials (BOD) are removed by bacterial oxidation of the waste to carbon dioxide and water. Bacterial synthesis of wastewater is enhanced by injection of oxygen.

Sediment Particulate organic and inorganic matter that accumulates in a loose, unconsolidated form on the bed of streams, rivers, lakes, bays, estuaries, and oceans.

Sediment oxygen demand (SOD) The solids discharged to a receiving water are partly organics, and upon settling to the bottom, they decompose anaerobically as well as aerobically, depending on conditions. The amount of oxygen consumed in the sediment bed during aerobic decomposition of particulate organic carbon deposited on the bottom of a waterbody; represents another dissolved oxygen loss or sink for the waterbody.

Significance level A statistical measure of the certainty that can be associated with the results of a statistical analysis.

Stabilization pond A large earthen basin that is used for the treatment of wastewater by natural processes involving the use of both algae and bacteria.

Standard Industrial Classification (SIC) Codes Four-digit codes established by the Office of Management and Budget (OMB) to classify commercial and manufacturing establishments by the principal type of activity. Example: 4952 = municipal wastewater treatment plant.

Station (monitoring) Specific location in a waterbody chosen to collect water samples for the measurement of water quality constituents. Stations are identified by an alphanumeric code identifying the agency source responsible for the collection of the data and a unique identifier code designating the location. Station measurements can be recorded from either discrete grab samples or continuous automated data acquisition systems. Station locations are typically sampled by state, federal, or local agencies at periodic intervals (e.g., weekly, monthly, or annually) as part of a routine water quality monitoring program to track trends. Station locations can also be sampled only for a period of time needed to collect data for an intensive survey or a special monitoring program.

Stoichiometric ratio Mass-balance-based ratio for nutrients, organic carbon, dry weight, and algae (e.g., nitrogen-to-carbon ratio of algae).

STORET U.S. Environmental Protection Agency (USEPA) national water quality database for STOrage and RETrieval (STORET). Mainframe water quality database that includes millions of data records of physical, chemical, and biological data measured in waterbodies throughout the United States.

Storm runoff Rainfall that does not evaporate or infiltrate into the ground because of impervious land surfaces or a soil infiltration rate lower than rainfall intensity, but instead flows onto adjacent land or waterbodies or is routed into a drain or sewer system.

Stream order A ranking, developed by A. N. Strahler, of the relative size of streams and rivers within a watershed based on the network of tributaries. The smallest, headwater stream is classified as an Order 1 stream. The stream formed by the confluence of two or more Order 1 streams is classified as an Order 2 stream. In the United States, the Mississippi River is an Order 10 river. See Strahler, A. N. 1960. *Physical Geography.* John Wiley and Sons, New York.

Streamflow Discharge that occurs in a natural channel. Although the term "discharge" can be applied to the flow of a canal, the word "streamflow" uniquely describes the discharge in a surface stream course. The term streamflow is more general than "runoff," as streamflow may be applied to discharge whether or not it is affected by diversion or regulation.

Submersed or submerged aquatic vegetation Rooted aquatic plants that grow in shallow clear water.

Surface waters Water that is present above the substrate or soil surface. Usually refers to natural waterbodies such as streams, rivers, lakes and impoundments, and estuaries, and coastal ocean.

Suspended solids or load Organic and inorganic particles (solids/sediment) suspended in and carried by a fluid (water). The suspension is governed by the upward components of turbulence, currents, or colloidal suspension.

Tenth percentile A low-end statistic based on ranking a data set from the minimum to the maximum value. The tenth percentile defines the value for which 10% of the ranked data set is less than the statistic and 90% of the data set is greater than the statistic. Used in this study to define "worst-case" conditions for evaluation of "before and after" trends of dissolved oxygen.

Tertiary treatment Waste treatment processes designed to remove or alter the forms of nitrogen or phosphorus compounds contained in domestic sewage.

Total coliform bacteria A particular group of bacteria that are used as indicators of possible sewage pollution.

Total Kjeldahl nitrogen (TKN) The sum total of organic nitrogen and ammonia nitrogen in a sample, determined by the Kjeldahl method.

Total Maximum Daily Load (TMDL) The sum of the individual wasteload allocations and load allocations. A margin of safety is included with the two types of allocations so that any additional loading, regardless of source, would not produce a violation of water quality standards.

Toxic substances Those chemical substances, such as pesticides, plastics, heavy metals, detergent, solvent, or any other material, that are poisonous, carcinogenic, or otherwise directly harmful to human health and to biota in the environment.

Transport of pollutants (in water) Transport of pollutants in water involves two main process: (1) advection, resulting from the flow of water, and (2) diffusion, or transport due to molecular and turbulent mixing in the water.

Tributary A lower order stream compared to a receiving waterbody. "Tributary to" indicates the largest stream into which the reported stream or tributary flows.

Trickling filter A wastewater treatment process consisting of a bed of highly permeable medium (e.g., gravel or stones) to which microorganisms are attached and through which wastewater is percolated or trickled over the biofilm that forms on the media.

Turbidity Measure of the amount of suspended material in water.

Ultimate biochemical oxygen demand (UBOD or BOD_u) Long-term oxygen demand required to completely stabilize organic carbon and ammonia in wastewater or natural waters; defined as the sum of ultimate carbonaceous BOD and nitrogenous BOD.

Urban drainage Water derived from surface runoff or shallow groundwater discharge from urban land use areas.

Waste load allocation (WLA) The portion of a receiving water's total maximum daily load that is allocated to one of its existing or future point sources of pollution.

Wastewater Usually refers to effluent from an industrial or municipal sewage treatment plant. See also **domestic wastewater**.

Wastewater treatment Chemical, biological, and mechanical processes applied to an industrial or municipal discharge or to any other sources of contaminated water in order to remove, reduce, or neutralize contaminants prior to discharge to a receiving water.

Water pollution Any condition of a waterbody that reflects unacceptable water quality or ecological conditions. Water pollution is usually the result of discharges of waste material from human activities into a waterbody.

Water quality Numerical description of the biological, chemical, and physical conditions of a waterbody. It is a measure of a waterbody's ability to support beneficial uses.

Water quality criteria (WQC) Water quality criteria include both numeric and narrative criteria. Numeric criteria are scientifically derived ambient concentrations developed by USEPA or the states for various pollutants of concern to protect human health and aquatic life. Narrative criteria are statements that describe the desired water quality goal.

Water quality standard (WQS) A water quality standard is a law or regulation that consists of the beneficial designated use or uses of a waterbody, the numeric and narrative water quality criteria that are necessary to protect the use or uses of that particular waterbody, and an antidegradation statement.

Watershed See **drainage basin**.

Zooplankton Very small animals (protozoans, crustaceans, fish embryos, insect larvae) that live in a waterbody and are moved passively by water currents and wave action. See **plankton**.

INDEX

301(h) (*see* Legislative and regulatory history)
7-day, 10-year flow (7Q10) (*see* Statistics, streamflow)
Accounting unit, definition of, (*see* Hydrologic Unit Code)
Activated sludge, 25, 26, 39, 44, 155, 369, 582
Advanced primary treatment, 42, 43, 57
  effluent characteristics of, 76–78
Advanced secondary treatment, 42, 45, 57, 298, 361, 369, 372, 374, 376–377, 379, 384, 385, 386
  effluent characteristics of, 76–78
Advanced treatment (AT), 182–184, 452, 454, 455, 582–584, 598
  effluent characteristics of, 76–78
Advanced wastewater treatment (AWT), 42, 45, 57, 59, 60, 64, 70, 97, 99, 154, 155, 273, 286, 453, 454, 598
Advisory, fish consumption, 248, 249, 281, 358, 389
Aerobic, 24, 25, 43, 598
Algae (*see also* Phytoplankton), 2, 6, 44, 189, 302, 315
  blooms of, 222, 249, 286, 290, 298, 382, 383
  blue-green, 294
  Microcystis aeruginosa, 294
  nuisance blooms of, 380
  respiration of, 6, 24, 43, 46, 106, 271
Albany, New York, 182, 223, 228
Allegheny River, 345, 346
American Heritage River, 308, 392, 411
Ammonia (NH3-N):
  degradation, of water quality, 334
  effluent load, long term trends, 374
  nitrification, in natural waters, and, 24, 26, 43, 46, 116, 118
  nitrification, in biological waste treatment, 45, 64, 376, 385
  nitrogenous BOD (NBOD), as component of, 48
  improvements, in water quality quality of, 155, 183, 210, 238, 248, 266, 269, 280, 354, 392
  removal, wastewater of, 279, 298, 339, 369, 376, 384, 390
  simulated distribution of, with water quality model, 271, 273, 384
  unionized, 369, 376, 383
Anadromous, fish, 204, 209, 221, 222, 275, 324
Anaerobic, 25, 598
Anoxic, 262, 384
Appomattox River, 312, 316, 318, 319
Aqueducts (*see* Water supply)
Ardern, E. and W.T. Lockett, 25
Arkansas-White, Hydrologic Region 11
  dissolved oxygen, before and after CWA, 164, 171
Arthur Kill, 184, 213, 228, 235, 246, 247
ASIWPCA (*see* Association of State and Interstate Water Pollution Control Administrators)
Assimilative capacity, 263, 316, 335, 367, 368, 405
Association of State and Interstate Water Pollution Control Administrators (ASIWPCA), 4, 182, 450, 451
AT (*see* Advanced treatment)
Atlanta, Georgia, 19, 280, 327, 329–342
Atlanta Regional Commission (ARC), 327, 333, 335, 337–339
Atmospheric:
  deposition, of heavy metals and organic chemicals, 75, 280, 389
  reaeration, of dissolved oxygen (*see* Dissolved oxygen)
Authority, for water pollution control:
  federal government, 5, 28, 30, 33, 34
  Clean Water Act of 1972, after, 27, 189
  Clean Water Act of 1972, before, 27, 96, 189
  local government, 19, 335
  state government, 27, 28, 33, 96, 316, 372, 373, 405
Autotrophs, 106
AWT (*see* Advanced waste treatment)

Bacteria, 2, 44, 249
  anaerobic, 25
  autotrophic, 6, 24, 43
  coliform, 189, 290, 301
  fecal, 37, 38, 76, 238, 300, 369, 370, 372, 379, 383, 391, 392
  total, 243, 244, 301, 368
  combined sewers, as source of, 206, 222, 237, 242, 300, 369
  combined sewers, effluent characteristics of, 73–74
  degradation, of ambient water quality, from water pollution, 205, 262, 290, 334, 345, 350, 358, 368, 370, 376, 404, 408

Bacteria (*continued*):
  decomposition, of organic matter and, 24, 25, 43, 46, 106, 116
  improvement, of ambient water quality, from water pollution control, 237, 243, 248, 249, 273, 279, 280, 300, 341, 369, 372, 379, 392
  model, water quality of, 383
  nitrifying, 6, 24, 43, 46, 118
  public health, and risk of exposure to, 368, 372
  secondary treatment, effluent characteristics, April, 1973 definition of, 36
  standards, water quality, for coliform bacteria, 37, 38, 206, 222, 243, 300, 351, 369, 391
    primary contact, 243, 300
    secondary contact, 298, 300
    urban storm water, effluent characteristics, 73–75
    wastewater, as source of, 25, 44, 262, 279, 300
Backwater, pools, 347, 363, 383, 389, 392
Beaches, recreational, 23, 225, 300, 301
  closure of:
  development, 260
  protection of public health, 73, 222, 242, 243, 245
  Enhanced Beach Protection Program, New York City, 245
  re-opened, 243, 248
Benefits:
  ecological, 392
  economic, 177, 183, 249, 280, 411
  environmental, 36, 177
  natural resources of, 28
  recreational, 280, 411
  spatial scale of, 9, 119, 154
  water pollution, control of, 4, 9, 28, 38, 194, 346
  water quality, of Clean Water Act, 194, 270, 271, 286, 294, 321, 383–384
  water quality, of secondary treatment, 9, 39, 183
  water resource users, 12
Best management practices (BMP), 75
Best practicable treatment (BPT), 32, 33, 38
  minimum level, of waste treatment, as, 33
Big Sioux River, 159, 162
  Lower, 143, 159, 162
Bioaccumulation, of toxic organic chemicals, 392, 411
  biota, benthic, 280
  birds, fish-eating, 246, 247, 388
  fish, 222, 246, 280, 358, 388, 390, 408
  mammals, fish-eating, 388
Biochemical oxygen demand (BOD):
  domestic sewage, decomposition of, 26
  effect of wastewater discharges, on dissolved oxygen of, 47, 116
  incubation period, 26, 46, 47
  indicator of wastewater strength as, 5, 26, 95
  industrial wastewater, decomposition of, in, 26
Biochemical oxygen demand (BOD), water quality:
  case study waterways, 334, 353
  spatial distribution of, in hypothetical river, 116–118
  trends, long-term:
    case study waterways, 206–207, 238, 248, 266, 377
    catalog units, (1961–1995), 143–144, 154–156, 159, 162
    RF1 reaches, (1961–1995), 154–156, 159, 162
Biochemical oxygen demand (BOD), effluent characteristics, of:
  combined sewer overflows, 73–74
  industrial wastewater, 71–73
  municipal wastewater, 76–78
  rural runoff, 73–75
  secondary treatment, April, 1973 definition of, 36
  urban storm water, 73–75
Biochemical oxygen demand (BOD), effluent loads, national, ca. 1995
  catalog units, distribution of, by, 97
  combined sewer overflows, 84, 97
  industrial wastewater, 76, 78, 82–83, 97
  municipal wastewater, 76–77, 80–81, 97
  rural runoff, 84–85, 88–89, 97
  urban storm water, 84–87, 97
Biochemical oxygen demand (BOD), effluent loads, case study waterways, 189–191, 263–265, 273, 276, 279, 292, 295, 298, 318, 320, 321, 323, 325, 332, 339, 352, 377, 378, 390, 406
Biochemical oxygen demand (BOD), effluent loads, national, for industrial wastewater, major river basins, for 1950, 1973, and 1995, 581–584
Biochemical oxygen demand (BOD), effluent loads, national, for municipal wastewater, 17, 42
  historical trends, (1940–1996) of, 57–62, 96–97, 453–454
  major river basins, for 1950, 1973, and 1995 of, 581–584
  projected trends, (1996–2016) of, 64–70, 97, 99, 454–455
Biochemical oxygen demand (BOD), influent loads, national, for municipal wastewater, 17, 42
  historical trends, (1940–1996) of, 50–57, 96, 453–454
  projected trends, (1996–2016) of, 64–70, 97, 99, 454–455
Biochemical oxygen demand, (BOD), removal efficiency, for municipal wastewater, 26, 37
  actual performance, 57
  design-based, 42, 65
  historical trends, (1940–1996) of, 59–64, 97, 453–454

projected trends, (1996–2016) of, 65–70, 97, 99, 454–455
Biochemical oxygen demand (BOD), types of:
BOD5, 5-day biochemical oxygen demand, 43
BODU, ultimate biochemical oxygen demand, 46, 48
CBOD5, 5-day carbonaceous biochemical oxygen demand, 43, 48
CBOD, carbonaceous biochemical oxygen demand, 43, 47
CBODU, ultimate carbonaceous biochemical oxygen demand, 43, 48
NBOD, nitrogenous biochemical oxygen demand, 43, 47, 48, 55, 60, 64, 116, 118, 319
NBODU, ultimate nitrogenous biochemical oxygen demand, 43, 48
Total BOD, 43
Biochemical oxygen demand (BOD), wastewater characteristics:
ratio, of ultimate BOD to 5-day carbonaceous BOD, 47, 48, 55, 295
raw, untreated, 76–78
reaction rate, 386
reaction rate of oxidation, and waste treatment level, 271, 295, 384, 386
Biogeochemical, processes in natural waters, 74
Biological:
indicators (*see* Indicators, biological)
productivity (*see* Productivity, biological), 258
resources (*see* Resources, biological)
Birds, 213, 246, 302, 304, 308, 315, 324, 325, 388
Birds, fish-eating:
Cormorants, 324, 388
Eagles, 222, 246, 324, 388
Mergansers, 304
Ospreys, 246, 248, 324
Peregrine falcons, 248, 388
Birds, wading:
Egrets, 213, 246, 247, 324, 388
Herons, 183, 213, 246–248, 324, 388
Ibises, 213, 246, 247
Birds, waterfowl:
Canada geese, 304
ducks, 246, 302, 304, 388
gadwalls, 302, 304
widgeons, 302, 304
Blackstone River, 22, 182, 417, 436
Blue crab (*see* Shellfish)
Blue green algae (*see* Algae)
BMP (*see* Best Management Practices)
Boating (*see* Recreation, water-based)
BOD (*see* Biochemical oxygen demand, type of)
BOD5 (*see* Biochemical oxygen demand, type of), 600
BODU (*see* Biochemical oxygen demand, type of)
Boise River, 183
Boston, Massachusetts, 77, 85, 183, 215
BPT (*see* Best Practicable Treatment)
Caddisflies (*see* Macroinvertebrates, aquatic)
California, Hydrologic Region 18, dissolved oxygen, before and after CWA, 164
Canada geese (*see* Birds, waterfowl)
Capital costs (*see* Expenditures, water pollution control)
Carbon:
cycle of, 24, 106
dioxide, 24, 25, 106
inorganic, 294
tetrachloride, 408
Carbon, organic:
decomposition of, 24, 43, 116
labile fraction of, 46, 271, 295
particulate (POC), municipal wastewater characteristics and loads, 449, 450, 452, 455
refractory fraction of, 46, 271, 295
removal, in wastewater treatment, 45, 176, 315, 320, 325, 582
spatial distribution of, in hypothetical river, 116–118
total (TOC) loads estimates for New York Harbor, 235
wastewater, as source of, 315
Carbonaceous BOD (CBOD) (*see* Biochemical oxygen demand)
Carson, Rachel, 33
Case study waterways:
literature review of, 182–183
Chattahoochee River, 327–344
Connecticut River, 199–212
Delaware estuary, 255–284
Hudson–Raritan estuary, 213–254
James estuary, 311–326
Ohio River, 345–360
Potomac estuary, 285–310
Upper Mississippi River, 361–397
Willamette River, 399–413
Categories, of waste treatment, municipal, 42, 44–45
Catalog Unit, definition of, (*see* Hydrologic Unit Code)
Cattaraugus Creek, 143
CBOD (*see* Biochemical oxygen demand, types of)
CBOD5 (*see* Biochemical oxygen demand, types of)
CBODU (*see* Biochemical oxygen demand, types of)
Chattahoochee-Flint-Apalachicola River basin, 327
Chattahoochee River, 12, 189, 327, 328, 330–334, 336, 337, 339–342
advanced secondary, upgrade to, 339
ammonia:
water quality, 334
removal, effluent of, 339
assimilative capacity, of river, 335
Atlanta Regional Commission (ARC), 335
bacteria, degradation, of water quality, from water pollution, 334

Chattahoochee River (*continued*):
bacteria, improvement, of water quality, from water pollution control, 341
biochemical oxygen demand (BOD),
effluent loads, municipal facilities, 339
water quality, 334
Clayton, R.M., municipal wastewater plant, 334, 335
improvements in effluent quality, 336, 338, 339
upgrades, 338, 339
Clean Water Act of 1972 (*see* legislative and regulatory history)
combined sewer overflows (CSOs), 334, 339
expenditures, for separation of, 340
remedial measures plan, 340
dams and reservoirs, 335, 336, 342
Buford Dam, 329, 332, 333, 336, 341
Morgan Falls Dam, 329, 333
regulation of streamflow (*see* streamflow)
West Point Dam, 327, 333, 340
dissolved oxygen, water quality:
degradation of:
from water pollution, 334, 335, 336
improvement of:
from water pollution control, 336
drainage area:
Chattahoochee-Flint-Apalachicola basin of, 327
Upper Chattahoochee River basin of, 328
effluent limits, 336
erosion, of soil, 335
expenditures, water pollution control, 335, 336, 341, 342
Clean Water State Revolving Fund (CWSRF), 336
Construction Grants Program, 336
fish kills, 334, 341
fishery resources, 340, 341
bluegills, 341
catfish, 341
redbreast sunfish, 341
recreational, 341
surveys, 340
wastewater loads, effect of, on, 341
floodplain, 335
Georgia Environmental Protection Division (EPD), 335, 336, 339, 340, 342
Georgia Department of Natural Resources (GADNR), 342
Index of Biotic Integrity (IBI), 340
Lake Sidney Lanier, 327, 333
land uses, 332
legislative and regulatory history, 335
Clean Water Act, of 1972, 336
Georgia Water Quality Control Act, of 1964, 335
Metropolitan River Protection Act (MRPA), of 1973, 335
Soil Erosion Act, of 1975, 335
macroinvertebrates, benthic fauna, 341
models, surface water quality, 336
National Pollutant Discharge Elimination System (NPDES), 336, 342
nonpoint sources, 335, 342
agricultural runoff, 340
construction practices, 340
sediment loading, 342
urban runoff, 334, 342
phosphorus:
ban, on phosphate detergents, 339
effluent loads, 338, 339
removal, biological of, 339
water quality, 339
point sources, impact on impairment of water quality, 342
pollution tolerant, biota, 341
population, 327, 331, 336
density, 331
Metropolitan Statistical Area (CMSA), counties, 331
trends (1940–1996), 331
primary treatment, 334, 336, 338
rank, national, of Chattahoochee River, by flow, 327
raw sewage, 334
Reach File Version 1 (RF1), reach, 337
recreation, water–based, 327, 333, 335, 341
secondary treatment, 336, 338, 339, 342
sludge beds, 341
stormwater runoff, 342
streamflow, 327, 329
drought, 334
gage, USGS monitoring, near Atlanta, Georgia, 329
low-flow, 329, 334
regulation of, by dams and reservoirs, 329, 336
mean, summer, trends, (1940–1995), 330
mean, monthly, 330
minimum, 329, 330, 336
suspended solids:
nonpoint source loads of, 340
water quality, 334
wastewater, collection system, 335
wastewater, industrial, 334, 340
wastewater, municipal, 340
effluent flow, 336
effluent loads, 336, 338, 339
land application of, 339
water quality:
criteria, 336
degradation, 334, 340, 341, 342
improvements, from wastewater upgrades, 336, 341, 342
issues, historical, 333, 340, 341
management, 335
monitoring, 342
standards, 336, 340
violations, 341
watershed:
characteristics, physiographic of, 328
management, 342
water uses, 332, 333

West Point Lake, 333
withdrawals, of water, for water supply, 327, 332, 333, 336, 339, 340, 342
intakes, 335, 342
municipal, 327, 332, 333, 335, 336, 339, 340, 342
power generation, 332, 333
projected increase of, 333, 342
Chemical Engineering Plant Cost Index, 90
Chesapeake Bay, 1, 8, 271, 285, 287, 301, 304–307, 311, 313, 316, 317, 324
Chicago, Illinois, 19, 26, 85
Chicago Sanitary District, 22
Chicago Ship and Sanitary Canal, 22
Chickahominy River, 318
Chlorides, 22, 179, 269, 598
Chlorination, of water supplies, (*see* Disease, waterborne)
Chlorophyll, 271, 294, 298, 321, 380, 384
Cholera (*see* Disease, waterborne)
Cincinnati, Ohio, 345, 347, 351–353, 355–358
Clams, hard (*see* Shellfish)
Clarity, water (*see* Transparency, water)
Clean Water Act of 1972, (PL 92-500) (CWA) (*also* Federal Water Pollution Control Act of 1972), (*see* Legislative and regulatory history)
Clean Water Action Plan, 85
Clean Water Needs Survey (CWNS) (*see* Wastewater, municipal, national inventory)
Clean Water State Revolving Fund (CWSRF) (*see* Expenditures, water pollution control)
Cloaca Maxima, 18
Closure, of shellfish beds (*see* Shellfish)
CMSA (*see* Consolidated Metropolitan Statistical Area)
Coastal plain, 138, 255, 258, 273, 311
Cold-water fish, and dissolved oxygen (*see* Dissolved oxygen, criteria and standards)
Coliforms (*see* Bacteria)
Columbia River, 399–401, 404
rank, national, of Columbia River by discharge, 399
Combined sewer overflows (CSOs), 17, 33, 41, 70, 205, 215, 220, 264, 281, 319, 334, 339, 358, 376
capture of, 238, 243, 249
closure, of beaches and shellfish beds, 73
coliform bacteria, and, 369, 391
CSO Abatement Program, New York City, 224
CSO controls, 243
CSO Strategy, national, 224
effluent characteristics of, 74
effluent load ca. 1995, BOD5, catalog unit distribution, 84
effluent load, from, 206, 235, 237
expenditures, for separation of, 340, 369
fecal coliform bacteria, as source of, 206, 222, 237, 242
fecal coliform bacteria, effects of, 300
inventory of facilities, national, 84
remedial measures plan, 340
secondary treatment, effect on compliance with effluent limits, 37
separation, of stormwater and raw sewage, 23, 340, 369, 372, 391
water quality, effect of, on, 73
Commercial fishery (*see* Fishery resources)
Commercial and industrial water use, as component of wastewater flow, 50, 54
Connecticut Department of Environmental Protection (CtDEP), 209
Connecticut River, 12, 183, 199–210, 417, 426, 432, 436, 447
bacteria, degradation, of water quality, from water pollution, 205
biochemical oxygen demand(BOD), water quality, 206, 207
Clean Water Act, of 1972 (*see* legislative and regulatory history)
combined sewer overflows (CSOs), 205
dams and reservoirs, 205
effluent load, 206
fecal coliform bacteria, as source of, 206
fish ladders, 207, 210
fishery resources, effect of, on, 204, 207, 208
regulation of streamflow (*see* streamflow):
development and growth, history of, 204
dissolved oxygen, water quality, 206, 207
before and after POTW upgrades, 207
drainage area, of Connecticut River basin, 199, 200
effluent limits:
technology-based, 206
water quality-based, 206
enforcement, interstate conference, for water pollution control, 205
eutrophication, 210
expenditures, water pollution control, 205, 210
Construction Grants Program, 205
land use, 204
legislative and regulatory history, 205
Clean Water Act, of 1972, 200, 205, 206
Federal Water Pollution Control Act, of 1956, 205
Lower Connecticut, 199, 201, 203, 207–210
nitrogen, water quality, 210
nonpoint sources, export coefficients, land use dependent, 206
phosphorus, water quality, 210
population:
Metropolitan Statistical Area (MSA), counties, 203
served, by municipal wastewater facilities, 206
trends (1940–1996), 203–204
primary treatment, 205
recreation, water-based:
boating, 199
fishing, 209
resources, fishery, 207–209
Atlantic salmon, 207–208
American shad, 209, 210

Connecticut River (*continued*):
riverfront:
deterioration of, and water pollution, 199
economic revitalization of, 199
secondary treatment, 205
standards, water quality, 206
bacteria, fecal coliform, 206
dissolved oxygen, 206
"fishable and swimmable", 206
streamflow, 201–203
gage, monitoring, at Thompsonville, Connecticut, USGS, 201
mean, summer, trends, (1940–1995), 201–203
mean, monthly, 201–203
regulation of, by dams and reservoirs, 201, 203
suspended solids, effluent discharge of, 205
wastewater, industrial, 205–206
effluent loads, 206
pulp and paper mills, 205
wastewater, municipal, effluent loads, 206
water quality:
issues, contemporary, 210
issues, historical, 204–206
water supply:
agricultural, 204
municipal, 204
Consolidated Metropolitan Statistical Area (CMSA), 218, 260
Construction Grants Program (*see* Expenditures, water pollution control)
Copper:
contamination, of sediment bed, 234, 248, 280
contamination, of water column, 280, 358
loads, historical trends of, 234
model, water quality of, 235
standards, water quality, 223, 408
trends, of improvement in water quality, 354
Cormorants (*see* Birds, fish-eating)
Corvallis, Oregon, 405
Criteria, water quality (WQC) (*see* Dissolved oxygen, criteria and standards)
Critical location, of oxygen sag (*see* Dissolved oxygen)
Crow River, 363
Cuyahoga River, 1, 182, 420, 437
CWA (*see* Legislative and regulatory history)
CWNS (*see* Clean Water Needs Survey)
CWSRF (*see* Clean Water State Revolving Fund)

Dams, 257, 259, 273, 335, 342
Buford Dam, Georgia, 329, 332, 333, 336, 341
canalization, of Ohio River, 347
degradation, of biological communities, effect on, 392
dredging, of locks and dams, 364
fish ladders, 207, 210, 324, 410, 411
fishery resources, effect on, 204, 207, 208, 274, 275, 405, 409, 411
flood control, 324, 363
flow augmentation, by reservoirs, of streamflow (*see* Streamflow)
Holyoke Dam, Connecticut, 205
locks and dams, 363, 367, 382, 383
Morgan Falls Dam, Georgia, 329, 333
navigation, and locks and dams, 363, 364, 382, 392
regulation of, streamflow, by locks, dams and reservoirs (*see* Streamflow)
water quality, effect of, on, 367
West Point Dam, 327, 333, 340
Data mining, 12, 106, 119–126, 162, 173–174, 177, 179
filtering, hydrologic, for dry, low-flow conditions, 121, 130, 143, 173
rules, for data screening, 7, 119–126, 173–174, 176
paired data sets, 174
performance measures, for water quality, 177
signal, of point source impacts on water quality, 9, 10, 105–107, 118–119, 121–122, 144, 154–156, 159, 162, 173
spatial scale, effect of, 7, 9–12, 105, 118, 122, 125–126, 144, 154–156, 159, 162, 172, 177
stations, water quality monitoring, selection of, 120–122, 125, 174
point source discharges, downstream location of, on RF1 reach, 122, 159–162, 174
streamflow ratio (*see* Streamflow, ratio, summer)
temporal, filtering, Before and After CWA years, 120–121, 126
worst-case, summary statistics:
tenth percentile, for dissolved oxygen, 121–123, 125, 127, 144, 155, 159, 172, 174
ninetieth percentile, for biochemical oxygen demand, 143, 156
DDT:
ban, on, 324, 388
bioaccumulation of, 246, 247
criteria, exceedance of, 408
mosquito control, spraying of, 246, 277
sediment contamination, 408
"Dead zone", of low dissolved oxygen
Delaware estuary, 275
Gulf of Mexico (*see* Gulf of Mexico)
Decomposition:
bacterial, of organic matter, 24, 25, 43, 44, 46, 106, 116
organic carbon of, 24, 43, 46, 116
organic matter of, 6, 24, 106, 116
rate of, 138
sewage sludge beds of, 368
wastewater of, 26
zone, of active, in a river, 116, 118, 155, 156, 173
DECS (*see* Delaware Estuary Comprehensive Study)
Deforestation, 269

Degradation zone, of dissolved oxygen (*see* Dissolved oxygen)
DEL USA (*see* Delaware Estuary Use Attainability Study)
Delaware Bay, 130, 133, 256–259, 261, 264, 276
Delaware estuary, 12, 184, 192, 193, 255, 256, 258, 260–281
Delaware Estuary Comprehensive Study (DECS), 262–263, 270
Delaware Estuary Use Attainability Study (DEL USA), 271
Delaware River, 189, 255–264, 266, 270, 273, 275, 277, 279–281, 417, 432, 437, 447
acidic conditions, in river, 262
advanced waste treatment, 273
advisory, fish consumption, 281
anoxic, 262, 279
bacteria, degradation, of water quality, from water pollution, 262
bacteria, improvement, of water quality, from water pollution control, 273, 279, 280
bacteria, fecal coliforms, effluent discharges from wastewater, 279
bioaccumulation, of toxic organic chemicals, 280
biochemical oxygen demand (BOD), water quality, 266
biochemical oxygen demand (BOD), loads,
nonpoint source, rural runoff, 264, 265
nonpoint source, urban runoff, 264, 265
municipal and industrial point sources, 263–265, 273, 276, 279
relationship, of effluent loads and dissolved oxygen, 189–191
ultimate BOD, and wasteload allocation (WLA), 270
birds:
shorebirds, 260
waterfowl, 261, 273
chlorides, 269
Clean Water Act, of 1972 (*see* legislative and regulatory history)
combined sewer overflows (CSOs), 264, 281
dams and reservoirs, 257, 259, 273
fishery resources, effect of, on, 274, 275
regulation of streamflow (*see* streamflow)
DDT, mosquito control, spraying of, 277
DECS (*see* Delaware Estuary Comprehensive Study)
DEL USA (*see* Delaware Estuary Use Attainability Study)
development and growth, history of,
colonial era, 261, 273
industrial revolution, 261
dissolved oxygen, water quality, 265, 266, 275, 280
"dead zone", low levels of, 275
minimum oxygen, 265, 266, 280
sag, critical location, 265, 273
drainage area:
Delaware River of, 256
Schuykill River of, 256
DRBC (*see* Delaware River Basin Commission)
economic benefits, of Clean Water Act, 280
effluent limits:
technology-based, 263
water quality-based, 263
fall line, at Trenton, New Jersey, 258, 263
fish consumption advisories (*see* advisory, fish consumption)
fish kills, 274
fishery resources, 260, 273–280
Alewives, 277
American shad, 260, 273, 275–277, 279, 280
Herring, 273, 277
Sheepshead, 273
Striped bass, 260, 275–277, 279
Sturgeon, 260, 273, 275–277, 279
White perch, 277
commercial, 277, 280
recreational, 277, 280
spawning, 274, 275, 277
wastewater loads, effect on, 276
"fishable standards," 281
habitat, destruction of, 279
heavy metals:
sediment contamination, 280
sources of, 280
water quality, 273
hydrogen sulfide, 262, 279, 368
hypoxic, 262
INCODEL (*see* Interstate Commission Delaware River Basin)
land use, 259, 260, 263, 269
agricultural practices, 269
deforestation, 269
legislative and regulatory history, 262, 263
Clean Water Act, of 1972, 263, 270, 271, 280
Lower Delaware, 257, 259, 261, 264
marshes, 260, 273
models, surface water quality, 262
Dynamic Estuary Model, (DEM), 184, 271, 272
hydrodynamic, 271
review, literature of, 270–271
evaluation, of wastewater upgrades, 189, 194, 271–273
nitrification, 269, 271
nitrogen, water quality, 266, 269, 273
ammonia-nitrogen, 266, 271, 273, 280
cycle, 271
nitrate-nitrogen, 269, 271
organic-nitrogen, 271
total, 266
nitrogen, effluent loads, 279
nitrogenous BOD (NBOD), 270
nonpoint sources, export coefficients, land use dependent, 263–265
phosphorus, water quality, 266, 269, 273, 280
phosphorus, effluent loads, 269, 279
ban, on phosphate detergents, 269
population, 281

DEL USA (*continued*):
Metropolitan Statistical Area (MSA), counties, 260
density, 260
trends (1940–1996), 259, 260
primary treatment, 262, 270, 271
productivity, biological, 258, 259
rank, national, of Delaware River by length and flow, 255
raw sewage, 262, 281
Reach File Version 1 (RF1), reach, 266
recreation, water-based, 280
riverfront, 279, 280
deterioration of, and water pollution, 279
economic revitalization of, 279, 280
salinity, 258
intrusion, 259
secondary treatment, 262, 263, 270, 271, 273
sediment toxicity, 281
shellfish, 260, 278
blue crabs, 277
commercial landings of, 277
hard clams, 277
horseshoe crabs, 277
lobsters, 277
oysters, 260, 277, 279
shipping, navigation, 258
stormwater runoff, 280
streamflow, 257–258, 273
gage, USGS monitoring, at Trenton, New Jersey, 257
drought, 259
during early 1960s, 257
low-flow, 257, 273
regulation of, by dams and reservoirs, 257, 259, 273
mean, summer, 1940–1995, 257
mean, monthly, 257
suspended solids, water quality, 273
tidal freshwater, 258
transition zone, 258, 259
toxic organic chemicals, 280
turbidity, 258, 259
waste load allocation (WLA), 262, 263, 270, 271, 281
wastewater, industrial, 261, 262, 273, 275, 279
effluent flow, 271
effluent loads, 264–266
wastewater, municipal, 261, 273, 275, 279
communities served, by, 262
effluent flow, 271
effluent loads, 264–266
water quality:
degradation, from water pollution, 261, 273
improvements, from water pollution control, 266, 273, 277, 280
issues, contemporary, 280–281
issues, historical, 261–262
water quality standards, 263, 270, 273
water quality zones, in Delaware estuary, 263
wetlands, 274
withdrawal, of water, for water supply:
industrial, 259, 260
intakes, 261, 269
municipal, 260
Delaware River Basin Commission (DRBC), 259, 262–263, 265, 269–271
Denitrification, 271
Department of Natural Resources, Georgia, 339
Depletion, of dissolved oxygen (*see* Dissolved oxygen)
Des Plaines River, 22
Dilution, in-stream, of wastewater, 26, 108, 110, 120, 179, 292, 319, 367
Dioxins, 408
DMRs (*see* Discharge Monitoring Reports)
NPDES (*see* National Pollutant Discharge Elimination System)
Disease, waterborne, 28
"Dark" Ages, during, 18
Great Sanitary Awakening, 19, 21, 96
Greek and Roman empires, and, 18
International Sanitary Conferences (1851–1938), 19
Snow, Dr. John, 19, 347, 364
Talmud, Old Testament, 18
Thames River, 19, 23
chlorination, of water supplies, and, 19, 20, 22, 237, 242, 243, 245, 300, 368
cholera epidemics, 19
household piped water supply, introduction of, 19
nineteenth century, during, 19
public health and, 18–20, 22, 96
public sanitation and, 18, 19
public water supplies and, 19
shellfish beds, closure (*see* Shellfish)
twentieth century, during first half of, 19
typhoid epidemics, 19, 20, 242, 368
Dissolved oxygen (DO):
analytical methods, 107
aquatic organisms, and survival, 110
data mining, Before and After CWA, example of methodology, for, 9–13, 123–125
decomposition zone, active, 116, 118, 156, 173
degradation zone, 116, 118, 156, 173
demand, for, 26, 43, 47, 95, 110, 118, 179, 406
indicator, of water quality, as, 23–25, 106, 120, 176, 179, 392
location, critical, for dissolved oxygen sag, 118, 154, 162, 298
natural factors, effect of, on, 6,
nitrification, effect of, on, 24, 116
population impacted, estimate of, by Clean Water Act, 177, 179
reaeration, atmospheric, 6, 116, 118, 120, 179
recovery zone, 116, 118, 156, 159
saturation (*see* Dissolved oxygen, saturation)
seasonal variation of, 110, 120, 123, 292
sediment oxygen demand (SOD), 271, 319

signal, effect of point source discharges on, 7, 9, 12, 105–108, 118–119, 144, 154–155, 159, 176, 181
spatial scale:
aggregation of data, 126,
effect of, on trends of, 7, 9–12, 105–107, 119, 125–126, 154–162, 173, 176
streamflow, effect of, on, 110, 377–379, 385, 391
wastewater discharges, effect of, on, 2, 7
water temperature, effect of, on, 110, 116, 118
worst-case conditions, for, 9, 12, 106–108, 118–119, 121, 125–126, 162, 164, 171, 181
case study waterways, summary of, 184
spatial dependence of, 116–119, 125–126, 172
statistics of, (*see* Statistics, dissolved oxygen)
temporal dependence of, 108, 110–111, 116
Dissolved oxygen (DO), water quality, 206, 241, 248, 356, 377–379
before and after Clean Water Act, changes in,
catalog units, 130, 138–139, 143, 144, 154, 175
major river basins, 162, 164, 171, 176
POTW upgrades, Connecticut River 207
RF1 reaches, 127–130, 154, 175
degradation of, from water pollution, 334, 335, 336, 379
depletion of, 26, 194, 205–206, 221–222, 273, 276, 285, 290, 292, 301, 306, 315, 324, 350, 367, 371, 376, 404, 406
drought, during early 1960s, effect of, on, 298
drought, during 1988, effect of, on, 352, 377, 379
historical, records of, 286, 290, 292, 298, 319, 325, 368, 376, 384
improvement of, from water pollution control, 336, 371, 372, 376, 377, 379, 385
low levels of, 275, 368, 370, 411
minimum, levels of, 265, 266, 280, 292, 294, 315, 320, 323, 324, 368, 377
models, water quality of, 292, 295, 298, 323, 324, 384–385,
sag, 118, 119, 154–156, 159, 162, 173, 189, 192, 193, 266, 275,
sag, critical location of, 144, 154, 265, 273, 292, 298, 319, 320, 325, 352, 377, 379, 385
spatial distribution of:
in hypothetical river, 116–118
in streams and rivers, 6, 105, 116–118, 265, 267, 268, 292, 298, 319, 353, 376, 406, 407, 409
STORET, as data source, 377
trends, long-term of:
case study waterways, 184, 188, 189, 206–207, 238, 247, 336, 265–266, 292, 294, 319–321, 352–353, 377–379, 406
catalog units, (1961–1995), 143–144, 154–156, 159, 162, 175
RF1 reaches, (1961–1995), 130, 133–135, 156, 159, 162, 175
Dissolved oxygen (DO), criteria and standards:
benchmark, 5 mg/L, 6, 12, 110, 120, 127–130, 139, 194
cold-water, fish and biota, 6
criteria, water quality (WQC), 6, 7, 31, 32, 189, 336, 351
saturation, as water quality criterion, 24, 120
standards, water quality (WQS), 27, 154, 273, 294, 298, 300, 317, 324, 385, 390
warm-water, fish and biota, 6, 120, 125, 127
Dissolved oxygen (DO), effluent characteristics, of municipal wastewater, 78
Dissolved oxygen (DO), saturation, 164, 179
degradation, of water quality, from water pollution, 222, 368
factors, influencing, in natural waters, 24, 108, 110, 116
improvement, of water quality, from water pollution control, 247
indicator, of water quality, as, 120, 179
trends, long-term of, in New York Harbor, 238, 247
wastewater effluent, levels of, 78, 108
DO (*see* Dissolved Oxygen)
Domestic wastewater, 54, 453
biochemical oxygen demand (BOD), 26
Drainage area (*see* also River basin name)
case study river basins of, 199, 200, 213, 256, 285, 286, 311, 312, 318, 319, 327, 328, 346, 363, 399
contributing to streamflow at USGS gage, 110, 364
Hydrologic Regions of, 199, 213, 255, 285, 327, 362, 345, 399
Mississippi River basin of, 361, 382
Drainage basin (*see* also Watershed):
Delaware River of, 269, 273
Hudson River of, 215, 249
nonpoint source runoff, and, 73, 74
Mississippi River of, 361
Ohio River of, 346
Upper Mississippi River, 361, 366
DRBC (*see* Delaware River Basin Commission)
Dredged channels, 382, 383
Drought (*see* Streamflow)
Ducks (*see* Birds, waterfowl),

Eagles (*see* Birds, fish-eating)
Eagle Creek, 156
East River, 225, 228, 229, 235, 238, 239, 243
Economic benefits, of Clean Water Act (*see* Benefits)
Efficiency, of wastewater treatment (*see* Wastewater, municipal, removal efficiency)
Effluent characteristics (*see* Wastewater, municipal; also Wastewater, industrial)
Effluent guidelines (*see* Wastewater, industrial)

Effluent limits (*see* Wastewater, municipal, effluent limits)
Effluent regulations, 119, 181
BPT (*see* Best practicable treatment)
Discharge Monitoring Reports (DMRs) (*see* Monitoring, of effluent quality)
discharge limitations, 33
discharge permits, 4
economic considerations, 36
minimum technology, 3, 4, 38, 93, 96, 171, 175, 189
National Pollutant Discharge Elimination System (NPDES), 4, 34, 38, 40, 42, 47, 57, 70, 85, 95–96, 206, 264, 336, 342, 370, 405
percent removal, 37, 40, 41
secondary treatment, as minimum requirement, 36, 189
Enforcement, interstate conferences, for water pollution control, 184, 415–446
Connecticut River, 205
joint state–federal, (1957–1971), 184, 419–423
potential enforcement actions, (1963), 184, 416–419
Upper Mississippi River, 361
Egrets (*see* Birds, wading):
Embayments, 221, 294, 311
Environmental Protection Division, Georgia (EPD) (*see* Chattahoochee River)
EPD (*see* Environmental Protection Division, Georgia)
Epidemics, of cholera and typhoid fever (*see* Disease, waterborne)
Estuarine:
case study waterways, 183, 184
circulation, 285
fishery (*see* Fishery resources)
models, water quality, 13, 184, 189, 242, 262, 270, 271, 294, 321, 325
Eugene, Oregon, 400, 403, 404
Eutrophication, 210, 242, 271, 292, 294
cultural, 182, 214
Lake Erie of, 182, 422, 424, 426, 427, 432, 447
Lake Pepin, of (*see* Upper Mississippi River)
nutrient enrichment (*see* Nutrients)
Expenditures, water pollution control, 4, 13, 90, 91, 93, 95, 98, 105, 194, 501
Chemical Engineering Plant Cost Index, 90, 93
Clean Water State Revolving Fund (CWSRF), 1, 3, 5, 13, 90–91, 95, 98–99, 501, 502
Construction Grants Program, 1, 3–5, 31, 13, 30–32, 34, 39, 70, 85, 90, 92, 93, 96, 98–99, 194, 205, 224, 229, 336, 501, 502
Grants Information and Control System (GICS), 90–92, 503
Gross Domestic Product (GDP), 93
Pollution Abatement Cost Expenditures (PACE), 4, 91
State Revolving Fund (SRF) (*see* Clean Water State Revolving Fund):
capital costs, 1, 4, 13, 93, 98
future, projected, for water pollution control, 13, 70, 93, 95, 98
operations and maintenance costs (O&M), 1, 4, 13, 98, 336, 501, 504–506
private sector, 4, 17, 90–94, 98, 194
public sector, 4, 17, 90–94, 98, 194

Fall Creek, 156
Fall line, 258, 263, 285, 286, 290, 315
Fecal coliforms (*see* Bacteria)
Federal Reserve Board of Governors, 593, 594
Federal Water Pollution Control Act, Amendments (*see* Legislative and regulatory history)
Federal Water Pollution Control Act, Amendments of 1972 (*see* Legislative and regulatory history, Clean Water Act of 1972)
Federal Water Pollution Control Administration (FWPCA), 31–33
monitoring program, water quality, 368
network, national surveillance, 138
Fin rot (*see* Fishery resources)
Fish (*see* Fishery resources)
Fish consumption advisories (*see* Advisory, fish consumption)
Fish kills, 6, 274, 290, 334, 341, 380, 383, 404
Fish ladders (*see* Dams)
"Fishable and swimmable", 4, 34, 194, 195
"fishable standards", 281
Fishery resources (*see also* Shellfish), 219, 273, 275–281, 301, 306–308
Alewife, 221
American lobsters, 277
American shad, 209, 245, 261, 273, 275–277, 279, 280, 290
Atlantic salmon, 204, 207, 210
Bass Master Tournaments, 208, 311, 324, 357, 358
Blue sucker, 389
Catfish, 6, 315, 340, 341, 389, 411
Common carp, 389
Crappies, 411
Largemouth bass, 308, 324, 340, 409
Pacific salmon, 405, 408, 410, 411
Paddlefish, 389
Shortnose sturgeon, 248, 273, 275–277, 279
Smallmouth bass, 183, 324
Smallmouth buffalo, 389
Steelhead, 409, 411
Striped bass, 221–223, 245, 248, 261, 275–279, 324
Sturgeon, 221, 248, 261, 273, 275–277, 279, 411
Walleye, 183, 368, 389
White perch, 277–278
Winter flounder, 245
commercial, 221, 245, 280
dams, effect on fish (*see* Dams)
estuarine, 221
fin rot, 245
habitat, for, 219, 221, 246, 247, 274, 279, 301, 302, 304, 306–308, 324, 356–358,

362, 380, 392, 399, 411
spawning of, 245, 273, 275, 277, 290, 302, 409–411
Fishing:
commercial (*see* Fishery resources)
sport (*see* Recreation, water-based)
Flint River, 183, 327, 437
Floods, 110, 287, 324, 363–365, 370, 380–383, 386, 392
Flow-augmentation (*see* Streamflow)
Fossil fuel, 269, 335
Fox River, 130
Fox River, lower, 143
Free-flowing, river, 122
Frequency distributions, of dissolved oxygen (*see* Statistics, dissolved oxygen)
FWPCA (*see* Federal Water Pollution Control Administration)

Gaging stations, USGS, streamflow (*see* Streamflow, gaging stations, USGS)
Galveston Bay, 79, 182, 424, 427, 437, 447
GAO (*see* General Accounting Office)
GDP (*see* Gross Domestic Product)
General Accounting Office (GAO), 4, 182, 194
Georgia EPD (*see* Environmental Protection Division, Georgia)
GICS (*see* Grants Information and Control System)
Grants Information and Control System (GICS) (*see* Expenditures, water pollution control)
Great Basin, Hydrologic Region 16
dissolved oxygen, before and after CWA, 164
Great Flood of 1993, Upper Mississippi River (*see* Streamflow)
Great Lakes, Hydrologic Region 4,
dissolved oxygen, before and after CWA, 164
Great Miami River, 152
Great Sanitary Awakening (*see* Disease, waterborne)
"Great Stink", London (*see* Wastewater, collection and disposal systems)
Greater than secondary (*see* Secondary treatment)
"Green Movement", 34
Gross Domestic Product (GDP), 93, 505–506
Gulf of Mexico, 84, 327, 329, 361, 433
Action Plan, NOAA and EPA, for Gulf of Mexico Initiative, 382
"Dead Zone", and algal blooms, 382
nitrogen load, of Mississippi River, to, 382

Habitat, for fish (*see* Fishery resources)
Habitat, destruction of, 221, 274
Hackensack River, 228
Hard clams (*see* Fishery resources)
Harlem River, 228
Hartford, Connecticut, 199–201, 203–208
Heavy metals (*see also* Copper and Lead), 370, 380, 391
contaminant, as, for water supply users, 2
deposition, atmospheric of, 75, 280, 389
water quality, 273, 354, 355, 370, 408
effluent load of, 235, 248
effluent load, controls, for, 44, 189, 234, 248, 370, 371, 391
issues, contemporary, for water quality management, as, 195, 222, 280, 358, 380, 391, 392, 408
pretreatment, industrial, 234, 391
sediment bed, contamination, by, 235, 280, 382, 383
sediment bed, cores, and, 370
sources of, 280, 370, 371, 392
standards, water quality, for, 222, 408
state variables, of water quality model, as, 223, 235
storm water, urban, runoff characteristics, and, 75
wastewater, industrial loads, and, 71, 392
Hepatitis, infectious, 221, 242
Herons (*see* Birds, wading)
Heterotrophs, 106
Hoosic River, 183, 417, 438
Houston Ship Channel, 1, 182, 447
HUC (*see* Hydrologic unit code)
Hudson River, 182, 213, 215, 217, 219, 221, 225, 226, 228, 229, 232, 235, 238, 239, 243, 245, 248, 422, 427, 432, 438, 447
Lower, 221, 226, 228–235, 243, 245, 591
Middle, 226, 228, 235
Upper, 591
Hudson-Raritan estuary, 12, 213, 216, 218, 223, 226, 235, 243, 246–249
advanced primary, 226, 232
advisory, for fish consumption, 248, 249
algae, blooms, 222, 249
bacteria, contamination of shellfish beds, 221, 222
bacteria, degradation, of water quality, from water pollution, 222
bacteria, improvement, of water quality, from water pollution control, 237, 243, 248, 249
beaches, recreational, 225, 242–243, 245, 248–249
closure of, 222, 242, 243, 245
Enhanced Beach Protection Program, 245
re-opening of, 243, 248–249
benefits, of water pollution control, 249
bioaccumulation, of toxic organic chemicals, 222, 246, 247
biochemical oxygen demand (BOD):
effluent loads, 232
per capita loading rate of, 232
water quality, 206, 207, 238, 248
birds:
Canada goose, 246
eagles, 222, 246
egrets, 213, 246, 247
herons, 213, 246–248
ibises, 213, 246, 247
ospreys, 246, 248
peregrine falcons, 247, 248

Hudson-Raritan estuary (*continued*):
chlorination, of wastewater effluent, 242, 243, 245
Clean Water Act (*see* legislative and regulatory history)
combined sewer overflows (CSOs), 215, 220, 222, 224, 238, 249
capture of, 238, 243, 249
CSO Abatement Program, New York City, 224
CSO controls, 243
CSO Strategy, national, 224
effluent load, 235, 237
fecal coliform bacteria, as source of, 222, 237, 242
data sources, for case study, 213
DDT, for mosquito control, spraying of, 246
debris, floatable, 245
development and growth, history of, 214–216, 218
disease, waterborne, 221, 242
gastroenteritis, 242
hepatitis, infectious, 221, 242
swimmer's itch, 242
swimmer's ear, 242
typhoid fever, 221, 242
dissolved oxygen, water quality, 238, 241, 248
trends, long-term, of saturation, 238
drainage area, of Hudson River basin, 213
dredge spoils, 221
expenditures, water pollution control, 223, 224, 249
Construction Grants Program, 224, 229
Pure Waters Program, New York State, 224
public works, during, Great Depression, 223
Federal Water Pollution Control Act (*see* legislative and regulatory history)
fish consumption advisory (*see* advisory, fish consumption)
fishery resources, 219, 221, 245
American shad, 221, 245
Alewife, 221
Shortnose sturgeon, 248
Striped bass, 221–223, 245, 248
Sturgeon, 221, 248
Weakfish, 245
Winter flounder, 245
commercial, 219, 221, 245, 249
fin rot, 245
recreational, 220
Harbor Estuary Program, (HEP), 222, 235, 242
Harbor Herons Complex, 246
heavy metals, effluent loading, 248
copper, 234, 235
lead, 234, 235
heavy metals, sediment contamination (*see* sediment bed, contamination)
indicators, biological, 248
Interstate Sanitation Commission (ISC), 223, 229
land use, 218
legislative and regulatory history, 223–224
Clean Water Act of 1972, 224, 225, 229, 243, 248
Federal Water Pollution Control Act of 1948, 223
Federal Water Pollution Control Act of 1956, 223
Federal Water Pollution Control Act of 1965, 224
marine borers, (*see* shipworms)
Metropolitan Sewerage Commission, 220, 221, 223
models, surface water quality, 222, 223, 242
heavy metals, 223, 235
hydrodynamic, 242
System Wide Eutrophication Model, (SWEM), 242
toxic organic chemicals, 235
monitoring, water quality, 220, 221, 248
twentieth century, early, 220, 221, 248
twentieth century, latter half, contemporary, 248
nitrogen, water quality, 238, 248
ammonia-nitrogen, 238, 248
nitrogen, total, effluent loads, 232, 233
nonpoint sources, 206, 226, 235, 249
urban runoff, 215, 235
watershed runoff, 237
phosphorus, total, effluent loads, 233, 234
ban, on phosphate detergents, 234
polychlorinated biphenyls (PCBs), 222, 223, 235, 246
population:
Metropolitan Statistical Area (MSA), counties, 216, 218
served, by municipal wastewater facilities, 226, 228, 229, 237
trends (1940–1996), 218
pretreatment, industrial (*see* wastewater, industrial)
primary treatment, 223, 232
rank, national, of Hudson River by length and flow, 213
raw sewage, 220–222, 225, 226
bypasses of, from failure of collection system, 226, 245
elimination, of discharges of, 226, 232, 237, 243, 249
recreation, water–based, 249
boating, 249
fishing, 220, 249
Sanitary Commission, 225
secondary treatment, 225, 226, 229, 232, 242
sediment contamination, 249
copper, 235
lead, 235, 248
polynuclear aromatic hydrocarbons (PAHs), 235
toxicity, and, 235
shellfish, 221, 222, 245, 248

beds, closure of, 221, 222
beds, re–opening of, 245, 248
clams, hard, 221
contamination, of beds, by raw sewage, 221
oysters, 221
scallops, bay, 221
shipping, navigation, 219
shipworms, (marine borers), 241, 248
sludge, disposal:
biosolids, and, 224
dewatering, and, 224
ocean dumping, as practice for, 224
standards, water quality:
bacteria, fecal coliform, 222, 243
heavy metals, 222
streamflow, 215
drought, early 1960s, 215, 224
gage, USGS monitoring, at Green Island, New York, 215
mean, summer, trends, (1940–1995), 215
mean, monthly, 215
suspended solids, 232, 237
effluent loads, 232
per capita loading rate, 232
watershed runoff, 237
toxic organic chemicals, 222, 223, 241, 246, 248
wastewater, industrial, 215, 221
effluent flow, 215
effluent loads, 235
pretreatment, 222
wastewater, municipal, 215, 226
effluent flow, 215, 226, 229, 232
effluent loads, 226, 235
water quality:
issues, contemporary, 249
issues, historical, 220–222
water uses, 219
withdrawals, of water, for water supply, 219
per capita use, rate of, 229, 232
wetlands, 214, 219, 246
Hydrogen sulfide, 25, 220, 262, 279, 367
Hydrologic region, definition of, (*see* Hydrologic Unit Code)
Hydrologic Unit Code (HUC), 8, 9
accounting unit, definition of, 8, 9
catalog unit, definition of, 9
hydrologic region, definition of, 8
major river basin, definition of, 8
subregions, definition of, 8
Hydroelectric, as source of power, (also hydropower), 358, 363, 366, 403
Hypoxia, 262, 382, 384

IBI (*see* Index of Biotic Integrity)
Ibises (*see* Birds, wading)
IFD (*see* Industrial Facilities Discharge database)
Illinois River, Upper, 183
INCODEL (*see* Interstate Commission Delaware River Basin)
Index of Biotic Integrity (IBI), 306, 307, 340, 357, 389
Indicators, biological, 248, 369
Industrial Facilities Discharge database (IFD), 122
Industrial pretreatment (*see* Wastewater, industrial)
Industrial Production Index (IPI), 591, 593, 594, 597
Industrial revolution, 261
Industrial wastewater (*see* Wastewater, industrial)
Industrial water use (*see* Withdrawals, industrial, self-supplied)
Infectious hepatitis (*see* Hepatitis, infectious)
Infiltration and inflow, 41
as component of wastewater flow, 50, 54
International Sanitary Conferences (*see* Disease, waterborne)
Interstate waters, 27, 29–32, 96, 405, 417
Interstate Commission Delaware River Basin (INCODEL), 262
Interstate Sanitation Commission, NY-NJ-Ct (ISC), 223, 229
Inventory, of municipal wastewater (*see* Wastewater, municipal),
Investment, in water pollution control (*see* Expenditures, water pollution control)
IPI (*see* Industrial Production Index)
ISC (*see* Interstate Sanitation Commission)

Jamaica Bay, 221, 228, 238
James estuary, 12, 184, 192, 193, 311, 312, 314–321, 325
James River, 184, 311–318, 321, 322, 324, 325, 432, 438
advanced primary, wastewater treatment, 318
advanced secondary, wastewater treatment, 324
algae, 315, 324
assimilative, capacity, of James River, 316
beaches (*see* recreation, water-based)
biochemical oxygen demand (BOD):
water quality, simulated, with James River Model(JMSRV), 322
upstream, ultimate BOD, 319
biochemical oxygen demand (BOD), effluent loads, 318, 320, 321, 323, 325
birds:
cormorants, 324
eagles, bald, 324
egrets, cattle, 324
herons, great blue, 324
ospreys, 324
vultures, turkey, 315, 324
carbon, effluent loads, 315
removal of, 320, 325
chlorophyll, 321
Clean Water Act, of 1972 (*see* legislative and regulatory history)
combined sewer overflows (CSOs), 319
dams and reservoirs, 324
fish ladders, 324

James River (*continued*):
flood control, 324
regulation of streamflow (*see* streamflow, regulation of)
dissolved oxygen, water quality,
depletion of, 315, 324
historical, records of, 319, 325
minimum, levels of, 315, 320, 323, 324
sag, critical location of, 319, 320, 325
simulated, with James River Model(JMSRV), 323, 324
spatial distribution of, 319
DDT, 324
drainage area, of James River basin, 311, 312, 318, 319
effluent limits, water quality-based, 324
expenditures, water pollution control, 316
fall line, at Richmond, Virginia, 315
"fishable", 316
fish kills, 315
fishery resources:
Bassmasters Fishing Tournament, 311, 324
catfish, 315
recreational (*see* recreation, water-based)
Smallmouth bass, 324
spawning, 324
Striped bass, 324
habitat, for wildlife, 324
land use, 311
legislative and regulatory history, 316
Clean Water Act, of 1972, 316, 321, 324
Federal Water Pollution Control Act, of 1956, 316
Section 208 Areawide Study, Richmond-Crater, 316
models, surface water quality, 316, 321, 325
evaluation, of phosphorus controls, 321
evaluation, of wastewater upgrades, 189, 194
James Estuary Model (JEM), 316
James River Ecosystem Model, 316
James River Model (JMSRV), 316, 321
nitrogen, effluent loads, 315
nitrogenous BOD(NBOD), 319
nonpoint source, runoff, loads, 319
nutrient:
effluent loads, 315, 321, 325
enrichment, 316, 317
water quality, 325
phosphorus, effluent loads, 315, 318
ban on phosphate detergents, 316–318, 321, 325
controls on, 317, 318, 321
population, 314, 315
growth, 316
Metropolitan Statistical Area (MSA), counties, 314
trends (1940–1996), 315
primary treatment, 317, 321–323, 325
rapids, whitewater, Class IV, 311, 325
raw sewage, 315
recreation, water–based, 315
boating, 311
fishing, sport, 311
kayaking, 325
rafting, whitewater, 325
Richmond, Virginia, 280, 311–317, 319–321, 323–325
riverfront:
Canal Walk, 325
deterioration of, and water pollution, 315
revitalization of, 325
secondary treatment, 316, 318, 320–322, 324, 325
greater than, 324, 325
State Water Control Board, 316
stormwater overflow basin, 316
streamflow, 311, 319
10–year, 7–day minimum flow (7–Q-10), 322, 324
gage, USGS monitoring, near Richmond, Virginia, 311
floods, 324
low-flow, 318, 322, 323, 324
mean, annual, 311
mean, summer, (1940–1995), 312
mean, monthly, 312
regulation of, by dams, 324
suspended solids, effluent loads, controls on, 316
tidal, 320
Upper James River, 324
Upper James River Wasteload Allocation Plan, 316
waste load allocation (WLA), 316, 318
wastewater, collection system, 315
wastewater, industrial, 316, 317
wastewater, municipal, 316, 317
water quality:
degradation, from water pollution, 315
improvements, from water pollution control, 320, 321, 324, 325
issues, historical, 315
plans, management, 316
water quality standards, 321
dissolved oxygen, 315, 324
nutrient enrichment, 316
withdrawals of water, for water supply:
industrial, 312
municipal, 311, 312

Johnson, Lyndon, B., President, 301

Kill van Kull, 221, 228, 246, 247
Kolmogorov-Smirnoff (KS) test (*see* Statistics, dissolved oxygen)

Ladder, water quality (*see* National Water Pollution Control Assessment Model)
Lake Dubay, 143
Lake Erie, 182, 422, 424, 426, 427, 432, 447
Lake Itasca (*see* Upper Mississippi River)
Lake Pepin (*see* Upper Mississippi River)
Lake Sidney Lanier (*see* Chattahoochee River)
Lake Washington, 182, 447

Land use:
agriculture, 1, 74, 154, 199, 204, 269, 290, 345, 350, 362, 366, 382, 399
characteristics of, 154, 204, 218, 259, 260, 263, 290, 311, 332, 345, 350, 361, 366, 403
deforestation, 269
export coefficients, for nonpoint source runoff, 75, 206, 264, 381
forest, 74, 75, 204, 290, 332, 366
industrial, 332, 345, 350
influence, on watershed runoff, 74, 106, 206, 264, 366, 380
transformation of, 75
urban, 154, 259, 290, 332, 345, 366
Lawrence Experiment Station, 5, 25
Lead:
combined sewer overflows (CSO), as source of, 73
loads, historical trends of, 234
contamination, of sediment bed, 234, 248
model, water quality of, 235
Legislative and regulatory history, of water pollution control, 17
Clean Water Act of 1972, (PL 92–500) (also Federal Water Pollution Control Act of 1972), 1, 27, 33, 36, 40, 263, 334, 390, 391, 415
goals and objectives of, 5, 6, 34, 176, 182, 194, 195
history of, 34
improvements, in water quality, attributed to, 4, 13, 99, 119, 324, 411
national policy of, 34, 195
secondary treatment, as minimum technology, 42, 118, 171, 181
Waste-water discharges, regulatory controls on:
Clean Water Act Amendments of 1977, (PL 95–217), 38
Clean Water Restoration Act of 1966
Clean Water State Revolving Fund (CWSRF) (*see* Expenditures, water pollution control)
Construction Grants Program (*see* Expenditures, water pollution control)
Delaware River Basin Commission(DRBC), establishment of, 262–263
Federal Water Pollution Control Act of 1948, (PL 80–845), 28, 30, 223, 350
Federal Water Pollution Control Act Amendments of 1952, (PL 82–579), 30
Federal Water Pollution Control Act Amendments of 1956, (PL 84–660), 30, 85, 205, 223, 290, 316
Federal Water Pollution Control Act Amendments of 1961, (PL 87–88), 30, 31
Federal Water Pollution Control Act Amendments of 1971, (PL 92–240), 32
Federal Water Pollution Control Act Amendments of 1972, (PL 92–500) (see Legislative and regulatory history, Clean Water Act of 1972)
Georgia Water Quality Act, of 1964, 335
International Sanitary Conferences (1851–1938), influence on, 21
Interstate commerce clause, of U.S. Constitution, 96
Interstate Commission Delaware River Basin (INCODEL), establishment of, 262
Interstate Sanitation Commission, NY-NJ-Ct (ISC), establishment of, 223
Interstate waters (*see* Interstate waters)
Metropolitan River Protection Act, 1973, of Georgia (MRPA), 335
Metropolitan Sewerage Commission, of New York, establishment of, 220
Municipal Wastewater Treatment Construction Grants Amendment of 1981, (PL 97–117), 39
National Municipal Policy of 1984 (49 FR 3832–3833), 40
National Pollutant Discharge Elimination System (NPDES) (see Effluent Regulations)
Nixon, President Richard M. (*see* Nixon, Richard M.)
Oil Pollution Act of 1924, 28
Ohio River Valley Sanitation Commission (ORSANCO), establishment of, 345, 346
Oregon State Sanitary Authority, establishment of, 405
Public Health Act of 1848, Great Britain, 21
Public Health Service Act of 1912, 28
Rivers and Harbors Act of 1890, amended 1899, 27, 28
Roosevelt, President Franklin D. (*see* Roosevelt, Franklin D.)
Secondary Treatment Regulations (*see* Secondary treatment)
Section 208, areawide basin plans, 271, 316, 383
Soil Erosion Act, of 1975, Georgia, 335
timeline, of federal water pollution control acts, 29, 35
waivers, Section 301(h), for marine wastewater discharges, 33, 38, 52
Water Pollution Control Act, of 1945, Minnesota, 373
Water Pollution Control Act Amendments of 1972 (*see* Legislative and regulatory history, Clean Water Act of 1972)
Water Quality Act of 1965, (PL 89–234), 31, 224, 300, 316, 351, 405
Water Quality Improvement Act of 1966, (PL 89–753), 32
Water Quality Improvement Act of 1970, (PL 91–224), 32, 33
Lehigh River, 182, 261, 438
Less than secondary (*see* Secondary treatment)
Locks (*see also* Dams), 363–365, 382, 383, 386, 387, 392
London, England, 22
Long Island Sound, 1, 199, 201, 204, 210, 215, 220, 228, 229, 242, 245, 424, 427, 432, 439

Louisville, Kentucky, 345, 347, 348, 352, 353, 355–357
Low-flow (*see* Streamflow)
Lower Connecticut River (*see* Connecticut River)
Lower East River, 225, 238, 239
Lower Hudson River (*see* Hudson River)
Lower Fox River (*see* Fox River), 143
Lower Great Miami, 143
Lower New York Bay, 238, 248
Lower Spokane River, 143
Lower Mississippi, Hydrologic Region 8
  dissolved oxygen, before and after CWA, 164
Lower White River (*see* White River)

Macroinvertebrates, aquatic, 341, 408
  benthic fauna, 341, 357, 369, 408
  blood-worms, 341
  caddisflies, 408
  marine borers (also shipworms, *Teredos*), 241, 248
  shipworms, 241
  mayflies (*Hexagenia*), 371, 372, 391, 408, 409
  pollution intolerant, 248, 306, 357, 408–409
  pollution tolerant, 341, 358, 408–409
  shipworms, 241, 248
  sludgeworms, 408
  stoneflies, 408, 409
  Teredos, 241, 248
Mammals, fish-eating, 388
Marine borers (*see* Macroinvertebrates, aquatic)
Marshes (*see also* Wetands), 214, 246, 260, 273, 277
Massachusetts State Board of Health, 21
Mathematical models, of water quality (*see* Models, surface water quality)
Maumee River, 130, 135
Mayflies (*see* Macroinvertebrates, aquatic)
Mercury:
  contamination, of water column, 280
  contamination, of sediment bed, 280, 370, 371, 372, 379
  dental amalgam, as source of, 392
  deposition rates, historical trends of, 371
  loading, 372, 391
  model, water quality of, 235
  sources of, from municipal and industrial wastewater, 371, 391
  standards, water quality, violations of, 223, 408
Merrimack River, 22, 417, 427, 432, 433, 439
Metropolitan River Protection Act, 335
Metropolitan Sewerage Commission, of New York, 23, 220, 223
Metropolitan Statistical Area (MSA), 184, 219, 260, 289, 314, 331, 349, 366, 403
Mid-Atlantic, Hydrologic Region 2, 164, 213, 255, 285, 311
  dissolved oxygen, before and after CWA, 164
  drainage area of, 213, 255, 285
Milwaukee River, 130, 134
MSA (*see* Metropolitan Statistical Area)
Minimum technology, for wastewater treatment (see Effluent regulations)
Minneapolis, Minnesota, 111, 144, 162, 189, 361–363, 366, 367, 369, 372, 388
Minnesota Department of Health, 389
Minnesota Pollution Control Agency, 383
Minnesota River, 9, 154, 363, 366, 373, 376–378, 380–382, 384, 386, 391, 392
  land uses, 369
  phosphorus yield of, 381, 382
  sediment yield of, 369
Minnesota River Assessment Project, 382
Minnesota-Wisconsin Boundary Area Commission, 388
Mississippi River, 22, 77, 78, 85, 97, 345–347, 361, 362
  annual mean flow, 361
  drainage area, of basin, 361, 382
  land use, agricultural, 382
  length of, 361
  rank, national and international, of Mississippi River, by length and flow, 361
Missouri, Hydrologic Region 10, 164, 171, 362
  dissolved oxygen, before and after CWA, 164, 171
  nonpoint source load, rural, of BOD5, ca. 1995, 84
Models, surface water, 13, 183, 184, 189, 194, 242, 270, 271, 294, 316, 321, 382, 383
  AESOP, Upper Mississippi River, 383
  Chattahoochee River, 336
  Delaware Estuary Comprehensive Study, (DECS), 262, 270–271,
  Dynamic Estuary Model, (DEM), Delaware estuary, 184, 271, 272
  James River, (JMSRVM), 184,
  NWPCAM (*see* National Water Pollution Control Assessment Model)
  QUAL-II, Upper Mississippi River, 383
  Potomac Eutrophication Model, (PEM), 184, 294
  System Wide Eutrophication Model, (SWEM), New York Harbor, 242
  WASP5–EUTRO5, Upper Mississippi River, 184, 383
  biochemical oxygen demand(BOD) of, 116–118, 270, 271, 383
  carbon, organic, 271
  dissolved oxygen of, 116–118, 242, 270, 271, 382, 383
  eutrophication of, 242, 271, 382, 383
  heavy metals of, 223, 235
  hydrodynamics of, 242, 271, 382
  nutrients of, 116–118, 271, 294, 382, 383
  sediment transport of, 382
  spatial distributions, simulated, of
    BOD, DO, nitrogen, phosphorus, algae, for case study waterways, 189, 384, 385
    carbon, nitrogen and oxygen, in a hypothetical river, 116–118
    toxic organic chemicals of, 235, 271
Monangahela River, 345, 346
Monitoring, of effluent quality, of wastewater dischargers, 336, 371, 376, 391

Discharge Monitoring Reports (DMRs), 47
Permit Compliance System (PCS), 76
Monitoring, of streamflow, at USGS gages (see Streamflow, gaging stations, USGS)
Monitoring, of water quality:
data, availability of historical, 179
evaluations, of long-term trends of, 248, 342, 354, 358, 369, 377
FWPCA, technical assistance, for, 224, 368
FWPCA, surveillance network, 107–108
programs, 13, 27, 39, 47, 105, 223, 248, 393, 406
programs, historical, during early twentieth century, 22, 24, 107, 248
stations, Before and After CWA changes, dissolved oxygen, 7, 9, 106, 119, 123, 125, 127, 154, 172, 174
stations, location of, 7, 9, 10, 106, 118–119, 155, 156, 173, 174, 206,
STORET, as national archive of data, 176, 177
MSX, disease, of oysters (*see* Shellfish)
Municipal wastewater (*see* Wastewater, municipal)
Mussels, freshwater (*see* Shellfish)

Narragansett Bay, 182, 183, 440
National Commission on Water Quality (NCWQ), 184, 383, 415, 446, 581
National Marine Fisheries Service (*see* National Oceanic Atmospheric Administration)
National Pretreatment Program, 73
NCWQ (*see* National Commission on Water Quality)
National Oceanic Atmospheric Administration (NOAA Action Plan, for Gulf of Mexico Initiative (*see* Gulf of Mexico)
National Marine Fisheries Service, 245
typical pollutant concentration (TPC) (*see* Wastewater, industrial)
NOAA (see National Oceanic Atmospheric Administration)
National Pollutant Discharge Elimination System (NPDES) (*see* Effluent regulations)
National Urban Runoff Project (NURP) (*see* Storm water, urban runoff)
National Water Pollution Control Assessment Model (NWPCAM), 6, 71, 155, 177
Clean Water Needs Survey (CWNS), as data source, 76
Industrial Facilities Discharge Database (IFD), as data source, 122
Permit Compliance System (PCS), as data source, 76, 122
Reach File Version 1 (RF1), as data source, 76, 122
Reach File Version 3 (RF3), as data source, 76
framework, 17, 34, 76, 271
water quality ladder, 76
Navesink River, 245
Navigation:
channels, 258, 364, 383
interstate waters, impedance of, in, 27, 28
locks and dams, and, 363, 364, 382, 392
maintenance of river, for, 382, 383
projects, 380, 392
water use of waterway, as, 204, 215, 219, 298, 345, 350, 363, 366, 382, 403
NBOD (*see* Nitrogenous biochemical oxygen demand)
Needs Survey (*see also* Clean Water Needs Survey), 45, 65–68, 76, 84, 97, 122
Neches River, 183
Neshaminy Creek, 182, 257
New England, Hydrologic Region 1, drainage area of, 199
New England Interstate Water Pollution Control Commission (NEIWPCC), 204
New York City, 19, 23, 107, 189, 213–215, 218, 220, 222–225, 228, 232, 233, 237, 242, 243, 245, 247, 248
New York City Department of Environmental Protection (NYCDEP), 224, 240, 245, 248, 249
New York Harbor, 23–24, 107, 184, 189, 214–216, 218–222, 227, 234, 237, 238, 241–244, 247, 248, 277, 280
New York State Board of Health, 21
Newark Bay, 221, 228, 235
Ninetieth Percentile (see Statistics, biochemical oxygen demand)
Nitrate (NO3–N):
consumption, of dissolved oxygen, during nitrification, 24, 43
conversion from ammonia-N, as end product of nitrification, 24, 43, 118, 210, 269
eutrophication, to coastal, as contribution, 210, 382
product, end, of organic matter decomposition, as, 24
spatial distribution of, in hypothetical river, 116, 118
removal, by denitrification, in advanced wastewater treatment of, 45, 64
state variable, of water quality model, as, 271, 384
trends, long-term, water quality of, 269
Nitrification:
advanced secondary treatment, as unit process, 369, 372, 376, 377, 379, 384–386, 390
advanced waste treatment, as unit process, 45, 64
BOD5, contribution to of, 47
BOD ultimate, contribution to, 26, 43, 46
consumption, of dissolved oxygen, and, 6, 43, 47, 48, 116
inhibition of, 46, 454
initiation of, 46, 118
laboratory samples, for BOD5 measurements, in, 46, 47
suppression of, in, 47
model, water quality, kinetic process, as, 271, 361
natural waters, in, occurrence of, 269

Nitrification (*continued*):
"seed" population, of nitrifying bacteria, for, 46, 118
sequence, of reactions, during, 24, 118
Nitrite (NO2–N):
consumption, of dissolved oxygen, during nitrification, 24, 43
conversion from nitrate-N, as intermediate product of nitrification, 43
state variable, of water quality model, as, 271, 384
product, intermediate, of organic matter decomposition, as, 24
Nitrobacter, 43
Nitrosomonas, 43
Nitrogen:
ammonia (NH3–N) (*see* Ammonia)
denitrification (*see* Denitrification)
effluent controls, on, 249, 292
effluent loading of, 242, 292, 298
eutrophication, of coastal, as cause, 210, 382
hydrolysis (*see* Organic nitrogen)
influent, characteristics, of municipal wastewater, 76–78
limitation, of algal bloom, 294
model, water quality, cycle of, 271
nitrification (*see* Nitrification)
nitrate (NO3–N) (*see* Nitrate)
nitrite (NO2–N) (*see* Nitrite)
organic nitrogen (*see* Organic Nitrogen)
ratio, stoichiometric, of oxygen to nitrogen (*see* Stoichiometry)
reactions, sequential, in cycle of, 24, 118, 271
removal, in advanced wastewater treatment of, 45
removal, and ultimate BOD, in wastewater treatment of, 176
spatial distribution of, in hypothetical river, 116–118
total Kjedhal nitrogen (TKN) (see Total Kjedhal nitrogen)
total nitrogen (TN) (*see* Total nitrogen)
wastewater, in, constituent species of, 43, 47, 48, 315
water quality, and, 47
Nitrogen, effluent characteristics of:
combined sewer overflows, 73–74
industrial wastewater, 71–73
municipal wastewater, 76–78, 454
rural runoff, 73–75
urban storm water, 73–75
Nitrogen, effluent, of municipal wastewater, national trends, as NBOD, 454
loads, trends (1940–1996), 59–60, 473
loads, projected trends (1996–2016), 64–69, 473
removal efficiency, trends (1940–1996) of, 63–64
removal efficiency, projected trends (1996–2016) of, 473
Nitrogenous biochemical oxygen demand (NBOD)(*see* Biochemical oxygen demand)
Nitrogenous BOD (NBOD) (*see* Biochemical oxygen demand)
"No discharge", 48, 49
Nixon, President Richard M., 34
Nonpoint sources (NPS):
best management practices (BMPs)(*see* Best management practices)
data mining, methodology for, and, 122
degradation, of water quality, from, 342
dissolved oxygen, on, impact of, 7
flow, effect of, on, 120, 380–382, 384
land uses, and, relationship of, 73–75, 249, 263
land uses, and export coefficients (*see* Land use)
load, as component of total point and nonpoint source load, 6, 17, 70, 76–79, 84, 85, 97, 98, 155, 181, 206, 226, 235, 264, 265, 319, 382, 383, 406
load, of BOD5, national, (ca. 1995), from, 84–85
load, of nutrients, 380, 392
load, of pesticides, 392
load, of phosphorus, 382
load, of sediments, 369, 392
management and control of, 74, 75, 177, 195, 335, 358
National Water Pollution Control Assessment Model, as input to, 71, 76
strategies, for control of, 75
watershed, in, origin of, 73, 74
NPS (*see* Nonpoint sources)
NPDES (*see* Effluent regulations)
NRC (*see* National Research Council)
NURP (*see* National Urban Runoff Project)
Nutrient(s):
benthic flux of, 271
cycle, 106
enrichment, 106, 195, 222, 316, 317, 411
improvement, of water quality, from water pollution control, 189, 325
loads, from point and nonpoint source of, 235, 237, 292, 298, 302, 308, 315, 321, 325, 363, 380, 392, 411
loads, reduction of, 316, 325
plant production, and, 106
plant uptake of, 304
pollutant, as, of natural waters, 2, 85
removal, in wastewater, by treatment processes of, 44, 45
removal, in Potomac River, by submersed aquatic vegetation (SAV) beds of, 307
state variables, of water quality model, as, 271, 294, 382, 383
NWPCAM (*see* National Water Pollution Control Assessment Model)
NYCDEP (*see* New York City Department of Environmental Protection)

O&M (*see* Expenditures, Operations and maintenance)
Oconto River, 143, 151
Office of Management and Budget (OMB), 203, 216, 260, 288, 314, 331, 348, 403

Ohio River, 2, 12, 84, 85, 164, 167, 184, 345–353, 355–358, 363
advisory, fish consumption, 358
bacteria, degradation, of water quality, from water pollution, 345, 350, 358
bioaccumulation, of toxic organic chemicals, 358
biochemical oxygen demand (BOD),
water quality, 353
municipal and industrial, point source loads, 352
biota of, 356
combined sewer overflows (CSOs), 358
dams and reservoirs
canalization, 347
regulation of streamflow by, (*see* streamflow)
wicket, submergible, 347
data sources, for case study, 345
development and growth, of river basin, 347
disease, waterborne, and epidemics, 345, 350
dissolved oxygen, water quality, 356
depletion of, 350
drought, 1988, effect of, on, 352
minimum, critical location of, 352
sag, 352
trends, long-term, 353
drainage area, of Ohio River, ORSANCO DistriCt, 346
effluent limits, 352
expenditures, water pollution control, 351
fish consumption advisory (*see* advisory, fish consumption)
fishery resources:
bass tournaments, 357, 358
commercial, 355, 357
dissolved oxygen, effect of, on, 356, 350
diversity, of species, 355, 356
sport fishing (*see* recreation, water-based)
surveys, 356
heavy metals, water quality, 354, 355, 358
hydropower, development of, 358
Index of Biotic Integrity (IBI), 357
land use:
agricultural, 345, 350
industrial, 345
mining, of coal, 350
legislative and regulatory history:
Bense Act, of 1908, 350
Federal Water Pollution Control Act, of 1948, 350
Federal Water Quality Act. of 1965, 351
ORSANCO CompAct, of 1948, 350
Lower Ohio, 347
nitrogen, water quality, ammonia-nitrogen, 354
nonpoint sources, controls on, 358
ORSANCO (*see* Ohio River Valley Sanitation Commission)
phosphorus, water quality, 354
population, 347, 352
density, 345
Metropolitan Statistical Area (MSA), counties, 348
trends (1940–1996), 348, 350
rank, national, of Ohio River, by length and flow, 345
raw sewage, 345, 352
recreation, water-based, 345, 350, 358
contact (i.e, swimming), support of, 358
sport fishing, 355, 357
secondary treatment, 351, 352, 358
equivalent, for industrial dischargers, as, 351
minimum for municipal dischargers, as, 351
shipping, navigation, 345, 350
streamflow, 346, 347
drought, 350
drought, of 1988, 352
gage, USGS monitoring, at Louisville, Kentucky, 347
low-flow, 347, 352
mean, monthly, 347
mean, summer, trends (1940–1995), 347
regulation of, by dams, 347
taste and odor, 350
toxic organic chemicals, 358
turbidity, 353
wastewater, industrial, 345, 351
wastewater, municipal:
effluent flow, 352
effluent loads, 352
population served, by, 351
standards, for treatment, 351
water quality:
degradation, 350, 358
improvements, from wastewater upgrades, 351, 352, 354, 358
issues, contemporary, 358
issues, historical, 350
management, regional of, 345, 346, 350, 351, 352, 358
monitoring, 354
water quality standards and criteria, 353, 354
bacteria, coliform, 351
dissolved oxygen, 352
watershed, characteristics, physiographic of, 346
water uses, 345, 350, 358
withdrawals, of water, for public water supply, 345, 350, 358
Ohio, Hydrologic Region 5, 345
dissolved oxygen, before and after CWA, 164
drainage area, of basin, 345
nonpoint load, rural, of BOD5, ca. 1995, 84
Ohio River Valley Sanitation Commission (ORSANCO), 2, 345–347, 350–354, 356–358
Operations & maintenance (O&M)(see Expenditures, water pollution control)
Organic matter, 2, 3, 24–26, 43, 44, 46, 47, 106, 116, 182, 455
breakdown, in wastewater, 25
carbon-to-dry weight ratio (C:DW)
decomposers of, 25
decomposition, and, 6, 24, 106, 116
degradation, and biochemical oxygen demand (BOD), 26, 116

Organic matter (*continued*):
dissolved organic matter (DOM),
labile, 46, 47, 271, 295, 384, 386
oxidation of, 26, 44, 47, 118, 295, 319, 384, 386, 450
particulate organic matter (POM), 455
production, 106
refractory, 46, 271, 295, 384, 386
wastewater, in, 25, 386
Organic nitrogen:
cycle, nitrogen of, as component, 24, 116, 118, 271
decomposition, of organic matter, as produCt, 43
hydrolysis of, 24
state variable, of water quality model, as, 271, 384
nitrification, and, 46
wastewater, in, form of nitrogen, as, 48
ORSANCO (*see* Ohio River Valley Sanitation Commission)
Ospreys (*see* Birds, fish-eating)
Oxidation ponds, 48, 49
Oxygen (*see* Dissolved oxygen)
demand (*see* Dissolved oxygen)
depletion (*see* Dissolved oxygen)
sag (*see* Dissolved oxygen)
saturation (*see* Dissolved oxygen)
Oysters (*see* Shellfish)

PACE (*see* Pollution Abatement Cost Expenditures)
Pacific Northwest, Hydrologic Region 17, 399
dissolved oxygen, before and after CWA, 164, 171
drainage area of, 399
PAHs (*see* Polynuclear aromatic hydrocarbons)
Pamlico-Albemarle Sound, 182
Passaic River, 22, 184, 220
Pasteur, Dr. Louis (*see* Disease, waterborne)
Pathogens, 2, 20, 44, 75, 85, 177, 242, 245
PE (*see* Population equivalent)
Per-capita water use, 41, 50, 97
Peregrine falcons (*see* Birds, fish-eating)
Permit Compliance System (PCS), 76, 122, 583
Persistence, of low-flow, dry conditions (*see* Streamflow)
Pesticides, 280, 363, 380, 392, 408
pH, 37, 38, 46, 262, 294, 353, 356
Philadelphia, Pennsylvania, 215, 216, 255, 256, 258–263, 265, 266, 269, 270, 273, 275, 276, 279, 280
Phosphorus:
water quality, 210, 266, 269, 273, 280 302, 308, 339, 354, 380
ban, on phosphate detergents, 234, 269, 316–318, 321, 325, 339, 380
benthic release of, 384
biological removal of, 339, 391
effluent controls, on, 292, 298, 302, 317, 318, 321, 383
export coefficients, for total phosphorus yield, 381, 382
influent, characteristics, of municipal wastewater of, 76–78
Lake Pepin Phosphorus Study, 366, 380
load, budget, of point and nonpoint sources of, 382
load, nonpoint sources of, 380–382, 384
load, point source effluent of, 233, 234, 269, 279, 315, 318, 338, 339, 380–382, 384
land use, and, nonpoint source loads, 366, 380, 381
model, water quality, case study waterways of, 189, 321, 384, 385
model, water quality, evaluation, of phosphorus controls, 321
phosphate, 24
sediment bed concentration, of phosphorus, 380
streamflow, effect of, on nonpoint source load of, 381, 382, 384
soluble reactive (SRP), 380
suspended solids, and interaction of, 384
total (TP), 45, 73, 210, 226, 233, 266, 268, 269, 273, 279, 280, 380–382
Phosphorus, effluent characteristics, of:
combined sewer overflows, 73–74
industrial wastewater, 71–73
municipal wastewater, 76–78
rural runoff, 73–75
urban storm water, 73–75
Photosynthesis, 44, 106, 298, 308
Phytoplankton (see also Algae), 294, 315, 321
Piedmont, 138, 255
Pittsburgh, Pennsylvania, 19, 345, 347, 350
Plants, aquatic, 106, 206
Eurasian watermilfoil, 301
Hydrilla, 307
macrophytes, 301
submersed (also submerged) aquatic vegetation (SAV), 189, 301–308
Water chestnut, 301
PMSA (*see* Primary Metropolitan Statistical Area)
POC (*see* Carbon, organic, particulate)
Point source (PS):
controls, effectiveness of, 7, 176, 177
controls, regulatory of, 99, 171, 189, 242, 321, 406
criteria, for case study selection, as, 183
data mining, screening rules, for, 119, 122, 173, 174
definition of, 70
degradation, of water quality, from, 342
effluent flow of, 237, 583
effluent load, as component of total point and nonpoint source load of, 76, 78, 85, 97, 98, 319, 381, 382, 384, 406
effluent load, of BOD5, for catalog units, ca. 1995, from, 155
effluent load, from, 12, 138, 264, 373, 386, 583
effluent load, industrial, decline of, 85
effluent load, reductions of, 189, 206, 263, 265
effluent permits, for discharge of, 34
expenditures, water pollution control, 1, 98

inventory, dischargers of, 122, 207, 263, 264, 317, 583
Lower White River, to, discharges of, 156
signal of, and link to dissolved oxygen, 7, 9, 10, 12, 106–108, 119, 121, 122, 154, 155, 159, 172, 173
strategies, for control of, 75
Pollution Abatement Cost Expenditures (PACE), 4, 91
Pollution intolerant (*see* Macroinvertebrates, aquatic)
Pollution tolerant (*see* Macroinvertebrates, aquatic)
Polychlorinated biphenyls(PCBs), 222, 223, 280, 388, 389, 390
ban, on, 389, 390
lipid-normalized, in fish tissue, 389
Polynuclear aromatic hydrocarbons(PAHs), 235, 280
Population, of case study counties, 184–185
Population, of United States:
impacted, as of 1990 census, by Clean Water Act of 1972, 177, 179
rural, long-term trends (1900–1990), 23–24
total, long-term trends (1940–1996), 42
urban, long-term trends (1900–1990), 23–24
Population density, 203, 260, 331
Population equivalent (PE), of untreated municipal wastewater, 54, 453, 581, 584
Population served, by municipal wastewater facilities, 63, 99, 105
case study waterways, and, 206, 228, 229, 232, 234, 291, 335, 350, 351, 358, 367, 374, 405
data sources, for historical trends of, 50, 450–452, 581–583
data sources, for projected trends of, 454–455
impacted, by Before and After CWA changes in dissolved oxygen, 127, 138, 175
measure, of performance, for water pollution control expenditures, as, 4
301(h) waivers, granted, with, 52
in 1950, 26
in 1968, 6, 52, 96–97
in 1972, 3
in 1996, 6, 52, 96–97, 127, 175
trends, historical, (1940–1996), 6, 50–52, 229, 449
trends, projected, (1996–2016), 64, 65, 70, 97, 99
wastewater treatment of, categories, by, 52, 64, 96, 229, 358
Population served, by public water supply systems, 591–593, 595
Population served, by wastewater collection systems, 22, 229
Portland, Oregon, 189, 280, 399, 401, 403–408, 410, 411
Potomac estuary, 12, 182, 184, 189–193, 285, 286, 288–292, 294–300, 303, 307, 308
Potomac River, 22, 189, 285–288, 290, 291, 293, 295, 298, 300–303, 305, 307, 308, 418, 420, 425, 429, 430, 441, 448
advanced secondary waste treatment, 295, 298
advanced waste treatment (also tertiary), 286, 295, 298
algae, 302, 304
blooms of, 286, 290, 298
blue-green, 294
Microcystis aeruginosa, 294
American Heritage River, designation as, 308
bacteria, degradation, of water quality, from water pollution, 285, 290
bacteria, effluent discharge from wastewater, 300
bacteria, improvement, of water quality, from water pollution control, 300
beaches (*see* recreation, water-based)
biochemical oxygen demand (BOD), water quality:
simulated, with Potomac Eutrophication Model (PEM), 294
biochemical oxygen demand (BOD), effluent loads:
municipal point sources, loads, 292, 295, 298
relationship, of effluent loads and dissolved oxygen, 189–191, 306
birds:
watching of, 298, 307
waterfowl, 301, 302, 304, 307, 308
Blue Plains, wastewater treatment plant, 290–292, 294, 295, 298, 300, 301
chlorination, of wastewater effluent, 300
chlorophyll, 294, 298
combined sewer overflows (CSOs), 300
fecal coliform bacteria, effects of, 300
clarity, of water (see transparency)
Clean Water Act, of 1972 (see legislative and regulatory history)
Choptank River, 304, 307
data sources, for case study, 285
development and growth, history of, colonial era, 301, 302
depletion of, 285, 290, 292, 301, 306
historical, records of, 286, 290, 292, 298
minimum, levels of, 292, 294
sag, critical location of, 292, 298
seasonal variation of, 292
simulated, with Potomac Eutrophication Model (PEM), 294, 295, 298
spatial distribution of, 292, 298
drainage area, of Potomac River basin, 285, 286
effluent limits, 294
technology-based, 294
water quality-based, 294
eutrophication, 292
fall line, at Little Falls, Virginia, 285, 290
fish kills, 290
fishery resources, 301, 302, 304, 307, 308
American shad, 290
Bassmasters Fishing Tournament, 208
commercial, 298
habitat, 301, 302, 304, 306, 307
Largemouth bass, 308
professional guide services, 308
recreational (*see* recreation, water-based)

Potomac River (*continued*):
spawning, 290, 302
species diversity, 307
surveys, 306, 307
wastewater loads, effect of, on, 290
Index of Biotic Integrity (IBI), 306, 307
industrial activity, 290
land use, 290
legislative and regulatory history, 290
Clean Water Act, of 1972, 294
Federal Water Pollution Control Act, of 1956, 290
Water Quality Act, of 1965, 300
Microcystis aeruginosa (*see* Algae)
models, surface water quality:
Potomac Eutrophication Model, (PEM), 184, 294
evaluation, of wastewater upgrades, 189, 294, 295, 298
nitrogen, water quality, 308
nitrogen, effluent loads, 292, 298
controls, on, 292, 298
nutrient, effluent loads, 292, 298, 302
phosphorus, water quality, 302, 308
phosphorus, effluent loads,
controls, on, 292, 298, 302
plants, aquatic:
Eurasian watermilfoil, 301
Hydrilla, 307
macrophytes, 301
submersed (also submerged) aquatic vegetation (SAV), 189, 301–308
Water chestnut, 301
population:
Metropolitan Statistical Area (MSA), counties, 288
growth, 286, 290, 292
served, by municipal wastewater, 291
trends (1940–1996), 288
primary treatment, 290, 295
primary productivity, of algae, 294, 298
rank, national, of Potomac River by length and flow, 285
raw sewage, 290, 292
recreation, water-based:
beaches, 300, 301
boating, 290, 298, 300
fishing, 290, 298, 300, 307
swimming, 300, 301
riverfront:
deterioration of, and water pollution, 300
revitalization of, 300
salinity, 285, 294
toxicity, to algae, 294
secondary treatment, 286, 292, 294, 295
greater than, 298
shellfish, harvest, 300
stormwater, runoff, 300
streamflow, 286, 287
gage, USGS monitoring, at Little Falls, Virginia, 286
drought, during early 1960s, 287, 298
floods, 287
low-flow, 287, 290, 291
mean, summer, 1940–1995, 287, 295
mean, monthly, 287
submersed (also submerged) aquatic vegetation (see plants, aquatic)
suspended solids, water quality, 308
suspended solids, effluent loads, 302
controls, on, 302
swimming (*see* recreation, water-based)
tertiary treatment(*see* advanced wastewater treatment)
influence, on circulation, 285
freshwater, 285
Potomac, 302, 304, 307, 308
river, 285, 290, 300, 301, 304
transition zone, 285, 294, 304
transparency, of water, 298, 302
turbidity, 302
Upper Potomac River, 285, 286
wastewater, collection system, 290
wastewater, industrial, 290
wastewater, municipal, 290, 306
effluent flow, 291
effluent loads, 292, 302
water clarity (*see* transparency, of water)
water quality:
degradation, 290
improvements, from wastewater upgrades, 286, 300, 302, 308
issues, historical, 285, 290, 308
management plans, 290
monitoring program, during early twentieth century, 107
water quality standards:
bacteria, fecal coliform, 300
dissolved oxygen, 294, 298
primary contact, for coliform bacteria, 300
secondary contact, for coliform bacteria, 298, 300
withdrawals, of water, for water supply, 290, 298
diversions, 290
municipal (also public), 290, 298
POTW (*see* Publically Owned Treatment Works)
Pretreatment, industrial (*see* Wastewater, industrial)
Primary contact, and water quality standards (*see* Bacteria)
Primary Metropolitan Statistical Area (PMSA), 203, 216, 288, 314, 331, 348, 403
Primary treatment, of wastewater, 23, 25, 26, 44, 116, 373
effluent characteristics of, 76–78
minimum level of treatment, as, 33
Productivity, biological, 258, 259
Public:
awareness, of water pollution issues, 34, 249, 280, 281
support, for water pollution control regulations, 390, 399, 410
Public Health Service (*see* U.S. Public Health Service)

Public Law (PL) (*see* Legislative and regulatory history),
Public sanitation (*see* Disease, waterborne)
Public supply, of water (*see* Withdrawals, public supply)
Publically Owned Treatment Works (POTWs), 1–2
Pulp and paper, mills, 205, 404, 405
Pure Waters Program (New York State), 224

Quinebaug River, 183, 418, 441

Raritan Bay, 221, 228, 245, 248, 418, 420, 425, 430, 434, 441
Raw sewage:
- assimilation, capacity of waterway, for disposal of, 5
- characteristics of, 54, 55, 453, 455
- combined sewer systems, and, 73, 222, 242, 281, 369
- decomposition rates of, 46
- degradation, of water quality, from disposal of, 222, 243, 261, 262, 290, 292, 315, 399, 404
- discharges, from malfunctions and bypasses of, 226, 242, 245, 369, 370
- effluent load, accounted for by disposal of, 62, 220, 225, 226, 229, 232, 233, 243, 262, 352
- elimination, of disposal of, 237, 243, 249
- fish kills, caused by disposal of, 334
- inventory, of municipal facilities, discharging, 49
- population served, by untreated wastewater, 3, 26, 52, 229, 451–452, 581
- public concern, about disposal of, 27, 220, 372
- recreation, water-based, and public health impact, of disposal of, 242
- shellfish, and public health impact, of disposal of, 221
- urban water cycle, as weak link, from disposal of, 2
- wastewater collection systems, and, 22, 372
- waterborne disease, public health impact, of disposal of, 5, 345, 368, 372

Reach File, Version 1 (RF1):
- biochemical oxygen demand (BOD), improvements, at scale of, (see Biochemical oxygen demand, water quality)
- Chesapeake Bay watershed, river network map of, 8
- definition of, 7–9, 122
- dissolved oxygen, improvements, at scale of, (*see* Dissolved oxygen, water quality):
  - case study waterways, and, 184
  - reaches, paired, for major river basins of, 164
  - spatial scale, signal of, 144, 154, 159, 162
- identification code, 11-digit, example of, 9–10
- National Water Pollution Control Assessment Model (NWPCAM), framework, as, 76
- point source discharge locations, linked to, 122, 123
- population, in 1990, linked to, 177
- spatial scale, hierarchal of, as component, 126, 144, 177
- water quality monitoring station locations, linked to, 122

Reach File, Version 3 (RF3), 7
- National Water Pollution Control Assessment Model (NWPCAM), framework, as, 76

Reaeration, atmospheric (*see* Dissolved Oxygen)
Recovery zone (*see* Dissolved Oxygen)
Recreation, water-based, 2, 181, 182, 194, 199, 207, 242, 332, 335, 341, 345, 350, 358, 399
- boating, 199, 219, 249, 290, 298, 300, 311, 325, 386
- fishing, 209, 220, 249, 277, 280, 290, 298, 300, 307, 355, 357, 366, 403, 409, 410
- kayaking, 325
- rafting, whitewater, 325
- swimming, 300, 301

Removal efficiency, of treatment processes (see Wastewater, municipal)
Reservoirs, 120, 203, 257, 259, 342, 402, 411
Residential water use, as component of wastewater flow, 50, 54
Resources, biological, 181–183, 189, 194, 308, 386
Respiration, of algae (see Algae)
RF1 (*see* Reach File, Version 1)
RF3 (*see* Reach File, Version 3)
Richmond, Virginia, 280, 311–317, 319–321, 323–325
River basin, definition of, for Clean Water Restoration Act of 1966, 32
River basins, major:
- definition, of (*see* Hydrologic Unit Code)
- list of, in conterminous United States, 9, 163
- municipal and industrial wastewater loads, of BOD5 and TSS, by, 581–584
- municipal and industrial water withdrawals, by, 593–594

Riverfront:
- deterioration of, from water pollution, 199, 279, 300, 315, 391, 411
- economic revitalization of, 183, 199, 279–280, 300, 325, 391, 411

River Raisin, 130
Roosevelt, Franklin, D., President, 28, 262
Root River, 130
Rum River, 363
Runoff, rural:
- coefficients, export, land use dependent, for, 75, 206, 263–265
- characteristics of, 73–75
- controls, on, 391
- data source, for catalog unit-based load estimates of BOD5, 6, 84
- distribution, geographic of, 84, 85
- issue, contemporary, for watershed management, as, 392

Runoff, rural (*continued*):
load, as component of total point and nonpoint source load, as, 85, 97, 155, 206, 264, 265
load, BOD5, catalog units, from (*see* Biochemical oxygen demand, effluent loads ca. 1995)
Runoff, watershed (*see* Nonpoint sources)
"Rust Belt", 85

Sag, of dissolved oxygen, in a river (*see* Dissolved Oxygen)
Salem, Oregon, 257, 402, 403, 405, 406, 408, 409
Salinity, 235, 258, 259, 285, 294
Salinas River, 143
St. Paul, Minnesota, 1, 110, 111, 162, 189, 361–377, 379, 380, 384, 386, 391
Sandy Hook Bay, 235
Sangamon River, 22
Schuylkill River, 182, 256, 265, 273
Secchi depth (*see* Transparency, water)
Secondary contact, and water quality standards (*see* Bacteria)
Secondary treatment, of municipal wastewater:
benefits, water quality of, 9, 39, 183
deadline, for implementation of, 38–40,
definitions, regulatory, 36, 37–39
dry weather baseflow, 41
effluent characteristics of, 76–78
"equivalent", as defined in PL 97–117, 39, 40
greater than (also better than), 42, 48, 298, 324, 325, 390
historical development of, 5, 25–27
infiltration and inflow, 41
less than, 42, 48, 57
minimum acceptable technology, as, 3–5, 17, 34, 36, 118, 171–172, 175, 176
regulations, for (*see* Legislative and regulatory history)
regulations, Clean Water Act, for, 36, 38–41
water quality standards, and, 30
wet weather flows from combined storm and sanitary sewers, 37, 41
Sediment bed, contamination (*see* Toxic organic chemicals; *see also* Heavy metals)
Sediment bed, toxicity (*see* Toxic organic chemicals)
Sediment oxygen demand (SOD) (*see* Dissolved oxygen)
Shellfish:
beds, closure of, 73, 222
beds, contamination, by pathogens of, 221, 222, 245
beds, re-opening of, 245, 248
bioaccumulation, of DDT, 246
blue crabs, 277, 279
consumption, as pathway for waterborne disease of, 242
hard clams, 221, 277
harvest, commercial of, 219, 277, 278
decline of, 24, 215, 219, 221, 277
resurgence of, 277, 278
mussels, freshwater, 356
oysters, 221, 261, 277, 279
MSX, disease of, 277
protection, 301(h) criteria for of, 39
resource, natural, as, 213, 277
scallops, bay, 221
standards, bacterial, water quality, for consumption of, 300
wastewater, disposal, effect of, on, 2, 73, 221
Shenandoah River, 22, 442
Shipping, as commercial use of waterways, 215, 218, 219, 241
Shipworms (*see* Macroinvertebrates, aquatic)
Shrewsbury River, 221
SIC (*see* Standard Industrial Classification)
Signal, of point source impact, on water quality (see Data mining)
Sludge:
bottom deposits of, in rivers, 341, 368, 404
disposal of, 224
sewage, 224, 290, 367, 368, 369
Sludgeworms (*see* Macroinvertebrates, aquatic)
SOD (*see* Sediment oxygen demand)
Snow, Dr. John (*see* Disease, waterborne)
SOD (*see* Dissolved Oxygen, sediment oxygen demand)
South Atlantic-Gulf, Hydrologic Region 3, 327
dissolved oxygen, before and after CWA, 164
drainage area, of basin, 327
Spatial distribution, in rivers, of dissolved oxygen (see Dissolved Oxygen)
Spawning, of fish (*see* Fishery resources)
Spokane River, 143
Sport fishery (*see* Fishery resources)
SRF (*see* State Revolving Fund; *see also* Clean Water State Revolving Fund)
St. Croix River, 9, 154, 363, 366, 367, 373, 377, 378
phosphorus yield of, 381
sediment load of, 380
Stabilization pond, wastewater, 39, 44, 48
Standard Industrial Classification (SIC), 583
Standards, water quality (*see* Water quality standards)
State Revolving Fund (*see also* Clean Water State Revolving Fund), 1, 90, 95, 98, 99, 336, 501, 502
State Water Control Board, (SWCB, Virginia), 316
Statistics, biochemical oxygen demand, "worst-case", before and after CWA,
ninetieth percentile of, 143, 155, 156
Statistics, dissolved oxygen, "worst-case", before and after CWA, 7, 9–10, 106, 119, 125, 126, 156, 164, 171, 172, 174, 175
frequency distributions, for major river basins of, 164–171
paired data, Before & After CWA, 7, 126, 162, 164, 171
samples, minimum number required, 121
Kolmogorov-Smirnoff (KS) test, 164, 171

significance level, statistical, of analysis of, 162, 164
Student's t-test, paired, 126, 162, 164, 171
tenth percentile of, 121, 122, 125, 144, 154, 159, 164, 174
Statistics, streamflow:
7-day, 10-year flow (7Q10), 154, 182, 203, 273, 287, 322, 365
mean, annual, 213, 255, 286, 311, 327, 330, 345, 346, 361, 365, 399
mean, monthly, long-term, 201, 215, 257, 273, 287, 312, 330, 347, 364, 365, 400, 402
mean, summer, trends, long–term, 76, 110, 201, 215, 257, 273, 287, 312, 330, 347, 364, 365, 400, 402
Stoichiometry, 48, 59
carbon to dry weight (C:DW), 452, 455
CBODU to BOD5 (CBODU:BOD5), 47, 48, 55, 77–78, 295
oxygen to nitrogen (O2:N), 48, 59, 454
Stoneflies (*see* Macroinvertebrates, aquatic)
STORET, USEPA water quality database, 176
archive, of historical water quality data, 106–107, 177
data source, for Before and After CWA trends, 12, 106–107, 181
potential bias of, for statistical analysis, 176
example application of methodology, 123–125
data source, for case studies, 184, 199, 213, 255, 285, 327, 345, 361, 399
inventory, of dissolved oxygen records, (1941–1995), 107, 109
Stormwater, urban runoff, 75, 370
coefficients, export, land use dependent, for, 75, 263–265
combined, with wastewater (*see* Combined sewer overflow)
characteristics, effluent of, 73–75
controls, on, 391
data source, for catalog unit-based load estimates of BOD5, 6, 84
distribution, geographic of, 84, 85
heavy metals, as source of, 75
issue, contemporary, for watershed management, as, 249, 280, 342, 358, 392
load, as component of total point and nonpoint source load, as, 85, 97, 155, 206, 264, 265
load, BOD5, catalog units, from (*see* Biochemical oxygen demand, effluent loads ca. 1995)
pathogens, as source of, 75
National Urban Runoff Project (NURP), 75
secondary treatment, effect on compliance with effluent limits, 37
sediments, as source of, 75
systems, collection, 75
toxic organic chemicals, as source of, 75
wastewater flow, as component of, 50, 54
Stratification, 120
Streamflow, 201–203, 215, 257–258, 273, 286, 287, 311, 319, 327, 329, 346, 347, 361, 364, 399, 400
10-year, 7-day flow (7Q10) (*see* Statistics, streamflow)
augmentation, by reservoirs of, 400, 405, 406, 410, 411
drought, 173, 220, 259, 292, 294, 334, 345, 350, 365
drought, during early 1960s, 111, 116, 155, 194, 215, 224, 257, 287, 298
drought, during 1987–1988, 111, 155, 352, 377, 379, 380, 382, 383, 385, 391
floods, 287, 324
gaging stations, USGS, 110, 201, 215, 256, 257, 286, 311, 329, 347, 364, 379, 402, 537
drainage area, contributing, to, 110, 364
Great Flood of 1993, Upper Mississippi River, 110, 364, 365, 381, 383, 386
high-flow, 368, 369
low-flow, 118, 154, 257, 273, 287, 290, 291, 318, 322–324, 329, 334, 347, 352, 364, 370, 378–380, 400, 402
dissolved oxygen, on, effect of, 108, 110, 377–379, 385, 391
persistence, of dry, conditions of, 110–111, 121, 123, 155, 173
mean, annual (*see* Statistics, streamflow)
mean, monthly (*see* Statistics, streamflow)
mean, summer (*see* Statistics, streamflow)
minimum, 329, 330, 336, 364
nonpoint source loads, on, effect of, 381, 382, 384
regulation, by locks, dams and reservoirs of, 201, 203, 257, 259, 273, 324, 329, 336, 347, 363, 365
seasonal variation of, 110, 120, 123
Streamflow, ratio, summer, 116, 365, 379
catalog unit maps, United States, (1961–1995), 537–539
catalog unit maps, United States, comparison of 1963 and 1988 conditions, 110–111
data mining, rules, for, 121, 173
definitions, for dry, normal, and wet conditions of, 110
example, Upper Mississippi River of, 110–111, 123
methodology, for calculation of, 110–111
runoff, normalized as cfs per square mile, for calculation of, 110, 537, 540
sliding window, for spatial interpolation of, 110–111
Student's T-test (*see* Statistics, dissolved oxygen)
Subregion, definition of, (*see* Hydrologic Unit Code)
Surveys, of water pollution, during early twentieth century (*see* Monitoring, of water quality)
Suspended solids, water quality, 189, 205, 273, 302, 308, 334, 411
data mining, as candidate for, 177
interaction, in river, with phosphorus, 384

Suspended solids, water quality (*continued*):
state variable, of National Water Pollution Control Assessment Model (NWPCAM), as, 76
Suspended solids, effluent characteristics:
combined sewer overflows of, 73–74
industrial wastewater of, 71–73
municipal wastewater of, 76–78, 455
rural runoff of, 73–75
secondary treatment, April, 1973 definition of, 36
urban storm water of, 73–75
Suspended solids, effluent loads, industrial wastewater, national estimates,
major river basins, for 1950, 1973, and 1995 of, 581–584
Suspended solids, effluent loads, municipal wastewater, 205, 226, 235, 237, 232, 302, 374
controls, on of, 273, 302, 308, 316, 336, 342, 376, 390
per capita loading rate of, 232
Suspended solids, effluent loads, municipal wastewater, national estimates,
major river basins, for 1950, 1973, and 1995 of, 581–584
trends, (1940–1996) of, 453–454
projected trends, (1996–2016) of, 454–455
removal efficiency, trends, (1940–1996) of, 453–454
removal efficiency, projected trends, (1996–2025) of, 454–455
Suspended solids, nonpoint source loads, 235, 237, 340, 380
watershed runoff, 237
Suspended solids, wastewater treatment processes, removal of, 44, 45
Susquehanna River, 150
SWCB (*see* State Water Control Board, Virginia)
SWEM (*see* System Wide Eutrophication Model)
Swimming beaches (*see* Beaches, recreational)
System Wide Eutrophication Model (SWEM) (*see* Models, water quality)

t-test (*see* Student's t-test)
Tennessee, Hydrologic Region *6, 8*
dissolved oxygen, before and after CWA, 164, 171
Teredos (*see* Macroinvertebrates, aquatic)
Tertiary treatment, 45, 295, 405
Texas-Gulf, Hydrologic Region 12,
dissolved oxygen, before and after CWA, 164
Thames River, 19, 21, 23
"Three-legged stool", 5, 7, 17, 95, 98, 105, 119, 154, 171, 181, 598
Tidal freshwater, 258, 259
TOC (*see* Carbon, organic, total)
Total coliform (*see* Bacteria)
Total Kjedhal nitrogen (TKN), 48, 55, 59, 60, 76, 374, 376, 450, 452, 454, 490
Total nitrogen (TN), 73, 118, 210, 226, 231, 266, 268, 292
Total suspended solids (*see* Suspended solids)
Toxic organic chemicals, 85, 241, 246, 248
bioaccumulation of, (*see* Bioaccumulation, toxic organic chemicals)
criteria, water quality, for, 408
data mining, candidate, as, 177
groundwater, contamination, by, 392
issue, contemporary, for water quality management, as, 195, 222, 223, 249, 280, 281, 358, 380, 391, 392, 408, 411
see DDT
see polychlorinated biphenyl (PCBs)
see polynuclear aromatic hydrocarbon (PAHs)
sediment bed, contamination, by, 195, 222, 249, 382, 383, 392
sediment bed, toxicity, 235, 281
state variables, of water quality models, 235, 271
storm water, urban runoff, as source of, 75
wastewater, industrial, and, 392
wastewater, primary treatment, and removal of, 44
TPC (*see* Wastewater, industrial, typical pollutant concentration)
Transition zone, in a tidal river, 258, 259, 285, 294, 304
Transparency, of water:
secchi depth, 303
turbidity, 258, 259, 302, 353, 369
water clarity, 298, 302–304
Travel time (in a river), 118, 400
Trenton, New Jersey, 255–261, 263, 273, 276, 280
Trickling filter, 5, 25, 26, 39, 45, 582
Turbidity (*see* Transparency, water)
Twin Cities, Minnesota, 144, 361, 363, 366–370, 372, 373, 378, 390, 391
Typhoid fever (*see* Disease, waterborne)
Typical pollutant concentration (TPC) (*see* Wastewater, industrial)

Ultimate BOD (*see* Biochemical oxygen demand)
U.S. Army Corps of Engineers (USCOE), 364, 382, 383, 386, 388, 402, 405
U.S. Census Bureau (USCB), 20, 21, 24, 65, 331, 454
U.S. Department of Commerce (USDOC), Bureau of Economic Analysis (BEA), 3, 91
U.S. Department of Interior (USDOI), 31, 184
U.S. Environmental Protection Agency (USEPA), 1
creation of, as federal regulatory agency, in 1970, 31
U.S. Fish and Wildlife Service, 388
U.S. Geological Survey (USGS):
Hydrologic Unit Code (HUC) of, developer, as, 8
streamflow, data source, for historical records, as, 110
streamflow, gaging stations, USGS (*see* Streamflow, gaging stations)
withdrawals, of water, data source, for municipal and industrial supply, 260, 591–594

U.S. Public Health Service (USPHS):
  assistance, technical, to states, for water pollution studies, 30, 223, 224
  inventories, of municipal wastewater facilities (*see* Wastewater, municipal, national inventory)
  surveys, of water pollution, 262, 290, 372
Upper Black Warrior River, 79
Upper Hudson River (*see* Hudson River)
Upper Illinois River (*see* Illinois River)
Upper Mississippi River, 144, 154, 162, 168, 184, 189–193, 361–374, 376–393, 421, 428, 448
  activated sludge, 369
  advanced secondary, with nitrification, 361, 369, 372, 374, 376–377, 379, 384, 385
  advisory, fish consumption, 389
  American Heritage River, designation as, 392
  assimilative capacity, of river, 367, 368
  bacteria, degradation, of water quality, from water pollution, 368, 370, 376
  bacteria, improvement, of water quality, from water pollution control, 369, 372, 379, 392
  bioaccumulation, of PCBs, 388–390
  biochemical oxygen demand (BOD):
    water quality, trends, 377
    effluent load, trends, municipal, from METRO facility, 374, 378, 390
    model simulation of, 384–385
    STORET, as data source for ambient trends, 377
  birds, 388
  channelization, of river, as contemporary issue, 392
  chlorination:
  failure, of public water supply, 368
    sewage effluent of, 368
  chlorophyll, 380
    model simulation of, 384–385
  combined sewer overflows(CSOs), 369, 376, 391
    coliform bacteria, and, 369, 391
    expenditures, for separation of, 369
    separation, of stormwater and raw sewage, 369, 372, 391
  dams (*see* locks and dams)
  data mining, examples of methodology:
  dissolved oxygen, for, 9–13, 123–125
    streamflow ratio, summer, for, 110–111
  data sources, for case study, 361
  DDT, 388
  development and growth, history of, 366, 371, 380
  disease, waterborne, and public health, 368, 372
    dissolved oxygen, water quality, 377–379
    depletion of, 371, 376
    degradation of, 379
    drought, 1987–1988, effect of, on, 377, 379
    historical records of, 368, 376, 384
    low levels of, 368, 370
    indicator, as, 392
    minimum oxygen, 368, 377
    model simulation of, 384–385
    improvements, in, 371, 372, 376, 377, 379, 385
    sag, 144, 154, 377, 379, 385
    saturation, 368
    spatial distribution of, 376
    standards (*see* water quality standards)
    STORET, as data source for ambient trends, 377
    streamflow, effect of, on, 377–379, 385, 391
    trends, long-term, 376–379
    typhoid, epidemic of, 368
  drainage area:
    Catalog Unit 07010206 of, 363
    USGS streamflow gage, at St. Paul, Minnesota, and, 364
  economy, of river basin, 362, 366
    agriculture, 362, 366
    forest products, 362, 366
    effluent limits, 373
    industrial wastewater, 373
    municipal wastewater, 373
  enforcement, interstate conference, for water pollution control, 361, 421
  eutrophication (*see* Lake Pepin)
  expenditures, water pollution control, 373, 379, 386, 390–392
  fish consumption advisory (*see* advisory, fish consumption)
  fish kills (*see* Lake Pepin)
  fishery resources: 366, 368, 389
    blue sucker, 389
    carp, 368
    catfish, 389
    commercial, 366
    electro-fishing, 389
    paddlefish, 389
    recreational, 366
    smallmouth buffalo, 389
    walleye, 368, 389
    wastewater loads, effect of, on, 368
  flood-control, 364, 380, 383, 392
  flood plain, 388
  grey Cloud, Minnesota, 369, 377, 379
  habitat:
    alteration, and loss of, 380, 383, 392
    fish and wildlife, for, 362
    invertebrates, benthic, for, 372
  heavy metals (*see also* mercury), 370, 380, 391, 392
    effluent load, reduction of, 370, 371, 391
    sediment cores, 370
    sediments, contaminated, 382
    sources of, 370, 371, 392
  hydropower, 363, 366
  Index of Biotic Integrity (IBI), 389
  Investments:
    ecological integrity of river, for restoration of, 392
    marinas, in, 386
    riverfront development, in, 386

Upper Mississippi River (*continued*):
  Lake Itasca, 361, 362
  Lake Pepin, 364, 388
    algal blooms, 380, 383
    eutrophication, 380–383
    fish kills, 380, 383
    mercury load, 391
    PCB levels, in fish, 389
    phosphorus, water quality, 380
    phosphorus load, 380, 381
    Phosphorus Study, 366, 380
    sediment bed contamination of, 370, 371, 379, 392
    sediment bed concentration, of mercury, 370, 371, 372, 379
    sediment bed concentration, of phosphorus, 380
    sediment deposition, to, 380
  land use, 361, 369
    agricultural, 366, 392
    forest, 366
    urban, 366, 392
  legislative and regulatory history:
    Clean Water Act, of 1972, 373, 383, 390–392
    Metropolitan Drainage Commission, 372
    Minnesota Board of Health, 368, 372
    Minnesota Pollution Control Agency, 373
    Minnesota Water Pollution Control Act, of 1945, 373
    Section 208, of Clean Water Act of 1972, 383
  locks and dams, 363, 367, 382, 383
    degradation, of biological communities, effect of, on, 392
    dredging, 364
    elevation, of river, 364
    fishery resources, effect of, on,
    flood control, for, 363
    navigation, for, 363, 364, 382, 392
    regulation of streamflow (*see* streamflow)
    water quality, effect of, on, 367
  mayflies (*hexagenia*), 371, 372, 391
    resurgence of, 372, 391
    scarcity of, 371
  mercury, 371–372, 379, 391
    dental amalgam, as source of, 392
    deposition rates, historical trends, 371
    loading, 372, 391
    sediment contamination, in Lake Pepin (*see* Lake Pepin)
    sources of from municipal and industrial wastewater, 371, 391
  Metropolitan Council Environmental Services (MCES), 361, 371, 373, 382, 390, 391
  Metropolitan Waste Control Commission (MWCC), 373, 374
  Metropolitan Wastewater Treatment Plant, St. Paul, Minnesota, (METRO), 123, 125, 144, 154, 361, 367–370, 373, 377, 381, 389, 391
    effluent load, of phosphorus, 381
    impact of on dissolved oxygen and un-ionized ammonia, 383
  model simulation, of effluent regulations, 384–386
    population served, by (*see* population)
    upgrades, 369, 372–374, 376, 377, 379, 385, 390
  models, surface water quality, 361, 382, 383
    advanced eutrophication of, 382, 383
    bacteria of, 383
    evaluation, of wastewater upgrades, 189, 194, 384–386
    hydrodynamic, 382, 383
    post-audit, 379, 383
    review, literature of, 383
    sediment transport of, 382, 383
    waste load allocation, for, 383
  Newport, Minnesota, 369, 379
  nitrogen, water quality,
    model simulation of, 384–385
    un-ionized ammonia, 369, 376, 383
  nitrogen, effluent load, trends (1940–2000), from METRO facility, 376
    ammonia-nitrogen, 369, 374, 376, 390
    total Kjedhal-nitrogen, 374, 376
  nonpoint sources, 369, 391, 392
    export coefficients, for total phosphorus yield, 381
    nutrients of, 363, 380, 381, 392
    pesticides of, 363, 380, 392
    rural runoff, 391, 392
    sediment of, 369, 380, 392
    streamflow, effect of, on phosphorus and sediment load, 381, 382, 384
    urban runoff, 391, 392
  phosphorus, 380, 381
      benthic release of, 384
      biological removal of, 391
      budget, of point and nonpoint source loads, 382
      effluent controls, on, 383
      nonpoint source loads, 380–382, 384
      point source loads, 380, 382, 384
      sediment concentration, in Lake Pepin (*see* Lake Pepin)
      ban, on phosphate detergents, 380
      land use, and, nonpoint source loads, 366, 380, 381
      water quality, in Lake Pepin (*see* Lake Pepin)
  Pigs Eye Lake, 373, 388
  pollution intolerant, 371
  pollution tolerant, 368
  Pool, of Upper Mississippi River
    2, 364, 369, 371, 372, 383, 386, 389
    3, 371
    4 (Lake Pepin), 364, 371, 389
  population, 361, 366, 368
    Metropolitan Statistical Area (CMSA), counties, 366
    served, by METRO facility, 367, 374
    trends (1940–1996), 367

pretreatment, industrial (*see* wastewater, industrial)
primary treatment, 361, 368, 373, 384, 385
productivity, biological,
rank, national, by length and flow (*see* Mississippi River)
raw sewage, 367–370, 372
Reach File Version 1 (RF1), reach, 377
recreation, water–based, 366, 386, 388
  boating, 386
  marinas, 386
resources, biological, 386
riverfront, 386, 391
  deterioration of, and water pollution, 391
  economic revitalization of, 391
  St. Paul Riverfront Corporation, 391
secondary treatment, 361, 369, 383, 384, 385
  better than, 390
sediment:
  contamination, 389
  deposition, to Lake Pepin (*see* Lake Pepin), 380
  disposal, of dredged materials, 382, 383
  load, and land use, 366, 380
shipping and navigation, 366, 380, 392
sludge, sewage, 368, 369
STORET, USEPA water quality database, 377
stormwater, 367, 369, 389, 391
  runoff, urban, 370
streamflow, 361, 364
  10–year, 7–day minimum flow (7–Q-10), 365
  drought, 365, 385
  drought, of 1988, 377, 379, 380, 382, 383, 385, 391
  gage, USGS monitoring, at St. Paul, Minnesota, 364, 379
  Great Flood of 1993, 110, 364, 365, 381, 383, 386
  high flow, 368, 369
  low-flow, 364, 370, 378–380
  mean, annual discharge, 365
  mean, summer, trends (1940–1995), 365
  mean, monthly, 365
  minimum, 364
  ratio, summer, 110–111, 365, 379
  regulation of, by locks and dams, 363, 365
suspended solids:
  effluent load, from METRO facility, 374, 376, 390
  interaction, in river, with ambient phosphorus, 384
toxic organic chemicals, 380, 391, 392
  bioaccumulation of, 392
  groundwater contamination, 392
  sediment bed contamination, 382, 392
turbidity, 369
wastewater, collection system, 366, 367, 373
wastewater, industrial, 367, 373, 389, 392
  effluent limits, 373
  heavy metals, load, 392
  pretreatment, 370, 372, 388, 391
  toxic organic chemicals, 392
wastewater, municipal, 367, 373, 374, 389, 392
  communities served, by, 373
  effluent flow, 374
  effluent limits, 373
  effluent load, 363, 373, 374, 376, 380, 381
  nutrient load, 392
  population served, by (*see* population)
water, drinking, 372
water quality:
  degradation, 367–368
  improvements, from wastewater upgrades, 368, 376, 377, 379, 392
  issues, contemporary, 380, 392
  issues, historical, 367–371, 376, 386
  locks and dams, and effect on (*see* locks and dams)
  monitoring stations, 369
  surveys, early twentieth century, 107, 372, 390
watershed, 361–364
  characteristics, physiographic of, 363, 364
water pollution control, effectiveness of, 391
water quality standards:
  dissolved oxygen, 369, 379, 383, 385, 390, 391
  fecal coliform bacteria, 369, 391
  un-ionized ammonia, 369, 376
water uses, 366, 388
wetlands, 366, 379, 380, 388, 392
withdrawals, of water, for public water supply, 366–368, 372
Upper Mississippi River, Hydrologic Region 7, 362
  dissolved oxygen, before and after CWA, 164
  drainage area, of basin, 362
  nonpoint load, rural, of BOD5, ca. 1995, 84–85
Upper Mississippi River Conservation Committee, 392
Upper White River, 154–156, 159, 162
Urban runoff, of storm water (*see* Storm water, urban, runoff)
Urban storm water (*see* Storm water, urban, runoff)
Urban water cycle (*see* Water cycle, urban)
USCB (*see* U.S. Census Bureau), 20, 21, 24
USDOC (*see* U.S. Department of Commerce)
USDOI (*see* U.S. Department of Interior)
  USEPA (*see* U.S. Environmental Protection Agency)
USGS (*see* U.S. Geological Survey)
USPHS (*see* U.S. Public Health Service)

Vermilion River, 373

Waivers, 301(h) (*see* Legislative and regulatory history)
Warm-water fish, and dissolved oxygen (*see* Dissolved oxygen, criteria and standards)
Washington, DC, 182, 189, 285, 286, 288–292, 295, 298, 300–302, 305, 307, 308

Waste Load Allocation (WLA), 27, 96, 262, 263, 270, 271, 281, 316, 318
Wastewater, collection and disposal systems, 17–19, 22, 219–220, 248, 298, 327, 366, 373, 391
  "Great Stink", London (1858), 21
  Greek and Roman empires, 18
  storm water, urban (*see* Storm water, urban, runoff)
  twentieth century, first half of, 22
Wastewater, industrial, 23, 181, 194, 224, 247, 255, 263, 317
  controls, regulatory, for disposal of, 13, 33, 37, 96, 176, 189, 206, 234, 263, 317, 351, 405, 406, 408, 411
  Discharge Monitoring Reports (DMRs), 47
  degradation, of water quality, from disposal of, 28, 204, 205, 221, 234, 261, 262, 275, 277, 316, 340, 345, 367, 389, 392, 399, 404
  dissolved oxygen, effect on, of disposal of, 106, 119, 138, 404
  effluent characteristics of, 26, 71–72
  effluent guidelines, for, 71, 73
  effluent limits, for, 47, 176
  effluent load, 6, 17, 70, 76, 386
  effluent load, as component of total point and nonpoint source load, 78, 79, 85, 97, 98, 122, 155, 206, 215, 235, 237, 264, 265
  effluent load, of BOD5, catalog units, ca. 1995, 78–79, 82–83
  effluent load, of BOD5 and TSS, major river basins, for 1950, 1973, 1995, 581–584
  effluent load, declines, from water pollution control, 85, 266, 273, 279, 371, 388
  heavy metals, source, as, 280
  Industrial Facilities Discharge database (IFD), 122
  municipal wastewater, to, contribution of, 41, 50, 54, 71, 73, 452, 453
  National Pretreatment Program, 73
  pretreatment of, 39, 73, 85, 222, 224, 234, 370, 372, 388, 391
  treatment of, 71
  typical pollutant concentration (TPC), for, 71–72, 583
  waste load allocation, for, 270, 271, 281, 383
  water use, as, disposal of, 31, 199, 204, 219, 403
Wastewater, municipal, effluent characteristics of, 72, 74, 77, 78, 295, 450, 488–494
  carbonaceous, 5-day BOD (CBOD5), 48, 77–78, 453, 454
  carbonaceous, ultimate BOD (CBODU), 48, 454
  categories, of wastewater treatment (also types) (*see* Categories)
  dissolved oxygen, saturation, 108
  nitrogen:
    ammonia-N, 77–78
    total Kjedhal nitrogen (TKN), 48, 55, 77–78
    total nitrogen (TN), 77–78
  nitrogenous BOD (NBOD), 48, 55
  organic carbon:
    particulate(POC), 455
    total (POC), 77–78
  per capita, wastewater flow rate, 41, 48, 55, 65, 97, 229, 232, 595
  per capita, pollutant loading rate,
    biochemical oxygen demand (BOD5), 54, 232
    nitrogenous biochemical oxygen demand (NBOD), 55
    suspended solids, total (TSS), 232
  phosphorus, total (TP), 77–78
  population equivalent (PE), 54, 453, 581, 584
  ratios, stoichiometric (*see* Stoichiometry)
  removal efficiency, 57, 59, 60, 61, 63, 77–78
    actual peformance, 57
    design-based, 42
  suspended solids, total (TSS), 77–78, 455
  ultimate BOD (BODU), 454
Wastewater, municipal, effluent limits, 32, 38, 71, 290, 336, 373
  ban, on phosphate detergents, 234, 269, 316–318, 321, 325, 339, 380
  technology-based, 3, 34, 38, 70, 96, 107, 119, 154, 176, 181, 184, 189
  water quality-based, 3, 34, 70, 96, 107, 119, 154, 176, 189, 206, 263, 324
Wastewater, municipal, effluent loads:
  component, of total point and nonpoint source load, as, 78, 79, 85, 97, 98, 122, 155, 206, 215, 235, 237, 264, 265
  Construction Grants Program Awards, and, 90
  BOD5, national (*see* Biochemical oxygen demand (BOD), effluent loads, national)
  BOD5, major river basins (*see* Biochemical oxygen demand (BOD), effluent loads, national)
  TSS, major river basins (*see* Suspended solids, effluent loads, national)
Wastewater, municipal, influent loads,
  BOD5, national (*see* Biochemical oxygen demand (BOD), influent loads, national)
Wastewater, municipal, inventories, national, of treatment plants, 4
  categories, of wastewater treatment (also types) (*see* Categories)
  Clean Water Needs Survey (CWNS), 48, 65–68, 76, 84, 97, 122, 454, 583
  Needs Surveys (*see* Clean Water Needs Surveys)
  trends, historical, (1940–1996) of, 48, 49
  trends, projected, (1996–2025) of, 65–68, 70
  U.S. Public Health Service, inventories, 3, 26, 42, 48, 50
Wastewater, municipal, population served (*see* Population served, by municipal wastewater facilities)
Water clarity (*see* Transparency, water)
Water cycle, urban, 2, 5, 3, 22, 23, 26, 31, 34, 75, 95, 96, 99
  piped water, in household, 19
  public health, and, 18